Modern Applications of EPR/ESR

From Biophysics to Materials Science

Springer
Singapore
Berlin
Heidelberg
New York
Barcelona
Budapest
Hong Kong
London
Milan
Paris
Santa Clara
Tokyo

Modern Applications of EPR/ESR

From Biophysics to Materials Science

APS'97

Proceedings of the
First Asia-Pacific EPR/ESR Symposium
Hong Kong
20 – 24 January 1997

C.Z. Rudowicz
(Editor)

K.N. Yu
H. Hiraoka
(Associate Editors)

Springer

Czeslaw Z. Rudowicz and K.N. Yu
Department of Physics and Materials Science
City University of Hong Kong
Tat Chee Avenue
Kowloon
Hong Kong

H. Hiraoka
Department of Chemisty
Hong Kong University of Science and Technology
Clear Water Bay
Kowloon
Hong Kong

Library of Congress Cataloging-in-Publication Data

Asia-Pacific EPR/ESR Symposium (1st : 1997 : Hong Kong)
Modern applications of EPR/ESR : from biophysics to materials science : proceedings of the the First Asia-Pacific EPR/ESR Symposium, Hong Kong, 20–24 January 1997 / Czeslaw Z. Rudowicz (editor), K.N. Yu, H. Hiraoka (editors).
p. cm.
Includes bibliographical references.
ISBN 9813083239
1. Electron paramagnetic resonance spectroscopy--Congresses.
2. Electron paramagnetic resonance spectroscopy--Industrial applications--Congresses. I. Rudowicz, Czesław. II. Yu, K. N., 1963– . III. Hiraoka, H., 1932– . IV. Title.
QC763.A78 1997
543'.0877--dc21

97-22627
CIP

ISBN 981-3083-23-9

Printed in Singapore

Typesetting: Camera-ready by authors
SPIN 10636243 5 4 3 2 1 0

FOREWORD

The First Asia-Pacific EPR/ESR Symposium organised by the Department of Physics and Materials Science, City University of Hong Kong, was held in January 20–24, 1997, at the City University of Hong Kong. The Symposium was attended by about 140 participants from 22 different countries, among whom 110 were from Asia-Pacific, Europe and North America whereas 30 were from Hong Kong, mainland China and Taiwan. There were 8 invited plenary lectures and 3 invited plenary talks, 114 oral presentations and 69 poster presentations, making a total of 183 presentations.

The interdisciplinary character of the EPR/ESR spectroscopy was reflected in the Scientific Programme, which consisted of the plenary sessions and four parallel sessions

- biology, life and medical sciences
- chemistry, earth and environmental sciences
- physics and materials science
- new developments and cross-disciplinary areas

The meeting featured the following plenary lectures :
Prof. S. H. Choh (Republic of Korea): *EPR study of paramagnetic defects in* $LiNbO_3$ *single crystals* • Prof. Richard Cammack (UK): *Bio-medical applications of EPR spectroscopy* • Prof. C.A.J. Ammerlaan (The Netherlands): *Magnetic resonance spectroscopy of rare-earth doped semiconductors* • Prof. J. M. Spaeth (Germany): *Electrical detection of EPR and ENDOR: New possibilities for the study of defects in semiconductors* • Prof. John R. Pilbrow (Australia): *Pulsed and CW EPR in materials science* • Dr. Richard Bramely (Australia): *EPR meeting the challenges of new materials* • Prof. M. Ikeya (Japan): *New applications of electron spin resonance — dating, dosimetry and microscopy* • Prof. Noboru Hirota (Japan): *Chemically induced dynamic electron polarization (CIDEP) studies of photochemical reactions at different magnetic fields*;

and plenary talks:

Prof. Mitsuhiro Motokawa (Japan): *Submillimeter wave ESR of magnetic materials in high magnetic fields* • Prof. Yves Servant (France): *EPR studies of spin transitions: A review* • Prof. A. Romanyukha (Russia): *Comparative study of different EPR materials for use in clinical dosimetry*.

The Symposium has provided opportunities for interaction between specialists in many different fields. We were fortunate that many distinguished scientists had joined the Symposium. The Symposium has provided also an opportunity to bring the state-of-the-art knowledge of EPR to Hong Kong and Southern China.

During the Symposium the Inaugural Meeting of the Asia-Pacific EPR/ESR Society was held. A Welcome Address from Prof. Jim Norris, the President of the International EPR/ESR Society was read at the Inaugural Meeting. The 62 researchers from Asia-Pacific region present have elected the Provisional Council as follows:

President	Prof. Czeslaw Rudowicz
Vice-President	Prof. Asako Kawamori (Japan)
Secretary	Dr Y. Y. Yeung (Hong Kong) [nominated later by the President]
Treasurer	Prof. Hiroaki Ohya-Nishiguchi (Japan)
Representatives	Prof. John Pilbrow (Australia/New Zealand) Prof. Asako Kawamori (Japan) [also Vice-President] Prof. R. Y. Zhan (Peoples' Republic of China) Prof. S. V. Bhat (India) Prof. S. H. Choh (Republic of Korea) Prof. N.T. Nguyen (Vietnam) Prof. A. Ziadtinov (Russia — Far East)

It appears that the Symposium was a great success judging by the positive feedback received from many participants. Moreover, the Asia-Pacific participants gathered at the Inaugural Meeting of the Asia-Pacific EPR/ESR Society unanimously voted a proposal that the second Asia-Pacific EPR/ESR Symposium be held, not in another country as suggested by the LOC Chairman, but again at CityU in Hong Kong in 1999. However, since continuing organisation of the Symposium would involve significant University resources, at this time the University does not feel it is able to support a second event in the next few years.

Efforts will be made to find alternative arrangements for the next Symposium. As President of the Asia-Pacific EPR/ESR Society, I will try to convince researchers from other institutions within the Asia-Pacific region to take over the organisation of the Symposium. Probably we should follow the pattern adopted by the European Federation of EPR Groups. The European EPR Symposium is to be organised every three years in a different European country with a national EPR Group in charge of the local organization being also responsible for handling the financial affairs.

On behalf of the Local and International Organizing Committee I would like to thank the organisations who financially supported the Symposium, namely, Lee Hysan Foundation, The Croucher Foundation and Beijing–Hong Kong Academic Exchange Centre.

This was the first international conference our local team and I had organised. The administrative and printing costs had been kept to a minimum. We had not hired a professional conference organiser. Instead we have provided as many travel and accommodation grants to our colleagues in need as possible as well as waived the registration fee for about 40% of the participants. We were helped in this regard by the caring attitude of our Invited Speakers who opted for a standard accom-modation instead of an upper class hotel. This Symposium may be compared to a "home-made cake". Thank you to all who have helped to successfully "bake the cake", especially the hardworking colleagues from the Local Organizing Committee.

This volume of Proceedings includes the papers submitted within the time limit set by the publisher. All papers have been refereed by the members of the Review Panel listed later in this volume, who are experts in their own areas. The efforts of all the reviewers and Springer-Verlag, Singapore in achieving high standard of scientific and technical quality of the Proceedings are gratefully acknowledged.

Czeslaw Rudowicz
Chairman
Local and International Organizing Committee
First Asia-Pacific EPR/ESR Symposium

Post scriptum (November 1997)

The Asia-Pacific EPR/ESR Society will provide the logistic support and coordination for the organization of the future Symposia. The President of the Asia-Pacific EPR/ESR Society has recently conducted a search for an Asia-Pacific EPR/ESR group willing to host the next Symposium. After consultations with the potential hosts it has just been decided that the 2nd Asia-Pacific EPR/ESR Symposium will be held at Zhejiang University, Hangzhou, P.R. China, 15–19 November 1999. Prof. Yuanzhi Xu, the Chairman of the Local Organizing Committee, is the member of Magnetic Resonance Special Committee, Chinese Physical Society.

The Second Announcement and Call for Papers will be circulated in January 1998. If your name was not on the name list of the participants of the 1st Asia-Pacific EPR/ESR Symposium please provide us with your name and full address. For more information contact:

Prof. Yuanzhi XU, Chairman, LOC
The 2nd Asia-Pacific EPR/ESR Symposium
Department of Chemistry, Zhejiang University,
Hangzhou 310027, P. R. China
Phone: +86 571 7984095 or +86 571 7951352;
Fax: +86 571 7984095 or +86 571 7951895;
E-mail: xyz@public.hz.zj.cn

or

Prof. Czeslaw Rudowicz, The President
The Asia-Pacific EPR/ESR Society
Department of Physics and Materials Science, City University of Hong Kong,
83 Tat Chee Avenue, Kowloon, Hong Kong SAR, P. R. China
Phone: (852) 2788-7787 Fax: (852) 2788-7830
E-Mail: APSEPR@CITYU.EDU.HK

CONTENTS

1st Asia-Pacific EPR/ESR Symposium
City University of Hong Kong, Hong Kong
January 20–24, 1997

SPONSORS & COMMITTEES

Organised by : Department of Physics & Materials Science
City University of Hong Kong

Co-sponsored by : Lee Hysan Foundation
The Croucher Foundation
Beijing–Hong Kong Academic Exchange Centre

International Organising Committee:

Chair:	Prof. C. Rudowicz	City University	(Hong Kong)
Co-chair:	Prof. S. H. Choh	University of Korea	(Korea)
Co-chair:	Prof. J. R. Pilbrow	Monash University	(Australia)
Co-chair:	Prof. O.N. Hiroaki	Yamagata Technopolis Foundation	(Japan)
Co-chair:	Prof. H. Takeuchi	Nagoya University	(Japan)
Co-chair:	Prof. J.L. Rao	Sri Venkateswara University	(India)
Co-chair:	Prof. R. Zhan	Changchun Institute of Applied Chemistry	(P.R. China)
Co-chair:	Prof. S.K. Misra	Concordia University	(Canada)
Co-chair:	Prof. L. Kevan	University of Houston	(USA)
Co-chair:	Prof. S.A. Marshall	Roosevelt University	(USA)

Local Organising Committee :

Honorary Chairman: Prof. E.C.M. Young, Vice-President,
City University of Hong Kong

Chairman :	Prof. Czeslaw Rudowicz	City University of Hong Kong
Secretary:	Dr Peter K.N. Yu	City University of Hong Kong
Treasurer:	Dr Ken G. G. Siu	City University of Hong Kong
Members:	Dr Dan D.S. Chiu	City University of Hong Kong
	Prof. H. Hiraoka	Hong Kong University of Science and Technology
	Dr Aaron H. P. Ho	City University of Hong Kong
	Dr C.C. Lam	City University of Hong Kong
	Dr P.L. Leung	City University of Hong Kong
	Dr Kevin W.P. Leung	Chinese University of Hong Kong
	Dr Sudhana B Madhu	City University of Hong Kong
	Dr Edwin Y.B. Pun	City University of Hong Kong
	Prof. S.K. Wong	City University of Hong Kong

EDITORIAL BOARD

Editor:

Associate-Editors:

REVIEW PANEL

Chapter 1

Biology, Life and Medical Sciences

Bio-medical Applications of EPR Spectroscopy

Richard Cammack and Jasvinder K. Shergill

Centre for the Study of Metals in Biology and Medicine, King's College, Campden Hill Road, London W8 7AH, U.K.

Abstract. EPR spectra of biological materials, from microbial cells to human tissues, show numerous signals which can be identified. These include various forms of free radicals and iron proteins. Examples will be given of the application of EPR methods to derive detailed molecular information about paramagnetic species in complex biochemical systems. Most free radical species are too short-lived to detect directly but a few are stable enough to be detected in whole cells. Notably the tyrosine free radical of ribonucleotide reductase in tissue-culture cells shows changes in response to the intracellular iron status. Nitric oxide has attracted considerable interest for its cytotoxic effects when produced, for example, by macrophages. EPR can help to determine the sites of such tissue damage. Nitrosyl-iron centers may be observed by EPR spectroscopy, generated both from naturally-occurring iron proteins and from spin-trap compounds. Such compounds may be important in neurodegenerative disorders such as Parkinson's disease.

More discriminating techniques, electron-nuclear double resonance (ENDOR) and electron spin-echo envelope modulation (ESEEM) spectroscopy, in one and two dimensions, have been used to observe hyperfine interactions with nuclei in the vicinity of paramagnetic centres in proteins. By judicious choice of sample preparation and measurement conditions, it is possible to observe the spectra due to individual paramagnetic species. An example is the Rieske iron-sulfur protein in the mitochondrial respiratory chain. New subjects for investigation are being constructed by genetic engineering techniques.

1. Introduction

In this paper we review some applications of continuous-wave and pulsed EPR spectroscopy to the study of biological systems, mainly using examples from our recent work. We will consider mainly naturally-occurring paramagnets in biological materials. Spin labels and most spin traps will not be discussed for reasons of space. We begin with some general EPR measurements of paramagnets related to medical investigations, then some detailed molecular studies of metal centres in enzymes.

EPR has a number of advantages over other spectroscopic methods for the investigation of biological systems. It is sensitive, compared, for example, with NMR; it is also quantitative. However not all redox states of proteins are EPR-detectable. Paramagnets that can be detected at room temperature EPR include stable free radicals such as spin traps and nitroxide spin labels, Mn(II) and Cr(V). Transition metals are normally measured at cryogenic temperatures, for example Fe(III), Cu(II), Mo(V).

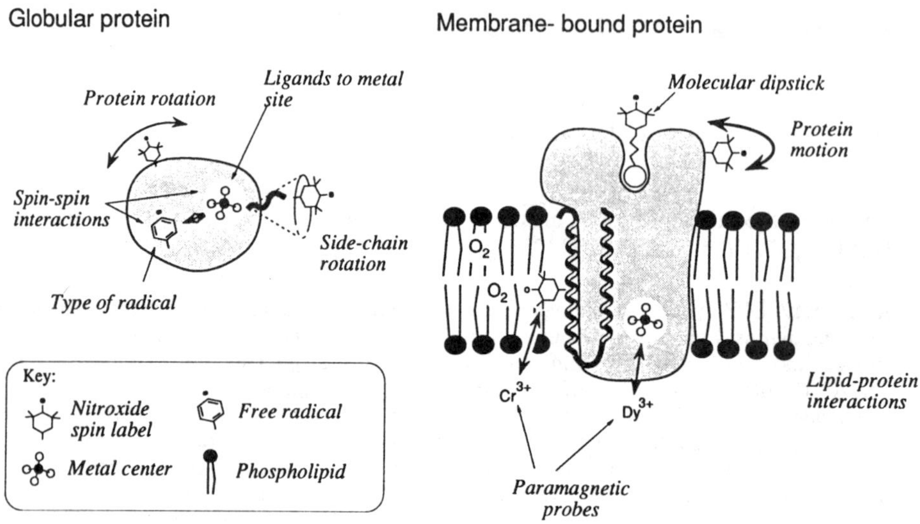

Figure 1. Information about proteins that can be derived by EPR spectroscopy

EPR can provide probably aa wide a range of information as any spectroscopic technique (Fig. 1). Examples in biological systems include:

- *Novel biochemical species* can readily be identified from their lineshapes and temperature dependence. For transition-metal centres, the EPR spectrum spin states and coordination geometry. The spectra of iron proteins show a particularly rich diversity and are prominent in low-temperature EPR spectra of living tissue. Iron-sulfur proteins in electron transfer chains of mitochondrial respiration and photosynthesis were first identified using EPR.
- *The kinetics of biological processes* which involve one-electron transfer and photochemical reactions.
- Effects of *molecular motion* (using nitroxide spin labels), for example motion of whole proteins or parts of protein in solution, or freedom of motion of a ligand exposed to solution (molecular dipstick).
- Hyperfine interactions allow the determination of *electron density distribution* in electron carriers, or identification of nuclei surrounding the paramagnet.
- Electron spin-spin interactions allow the estimation of distances between paramagnetic centres; oxygen concentration in the vicinity of a spin probe (EPR oximetry), and lipid-protein interactions. Paramagnetic relaxation probes allow the estimation of distances of paramagnetic centres from the macromolecule surface, and, for membrane proteins, on which side of the membrane they are located. Electron transfer proteins or protein complexes usually contain several paramagnetic centres, so that the effects of spin-spin interactions between them can be observed.

The environment of the paramagnet may be explored in more detail by hyperfine couplings, for example using ENDOR and ESEEM. An example of this is interactions with solvent protons may also be observed by exchange with deuterium.

Sample preparation is important in order to obtain optimum results from EPR. Requirements include adequate concentration; detectable oxidation state; and minimization of interfering signals.

2. Iron and Radical in Ribonucleotide Reductase.

Conditions of iron overload are widespread and include hemochromatosis and β-thalassemia. In β-thalassemia, an inherited recessive disease, a faulty hemoglobin is produced. To allow respiration, blood transfusions are used therapeutically. However these lead to a build-up of iron in the tissues and early mortality, usually due to heart failure. Iron may be removed from the body by treatment with iron chelators. The one in present use is desferrioxamine (DFO), which is expensive and requires frequent injections. New iron chelators, the hydroxypyridones, are being developed. These are orally active and charge-neutral in their free and iron-bound forms. However the chelators have side-effects, notably inhibition of ribonucleotide reductase, and depletion of the low-molecular weight intracellular iron pool [1].

Ribonucleotide reductase is an enzyme required for the synthesis of deoxyribonucleotides of DNA. The enzyme depends for its activity on the presence of a tyrosyl cation radical in the protein, formed in the enzyme by the action of oxygen on a dinuclear iron centre.

The radical in ribonucleotide reductase may be observed by EPR and shows a characteristic hyperfine splitting due to interaction with the β-CH_2 protons of the tyrosine residue. This stable radical species may be detected by EPR spectroscopy in cultures of animal and human cells [2]. We studied the human leukemia K562 cells in which the radical signal was observed quite prominently, at temperatures around 30 K [1]. The amplitude of the signal correlated with ribonucleotide reductase activity. In the same samples, chelated Fe(III) could be observed at $g = 4.3$. This arises from the intracellular iron pool and not just from ribonucleotide reductase.

We have studied the effect of the iron chelators on the level of the ribonucleotide reductase radical and iron chelates in K562 cells. It was found that formation of iron chelates by the hydroxypyridone CP19 occurred within minutes, much more rapidly than with DFO, and that loss of the ribonucleotide reductase radical paralleled the appearance of Fe(III)-chelate complexes. Moreover, on washing the CP19 from the cells, the ribonucleotide reductase radical was restored very rapidly. These results have implications for the therapeutic use of these chelators.

3. Nitric Oxide Biochemistry

Nitric oxide (nitrogen monoxide, NO) has recently been recognized as an important metabolite in the body [3]. It is produced by oxidation of the amino-acid arginine by nitric oxide synthases (NOS). In blood vessels it is produced by endothelial nitric oxide synthase (eNOS), and causes decreased blood flow and suppresses blood clotting by platelets. In brain it is produced by neuronal nitric oxide synthase (nNOS), where it is involved in nerve potentiation and short-term memory. In these signalling processes NOS is constitutively expressed, and the levels of NO involved are very low, less than 1 micromolar. At these levels NO is relatively stable and non-toxic. NO is produced in much higher amounts by certain cells of the immune system, which act to destroy foreign cells such as bacteria or cancer cells. The enzyme involved is inducible nitric oxide synthase (iNOS), which is synthesized in response to an immune challenge, in the blood by macrophages, and in the brain by microglial cells. This produces NO at higher levels, of the order of 1 - 10 μM.

From the EPR point of view NO is an interesting molecule in having an unpaired electron; complexes of ions such as Fe(II) become converted to EPR-detectable states by complexation with NO. Thus EPR is a viable method of detection of NO in biological materials.

The mechanism of damage to cells by NO is not certain, though it has been demonstrated in recent studies that NO reacts with metalloproteins of the mitochondrial respiratory chain [4]. Certainly NO can react with model iron-sulfur complexes, forming iron-sulfur-nitrosyl (Fe-NO-S) complexes, some of which are detectable by EPR. However our studies on iron-sulfur proteins indicate that they are relatively resistant to NO itself, at concentrations around 1 mM. A more potent compound, peroxynitrite, may be formed by the reaction of NO with superoxide ion, which is also a product of macrophages during the immune response. EPR has shown that iron-sulfur clusters in mitochondrial membranes are depleted by treatment with peroxynitrite.

3.1 Nitric Oxide in Endotoxic Shock

In response to invasion by bacteria, the cells of the immune system such as macrophages are stimulted to release NO and superoxide. The damage to the body by these toxic species can lead to a condition known as endotoxic shock, which is potentially fatal. This condition can be modelled in animal systems by injection of bacterial lipopolysaccharide (LPS), which stimulates the immune system in the same way.

The production of nitric oxide in the body leads to the formation of nitrosyl iron complexes, notably nitrosyl hemoglobin (Hb-NO). Nitrosyl hemoglobin builds up in the blood, and can be observed by EPR. A component of its spectrum of Hb-NO has a prominent three-line hyperfine pattern from the ^{14}NO. This pattern is typical of pentacoordinate nitrosyl heme proteins in which the proximal histidine ligand has been dissociated from the iron. The existence of this signal is an indication of the presence of high concentrations of nitric oxide.

Another method to detect NO is to use a spin-trapping compound such as iron(II) - diethyldithiocarbamate (Fe-DETC) [5]. This compound associates with cell membranes and persists in the body for about 1 day, during which time it accumulates the NO as a nitrosyl complex. This can readily be detected by EPR of tissues, with a triplet hyperfine splitting due to the ^{14}N nucleus (Fig. 2).

We have exploited these EPR signals to investigate various treatments for endotoxic shock (Fig. 2). One of these is to deplete the arginine from the blood by means of the enzyme arginase, thereby depriving iNOS of its substrate. This was shown to decrease the level of Fe-DETC-NO in various tissues of rats after treatment with LPS and Fe-DETC [6]. Another method was to use 2,4-diamino-6-hydroxypyrimidine (DAHP), an antagonist of tetrahydrobiopterin, a component of NOS. The levels of Hb-NO in kidneys of rats after treatment with LPS were considerably decreased by treatment with these compounds [7]. These results indicate the utility of EPR for observing the effects of chemotherapy for endotoxic shock.

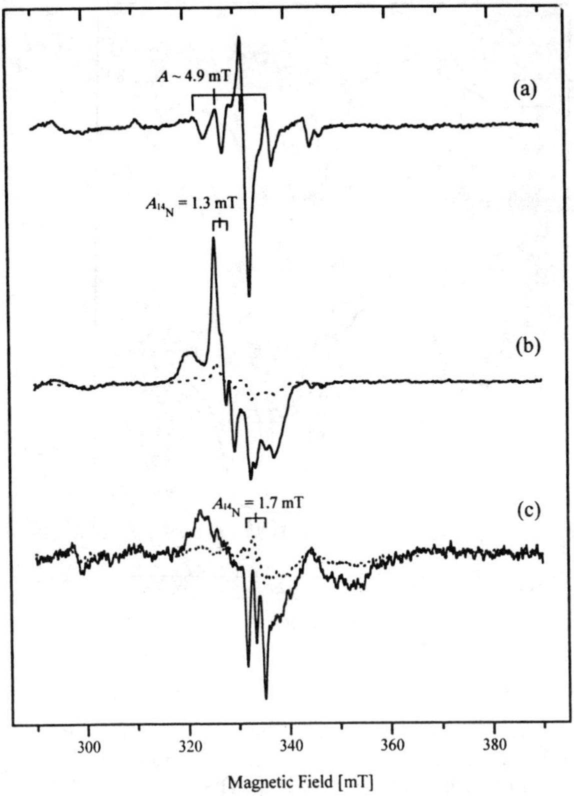

Figure 2. EPR spectra from LPS-treated rats [6,7].(a) EPR spectrum, recorded at 15 K and 9.34 GHz, of a homogenate of spleens from rats injected with DETC and Fe-citrate; the part of the spectrum indicated with $A = 4.9$ mT arises from the Cu-DETC complex. (b) Similar spectra from rats additionally administered with LPS; the part of the spectrum indicated arises from the Fe-DETC-NO. For the spectrum with the dotted line, arginase had also been administered. (c) EPR spectrum, recorded at 40 K and 9.375 GHz, of a homogenate of kidneys from rats treated with LPS; the part of the spectrum indicated arises from the g_z region of 5-coordinate nitrosylhemoglobin.. For the spectrum with the dotted line, the biopterin antagonist DAHP had also been administered.

3.2 Parkinson's Disease

Idiopathic Parkinson's disease is a common neurodegenerative disease which affects about 1 in 40 of the population. It is characterized by a degeneration of a specific area of the brain known as the substantia nigra, which is required for synthesis of dopamine. It was proposed that the condition was associated with damage to the mitochondria of the substantia nigra. By EPR, we examined samples of this area of the brain from patients who had died from Parkinson's disease, and compared them with samples from an adjacent area of the brain, the globus pallidus, which is not affected by the disease [8]. As controls, brain samples were taken from patients who had died from unrelated causes, and matched for age and post-mortem delay. Iron-sulfur clusters of the mitochondrial respiratory chain were observed by EPR in globus pallidus, and no effects were observed in Parkinson's disease patients.

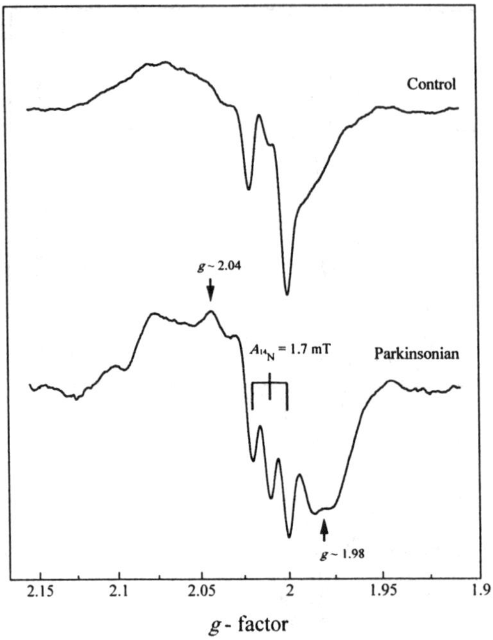

Figure 3. EPR spectra of postmortem samples of substantia nigra from control and Parkinson's disease human brain [8]. Measurement conditions were: temperature 13 K, microwave power, 20 mW, frequency, 9.44 GHz. The hyperfine splitting assigned to 5-coordinate nitrosylhemoglobin is indicated.

However in the substantia nigra, the spectra were dominated by nitrosyl heme signals resembling those of Hb-NO (Fig. 3). The samples from Parkinson's disease brain showed a characteristic form which indicated a further nitrosyl-heme species. These results were attributed to the production of NO by microglial cells. Further work should establish whether this is a cause or effect of the disease process.

4. Biochemical Applications of Pulsed EPR Spectroscopy

- Relaxation times - determination of T_1 and T_2. These provide exact estimates of the spin-label motion. Distances between paramagnets in proteins can also be estimated [9]. (Fig. 1).
- Flash photolysis - investigations of photosynthetic electron transfer. The spectra of the intermediates consist mainly of spin-correlated radical pairs.
- Electron spin-echo envelope modulation (ESEEM): an effective method to measure hyperfine interactions with quadupolar nuclei such as ^2H and ^{14}N.

ESEEM provides detailed information about the hyperfine couplings from nuclei in the environment of a specific centre. Generally the method has been applied to simple chemical systems or purified biochemical materials, but we investigated whether it can be applied to individual species in a complex biological system. We chose the iron-sulfur clusters of the mitochondrial respiratory chain, a system of at least 20 electron carriers which can be paramagnetic in certain oxidation states.

Overlap between spectra complicates the use of ESEEM and ENDOR on unpurified systems. It is more difficult to select the paramagnet of interest than in continuous-wave EPR since it is not possible to use power saturation of signals.

Careful control of sample preparation and measurement conditions can help to focus on the spectrum of the species of interest. By choice of the source of material, or by genetic engineering methods, the sample can be enriched in the protein of interest. Control of the redox state of the sample may render the centre of interest selectively paramagnetic. Choice of the characteristic g-factor of the cluster can select one species in a sample, and, in fact a sub-set of orientations for an anisotropic species.

We used these methods to observe ESEEM spectra of the Rieske clusters of the mitochondrial respiratory chain. These clusters were believed, on spectroscopic evidence, to have co-ordination by two histidine and two cysteine residues instead of four cysteines. This proposed structure has recently been supported by x-ray crystallography. .

4.1 Rieske clusters in Complex III of the respiratory chain

Rieske [2Fe-2S] clusters are involved in the interaction of Complex III (ubiquinol: cytochrome *c* reductase) with quinones in the membrane. Quinone inhibitors such as 5-*n*-undecyl-6-hydroxy-4,7-dioxobenzothiazole (UHDBT) alter the EPR signal of the Rieske protein, and the question arose as to whether this implied a change in the co-ordination environment of the cluster. The effect of quinones on the EPR spectrum is lost on isolation of the proteins, so purified protein could not be used in the investigation.

The ESEEM spectrum of the Rieske protein was detected by using concentrated mitochondrial membranes from beef heart, treated with ascorbate so that the Rieske clusters were selectively reduced [10]. ESEEM spectra of the Rieske protein (Fig 4Aa) revealed a number of nuclear frequencies in the range 1 - 8 MHz, attributed to ^{14}N nuclei weakly coupled to the cluster. Two pairs of resonances could be attributed to two nitrogens which were potential ligands to the cluster. The spectra were similar to those previously observed in purified Rieske proteins [11].

Addition of UHDBT to the membranes did not significantly alter the ESEEM pattern (Fig 4A(a), dotted line), indicating that the quinone compound does not distort the nitrogen co-ordination to the clusters.

4.2 Benzene Dioxygenase

This enzyme, from *Pseudomonas putida*, is important in bioremediation of polluted environments. It also has potential applications in asymmetric synthesis and in biosensors. Its EPR spectrum is characteristic of Rieske iron-sulfur proteins, although the redox potentials are quite different. 1-dimensional ESEEM of the catalytic iron-sulfur protein of benzene dioxygenase (ISP_{BED}) showed the presence of ^{14}N resonances similar to those in the mitochondrial Rieske proteins [12] (Fig. 4Ab).

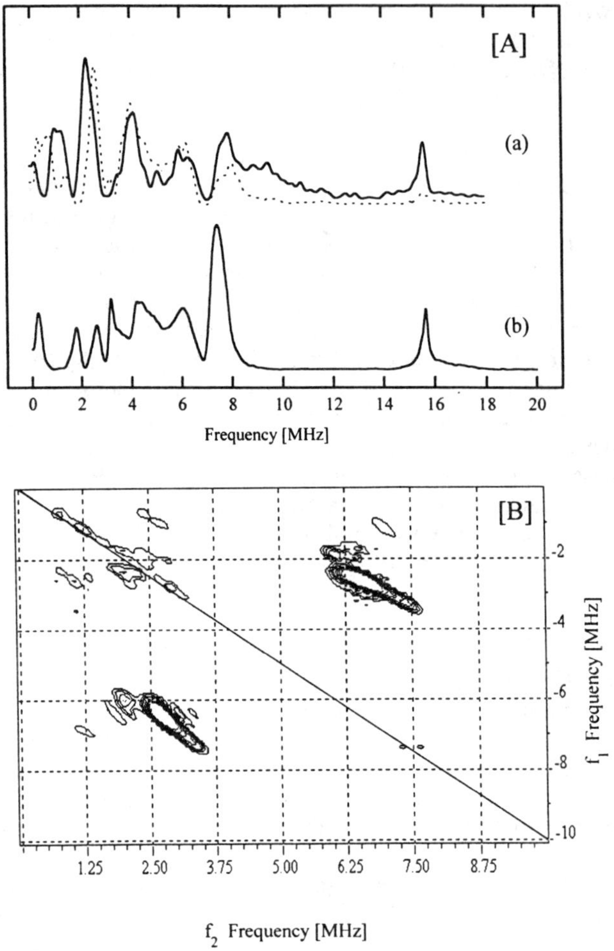

Figure 4. ESEEM spectra of ^{14}N in Rieske-type [2Fe-2S] clusters [10, 14].
[A] Fourier-transformed 3-pulse ESEEM spectra, recorded by the three-pulse stimulated-echo sequence. (a) the Rieske protein in bovine heart mitochondrial membranes. The spectrum with the dotted line is from membranes treated with UHDBT. Measurement conditions: temperature 3.7 K, B_0 field 367.61 mT, microwave frequency, 9.72 GHz, τ - value, 112 ns, pulse width for a $\pi/2$ pulse, 16 ns. (b) purified ISP_{BED} from *P. putida*. The B_0 field was 365.1 mT, microwave frequency, 9.77 GHz.
[B] 2-dimensional HYSCORE spectrum of the [2Fe-2S] cluster of ISP_{BED}, recorded at $g_y = 1.914$. 256 steps of f_2, f_1 were taken at 16 ns intervals, and Fourier transformed in both dimensions. The contour representation is from the $[+f_2, -f_1]$ quadrant. The two pairs of crosspeaks at [(2.7),(-6.4), (6.5, -2.6)] and [(3.3),(-7.3), (7.2, -3.3)] MHz correspond to the $\nu_{dq}^{\pm}$ frequencies of the two ^{14}N nuclei coordinating the [2Fe-2S] cluster.

The 2-dimensional EPR spectroscopic technique, hyperfine sublevel correlation spectroscopy (HYSCORE) provides a means of assigning resonances which are associated with a particular nuclear hyperfine interaction [13].

For the ESEEM spectra of the Rieske proteins there is an uncertainty about the assignment of the resonances to individual nitrogen nuclei, and this information is needed to determine the coupling parameters of the nuclei.

2D- spectra of the Rieske protein in ISP_{BED} (Fig. 4B) indicated the hyperfine couplings of three nitrogen nuclei to the cluster, two with strong couplings (4.8 and 3.6 MHz) and quadrupole couplings typical of imidazole nitrogens (2.3 and 2.5 ± 0.2 MHz, respectively) and one with a weaker coupling (1.1 MHz) and quadrupole coupling (3.3 MHz) typical of a distant peptide nitrogen [12].

The amino-acid sequence of the benzene dioxygenase iron-sulfur protein has been determined [12]. In this sequence, cysteines 96 and 116, and histidines 98 and 119 have been identified as likely ligands to the [2Fe-2S] cluster. The protein has been cloned and modified by sequence-directed mutagenesis. Substitution of histidine 119 by cysteine led to loss of the Rieske cluster. Substitution of histidine 98 by cysteine yielded a protein with a highly modified EPR spectrum. In the ESEEM spectrum of the H98C mutant the typical pattern of high-frequency modulations typical of Rieske clusters was absent, indicating a novel coordination environment for the cluster.

5. Concluding Remarks

The exploration of new bio-medical problems presents new challenges to extend the limits of EPR spectroscopy. As has previously occurred in physical chemistry, new principles begin to emerge, and one would like to construct new molecules in order to investigate these principles. In biochemistry of proteins we now have the possibility to construct such subjects for study, by the techniques of genetic engineering.

Acknowledgements

We thank the many colleagues who have contributed to this work, in particular Dr A. Bune, Dr C.S. Butler, Dr H.T. Cook, Dr C.E. Cooper, Prof R.C. Hider, Prof M.N. Hughes, Ms G.R. Lynagh, Dr J.R. Mason, Dr J.B. Porter and Prof A.H.V. Schapira. This work was supported by the Biotechnology and Biological Sciences Research Council the National Kidney Research Council, and Bruker Spectrospin Ltd.

References

1. Cooper, C.E., Lynagh, G.R., Hoyes, K.P., Hider, R.C., Cammack, R. and Porter, J.B. The relationship of intracellular iron chelation to the inhibition and regeneration of human ribonucleotide reductase as studied by EPR spectroscopy. *J. Biol. Chem.*, **271** (1996) 20291-20299.
2. Nyholm, S.; Mann, G. J.; Johansson, A. G.; Bergeron, R. J.; Graslund, A.; Thelander, L. Role of ribonucleotide reductase in inhibition of mammalian cell growth by potent iron chelators. *J. Biol. Chem..* ; **268** (1993) 26200-26205.
3. Moncada, S.; Higgs, E. A. Molecular mechanisms and therapeutic strategies related to nitric oxide. *FASEB J.* 1995; 9(13): 1319-1330.
4. Henry, Y.; Lepoivre, M.; Drapier, J. C.; Ducrocq, C.; Boucher, J. L.; Guissani, A. EPR characterization of molecular targets for NO in mammalian cells and organelles. *FASEB J.* **7** (1993) 1124-1134.

5. Kubrina, L. N.; Caldwell, W. S.; Mordvintcev, P. I.; Malenkova, I. V.; Vanin, A. F. EPR Evidence for Nitric Oxide Production from Guanidino Nitrogens of L-Arginine in Animal Tissues Invivo. *Biochim. Biophys.Acta.* **1099** (1992) 233-237.
6. Bune, A.J., Shergill, J.K., Cammack, R. and Cook, H.T. L-arginine depletion by arginase reduces nitric oxide production in endotoxic shock: an electron paramagnetic resonance study. *FEBS Lett.* **366** (1995) 127-130.
7. Bune, A.J., Brand, M.P., Heales, S.J.R., Shergill, J.K., Cammack, R. and Cook, H.T. Inhibition of tetrahydrobiopterin synthesis reduces in vivo nitric oxide production in experimental endotoxic shock. *Biochem. Biophys. Res. Commun.* **220** (1996) 13-19.
8. Shergill, J. K., Cooper, C.E., Cammack, R., Mann, V., Cooper, J.M. and Schapira, A. H. V. Detection of nitrosyl complexes in human substantia nigra, in relation to Parkinson's Disease. *Biochem. Biophys. Res. Commun.* **228** (1996) 298-305.
9. Cammack, R., Williams, R., Guigliarelli, B. More, C. and Bertrand, P. (1994) Long-distance spin-spin interactions with Fe-S clusters as observed by EPR spectroscopy. *Biochem. Soc. Trans.* **22**, 721-725.
10. Shergill, J.K. and Cammack, R. ESEEM and ENDOR studies of the Rieske iron-sulphur protein in beef heart mitochondrial membranes. *Biochim. Biophys. Acta,* **1185** (1994) 35-42.
11. Britt, R. D.; Sauer, K.; Klein, M. P.; Knaff, D. B.; Kriauciunas, A.; Yu, C. A.; Yu, L.; Malkin, R. Electron spin echo envelope modulation spectroscopy supports the suggested coordination of two histidine ligands to the Rieske Fe-S centers of the cytochrome b_6f complex of spinach and the cytochrome bc_1 complexes of *Rhodospirillum rubrum, Rhodobacter sphaeroides* R-26, and bovine heart mitochondria. *Biochemistry.* 30 (1991) 1892-1901.
12. Mason, J. R., Butler, C.S., Cammack, R. and Shergill, J.K. Structural studies on the catalytic component of benzene dioxygenase from *Pseudomonas putida. Biochem.Soc. Trans.* **25** (1996) 90-95.
13. Höfer, P. Distortion-free electron spin-echo envelope modulation spectra of disordered solids obtained from 2-dimentsional and 3-dimensional HYSCORE experiments. J. Magn. Reson. Ser. A **111** (1994) 77-86.
14. Shergill, J.K.; Joannou, C.L., Mason, J.R. Cammack, R. Coordination of the Rieske-type [2Fe-2S] cluster of the terminal iron-sulfur protein of *Pseudomonas putida* benzene-1,2-dioxygenase, studied by 1D- and 2D-ESEEM spectroscopy. *Biochemistry,* **34** (1995) 16533-16542.

Comparison Study of Different EPR Dosimeter Materials

D. Regulla [1], A. A. Romanyukha [2] and E.A. Ignatiev [2]

[1] GSF-National Research Center for Environment and Health, Institute of Radiation Protection, D-85764 Neuherberg, Germany
[2] Institute of Metal Physics, Russian Academy of Sciences, Ekaterinburg, 620219, Russia

Abstract. The main dosimetric properties of EPR dosimetry using alanine, synthetic hydroxyapatite with different carbonate content, and polytetrafluoroethylene are compared, problems and perspectives of the EPR dosimetry discussed.

Introduction

To the present time two types of dosimetric methods have achieved wide spread acceptance in the world, i.e. the conventional ionization chamber dosimetry and, more recently, the thermoluminescence dosimetry (TLD). TLD has certain perspectives for reasons of a small detector volume down to 1 mm^3 or layers of 0.1 mm thickness, precision of about 0.3 % (for one relative standard deviation), the remarkable dynamics of the dose-effect relationship from 0.5 μGy to more than 100 Gy and easy automation of the readout process. Other available dosimetry methods, such as Fricke dosimetry and calorimetry, have more limited application and serve for special tasks, e.g. calibration and reference dosimetry. EPR dosimetry as presented here is one of the modern alternatives in dosimetry, with potential spin-offs for application in high-dose standardization, routine high-level dosimetry, radiation processing dosimetry, therapy level dosimetry and retrospective dosimetry. The main advantages of this new technique are an also dynamic dose range but shifted to the upper part of the dose scale, till 10^9 Gy (Wieser *et al.*, 1990), the archival character for dose information based on a non-destructive readout that allows for repeated EPR measurements, and the tissue equivalency of certain EPR detector materials, e.g. alanine (Regulla and Deffner, 1982); the latter property makes the method applicable also in high-energy electron dosimetry without the impact by perturbation and displacement effects as known from ionization chamber dosimetry. The relatively high costs of the EPR spectrometer must be considered as a disadvantage of EPR dosimetry. However, during the recent past new and relatively lower-cost EPR spectrometers became available or are in view, with specifications well acceptable for the dosimetric purposes (Maier and Schmalbein, 1993, Nakanishi *et al.*,1993, Kojima *et al.*, 1993).

As a consequence EPR dosimetry with alanine detector samples became a standard dosimetry method in Europe, USA, Japan and some other countries world-wide. It is one of the very few examples where EPR spectroscopy has been accepted outside of its pure scientific application. The fields of application of alanine/EPR dosimetry became wide-spread, for instance in quality assurance and transfer dosimetry, organized by national and international bodies, e.g. by the International Atomic Energy Agency, Vienna (Nam and Regulla, 1989).

In spite of the above mentioned attractive properties of alanine as a detector material, the search for materials that provide a methodological progress and expansion is continued, especially for the range of radiotherapy-level doses. The scope of the present work is therefore to investigate the dosimetric properties and particularly the lower detection threshold of (i) synthetic hydroxyapatite with different carbonate content, known from tooth enamel dosimetry (Romanyukha *et al.*, 1996) and (ii) easily available polytetrafluoroethylene, and compare them with the ones of alanine.

Alanine dosimetry

The chemical formula of alanine is $CH_3{\cdot}CH(NH_2){\cdot}COOH$; the stable free radicals result from the induced by radiation loss of NH_2 groups. Bradshaw *et al.* first suggested the use of alanine as an EPR radiation detector in 1962. The alanine is sensitive to different types of radiation, e.g. alpha, beta and gamma radiation, as well as electrons, x-rays, neutrons and protons. The radiation induced effect has been developed to a method of high precision dosimetry at the GSF since the late seventieth (Regulla and Deffner, 1982), and later at other laboratories.

The alanine/EPR dosimetry is a relative method providing calibration. For correlation with the dose quantity used for calibration, the peak-to-peak amplitude of the central peak of the first derivative of the EPR absorption spectrum of alanine is used. The resulting calibration function of the EPR amplitude-versus-dose is linear up to almost 10 kGy and can be transfered between different EPR laboratories (Regulla *et al.*, 1993). Most of the GSF developments on alanine/EPR dosimetry has successfully been performed using the X-band EPR spectrometers Varian E9 as well as later Bruker ESP 300 and EMS 104 (Maier and Schmalbein, 1993).

During roughly 35 years passed quite a number of publications dealt with alanine dosimetry were published and different sample compositions proposed. Since the alanine is taken analytical-grade from Fluka, Merck or Serva, the difference consists mainly in the binder (paraffin, polyolefins, polyethylene etc. which adds to alanine), in the shape of dosimeter (cylinder, pellet, film etc) and in size. High quality alanine dosimeters were developed by Regulla and Deffner in 1982 with the samples consisting of a blend of, by mass, 80 % commercially available L-alanine and 20 % paraffin. Its shape was cylindrical with 4.9 mm in diameter and 10 mm in length, its mass 250 mg. A pill-press is used to produce such dosimeters. The GSF alanine/EPR dosimetry method became subject of a standard of the American Standards for Testing and Materials (ASTM, 1994)

A number of intercomparison studies were carried out with the GSF alanine detectors to determine the reliability, uncertainty and confidence level (Regulla *et al.*, 1993; Feist *et al.*, 1993). As a result of comparative measurements with ionzation chamber and thermoluminescence dosimeters it could be shown that a coefficient of variation of lower than 0.5 % is achievable with the alanine/EPR dosimetry in the operational dose range; here 30 Gy absorbed dose in water were chosen. The standard operational dose range of alanine/EPR dosimetry is about 5 Gy - 50 kGy. As was estimated by Wieser *et al.* in 1993, the lower limit of detection of the alanine samples is expected to be roughly 1 mGy for EPR spectrometers with a standard sensitivity of $8 \cdot 10^{13}$ spins/T. Unfortunately, there are non-radiation induced components in the EPR spectrum of alanine and their contributions are equivalent to doses of about several hundred mGy. An application of alanine/EPR dosimeters at lower doses was shown to be possible but required fast Fourier transform signal treatment; this technique allowed for a detection limit as low as about 50 mGy, at the present state of development (Ruckerbauer *et al.*, 1996).

Hydroxyapatite dosimetry

The idea to use carbonated hydroxyapatite, $Ca_{10}[(PO_4)_{6-x}(CO_3)_x][(OH)_{2-2y}(CO_3)_y]$ as an EPR dosimeter results from retrospective EPR dosimetry with tooth enamel and bones. Carbonated hydroxyapatite is the mineral component of above biological tissues. The indices x and y in the chemical formula of carbonated hydroxyapatite characterize the carbonate concentration. As known from retrospective dosimetry (see for instance, Romanyukha and Regulla, 1996) tooth enamel is sensitive to ionizing radiation such as photons, electrons, α-particles and heavy ions in a dose range from about 100 mGy to above 100 Gy, with constant EPR response. The radical involved in EPR dosimetry with hydroxyapatite is CO_2^-; it is long-lived with a life time of 10^7 years, for storage at 25°C. These dosimetric properties have been evaluated for natural hydroxyapatite. Its synthetic production is difficult. Only few publications report on the EPR properties of synthetic apatites (Callens *et al.*, 1989, 1991; Ishii and Ikeya, 1993; Romanyukha *et al.*, 1996).

Romanyukha *et al.*, 1996 prepared synthetic apatites with different carbonate content by hydrolyzing monetite, $CaHPO_4$, in a $NaCO_3$ solution at a temperature of 95°C. The carbonate concentration in apatite was regulated by variation of the $NaCO_3$ content in the solution. Three different concentrations of $NaCO_3$ were examined in this investigation, i.e. 0.01, 0.1 and 0.25 mol/l. For irradiation different types of radiation sources were used: A ^{60}Co source from medical therapy and an x-ray tube from diagnostics (60 kV, 40 mA) at the Secondary Standard Dosimetry Laboratory of GSF (Nahrstedt, 1995) as well as an ultrviolet (UV) light source (NK 6/12, produced by Heraeus). The UV lamp is a low-pressure mercury lamp with an electric power of 6 W and the major spectral line at 254 nm. The hydroxyapatite samples were investigated in the physical shape of powder and pellets.

Fig. 1 depicts the EPR spectra for the sample with 0.1 mol/l of $NaCO_3$. The ^{60}Co gamma dose was 5 Gy, the x-ray dose 2 Gy. The dose from the UV lamp could not be determined, but irradiation was continued during 2 hours in 4 cm distance. As shown in Fig. 1, ^{60}Co gamma radiation produces the same type of radicals in synthetic hydroxyapatite as it does in tooth enamel ($g = 2.0008$), whereas UV and x-rays induce two signals which have an isotropic shape and are distinguished from ^{60}Co by the g-value and linewidth. The signal of UV-exposed samples at the lower magnetic field ($g = 2.0117$) decayed completely after two weeks at room temperature. For x-ray irradiated samples this decay period may even be shorter. The signal with $g = 2.0008$ however was stable during one month of observation, under the same temperature condition.

The EPR spectra of synthetically carbonated apatites, after x-ray irradiation, have been studied in details by Callens *et al.*, in 1989 and 1991. They found five different EPR signals, caused by five different radicals: (1) An anisotropic signal with $g_x = 2.0406$, $g_y = 2.0330$, $g_z = 2.0276$, $\Delta B = 4.9$ G (surface O^-), denoted by A1; (2) an isotropic signal with $g_x = 2.0115$, $\Delta B = 3.3$ G (CO_3^- at phosphate lattice site), denoted by I2; (3) an anisotropic signal with $g_x = 2.00170$, $g_y = 2.00084$, $g_z = 2.0060$, $\Delta B = 3.4$ G (surface CO_3^-), denoted by A2; (4) an isotropic signal with $g = 2.0006$, $\Delta B = 3.0$ G (CO_2^- at phosphate lattice site), denoted by I3; (5) an anisotropic signal with $g_x = 2.0030$, $g_y = 2.0015$, $g_z = 1.9970$, $\Delta B = 1.8$ G (surface CO_2^-), denoted by A3. In this terminology, irradiations with x-rays and UV light mainly produce I2 and I3 signals, whereas irradiations with ^{60}Co gamma rays produce A3 type signals, in the presently investigated synthetic hydroxyapatite. The latter means that UV and x-rays generate effects on phosphate lattice sites, whereas ^{60}Co gamma rays generate effects on the surface. This finding is probably due to the different ionization energies of the two defects (bulk and surface) and needs additional studies. The question rises why such difference in effects of UV, x-ray and ^{60}Co is not observed in tooth enamel. The difference can probably explained by the crystalline structure of hydroxyapatites in tooth enamel and dentine, as compared with the synthetic powder samples.

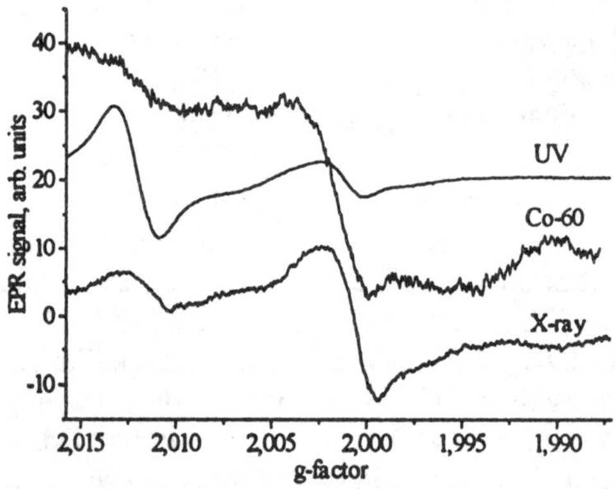

Fig. 1 EPR spectra of synthetic carbonated hydroxyapatite irradiated with different radiation types

In spite of the problems with the interpretation of the origin of radiation effects, the synthetically carbonated hydroxyapatites turned out suitable for EPR dosimetry. The lower detection limit is close to 1 Gy, for ^{60}Co gamma irradiation. This figure can be improved by optimizing the conditions of chemical synthesis especially due to the carbonate content. The radiation sensitivity of synthetic hydroxyapatite is considerably higher for x-rays and UV radiation. The finding of different EPR spectra in synthetic apatites after irradiation with ^{60}Co, x-rays and UV allows to distinguish between the type of radiation source. Moreover, the materials investigated seem to be very promising for UV dosimetry. The existence of a non-stable line in the spectrum of synthetic hydroxyapatite after irradiation with x-rays and UV can be used to estimate the time elapsed between irradiation and measurement. Concluding synthetically carbonated hydroxyapatites is a potential candidate for ESR dosimetry and may serve as a model substance for center structures and mechanisms in biophysical tooth and bone dosimetry.

Polytetrafluoroethylene dosimetery

Polytetrafluoroethylene (PTFE) has intensively been studied by EPR spectroscopy gamma ray and electron dosimetry since the late sixtieth (Judeikis et al., 1968; Siegel and Hedgpeth, 1967; Tamura *et al.*, 1970). When PTFE is irradiated and analyzed by EPR spectroscopy in vacuum and at room temperature, chains of $RCF_2C{\cdot}FCF_2R$ and propagating $RCF_2CF{\cdot}$ radicals are formed. When irradiation is performed in air, chain peroxide radicals $RCF_2CFOO{\cdot}CF_2R$ and propagating peroxide radicals $RCF_2CF_2OO{\cdot}$ are stabilized. The relative concentrations of the two radicals depend on the dose, with the chain peroxide radical dominating at doses $< 10^5$ Gy (Judeikis *et al.*, 1968). The EPR spectrum of peroxide radicals exhibits an asymmetric singlet (Fig. 2) with a g-value $g \approx 2.016$ that is close to the g-value of the propagating radical signal. The radical yield (G-value) of peroxide radicals is about 0.18 (Tamura *et al.*, 1970), whereas the G-value of alanine is about 1 (Ikeya, 1993). Some authors observed an essential radical yield decrement from about 0.22 to about 0.12 with the dose increasing from 10^3 to 10^5 Gy (Judeikis *et al.*, 1968).

The applicability of PTFE dosimeters was evaluated in the range 1 Gy - 10 MGy (Judeikis *et al.*, 1968). Above 1 MGy the concentration of the radicals is approaching saturation. Linear fitting of amplitude dependence is permissible for doses below 1 MGy and powder samples (Tamura *et al.*, 1970). Non-linear radical concentration-versus-dose dependence ($R \sim D^{0.88}$) was observed for PTFE pieces (Teflon 7) (Judeikis *et al.*, 1968).

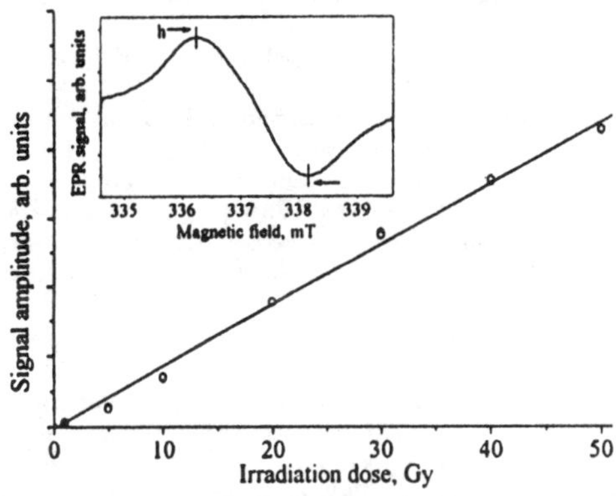

Fig. 2. Dose dependence of the EPR signal of PTFE (Bruker ESP 300, microwave power 8 mW, modulation field 1 mT (100 kHz). Insert: EPR spectrum of PTFE irradiated with a dose of 50 Gy.

The applicability of PTFE dosimeters was studied for the therapy dose range from 1 to 50 Gy. The PTFE samples (Teflon 4) were in the form of rods with 5 mm diameter and 10 mm length. The dosimeters were irradiated with at the medical ^{60}Co gamma-ray source. Fig. 2 depicts the dependence of the EPR signal amplitude versus absorbed dose. Square deviation of points from the straight fitting line is about 1.3 Gy, whereas that for alanine dosimeters investigated is about 0.2 Gy for the same dose range and irradiation conditions. It may be explained both by non-linear amplitude-versus-dose dependence for PTFE samples and by their relatively low radiation sensitivity.

The advantages of PTFE over alanine are the high mechanical and chemical rigidity as well as the broad temperature range of operation (up to 150°C). Both materials are characterized by a relatively low energy dependence of response (Tamura *et al.*, 1970). However PTFE, during the first days after exposure, shows an essential fading of the radiation induced EPR signal which effect is inversely proportional to the dose applied. Therefore it is necessary to pre-anneal PTFE at 50-100°C before measurement or wait up to one week before readout (Judeikis *et al.*, 1968). Another disadvantage of polytetrafluoroethylene is a non-negligible temperature dependence of the radical generation rate (Tamura *et al.*, 1970) and an essential decrement of radical yield at irradiation for some kinds of PTFE (Judeikis *et al.*, 1968).

References

ASTM-Standard E10.01-F. Standard practice for use of the alanine / ESR dosimetry system. (1994)

W.W. Bradshaw, D.G. Cadena, G.W. Crawford and H.H.W. Spetzler. The use of alanine as a solid dosimeter. *Radiat. Res.* **17**, (1962) 11-21.

F.J. Callens, R.M.H. Verbeeck, D.E. Naessens, P.F.A. Matthys and E.R. Boesman. Effect of carbonate content on the EPR spectrum near g = 2 of carbonated calciumapatites synthesized from aqueous media. *Calcif. Tissue Int.* **44**, (1989) 114-124.

F.J. Callens, R.M.H. Verbeeck, D.E. Naessens, P.F.A. Matthys and E.R. Boesman. Effect of carbonate content and drying temperature on the EPR spectrum near g = 2 of carbonated calciumapatites synthesized from aqueous media. *Calcif. Tissue Int.* **48**, (1991) 249-259.

H. Feist, D. Regulla and A. Wieser. Is alanine/ESR dosimetry now an alternative to ferrous sulphate dosimetry? *Appl. Radiat. Isot.* **44**, (1993) 47-51.

M. Ikeya. New Applications of Electron Spin Resonance. Dating, Dosimetry and Microscopy. (World Scientific, Singapore, 1993).

H. Ishii and M. Ikeya. Defects in synthetic apatite powder and sintered materials. *Appl. Radiat. Isot.* **44**, (1993) 95-100.

H. S. Judeikis, H. Hedgpeth, and S. Siegel. Free radical yields in polytetrafluoroethylene as the basis for a radiation dosimeter. *Radiat. Res.* **35**, (1968), 247-262

T. Kojima, Y. Haruyama, H. Tachibana, R. Tanaka, J. Okamoto, H. Hara and Y. Yamamoto. Development of a portable ESR spectrometer as a reader for alanine dosimeters. *Appl. Radiat. Isot.* **44**, (1993) 361-366.

D. Maier, D. Schmalbein. A dedicated EPR analyzer for dosimetry. *Appl. Radiat. Isot.* 44, (1993) 345-350.

U. Nahrstedt. The GSF Secondary Standard Dosimetry Laboratory for Photon and Beta Radiation. GSF-National Research Center for Environment and Health, Neuherberg, Germany, GSF-Report 9/95 (1995).

A. Nakanishi, N. Sugahara and A. Furuse. Portable ESR spectrometer with Neomax (Nd-Fe-B) permanent magnet circuit. *Appl. Radiat. Isot.* **44**, (1993) 357-360.

W.J. Nam, D.F Regulla. The significance of the International Dose Assurance Service for radiation processing. *Appl. Radiat. Isot.* **40**, (1989) 953-956

D. Regulla, U. Deffner. Dosimetry by ESR spectroscopy of alanine. *Appl. Radiat. Isot.* **33**, (1982) 1101-1114.

D.F. Regulla, A. Bartolotta, U. Deffner, S. Onori, M. Pantoloni and A. Wieser. Calibration network based on alanine/ESR dosimetry. *Appl. Radiat. Isot.* **44**, (1993) 23-31

A.A. Romanyukha, A. Wieser and D. Regulla. EPR dosimetry with different biological and synthetic carbonated materials. *Radiat. Prot. Dosim..* **65**, (1996) 389-392

F. Ruckerbauer, M. Sprunck and D.F. Regulla. Numerical signal treatment for optimized alanine/ESR dosimetry in the therapy-level dose range. *Appl. Radiat. Isot.* **47**, (1996) 1263-1268

S. Siegel and H. Hedgpeth. Chemistry of irradiation induced polytetrafluoroethylene radicals: I. Re-examination of the EPR spectra. *J. Chem. Phys.* **46**, (1967) 3904-3912

N. Tamura, Y. Oshima, K. Yotsumoto and H. Sunaga. Radiation dosimeter using peroxy radicals in irradiated polytetrafluoroethylene. *Japan. J. Appl. Phys.* **9**, (1970) 1148-1153

A. Wieser and D.F. Regulla. Ultra high-level dosimetry by ESR spectroscopy of crystalline quartz and fused silicate. *Radiat. Prot. Dosim.* **34**, (1990) 291-294

A. Wieser, C. Lettau, U. Fill and D.F. Regulla. The influence of non-radiation induced ESR signal from paraffin-alanine probes for dosimetry in the radiotherapy dose range. *Appl. Radiat. Isot.* **44**, (1993) 59-65.

Comparison of the Effect of High-spin and Low-spin Fe(III) on Nitroxyl T_1 in a Spin-labeled Porphyrin

Margaret H. Rakowsky, Gareth R. Eaton, and Sandra S. Eaton

Department of Chemistry and Biochemistry, University of Denver, Denver, Colorado 80208 USA

Abstract. Long-pulse saturation recovery was used to measure spin-lattice relaxation rates, $1/T_1$, for the nitroxyl radical in a metal-free spin-labeled porphyrin, and in complexes of that porphyrin with low-spin and high-spin Fe(III). Between about 10 and 120 K high-spin Fe(III) (S = 5/2) had a greater impact than low-spin Fe(III) (S = 1/2) on the nitroxyl relaxation rate, but the difference was not as large as would be predicted if *only* the larger value of S and the faster electron spin relaxation rates for the high-spin Fe(III) than for low-spin Fe(III) were considered. The discrepancy is attributed to the importance of terms with the difference in resonance frequency for the fast-relaxing (iron) and slowly-relaxing (nitroxyl) spins in the denominator. For most of the metal transitions this difference is much smaller for low-spin iron than for high-spin iron, which enhances the effect of the low-spin iron relative to high-spin iron at the same metal relaxation rate.

1 Introduction

One of our long-term goals is development of methods to determine distances between paramagnetic centers by EPR. Electron spin relaxation rates are sensitive to spin-spin interaction over longer distances than are line shapes, which provides, in principle, the possibility to determine longer interspin distances. To test methods for determining interspin distances from changes in relaxation rates in systems with inequivalent paramagnetic centers we are first examining systems in which interspin distances are well-defined.

In the early NMR literature Bloembergen [1] and Abragam [2] presented expressions for the impact of a rapidly-relaxing spin on the relaxation of a more slowly-relaxing spin to which it is spin-coupled via dipolar and/or exchange (scalar) interaction. Their approach was applied to EPR of immobilized samples by Likhtenshtein and co-workers [3] and by Hyde and Rao [4], using eq. (1) in which the labels B, C, and E refer to the components of the spin-spin interaction Hamiltonian that give rise to each term.

$$\frac{1}{T_{1s}}=\frac{1}{T_{1s}{}^{o}}+S(S+1)[\underset{\text{B term}}{\frac{b^2T_{2f}}{1+(\omega_f-\omega_s)^2T_{2f}{}^2}}+\underset{\text{C term}}{\frac{c^2T_{1f}}{1+\omega_s{}^2T_{1f}{}^2}}+\underset{\text{E term}}{\frac{e^2T_{2f}}{1+(\omega_f+\omega_s)^2T_{2f}{}^2}}]$$

$$b^2=\frac{8}{3}[-\frac{J}{2}-\frac{1}{4}g_sg_f\beta^2\frac{(1-3\cos^2\theta)}{\hbar r^3}]^2$$

$$c^2=3g_s{}^2g_f{}^2\beta^4\frac{\sin^2\theta\cos^2\theta}{\hbar^2r^6}$$

$$e^2=3g_s{}^2g_f{}^2\beta^4\frac{\sin^4\theta}{\hbar^2r^6} \qquad (1)$$

where "f" and "s" denote the fast- and slowly-relaxing spins, respectively, $T_{1s}{}^{o}$ is T_1 for the slowly-relaxing spin in the absence of spin-spin interaction, T_{1s} is T_1 for the slowly-relaxing spin perturbed by the fast-relaxing spin, S is the electron spin on the faster-relaxing center, ω_f and ω_s are the resonant frequencies for the fast- and slowly-relaxing spins, respectively, r is the interspin distance, J in units of s^{-1} is the electron-electron exchange interaction for the Hamiltonian written as $-JS_1 \cdot S_2$, and θ is the angle between the interspin vector and the external magnetic field.

Subsequent applications in EPR have been reviewed [5]. Early work used continuous wave (CW) power saturation curves to determine the relaxation times for the slowly relaxing partner. More detailed information can be obtained via analysis of the full shape of a saturation recovery curve than by power saturation. In an analysis of the effect of low-spin Fe(III) on nitroxyl saturation recovery curves we proposed that the B term should be replaced by eq. (2) to avoid a singularity that could otherwise occur when $\omega_f - \omega_s$ goes to zero and to ensure that in the limit of strong interaction this contribution approaches $1/T_{1f}$.

$$\frac{b^2T_{2f}}{1+(\omega_f-\omega_s)^2T_{2f}{}^2+b^2T_{1f}T_{2f}} \qquad (2)$$

The use of this modified version of eq. (1) with experimental values of T_{1f} and T_{2f} and inclusion of known g anisotropy permitted determination of interspin distances for spin-labeled low-spin Fe(III) porphyrins [6] and spin-labeled methemoglobin [7] that were in good agreement with known structures of the compounds.

We now seek to compare the effects of high-spin Fe(III) with those of low-spin Fe(III) for approximately the same iron-nitroxyl distance. The comparison samples were obtained by changing the axial ligand on an Fe(III) porphyrin from chloride, which gives a 5-coordinate high-spin complex with zero-field splitting

(ZFS) about 7.0 cm^{-1} (see references in [6]), to 1-methyl-imidazole (MeIm) which gives a 6-coordinate low-spin complex. Since the porphyrin provides a relatively rigid molecular geometry we anticipate that the iron-nitroxyl distance will be approximately the same in the high-spin and low-spin complexes, thereby permitting examination of the effect of changing properties of the metal at constant interspin distance. Data were obtained in ca. 1 mM solutions that formed glasses upon cooling, so average intermolecular spin-spin distances were much longer than the pairwise interspin distance within the complex, and consequently the analysis considers only the pairwise intramolecular interaction.

2 Experimental Section

FeTTPCl (iron(III) tetratolylporphyrin chloride), H_2(*trans*-TPP-SL) (5,10,15-20-tetraphenyl-2-(3-(N-(2,2,6,6-tetramethyl-4-piperidinyl-1-oxy)acrylamido))-porphyrin), Fe(*trans*-TPP-SL)$(MeIm)_2^+$, and Fe(*trans*-TPP-SL)Cl were prepared as reported previously [6]. Samples for EPR study were ca. 1 mM in toluene or 2:1 toluene:chloroform and were degassed by multiple cycles of freeze-pump-thaw degassing.

A locally-constructed spectrometer [8] operating at 9.2 GHz was used for the saturation recovery (SR) experiments. Temperatures between 5 and 70 K were obtained with liquid helium and an Oxford ESR900 flow cryostat. Temperatures between 80 and 120 K were obtained with liquid nitrogen and a Varian dewar flow assembly. To test for spectral diffusion the pump time was increased until further increase gave no detectable change in the recovery time constant. Pump times used for data collection were $>T_1$. Typically 5000 to 10000 recovery curves were signal-averaged in a LeCroy digital oscilloscope or with an EG&G 9825 digitizer and stored in a PC. SR data for the spin-labeled porphyrins were recorded in the center of the nitroxyl spectrum.

Direct measurements of T_1 and T_2 for the low-spin Fe(III) by inversion recovery or electron spin echo (ESE) at temperatures up to about 20 K were reported by Rakowsky et al. [6]. Analogous measurements for the high-spin Fe(III) were performed at temperatures between 4.5 and 6 K at several magnetic fields including fields that correspond to the principal axes. The observed temperature dependence of T_1 and T_2 below 6 K suggested that above about 15 K $T_1 \sim T_2$ for the high-spin Fe(III). At higher temperatures the iron relaxation rates were determined by simulation of the CW line shapes as a Gaussian distribution of Lorentzian spin packets characterized by T_2. The width of the Gaussian distribution was assumed to be independent of temperature. The relaxation times obtained from the CW lineshapes were the same for the parallel and perpendicular axes.

3 Results and Discussion

Due to the orientation dependence of $\omega_f - \omega_s$ and of the spin-spin interaction (equation (1)) the SR curves for the nitroxyl in the spin-coupled systems are not single exponentials. However, to give a qualitative comparison of the effects of the iron on the nitroxyl relaxation rate, values of T_1 displayed in figure 1 were obtained by single-exponential fits to the experimental SR data. Throughout the temperature interval 10 to 120 K the nitroxyl relaxation rates were faster for high-spin Fe(*trans*-TPP-SL)Cl than for low-spin Fe(*trans*-TPP-SL)(MeIm)$_2^+$. The ratio of the nitroxyl relaxation rates for nitroxyl in the high-spin and low-spin complexes was about 20 at 20 K and about 2 at 100 K. If the factor of S(S+1) in eq. (1) were the dominant difference between the high-spin (S = 5/2) and low-spin (S = 1/2) cases one would

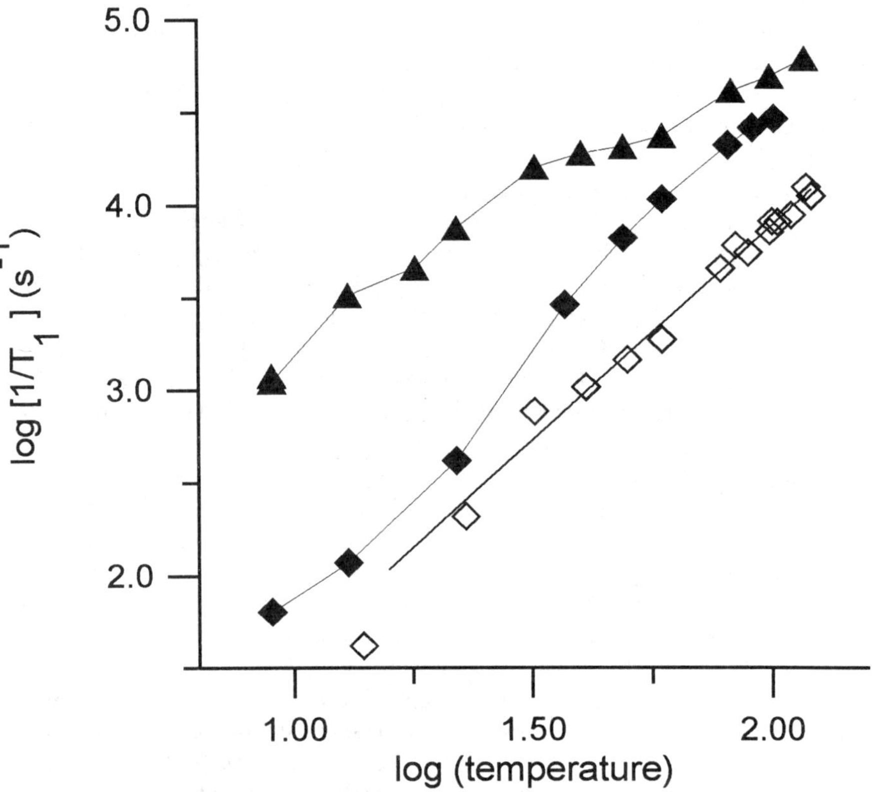

Fig. 1. Nitroxyl relaxation rates as a function of temperature for (◇) H_2(*trans*-TPP-SL) in toluene solution, the least-squares fit line is $\log(1/T_1) = 2.33 \log(T) - 0.76$, (◆) low-spin Fe(*trans*-TPP-SL)(MeIm)$_2^+$ in 2:1 toluene:chloroform, and (▲) high-spin Fe(*trans*-TPP-SL)Cl in toluene. The lines through the data for the Fe(III) complexes connect the data points. Analysis of the data for Fe(*trans*-TPP-SL)(MeIm)$_2^+$ gave an iron-nitroxyl interspin distance of 13 Å [6].

Table 1. Relaxation Times for High-spin and Low-spin Fe(III) in Fe(TPP)X

X	20 K	100 K
$(MeIm)_2$, low-spin	$T_1 = 3 \times 10^{-6}$ s $T_2 = 3 \times 10^{-7}$ s	$T_1 \sim T_2 = 2 \times 10^{-9}$ s
Cl, high-spin	$T_1 \sim T_2 = 5 \times 10^{-9}$ s	$T_1 \sim T_2 = 2 \times 10^{-10}$ s

have expected a temperature-independent ratio of 11.7. However, the experimental data (Figure 1) exhibit a temperature dependence of the relative effects of high-spin and low-spin Fe(III).

A major contribution to the difference in effects of the high-spin and low-spin Fe(III) is the difference in iron relaxation times. Relaxation time data for FeTTPX at 20 and 100 K are summarized in Table 1. Relaxation times for the iron in the spin-labeled complexes were the same, within experimental uncertainty, as in the iron porphyrin without the spin label. Since the ZFS for the high-spin complexes is much larger than the X-band EPR quantum, only the transitions between the $m_s = \pm 1/2$ levels are observed at X-band so the measured relaxation times are for these transitions.

At 20 K the high-spin Fe(III) relaxes 10^2 to 10^3 times faster than low-spin Fe(III) and at 100 K the high-spin Fe(III) relaxes about 10 times faster than the low-spin Fe(III). If the C or E terms in eq. (1) dominated for both spin states, the product of S(S+1) times T_{1f} or T_{2f} would predict 10^3 to 10^4 times larger effect for high-spin than for low-spin iron at 20 K and about 100 times larger at 100 K. The larger difference between the effects of the two spin states at 20 K than at 100 K is consistent with the data in Figure 1, but the magnitude of the observed difference is smaller at both temperatures than predicted by the C and E terms. Therefore, the effect of the B term is examined in the following paragraph.

For Fe(*trans*-TPP-SL)$(MeIm)_2^+$ the iron g values and corresponding resonance fields at 9.2 GHz are $g_x = 1.50$ (4525 G), $g_y = 2.13$ (3187 G), and $g_z = 2.615$ (2596 G). Since the nitroxyl resonance is centered at 3385 G all of the low-spin iron transitions are within 800 to 1200 G of the nitroxyl in a field-swept experiment. Although the separations would be somewhat different when calculated at constant field, it can be seen that $\omega_f - \omega_s$ in the denominator of the B term is substantially less than ω_s, which appears in the denominator of the C term. Thus, the dependence on $\omega_f - \omega_s$ in the denominator of the B term enhances the contribution from the B term relative to that for the C term for the low-spin complexes. For high-spin Fe(III), calculation of $\omega_f - \omega_s$ for the $m_s = \pm 1/2$ transitions using fictitious $S' = 1/2$ and effective g-values of 2 to 6, gives some orientations with relatively small values of $\omega_f - \omega_s$ and therefore enhanced contributions of the B term relative to the C term although many orientations give large values of $\omega_f - \omega_s$. Furthermore, for the $m_s = \pm 1/2$ to $\pm 3/2$ and $\pm 3/2$ to $\pm 5/2$ transitions $\omega_f - \omega_s$ is of the order of the zero-field splitting which is much larger than the X-band EPR quantum. Thus, the denominator of the B term is much larger than for the C term and the B term for these transitions makes very little contribution to enhanced nitroxyl relaxation. The smaller denominator for the B term for many of the transitions for low-spin Fe(III) than for high-spin Fe(III) partially compensates for the larger values of S(S+1) and T_{2f} for high-spin Fe(III) than for low-spin Fe(III). These differences in the denominator of the B term appear to be the dominant contribution to the smaller difference in relaxation times for nitroxyl in the presence of high-spin Fe(III) than for low-spin Fe(III) than predicted based only on magnetic moments and iron relaxation times.

Although equation (1) is written in a general form that includes S, it does not explicitly include the effects of ZFS on the contributions for various values of the metal m_s for S>1/2. Work is underway to include the effect of ZFS greater than the EPR quantum on the analysis of the effect of high-spin Fe(III) on the nitroxyl relaxation to determine the interspin distance.

Acknowledgments

The financial support of this work by National Institutes of Health Grant GM21156 (G.R.E. and S.S.E.) and National Science Foundation grant CHE8908436 (M.H.R.) is gratefully acknowledged.

References

1. N. Bloembergen, On the interaction of nuclear spins in a crystalline lattice, *Physica* **15** (1949) 386-426.
2. A. Abragam, Overhauser effect in nonmetals, *Phys. Rev.* **98** (1955) 1729-1735.
3. A. V. Kulikov and G. I. Likhtenstein, The use of spin relaxation phenomena in the investigation of the structure of model and biological systems by the method of spin labels, *Adv. Mol. Relax. Interact. Proc.* **10** (1977) 47-79.
4. J. S. Hyde and K. V. S. Rao, Dipolar-induced electron spin-lattice relaxation in unordered solids, *J. Magn. Reson.* **29** (1978) 509-516.
5. S. S. Eaton and G. R. Eaton, Interaction of spin labels with transition metals. Part 2, *Coord. Chem. Rev.* **83** (1988) 29-72.
6. M. H. Rakowsky, K. M. More, A. V. Kulikov, G. R. Eaton, and S. S. Eaton, Time-domain electron paramagnetic resonance as a probe of electron-electron spin-spin interaction in spin-labeled low-spin iron porphyrins, *J. Am. Chem. Soc.* **117** (1995) 2049-2057.
7. V. Budker, J.-L. Du, M. Seiter, G. R. Eaton, and S. S. Eaton, Electron-electron spin-spin interaction in spin-labeled low-spin methemoglobin, *Biophys. J.* **68** (1995) 2531-2542.
8. R. W. Quine, S. S. Eaton, and G. R. Eaton, Saturation recovery electron paramagnetic resonance spectrometer, *Rev. Sci. Instrum.* **63** (1992) 4251-4262.

EPR Tooth Enamel Dosimetry and its Application to Study Radiological Consequences of the Chernobyl Accident

A. Bougai, V. Baryakhtar, M. Zacharash, V. Gaichenko

Institute of Semiconductor Physics, National Academy of Sciences of Ukraine, Ukraine

Abstract. EPR tooth enamel dosimetry is the only experimental method of the retrospective individual dosimetry in accident cases. The international experimental collaboration project on retrospective dosimetry and dose reconstruction in 1993-1994 supported by European Commission resulted in improvement of dosimetry techniques. Advanced technique was used for retrospective dosimetry of Chernobyl victims. The same technique was used for individual dosimetry and radiological studies of wild animals in Chernobyl area.

1 EPR Dosimetry

Dosimetry is an outstanding application of the EPR technique. It is based on the paramagnetism of many radiation induced centers in the solid state. The concentration of such centers usually is a measure of the irradiation dose. Many solid-state substances potentially can be used for EPR dosimetry.

After the 4th International Symposium on ESR Dosimetry and Applications in Munich in 1995 it is clear that the most widely and effectively used are two types of EPR dosimetry: reference alanine EPR dosimetry and accident tooth enamel EPR dosimetry. Alanine EPR dosimetry was first elaborated in Germany by Regulla and Deffner (1982). The use of tooth enamel EPR for the accident dosimetry was first proposed in Japan by Ikeya *et. al.* (1984). Both ideas have been powerfully developed. Alanine EPR dosimetry has been selected by IAEA as a transfer dosimetry system and is widely used through the world for the control of industrial and medical radiation sources. Tooth enamel EPR dosimetry is currently the only reliable accident dosimetry system for the retrospective individual dosimetry.

2 Technique of EPR tooth enamel dosimetry

EPR dosimetry technique involves some important points: sample preparation, choice of optimal EPR spectrometer parameters, calibration of radiation sensitivity of the enamel, EPR spectrum processing, and choice of line shapes for radiation-induced and background spectral lines.

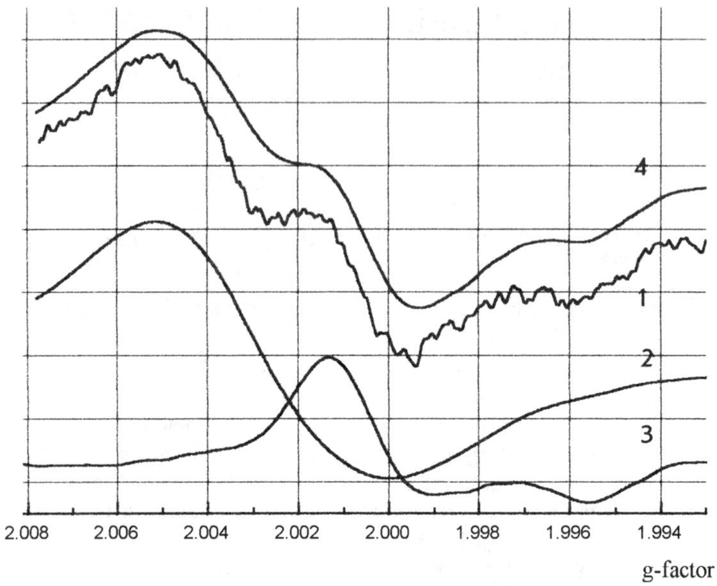

Fig. 1. 1 - experimental EPR spectrum of irradiated tooth enamel; 2 - radiation induced spectrum; 3 - background spectrum; 4 - superposition of 2 and 3 spectra. must coincide with 1.

The spectrum processing is illustrated at the Fig. 1. The experimental spectrum of irradiated tooth enamel 1 is the superposition of a radiation induced spectrum 2 and background spectrum 3.

At present there are no standard conditions of the EPR tooth enamel dosimetry, and each laboratory has its own technique which may differ in details from those of other laboratories. The Chernobyl accident has promoted the interest in dosimetry, and European Commission have funded work on inter-comparison and intercalibration of the existing techniques in laboratories of different countries of former Soviet union and Europe, for development of the optimal EPR tooth enamel technique. This work was performed in 1994-1995 within the framework of the Experimental Collaboration Project № 10 (ECP 10) "Retrospective dosimetry and dose reconstruction" (1996). This work was discussed at the 4th International Symposium on ESR Dosimetry and Applications (1995) in Munich and at the First International Conference "The Radiological Consequences of the Chernobyl Accident" (1996) in Minsk.

2.1 Summary of ECP10 Project on intercomparison

Eight laboratories from Ukraine, Russia, Belarus, Estonia, USA and Germany have taken part in the work on the intercomparison of EPR tooth dosimetry techniques, and each laboratory has presented its own version of the technique.

In the course of the intercomparison work each laboratory have obtained a set of 15 standardized tooth enamel samples irradiated by a gamma source with different doses unknown to participants. The results obtained in laboratories were compared with genuine values at the special workshop. The publication of results was made on behalf of all coworkers of all involved laboratories (Wieser *et. al.*, 1996). Below , the particular estimations of authors of this paper are presented.

The work has allowed participants to decide on the optimal modern EPR tooth dosimetry technique. The main steps are: (a) careful preparation and purification of samples using mechanical, chemical and ultra sonical treatments; (b) using of highly sensitive EPR spectrometers and defined parameters of equipment, such as microwave power ($\geq$ 10 mW) and magnetic field modulation amplitude (0.3 mT); (c) use of the suitable computer software distinguish the radiation induced part of the EPR spectrum; (d) use of the correct background line shape.

Employment of a correct procedure and adequate equipment allows measurement of irradiation doses (for about 90% of experimental cases) from 300 mGy and higher with an error of about $\pm$ 20-30 %. The lower limit of the sensitivity is 100-200 mGy depending on the sample quality and weight, but the error in this case is up to 50% and higher. In some cases (about 10% of experimental samples) there are some difficulties in tooth dosimetry because of individual peculiarities of the teeth.

2.2 Background line shape

The background line shape is an important parameter of EPR tooth dosimetry. The physical principle of the method is the generation of paramagnetic centers in the crystal lattice of enamel during the irradiation. The concentration of radiation- induced paramagnetic centers is proportional to the absorbed irradiation dose and can be measured with the ESR. One of the problems of correct measurement of radiation induced paramagnetic centers is the presence of non radiation-induced paramagnetic centers in enamel. Usually the ESR spectrum of such centers is superimposed at the radiation induced spectrum. The non radiation-induced ESR line in tooth enamel is called the background line. To measure the radiation induced part one must subtract the background spectrum from the experimentally observed line.

So there are questions about the correct line shape of the background spectrum, its possible variations from one sample to another and from one laboratory to another, and the possible influence of the background line shape on the obtained dose value.

Five background lines used in different laboratories were compared together and with Gaussian (G) and Lorentzian (L) line shapes. The averaged background line) together with G and L is presented at Fig.2.

The important characteristic of the background line is its asymmetry compared to Gaussian and Lorentzian. It means that the use of symmetric background lines should give considerable overestimation of results.

The solid line at Fig. 2 can be considered as an idealized background line. It can be presented also in digital form (Table 1) for a computer code.

Application of the background line shown at Fig.2 and Table 1 is one of steps in the sandard procedure of the ESR/tooth dosimetry.

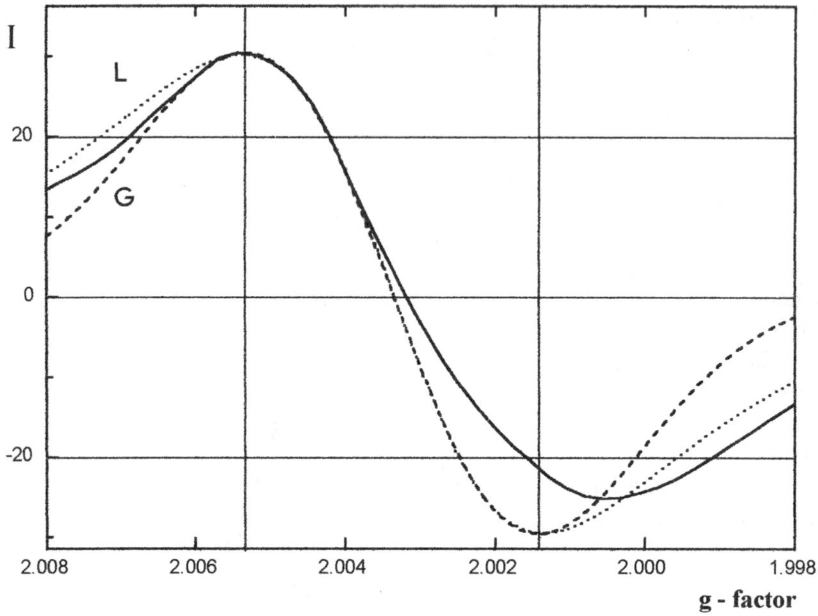

Fig. 2. Background (solid), Gaussian (G) and Lorentzian (L) lines .
I is intensity of line at given value of g in arbitrary units.

3. Retrospective individual dosimetry of Chernobyl liquidators

The most dramatic consequence of the Chernobyl accident was irradiation of many people - nuclear power plant workers, inhabitants of contaminated territories, liquidators. Health consequences depend on irradiation dose. There is a serious problem with dosimetry of Chernobyl survivors in Ukraine, Belarus and Russian

Table 1. The digital code of ithe idealized background line shape. I is intensity of line at given value of g an arbitrary units.

g	1.9980	1.9982	1.9985	1.9987	1.9990	1.9992	1.9995
A	-16.89	-18.35	-19.91	-21.33	-22.67	-23.71	-24.41
g	1.9997	1.9999	2.0002	2.0005	2.0007	2.0009	2.0012
A	-24.93	-25.09	-24.77	-23.80	-22.35	-20.45	-18.39
g	2.0014	2.0016	2.0019	2.0022	2.0024	2.0026	2.0029
A	-16.14	-13.55	-10.57	-7.15	-3.35	0.99	5.35
g	2.0031	2.0033	2.0036	2.0039	2.0041	2.0043	2.0046
A	9.99	15.04	20.11	24.26	27.32	29.18	30.21
g	2.0048	2.0051	2.0053	2.0056	2.0058	2.0060	2.0063
A	30.54	29.87	28.19	26.26	24.34	22.24	20.14
g	2.0065	2.0068	2.0070	2.0073	2.0075	2.0077	2.0080
A	18.29	16.70	15.35	14.17	12.96	12.96	12.96

Federation. The National Registry of Chernobyl survivors in Ukraine contains the names of 350 000 liquidators. Only for about 20% of them there is any information on individual doses. EPR tooth enamel dosimetry allows us to obtain experimental values of individual doses for some of the survivors, and verify theoretical estimations of collective doses.

We have used the EPR dosimetry for the retrospective reconstruction of individual doses of a group of servicemen liquidators, who are under the regular medical control at the Military Medical Board. Extracted bad teeth of these people were used for dosimetry. For some of these people the "official" irradiation doses are known from records in their medical cards. As a rule the records on individual doses were made by the dosimetry board using data of external dosimetry at the place of their work at the Chernobyl area in 1986-1987. Results of our measurements are shown in Table 2.

4 EPR dosimetry of wild animals in the Chernobyl area

The 30-km area around the Chernobyl Nuclear Power Plant (the Chernobyl Zone) is heavily contaminated with radioactive materials.

Table 2. Individual doses of liquidators measured by EPR tooth enamel techniques

Case No	Sample No	"Official" Dose	EPR Dose, Gy
1	3213	0.10	0.25
2	4824	0.019	0.80
3	4994	0.013	0.60
4	5706	0.009	0.030
5	6170	-	0.20
6	6662	0.035	0.20
7	6826	0.03	0.19
8	7747	-	0.25
9	8474	-	0.15
10	8797	-	0.30
11	9436	-	0.15
12	11956	-	0.15
13	NN 1	-	0.20
14	NN 2	-	1.75
14 cases	------	-	$\leq$ 0.10

Civilian people were evacuated at early stage of the accident and since that time this area became a reserve territory. As a result the wild animal population became large. Thousands of wild boars, several hundreds of red deer and elks and many small animals live here now. Wild boars often occupy deserted country houses and feed on fruits from deserted orchards.

Wild animals live in the very contaminated territories. Now the averaged values of gamma-dose rate in some places of the Chernobyl Zone are up to 10 mR/h (0.1 mGy/h). However the local level of contamination is very variable. Over distances of hundreds of meters it can change by two orders of magnitude.

The animals are irradiated during their life (usually several years) and their cumulative dose can be large. The wildlife at Chernobyl is, unintentionally, involved in the enormous experiment on the influence of different levels of radiation on living beings. Animals in the Chernobyl Zone are objects of scientific research of Ukrainian scientists, and the question arises about the individual doses of these animals. Now the only possibility to answer this question is EPR tooth enamel dosimetry.

Table 3. Individual doses of wild boars at Chernobyl area

Case No	Date of hunt	Place of hunt	Age, years	Activ. of muscle, Bq, ^{137}Cs	Dose, Gy
1	26.02.91	Ilyinysy	2	59000	0.60
2	21.02.91	Zimowistche	2	4700	0.40
3	21.02.91	Maschevo	2	83000	1.10
4	05.06.92	Opachichi	3	2600	0.35
5	06.05.92	Kriva Gora	4	40500	0.95
6	08.06.92	Opachichi	5	270	0.95
7	19.05.93	Jampol	2	23000	1.50
8	21.05.93	Korogod	1	4400	0.60
9	21.05.94	Zimowistche	3	8000	5.60
10	21.05.94	Usov	4	201000	16.40
11	22.05.94	Opachichi	?	1400	2.70
12	14.10.94	Ivanovka	1	?	<0.10
13	30.10.94	Maschevo	2-3	?	6.10
14	27.02.95	Krasno	-	?	7.70
15	13.01.96	Buda	2	?	0.80
16	09.02.96	Kriva Gora	1	?	0.50
17	16.02.96	Opachichi	3	?	0.20

Results of our measurements are shown in the Table 3. We can see that all animals in the Chernobyl area have received significant irradiation doses and appreciable part of them obtain very large doses (at the level of several Gy) of external irradiation. The left bank floodplain of the Pripyat River, where many animals live, is especially contaminated. Our results show that irradiation doses are especially large for animals which live at this territory.

The results of EPR dosimetry have been used zoologists in study of irradiation impact on animals.

References

1. 4th International Symposium on ESR Dosimetry and Applications, 15-19 May, 1995, GSF, Munchen/Neuherberg. The proceedings will be published in *Appl. Radiat. Isotopes.*
2. D.F.Regulla and U.Deffner, Dosimetry by ESR spectroscopy of alanine, *Int. J. Appl. Radiat. Isot.* **33** (1982) 1101-1114.
3. M.Ikeya, J.Miyajima and S.Okajima, ESR dosimetry for atomic bomb survivors using shell buttons and tooth enamel, *Japan J. Appl. Phys.* **23** (1984) 697-699.
4. Retrospective dosimetry and dose reconstruction. *Experimental collaboration project No 10. Final Report.* EUR 16540 EN. Editors I.K.Bailiff and V.Stepanenko (1996) pp. 57-76.
5. V.Chumak, A.Bougai, A.Wieser *et. al.,* The first International intercomparison of EPR-dosimetry with teeth: first results, in *Proceedings* of the 4th International Symposium on ESR Dosimetry and Applications, 15-19 May, 1995, GSF, Munchen/Neuherberg , will be published in *Appl. Radiat. Isotopes.*
6. A.Wieser, A.Bougai *et. al.*, International intercomparison of dose measurements using EPR spectrometry of tooth enamel, in *The Radiological Consequences of the Chernobyl Accident,* Proceedings of the First International conference, Minsk, Belarus (1996) pp. 957-964.
7. V.G.Baryachtar, A.A.Bugai *et al.*, The EPR Investigation of Tooth Enamel for Measurement of Absorbed Gamma Doses of People Irradiated in Chernobyl Accident. *Reports of the Ukrainian Academy of Sciences* (1993) 170-172.
8. V.G.Baryachtar, A.A.Bugai *et al.*, Experimental Retrospective Dosimetry and Chernobyl Problems. *Reports of the Ukrainian Academy of Sciences* (1994) 67-72.
9. V.G.Baryachtar, , A.A.Bugai *et al.*, External Irradiation Doses of Wild Animals in 30-km Area of Chernobyl Power Plant. *Reports of the Ukrainian Academy of Sciences* (1994) 149-152.

The Nitric Oxide Cycle — a New Metabolic Cycle of Intracellular Signalling System

Valentin P. Reutov, and Elena G. Sorokina

Institute of Higher Nervous Activity and Neurophysiology Russian Academy of Sciences, Butlerova Str., 5a, Moscow 117485 Russia

Abstract. The method of electron paramagnetic resonance (EPR) has been applied to study the mechanism of reduction of nitrite ions (NO_2^-) to nitric oxide (NO).The concept of the nitric oxide cycle was suggested, which helps explain numerous experimental data indicating a sharp increase in NO synthesis under hypoxic/ischaemic conditions. We believe that activation of NO formation is based on the two mechanisms, i.e., the intensified of L–arginine utilisation evidenced by its decrease in blood serum in hypoxia, registered by the authors, and transition of a number of heme–containing proteins (hemoglobin, myoglobin, cytochrome oxidase and cytochrome P–450) to the deoxy–form, in which these proteins can reduce NO_2^- to NO.

1 Introduction

Recently it has been shown that NO performs one of the key functions in the regulation of intracellular signalling system in the mammalian organism [1—5]. Researchers assign NO to the secondary messengers, along with Ca^{2+}, cAMP, cGMP, inositoltrisphosphate, diacylglycerol and cADP–ribose [6—10]. Nitrates, nitrites and NO were considered xenobiotics up to 1980. However, in the last decade this opinion has changed considerably due to the discovery in NO of the ability to regulate intracellular concentration of Ca^{2+} [11—15].

Study of the mechanisms of the toxic action of nitrates and nitrites has led us to the problem of the effect of NO on the regulation of cellular metabolism of mammals. It was shown earlier that nitrites could exert their toxic effects on the mammalian organism at practically all levels: the whole body, single organs, cells, subcellular and membrane structures, as well as at the level of macromolecules and single molecules [16—19]. At the same time, it was difficult to understand why simple molecules such as NO_3^- and NO_2^- ions with their weak or moderate oxidative properties have such a wide spectrum of regulatory, pharmacological and toxic action. The character of this action (from regulatory to toxic) largely depended on the dose, and was determined less by the individual characteristics of the mammalian organism [18, 19].

2 The Mechanisms of Regulatory, Pharmacological, and Toxic Effect of Nitrogen Compounds

The doses of nitrites from minimally effective to lethal were used for a more full determination of the spectrum of their action. This approach turned out to be fruitful and made it possible to establish that the universal mechanisms of the toxic action of nitrites were based on the NO_2^- ion transformation into NO (Fig. 1).

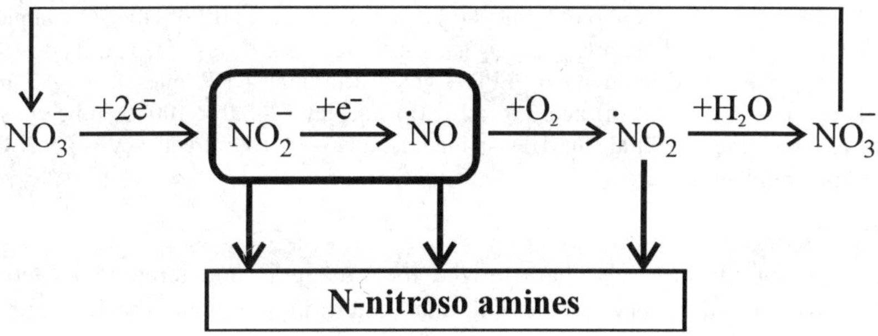

Fig. 1. Main products of NO_2^- and NO_3^- ion transformation

It is exactly the step of NO_2^- ions into NO that determines the multiple toxic properties of nitrites: the ability of NO_2^- ions to oxidize hemoglobin and form Hb–NO complexes, to effect the activity of soluble guanylate cyclase and the cGMP level as the intracellular concentration of Ca^{2+} ions [16—20]. Cellular metabolism is regulated by numerous enzymatic systems as well as by secondary messengers through changes in the intracellular Ca^{2+} concentration [1—5, 7]. Since the secondary messengers function in the systems with the amplification coefficient of 10^6—10^8 [20—22], the role of this link in the mechanisms of toxic effects of NO_2^- could hardly be overestimated. It is known that NO, being oxidized to nitrogen dioxide (NO_2), can react with proteins (binding to SH–groups of cysteine and OH–groups of tyrosine residues) and with unsaturated fatty acids included in the membranes of cells and subcellular structures [16—18].

Free–radical products, NO and NO_2, can damage proteins and unsaturated fatty acids, to decrease the activity of most of the enzymatic systems (except soluble heme–containing guanylate cyclase and ADP–ribosyl transferase), to destroy the integrity of cellular and subcellular structures, to interrupt oxidative phosphorylation, to inhibit electron transport through the respiratory cycle of mitochondria, to decrease the ATP level in mammalian blood and tissue cells, to oxidize blood pigment, and to participate in the formation of R– and T–conformers of Hb–NO complexes (Fig. 2) [18, 19, 23, 24].

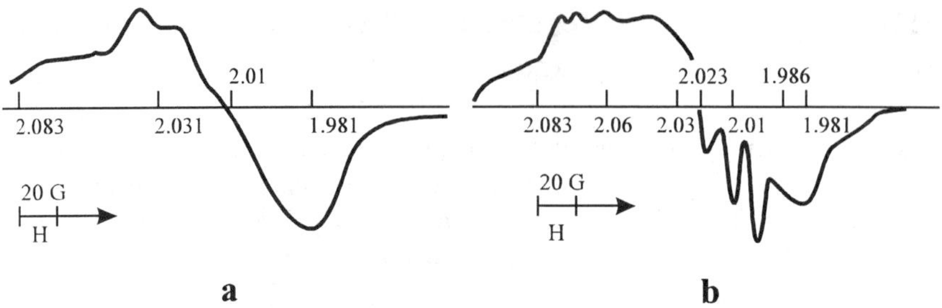

Fig. 2. EPR spectra for the R–conformers (a) and T–conformers (b) of Hb–NO complexes formed in the animal blood after $NaNO_2$ administration in a dose 7 (a) and 3 mg/100 g (b) of body weight. Conditions of EPR signal recording: UHF intensity — 2 mW; amplification — 5.0×10^2 (a) and 3.0×10^3 (b); display — 8.0×100 G, time constant $\tau = 0.128$ sec; the amplitude of HF — modulation — 4.0 G; frequency — 9.3 Ghz; temperature of recording T = 77 K

These compounds also are characterized by mutagenic and teratogenic activity [25]. Thus, the discovery of the link that determines a wide spectrum of the regulatory, pharmacological, and toxic effects proved to be very important for understanding this phenomenon. It should be noted that, along with its toxicity, carcinogenic N–nitroso–compounds are formed at the stage of NO_2^- ion transformation into NO (see a Fig. 1).

The ability to control processes of transformation of nitrites to NO could help pathophysiologists and pharmacologists to correct functional disorders in patients. Knowing the mechanisms of reduction of NO_2^- ions to NO and the ways for changing the nitrite reductase properties of enzymatic systems and low–molecular compounds, toxicologists could reduce the unfavourable effects of nitrites on the human and animal organism. On the basis of these mechanisms, hygienists and oncologists would be able to determine more accurately the permissible concentrations of nitrites and nitrates for the population and especially for subjects genetically predisposed to oncological diseases and patients in the rehabilitation period.

In connection with the important step of the transformation of nitrites to NO (see a Fig. 1.), three tasks were set up: 1) to localize the sites of reduction of NO_2^- ions to NO in the mammalian organism; 2) to establish the nature of enzymes and low–molecular–weight organic compounds that can perform the nitrite reduction function in mammals; and 3) to analyse the factors affecting the activation and inhibition processes of reduction of nitrite to NO.

3 The Nitrite Reductase Activity of Heme–Containing Proteins

It was shown that blood hemoglobin in the deoxy–form can reduce nitrites to NO [19, 26]. Oxygen can inhibit the nitrite reductase activity of hemoglobin [26]. It was also found that myoglobin in the deoxy–form can function in a similar way

[19, 27]. The reduction of nitrites to NO in mammalian tissues can take place in mitochondria [19, 27] and microsomes [28]. The nitrite reductase function in mitochondria is performed by cytochrome oxidase [19, 27], and in microsomes by cytochrome P–450 [28]. Therefore, heme–containing proteins such as hemoglobin, myoglobin, cytochrome oxidase and cytochrome P–450, which usually react with oxygen in the deoxy–form, could reduce nitrites to NO and thus close the chain of transformation of L–arginine → NO → NO_2^-/NO_3^- to a single NO cycle (fig. 3).

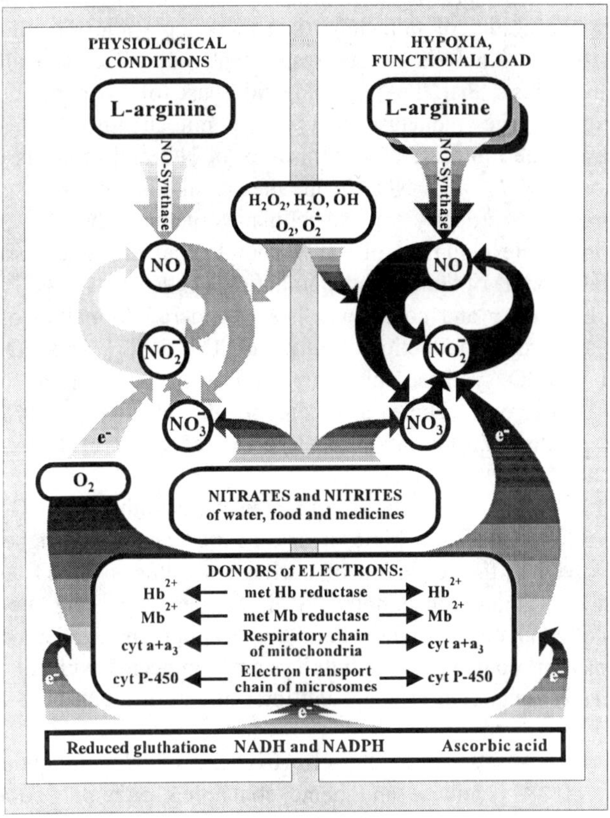

Fig. 3. The nitric oxide cycle

In view of the fact that only reduced forms of heme–containing proteins can reduce NO_2^- to NO, one could not ignore the role of the enzymatic and non–enzymatic systems involved in the transfer of electrons to these heme–containing proteins. They primarily include the metHb and metMb reductases, as well as the electron transport chains of mitochondria and endoplasmatic reticulum [19, 27, 28]. The role of ascorbic acid and reduced glutatione should be noted among the low–molecular–weight systems [19, 27, 30].

4 Physiological Role of Nitric Oxide Cycle in Mammals

What is the biological sense of the nitrite reductase properties? Many biologically important functions in the living organisms are mutually interconnected. The contractile and ATP–ase activity of myosin or processes of oxidation and phosphorilation in mitochondria can serve as an example. The role of mitochondria and endoplasmic reticulum in the regulation of intracellular Ca^{2+} is well known as being important in the processes of intracellular signalling[1—8]. Much less is known about the ability of mitochondria and endoplasmic reticulum to reduce NO_2^- ions to NO. Analysis of our results and published data allowed us to advance a hypothesis that the physiological sense of coupling of the Ca^{2+}–mobilising and nitrite reductase activity in mitochondria and endoplasmic reticulum lies in the more effective utilisation of NO and products of metabolic transformation (NO_2^- and NO_3^-) in the regulation of intracellular Ca^{2+} concentration. This property of subcellular structures can be of particular importance in the development of many pathological processes associated with ischaemia and hypoxia of human and animal tissue cells.

Recently J. L. Zweier and co–authors has considered formation of NO in the ischaemic heart by direct reduction of nitrite to NO [15, 32]. This NO formation is not blocked by NO synthase inhibitors, and with long periods of ischaemia progressing to necrosis. These data support the existence of NO–synthase–independent mechanism formation of NO by reduction of NO_2^- ions during hypoxia/ischaemia.

Hypoxia and ischaemia is associated with intensified synthesis of NO, nitrites, and nitrates [13, 32]. At the same time, it is accompanied by the higher activity of enzymes involved in the reduction of nitrites and nitrates to NO and decreased blood serum L–arginine concentration [31]. The available data suggest that the mechanisms of adaptation to hypoxia and the development of tolerance in animals to reduced oxygen content can include systems connected with nitrite reductase activity of heme–containing proteins and intensified utilisation of L–arginine [26, 30, 31].

The question arises: what is the effectiveness of NO formation with the participation of NO synthase and heme–containing proteins? To evaluate the potency of mammalian NO synthases and the nitrite reductase activity of heme–containing proteins and of enzymatic and nonenzymatic systems involved in the reduction of heme–containing proteins and thus directly or indirectly reducing NO_2^- ions to NO, it is sufficient to compare the constant–state concentrations of nitrites, nitrates and NO under common physiological conditions and in nitrite loading. The physiological concentration of L–arginine in the cell of mammalian tissues and blood serum is within 100—500 μM, as a rule. The concentration of NO does not exceed 10^{-8} — 10^{-9} M, that of NO_2^- ions 10^{-6} M, and of NO_3^- ions 10^{-5} M. These concentrations are nothing other than the maximum allowable concentration (MAC) developed in the mammalian organism and possibly in eukaryotic cells in the process of evolution. When nitrites are administered in a dose of 50—70 mg/kg of body weight, the blood serum concentration of nitrites is

40—100 mg/ml in the first hour after their absorption. The concentration of Hb–NO complexes in the blood in this case approaches 10—15% of the total hemoglobin content. Since hemoglobin concentration in the mammalian blood is ~ 12—16 g% or about 2×10^{-3} M, the NO concentration in these cases can reach at least 10^{-4} M and that of nitrites 10^{-3} M. These data indicate that formation of NO, nitrites and nitrates from L–arginine in the mammalian cells is limited. At the same time, when nitrites enter the animal organism, the level of NO can exceed the normal physiological concentration by 1000 times and more.
Thus, it is obvious that the principle of hygienic regulation of nitrate, nitrite and NO concentration in the mammalian organism are observed quite strictly. During the whole process of evolution, eukaryotes, invertebrates, vertebrates, and mammals did not experience such an intensive nitrate–nitrite pressure as that to which they were exposed in the 20th century and especially in its second half. The scientific and technological progress and hence the intensified production of mineral nitrogen fertilizers, and wide use of nitrates and nitrites in various fields of industry, in agriculture, and pharmacology have led to the situation where the organism's "resistance reserve", which was provided by the evolutionary process and which is several orders of magnitude higher than the MAC of nitrates nitrites, and NO, turned out to be easily overcome by *Homo sapiens* conscious (or rather unconscious) activity.

5 Conclusions

Thus, our concept of activation of NO formation under hypoxia is based on two mechanism: first, activation of NO synthase and intensified L–arginine utilization evidenced by a decrease in L–arginine concentration in blood serum and an increase in the level of NO complexes with heme or non–heme iron; and second, the transition of a number of heme–containing proteins such as hemoglobin, myoglobin, cytochrome oxidase, and cytochrome P–450 to the deoxy–form, which make them capable of reducing NO_2^- ions to NO.
We believe that the nitric oxide cycle emerged and was preserved during evolution mainly as a regulatory mechanism of intracellular Ca^{2+} concentration. Its effectiveness sharply increases under functional loads due to intensified oxygen utilization, in brain and myocardial ischaemia, and in many pathological processes taking place under hypoxia. This specific feature of NO cycle functioning is explained largely by the fact that the heme–containing proteins being transformed into the deoxy–form begin to actively reduce NO_2^- ions. A higher functional load is associated, as a rule, with intensified oxygen utilization and a higher probability that heme–containing proteins will transform to the deoxyform. That is why the role of reduction of nitrites to NO and the whole NO cycle will sharply increase under such conditions. Only in the cases when the oxygen use is completely compensated by its supply, the role of the nitric oxide cycle can remain the same as it fulfils in usual physiological conditions.

Acknowledgements
The research is supported by Russian Foundation for Fundamental Studies.

References

1. C.J. Zong, and T.W. Stone, EDRF: a Ca — dependent chemical noiety, *Trends Pharmacol Sci.* **6**, 7 (1985) 285—290.
2. H. Maeda, T. Akaike, M. Yoshida, K. Sato, and Y. Noguchi, A new nitric oxide scavenger, imidazolineoxyl N–oxide derivative, and its effects in and microbiology, in *The Role of Nitric Oxide in Physiology and Pathophysiology,* edited by H. Koprowski and H. Maeda, Springer–Verlag Berlin (1995) pp. 37—50.
3. M.J. Berridge, and R.F. Irvine, Inositol phosphates and cell signalling, *Nature,* **341,** (1989) 197—205.
4. J.S. Stamler, S–Nitrosothiols and the bioregulatory action of nitrogen oxides through reactions with thiol groups, in *The Role of Nitric Oxide in Physiology and Pathophysiology,* edited by H.Koprowski and H.Maeda, Springer–Verlag, Berlin (1995) pp. 19–36.
5. D. Bredt, P. Hwang, and S. Snyder, Localization of nitric oxide synthase indicating a neural role for nitric oxide, *Nature*, **347** (1990) 768—770.
6. B. Brune, and E. Lapetina, Activation of a cytosolic ADP–ribosyltransferase by nitric oxide–generative agents, *J. Biol. Chem.*, **264** (1989) 8455—8458.
7. A.Galione, Ca^{2+}–induced Ca^{2+}–release and its modulation by cyclic ADP–ribose, *Trends Pharmacol. Sci.*, **13**, 8 (1992) 304—306.
8. J. Gartwaite, S.L. Charles, and R. Chess–Williams, Endothelium — derived relaxing factor release on activation of NMDA receptors suggests role as intracellular messenger in the braine, *Nature,* **336** (1988) 385—388.
9. J.Garthwaite, Nitric oxide synthesis linked to activation of excitatory neurotransmitter receptors in the brain, in *Nitric Oxide from L–Arginine: a Bioregulatory System. Proc. of the Symposium Held at the Royal Society London, September 14—15, 1989,* edited by S. Moncada and E.A. Higgs, Excerpta Medica, Amsterdam (1990) pp. 115—138.
10. R.G.Knowles, M. Palacios, R.M.J. Palmer, and S. Moncada, "Formation of nitric oxide from L–arginine in the central nervous system: a transduction mechanism for stimulation of the soluble quanylate cyclase", *Proc. Nat. Acad. Sci. USA*, **89** (1989) 5159—5162.
11. R.Busse, J. Fleming, and V.B. Schini, Nitric oxide formation in the vascular wall: regulation and functional implication, in *The Role of Nitric Oxide in Physiology and Pathophysiology*, edited by H. Koprowski and H. Maeda, Springer–Verlag, Berlin (1995), pp. 7—18.
12. I.Yu. Malyshev, A.V. Malugin, L.Yu. Golubeva, T.A. Zenina, E.B. Manuchina, V.D.Mikoyan, and A.F. Vanin, Nitric oxide donor induces HSP70 accumulation in the heart and in cultured cells, *FEBS Letters*, **391** (1986) 21—23.
13. E.B. Manukhina, A.V. Lapshin, F.Z. Meerson, Physical training limits the fall of blood pressure and the endothelium overactivation in acute myocardial infarction, *Physiol. Res.*, **45** (1996) 261—266.
14. A.F. Vanin, R.A. Stukan, and E.B. Manukhina, Physical properties of dinitrosyl iron complexes with thiol — containing ligands in relation with their vasodilator activity, *Biochimica et Biophysica Acta*, **1295** (1996) 5—12.
15. J.L. Zweier, P. Wang, A. Samouilov, and P. Kuppusamy, Enzyme — independent formation of nitric oxide in biological tissues, *Nature Medicine*, **1**, 8 (1995) 804—809.

16. V. P. Reutov, Ya. I. Azhipa, and L.P. Kayushin, The paramagnetic resonance study of products of nitrogen oxide reactions with some organic compounds, *Bull. of Experimen. Biol. and Med.*, 9 (1978) 299 — 301.
17. V.P. Reutov, Ya. I. Azhipa, and L.P. Kayushin, The paramagnetic resonance study of products during interaction of nitrogen dioxide with oleic acid and tyrosine, *Dokl. Akad. Nauk SSSR*, **241**, (1978) 1375—1377.
18. V.P. Reutov, E.G. Sorokina, V.G. Pinelis, T.S. Korshunova, A.A. Rodionov, V.B. Koshelev, S.M. Strukova, L.P. Kayushin, P. Braque, and L.Kh. Komissarova, Compensatory — adaptive mechanisms during nitrite hypoxia in rats, *Bull. of Experimen. Biol. and Med.*, 11 (1993) 1387—1390.
19. Ya.I. Azhipa, V.P. Reutov, and L.P. Kayushin, Ecological and medico–biological aspects of the problem of environmental pollution with nitrates and nitrites, *Human Physiol.* **16**, 3 (1990) 220—236.
20. V.P. Reutov, and S.N. Orlov, The physiological significance of guanilate cyclase and the role of nitric oxide and nitrogen compounds in the regulation of activity of this enzyme, *Human Physiology*, **19**, 1 (1993) 71—79.
21. P.G. Kostyuk, and E.I. Chazov, Intracellular signalling: biological and medical aspects of the issue, *Usp. Fiziol. Nauk*, **19**, 4 (1988) 3—11.
22. G.N. Kryzhanovskii, The pathology of regulatory mechanisms, *Pathol. Physiol. Experimen. Terap.*, 2 (1990) 3—11.
23. A.M. Mansouri, Methemoglobinemia, *Am. J. Med. Sci.*, 289, 5 (1985), 200—209.
24. E. Hegesh, and J. Shiloah, Blood nitrates and infantile methemoglobinemia, *Clin. Chim. Acta*, **125**, 2 (1982) 107—115.
25. S.R. Tannenbaum, V.R. Young, L. Green, and K. Ruiz de Luzuriada, Intestinal formation of nitrite and N–nitroso compounds, *JARS Sci. Publ.*,31 (1980) 281—287.
26. V.P. Reutov, Ya.I. Azhipa, and L.P. Kayushin, Oxygen as an inhibitor of hemoglobin nitrite reductase activity, *Izv. Akad. Nauk SSSR, Ser. Biol.*, 3 (1983), 408—418.
27. V.P. Reutov, *Study of the mechanisms of NO_2^- ion reduction in the blood and cell mithochondria of mammalian tissues*, Moscow University Press, (1988) pp. 1—24.
28. G.S. Duthu, and H.G. Shertzer, Effect of nitrite on rabbit liver mixed function oxidase activity, *Drug. Metab. Disposit. Biol. Fate Chem.* **7**, 5 (1979) 263—269.
29. Ya. I. Azhipa, V.P. Reutov, and L.P. Kayushin, Conformational isomers of hemoglobin complexes with nitric oxide which arise in the blood under the effect of sodium nitrite, *Izv. Akad. Nauk SSSR. Ser. Biol.*, 2 (1983) 240—250.
30. V.P. Reutov, L.P. Kayushin, and E.G. Sorokina, Physiological role of nitric oxide cycle in human and animal organism, *Human Physiology*, **20**, 3 (1994), 219—225.
31. V.P. Reutov, E.G. Sorokina, A.A. Rodionov, and T.S. Korshunova, Peptides and amino acid in blood of mammals under the action of nitrites, in *Structure and Function of Regulatory Polypeptides*, edited by E. Frolova, Russian Academy Press (1992) pp 67—68.
32. J.L. Zweier, P. Wang, and P. Kuppusamy, Direct measurement of nitric oxide generation in the ischemic heart using electron paramagnetic resonance spectroscopy, *J.Biol. Chem.*, **270** (1995), 304—307.

Analysis of Slow Motions in Serum Albumin Solution and Gel by Vector EPR

Tomoya HAYASHI, Yuhei SHIMOYAMA[†] *and Seiichi ERA*

Department of Physiology, Gifu University School of Medicine, Gifu 500 and
[†] Department of Physics, Hokkaido University of Education, Hakodate 040, Japan

Abstract. Serum albumin has one reactive sulfhydryl (Cys-34) which is one of many important binding sites. Cys-34 is located in a molecular crevice and undergoes a restricted motion. We studied the molecular motion of Cys-34 of bovine serum albumin (BSA) in solution and gel states by vector electron paramagnetic resonance (EPR) using a maleimide spin label.

The vector EPR spectra of BSA solution and gel states showed two kinds of signals; slow and fast motional signals. The former one indicated a slow motion from the spin label bound to Cys-34 which undergoes restricted motion in the crevice of BSA. The latter one might originate from the spin label bound to ε-NH_2 of lysine residues on the surface of BSA. The integrated value obtained from the vector EPR in gel state was much larger than that in solution state, indicating that the molecular motion of the spin label bound to Cys-34 in gel state is considerably immobilized.

1 Introduction

It is well known that serum albumin possesses a free sulfhydryl (Cys-34). Cys-34 can bind cysteine, glutathione and various metals such as Cd, Au, Hg and Ag[1]. Hull *et al.*[2] found that Cys-34 exists in a molecular crevice of approximately 10 Å in depth by conventional electron paramagnetic resonance (EPR) method. Using X-ray crystallographic method, Carter and Ho[3] also revealed that Cys-34 is located in a molecular crevice on the surface of the protein and it is protected by several residues. Therefore, a molecular motion of Cys-34 might be anisotropic and slow motions. However, controversy remains about the detailed micro-environment for Cys-34 in the crevice.

Commercial bovine serum albumin (BSA) Fraction V contains small amounts of a proteolytic enzyme. The BSA results in the formation of partially hydrolyzed one by the enzyme in the F-form (ca. pH 3.8), and it is called BSA*[4]. The BSA* has a tendency to form a transparent gel at pD

4.0 in D_2O at concentrations above 8%, and it is called BSA*-gel[5]. Detailed knowledge of physicochemical properties in BSA*-gel is still unknown.

Saturation transfer EPR (ST-EPR)[6] or vector EPR[7] using spin label is suitable for measuring anisotropic and slow motions. In the present study, we studied more accurately molecular motion of Cys-34 in BSA solutions (the N-form (pH 5.7) in H_2O and the F-form (pD 4.1) in D_2O) and BSA*-gel by vector EPR using a maleimide spin label. We were also able to demonstrate the power of the vector EPR method for the characterization and detection of the slow molecular motion which is outside the limits of measurement by the conventional EPR method.

2 Materials and Methods

2.1 Preparation of Spin Labeled BSA Solution (N-form)

Crystallized BSA (Armour Pharma. Co.) was defatted according to the procedure of Sogami and Foster[8]. 3 - maleimido - 2, 2, 5, 5 - tetramethyl - 1 - pyrrolidinyloxyl (Aldrich Chem. Co.) was used in this study. To label Cys-34, BSA in 0.20 M KCl was allowed to react with each of spin labels (1.0 mol spin label per mol BSA) with stirring for 48 hours at 4℃. The solution was dialyzed exhaustively against distilled water at 4℃ or performed gel filtration at 4℃. The solution was filtered through a Millex filter (0.45 μm) and then was put into capillaries (0.5 mm ϕ) in an EPR tube (5 mm ϕ) to avoid dielectric loss in an EPR cavity.

2.2 Preparations of Spin Labeled BSA Solution (F-form) and BSA*-gel

BSA Fraction V (Sigma Chem. Co.) in 0.20 M KCl was allowed to react with each of above spin labels (1.0 mol spin label per 1.0 mol BSA) with stirring for 48 hours at 4℃. The mixture was dialyzed exhaustively with distilled water at 4℃ and underwent ultrafiltration at 4℃. The solution was then lyophilized. For making 10% spin labeled BSA solution, a lyophilized sample was dissolved in D_2O, and pD was adjusted to 4.10 (pD range of the F-form) by adding 0.10 M DCl. The solution was filtered through a Millex filter (0.45 μm) and then was put into sterilized capillaries (0.5 mm ϕ) in an EPR tube (5 mm ϕ) to avoid dielectric loss in an EPR cavity. These EPR tubes containing spin labeled BSA solution in D_2O were kept at 35±0.3℃, resulting in the formation of a transparent spin labeled BSA*-gel.

2.3 Measurements of Vector EPR Spectra

We made a system for measurement of vector EPR; a lock-in amplifier (NF Electronic Instr., 5610) and a personal computer (NEC, PC-9801 DA) were attached to an EPR spectrometer (JEOL, JES-FE1X). The vector EPR spectra were recorded with a microwave power of 121 mW, 100 kHz field modulation (0.2 mT), in the second harmonic mode, that is, 100/200 kHz of modulation/demodulation scheme. The phases of vector EPR spectra were measured with both in-phase (V_2) and out-of-phase (V_2') displays.

3 Results and Discussion

3.1 Spectral Features of BSA Solution (N-form)

The V_2' spectrum of spin labeled BSA solution (N-form) is shown in Fig. 1. This spectrum showed two kinds of signals; the one corresponded to strongly immobilized signals (L, L′, H and H′) and the other to weakly immobilized ones (F_L and F_H). The former ones indicated a slow motion from the spin label bound to Cys-34 which undergoes restricted motion in the crevice of BSA. The latter ones might originate from the spin label bound to ε-NH_2 of lysine residues on the surface of BSA, as considered by Griffith and McConell[9)], because free or unbound spin label was removed from the BSA solution by exhaustive dialysis or gel filtration in this study.

3.2 Spectral Features of BSA Solution (F-form) in D_2O and BSA*-gel

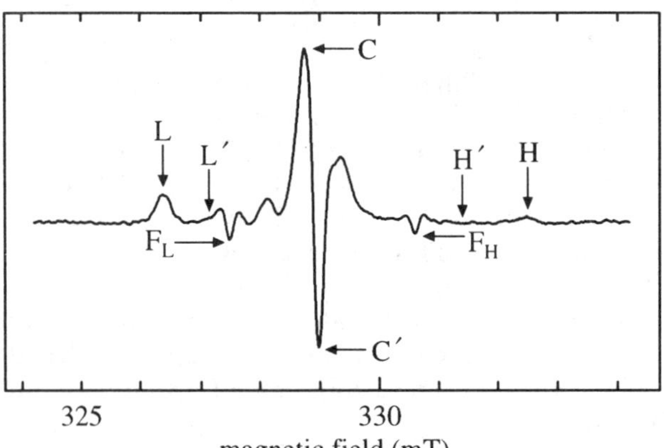

Fig.1. The out-of-phase (V_2') spectrum of spin labeled BSA solution (N-form) by vector EPR. Experimental conditions are as follows; 100 kHz/200 kHz of modulation/demodulation scheme, microwave power of 121 mW, magnetic modulation amplitude of 0.2 mT.

The EPR spectra of spin labeled BSA solution (F-form) in D_2O and BSA*-gel showed the same two kinds of signals as those in BSA solution (N-form) in H_2O; the signals originated from the spin label bound to Cys-34 in the crevice and the spin label bound to ε-NH_2 of lysine residues, indicated slow and fast motions, respectively. The V_2' spectral shapes between BSA solution (F-form) and BSA*-gel were considerably different. The signal intensities of L′, C′ and H′ in BSA*-gel were larger than those in BSA solution (F-form), suggesting that the molecular motion of the spin label bound to Cys-34 in the BSA*-gel was more strongly immobilized than that in thc BSA solution (F-form).

3.3 Phase Behaviors of Slow and Fast Motional Signals in Solution and Gel States of BSA

Shimoyama and Watari[7] demonstrated that a peak-to-peak height on the second harmonic mode of TEMPOL in the glycerol-water solution varies sinusoidally as functions of the modulation phase. We obtained two kinds of signals in both solution and gel states of BSA, which are slow and fast motional signals. Therefore, we investigated whether the signal phases between slow and fast motional signals in BSA solution (N-form) and BSA*-gel are different. We measured 36 spectra on the second harmonic mode with increments of 10 degree at instrumental phase angles. Signal phases between slow and fast motional signals in BSA solution (N-form) were almost no difference (Fig. 2-A). However, signal phase of slow motional signal in BSA*-gel was shifted from that of fast one (Fig. 2-B). It suggests that such a phase shift observed in BSA*-gel is related to the mobility of spin label.

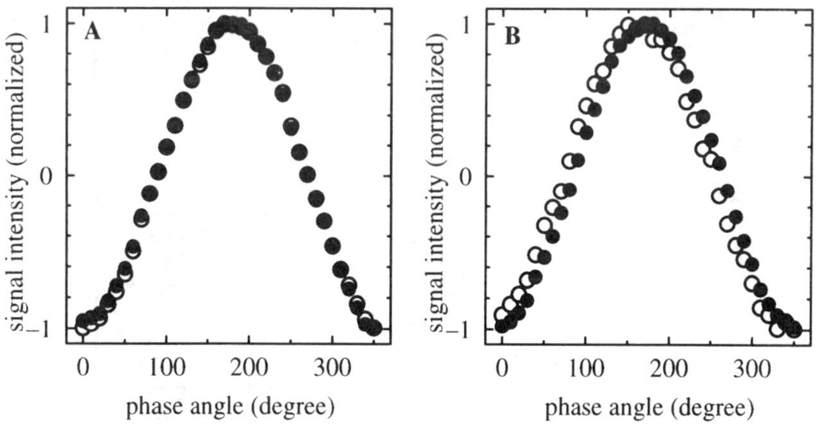

Fig.2. The normalized signal intensity of slow motional spin label (○) and fast one (●) in the vector EPR spectra, as a function of phase angle (degree). A): BSA solution (N-form), B): BSA*-gel. See Fig.1 for experimental conditions.

3.4 Molecular Motion of Cys-34 in BSA*-gel

Evans[10] demonstrated that there is a correlation of the integrated values (I) obtained from ST-EPR spectra with the rotational correlation times (τ_R) of a slow motional spin label. He also indicated that such a correlation may be useful in measuring an accurate τ_R value of slow motional spin label in the presence of fast motional one[10]. As the integration of fast motional signals in the V_2' spectrum becomes a differential form, fast motional ones have no influence upon I value obtained from the V_2' spectrum. Thus, I is only related to the amplitude for slow motional signals in the V_2' spectrum which is correlated with τ_R for the molecule. Consequently, it may be worth noting that I obtained from the V_2' spectra in the present study indicates directly the molecular motion of the spin label bound to Cys-34 in the crevice of the BSA molecule. The integrated spectra of BSA solution (F-form) and BSA*-gel are shown in Fig. 3. I values for BSA solution (F-form) and BSA*-gel were 19.0 and 86.7, respectively.

These values together with the results shown in Fig.1, 2-A and 2-B indicate that the molecular motion of the spin label bound to Cys-34 in BSA*-gel is considerably immobilized during the gelation of BSA*.

References

1. T. Peters, Jr., Serum albumin, *Adv. Protein Chem.* **37** (1985) 161-245.
2. H. H. Hull, R. Chang and L. J. Kaplan, On the location of the sulfhydryl group in bovine plasma albumin, *Biochim. Biophys. Acta* **400** (1975) 132-136.

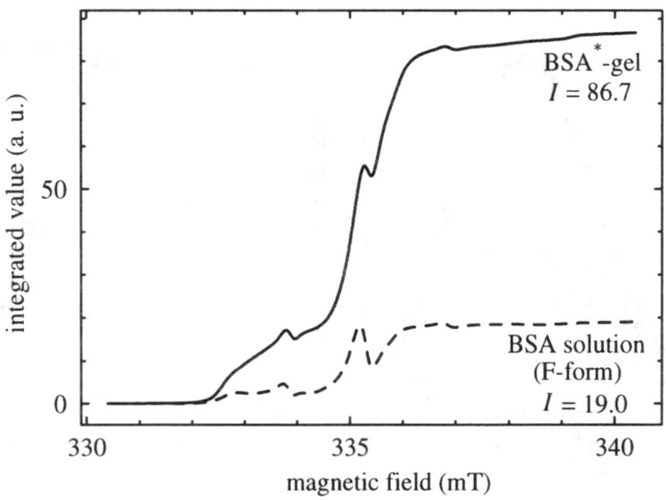

Fig.3. Integrated spectra of BSA solution (F-form) (dotted line) and BSA*-gel (solid line) obtained from V_2' spectra. See Fig.1 for experimental conditions.

3. D. C. Carter and J. X. Ho, Structure of serum albumin, *Adv. Protein Chem.* **45** (1994) 153-203.
4. W. D. Wilson and J. F. Foster, Conformation-dependent limited proteolysis of bovine plasma albumin by an enzyme present in commercial albumin preparations, *Biochemistry* **10** (1971) 1772-1780.
5. M. Sogami, S. Nagaoka, S. Era, H. Inouye, K. Kuwata, K. Kida, H. Watari, K. Shiga and K. Akasaka, Water structure in modified bovine plasma albumin (BPA*) gel and bovine mercaptalbumin solution, *Int. J. Peptide Protein Res.* **28** (1986) 130-140.
6. J. S. Hyde and D. D. Thomas, New EPR methods for the study of very slow motion : Application to spin-labeled hemoglobin, *Ann. N. Y. Acad. Sci.* **222**, (1973) 680-692.
7. Y. Shimoyama and H. Watari, Phase behaviors of saturation transfer ESR signals at high microwave fields, *Appl. Spectrosc.* **39** (1985) 170-173
8. M. Sogami and J. F. Foster, Isomerization reactions of charcoal-defatted bovine plasma albumin. The N-F transition and acid expansion, *Biochemistry* **7** (1968) 2172-2182.
9. O. H. Griffith and H. M. McConnell, A nitroxide-maleimide spin label, *Proc. Natl. Acad. Sci. USA* **55** (1966) 8-11.
10. C. A. Evans, Use of the integral of saturation transfer electron paramagnetic resonance spectra to determine molecular rotational correlation times. Slowly tumbling spin labels in the presence of rapidly tumbling spin labels, *J. Magn. Reson.* **44** (1981) 109-116.

AN ESR STUDY OF SUPEROXIDE GENERATION IN STIMULATED NEUTROPHILS FROM A CALF WITH BOVINE LEUKOCYTE ADHESION DEFICIENCY (BLAD)

Mikinori Kuwabara, Noriko Gouji, Osamu Inanami, Hidetoshi Higuchi[@] *and Hajime Nagahata*[@]

Graduate School of Veterinary Medicine, Hokkaido University, Sapporo 060, [@]Faculty of Veterinary Medicine, Rakuno Gakuen University, Ebetsu 069, Japan

Abstract. The ability of neutrophils from a Holstein calf with leukocyte adhesion deficiency (BLAD) (the proband with a genetic deficiency of the CR3 (Mac-1, CD11b/CD18) glycoprotein corresponding to the receptor of complement iC3b) to generate oxygen radicals was examined with electron spin resonance spectrometry (ESR) combined with a spin-trapping method. When neutrophils from BLAD were stimulated by serum-opsonized zymosan (s-OPZ) and IgG-opsonized zymosan (IgG-OPZ), the generation of superoxide anions (O_2^-) was about 15% that of normal neutrophils, whereas when they were exposed to protein kinase C stimulator, phorbol 12-myristate 13-acetate, the O_2^- generation was normal, indicating that (1) NADPH oxidase activity was intact and (2) the signal transduction pathways for superoxide generation from FcR to NADPH oxidase are affected by the presence of the receptor CR3. To clarify the roles of CR3, neutrophils from BLAD as well as normal calves were exposed to those stimuli in the presence or absence of inhibitors of several protein kinases affecting signal transduction pathways (H-7 and staurosporine as PKC inhibitors, genistein as a tyrosine kinase inhibitor, and wortmannin as a phosphatidyl inositol (PI) 3-kinase inhibitor). The results showed that genistein depressed the generation of superoxide in both neutrophils regardless of the difference of stimuli, whereas PKC inhibitors H-7 and staurosporine largely affected the superoxide generation in s-OPZ-stimulated neutrophils, but in s-OPZ-stimulated BLAD neutrophils. Similarly, wortmannin inhibited the superoxide generation in normal neutrophils but did not in BLAD ones when they were stimulated by s-OPZ. However, this did not inhibit the O_2^- generation in both neutrophils when they were stimulated by IgG-OPZ. These results indicated that the signal transduction pathways for the O_2^- generation in BLAD neutrophils, though its generation was low, were sensitive to tyrosine kinase inhibitor but insensitive to PKC and PI 3-kinase inhibitors. Since the FcR is conjugated with tyrosine kinase, it was inferred that the activation processes of NADPH oxidase via IgG-FcR interaction might be amplified by the activation of PI 3-kinase and PKC associating with the CR3 in normal neutrophils.

1 Introduction

Leukocyte adhesion deficiency is an autosomal recessive disease characterized by greatly reduced expression of the heterodimeric β2 integrin adhesion molecules on leukocytes. The β2 integrins include LFA-1, Mac-1 and p150,95 which consist of unique α subunits, CD11a, CD11b, and CD11c, respectively, and a common β subunit, CD18. Because β2 integrin expression requires intracellular association of the CD11 and CD18 subunits, defects in CD18 prevent expression of all β2 integrins (1). Mac-1 is a main β2 integrin in neutrophils and is a receptor of component iC3b (CR3). The generation of O_2^- in stimulated neutrophils was conducted by oxidation of NADPH and reduction of oxygen by NADPH oxidase. The binding of Fc region of IgG to its receptor (FcR) is an initial step to activate NADPH oxidase, and several protein kinases are thought to be concerned in the signal transduction pathways starting from the IgG-FcR binding and ending by the activation of NADPH oxidase (2). However, not only FcR but also other receptors such as CR1 (C3 receptor), LFA-1 (ICAM-1 receptor) and CR4 (p150,95 receptor) which can bind with opsonin factors are present on the surface membrane of neutrophil. The activation of these receptors by opsonin factors may give some effects on the signal transduction pathways from FcR to NADPH oxidase. A Holstein calf with leukocyte adhesion deficiency (BLAD) was a proband with a genetic deficiency of the CR3 (Mac-1, CD11b/CD18) glycoprotein corresponding to the receptor of complement iC3b in neutrophils (3). Our previous experiments showed that neutrophils obtained from BLAD were quite low in generating O_2^- when stimulated with serum-opsonized zymosans (s-OPZ) compared to those from normal calves (4,5). This result suggested that the defects of CR3 might affect the ability of neutrophils to generate O_2^-. In the present study, we applied the ESR method combined with spin-trapping to get information about the special interactions of the IgG Fc receptor (FcR) and the complement iC3b receptor CR3 leading to the generation of O_2^-.

Spin-trapping is a method designed to trap free radical intermediates like reactive oxygen radicals (O_2^-, ·OH, OCl^- and ·NO) (2). ESR combined with this method is especially useful to study oxygen-radical generation in biological systems (6,7). Several spin-trapping agents have been developed, and 5,5-dimethyl-1-pyrroline N-oxide (DMPO) is most suitable for trapping of reactive oxygen radicals among them (7). When the spin trap, DMPO, is present in an aqueous solution in which reactive oxygens such as O_2^- and ·OH are generated, the reaction shown in Fig. 1 take place to form the spin adducts (so-called spin-trapping reactions). The resulting nitroxide radicals

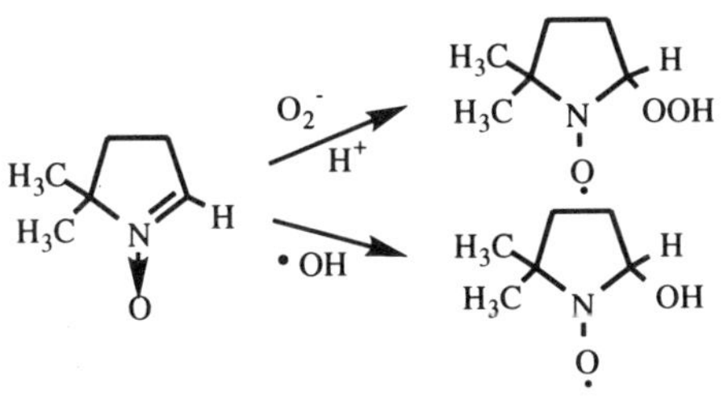

(1)

(2)

Fig. 1. Spin-trapping reactions of O_2^- (1) and OH radical (2) by DMPO are stable, so that these radicals are readily observed by conventional ESR spectrometry.

Since both DMPO-OOH and DMPO-OH adducts give their own characteristic ESR spectra, one can identify which kind of reactive oxygen is generated in biological system in question. However, DMPO-OOH adducts are automatically converted to the form of DMPO-OH adducts depending on the rate of O_2^- generation and post-reaction incubation time, and the ESR spectrum due to DMPO-OH adducts become dominant. Therefore, it is important to distinguish between the ESR spectrum due to DMPO-OH adducts which are originally produced in biological system and that due to DMPO-OH adducts which are derived from DMPO-OOH adducts. For this purpose, dimethyl sulfoxide (DMSO) is added to the reaction system as a OH radical scavenger. The reaction of OH radical with DMSO results in the formation of methyl radical ($\cdot CH_3$). The methyl radical is subsequently trapped by DMPO to form methyl spin adduct (Fig. 2). This adduct also gives its own characteristic ESR spectrum. Thus, observation of the ESR spectrum due to DMPO-CH_3 adducts instead of the DMPO-OH adducts means that OH radicals are surely produced in the biological system in question and not derived from the DMPO-OOH adducts.

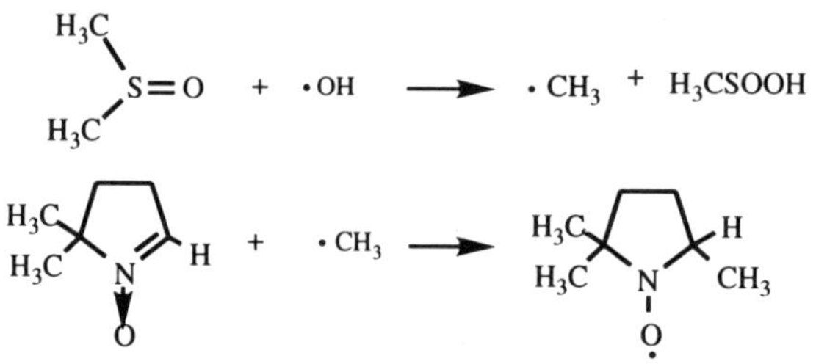

(3)

(4)

Fig. 2. OH radical scavenging by DMSO and the subsequent reaction between methyl radical and DMPO.

The ESR spectrum obtained from stimulated neutrophils consisted of two components corresponding to the DMPO-OOH and DMPO-OH adducts, respectively, as described later. Addition of DMSO to the reaction solutions gave no effects on the relative amounts of two adducts, suggesting that OH radicals were little directly produced in stimulated neutrophils.

2 Materials and Methods

2.1 Animals

A Holstein-Friesian calf (20 months old) referred to the Veterinary Teaching Hospital of Rakuno Gakuen University and diagnosed as bovine leukocyte adhesion deficiency (BLAD) because of greatly reduced expression of β subunit CD18 in CR3 receptor

(Mac-1). The clinical signs and hematological, biochemical and pathological findings were reported in previous papers (4,5,8). Five healthy calves were used for establishing normal response. Blood was collected from the jugular into evacuated tubes containing heparin as an anticoagulant.

2.2 Chemicals

Phorbol 12-myristate 13-acetate (PMA), zymosan A (from *S. Cerevisiae)*, Ficoll, wortmannin, 1-(5-isoquinolinylsulfonyl)2-methylpiperazine (H-7) and DMPO were purchased from Sigma Chemical Company. Bovine IgG and Conray 400 were obtained from ICN Biochemicals and Daiichi Pharmaceutical Company (Tokyo), respectively. Genistein and staurosporine were from Wako Pure Chemicals (Tokyo).

The buffer system used in this work was Hanks' balanced salt solution, pH 7.4, which is free from Ca^{++} and Mg^{++}, HBSS(-). PMA was dissolved in dimethyl sulfoxide (DMSO). DMPO was dissolved in pure water at the concentration of 0.5 M.

2.3 Isolation of Neutrophils

Isolation of neutrophils from normal and BLAD calves were carried out essentially according to the method previously described (5,9). In brief, the peripheral blood (10 mL) which was diluted twice with 30 mL of HBSS was layered onto 10 mL of Ficoll-Conray solution (density: 1.078) and centrifuged at 1600 g for 20 min at 20°C. The cells in the lower layer were treated by adding 0.85% Tris-HCl solution and incubating for 5 min at 4°C to lyse the red cells. The solution was again centrifuged at 500 g for 5 min. The remaining cells were washed twice with cold HBSS(-) and resuspended in the same solution at a concentration of about $5x10^7$/mL. The purity of the cell preparation was more than 94%.

2.4 Preparation of s-OPZ and IgG-OPZ

Zymosan A was suspended in physiological saline and boiled at 100°C for 30 min. After centrifugation, the zymosan A was washed with physiological saline and resuspended in HBSS(-) solution at a concentration of 10 mg/mL. Opsonization was carried out by mixing the zymosan A solution and cow fresh serum or IgG solution (20 mg/mL) at a ration of 1:3 and incubating for 30 min at 37°C with shaking.

PMA was dissolved in DMSO at a concentration of 500 μg/mL and diluted to a concentration of 50 μg/mL in HBSS.

2.5 Spin Trapping of O_2^- From Simulated Neutrophils in the Presence or Absence of Protein Kinase Inhibitor.

Twenty-five μL of a solution with each protein kinase inhibitor (genistein (tyrosine kinase inhibitor), wortmannin (PI 3-kinase inhibitor) and H7 or staurosporine (PKC inhibitor)) was added to 150 μL of HBSS containing $2x10^7$ neutrophils followed by a incubation for 10 min at 37°C. Then 25 μL of HBSS containing s-OPZ or IgG-OPZ or 25 μL of DMSO solution containing PMA was added to the solution followed by a further incubation for 10 min in the case of s-OPZ and IgG-OPZ, and for 5 min in the case of PMA. Finally, 25 μL of aqueous DMPO solution (0.5 M) was added to the solution. Immediately after 1-min incubation, the reaction mixture was transferred to a flat ESR cell (10 x 40 x 0.2 mm) and fitted in the cavity of the X-band ESR spectrometer (JEOL-RE-1X).

The ESR spectrum was recorded at 25°C. Control experiments were done in the absence of stimulant. The ESR conditions were as follows. Microwave power was 20 mW, modulation frequency was 100 kHz with an amplitude of 0.1 mT, sweep time was 10 mT/min, and the receiver gain was $1x10^3$ with response times of 0.5 sec.

3 Results and Discussion

3.1 Quantitative Measurements of O_2^- Generated in Stimulated Neutrophils From Normal and BLAD Calves by the ESR-spin Trapping Method

Fig. 3 shows the first derivative ESR spectra obtained from normal and BLAD neutrophils when stimulated with s-OPZ, IgG-OPZ and PKC activator PMA. ESR spectra obtained from s-OPZ- or IgG-OPZ-stimulated neutrophils were similar to an ESR spectrum of the DMPO-OH adduct. Therefore, OH radicals seemed to be generated in neutrophils which were stimulated with them (reaction 2). However, since addition of OH radical scavenger DMSO gave no effects on these ESR spectra (reactions 3 and 4), these ESR spectra were regarded as the results of trapping of O_2^-. Stimulation of neutrophils with PMA gave a typical ESR spectrum of DMPO-OOH, indicating the generation of O_2^- (reaction 1).

When BLAD neutrophils were stimulated with s-OPZ and IgG-OPZ, ESR spectra obtained were smaller than those of normal neutrophils (about 15% of normal neutrophils), whereas when BLAD neutrophils were stimulated with PMA, the generation of O_2^- was similar to that of normal neutrophils. O_2^- is known to be generated during the oxidation of NADPH by NADPH oxidase. Therefore, these results indicated that the activity of NADPH oxidase was intact in BLAD as well as normal neutrophils and the lack of complement iC3b receptor, CR3, were responsible for the abnormal generation of O_2^- in BLAD neutrophils.

Figure 4 shows the results obtained by quantitative measurements of O_2^- generated in normal and BLAD neutrophils when they were stimulated with s-OPZ and IgG-OPZ in the presence or absence of protein kinase inhibitor. The amount of O_2^-

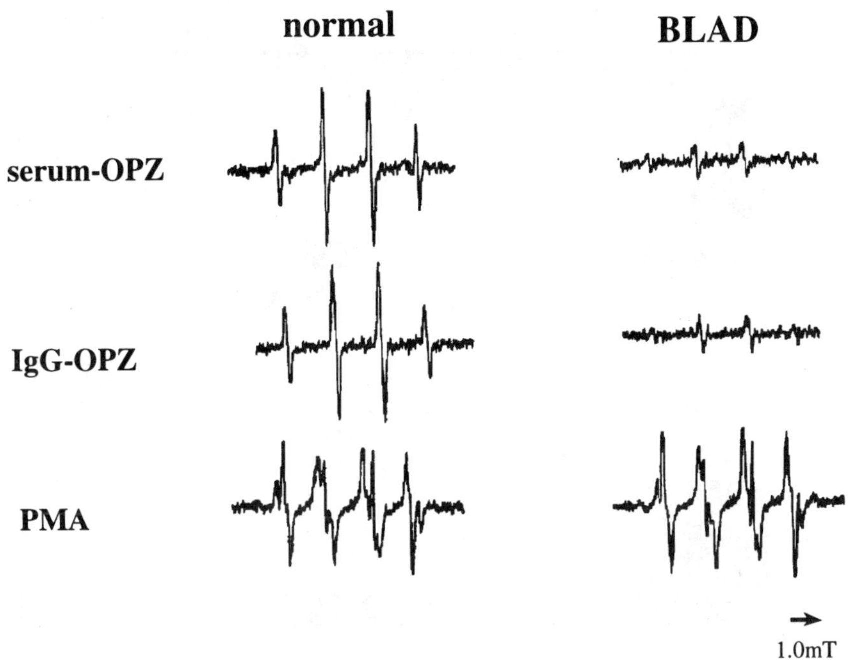

Fig. 3. ESR spectra obtained from stimulated neutrophils. Left, normal and right, BLAD neutrophils. Upper, neutrophils stimulated with s-OPZ, middle, those with IgG-OPZ, and those with PKC activator PMA.

generated in BLAD neutrophils was about 15% that of normal ones regardless of the difference of stimuli. This means that the lack of CR3 corresponding to the receptor of complement iC3b largely affected the ability of neutrophils to generate O_2^-. The interesting facts were that (1) tyrosine kinase inhibitor genistein showed the inhibiting effects for the O_2^- generation in both types of neutrophils regardless of the difference of stimuli, (2) PI 3-kinase inhibitor wortmannin did not suppress the ability of BLAD neutrophils to generate O_2^-, but suppressed that of normal neutrophils to the level of BLAD ones when they were stimulated with s-OPZ, and (3) this inhibitor gave no effects on both normal and BLAD neutrophils when stimulated with IgG-OPZ. In IgG-OPZ-stimulated neutrophils, no ligands to bind to the CR3 are present. Only the pathway from IgG-bound FcR to NADPH oxidase is activated. Therefore, these results indicated that (1) the pathway from IgG-bound FcR to NADPH oxidase was essential for generation of O_2^- and tyrosine kinase mediated the pathway between FcR and NADPH oxidase, and (2) PI 3-kinase associated with the CR3 and the PI 3-kinase activation through the CR3 acts on the FcR to promote its signal transduction to NADPH oxidase.

The treatments of neutrophils with PKC inhibitor H-7 (or staurosporine) largely affected the generation of O_2^- in normal but not in BLAD. This meant that PKC did

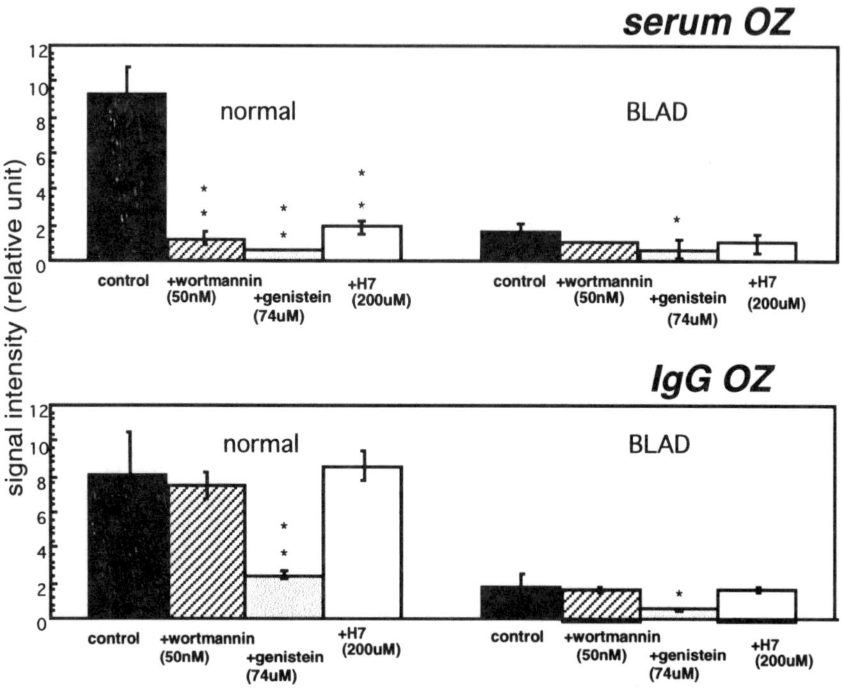

Fig. 4. Quantitative measurements of O_2^- generation in normal (left) and (right) neutrophils when stimulated with s-OPZ (upper) and IgG-OPZ (lower) in the presence or absence of protein kinase inhibitor.

not directly participate in the signal transduction pathway from FcR to NADPH oxidase. However, the generation of O_2^- in neutrophils directly stimulated with PMA was largely suppressed by the treatments with all inhibitors (genistein, wortmannin and H-7) regardless of the difference of neutrophils (data not shown), suggesting that direct activation of PKC by PMA leads to the generation of O_2^- through the signal transduction pathways other than that from FcR to NADPH.

3 Conclusions

From these results it was concluded that (1) the ability of neutrophils from BLAD to generate O_2^- was found to decrease to 15% that of neutrophils from normal calves, (2) all inhibitors used in the present experiments suppressed the O_2^- formation of neutrophils from normal calves, but only tyrosine kinase inhibitor genistein was effective to suppress that of neutrophils from BLAD, suggesting that the binding of iC3b to CR3 amplifies the signal transduction pathway from FcR to NADPH oxidase through tyrosine kinase was essential for generating O_2^-, (3) the ability of neutrophils from normal calves to generate O_2^- after inhibiting PI 3-kinase by wortmannin was quite similar to that of neutrophils from BLAD, suggesting that the binding of iC3b to CR3 amplifies the signal transduction pathway from FcR to NADPH oxidase through activation of PI 3-kinase, and (4) PKC inhibitor H-7 effectively suppressed the O_2^- formation of normal neutrophils but less effectively suppressed it of BLAD neutrophils as well as IgG-stimulated normal neutrophils, suggesting that PKC also participates in the modification of the signal transduction pathway form FcR to NADPH oxidase.

References

1. S. M. Francisco, A. N. Janice, E. Robbins, P, Simon, and T. A. Springer , A human leukocyte differentiation antigen family with distinct α-subunits and a common β-subunit, *J. Exp. Med.* **158**(1983)1785-2803.
2. V. D. Bianca, M. Grzeskowiak, S. Dusi, and F. Rossi, Transmembrane signaling pathways involved in phagocytosis and associated activation of NADPH oxidase mediated by FcγRs in human neutrophils, *J. Leukocyte Biol.* **53**(1993)427-438.
3. D. E. Shuster, M. E. Kehrli, M. R. Ackermann, and R. O. Gilbert, Identification and prevalence of a genetic defect that causes leukocyte adhesion deficiency in Holstein cattle, *Proc. Natl. Acad. Sci.* **89**(1992)9225-9229.
4. H. Nagahata, H. Noda, K. Takahashi, T. Kurosawa, and M. Sonoda, Bovine granulocytopathy syndrome: Neutrophil dysfunction in Holstein Friesian calves, *J. Vet. Med.* **34**(1987)445-451.
5. M. Kuwabara, Y. Nishimura, F. Sato, T. Takahashi, M. Tajima, K. Takahashi, and H. Nagahata, Spin-trapping and chemiluminescence studies of neutrophils from a Holstein-Friesian calf with bovine leukocyte adhesion deficiency, *Free Rad. Res.* **18**(1993)309-318.
6. E. G. Janzen, Spin trapping, *Acc. Chem. Res.* **3**(1971)31-40.
7. E. Finkelstein, G. M. Rosen, and E. J. Rauckman, Spin trapping of superoxide and hydroxyl radical: Practical aspects, *Archiv. Biochem. Biophys.* **200**(1980)1-16.
8. K. Takahashi, K. Miyagawa, S. Abe, T. Kurosawa, M. Sonoda, T. Nakade, H. Nagahata, H. Noda, Y. Chihaya, and E. Isogai, Bovine granulocytopathy syndrome of Holstein-Friesian calves and heifers, *Jpn. J. Vet. Sci.* **49**(1987)733-736.
9. H. Nagahata, A. Yatsu , and H. Noda, The evaluation of a quantitative assay for estimating the bactericidal activity of bovine neutrophils by nitroblue tetrazolium reduction, *Br. vet. J.* **142**(1986)578-583.

Redox Behavior of 5,5-Dimethyl-1-pyrroline-N-oxide (DMPO) as Studied by Electrochemical-ESR Measurement

Kunihiko TAJIMA, Nobuyuki ENDOH, Kenji KANAORI, and Keisuke MAKINO

Department of Polymer Science and Engineering,
Kyoto Institute of Technology, Matsugasaki, Sakyo-ku, Kyoto 606, Japan

Abstract. Redox behavior was explored for 5,5-dimethyl-1-pyrroline-N-oxide (DMPO) by electrochemical-ESR measurements. ESR signals ascribable to 5,5-dimethyl-1-pyrrolind-2-one-N-oxide radical (DMPOX) were observed by applying 400 mV (vs. Ag/Ag^{+}) to aqueous DMPO solutions. The DMPOX formation was supported by reversed-phase HPLC. The major precursor was found to be 1-hydroxy-5,5-dimethyl-1-pyrrolid-2-one.

1 Introduction

Nitrone spin traps are widely used to provide evidence for free radicals involvement in chemical and biological reactions.[1] 5,5-Dimethyl-1-pyrroline-N-oxide (DMPO) has been well established as one of the most useful spin traps to identify oxygen-centered radicals. Previously, however, we have detected 5,5-dimethyl-2-hydroxypyrroline-N-oxyl (DMPO/OH) and 5,5-dimethyl-pyrrolid-2-one-N-oxyl (DMPOX) radicals in aqueous DMPO solutions containing ferric ion,[2] and found that this radical formation does not involve free radical generation. In this reaction system, DMPO/OH is produced at the beginning of the reaction, and gradually converted into DMPOX through the formation of 2-hydroxy-5,5-dimethyl-1-pyrroline-N-oxide and its tautomer 1-hydroxy-5,5-dimethyl-1-pyrrolid-2-one (HDMPN) as intermediate species. The conversion of DMPO into DMPO/OH and further into DMPOX was interpreted as the nucleophilic addition of water molecules to DMPO chelating with ferric ion. Since the coordination and redox reactions proceed simultaneously in this reaction system, the accurate reaction mechanism for the generation of the DMPO/OH and DMPOX radicals were equivocal until present. To see if such species are also formed by electrochemical oxidation of DMPO, we applied electrochemical-ESR and reversed-phase (RP) HPLC to aqueous DMPO solutions.

2 Experimental

DMPO was obtained from Labotec (Tokyo) and used without further purification. 3,3,5,5-Tetramethylpyrroline-N-oxide (M4PO) and HDMPN were synthesized and purified according to the ordinary methods.[3] ESR spectra were recorded using a TE-300 X-band spectrometer (JEOL, Tokyo) coupled with and electrochemical-ESR cell (EL-3, JEOL), which consisted of gold-helix, gold-wire, and silver-wire as working-, counter- and reference-electrodes, respectively. Potassium perchlorate was used as a supporting electrolyte (0.1M). Products analysis was conducted by RPHPLC: monitoring wavelength, 220 nm; column, ODS; eluent, CH_3CN-H_2O, CH_3CN concentration gradient, 1-50 % (linear).

3 Results and Discussion

3.1 Electrochemical ESR of DMPO Derivatives

On supplying 400 mV (*vs.* Ag/Ag^+) to aqueous DMPO (1.0 mM) solutions containing $KClO_4$ (0.1 M), a triplet of triplet ESR hyperfine structure (hfs) was recorded (Fig.1-b) in an electrochemical-ESR cell. The hyperfine coupling constants (hfcc) were estimated as a^N=0.35 and a^H=0.71 mT by computer simulation. The small a^N hfcc value is characteristic of the nitroxide radical of pyrrolidone derivatives, where the carbonyl group reduces the spin density on the nitroxide moiety, and the observed ESR hfcc's were consistent with the values reported for DMPOX, thus allowing us to assign this radical as DMPOX. The signal intensity increased depending on the increase of applied potential. Unexpectedly, however, DMPO/OH ESR signals were not detected although it is one of the precursors for DMPOX in conventional chemical oxidation of DMPO. The lowest oxidation potential of 400 mV was obtained for DMPOX formation. The oxygen atom introduced to form DMPOX was possibly from water molecules, since identical ESR signals were equally detected both under the aerobic and anaerobic conditions.

Similar electrochemical-ESR measurements were also carried out for the precursor of DMPOX, HDMPN. No significant ESR signal was detected for aqueous HDMPN (1.0 mM) containing $KClO_4$ (0.1 M). On supplying 200 mV, which is lower than the oxidation potential for DMPO, a triplet of triplet ESR hfs attributable to DMPOX was observed (data not shown). This result implies that on the electrode applied 400 mV, DMPO is converted first into HDMPN and further oxidized to DMPOX at this potential.

Analogously, for M4PO, DMPOX-type derivative, 3,3,5,5-tetramethyl-pyrrolid-2-one-N-oxyl (M4POX, a^N=0.65 mT), was obtained by electrochemical conversion (Fig.2-b). The lowest oxidation potential for M4POX formation was 500 mV.

3.2 RPHPLC Products Analysis

Species produced from DMPO by electrolysis were also analyzed by RPHPLC. As shown in Fig.3-a and -b, DMPO eluted at 10 min and HDMPN at 18.6 min, indicating that DMPO used was free from HDMPN. After electrolysis of DMPO at 600 mV for 6 min, a pair of new peaks appeared at 1.8 and 2.2 min (Fig.3-c and the inset-c). The first peak was due to the inorganic salts present. In order to assign the second peak, HDMPN was chemically oxidized with PbO_2 and a peak with the retention time of 2.2 min appeared in the chromatogram, as represented in Fig.3-d and the inset-d. This reaction mixture gave strong ESR signals due to DMPOX, thus indicating that the chromatographic peak observed at 2.2 min is ascribed to DMPOX.

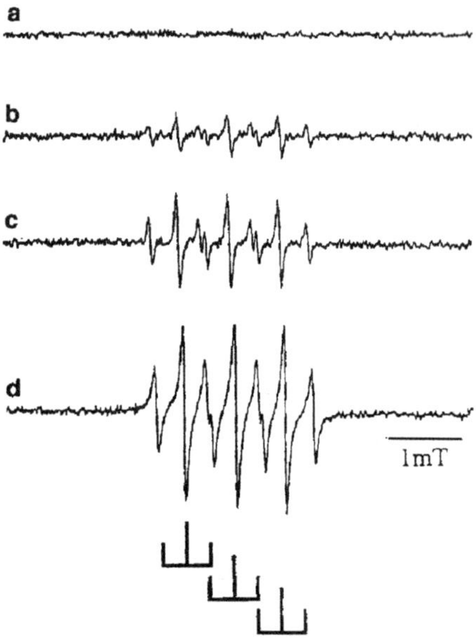

Fig.1 Electrochemical-ESR spectra observed at 300 K for an aqueous DMPO (10.0mM) solution containing 0.1M $KClO_4$. Applied potentials; a) 0 mV, b) 400 mV, c) 600 mV, and d) 1200 mV (*vs.* Ag/Ag^+).

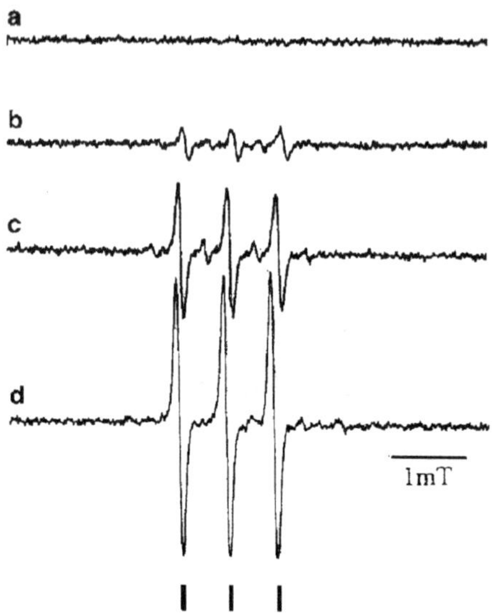

Fig.2 Electrochemical-ESR spectra observed at 300 K for an aqueous M4PO (10.0 mM) solution, containing 0.1M $KClO_4$. Applied potentials; a) 0 mV, b) 500 mV, c) 700 mV, and d) 1200 mV (*vs.* Ag/Ag^+).

Further RPHPLC analysis was conducted to see if DMPO/OH is generated in the electrochemical DMPO oxidation. For obtaining DMPO/OH, DMPO was added to Fenton reaction system composed of ferrous sulfate, hydrogen peroxide, and diethylenetriaminepentaacetic acid (DTPAC). As shown in Fig.3-e and its inset-e, an unknown peak was detected at 1.6 min in addition to the DMPO peak. The retention time of this unknown peak was close to that of DMPOX (2.2 min). Taking into account the fact that DMPO/OH is a mojor product of the DMPO-Fenton reaction system, this unknown peak may be due to DMPO/OH. To see if this is the case, electrochemical-ESR measurements were conducted for analogous system, and however, DMPO/OH was not detected, as depicted in the inset-c, indicative no DMPO/OH peak in the RPHPLC chromatograms. Consequently, DMPOX is only the major product of DMPO electrolysis at 400 mV (Scheme 1).

In summary, detection of DMPOX ESR signals can be an indicator for some redox process occurring in chemical and biological systems.

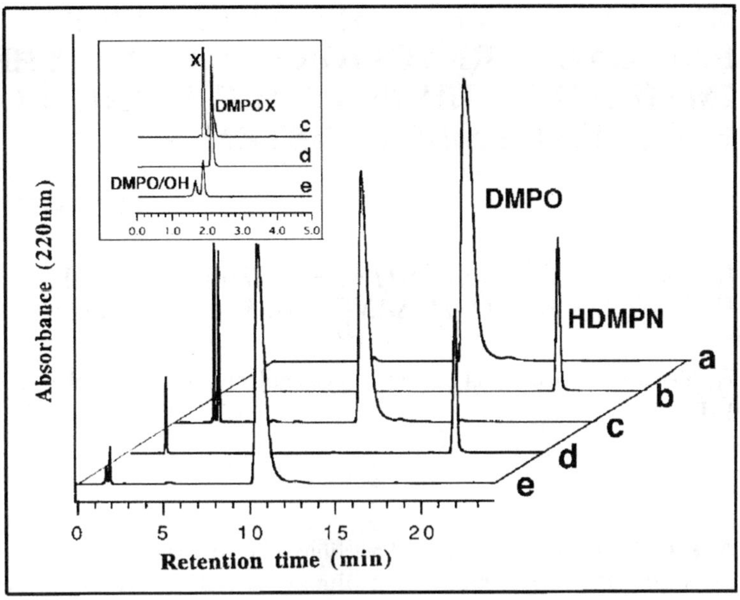

Fig.3 RPHPLC chromatograms observed for the solutions of a) DMPO, b) HDMPN, c) DMPO after electrolysis at 600 mV for 6 min, d) HDMPN oxidized by PbO_2, and e) DMPO of Fenton system composed of ferrous sulfate, DTPAC and hydrogen peroxide. Insets show enlarged chromatograms of Fig. 3-c, -d, and -e. X is due to salts in the buffer.

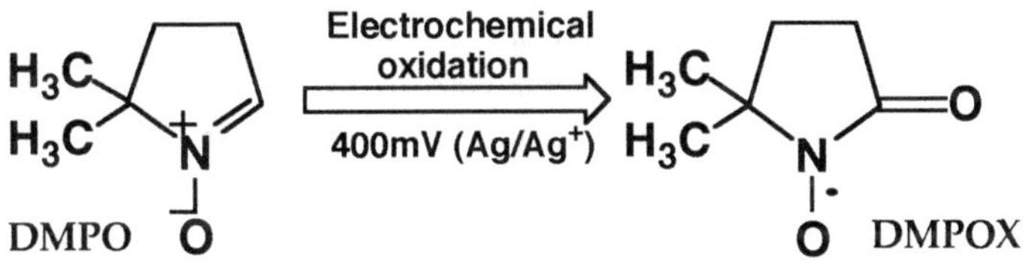

Scheme 1 Reaction mechanism for the electrochemical oxidation of DMPO at 400 mV (*vs.* Ag/Ag^+).

References

1. E. G. Janzen, *Acc Chem. Res.*, **4**, 31 (1971).
2. K. Makino, et al., *Can. J. Chem.*, **70**, 2818 (1992).
3. R. Bonnet, et al., *J. Chem. Soc.*, 2102 (1959); M. C. Kloetzel, et al.,

NON-INVASIVE MEASUREMENT OF RADICAL REACTIONS AND THEIR IMAGING IN LIVING MICE USING LOW-FREQUENT ESR-CT SYSTEM

H. UTSUMI, K. TAKESHITA, K. ICHIKAWA, J.-Y. HAN, K.-I. MATSUMOTO, H. SANO, and K.-I. YAMADA

Faculty of Pharmaceutical Sciences, Kyushu University, Fukuoka 812-82, JAPAN

Abstract. The *in vivo* free radical reactions were estimated in living mice with an *in vivo* ESR spectrometer using nitroxyl radicals as probes. One of the following nitroxyl radicals, 2,2,6,6-tetramethylpiperidine-1-oxyl (TEMPO), 2,2,5,5-tetramethylpyrrolidine-1-oxyl (PROXYL), 4,4-dimethyloxazolidine-1-oxyl (OXANO), and their derivatives, was dissolved in isotonic buffer and was intravenously, intramuscularly, transstracheally or intraperitoneally injected to female ddY mice. The ESR signal of nitroxyl radical in living mice was gradually decreased by reducing to the corresponding hydroxylamine, and the reduction rate depended on physiological and pathological conditions. Pre-treatment with antioxidants reduced the enhancement of signal decay by oxidative stress. The present paper demonstrates that the *in vivo* ESR technique using nitroxyl radical as a probe is very useful to estimate *in vivo* free radical reactions and to evaluate their relation to physiological and pathological phenomena.

1 Introduction

Free radicals such as active oxygen species and nitric oxide are believed to be very essential and functional compounds in various biological systems. Non-invasive measurement of *in vivo* free radical reactions is very important to understand the role of free radicals in the living body, since oxygen concentration in tissues is much lower than that in *in vitro* experiments and there are many substances and reaction pathways which influence *in vivo* radical reactions. Recently, L-band ESR spectroscopy has developed and enabled non-invasive *in vivo* measurement of radicals in whole animals (Bacic et al. 1989, Ferrari et al. 1990, Ishida et al. 1989, Nishikawa et al. 1985, Subczynski et al. 1986, Utsumi et al. 1990). ESR measurement of *in vivo* radical generation in living animals has rarely been reported because of poor sensitivity of the L-band ESR spectrometer. Nitroxyl radicals are susceptible to biological redox systems, indicating that combination of an L-band ESR spectrometer with nitroxyl radicals provide valuable information about the biological function of free radical reactions. Thus, we have studied *in vivo* free radical reactions using nitroxyl radicals which have been administered intravenously (Utsumi et al. 1990, 1993, Miura et al. 1992, 1995), intramuscularly (Masuda et al. 1991, Utsumi et al. 1993), intraperitoneally (Gomi et al. 1993), and transtracheally to mice (Takeshita et al. 1991, 1993). The results were that the reduction of nitroxyl radical depended on the physiological and pathological conditions such as aging, oxidative stress, etc.

In the present paper, we review our recent results about non-invasive ESR measurement of *in vivo* free radical reactions in living mice using nitroxyl radicals as probes and discuss the utilization of this technique for clarifying their role in biological responses.

2 Materials and Methods

Most TEMPO and PROXYL derivatives were purchased from Aldrich Chemical Co.. Glutaramide- and methylglutaramide-TEMPO and OXANO derivatives were synthesized in our laboratory. The water soluble probes were dissolved in isotonic solution, and the lipid probes were suspended with the other auxiliary lipids. Osmolarity of the solution was checked by freezing point depression. Pentobarbital sodium was purchased from Dainabot Co. Ltd. Other reagents were of the highest purity commercially available.

Female ddY mice (3-4 weeks old, 15-20 g body weight) were used throughout this study. Mice were anesthetized by intramuscular or intraperitoneal injection of pentobarbital (50 mg/kg) and fixed on a hand-made Teflon holder. Immediately after administration of isotonic solution of spin-probes through various routes, ESR spectra were measured as follows. The microwave frequency was 1.1-1.3 GHz and the power was 1.0-5.0 mW. The amplitude of the 100 KHz field modulation was 0.2 mT. The external magnetic field was swept at a scan rate of 5 mT/min.

3 Results and Discussion

Figure 1a, b and c show typical ESR spectra at the hepatic domain of a mouse with carbamoyl-PROXYL, hydroxy-TEMPO and OXANO administered into the tail vein, respectively. PROXYL and TEMPO derivatives gave three sharp-lines were observed with regular noise due to respiration. The hyperfine structure and the peak height ratios of carbamoyl-PROXYL coincided with those of the probe dissolved in saline without significant spin-exchange interaction. The quite similar

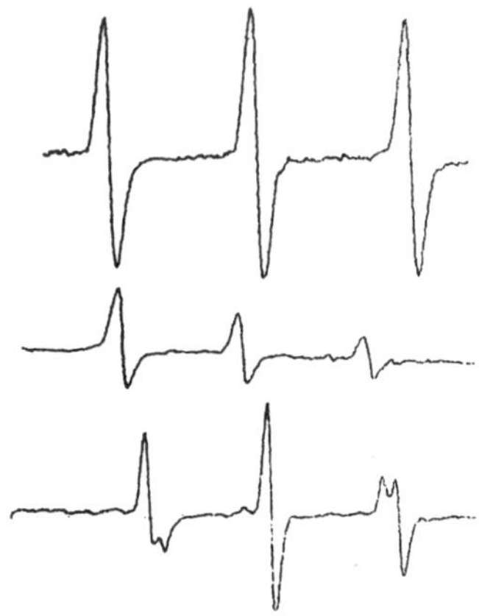

Fig. 1 ESR spectra of carbamoyl-PROXYL (a), hydroxy-TEMPO (b) and OXANO (c) in the hepatic domain of mice. One hundred microliters of an isotonic solution of spin-probes (280 mosM) were injected into a tail vein of a female ddY mouse, and ESR spectrum at the hepatic domain of mouse was observed with an L-band ESR spectrometer.

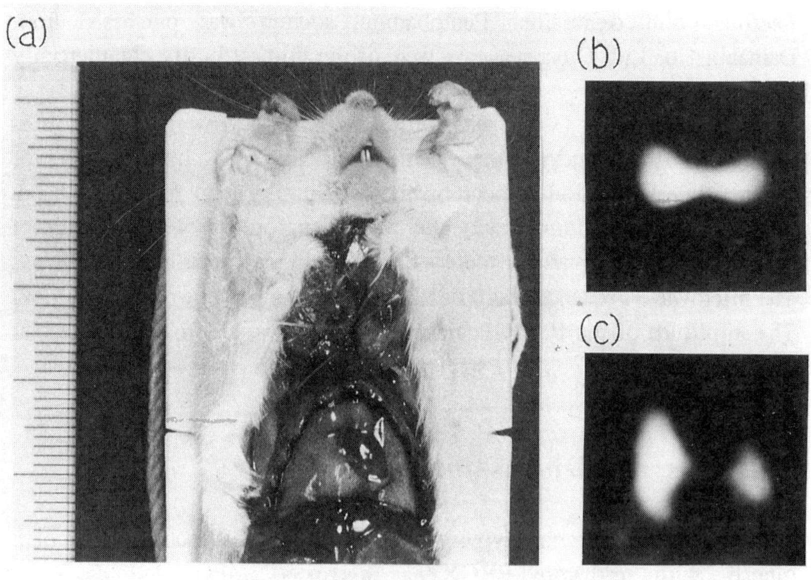

Fig. 2 ESR image of nitroxyl-probe injected into mouse lung (b, c) and an anatomical picture of the mouse used for ESR image (a). ESR image was performed from two directions after transtracheal injection of carboxy-PROXYL (15 mM, 0.9 ml) to mouse lung; from the ventral side (c) and from tail along the body axis (b). The scale of ESR imaging is adjusted to the anatomical picture.

spectra were also observed at head, chest, and lower abdomen after intravenous administration (Utsumi et al. 1990, Miura et al. 1992). OXANO gave a quite different spectrum having two set of triplet lines. One is due to probe in water phase and the other in oil phase.

Spin-probes injected transtracheally into mice lung showed the same triplet lines as those in water. Fig. 2 demonstrates the ESR-CT imaging of carboxy-PROXYL injected into mouse lung. The scale and position of the image agreed well to those of the lung, suggesting that the spin-probe distribute all over the lung (Takeshita et al. 1991).

Intramuscular administration of high concentration of spin-probe gave different spectrum at injected domain from those shown in Fig. 1. The shapes of the spectra depended on the period after injection, and the signals seemed to consist of several components. Computer simulation of the spectra indicated the diffusion of spin-probes in muscle (Masuda et al. 1991).

ESR signals of nitroxyl radicals decreased gradually in living mice. The semilogarithmic plot of the peak heights at hepatic domain after intravenous administration of spin-probes gave a straight line till at least 10 min, indicating that signal decay after i.v. injection should obey first order kinetics. However, the decay curve of at head after intraperitoneal administration was quite different and needed ca. 10 min for reaching the maximum peak height level (Gomi et al. 1993).

Table 1 demonstrates the decay constants of various spin-probes at head and breast after intravenous injection. The decay constants for TEMPOs were several times larger than those for the PROXYLs, and the probes having large n-octanol/water partition coefficients were reduced more quickly. This tendency agreed well to that obtained by *in vitro* experiments. The decay constants at head was significantly smaller than those at breast. The contribution of blood components after the collection to nitroxyl reduction must, however, be very small, since the decay constants of spin-probes in the collected blood were about 0.1 times those observed *in vivo* measurement (Utsumi et al. 1990). Interaction of spin-probes with blood vessels, endothelial cells or parenchymal cells of tissues may be important in the *in vivo* reduction of nitroxyl radicals. The difference of the decay constants among observing domains indicates that the nitroxyl reduction occur not only during blood circulation but also by the interaction with tissues.

There are several different mechanisms in spin-clearance *in vivo*, involving direct reduction to the corresponding hydroxylamine in blood circulation or cells, decomposition to the other non-paramagnetic compounds, excretion through kidney or liver to urine or feces, the binding to macromolecules, etc. However, signal decay in the early stages may be considered to arise from the reduction of nitroxyl radicals to the corresponding hydroxylamine. Small amount of blood was collected from the mouse pre-loaded with spin-probes, and the signal intensity in

Table 1. *In vivo* spin-clearance constants of nitroxyl radicals at head and breast of mice after intravenously injection (1/min, mean ± S.D)

Nitroxyl Radicals	Breast	Head
Hydroxy=TEMPO	0.97 ± 0.33	0.55 ± 0.13
Amino-TEMPO	1.53 ± 0.54	0.92 ± 0.14
CAT-1	0.21 ± 0.08	0.07 ± 0.05
Carboxy-TEMPO	0.46 ± 0.10	0.25 ± 0.02
Carboxy-PROXYL	0.04 ± 0.01	0.04 ± 0.01
Carbamoyl-PROXYL	0.10 + 0.05	0.05 + 0.01

collected blood was compared between before and after oxidation with ferricyanide. The result was the stoichiometric restoration of signal intensity to the amount of injected one after oxidation (Gomi et al. 1993). The presence of the corresponding hydroxylamine was also confirmed by thin-layer chromatography of the metabolic products in mice lung (Takeshita et al. 1993).

Mouse lung has also reduction system against nitroxyl radical, which is unwashable and inactivated by homogenization (Takeshita et al. 1991, 1993). Nitroxyl radicals without any charges were reduced much more quickly than those with charges. The reduction was inhibited strongly with membrane-soluble SH-blocker, N-ethylmaleimide, and amphiphilic ones whose maleimide groups locate within membrane hydrophobic region, but not with membrane-insoluble one, p-chloromercuriphenylsulfonic acid (Takeshita et al. 1993). The membrane-soluble reducing SH reagent, dithiothreitol, enhanced the reduction of nitroxyl radicals, although it showed no direct reactivity towards the radicals. These results indicate that the present reduction system in mouse lung is located within membrane hydrophobic region and that SH-modifying lipophilic compounds regulate the activity of reduction.

Table 2 represents the decay constants of carbamoyl-PROXYL at head of 6- and 30-month-old, ad libitum-fed, and 39-month-old, food-restricted mice after intraperitoneal administration (Gomi et al. 1993). The decay-constant for 30-month-old ad libitum-fed mice was significantly smaller than that for 6-month-old ad libitum-fed mice. The reduction for 39-month-old, food-restricted mice was significantly faster than that for 30-month-old ad libitum-fed mice and close to that for 6-month-old ad libitum-fed mice. These results suggest that reducing capacity in the cardiovascular system declines in old mice and that a food-restriction prevents its age-dependent retardation.

Oxidative stress caused an enhancement of spin clearance. Decay constant of carbamoyl-PROXYL in the abdomen under 80% oxygen was significantly greater than that under 20% oxygen ($p < 0.001$) (Miura et al. 1992). The pre-load of antioxidants such as Trolox, uric acid, and

Table 2 Effect of age and feeding on spin-clearance constants of carbamoyl-PROXYL at mouse head. (Gomi F. et al. (1993))

Age(month)	No. of Mice	Feeding	Clearance-Conctants(/min)
6	10	*ad libitum*	0.038 ± 0.006
30	4	*ad libitum*	0.026 ± 0.004
39	4	restricted	$0.035 + 0.004$

glutathione retarded the enhancement of signal decay under hyperoxia (Miura et al. 1995), and their retardation were corresponded to those estimated with TBA-reactive substances. Ascorbic acid did not show any effect *in vivo*, although it can reduce nitroxyl radical quickly *in vitro*. Ischemia-reperfusion also enhanced the *in vivo* signal decay (Utsumi et al. 1993). The decay at mouse thigh treated with ischemia-reperfusion was significantly faster than that without ischemia-reperfusion ($p < 0.05$), and the increment of the ratio was inhibited by the treatment of SOD or allopurinol.

The present paper strongly suggests that combination of *in vivo* ESR spectroscopy with nitroxyl radicals as a probe is very useful technique to estimate the free radical reactions under various physiological and pathological conditions and to evaluate the *in vivo* effects of antioxidants on biological radical reactions. Recently, drug delivery systems (DDS) are being actively investigated using radio-active compounds in the field of pharmaceutical sciences. *In vivo* ESR technique is also a very attractive method for DDS investigations, if nitroxyl radicals are used as markers of DDS (Yamaguchi et al. 1996). Imaging picture of DDS can also be constructed with an ESR-CT system, demonstrating the advantage of *in vivo* ESR-CT technique to investigation of pharmacodynamics of DDS in living body.

Acknowledgements

This work was in part supported by Grant-in-Aids for General Scientific Research, Developmental Scientific Research, and Scientific Research on Priority Areas from the Ministry of Education, Science and Culture of Japan, and Special Coordination Funds of the Science and Technology Agency of the Japanese Government.

Refereces

1. Bacic, G., Nilges, M. J., Magin, R. L., Walczak, T., & Swartz, H. M. (1989) *In vivo* localized ESR spectroscopy reflecting metabolism. *Magn. Reson. Med.* 10, 266-272
2. Ferrari, M., Colacicchi, S., Gualtieri, G., Santini, M. T., & Sotgiu, A. (1990) Whole mouse nitroxide free radical pharmacokinetics by low frequency electron paramagnetic resonance. *Biochem. Biophys. Res. Commun.* 166, 168-173.
3. Gomi, F., Utsumi, H., Hamada, A., & Matsuo, M. (1993) Aging retards spin clearance from mouse brain and food restriction prevents its age-dependent retardation, *Life Science*, 52(25) 2027-2033
4. Ishida, S., Kumashiro, H., Tsuchihashi, N., Ogata, T., Ono, M., Kamada, H., & Yoshida, E. (1989) *In vivo* analysis of nitroxide radicals injected into small animals by L-band ESR technique. *Phys. Med. Biol.* 34, 1317-1323
5. Masuda, S., Utsumi, H., & Hamada, A. (1991) *In vivo* ESR study on diffusion of spin labeled compounds in femoral muscle. *Jap. Mag. Reson. Med.* (in Japanese) 2, 69-74
6. Miura, Y., Utsumi, H., & Hamada, A. (1992) Effect of inspired oxygen concentration on *in vivo* redox reaction of nitroxide radicals in whole mice, *Biochem. Biophys. Res. Commn.*, 182, 1108-1114
7. Miura, Y., Hamada, A., & Utsumi, H. (1995) *In vivo* ESR studies of antioxidant activity on free radical reaction in living mice under oxidative stress; *Free Rad. Res.*, 22, 209-214
8. Nishikawa, H., Fujii, H., & Berliner, L. J. (1985) Helices and surface coils for low-field *in vivo* ESR and EPR imaging applications. *J. Mag. Res.* ; 62, 79-86
9. Subczynski, W. K., Lukiewica, S., & Hyde, J. S. (1986) Murine *in vivo* L-band ESR spin-label oximetry with a loop-gap resonator. *Magn. Reson. Med.* 3, 747-754

10. Takeshita, K., Utsumi, H., & Hamada, A. (1991) ESR measurement of radical clearance in lung of whole mouse; *Biochem. Biophys. Res. Commn.*, 177, 874-880
11. Takeshita, K., Utsumi, H., & Hamada A. (1993) Whole mouse measurement of paramagnetism-loss of nitroxide free radical in lung with L-band ESR spectrometer, *Biochem. Mol. Bio. Int.*, 29, 17-24
12. Utsumi, H., Muto, E., Masuda, S., & Hamada, A. (1990) *In vivo* ESR measurement of free radicals in whole mice. *Biochem. Biophys. Res. Commun.* 172, 1342-1348.
14, Utsumi, H., Takeshita, K., Miura, Y., Masuda, S., & Hamada, A. (1993) *In vivo* ESR measurement of radical reaction in whole mice. -Influence of inspired oxygen and ischemia-reperfusion injury on nitroxide reduction-; *Free Rad. Res. Commun.*, 19, s219-225
15. Yamaguchi, T., Itai, S., Hayashi, H., Soda, S., Hamada, A., & Utsumi, H. (1996) Non-invasive analysis of parenteral lipid emulsion in mice by *in vivo* ESR spectroscopy, *Pharmaceu. Res.*, 13, 729-733

In-vivo Application of Electron Paramagnetic Resonance (EPR) in Human Skin

Th. Herrling, J. Fuchs, L. Zastrow**, B. Malenke[#], N. Groth*

Technische Fachhochschule Berlin, Rudower Chaussee 6, 12489 Berlin, Germany
* Klinikum der J.-W.-Goethe Universität, Zentrum der Dermatologie und Venerologie, Frankfurt
** Lancaster Group, International Research Center, Monaco
# Galenus Inc., Berlin

Abstract. EPR is currently being explored for the study of living biological systems. EPR Imaging of spin labels is a powerful method for investigating skin and can give information about biophysical and biochemical processes. Among this processes the study of nitroxide radical interactions with tissue oxidants and antioxidants is of growing interest. We investigated the piperidine-type nitroxides Tempo and Tempol and their reaction in skin biopsies of hairless mices and in human skin.

1 Introduction

EPR spectroscopy and imaging techniques are currently being explored for biological in-vivo investigations (Berliner 1989, Swartz and Walczak 1993). Presently the application seems to be restricted to the study of small animals because microwaves have a limited tissue penetration. Skin is a target organ of the in-vivo EPR technique, because microwave penetration is not a limiting factor in this tissue (Fuchs et al. 1991, 1992, 1994). The potential applications of EPR in conjunction with nitroxide radicals in skin research include biophysical and biochemical studies (Herrling et al. 1992) such as oximetry, analysis of membrane fluidity and polarity, free radical detection and measurement of redox interactions with antioxidants and oxidants.

Although nitroxides are among the most stable free radicals known, they are chemically not inert and lose their paramagnetic properties in biological tissues. This feature can be exploited for biochemical investigations. Nitroxides can act as electron acceptors, forming hydroxylamines, and as electron donors, giving oxoammonium cations. Although the stability of nitroxide free radicals in solution, cell suspension, tissue homogenates of various origin and in animals (Eriksson et al. 1987, Swartz et al. 1986, Berliner and Wan 1989, Keana et. al 1989, Kveder et al. 1988) has been investigated in detail, only few studies on nitroxide redox reactions in skin have been performed. In previous investigations Fuchs et al. (1989) have shown that nitroxide radicals are readily metabolized in mouse skin and in mouse skin homogenate. The nitroxide ring structure and to a minor extent the ring substituents determine nitroxide stability in skin and skin cells (Fuchs et al. 1993). In-vivo EPR spectroscopy was successfully performed in human skin using S-band frequency and rapid nitroxide reduction was observed (Fuchs 1995). The reduction of nitroxides could be useful as an indicator of redox metabolism in-vivo. To explore and to understand the complex mechanisms of nitroxide radical penetration and scavenging in mouse and human skin, we have analyzed the reactions of different nitroxides like Tempo or Tempol. Furthermore we have studied the metabolism of Tempo in human skin before and after modulation of cellular antioxidant status.

1.1 Materials and Methods

Nitroxide Assay

The nitroxide assay must posses special properties including sufficient reduction potential, correct polarity and good permeability through the cell membranes and the horny layer of the skin.

The proper solubility of the test assay is a particularly important factor for rapid penetration. The test assay is structured as a microemulsion in order to address the unique character of the skin, in which the horny layer is more lipophilic while the epidermis contains is more hydrophilic. Both the lipophilic and amphiphilic piperidine nitroxides Tempo (2,2,6,6-tetramethyl-1-piperidinyloxy) and Tempol (2,2,6,6-tetramethyl-1-piperidinyloxy-4-ol) were used to accommodate the reduction by the various enzymatic and nonenzymatic antioxidants.

1.2 Skin Preparation

Mouse skin sheets were obtained from hairless mice. Total skin thickness varied between 400 and 800 μm depending on the age of the animals (12 - 24 weeks). Epidermal thickness at the back of the animals was 35 ±5 μm, and for follicular epidermis 160 ±10 μm. To prevent a rapid nitroxide reduction the punch biopsies were preincubated for 10 min in a 50 mM solution of N-ethyl-maleimide (NEM). Tempo or Tempol 100mM in microemulsion was placed on the epidermal surface of a freshly excised skin biopsy for tomographic measurements.

Further more spectroscopic measurements were performed at the inside of the forearm of human subjects (Herrling 1996). The skin was cleaned by isotonic sodium chloride immediately before the measurement. The concentration of the solved nitroxide assay in the microemulsion was 1.0 mM. Ten microliters of the nitroxide stock solution of Tempo were applied on the epidermal surface of the skin incubated at normal room temperature for 240 seconds and washed off with isotonic sodium chloride.

1.3 Methods for Skin Measurements

Four methods (Herrling 1997) were developed for collecting 2D EPR images; 2D Spatial Imaging TO (x,y), 2D Spatial-Spectral Imaging TO (*r*, B), 2D Spatial-Time Imaging TO (*r*, t) and 2D Spectral-Time Imaging S (B,t). We used 2D Spatial-Spectral Imaging TO (*r*,B) and 2D Spectral-Time Imaging S (B,t) for skin investigations. The 2D Spatial-Spectral Imaging is defined by one spatial axis *r* and one spectral axis B. Images with one spatial and one spectral dimension can be obtained using a MOdulated field gradient with Simultaneous superimposed magnetic field Scan (MOSS) which is described in detail by Herrling (1992). A great advantage of MOSS is the fact that it is possible to measure only the spectrum from the volume of interest, and not the whole sample or part of the sample, as is the case which the method using a series of different magnetic field gradients (Maltempo et al. 1988). So the EPR spectrum was only recorded from the epidermis as a function of time, determining the penetration process of stable nitroxides through the horny layer of the skin.

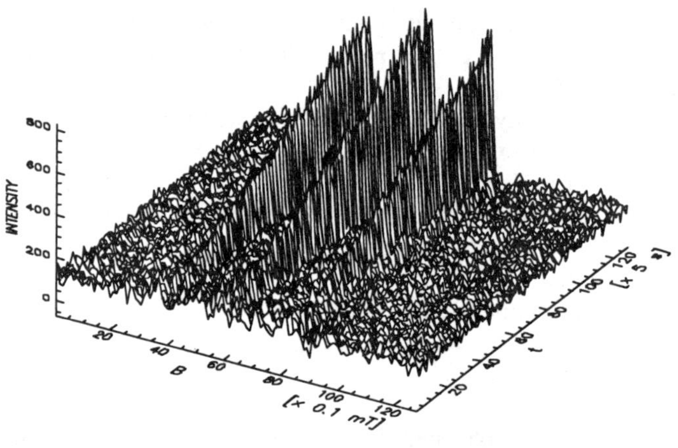

Fig. 1. Increase of Tempo in the epidermis of a hairless mouse

The representation of spectra measured at different times produces a 2D spectral-time image S (B,t) which provides a simple method for visualising the reduction of the spin label.

2 Results

We used Tempo and Tempol, chemically reactive nitroxides to measure their penetration in the epidermis of skin biopsies of hairless mice. To prevent nitroxide reduction the biopsies were treated by NEM for 10 min. The increase of Tempo in the epidermis as a function of time can be seen in Fig. 1. The second derivative of the Tempo spectrum shows that there is no changing in the line shape, giving information that there is no immobilisation of Tempo in the epidermis. Figure 2 represents the cross section for the mid field peak which shows the increase of the nitroxide concentration in the epidermis. Contrary to Tempo, there is no visible Tempol in the epidermis for the same treatment time if we look at the 2D image in Fig. 3. An exact analysis of the 2D image using a higher amplification results in the cross section shown, measured for the mid field peak. Figure 4 shows much smaller increase for Tempol. This image shows that it is much more harder for Tempol to overcome the barrier (horny layer) of the skin. So we have used Tempo for our measurements on human skin. The 2D Spatial-Spectral Image TO (r,B) measured 30 min after treatment of Tempo shows a constant distribution of the spin label which is extended over a range of 800μm from the horny layer to the dermis. This 2D image is shown in Fig. 5 for the second derivative of the Tempo spectrum. The spin label reduction S (B,t) as a function of time was recorded and is represented in Fig. 6. It demonstrates the intensity decrease of the mid field resonance (first derivative) of the three line ^{14}N spectrum of Tempo measured in the skin of the forearm of a human subject. The figure shows ten spectra measured in a time interval of 60 s. Every spectrum is recorded with 1000 data points. The signal intensity I(t) is measured in volts. The reduction curve is nomalized to the intensity of the first spectrum I(0) at time t=0. This reduction curve is fitted to an exponential function (e-curve). The exponent x of the curve function

$$I(t) / I(0) = e^{-xt}$$

describes the rate of the clearance process of Tempo in the skin of the subject. The slope of the reduction curve of the measured spectra in Fig. 6 and the fitted e-function are represented in Fig. 7. Both the exponent or the half life time $t_{1/2}$ give an information about the antioxidative potential of the skin against free radicals.

3 Conclusion

EPR spectroscopy and imaging seems to be a useful method for the in-vivo investigation of mammalian skin especially of human skin. It is a unique method to study paramagnetic centres, such as free radicals, in biological samples. EPR imaging allows the visualization of free radicals. EPR imaging with modulated field gradients allows spatial resolution of paramagnetic centres in different tissue planes, and the possibility of obtaining an EPR spectrum in a selected volume part. A spatial resolution (actual point distinction) between 10 and 100μm can be obtained. X-band microwaves can be used for imaging full-thickness skin biopsies of hairless mice in a rectangular cavity. In vivo skin spectroscopy is possible by employing surface coils at S-band frequency (Herrling else where). The determination of the antioxidative potential (AOP) of human skin by means of the reduction kinetic containing a semistable nitroxide radical can be a safe and feasible non-destructive method for the characterisation of the safety potential against free radicals. Various measurements (Herrling 1996) have shown that oxidative stress like UV-irradiation influences the AOP of the skin and generates a slower reduction of the nitroxide radical.

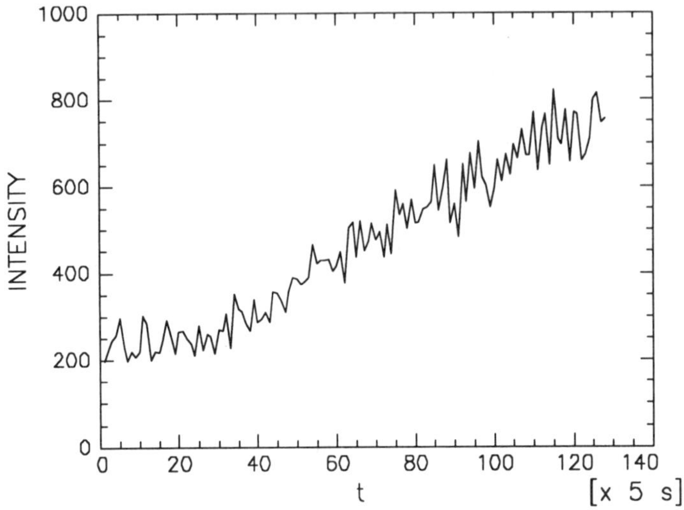

Fig. 2. Penetration profile of Tempo - cross section of the mid field peak

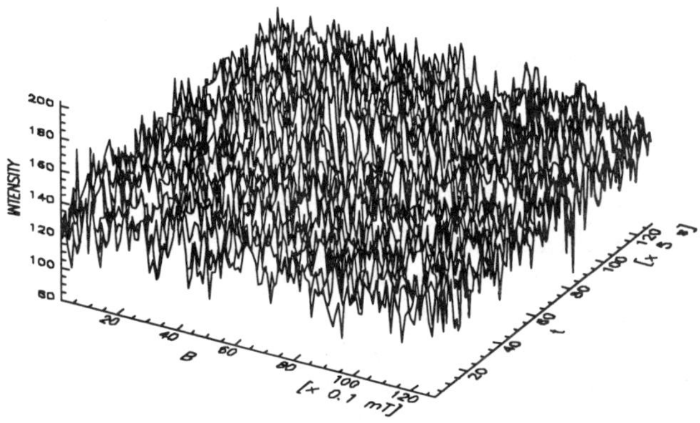

Fig. 3. Spectral-time image for Tempol measured in the plane of the epidermis

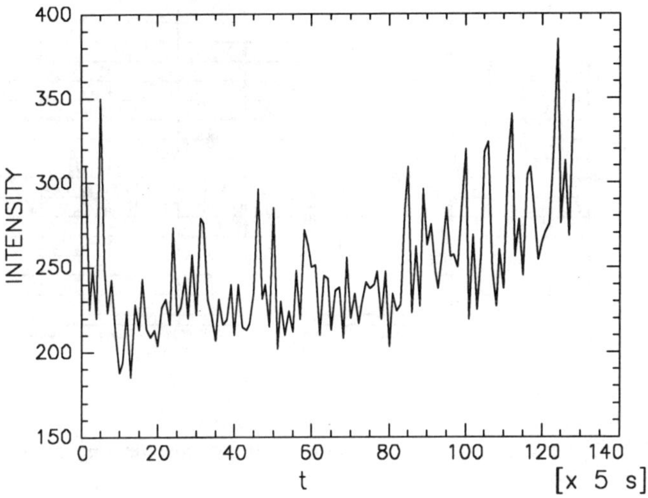

Fig. 4. Cross section of the mid field peak of Tempol measured at a higher amplification

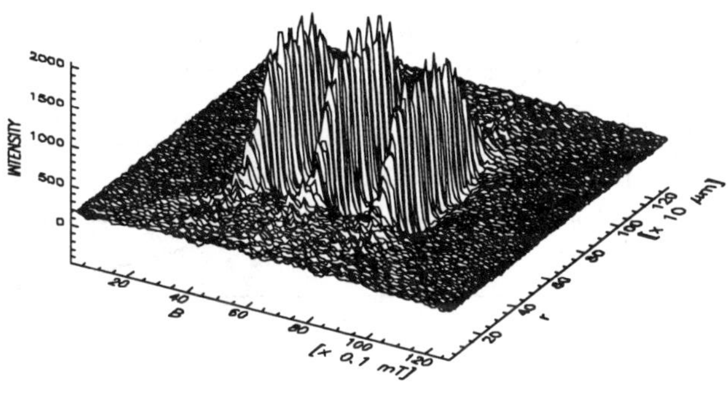

Fig. 5. Spatial-spectral distribution of Tempo in the skin biopsy of a hairless mouse

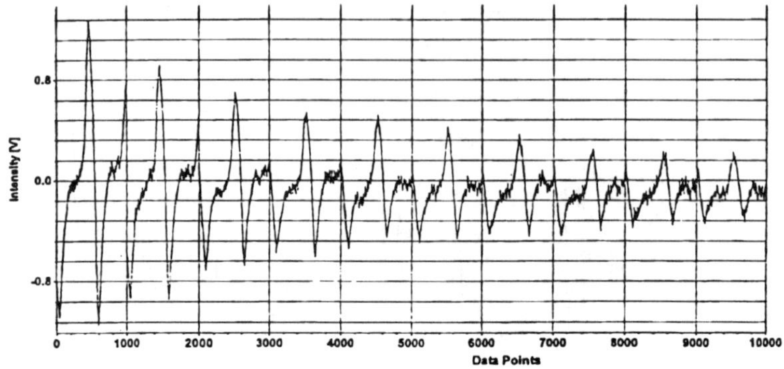

Fig. 6. Ten spectra measured in a time interval of 60s, charaterizing the reduction process of Tempo in human skin

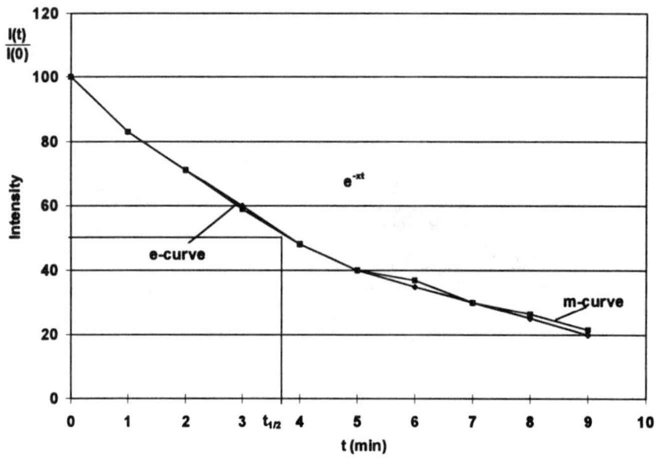

Fig. 7. Fitting of the measured reduction curve by an e-function (e-curve)

References

1. Fuchs J, Milbradt R, Groth N, Herrling T, Zimmer G, and Packer L. (1991) *One and two dimensional EPR (Electron Paramagnetic Resonance) imaging in skin.* Free Radical Research Communication, **15**, 245-253
2. Fuchs J, Milbradt R, Groth N, Herrling T, Zimmer G, and Packer L. (1992) *Electron Parainagnetic Resonance (EPR) imaging in skin: biophysical and biochemical microscopy.* J Invest Dermatol **98**, 713-719
3. Fuchs J, Groth N, Herrling T, and Packer L. (1994) In vivo electron paramagnetic resonance imaging of skin. Methods Enzymology **233**, 140-149
4. Herrling, Th., Groth, N., Thiessenhusen, K.-U., Fuchs, J., Ewert, U. in: *Magnetic Resonance Microscopy*, Blümich, B., Kuhn, W., Eds., pp. 563-572, New York: VCH Verlagsgesellschaft, 1992.
5. Eriksson UG, Ogan MD, Peng CT, Brasch RC, and Tozer TN. (1987) *Metabolic fate in the dog of the nitroxide moiety in a compound with potential utility as a contrast agent in MRI.* Mag Res Med **5**, 73-77
6. Swartz HM, Sentjurc M, and Morse II, PD. (1986) *Cellular metabolism of water soluble nitroxides: effect on rate of reduction of cell / nitroxide ratio, oxygen concentration and permeability of nitroxides.* Biochem Biophys Acta **888**, 82-90
7. Berliner LJ, and Wan X. (1989) *In vivo pharmacokinetics by Electron Magnetic Resonance Spectroscopy.* Magnetic Resonance Medicine **9**, 430-434
8. Keana JFW, Pou S, and Rosen GM. (1987) *Nitroxides as potential contrast enhancing agents for MRI application: influence of structure on the rate of reduction by rat hepatocytes, whole liver homogenate, subcellular fractions and ascorbate.* Magnet Res Med **5**, 525-536
9. Kveder M, Sentjurc M, and Schara M. (1988) *Spin probe reduction in cells and tissues.* Mag Res Med **8**, 241-247
10. Fuchs J, Mehlhom RJ, and Packer L. (1989) *Free radical reduction mechanisms in mouse epidermis and skin homogenates.* J Invest Dermatol **93**, 633-640
11. Fuchs J, Freisleben HJ, Podda M, Zimmer G, Milbradt R, and Packer L. (1993) *Nitroxide radical biostability in skin.* Free Rad Biol Med, **15**, 415-423
12. Fuchs J, Groth N, and Herrling T. (1995) *In-vivo Electron Paramagnetic Resonance in Human Skin.* II International Workshop on In-vivo ESR and ESR imaging. L'Aquila, Italy, September 10-14
13. Herrling, Th., Zastrow, L., Groth, N., Fuchs, J., Stanzl, K.: *Detection and Influencing of the Antioxidative Potential (AOP) of Human Skin*, SÖFW Journal 7/1996, 472-476
14. Herrling, Th., Thiessenhusen, K.-U., Ewert, U.: J. Magn. Reson., **100**, 123, (1992)
15. Maltempo, M.M., Eaton, S.S., Eaton, G.R.: J. Magn. Reson., **77**, 75, (1988)

Study on Oxygen Radicals and Natural Antioxidants with ESR Technique

B-L Zhao, Q.Guo, J-W Hou ,W-J Xin

Institute of Biophysics, Academia Sinica, Beijing, 100101, China

Abstract. We have studied the mechanisms of production of active oxygen radicals from stimulated PMN with PMA by ESR spin trapping technique. It was found that O_2^- and .OH were released from a site near the outside membrane of the stimulated PMN. The scavenging effects of natural antioxidants on oxygen-free radicals have been developed in several model systems including aqueous, enzyme, cell and tissue systems and about 20 kinds of Chinese traditional medicines and natural antioxidants were investigated.

1 Introduction

In biological system, electron transfer is one of basic change. Oxygen is the most important acceptor of electron. Different oxygen radicals are obtained with the accommodation of a different number of electrons. Addition of one electron generates superoxide anion (O_2^-), whereas the addition of three electrons produces hydroxyl radical (.OH). Oxygen free radicals play an important role in biological system and have close relationship with several serious diseases. Antioxidants, special natural antioxidants can protect the human body against the diseases. Using ESR technique, we have studied the oxygen radicals generated from stimulated PMN (Polymorphonuclear luekocytes) with PMA (phorbol myristate acetate) and the scavenging effects of natural antioxidants on the oxygen-free radicals.

2 Studies on the Mechanism of Production of Oxygen Free radicals in Biological Systems.

The ESR spectrum of spin adducts resulting from a mixture containing PMN stimulated with PMA and spin trapped by DMPO (5,5'-dimethyl-1-pyrroline-1-oxide) is shown in Fig.1A. This is a composite of two distinct products: DMPO-OOH (g=2.0055, a_N=14.3G, a_H^β =11.3G, a_H^γ=1.26G) and DMPO-OH (g=2.0053, a_N = a_H =14.9G). In order to identify the source of superoxide anion and hydroxyl radicals in the ESR spectrum, catalase was added into the system. It was found that the signals decreased but still contained two components, DMPO-OOH and DMPO-OH (Fig.1B). After SOD was added to the system, both DMPO-OOH and DMPO-OH were eliminated from the spectrum (Fig.1C), indicating that the ESR spectrum in Fig.1A came mainly from superoxide anion and hydroxyl radicals resulting from the superoxide anion produced from PMA stimulated PMN.

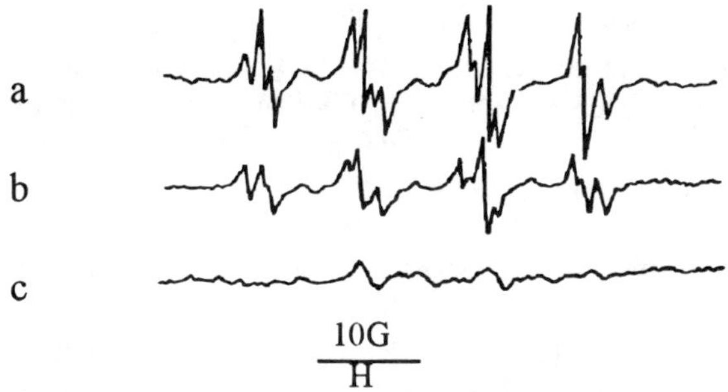

Fig. 1. A. The ESR spectrum resulting from 1.5x10^7 PMN/ml stimulated with 100ng/ml PMA in the presence of 0.1 mM DETAPAC (diethylenetriaminepentacetic acid) and 0.1 M DMPO. Conditions were the same as those for A, except that 300 U/ml catalase (B) and 300U/ml SOD(C) were present respectively.

The broad peak in Fig. 2B arises from hyperfine splitting of nitrogen of spin probe CTPO (3-carbamoyl-2,2,5,5-tetramethyl-3-pyrroline-1-yloxy). Because of interaction of the spin magnetic moment of nitroxide with the spin magnetic moment of oxygen, no obvious change in the spectrum even after 20 minutes., indicated that the respiration of PMN was not strong enough to cause the change of the spectrum in this condition. After stimulation with PMA, the spectrum in Fig. 2A became different to that in Fig. 2B. It was split into over 10 hyperfine peaks. That is because after stimulation with PMA, the respiratory burst of PMN had produced oxygen free radicals and caused oxygen consumption. The spin-spin interaction between oxygen and spin probe had decreased and the superhyperfine structure from hydrogens of 4 methyl groups in the probe had appeared. CTPO can not penetrate the cell membrane during the experiment. So these results reflect the oxygen consumption in the intercellular medium during the production of active oxygen radicals by PMA stimulated PMN. But it can not exclude intracellular oxygen consumption.

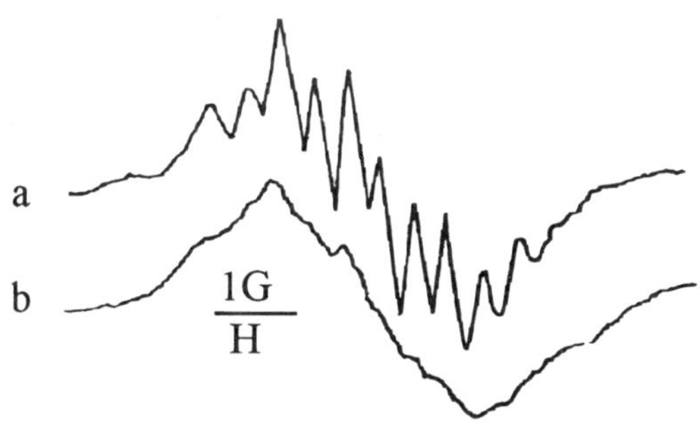

Fig. 2. The ESR spectrum resulting from 10^{-4}M CTPO in 2.5x10^7 PMN/ml stimulated with (A) and without (B) 100ng/ml PMA.

The superhyperfine structure from hydrogens of methyl groups appeared in the ESR spectrum of TEMPOL (2,2,6,6-tetramethyl piperidine-1-yloxy) in PMA stimulated PMN

(Fig.3A). But only a wide peak without any splitting appeared in Fig. 3B. The spin probe TEMPOL can penetrate the cell membrane and is present in both intercellular and intracellular media. The ESR spectrum in Fig. 3A should be the overlap of the contribution from the probe in both intercellular and intracellular media. The transition element ion Cr^{3+} with spin magnetic moment can widen the ESR spectrum of the spin probe. The straight line in Fig. 3C was the result of TEMPOL mixed with CrOX (Potassium trioxatochromate) in aqueous solution. However, CrOX can not penetrate the cell membrane and can only widen the ESR spectrum of TEMPOL in the intercellular medium. Furthermore, addition of CrOX to the mixture of CTPO and PMN resulted a loss of the ESR spectrum showing that both CTPO and CrOX are only in the intercellular medium. So the ESR spectrum in Fig. 4B without any splitting came from the contribution of TEMPOL in the intracellular medium. This indicated that the superhyperfine splitting in Fig. 3A coming from the contribution of TEMPOL in the intercellular medium was due to production of active oxygen free radicals and oxygen consumption by PMA stimulated PMN, and suggested that no oxygen consumption was found in the intracellular medium during production of active oxygen free radicals in the PMA stimulated PMN. This means that the location of production of active oxygen free radicals is in the cell mambrane and may be near the outside layer of the cell mambrane. SOD is macromolecular and can not penetrate cell membranes. When it was added into the system, the oxygen-free radicals were eliminated further indicating that the oxygen radicals were out of the cells.

Fig. 3. The ESR spectrum resulting from 10^{-4}M TEMPOL in 2.5×10^7 PMN/ml stimulated with 100ng/ml PMA without (A) and with (B) 37mM CrOX; (C) ESR spectrum of 10^{-4}M TEMPOL and 37mM CrOX in PBS.

3 Scavenging Effects of Natural Antioxidants on Oxygen-free Radicals.

Measurement method for the scavenging effects of natural antioxidants on oxygen free radicals have been developed in several model systems including aqueous, enzyme, cell and tissue systems and about 20 kinds of Chinese traditional medicines and natural antioxidants were investigated. Here we just give an example of the scavenging effects and molecular mechanism of green tea polyphenols (GTP) on oxygen free radicals.

3.1 Scavenging Effects of GTP on Oxygen-free Radicals Generated from PMA stimulated PMN.

Fig. 4 shows the ESR spectra of scavenging effects of GTP on oxygen radicals generated from PMA stimulated PMN and compared with RA (rosmary antioxidants), Cur (curcumin), vitamin C and vitamin E. It can be found that F6 (a frantion obtained from purification of green tea) and GTP have very strong scavenging effects on oxygen radicals in this sysytem. They almost eliminate all active oxygen free radicals, which is even stronger than that of Vit.C and Vit.E. RA and Cur have weak scavenging effects on oxygen radicals.

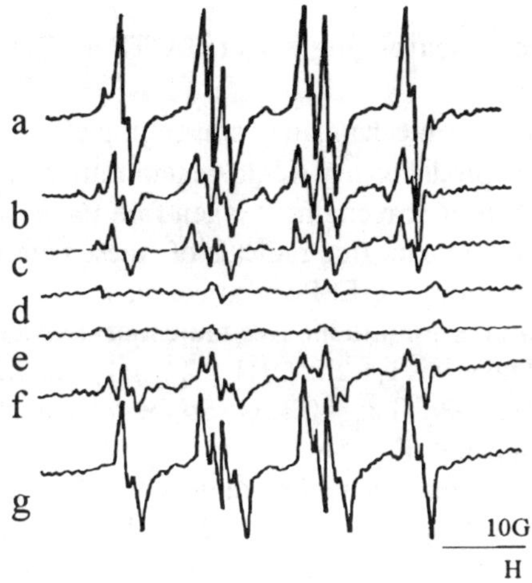

Fig 4. (A)The ESR spectra resulting from a mixture containing 10^7 Cells/ml PMN, 0.1 mM DETAPAC, 100ng/ml PMA, 0.1 M DMPO without (A) and with 200 μg/ml F6 (B),GTP (C), RA (D), Cur (E), Vit.C (F) and Vit.E (G) respectively.

3.2 Scavenging Effects of GTP on O_2^- and .OH Generated from Aqueous Solution

In order to measure and compare the scavenging effects of the antioxidants on superoxide and hydroxyl radicals quantitatively, two aqueous solution models were used. The scavenging effects of GTP and other antioxidants on O_2^- generated from irradiation of riboflavin/EDTA was showed in Table 1. The scavenging effects of GTP and other antioxidants on .OH generated from Fenton's Reaction was showed in Table 2.

Table 1 Scavenging effects of GTP and other anitoxidants on O_2^- generated from irradiation of riboflavin/EDTA

	PBS	GTP	F_6	VitC	VitE	Ra	Cur
E	0	72	74	96	23	23	17
δ(+)	0	7.4	4.7	6.4	6.1	5.1	5.5

Table 2 Scavenging effects of GTP and other anitoxidants on .OH generated from Fenton's Reaction

	PBS	GTP	F6	VitC	VitE	RA	Cur
E	0	12	13	56	35	16	69
δ(+)	0	8.6	16.8	5.9	13.5	19.0	11.7

3.3 Molecular Mechanism of Scavenging Effect of GTP on Oxygen-free Radicals

In order to study the molecular mechanism of scavenging effects of GTP on oxygen free radicals, we studied what products they would generate after oxidation of GTP and where was the active position of scavenging oxygen-free radicals in the GTP molecules. Relatively stable semiquinone free radicals of these compounds were found after oxidation in alkaline solution. The ESR spectra of the radicals were shown in Fig. 5. Fig. 5.A is the spectrum of semiquinone free radical of EGCG. It was found that the intensity ratio of ESR peaks was 1:1:2:2:1:1:1:1:2:2:1:1, and the hyperfin splitting constants were a_{H1}=3.71G, a_{H2}=0.96G, a_{H3}=0.32G. So we inferred that its reaction and active position may be as shown in Fig. 5A.

Fig. 5.B is the spectrum of semiquinone free radical of ECG. The spectrum is 3x2 line style, the intensity ratio of the peaks was 1:1:2:2:1:1, and the hyperfin splitting constants were a_{H1}=0.19G, a_{H2}=1.09G. It indicated that its reaction and active position may be as shown in Fig. 5B.

The ESR spectrum of semiquinone free radical of EC was shown in Fig.5.C. From the spectrum, we supposed that two free radicals were produced in one molecule of EC. In an active position, its peak intensity ratio was 1:1:1:2:1:1:1:1:1:1:2:1:1:1, and the hyperfin splitting constants were a_{H1}=0.19G, a_{H2}=0.32G, a_{H3}=0.64G, a_{H4}=3.39G, and in another active position, its ratio was 1:1:1:1:1:1:1:1, and the hyperfin splitting constants were a_{H1}=1.02G, a_{H2}=0.16G, a_{H3}=0.48G. So we speculated that it had two active positions and may be like this:

The ESR spectrum of semiquinone free radical of EGC was shown in Fig.5.D. This spectrum was formed and stable 60 min after the addition of EGC to the alkaline solution. Before that, the spectrum was varying with time. From the changes, it was inferred that the changes of active position of semiquinone free radical of EGC could be as follows.

From these results, it is obvious that these compounds exhibit different antioxidant active positions during scavenging oxygen free radicals.

A variety of clinical pathological events have been found to be linked to oxygen radical injury. Usually the generation and scavenging of oxygen free radicals are balanced in body. If there is imbalance in the regulatory metabolism of some enzymes, such as SOD or catalase, there will be excessive amount of oxygen free radicals in the body, which can damage the components of cells or even kill normal cells causing aging and some other serious diseases, for example, cancer, heart disease, stroke and emphysema ect. In this case, addition of antioxidants to body, special natural antioxi-

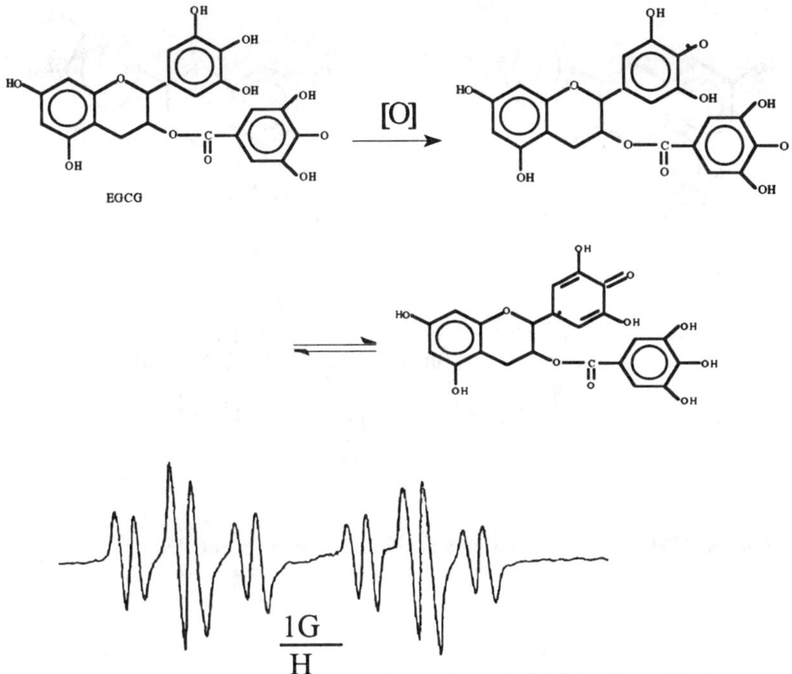

Fig. 5.A. The ESR spectrum of semiquinone free radical of EGCG after oxidation.

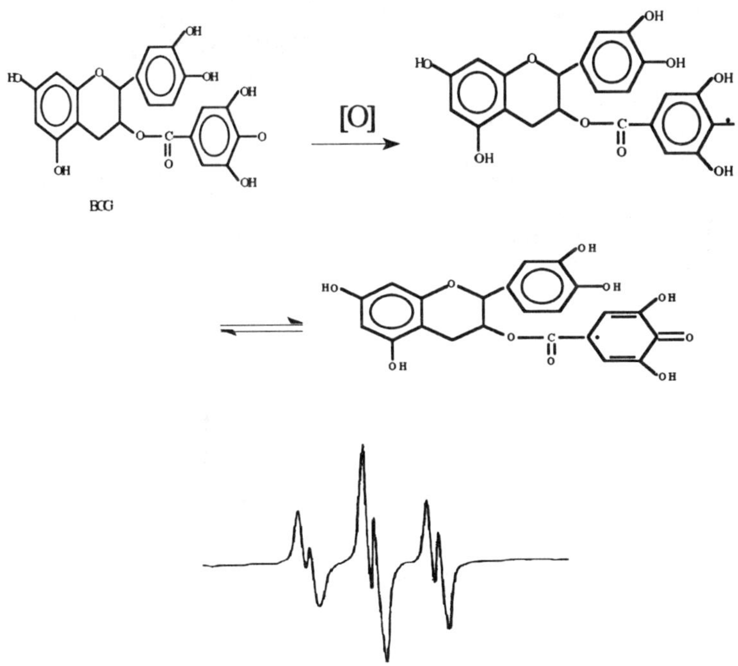

Fig. 5.B. The ESR spectrum of semiquinone free radical of ECG after oxidation.

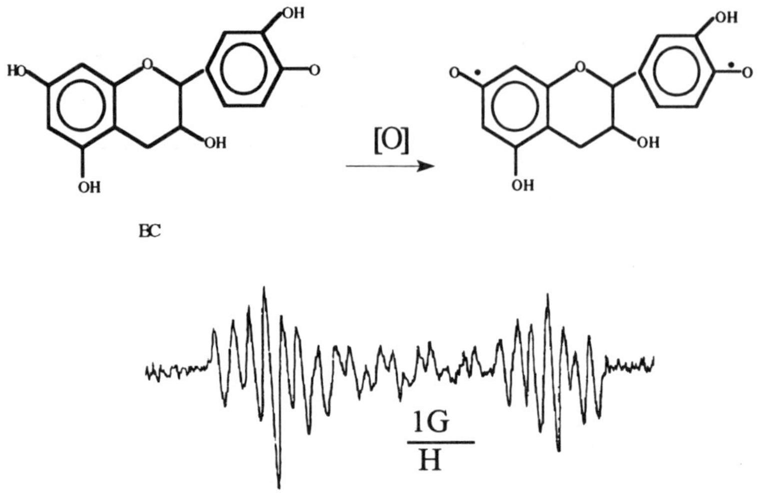

Fig. 5.C. The ESR spectrum of semiquinone free radical of EC after oxidation.

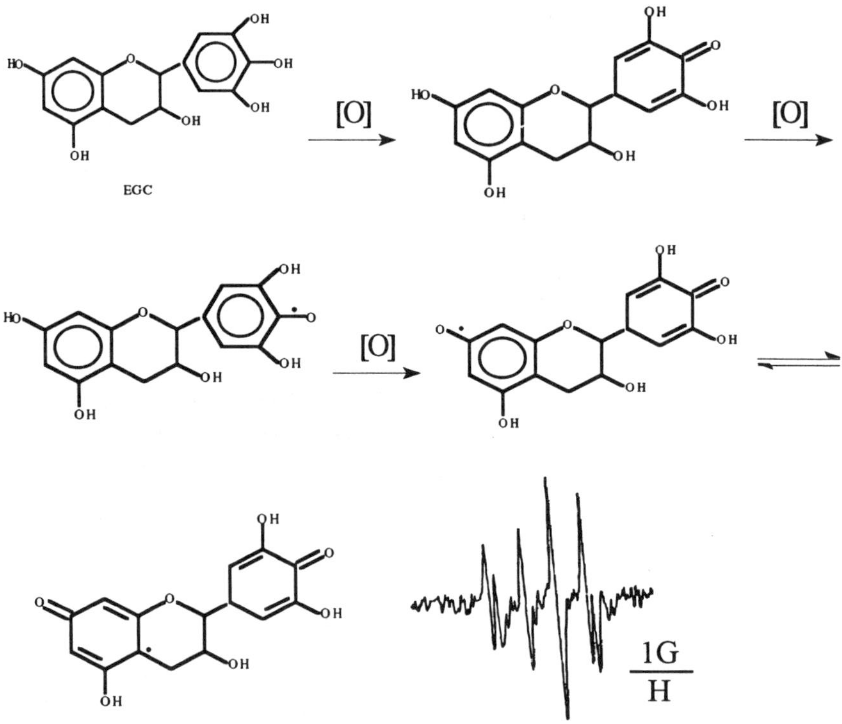

Fig. 5.D. The ESR spectrum of semiquinone free radical of EGC after oxidation.

dants, will be beneficial for prevention of such diseases and for human health. ESR technique can provide many very important information in this field.

Acknowledgements

This investigation was supported by a grant from the National Natural Science Foundation of China.

References

1. Zhao,B-L, Li,X-J & Xin, W-J: ESR studies on oxygen consumption in respiratory burst of human polymorphonuclear luekocytes. Cell Biol.Intern.Report 13(1989)317-323.
2. Zhao,B-L, Li,X-J & Xin W-J: ESR studies on active oxygen radicals produced in respiratory burst of human polymorphonuclear luekocytes stimulated by PMA. Cell Biol. Intern. Re port 13(1989) 529-535.
3. Zhao,B-L, Li,X-J & Xin,W-J: Scavenging effects of extracts of green tea and natural antioxi dants on active oxygen radicals. Cell Biophysics 14(1989) 175-185.
4. Zhao,B-L, Li,X-J & Xin,W-J: Studies on the effects of schizandra on active oxygen radicals by spin trapping technique. Cell Biol. Intern. Report 14(1990) 95-105.
5. Ju,H-S, Li,X-J, Zhao,B-L & Xin,W-J: The scavenging Effects of Berbamine on active oxygen radicals in phorbol ester-stimulated human polymorphonuclear luekocyte. Biochem. Pharmol. 39(1990) 1673-1678.
6. Li,X-J, Zhao,B-L, Xie,R-F & Xin,W-J: Studies on the active oxygen radicals produced by the luekocytes of malignant lymphoma patients. Nal. Med. J. China 70(1990)334-337.
7. Ju,H-S, Li,X-J, Zhao,B-L & Xin,W-J: Scavenging Effects of total flavonoid of hippophae on active oxygen radicals. Chinese Pharm. Bulletin 6(1990) 97-101.
8. Li,X-J, Zhao,B-L, Liu G-T & Xin,W-J: Scavenging effects on active oxygen radicals by schizandrins with different structures and configurations.Free Radical Biol. Med.9 (1990)99-104.
9. Wang,Y-Y, Zhao,B-L, Li,X-J, Hou,J-W and Xin,W-J: Studies on active oxygen radicals from human PMN respiratory burst induced with fluoride by spin trapping technique. Acta Sci Circums 10(1990)304-310.
10. Zhao,B-L, Duan,S-J, & Xin,W-J: Lymphocytes can produce respiratory burst and oxygen radicals as polymorphonuclear leucocyte. Cell Biophys, 17(1990)205-211.
11. Zhao,B-L, Yang,W-D, Zhu,H-L & Xin,W-J: Direct measurement of active oxygen radicals following reperfusion of ischemic rabbit myocardium with ESR technique Chinese Sci Bul. 35(1990) 56-61.
12. Yang,W-D, Zhu,H-L & Zhao,B-L: Direct measurement of active oxygen radicals produced from ischemia reperfusion rabbit myocardium and scavenging effect of salvia miltiorriza on the oxygen radicals by ESR technique. J. Myo. Vasc Sinica 17(1989) 178-180, 1989.
13. Zhao,B-L, Zhang,J-C & Xin,W-J: Studies on the free radicals produced from photolysis of hydroperoxided linoliec acid by ESR spin trapping technique. Chinese Sci. Bul. 36(1991)774-777.
14. Zhang, J-C, Zhao,B-L, & Xin, W-J: Evidence for L. against LOO. being spin-trapping by 4-POBN during the reaction of Fe^{2+}-induced lipid peroxidation. Appl.Magn.Reson .2 (1991) 251-250.

15. Shen,S-R, Yang,X-Q, Yang,F-J, Zhao,B-L & Xin,W-J: Scavenging effects of tea polyphenol compound and (-)-EGCG on oxygen free radicals. J.Tea Sci.12(1992)59-64.
16. Huang,K-T, Zhao,B-L: Scavenging effects of Radix ophiopogonis on free radicals. Chinese Herbs. 23(1992):538-539.
17. Shen,S-R, Yang,F-J, Zhao,B-L, & Xin,W-J: Coordinatingly Synergic effect of catechins during their antioxidation. J.Tea Sci. 13(1993):141-146.
18. Zhao,B-L, Xin,W-J, Chen,Y-T, Di,H, Guan,L. & Liu,W: Studies on the free radicals generated from ischemia, transplantation and reperfusion of kidney. Acta Biophys Sinica 10(1994) 170 -173.
19. Shi,H-L, Zhao,B-L,Xin,W-J: Scavenging effect of Baicalin on free radicals and its protection on erythrocyte membrane from radical injury. Biochem. Mole. Biol. Intern. 35(1995) 981-994.
20. Shi,H-L, Yang,F-J, Zhao,B-L & Xin,W-J: Effects of r-HuEPO on the biophysical characteristics of erythrocyte membrane in patients with anemia of chronic failure. Cell Res. 4 (1994):57-64.
21. Jiang,W.,Zhao,Y. Zhao,B-L,Xin,W-J: Studies on the scavenging effect of Tanshonon on lipid free radical of cardiac sarcoplasmic reticulum during peroxidation. Acta Biophysc Sinica 10 (1994) 685-689.
22. Zou,X-L,Wan,Q, Li,M-F, Zhao,B-L, Xin, W-J: Scavenging effect of green tea polyphenols on the oxygen free radicals generated from isolated ischemic reperfused rat myocardium. Chinese J. Magn. Res. 12(11995) 237-244.
21. Guo,Q.,Zhao,B-L, Li,M-F, Shen,S-R & Xin,W-J: Studies on protective mechanisms of four components of green tea polyphenols (GTP) against lipid peroxidation in synaptosomes. Biochem.Biopphys.Acta (in press) 1996.

EPR Instrumentation for In-vivo Investigation in Human Skin

Th. Herrling, J. Rehberg, F. Klein*, N. Groth*

Technische Fachhochschule Berlin, Rudower Chaussee 6, 12489 Berlin, Germany
* Galenus Inc., 12489 Berlin, Germany

Abstract. Two types of EPR spectrometers suitable for in-vivo studies on skin biopsies and skin are described. An X-band EPR spectrometer is combined with an imaging accessory designed for measurements on skin biopsies. An S-band EPR spectrometer enables in-vivo measurements on the skin of hairless mice and humans.

Key words: EPR imaging, in vivo spectroscopy, surface coil

1 Introduction

There is currently much interest in the design of instrumentation for electron paramagnetic resonance imaging studies of living systems. Although it is unlikely that images of large animals with resolution comparable to those currently obtained using NMR imaging will be achieved, it is nevertheless hoped that useful, complimentary data will ultimately be obtained using EPR techniques (Fairhurst 1990, Eaton 1988). The potential of using both NMR and EPR for localized in-vivo studies was suggested by Paul Lauterbur in an early paper on imaging (Lauterbur 1973). However, while NMR imaging and spectroscopy have progressed into clinical tools, EPR methods capable of studying processes in isolated tissues are only now beginning to be developed (Berliner and Fuji 1985, Bacic et al. 1988). There are several areas of physiological and pathophysiological importance for which EPR spectroscopy has significant advantages over NMR techniques; these include the measurement of oxygen concentrations and redox metabolism (Swartz 1987 and 1989). The main obstacle to extending EPR studies to in-vivo systems has been the severe dielectric loss and consequent heating that occurs in aqueous samples at conventional EPR frequencies. Commercial X-band EPR spectrometers, which operate at about 9 GHz, are unsuitable for the study of large biological samples. With aqueous samples, losses in the electric and magnetic fields are so severe at these frequencies that only small, thin samples, carefully positioned in the nodal plane of E-field, can be observed. Larger samples are then more easily accommodated, especially when alternative resonator structures designed to offer more suitable E-field geometries, are used. The extent to which the frequency must be reduced depends on the sample size. It must be low enough to give complete tissue penetration, yet not so low that there is a prohibitive, Boltzman factor, loss of sensitivity. Frequencies of about 1 GHz are considered a good compromise for small animals (Hutchison 1971, Subczynski et al. 1986). Loop-gap resonators, re-entrant cavities, and surface coils have recently been used to study samples ranging from plant stems to small mice at frequencies between 500 MHz and 2 GHz (Ogata et al. 1986, Sotgiu and Gualtieri 1985, Berliner et al. 1987). However, estimates from the data (Bottomley 1978) indicate that for larger samples, magnetic losses will indicate the use of lower frequencies. A whole body imaging spectrometer which works at a microwave frequency of 250 MHz was constructed by Symons (1995). Several resonators, ranging in diameter from 6.5 to 45 cm have been developed for this spectrometer. A microwave scanning microscope using a cavity at 9.5 GHz having an aperture over which a sample with a flat surface is moved with a computer-controlled X-Y stage has been developed by Ikeya et al. (1995) for measuring some biological materials.

1.2 Experimental Design

The experimental design is determined by the object under investigation. For skin biopsies we could use conventional X-band spectroscopy. The thickness of skin biopsies is smaller than 1 mm. Because of the fat and muscle content in skin tissues the dielectric loss enables X-band measurements for skin biopsies. The penetration depth of microwaves of 9.5 GHz is also sufficient for the measurement of biopsies.

The skin depth (DS) in muscle tissue as a function of microwave frequency is shown by Cheung (Cheung 1987). The skin depth is larger for fatty tissues. The skin contains more fatty tissues than muscle tissue and consequently the microwave attenuation is smaller for skin than other body organs. The high sensitivity for X-band spectroscopy enables EPR imaging of the skin with a high spatial resolution between 10 and 100μm which is necessary to distinguish the several skin layers.

The application of EPR to intact animals and humans requires two modifications.

The frequency of 9.5 GHz does not permit the investigation of large objects like biological samples, due to the high dielectric loss. The application of a low microwave frequencies of 3.0 GHz can solve this problem.

A low frequency system of 3 GHz was developed for the *in vivo* investigation of skin from a small animal (mouse) or human forearm. This microwave frequency requires a magnetic field of 0.11 T.

Such in vivo experiments demand a larger distance between the pole faces. The distance between pole faces of the conventional X-band EPR spectrometer ERS 221 is 60 mm and is much too small for such measurements. Therefore it was increased to 120 mm by decreasing the thickness of the pole pieces. The principal arrangement of the combined EPR apparatus for X-band EPR imaging and S-band in vivo spectroscopy is shown in Fig. 1. For this two possibilities of measurements an EPR spectrometer ERS 221 (build in the Academy of Science of GDR) was reconstructed and equipped with an imaging system and a rapid scan unit.

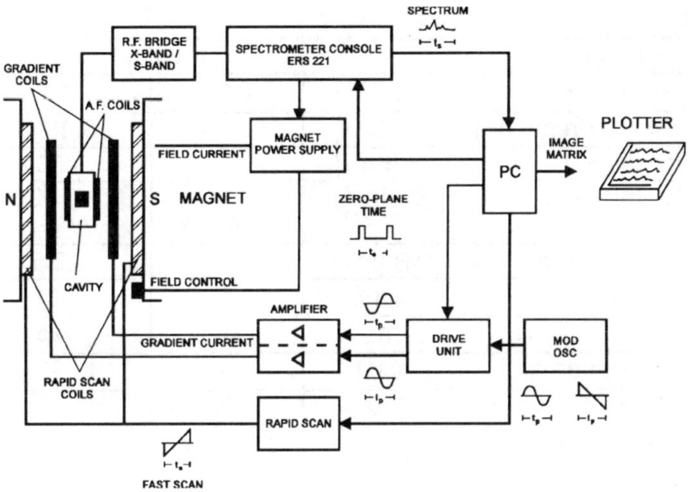

Fig. 1. Block diagram of the experimental apparatus for S-band spectroscopy and X-band imaging

The experimental arrangement for EPR imaging based on the method called MOSS (Modulated Gradient with Simultaneous Field Scan) is described in more detail by Fuchs et al (Fuchs 1994) and Herrling et al (Herrling 1992).

Because of movements of living samples it was necessary to enable a short measuring time. The combination with a rapid scan unit allows the registration of a spatial resolved spectrum in 200 ms. To increase the signal to noise ratio S/N 25 accumulations were measured. All measurements on living samples like on human skin or mouse skin were done by S-band spectroscopy with a special designed probe head like it is seen in Fig. 2. The probe head is based on a 90° bended surface coil with a diameter of 8 mm, and with an electronically matching system. A quartz plate is mounted on the lower sample site of the surface coil to accomplish a plane-parallel measuring area in the skin. Matching is accomplished by approaching the transmission lines in a $\lambda/_4$ region outside the loop via a piezo element. The two 100 kHz-modulation coils are arranged near and above the surface coil to generate a modulation field B_m in the skin layer. The rapid scan coils are mounted on the surface of the pole plates. For in vivo measurements the probe head was mounted on the forearm of a human which was subsequently placed in the magnet.

EPR imaging needs a high sensitivity if a spatial resolution in the μm-region is necessary. The thickness of the skin slices like horny layer and the epidermis are in the region of 50 μm to 200 μm. This spatial resolution demands a high signal intensity and high field gradients in the range of grad B > 1 T/m. Such measurements can only performed today on skin biopsies which are investigated in an X-band microwave cavity. The magnetic field gradient for spatial resolution is oriented in the Z-direction (direction of the main field B_0). The skin biopsy inside a tissue cell is arranged in the x-y plane in the midside of a rectangular H102 cavity. The parameters for the experimental arrangement are shown in Table 1.

Parameter		Value	
Magnetic Field Strength	B_0 [T]	0.34	0.11
Microwave Frequency	υ [GHz]	9.5 (X)	3.08 (S)
Microwave Power	P [mW]	50	60
Modulation Amplitude	B_m [mT]	0.15 (50kHz)	0.1 (100kHz)
Field Sweep	ΔB [mT]	12.5	2.5
Sweep Time	t_B [s]	5	0.2
Accumulations	n_A	1	25
Time Interval	t_i [s]	0	50
Projections	n_p	128	16
Image Matrix		128x128	16x128
Total Acquisition Time	t_a [min]	10.6	15.0
Field Gradient	T/m	4	0

Table 1. Comparison of spectrometer settings for X-band imaging and S-band spectroscopy

Another technical solution for using a compact probe head is described by Herrling et al (Herrling 1996).

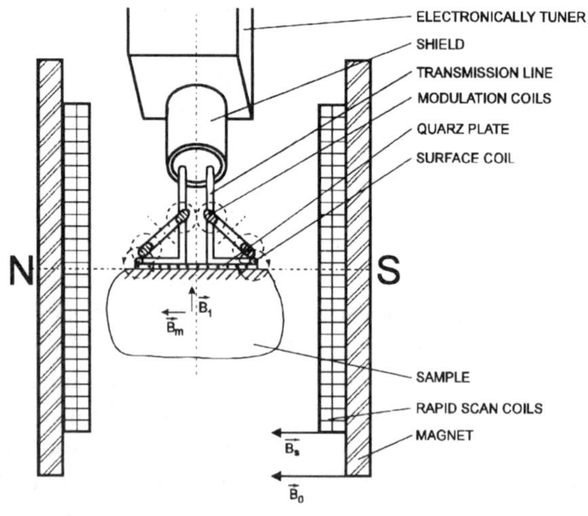

Fig. 2. Experimental arrangement consisting of the surface coil and the modulation coils for S-band spectroscopy in the pole gap of the electromagnet

2 Conclusion

Instrumentations for investigations in human skin can be subdivided in two categories. Measurements on skin biopsies of humans, mice or pigs can be performed in the X-band frequency range with a normal EPR spectrometer. EPR imaging of skin biopsies demands a special tomography accessory unit. The necessary signal intensity for EPR imaging requires the application of a semistable radical in the concentration range of 10 mM to 100 mM.

Measurements on a living subject need two preconditions:

1. sufficient skin depth of the microwaves with a small microwave attenuation,
2. fast measurements without influence of movement artifacts.

This two problems were solved by using the S-band frequency range and by application of Rapid Scan technique. Using Rapid Scan technique a high number of accumulations are necessary getting a sufficient signal to noise ratio.

References

1. Bacic G., Demsar F., Zolnai Z., and Swartz H.M., *Mag. Reson. Med. Biol.* **1,** 55 (1988)
2. Berliner L.J. and Fuji H., *Science* **227,** 517 (1985)
3. Berliner L.J., Fuji H., Wan X., and Lukiewicz S.J., *Magn. Reson. Med.* **4,** 380 (1987)
4. Bottomley P.A., *J. Phys. E* **11,** 413 (1978)
5. Cheung A.Y., Al-Alrash J.: *IEE Proceedings* Vol. 134, 6, 493 (1987)
6. Eaton S.S. and Eaton G.R., *Spectroscopy*, 1, 32 (1988)
8. Fairhurst S.A., Gillies D.G., and Sutcliffe L.H., *Spectrosc.* World **2**, 14 (1990)
9. Fuchs J., Groth N., Herrling Th., Packer L., *Methods in Enzymology, Vol.* **233,** *Oxygen Radicals in Biological Systems (Part C., Packer L., eds.),* pp. 140-149
10. Herrling Th., Groth N., Thiessenhusen K.-U., Fuchs J., Ewert U., *Magn. Res. Microscopy (Blümich B., Kuhn W., eds.), pp.* 563-572, (1992)
11. Herrling T.E., Groth N.K., and Fuchs J., *Appl. Magn. Reson.* **11**, 471-486 (1996)
12. Hutchison J.M.S., *J. Phys. E* **4,** 703 (1971)
13. Ikeya M., Hosokawa K., Hochi A., Yamamoto M., Sumimoko H., Ikeda S., *Magnetic Resonance in Medicine (H.Kamada Ed)* p 68, Vol. 6, Nikon-Igakukan 1995
14. Lauterbur P.C., *Nature (London)* **242,** 190 (1973)
15. Ogata T., Ono M., Fujisawa T., Yoshida E., and Kamada H., *Chem. Soc. Jpn. Chem. Lett.* 1681 (1986)
16. Sotgiu A. and Gualtieri G., *J. Phys. E* **18,** 899 (1985)
17. Subczynski W., Lukiewicz S., and Hyde J.S., *Magn. Reson. Med.* **3,** 747 (1986)
18. Swartz H.M., *Acta Biochim. Biophys. Hung.* **22,** 277 (1987)
19. Swartz H.M., *in „Advances in Magnetic Resonance Imaging"* (E.Feig, Ed.), Ablex Pub. o., Norwood, NJ, 1989
20. Symons M.CR., *Magnetic Resonance in Medicine (H.Kamada Ed)* p 56, Vol. 6, Nikon-Igakukan 1995

ESR Studies of NO Radicals Generated from Ischemia-Reperfusion Tissue and Polymorphonuclear Leukocyte

B-L Zhao, J-G Shen, J-W Hou, W-J Xin

Institute of Biophysics, Academia Sinica, Beijing, 100101, China

Abstract. NO free radicals bound to haemoglobin and generated from ischemia-reperfusion rat kidney, heart in vitro and in vivo and phagocytes were studied with low temperature ESR technique. An ESR signal which has a maximum peak at g=2.0790 and a minimum at g=2.0237, each of them was split into triplet a=16-17G and a signal at g=2.040 was obtained from the vein blood of ischemia-reperfusion kidney and heart in vitro respectively. A signal at g=2.035 with triplet a=13G of DETC-Fe-NO complex was obtained in ischimia-reperfused heart in vivo. The NO free radicals generated from PMA stimulated PMN measured.

1 Introduction

The free radical NO has many physiological functions, such as endothelium-derived relaxing factor (EDRF), inhibiting coagulation of platelet, reverse messenger for neuron transmission and plays an important role in immunity. It is also a free radical and plays "double edges" in body, e.g. in addition to its biological functions, it can also react with the components of cells and be poisonous to the cells. Because of the characteristic paramagnetic properties of iron-nitrosyl complex and its ability to detect such complexes against a large background of diamagnetic species in whole cells and tissues, ESR spectroscopy was used in this work.

2 ESR Spectra of NO Bound to Haemoglobin (Hb) Recorded at 123K

Figure 1 shows the different ESR spectra of NO bound to haemoglobin (Hb) recorded at 123K. The peaks at g=2.080 and g=2.010 with three hyperfine splitting (a_N=1.57 mT) is predominantly from NO bound to α–subunit of hemoglobin (α-NO complex) and the peaks at g=2.040 and g=2.015 are attributable to NO bound to β subunit of hemoglobin (β-NO complex) and nonheme iron-proteins. The ESR spectrum in Fig. 1a mainly comes from the NO bound to the α-subunit of Hb (90%α/10%β), the ESR spectrum in Fig. 1b partly comes from the NO bound to the α-subunit of Hb and partly to β-subunit of Hb (30%α/70%β) and the ESR spectrum in Figure 1c mainly comes from the NO bound to the β-subunit of Hb (10%α/90%β).

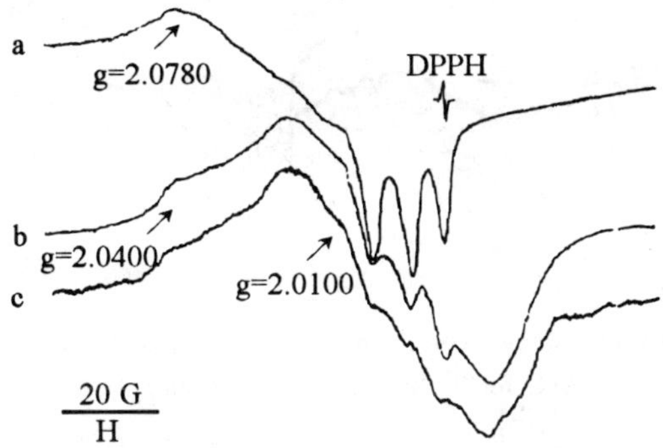

Fig.1. The ESR spectra of NO bound to different subunits of hemoglobin. (a) The ESR spectrum mainly comes from the NO bound to the α-subunit of Hb (90%α/10%β); (b) the ESR spectrum partly comes from the NO bound to the α-subunit and partly to β-subunit of Hb (30%α/70%β) and (c) the ESR spectrum mainly comes from the NO bound to the β-subunit of Hb (10%α/90%β).

3. NO-free Radicals Generated from Ischemia, Transplantation and Reperfusion of Rat Kidney

Figure 2 shows the ESR signal which has a maximum peak at g=2.0790 and a minimum at g=2.0237, each of them was split into triplet a=16-17G, was obtained from the vein blood of ischemia for 1 hour and reperfusion for 2 minutes in vivo and measured at 100K from the ischemia-reperfused kidney.

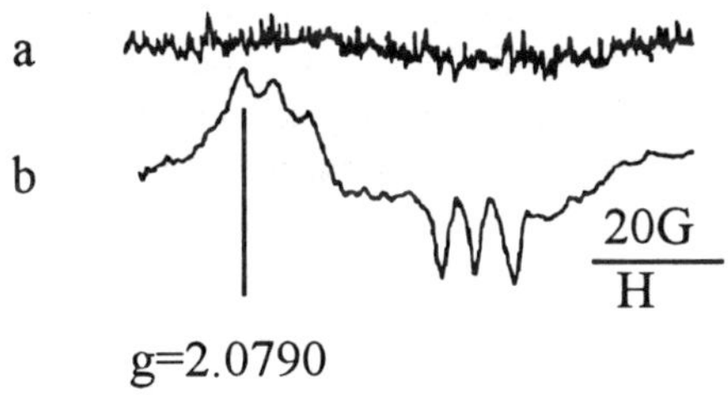

Fig.2. The ESR spectra of (a) Normal kidney vein blood (x1000) and (b) kidney vein blood of ischemia for 1 hour and reperfusion for two min (x100) recorded at 100K.

Figure 3 shows the ESR signal of kidney tissue kept in reperfusion solutions for 24 hours recorded at 100K, which is similar to that in Fig.2b. They are simillar to the NO bound to α-subunit of Hb.

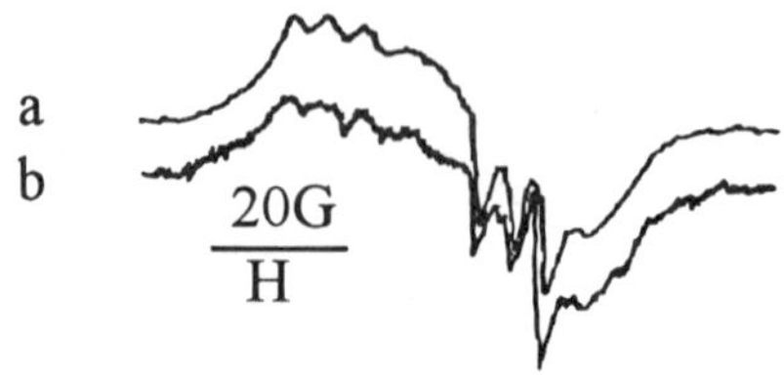

Fig.3. The ESR spectra of kidney tissue kept in (a) L-R reperfusion solution (x125) and (b) sacks reperfusion solution at 4° C for 24 hours recorded at 100K.

4. NO-Free Radicals Generated from Ischemia-Reperfusion Myocardium in vitro

Figuge 4 shows the ESR spectra recorded at 123K from the myocardium of normal (Fig. 4a), ischemia -reperfusion (Fig.4b), ischemia reperfusion with 100 mmol/L L-arginine (Fig. 4c) and ischemia-reperfusion with 100 mmol/L NAME (N^G-nitro-L-Arginine) (Fig. 4d). Only two peaks can be found at g=2.0050 and g=1.9372 respectively on the spectrum of normal myocardium, which correspond to the semiquinone free radicals and transition metal cation. Even after addition of L-arginine, there is no change in the spectra. In addition to these two peaks, there are other two peaks at g=2.040 and 2.030 on the spectrum of the ischemia-reperfused myocardium. The peak at g=2.040 is similar to that of the β-NO complex and g=2.030 has tentatively been reported to attribute to oxygen or alkyl peroxide radicals. When analogue of L-arginine, NAME, the inhibitor of NO synthesase, was added into the reperfusion solution, the signal at g=2.040 decreased, indicating that the signal connects with NO in the ischemia-reperfused myocardium. No NO free radicals can be detected in the normal myocardium even in the presence of L-arginine. The NO free radicals can be detectedin the ischemia-reperfused myocardium and are dose dependent with L-arginine. At the same time, the oxygen free radicals in the ischemia-reperfused myocardium also increased with the concentration of L-arginine added in the reperfusion solution. Figure

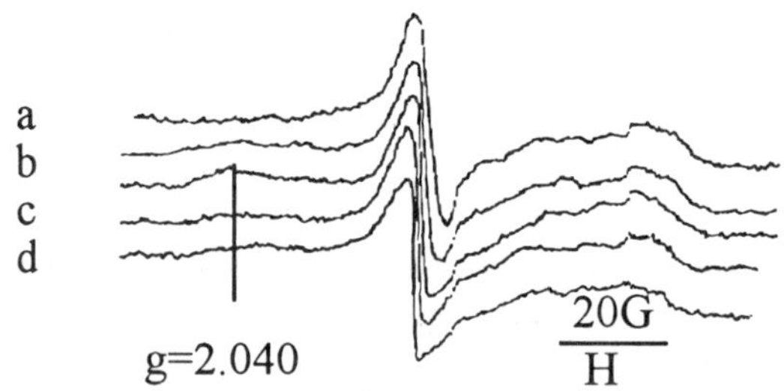

Fig. 4. ESR spectra measured at 123K from the myocardium of control (a); ischemia (120 min)-reperfusion (10 min) (b); ischemia (120 min)-reperfusion (10 min) with 100 mmol/L L-arginine (c); ischemia (120 min)-reperfusion (10 min) with 100 mmol/ L-NAME (d). Ischemia-reperfusion with 100μM Chinonin (e).

ure 4e shows the spectrum from the myocardium of ischemia-reperfusion with 100μM chinonin, which is similar to that of the myocardium with NAME.

5. NO-Free Radicals Generated from Ischemia-Reperfusion Myocardium in vivo.

Figure 5 shows the ESR spectra of NO free radicals from ischemia-reperfusion heart trapped with Fe-DETC (diethyldithiocarbamate) in vivo. It can be found that there is a maximum signal at g=2.035 with three hyperfine splitting (a_N=13G) and a small signal at g=2.02. After injection of L-arginine, the signal at g=2.035 was increased and after injection of NAME, the signal was inhibited. The NO free radicals generated from ischemia-reperfused heart at different conditions in vivo were measured and scavenging effects of antioxidants EGB761 (extracts from Ginko Biloba) and Chinonin on the free radicals was studied using this signal.

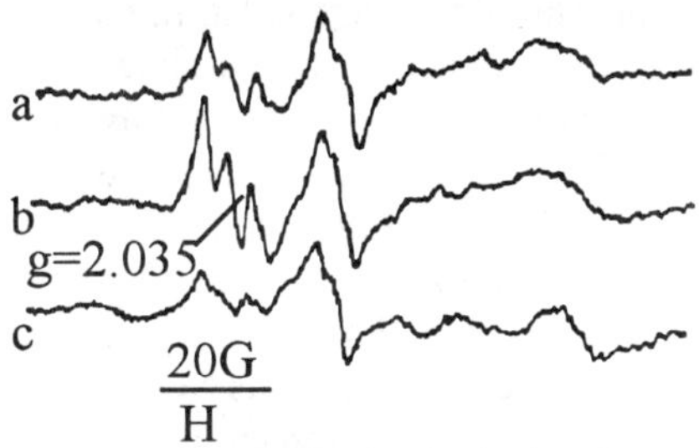

Fig.5. ESR spectra of DETC-Fe-NO complex in the ischemia-reperfused myocardium. (a) ischemia-reperfused heart; (b) ischemia-reperfused heart with L-arginine (5mg/kg,i.p.); (c) ischemia-reperfused heart with NAME (50mg/kg, i.p.).

6. NO-Free Radicals Generated from PMA Stimulated PMN

Recently it was found that stimulated PMN (Polymorphonuclear luekocytes) could release not only active oxygen but also NO free radicals. The spin trapping technique with nitroxide compounds has been successfully used for detecting oxygen free radicals. When these spin trapping agents were used in cell and biological systems, they encountered many difficulties. We have developed a novel method with nitroxide spin trap, DMPO (5,5-dimethyl-1-pyrroline-1-oxide), to detect NO free radicals generated from PMA (phorbol myristate acetate) stimulated PMN.

The ESR spectra of O_2^- generated from PMA stimulated PMN and spin trapped by DMPO and the effects of L-arginine and NGMA on the spectrum were shown in Fig.6.a, b and c. It can be found that addition of L-arginine could decrease significantly the ESR signal of spin trapped O_2^- in this system and it is dose dependent (table 1). Addition of NGMA could increase significantly the ESR signal trapped O_2^- in this system.

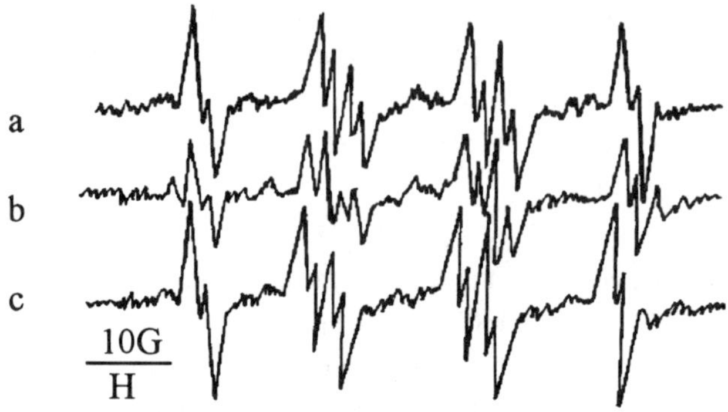

Fig. 6. ESR spectra of O_2^- spin trapped by DMPO (0.1 mol/L) generated from PMA (100 nmol/L) stimulated PMN (10^7 cells/ml) (a), and effect of L-arginine (b) and NGMA (c) on the spectrum.

To verify if L-arginine can scavenge the O_2^-, the scavenging effect of L-arginine on O_2^- generated from xanthine/xanthine oxidase was measured and no any scavenging effect was found in the concentration range using in above experiment. The enhancement effect of NGMA on the signal was also measured and no such effect was found. In order to measure how many NO was generated and reacted with O_2^- and formed $ONOO^-$ in above experiment, the synthesized NO was added directly into the PMA stimulated PMN system. It was found that the signal of DMPO-OOH was decreased while the signal of DMPO-OH was increased when the concentration of exogenous NO was lower than about 15 μmol/L and the signals of both DMPO-OOH and DMPO-OH decreased when the concentration of exogenous NO was higher than about 15 μmol/L. It was too complex to be used for measurement of NO generated in PMA stimulated PMN system. So the interaction of O_2^- generated from irradiation of riboflavin/EDTA with synthesised NO was investigated and it was found that the ESR signal of O_2^- trapped by DMPO was decrease with the increase of concentration of NO, which can give a standard curve. From this curve, the NO concentration generated from PMA stimulated PMN in the case of addition of L-arginine can be calculated (Table 1).

Table 1. Effect of L-arginine on the ESR signal of O_2^- and NO generated from PMA stimulated PMN.

L-A (mM)	1.8	1.0	0.1	0
NO (μML)	10.2±0.05	4.2±0.02	1.1±0.03	0.1±0.02

Acknowledgements

This investigation was supported by a grant from the National Natural Science Foundation of China.

Reference

1. Zhao, B-L, & Chen, W-C: Properties and biological function of NO. free radicals. Proc.Biochem.Biophys. 20(1993) 409-412.
2. Zhao,B-L, Xin, W-J, Chen, Y-T, Di, H, Guan, L. & Liu, W: Studies on the free radicals generated from ischemia, transplantation and reperfusion of kidney. Acta Biophys Sinica 10(1994) 170 -173.
3. Zhao, B-L, Shen, J-G, Li, M, Li, M-F, Wan, Q.& Xin, W-J: Scavenging effect of Chinonin on NO and oxygen free radicals and its protective effect on the myocardium from the injury of ischemia-reperfusion. Biochem. Biophys.Acta 1315(1996) 131-137.
4. Zhao, B-L, Shen, J-G, Hu, J-G, & Xin, W-J: Synergic effects of NO and oxygen free radicals in the injury of ischemia-reperfused myocardium -ESR Studies on NO free radicals gener ated from ischemia-reperfused myocardium. Science China (Series C) 26(1996) 331-338.
5. Zhao, B-L, Wang, J-C, Hou, J-W & Xin, W-J: Studies on nitric oxide free radicals generated from polymorphonuclear leukocytes (PMN) stimulated by phorbol myristate acetate (PMA). Cell Biol.Intern. 20,(1996) 343-350.
6. Shen, J-G., Zhao, B-L, Li, M-F, Wan, Q. & Xin, W-J: Inhibitory effects of Ginkgo biloba ex tract (EGB761) on oxygen free radicals, nitric oxide and myocardial injury in isolated ischemic- reperfusion hearts. In Proceedings of the international symposium on natural anti oxidants Molecular mechanisms and health effects (Eds. Packer, L. Traber, M.G. & Xin, W.)Raber,M.G. & Xin,W.), AOCS Press Champaign, Illinois (1996).

New Approach to EPR Dosimetry of Tooth Enamel

A. Bougai, V. Baryakhtar, S. Kolesnik

Institute of Semiconductor Physics, National Academy of Sciences, Ukraine

Abstract. The routine EPR tooth enamel dosimetry technique is discussed. The main shortcoming of it is the presence of a comparatively strong background spectrum of non irradiation origin. A new variant of dosimetry using a tooth as sensing radiation element is proposed. The new technique is based on the Fourier transformation of the EPR spectrum. The transformation leads to separation of radiation induced and background spectra. It allows improvement of sensitivity and accuracy of dose measurements.

1 Routine EPR tooth enamel dosimetry

EPR dosimetry is one of the most important applications of the EPR technique. It is based on EPR detection of paramagnetic centers which are introduced in a substance by irradiation. The use of a tooth enamel as a radiation sensing element now makes EPR dosimetry the only experimental method of the individual dose reconstruction in accident cases. EPR tooth enamel dosimetry was first used in Japan by Ikeya *et. al.* (1984) for the retrospective dosimetry of Hiroshima survivors.

Recently interest in EPR tooth enamel dosimetry was stimulated by Chernobyl accident. The European Commission has funded work on intercomparison of techniques in laboratories of different countries of the former Soviet Union and Europe for estimation of its possibilities, and for an elaboration of the optimal EPR tooth enamel dosimetry procedure. This work was carried out in 1994-1995 within the framework of the Experimental Collaboration Project №10 (ECP 10) “Retrospective dosimetry and dose reconstruction” (1996) and was discussed at the 4th International Symposium on ESR Dosimetry and Applications in Munchen by V.Chumak *et. al.* (1995) and at the First International Conference “The Radiological Consequences of the Chernobyl Accident” in Minsk by A.Wieser *et. al.* (1996).

The optimal dosimetry procedure includes sample preparation, the most important part of which is careful separation of dentine from the tooth enamel by means of the mechanical treatment and chemical etching processing in ultrasonic bath. This stage is necessary to reduce the non radiation part of the EPR spectrum (so called background spectrum). The detection of the EPR spectrum is carried out at a frequency of 9 GHz at room temperature. The microwave power level and the magnetic field modulation amplitude are very important. The EPR spectrum is registered in a digital form for its processing using computer software.

The EPR spectrum of a typical sample irradiated with the dose of 0.20 Gy has the form presented at Fig.1 (1). It consists of the non irradiated enamel spectrum

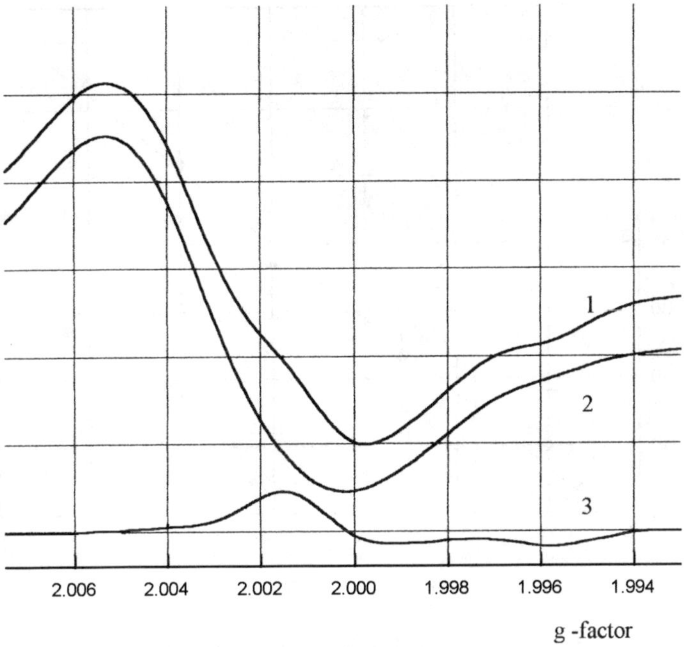

Fig. 1. EPR spectra of tooth enamel for a dose 200 mGy: (1) experimental; (2) background; (3) radiation -induced.

(this is so-called background signal) shown by the curve (2) and the radiation induced spectrum shown by the curve (3). The intensity of spectrum (3) is proportional to the irradiation dose. Spectral positions of the (2) and (3) spectra coincide. The small radiation induced spectrum (3) can be obtained as a difference of two comparatively large spectra: the experimental (1) and the background one (2). To carry out the subtraction and obtain the radiation induced spectrum a computer is generally used.

The calibration of sensitivity of each sample studied to irradiation must be done because of a large individual variability of the enamel. After the initial EPR signal measurement the sample is additionally irradiated 5 to 9 times by a calibrated source of gamma-radiation, and the EPR spectrum each time is registered. This procedure gives the dose dependence of the EPR signal which in general is linear (Fig.2) and allows us to find the initial dose of the sample. This procedure also increases the accuracy and sensitivity of the method. The real sensitivity limit is 0.1 - 0.2 Gy depending on the quality of the sample.

We can conclude that the main shortcomings of the usual technique are due to the presence of the comparatively strong background spectral line. In the absence of absence of a background line the sensitivity can be raised by an order of a magnitude.

Many efforts were undertaken to reduce the background spectrum. Below a new approach to the problem is proposed.

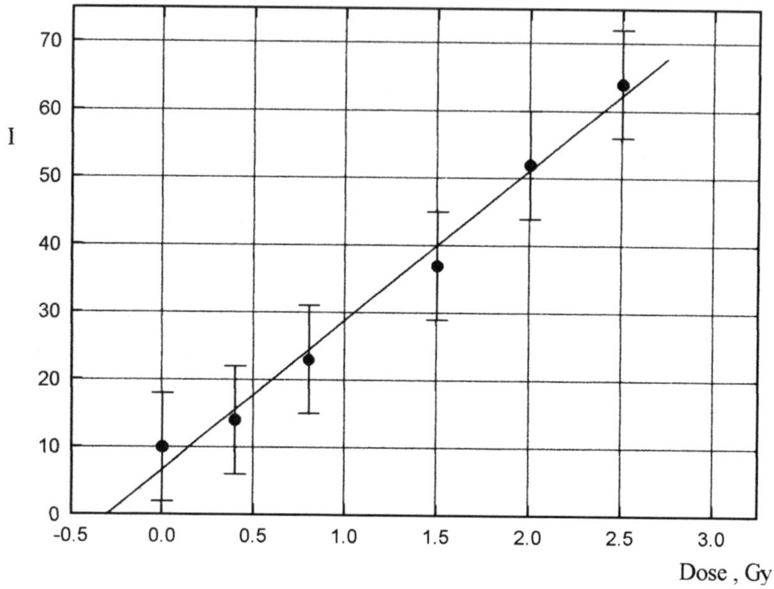

Fig. 2. Dose dependence of radiation induced spectrum

2 New approach to tooth enamel dosimetry

The main shortcoming of the routine EPR tooth enamel dosimetry is presence of background spectrum. The intensity of background spectrum of dentine is much higher than that of enamel. Because of this it is necessary to use extracted teeth and to separate the dentine from enamel. Thus the method in principle cannot be applied to *in vivo* measurements, and thus for inspection of most of nuclear incident victims. The necessity to separate the dentine also makes dosimetry very expensive.

We propose a new approach to retrospective dosimetry using tooth as a sensing element. The objective of efforts must be either elimination of the background spectrum, or obtaining of some new representation of the EPR spectrum where the radiation induced and background spectra are well resolved and there is a possibility of direct measurement of the radiation induced part of the spectrum without a necessity to subtract a background one.

We have found that the problem can be solved by the known mathematical method of Fourier transformation of the EPR spectrum. As we have found, this transformation leads to a separation of the radiation induced and the background parts of the spectrum.

We have used modern computer software for the fast Fourier transformation of the EPR spectrum.

At Fig.3, (1`) is a Fourier transformation of the experimental EPR spectrum, (2`) is a Fourier transformation of the background spectrum and (3`) is the radiation induced part of the spectrum. One can see that at the frequency of 0.22 Hz there is a radiation-induced line separated from the background spectrum. The

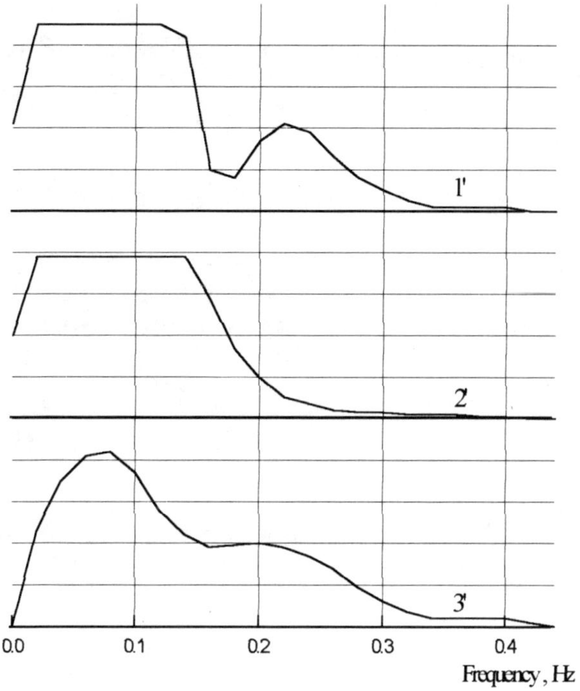

Fig. 3. Fourier transformation of EPR spectra shown at Fig. 1: (1') experimental; (2') background; (3') radiation induced

intensity of this line is linear function of the irradiation dose. It means that this part of the Fourier transformed spectrum is proportional to the number of radiation induced paramagnetic centers in the sample and can be used for the tooth enamel dosimetry.

The method was applied to the analysis of EPR spectra of Chernobyl liquidator teeth. The analysis of Fourier spectra allows us to obtain irradiation doses of an order of 0.20 Gy *without study of dose dependence.*

The advantage of the proposed method is a possibility to rise the sensitivity by an order of magnitude and to make measurements using teeth with dentine. This feature can be used to develop *in vivo* dosimetry techniques.

References

1. M.Ikeya, J.Miyajima and S.Okajima, ESR dosimetry for atomic bomb survivors using shell buttons and tooth enamel, *Japan J. Appl. Phys.* **23** (1984) 697-699.
2. Retrospective dosimetry and dose reconstruction. *Experimental collaboration project No 10. Final Report.* EUR 16540 EN. Editors I.K.Bailiff and V.Stepanenko (1996) pp. 57-76.
3. V.Chumak, A.Bougai, A.Wieser *et. al.*, The first International intercomparison of EPR-dosimetry with teeth: first results, in *Proceedings* of the 4th International Symposium on ESR Dosimetry and Applications, 15-19 May, 1995, GSF, Munchen/Neuherberg , will be published in *Appl. Radiat. Isotopes.*
4. A.Wieser, A.Bougai *et. al.*, International intercomparison of dose measurements using EPR spectrometry of tooth enamel, in *The Radiological Consequences of the Chernobyl Accident,* Proc. of the First International conference, Minsk, Belarus (1996) pp. 957-964.

Chapter 2

Chemistry, Earth and Environmental Sciences

EPR Meeting the Challenges of New Materials

Richard Bramley

Research School of Chemistry, Australian National University, GPO Box 414,
Canberra ACT 2601, Australia

Abstract. Following the discovery of EPR in bulk matter 50 years ago the initial phase emphasised the phenomenon as one of spectroscopic interest in its own right as much as for the insight it provided into the electronic structure of compounds of ions of the transition metals. Since then, EPR as a spectroscopic tool has matured and its use is driven now more by the challenges provided by materials. No longer do EPR spectroscopists choose materials suitable for the technique, rather EPR must rise to meet the demands of materials science. Some of the important landmarks on this journey will be visited beginning with detecting the EPR of electronically excited states. Some of the wealth of structural, kinetic and dynamic information which can be revealed will be described. In our own studies, a discovery connecting the Zeeman and fine-structure parts of the Hamiltonian meant that we could predict one knowing the other and thus tell how reactive an aromatic carbonyl compound would be in its lower excited state. A logical consequence is that one can dispense with the magnet (Zeeman part) and scan the frequency. This provided our entry into optically detected zero-field EPR (ZFEPR) and, in several labs, this matured into a versatile and often simple variant of EPR spectroscopy applicable to a wide variety of materials in both their ground and excited states. Pulsed EPR coupled with laser irradiation complement such studies by allowing more direct access to the kinetic and other time dependent properties of challenging materials. Changing back from optical to normal microwave detection, we are able to study ordinary materials in their ground states at zero field by scanning frequency. This technique further enhances the power of EPR spectroscopy. Advantages of ZFEPR as a procedure complementary to normal EPR will be illustrated. Some of our recent studies using home-built spectrometers will conclude the case that EPR can meet the challenge of new materials.

1 Introduction

Most branches of electromagnetic spectroscopy have their roots grounded in curiosity in the spectroscopy itself and not in our curiosity about materials, be they molecules in the gas phase or advanced technology composites. The substances used in these early growth phases are chosen to prove the spectroscopic point. Thus begun, the potential for aiding our understanding of materials is quickly recognised. In this context the term 'chemical shift' in NMR must have been used somewhat pejoratively by physicists for it meant that otherwise precise, or relatively precisely calculable, values for nuclear magnetic moments were not so precise after all. Nevertheless, the place that this radio-frequency branch of spectroscopy now holds in molecular and materials science must be second to none.

Physical techniques such as NMR and its microwave relative, EPR, may enjoy a simple applications phase where basic spectrometer design can satisfy most needs. This phase was somewhat shorter in NMR than in EPR, but there is no doubt that in the last decade or so, EPR spectroscopy has been driven by the challenges posed by materials and their chemistry; the latter on a nanosecond time scale, rather than by 'we should run the EPR spectrum anyhow'. My own research using EPR spectroscopy has been driven by materials curiosity, and the developments I have been involved in along with those of many others, have been made to meet the challenge posed by some material. I will describe some of them, in particular those less well covered by this symposium.

2 EPR in Excited States

Let's take basic EPR spectroscopy as read and note that all such studies in the first decade and a half from its discovery in bulk matter were on materials in their electronic ground states or on thermal averages including near-ground states. That electronically excited states could be equally well, and even better (kinetically), elucidated did not escape some of the EPR pioneers, notably Hutchison and his co-workers. Their triumph in this area was to report the EPR of the lowest electronic excited state of napthalene molecules isolated in a durene matrix.[1] In similar studies being undertaken in London,[2] it was noted that the signal intensity in similar systems was not a simple function of the triplet state lifetime. This signalled that there were kinetic factors operating other than just lifetime. What were they and could they be quantified? It was also apparent that many chemically interesting materials would have lowest excited states with lifetimes too short to allow an EPR observable population to be built up by photo-excitation with sensible light sources.

2.1 Optically Detected EPR

Independently, several groups set out to answer these questions, probably all remembering Brossel and Bitter's experiment on the optical detection of magnetic dipole allowed transitions in excited mercury atoms.[3] In this experiment, Hg atoms excited by polarised resonance light from a Hg lamp, were simultaneously irradiated at a fixed radio-frequency while a magnetic field scanned for changes in the polarisation of light transmitted through the optical/rf resonance cell.

This suggested that for organic molecular crystals, in which I was interested as basic models of materials photophysics and photochemistry, it might also be possible to detect EPR optically. A standard EPR spectrometer modulates the magnetic field and detects any resonance by demodulating the microwaves after their interaction with the material of interest using a high frequency diode and then looking for any signs of a response synchronous with the field modulation using a phase detector or lock-in amplifier. Many materials have optical properties which can be measured and which could thus act as detectors, such as optical absorption, fluorescence and phosphorescence, provided we can modulate them in some way. If any of these processes involves an excited state, commonly a spin-triplet, even indirectly, then amplitude modulation of the microwaves rather than the Zeeman field, could lead to a modulation of absorption or luminescence by alternate saturation and relaxation cycles which in turn could be detected with a photomultiplier rather than by a microwave diode (Figure 1). This optical signal could be decoded by a lock-in amplifier or box-car integrator synchronised to the microwave modulation frequency.

Three groups were quickly into print with their successes.[4,5,6] Such were the riches to be found in this new form of EPR spectroscopy that we spent several years exploiting them. The following specifics illustrate the still general power of this form of EPR. The triggering challenge for us at the time was to characterise the electronic structures and dynamics of materials so that one could for example design a polymer or dyestuff which would degrade in sunlight or be photo-stable as required. In this context it was important also to answer the question: is an excited triplet state $^3n\pi^*$ or $^3\pi\pi^*$ in character? The importance of this question was twofold. Firstly, there were ambiguities in the optical spectroscopic attempts to answer this question and, secondly, the photochemical proton extracting capability of excited aromatic carbonyl compounds[7] depended critically on their being $^3n\pi^*$, that is, with the excitation localised mainly on the carbonyl group. We chose 2-benzoylpyridine as our test case because its optical spectroscopic properties,[8] such as triplet lifetime, quantum yield for phosphorescence and phosphorescence polarisation, were known and it therefore looked like a good candidate.

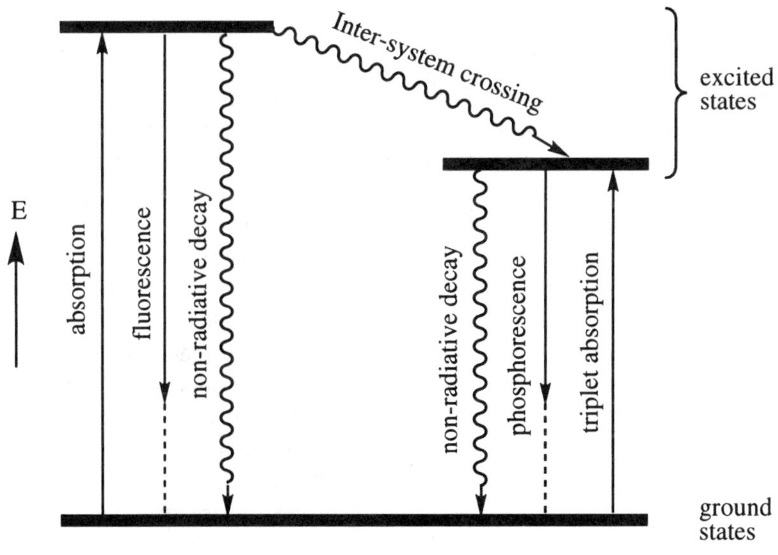

Fig. 1. Jablonski diagram showing some of the essential processes which may be influenced by microwave modulation of the triplet sub-level proportions and thus provide detection routes to other processes or the properties of other states.

2.2 Spectrometer Modifications for Optical Detection

Simple modifications to the basic spectrometer were needed. At first we used square wave field modulation applied to the normal cavity modulation coils by a modulation/lock-in amplifier system, thus bringing transitions into and out of resonance, saturating and relaxing. A mercury and/or xenon arc illuminated the cooled (4·2 K) sample through one port of the optical cavity and a photomultiplier collected the phosphorescence from the opposite port. The photomultiplier output was fed to the lock-in amplifier. (Appropriate optical filters were used on the excitation and observation sides.) The next development was to replace field modulation by amplitude modulation of the microwaves. A PIN diode switch was most conveniently inserted between the klystron and the bridge. A signal from the PIN diode driver was now used as a reference for the lock-in amplifier which received the photomultiplier current. This crude arrangement was improved by the insertion of a 10 KHz band-pass AFC (automatic frequency control) filter to block the effects of the PIN modulation on the AFC circuit of the (Varian) spectrometer. Optically detected EPR spectra could then be recorded in the usual way by scanning the magnetic field.[9] This gave simpler lineshapes than those obtained using square wave field modulation. Different spectrometers require the same approach in principle but will dictate differences in detail, for example, light pipe access to the sample might be more appropriate than a lens system. All that is required optically is efficient light delivery and collection. No AFC filter was needed in this case. Some other, more sophisticated arrangements have been reviewed.[10]

2.3 Photophysics and Photochemistry

The answers to our questions came as follows. 2-benzoylpyridine and $^3n\pi^*$ carbonyls generally, have low temperature triplet lifetimes of the order of a millisecond, too short for an adequate steady state triplet population and standard microwave detection. However, strong optically detected signals were seen (Figure 2). The Zeeman and fine-structure parameters of the spin-Hamiltonian were obtained in the usual way from rotation patterns (road maps) for pure single crystals. How could we be sure that the spectra we saw were coming from the host molecules and not from some guest impurity? The answer was provided by a simple experiment. By placing a scanning monochromator between the luminescing crystal and the photomultiplier and setting the field to be on a resonance, a microwave-modulated, phosphorescence spectrum was recorded. The electronic origin of this corresponded with the origin of the intrinsic phosphorescence spectrum recorded separately, thus establishing the identity of the optically detected EPR species as the host molecules.

Four sub-spectra were seen, one for each of the four magnetically distinct but crystallographically identical species, thus establishing the absence of truly excitonic behaviour in this particular triplet state. Closer examination of the intensity information in the road maps showed that the intensities were less than additive in the cross-over regions of resonances from pairs of molecules. This was successfully analysed in terms of a slow hopping model for the excitation, that is, polaronic rather

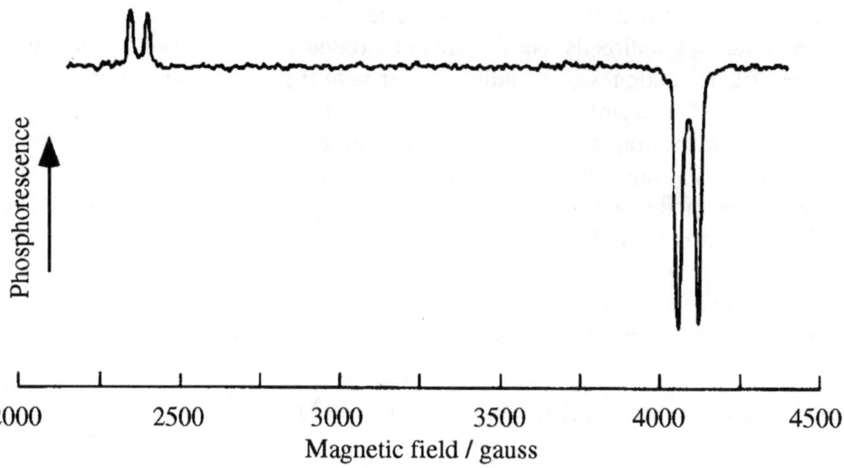

Fig. 2. Optically detected EPR spectrum of a 2-benzoylpyridine single at 4·2 K and 9·22 GHz. The crystal is so oriented in the field that resonances from pairs of molecules in the cells overlap.

than excitonic, with the vibrationally relaxed but electronically excited molecule transferring its excitation in a vibrationally accommodating lattice. The result of this analysis showed that this energy transfer was anisotropic, *least* preferring the crystal direction with the shortest C = O to C = O distance. Given that the excitation was $^3n\pi^*$ and largely localised on the C = O group, this was at first surprising but made sense in terms of the carbocyclic ring orientations and distances. Parallel rings and high overlap is a feature of excitonic aromatic systems.[11] There may, however, be phonon assisted processes, thus complicating this simple picture.

2.4 Kinetic Parameters

Optical microwave double resonance methods, as do all 2D spectroscopic procedures, provide a wealth of detailed information that cannot be approached by that from 1D techniques. For example, along with the host polaronic species, four additional sets of crystal symmetry related resonances were detected optically. On microwave resonance with any one of these signals the microwave modulated phosphorescence could be recorded by scanning a monochromator on the exit side. From the progression of bands in the ground state of this species, it was clearly an aromatic carbonyl compound of lower excited energy (deep trap) than the host species. Its fine-structure parameters were similar to those of the host material, so we believe that this was probably a substitutional impurity in which the benzene and pyridine rings were reversed or, less likely, where the pyridine ring was rotated 180°. Spectroscopically speaking, it is a deep trap. Further, by sitting on any one of these eight microwave (EPR) resonances and changing the monochromator over to the excitation side of the cavity, information could be obtained about their photoexcitation routes. The host (polaron) resonances could be excited either directly into the triplet state or indirectly via the singlet excited state followed by intersystem crossing. The excitation wavelength dependence of the deep trap and intrinsic (host) spaces (zero-field, *v. infra*) resonances were identical in the triplet-triplet region, thus confirming that the trap species received its energy from the host (hence the trap description). By using a box-car integrator rather than a lock-in amplifier to detect the microwave modulated phosphorescence, it was found that the singlet and triplet excitation patterns were different. This is not surprising given that the intersystem crossing rates are spin sub-level (symmetry) dependent.

The kinetic complexity of even a single isolated molecule can be appreciated from the simplest equation describing one sub-level of an excited triplet,

$$\dot{n}_i = F_i - k_i^r n_i - k_i^{nr} n_i + k_{ij}^{\mu}(n_j - n_i) + k_{ik}^{\mu}(n_k - n_i) + k_{ij}^{s}\left[(n_j - n_i) - c_{ij}(n_i - n_j)\right]$$
$$+ k_{ik}^{s}\left[(n_k - n_i) - c_{ik}(n_i - n_k)\right], \quad i = 1,2,3, \tag{1}$$

where n_i is the population of state $|i\rangle$, the rate coefficients for phosphorescence, non-radiative decay, microwave power and spin-lattice relaxation being k_i^r, k_i^{nr}, k_{ij}^{μ} and k_{ij}^{s}, and where $c_{ij} = [1 - \exp(|E_j - E_i| / kT)] / [1 + \exp(|E_j - E_i| / kT)]$ (see Figure 3).

One such equation is needed for each state in the excitation and decay loop. In simple situations, analytical solution is possible and numerical methods can clearly be

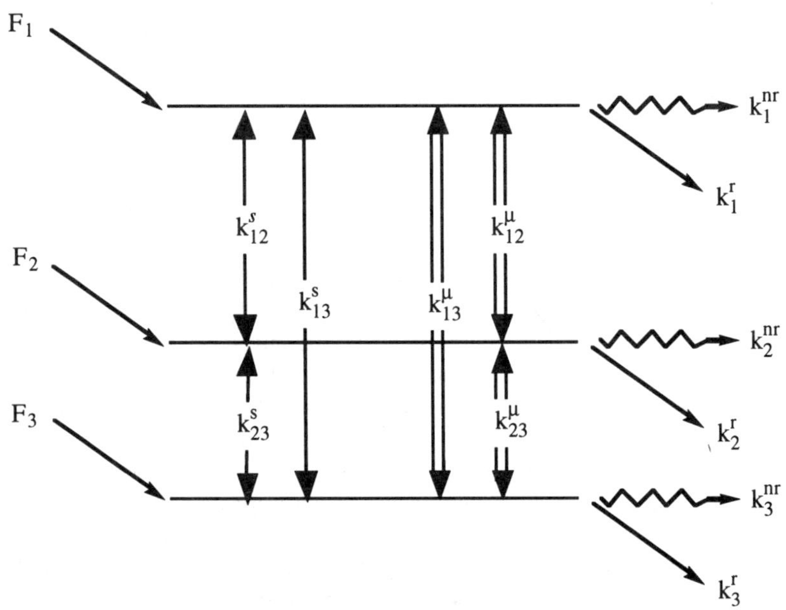

Fig. 3. Triplet sub-level energy diagram and processes occurring in an optical microwave double resonance experiment.

applied. In more complex situations there is fortunately a great deal of experimental control of the individual terms, for example, microwave power levels are totally under the experimenter's control, lower temperatures (pumped helium or helium 3) can exert powerful control over the spin-lattice relaxation times (even switching them off relatively), optical polarisation in the excited state can control sub-state feeding rates, or orientation selection can be used to the same end.

2.5 Optically Detected, Zero-Field EPR

The answer to settle the $^3n\pi^*$ or $^3\pi\pi^*$ question came principally from the fine-structure term of the spin-Hamiltonian with a correction due to spin-orbit coupling available from the Zeeman term.[12] With the emphasis on this term and an interest in the proximity of the two states perhaps thermally accessible, a magnet was not required. One could perhaps scan the frequency directly over the three resonances of a triplet. This provides a particularly simple EPR spectrometer comprising an amplitude modulated, variable frequency source, some arrangement to couple this radiation to the sample and in the case of luminescent materials, a photomultiplier.[9] For organic samples at liquid helium temperatures relatively low power sources can be used with surprisingly crude coupling of B_1 to the sample. Often a single loop short circuit at the end of a piece of coaxial cable would suffice, with an isolator to protect the source from reflected power. Just as crude were devices which were essentially coaxial airline sections with a hole for sample insertion, dimensioned to preserve characteristic impedance of the transmission line. In this case an isolator is not needed if the line is terminated properly. With such devices we could study a range of these materials (up to 26 GHz).

These zero-field splittings can be very large when the states are close, so large in fact that they can be optically resolved. So part of the armoury of the EPR spectroscopist is an optical spectrograph.[13,14]

2.6 Pulsed EPR and Other Methods for Non-Luminescing Excited States

What if one's triplet state does not luminesce? All is not lost, there are ways to get around this problem. If the the triplet lifetime approaches 100 ms or longer then standard methods apply; the early work on triplet napthalene in durene used standard Zeeman modulation and microwave detection.[1] For shorter lifetimes, pulsed light sources (flash or laser) can produce large instantaneous populations, almost always enhanced by spin polarisation (selective sub-level feeding). By increasing the bandwidth of the lock-in amplifier (post-phase-detector) and using a high modulation frequency, non-luminescing triplets can be studied successfully.

There are numerous lectures and posters describing pulsed EPR. It will therefore suffice to tell what we have done and are doing. We needed a spectrometer before commercial ones became available, so built it using sweepers from the ODZFR program as frequency sources and an old Varian magnet for the field, computer controllable. The rest we did ourselves. Our first studies were on potentially catalytic layered materials.[15] Our findings have been largely structural at this stage, such is the complexity. The power of pulsed EPR methods for such structural studies lies in the ability to observe magnetic nuclei beyond the first coordination sphere, out to about 5 Å, thus paralleling ENDOR studies. Other studies have been on high temperature superconductors.[16] Our current pulsed EPR work is with photochemistry looking at short timescale photophysics and photochemistry in organic molecules principally to elucidate both photophysical and photochemical mechanisms.

2.7 Zero-Field EPR on Ground States

Optically detected zero-field experiments offer another benefit over conventional EPR in that the resonant frequency is orientation independent. Only the intensity (polarisation) varies with orientation of B_1 in the molecular frame. This means that polycrystalline samples have the same frequency spectrum as a single crystal. The extension to amorphous materials and glasses is obvious. Now zero-field techniques could be particularly useful for ground states of the wide range of materials which can not be obtained in large single crystal form. All we have to do is dispense with optical detection and go back to microwave diode detection. By this time we had some variable frequency sources so we tried looking for the zero-field EPR spectrum in a ruby with a clockwork motor driven Varian X13B klystron, waveguide with a hole for sample insertion, and a waveguide crystal detector. The usual field modulation was applied. The result[17] was the expected single resonance at 11·5 GHz with negative Zeeman wings from which the g value could be recovered if required (with low precision). Being able to read a spin-Hamiltonian parameter off a frequency counter has a special appeal. It is emphasised that this is not a second derivative spectrum even though it looks like one. In principle, it is the difference between the zero-field absorption spectrum and the EPR absorption spectrum at peak ± field modulation amplitude. In the early phases of our work we used a second derivative Gaussian or Lorentzian simulation to assist location of lines in complex spectra, but only for assignment purposes. In some cases the lineshape can be far from second derivative in appearance, even one sided, and sometimes almost vanishes when the Zeeman component falls under the zero-field absorption.[18]

We set out to explore the potential of ZFEPR in the early 80's and published our results in part in a review.[17] We saw ZFEPR as a procedure complementary to normal (high variable field, fixed frequency) which had advantages in some circumstances more likely to turn up in the study of advanced materials than with canonical single crystals.

2.7.1 ZFR - The Experiment

Papers on ZFEPR would probably number a few hundred at most, even including the dominant fraction concerned with simple organic species in their excited triplet states.[19] Not surprisingly there is considerable variety in experimental approach as each group has engineered its experiment for a specialised materials problem. While the absence of a commercial spectrometer, or even a favoured variant among the lab-built may seem a disadvantage, the variety of approaches will offer a first solution for most new studies. The essential element of zero-field spectrometers is something which brings B_1 and the sample together. A few possibilities are listed in Table 1.

Table 1. Some environments for coupling B_1 to the sample.

Sample Irradiator	Resonant?	Remarks
Waveguide	No	Range limited to waveguide band, ridged waveguide best.
Co-axial line	No	Inhomogeneous B_1. Sample in air line or replacing dielectric. Size/mode problems above about 20 GHz.
Tuneable cavities and tuned coils	Yes	High B_1. Limited frequency range without (capacitive) coupling problems.
Loop-gap resonator (various tuning schemes)	Yes	High B_1. Full frequency range, acceptable coupling variation with inductive schemes.
Helix	Yes or No	Variety of approaches based on different propagation modes on helices. High optical access and easiest to use cryogenically.
Fabry-Perot	Yes	Suitable size for sub-millimetre work.

In addition, most nominally fixed-frequency resonators can be tuned magnetically or electrically to some extent. A tuning range of 10 per cent may be adequate for many experiments. Further details can be found in the literature. Magnetic field modulation is needed, preferably driven by a bi-directional squarewave. Phase-detecting the microwave output at twice this modulation frequency produces a spectrum which is the absorption at zero-field minus the absorption at peak modulation field.

2.7.2 The Scope for ZFEPR

ZFEPR is not a substitute for EPR, one is after all discarding a term in the Hamiltonian (except as a small perturbation from the modulating field). It is best seen as a facet of EPR or as a complementary experiment. As such it does offer benefits that are at times useful.

The absence of a field confers a number of benefits immediately. One does not have to worry about the accuracy of field measurement at the sample, nor of the orientation of the sample with respect to the field; there isn't one. Further, removing the Zeeman term in the Hamiltonian also removes the angular relationship of this term to other terms, a frequent source of error or uncertainty in spectral analysis. The nuclear Zeeman term also vanishes but nuclear quadrupole effects can now turn up in first order for some spin systems. The above features of ZFEPR all lead to increased precision of measurement for the parameters that do remain. This may be particularly important for higher order terms.

At this point it is worth emphasing that care needs to be exercised in comparing parameters derived from ZF and high-field EPR. Sweeping field and sweeping frequency are not formally equivalent. This has been discussed in detail by Pilbrow.[20,21] It is a difference in principle, but nevertheless one that correctly handled should lead to identical Hamilitonian parameters from the two experiments. The fact that results from the two experiments don't agree sometimes has in our case been the indication of the extra errors that can creep into a high field experiment (*v. supra*). The cases we studied involved narrow lines where the effects are smaller than the differences between ZFEPR and published EPR spectra.

A second group of benefits derives from the spectral equivalence of single crystals and randomly oriented samples, glasses, amorphous materials, or powdered crystals. This applies to the frequency axis, but not to the intensity axis. The latter polarisation factor has to be taken into account in simulations and spectral fitting.[18] Thirdly, both dispersion and resolution can be significantly improved at zero field. In high fields, the non-Zeeman terms are often perturbations on the Zeeman one, leading to overcrowding of transitions around $g = 2$. The often better resolution comes from the fact that the Zeeman effect is frequently non-linear or asymptotically zero at zero field, so dipolar broading is less of a problem in this case. Multiple sites may therefore be seen in ZFEPR that are convoluted in high field studies.[22]

ZFEPR has its limitations too. Sometimes the spectrum is so simple that information cannot in principle be obtained. For example $S = 3/2$ has one zero-field transition, but there are two fine-structure parameters (D and E). Only a combination of these two parameters can be obtained from the zero-field transition frequency.

2.8 Advances in ZFEPR

This recent phase of our work was not concerned with attempting to develop a spectrometer which was as sensitive as a normal EPR spectrometer, except that one varied frequency at zero field instead of field at fixed frequency. Our interest was rekindled because it became clear that the complexity and variety of modern materials were demanding something approaching that, particularly where small amounts of material were available. Our recent interest was triggered by needing to know something about the structure of partially stabilised zirconia (PSZ). Zirconia is ZrO_2. It has a number of desirable properties as a ceramic but one disadvantage in that it undergoes a phase change at high temperatures. This means that one could be driving a car with a ZrO_2 engine only to have it disintegrate as it warmed up to operating temperature. The incorporation of some yttria (Y_2O_3) or calcium oxide produces a stable material of widespread use, but one having fundamental questions remaining about its detailed structure. Since replacing Zr by Y or Ca requires charge compensation, this creates vacancies as oxygen is removed to maintain electrical neutrality. ZrO_2 is cubic with eight oxygens at the corners of a cube surrounding each zirconium atom. What is the distribution of oxygen vacancies; one per cube, two per cube, opposite corners on a cube diagonal, opposite on a cube face, or adjacent on a cube edge? Given the level of substitution or vacancy concentration, how are the vacancies on adjacent cubes arranged, randomly or with some correlation? We substituted a probe amount of Mn^{2+} for Ca^{2+} and Cr^{3+} for Y^{3+} and ran EPR spectra and hoped to answer these questions by superposition theory based on the spin-Hamiltonian parameters from EPR spectra. It was clear at the outset from the broad EPR spectra obtained in both cases, that there was significant disorder in the lattice. We added g as well as D and E strain to the model but the spectra defied our best efforts to crack the problem, hence the idea of running the ZFEPR spectra to sort out the fine-structure separately, and then using these results for a total simulation of the full EPR spectrum. For this and other studies we paid some attention to a broader range, higher sensitivity ZFEPR spectrometer which was also fully cryogenic. The results are not yet complete on the PSZ,[23] however we have begun to look at some exchange coupled transition metal systems.[18]

Fifteen years ago I had recorded the ZFEPR spectrum of anhydrous copper benzoate,[24] chosen because of all the copper carboxolate type dimers that were easily prepared, it had the highest deviation from axial symmetry (highest E value). The EPR spectra of this material were also broad, indicating disorder. A search for better EPR spectra suggested that the benzoic acid adduct (in the two axial positions of the dimer) would be a good candidate for a thorough analysis. This we have recorded and analysed, using full Zeeman subtraction in the simulation.[18]

A detail of the spectrometer may be interesting. Our resonator is a two gap single loop, loop-gap resonator driven by two (same sense) rotating shafts from the top. This can provide over two octaves of frequency coverage with one resonator pair and cover from about 1 – 14 GHz by changing the resonators. Other variants we have used include variable bridged loop-gap resonators and non-rotating but otherwise variable gap models. Making such devices is in principle simple; in practice there are many pitfalls. The most useful single rule is to keep structures inside the screen as small as possible, generally having dimensions no greater than one fifth of the shortest effective wavelength of radiation used.

2.9 Variable Frequency EPR

The ability to scan frequency has an obvious extension to normal EPR and one can combine our ZFEPR developments with a normal EPR spectrometer to get what might be called a VFEPR spectrometer. Apart from the obvious benefit of being able to check the validity of spin-Hamiltonians by recording spectra at two well separated frequencies, the principal gain is that it enables non-crossing regions of the Zeeman diagram to be precisely mapped, thus providing reliable values for higher order terms in the Hamiltonian with sufficient accuracy for them to be used for structural and theoretical considerations.[25] This is a wide open field and it is only with these methods or alternatively standard methods with great care, that one can have some real confidence in the experimental values. It is a form of 2D spectroscopy which promises useful results.

2.10 Pulsed ZFEPR

Finally, one can combine pulsed EPR spectroscopy with zero-field methods, initially by tuning the spectrometer to match a resonance at zero field. This has been recently demonstrated by Jeschke & Schweiger.[26] Note that this is a fixed frequency experiment. Scanning frequency rather than field in the pulsed environment poses some interesting problems which could no doubt be solved one by one, or the output given over to numerical methods for deconvolution.

One should not forget that magnetic resonance spectrometers are not aware that they are looking at nuclear or electronic paramagnetic samples. So one can study NMR, or zero-field NMR (NQR) in an EPR spectrometer and vice versa, particularly if they are of the variable frequency type. Thus, in a ZFEPR spectrometer designed for the VHF – UHF region (100–1000 MHz, say) one can study not only $S = {}^1/_2$, $I \geq {}^1/_2$ systems, but NQR as well. We have conversely recorded EPR spectra of spin-labels on a VHF spectrometer designed for NQR.

3 Conclusions

There is no doubt that EPR spectroscopy is alive and well and promises to be well up to meeting the challenges posed by modern materials science. The two main advances in EPR spectroscopy that have been discussed here have come about by adding new dimensions to basic, one-dimensional EPR.[27] Dimensions such as time and radio-frequency radiation have either created new spectroscopies or enriched one-dimensional forms. Examples are pulsed EPR and ENDOR.[28] By adding further dimensions to our growing library, and combining them in multidimensional spectroscopies, the power of EPR will continue to flourish.

References

1. C. A. Hutchison Jr., and B. W. Mangum, Paramagnetic resonance absorption in naphthalene in its phosphorescent triplet state, *J. Chem. Phys.* **34** (1961) 908-922.
2. H. P. Cleghorn, Ph.D. Thesis (1965) University of London, and personal communication.
3. J. Brossel, and F. Bitter, A new "double resonance" method for investigating atomic energy levels. Application to $Hg^3P_1{}^*$, *Phys. Rev.* **86** (1952) 308-316.
4. A. L. Kwiram, Optical detection of paramagnetic resonance in phosphorescent triplet states, *Chem. Phys. Lett.* **1** (1967) 272-275.
5. J. Schmidt, I. A. M. Hesselmann, M. S. de Groot, and J. H. van der Waals, Optical detection of electron resonance transitions in phosphorescent quinoxaline, *Chem. Phys. Lett.* **1** (1967) 434-436.
6. M. Sharnoff, ESR-produced modulation of triplet phosphorescence, *J. Chem. Phys.* **46** (1967) 3263-3264.
7. J. C. Scaiano, Solvent effects in the photochemistry of xanthone, *J. Am. Chem. Soc.* **102** (1980) 7747-7753.

8. M. Batley, R. Bramley, and K. Robinson, Photophysics of the lowest triplet state in 2-benzoylpyridine crystals. Part 1. Optical spectra, *Proc. Roy. Soc. Lond.* **A369** (1979) 175-185.
9. M. Batley, R. Bramley, F. Poldy, and K. Robinson, Photophysics of the lowest triplet state in 2-benzoylpyridine crystals. Part 2. Optically detected electron paramagnetic resonance in zero and high magnetic field, *Proc. Roy. Soc. Lond.* **A369** (1979) 187-206.
10. S. Geschwind, Optical techniques in EPR in solids, *Electron Paramagentic Resonance*, edited by S. Geschwind, Plenum Press, London (1972) pp. 353-425.
11. J. Jortner, S. A. Rice, J. L. Katz, and S.-I. Choi, Triplet excitations in crystals of aromatic molecules, *J. Chem. Phys.* **42** (1965) 309-323.
12. M. Batley, and R. Bramley, Spin-orbit coupling effects in EPR spectra of lowest triplet states of aromatic carbonyl compounds, *Chem. Phys. Lett.* **15** (1972) 337-342.
13. H. J. Griesser, and R. Bramley, Highly resolved dual phosphorescence of xanthone in several hosts, *Chem. Phys. Lett.* **83** (1981) 287-291.
14. J. D. Laposa, and R. Bramley, Temperature-dependent phosphorescence of deuterium-substituted xanthones, *J. Phys. Chem.* **88** (1984) 4641-4647.
15. V. Luca, D. J. MacLachlan, R. F. Howe, and R. Bramley, Synthesis and characterisation of a (Zn,Ti)-substituted layered silicate, *J. Mater. Chem.* **5** (1995) 557-564.
16. K.-P. Dinse, D. M. Wang, R. Bramley, and J. W. White, Determination of the London penetration length for Rb_3C_{60} with pulsed EPR, *Physica C* **218** (1993) 341-347.
17. R. Bramley, and S. J. Strach, Electron paramagnetic resonance spectroscopy at zero magnetic field, *Chem. Rev.* **83** (1983) 49-82.
18. C. D. Delfs, and R. Bramley, The zero-field EPR spectrum of a copper dimer, *Chem. Phys. Lett.* in the press (1997).
19. S. K. Lomer, and M. A. El-Sayed, The triplet state and molecular electronic processes in organic molecules, *Chem. Rev.* **66** (1966) 199-241.
20. J. R. Pilbrow, Lineshapes in frequency-swept and field-swept EPR for spin $^1/_2$, *J. Mag. Reson.* **58** (1984) 186-203.
21. J. R. Pilbrow, *Transition Ion Electron Pagmagnetic Resonance*, Clarendon Press, Oxford (1990).
22. R. Bramley, and S. J. Strach, Zero-field EPR of the vanadyl ion in ammonium sulfate, *J. Mag. Reson.* **61** (1985) 245-253.
23. K. J. Edler, D. J. MacLachlan, and R. Bramley, unpublished results.
24. Reference 17, p. 66.
25. R. Bramley, Variable-frequency electron paramagnetic resonance, *Int. Rev. Phys. Chem.* **5** (1986) 211-217.
26. G. Jeschke, and A. Schweiger, Zero-field electro spin echo envelope modulation spectroscopy, *Chem. Phys. Lett.* **259** (1996) 531-537.
27. R. Bramley, Electron Spin Resonance, in *Encyclopaedia of Inorganic Chemistry,* edited by R. B. King, Volume **3**, John Wiley & Sons (1993) pp. 1062-1081.
28. H. Barkhuijsen, R. de Beer, E. L. de Wild, and D. van Ormondt, Measurement of hyperfine interactions with electron spin-echo spectroscopy. Application to F centres in KCl, *J. Mag. Reson.* **50** (1982) 299-315.

CIDEP Studies of Photochemical Reactions at Different Microwave Frequencies, ST_0M and $ST_{-}M$

N. Hirota, Y. Miura, K. Ohara, and M. Terazima

Department of Chemistry, Graduate School of Science, Kyoto University, Kyoto 606 Japan

Abstract. Frequency dependence of the CIDEP spectra of the acetone ketyl radical and the *t*-butyl radical produced photochemically in 2-propanol were investigated by using cw time-resolved EPR spectrometers in the L-, S- and X-bands. The dependence of the ST_0M polarization on Q (half the difference of the resonant frequencies of two radicals of the pair) was examined in the acetone ketyl radical. It was found that the usual $Q^{1/2}$ dependence is valid at higher temperatures (> -10°C), but the spectra are better explained by the Q^0 dependence at lower temperatures (< -40°C). The ratio between the $ST_{-}M$ polarization and the ST_0M polarization increases at lower temperatures, but the extent of the increase is less at lower frequencies than at the X-band. On the other hand, the spectrum of the t-butyl radical is explained well by the $Q^{1/2}$ dependence even at very low temperatures. The difference in these two systems is ascribed to the different diffusional motions of the two systems.

1 Introduction

Transient EPR spectra of radicals obtained just after photochemical reactions reflect a non-Bolzmann distribution of spin states and show electron polarization called CIDEP (Chemically Induced Dynamic Electron polarization). CIDEP spectra of radicals and radical pairs produced in photochemical reactions in solution often provide valuable information about the details of the initial stages of reactions that cannot be obtained by other method. Both cw and pulsed time-resolved EPR techniques have been used extensively to study CIDEP phenomena in photochemical reaction systems in the last fifteen years. These studies have provided much valuable information concerning reaction mechanisms, structures and interactions in radical pairs, and spin and reaction dynamics of intermediate radicals and radical pairs. Yet, so far CIDEP studies have been mostly performed in the X-band (9.3 GHz, 330 mT), though there have been some attempts at different microwave frequencies such as L-band (Terazima,1991) and Q-band (Forbes, 1992,1993). Recently, we have made CIDEP studies in the L- (1.5 GHz, 50 mT) and S-band (3.0 GHz, 100 mT) as well as the X-band (Terazima,1994 and Ohara,1995). The main motives of our study are,

(1) Measurements at different magnetic field would be useful in examining the validity of CIDEP theories and sorting out correct CIDEP mechanisms.

(2) Further information about dynamics of precursor molecules and intermediate radicals and radical pairs may be obtained.

(3) Magnetic field dependence of CIDEP is interesting in connection with the magnetic field effects on chemical reactions and other related phenomena.

In previous papers we have reported magnetic field dependence of CIDEP due to triplet mechanism(TM) and spin-correlated radical pairs (SCRP). In the present paper we discuss the CIDEP spectra due to radical pair mechanism (RPM), ST_0M and ST_-M. measured at different microwave frequencies. Particularly, we focus our attention on the relation between CIDEP intensities and radical diffusion.

Basic theories of ST_0M and ST_-M have been worked out in 1970s mainly by Adrian (1991, 1992), Pedersen-Freed (1973), and Adrian-Monchick (1978). Although qualitative features of many CIDEP spectra have been successfully explained in terms of RPM and TM, quantitative comparisons of the experimental spectra and the theoretical predictions have been scarce. However, since CIDEP intensities are strongly dependent on radical diffusion, one might hope to obtain useful information about the diffusion of radicals in solution from a detailed analysis of CIDEP intensities. Previously Tominaga et al. (1990) made an attempt to calculate absolute magnitudes of the polarization observed in the photolysis of acetone in 2-propanol by solving the stochastic Liouville equation (SLE) following the procedure originally used by Pedersen and Freed (1973). It was found that in order to explain the observed large polarization it was necessary to assume slow diffusion of the radicals comprising the pair. This slow diffusion was ascribed to the microscopic solvent structure in 2-propanol that stabilizes the radical pair. Meanwhile, Shushin developed an effective and accurate method for an analytical solution of SLE and gave a simple expression for CIDEP intensities (1990). More recently he also considered CIDEP intensities due to RPM at low magnetic fields (1995) and (1997). It appears worthwhile to compare the experimental results with these theoretical predictions.

In the present work we have studied the CIDEP spectra of 2-hydroxylpropan-2-yl (acetone ketyl radical) produced by the hydrogen abstraction reaction of acetone and t-butyl radical obtained by the photolysis of di-t-butylketone(DTBK) at different magnetic fields. The experimental results will be compared with the predictions of the theories based on various models of radical diffusion.

2 Experimental

The X- and L- band time-resolved EPR spectrometers are the same as reported elsewhere (Terazima,1995). Briefly, the L-band spectrometer is based on a modified JEOL ES-LB2SX microwave unit. For S-band measurements modifications were made to the microwave unit. They included replacements of the oscillator, mixer and resonator to appropriate ones. Loop gap resonators of 15 mm diameters with slots for light irradiation were used for L-band measurements. The sensitivities of the S-band and L-band spectrometers were low because of low Q factors of the resonators and poor filling factors. The dc EPR signals were detected without field modulation and amplified by a wide band preamplifier. Signals were collected by a boxcar integrator (PAR 160) whose gate width was 0.2 μs, and the delay time was usually kept to be 1.0 μs after the laser excitation. The output was digitized by an A/D converter and stored in a personal computer.

The sample solutions were deoxygenated by bubbling nitrogen gas before and/or during the experiments and flowed into flat quarts cells (7mm width and 1mm path length) in the L-and S-band measurements and quartz tubings in the X-band measurements. Commercially available chemicals were used without further purification. 2-propanol was used as solvent.

An excimer laser with XeCl operation (Lumonics Hyper 400, 308nm, 7.5Hz) was used for photoexcitation of the sample.

3. Results and Discussion

3.1. *Q* Dependence of the ST_0M Polarization of the Acetone Ketyl Radical

The reaction scheme to produce the acetone ketyl radical in 2-propanol is considered to be given by,

$$(CH_3)_2C{=}O \;+\; (CH_3)_2CHOH \;\longrightarrow\; 2\,(CH_3)_2\dot{C}OH$$

In previous work Tominaga et al. (1990) found that the relative intensities of various hyperfine peaks of the acetone ketyl radical are temperature dependent. The reencounter model of Adrian (1971) predicts that the ST_0M polarization is proportional to $Q^{1/2}$, where Q is half the difference in the resonance frequencies of the two radicals of the

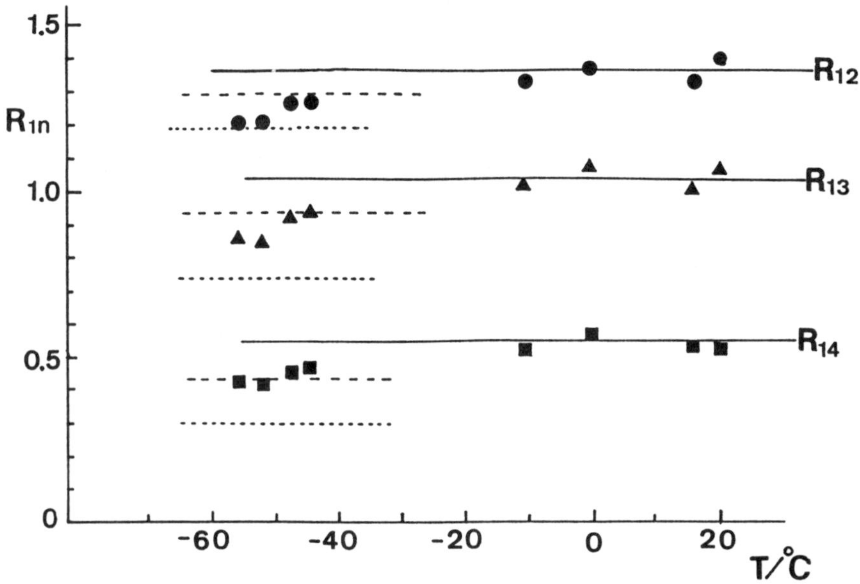

Fig. 1 Temperature dependence of R_{1n}, predicted by a) ——— $Q^{1/2}$ dependence b) – – – – by Q^0 dependence, and c) ------- Q^{-1} dependence. The calculated values were taken from Shushin (1990).

pair. Shushin (1990) considered Q dependence of the ST_0M polarization and predicted considerably different Q dependence for different types of diffusional motion of the radical in the case of the acetone ketyl radical.

In Fig.1 we show the temperature dependence of the relative intensities of different hyperfine lines of the deuterated acetone ketyl radical produced by the photolysis of acetone-d_6 in 2-propanol-d_8 in the X-band. Here, $R_{1n} = I_n/I_1$ is the ratio between the intensities of the hyperfine lines corresponding to $M_I = n$ and $M_I = 1$, respectively. It is seen that at higher temperatures the intensity ratios are in very good agreement with those predicted by the $Q^{1/2}$ dependence of the standard CIDEP theory. However, they deviate substantially from the $Q^{1/2}$ dependence at low temperatures. Shushin (1990) obtained analytical solutions of the sochastic Liouville equation for different model of diffusion. The results are summarized as follows.

a) Free diffusion model (FDM): Radicals diffuse freely, but two different cases can be distinguished depending on the diffusion constant. In the limit of $Qd^2/D<<1$, P_{ST0M} is proportional to $Q^{1/2}$. Here D is the diffusion constant and d is the distance of the closest approach. In the case of slow diffusion $Qd^2/D>1$, P_{ST0M} is independent of Q, or P_{ST0M} is proportional to Q^0.

b) Cage diffusion model: Radicals experiences strong cage potential due to strong interactions. Solutions for two limiting cases were also obtained. When $Q<<W_0$, P_{ST0M} is proportional to Q, whereas in the case of $Q>>W_0$, P_{ST0M} is inversely proportional to Q. Here W_0 is the sum of the relaxation rates of two radicals..

The relative peak intensities predicted by his analytical solutions are also indicated in Fig.1. It is seen that the relative intensities of the peaks at low temperatures (-40 - -60°C) are better described by the slow diffusion model. However, the value of Qd^2/D is 0.23 and 0.5 at -45°C and -60°C, respectively. In the above criterion it is still not in the slow diffusion limit. This seems to mean that the actual diffusion of the radical is slower than expected from the solvent viscosity, as suggested in our previous work (Tominaga,1990). Unfortunately it is not possible to determine the relative intensity ratios at further lower temperatures because of the superposition of the strong SCRP signals.

The Q dependence of the ST_0M polarization can also be studied from the L-band spectrum. We have simulated the L-band spectra of the acetone ketyl radical with different Q dependence of the ST_0M polarization. The experimental spectra at -30 - -60°C are compared with the simulated spectra with ST_0M polarization of different Q dependence. Better agreements are obtained with $P_{ST0M} \propto Q^r$ dependence, r being between 0 and 0.5. An example of the comparison between the experimental spectrum and the simulated one is shown in Fig. 2. Though the difference is small, it is seen that a better agreement is obtained for the spectrum at -60°C with Q^0 dependence. For example, the observed intensity ratio between the second and the third peaks from the right is in better agreement with that in c). This again indicates that the slow diffusion case is already realized, even though Qd^2 is still smaller than 1. At -75°C the experimental spectrum approaches more the simulated spectrum based on the cage diffusion model of $Q>>W_0$. This is consistent with the observation that SCRP spectrum of relatively long-lived radical pair is actually observed at very low temperature (Tominaga,1990).

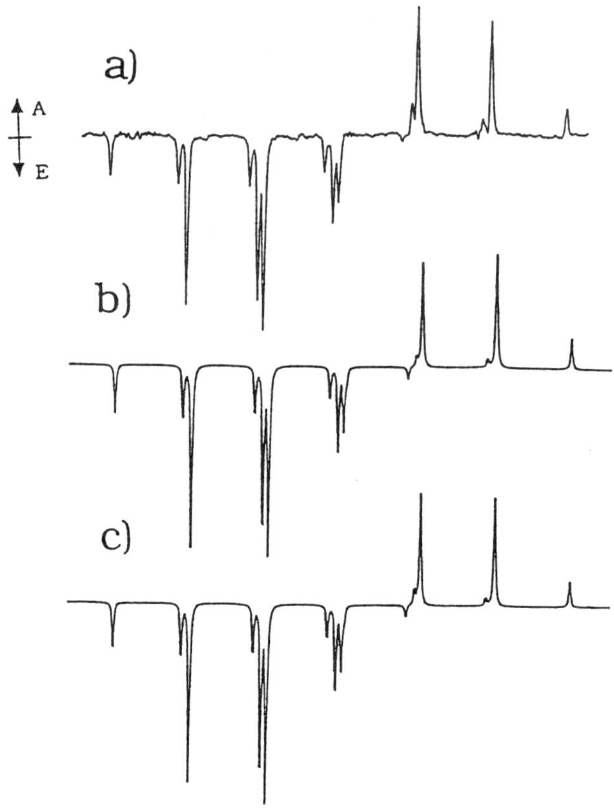

Fig. 2 L-band spectrum of the acetone letyl radical at -60 °C: a) experimental and simulated based on b) $Q^{1/2}$ dependence and c) Q^{0} dependence

3.2 Temperature Dependence of the Spectra at Different Microwave Frequencies: ST_0M vs. $ST_{-}M$

Temperature dependencies of the CIDEP spectra of the acetone ketyl radical in 2-propanol in the X- and L-band are shown in Fig. 3. The x-band spectrum shows splittings due to six equivalent methyl protons. The L-band spectrum shows lines due to the second order splittings by methyl protons. The X-band spectrum at higher temperature is of E/A* type presumably because of the contribution due to the reaction coming from the polarization of the thermalized triplet state of acetone (Levstein,1991). At lower temperature the spectrum changes to E*/A type. The net emissive component is believed to arise from the $ST_{-}M$ polarization because of the increased viscosity of the solvent. From Fig. 3 it is clearly seen that the net emissive component is greatly enhanced in the L-band spectrum compared with the X-band spectrum at the same temperature. According to the formula given for the $ST_{-}M$ polarization by Adrian and Monchick (1978), the magnitude of the $ST_{-}M$ polarization is predicted to be inversely proportional to the microwave frequency. Thus the

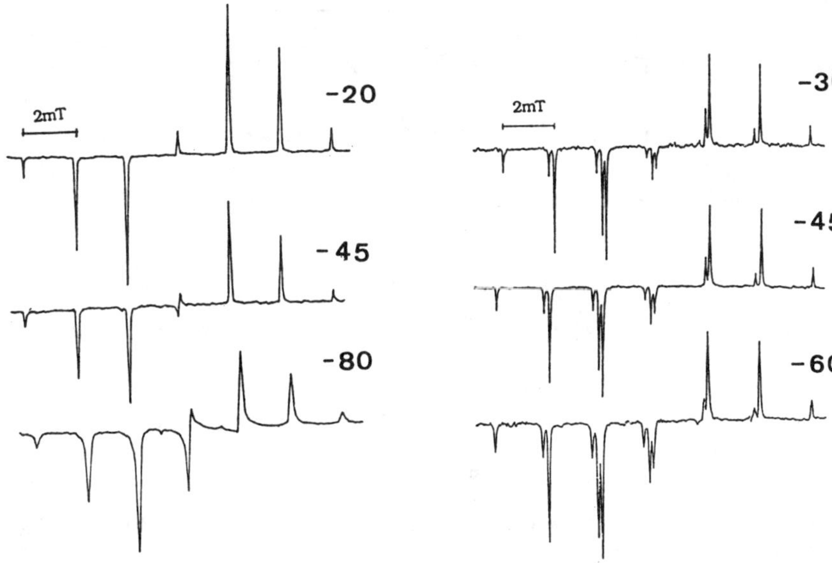

Fig. 3 Temperature dependence of the CIDEP spectra of the acetone ketyl radical: a) X-band, and b) L-band

observed result is qualitatively in agreement with the theoretical prediction. We have made a more detailed analysis in the following way.

We assume that the observed polarization mainly comes from the two contributions of RPM, namely ST_0M and ST_-M:

$$P = P_{STOM} + P_{ST\text{-}M} \tag{1}$$

P_{STOM} and $P_{ST\text{-}M}$ were estimated for the X-, S- and L-band spectra at different temperatures from the simulated spectra. Then the ratios of the two contributions, $R = P_{ST\text{-}M}/P_{STOM}$ were calculated for the X-, S- and L-bands. They are given in table 1.

Table 1. $R = P_{ST\text{-}M}/P_{STOM}$ at different temperatures for different bands

	L-band	S-band	X-band
-30°C	0.45	0.2	--
-45	0.60	0.3	--
-60	0.85	0.5	0.15
-75	1.2	0.9	0.40
-90	2.5	1.5	1.2

In Fig. 4 -ln$R$ is plotted against -lnD. Here D is the diffusion constant calculated from the Stokes-Einstein equation, $D = kT/3\pi\eta d$. The viscosity of 2-propanol is assumed

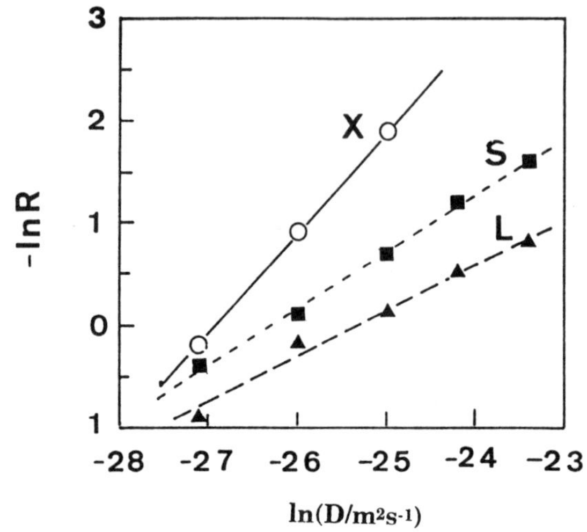

Fig. 4 Plot of ln$R$ for acetone ketyl radical versus ln$D$ in the X-, S- and L-bands. X, S and L stand for the X-, S- and L-bands, respectively.

to be given by η = 4.47x10^{-4}exp(2532/T)cp (Tonomura,1933). The hydrodynamic diameter d is taken to be 0.349nm. It is seen that the plots give straight lines in the all three bands, but slopes are quite different. It has been suggested that in a solvent of high viscosity the polarization due to ST_0M does not depend on D. ($P_{ST0M} \propto D^0$). Since $P_{ST\text{-}M}$ is proportional to D^{-1}, the slope is expected to be 1. This is in fact observed in the X-band. On the other hand the slope is more close to 0.5 at lower frequencies. If we assume that $P_{ST\text{-}M}$ is inversely proportional to the microwave frequency, $R(S)/R(X)$ = 3 and $R(L)/R(X)$ = 6, but the observed values are much smaller. This means that the prediction of the standard CIDEP theory is no longer valid at low magnetic fields. A theoretical analysis of the CIDEP phenomena at low magnetic fields was recently undertaken by Shushin (1955, 1997). His result indicates that separation of ST_0M and ST_- M in the conventional CIDEP theory is no longer valid at a low magnetic field. His analysis indeed predicts that R should be proportional to $D^{-0.5}$ as observed in our experiment in the S- and L-bands at temperature lower than - 50 ^oC.

3.3 *Q* Dependence of the L Band *t*-butyl Radical Spectrum

The Q dependence becomes quite drastic when only one of the radicals comprising the geminate pair has large hyperfine splittings (Shushin,1990)). This situation is realized in the case of the photodecomposition of di-t-butylketone (DTBK). Since the contribution of the TM polarization is very small in the L-band, the L-band spectrum is suited for examining the RPM polarization in detail. Furthermore, in the L-band spectra many peaks appear due to the second order hyperfine splitting, and an accurate comparison between the experimental and simulated spectra can be made. The photodecomposition is known to proceed by the following scheme (Jent,1987):

$$\text{R-CO-R} \rightarrow \text{R-CO} + \text{R}$$

$$\text{R-CO} \rightarrow \text{R} + \text{CO} \qquad (\text{R=C(CH}_3)_3)$$

The L-band spectra observed at different temperatures are shown in Fig.5. Only the spectrum of the *t*-butyl radical was observed, probably because the spin-lattice relaxation time of the acyl radical is very short. However, the lifetime of the acyl radical is considered to be relatively long-lived at low temperatures, making the acyl/t-butyl pair the only probable species involved in the ST_0M polarization (Jent,1989). As shown in the figure simulated spectra are strongly dependent on the Q dependence, because the acyl radical has negligibly small hyperfine splitting. The observed spectra can be simulated quite well by the $Q^{1/2}$ dependence of the ST_0M polarization with a small contribution of the ST_-M polarization even at very low temperatures. The Q dependence of this system is in a remarkable contrast to the acetone/2-propanol system, though the same solvent, 2-propanol was used in both cases. It is considered that different diffusional motions of radicals give rise to such a striking difference. In the case of DTBK the radicals in the geminate pair are not trapped in the cage and diffuse more freely in 2-propanol. Presence of the OH group in the acetone ketyl radical seems to be crucial for the slow diffusion of the radical in 2-propanol

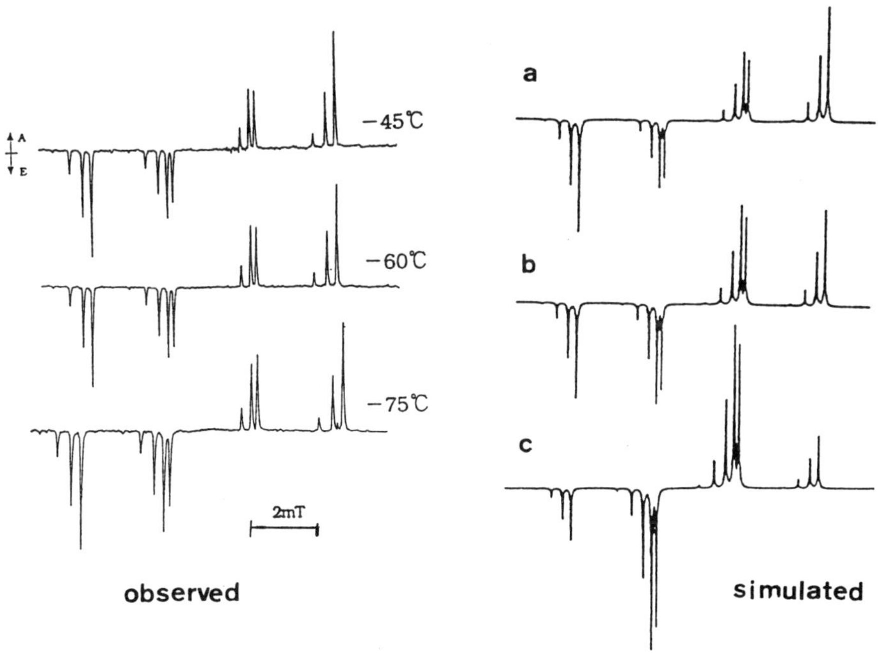

Fig. 5 L-band spectrum of the *t*-butyl radical observed and simulated with a) $Q^{1/2}$ dependence, b) Q^0 dependence and c) Q^{-1} dependence

4 Conclusion.

We have studied some CIDEP spectra of radicals produced by photochemical reactions at different microwave frequencies, paying attention to the relation between the CIDEP spectra and radical diffusion. The Q dependence of the CIDEP spectra is useful in clarifying the diffusional motion of the radicals in the initial stage of the reaction. It was found that the acetone ketyl radical created by the hydrogen abstraction reaction in 2-propanol is trapped in a solvent cage and its diffusional motion is strongly restricted by the interaction with the solvent. On the other hand, the motion of the *t*-butyl radical created by the photodecomposition of DTBK is well described by the free diffusion model even at very low temperatures.

Acknowledgments

We are indebted to Dr. K. Tominaga for some of the data discussed here. We thank Dr. A. I. Shushin for very helpful discussions. The work was supported in parts by a grant in aid for the priority area "Molecular Magnetism" by the Ministry of Education and Culture.

References

1. M. Terazima, S. Hayakashi, and T. Azumi, J. Phys. Chem. 95 (1991) 4297
2. M. D. E. Forbes, J. Phys. Chem., 96 (1992) 7386
3. M. D. E. Forbes and S. R. Ruberu, J. Phys. Chem. 97 (1993) 4297
4. M. Terazima, Y. Miura, K. Ohara, and N. Hirota, Chem. Phys. Lett., 195 (1994) 224
5. K. Ohara, M. Terazima, and N. Hirota, J. Phys. Chem. 99 (1995) 17814
6. F. J. Adrian, J. Chem. Phys. 54 (1971) 3918, 57 (1972) 5107
7. J. B. Pedersen and J. H. Freed, J. Chem. Phys. 58 (1973) 2746, 59 (1973) 2869
8. F. J. Adrian and L. Monchick, J. Chem. Phys. 71 (1979) 2600
9. K. Tominaga, S. Yamauchi, and N. Hirota, J. Chem. Phys. 92 (1990) 5172
10. A. I. Shushin, Chem. Phys. 144 (1990) 201, 223
11. A. I. Shushin, Chem. Phys. Lett. 170 (1992) 78
12. P. R. Levstein and H. van Willigen, J. Chem. Phys. 95, 900 (1991)
13. T. Tonomura, Sci. Rep. Tohoku Imp. Univ. Ser. 1 22 (1933) 104
14. A. I. Shushin, Chem. Phys. Lett. 237 (1995) 177
15. A. I. Shushin, Chem. Phys. in press
16. F. Jent and H. Paul, Chem. Phys. Lett., 160, 632 (1989)

The Late Pleistocene ESR Chronology of Cores ZQ_2 and ZQ_4 in the South China Sea

Yu-Guang Ye[1] *Shao-Bo Diao*[1] *and Jun-Cheng Gao*[2]

Institute of Marine Geology, MGMR, Qingdao 266071, China
National Institute of Metrology, Beijing 100013, China

Abstract. The Ages of 18 samples which were taken from cores ZQ_2 and ZQ_4 in Zhujiangkou Basin, the South China Sea, were obtained using Ge centres in quartz of sediments by the ESR technique. The correlation of ESR ages with ^{14}C ages, TL ages, ultrapaleontology, palynostratigraph, magnetostratigraph and oxygen isotope stratigraph, shows a correspondence with the horizons and also with the geological background.

1 Introduction

The Zhujing River Mouth Basin is in the continental shelt and upper stope area, the northen South China Sea. The National Offshore Petroleum Coporation of China once hoped that the Ministry of Geology and Minearl Resources would strengthen its investigation work in the fields of marine hazard geology and engineering geology. Guangzhou Marine Geological Survey, MGMR finished the whole investigation and study employing various advanced survey methods from 1986 to 1991. Core ZQ_4 and ZQ_2 were major cores in the survey and the

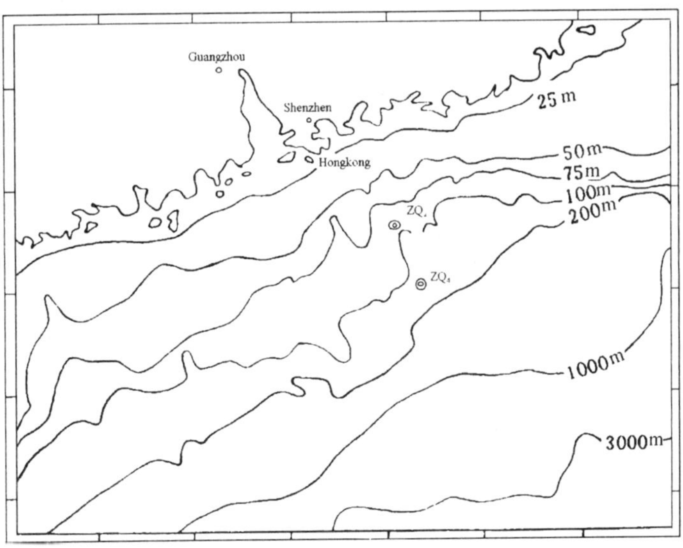

Fig. 1. Sketch map of locations of cores ZQ_2 and ZQ_4

locations are shown in Fig. 1. The core ZQ_4 is the standard profile in the area. We selected Ge centres as the dating signal in this work. The effect of natural light irradiation on Ge centres in quartz indicat that the ESR signals of Ge centres in quartz may satisfy the dating premise that the ESR signals were reset to zero after exposure to sunlight for one hour (Tanaka et al.,1985,Ye et al.,1993a,1993b,1995).

2 Samples

The position and lithology of the samples from core ZQ_4 and core ZQ_2 are shown in Table 1. As samples were taken several years after these cores were drilled, we did not determine water content in samples.

Table.1 Horizons, ESR ages and relevant parameters of samples

No.	Core	Depth(m)	Lithology	U(ppm)	Th(ppm)	K_2O(%)	D_E(Gy)	Age(ka)
373	ZQ_4	3.00-3.15	clayey silt	3.34	21.0	2.63	63.44	12.9
374	ZQ_4	4.00-4.15	clayey silt	3.60	20.5	2.36	72.57	15.5
376	ZQ_4	10.00-10.15	clayey silt	2.50	14.0	2.34	140.9	36.1
377	ZQ_4	11.00-11.15	clayey silt	4.50	10.1	1.93	128.7	34.6
379	ZQ_4	19.00-19.15	Silt	2.79	13.6	2.28	246.4	63.9
380	ZQ_4	21.00-21.15	clayey silt	3.21	13.6	2.69	271.1	62.3
381	ZQ_4	23.00-23.15	Sand-silt-clay	3.11	13.2	2.28	327.7	79.0
382	ZQ_4	25.00-25.15	Sand-silt-clay	3.58	13.5	2.74	382.0	85.4
383	ZQ_4	31.00-31.15	Sand	3.96	12.3	2.03	366.5	95.5
384	ZQ_4	35.00-35.20	Sand	3.14	6.5	1.68	387.9	136
385	ZQ_2	8.00-8.20	Clayey sand	1.57	5.8	2.34	33.9	11.4
386	ZQ_2	11.00-11.20	Sand-silt-clay	3.01	20.6	2.58	83.0	17.5
387	ZQ_2	13.00-13.20	Sand	2.49	13.7	2.50	121.4	30.6
388	ZQ_2	19.00-19.20	Sand	1.10	10.2	2.09	151.5	51.0
389	ZQ_2	21.00-21.20	Sand	1.44	11.8	2.28	315.7	88.7
390	ZQ_2	23.00-23.10	clayey silt	1.99	14.4	3.17	471.7	104.6
391	ZQ_2	31.00-31.10	clayey silt	2.44	13.4	3.13	603.5	133.8
392	ZQ_2	33.00-33.20	Sand	0.95	14.8	2.08	673.5	205.3

3 Experiments and Results

Samples were washed by water, sieved to obtain the 0.06-0.125mm grain fractions and treated with H_2O_2 to remove organic material. After cleaning, the separated grains were soaked in 6 mol/L hydrochloric acid for more than one day to dissolve carbonate. Then, they were etched in concentrate hydrofluoric acid

(40%) for about one hour to remove the outer layers of the quartz grains. The samples were thoroughly washed in water and dried at 40°C. Finally, any magnetic minerals in these samples were removed by a magnetic separater. Samples were divided into 9 aliquots (about 250 mg each), and were artificially irradiated with ^{60} Co- source to different doses. The irradiated samples were laid more than one week, then measured with a JES-FE1XG (JEOL Co.) spectrometer under the measurement condition of room temperature, X-band, mircowave power 2 mW, modulation amplitude 0.08 mT and magnetic field scanning range 337±5 mT. Fig. 2. shows a typical ESR spectra of a sample. The concentrations of U,Th and K_2O were determined by laser fluorescenec, colorimetric spectrophotometry and atomic absorption techniques respectively.

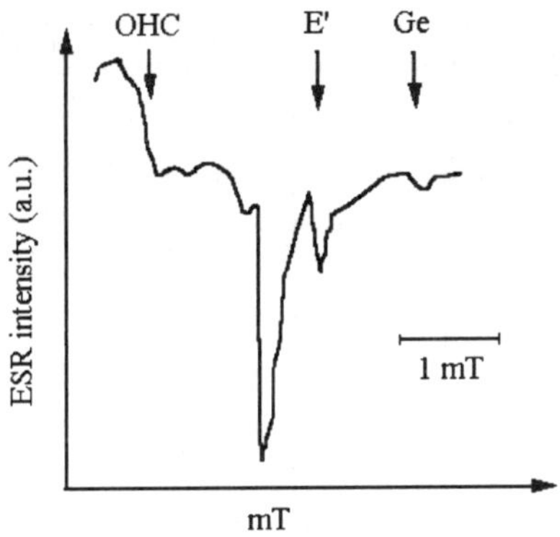

Fig. 2. A typical of ESR spectrum of the sample

We used linear fitting to determine D_E, and selected the value with large linear correlation coefficient, many data points and the smallest dose value as D_E(Ye et al., 1993c). The concentrations of U,Th, K_2O, D_E values and ESR ages of samples are show in Table 1.

4 ESR Chronologic Succession and Stratigraphic Correlation

Based on the results of ESR ages, ^{14}C ages, TL ages, ultrapaleontology, palynostratigraph, magnetostratigraph and oxygen isotope stratigraph, a chronologic succession is built up and show in Fig. 3. (Feng et al., 1996).The sum error of ESR ages is estimated to be 15-20%.

5 Reliability of ESR Ages Using Ge Centres

We can see from Fig.3 and Fig.4 that the ESR ages well correspond with stratigraphy, and thus ESR dating can serve as a good means of stratigraphic

division. Based on the oxygen isotope curve of core ZQ_4, the boundary of oxygen isotope stage 5/6 was at the core ZQ_4depth of 34.00 m, 128.0ka, under the boundary the ESR age was 136.1 ka at the core ZQ_4 depth of 35.00-35.15 m. The age at the reverse face between ultramicrofossil *Gehyrocapsa caribbeasmica* and *Emiliana huxleyi* should be 85.0 ka, the ESR age was 85.4 ka near the top boundary of the Nanwei Formation in core ZQ_4, TL age was 11.0 ka, ESR age was 12.9 ka, ^{14}C age was 13.9 ± 0.5 ka.
Based on ultramicrofossil assemblage and $CaCO_3$ content curve of core ZQ_2, the boundarf 31.0y of oxygen isotope stage 5/6 was at the core ZQ_2 depth of 30.00 m, -12.8 ka, under the boundary the ESR age was 133.8 ka at the core ZQ_2 depth o0-31.10 m. These datum show the reliability of ESR dating.

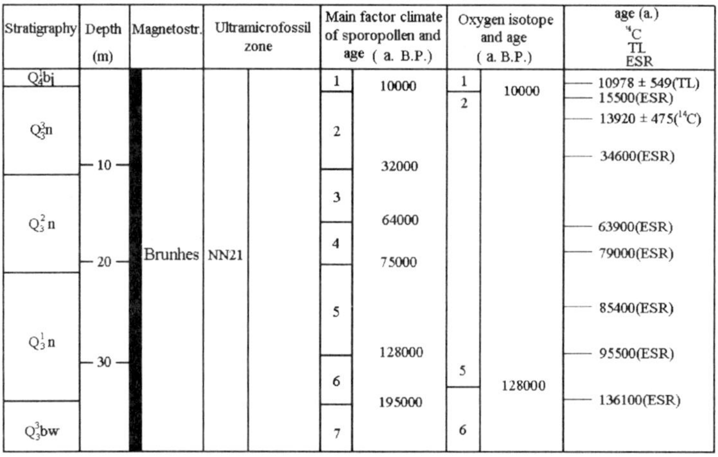

Fig. 3. A correlation of ESR ages with other stratigraphic methods ic core ZQ_4 (After Feng et al., 1996)

6 Summary

The effect of natural light irradiation on Ge centres of quartz indicates that the Ge centres of quartz may satisfy the dating premise that the ESR signals were reset after exposure to sunlight. We can obtained the sedimentary ages of marine sediments by using the Ge centres of quartz in sediments.

Acknowledgment

We are grateful for the financial support provided by the National Nature Science Foundation (No.49173177) and the Chinese Foundation for Development of Geological Sciences and Techniques (No. 90053).

Refenences

1. Feng Zhiqiang, Feng Wenke, Xue Wanjun, Liu Zonghui, Chen Junren and Li Weifeng, Evaluation of Marine Geologic Harzard and Engineering Geological Conditions in the northern South China Sea. Hehai University Publishing House, Nanjing, 1996, 50-81.
2. Tanaka T, Sawada S and Ito T., ESR dating of late Pleistoncene nearshore and terrace sands, in southern Kanto Japan. *ESR Dating and Dosimetry,* 1985, 275-280.
3. Ye Yuguang, He Jie, Diao Shaobo et al., Preliminary study on the Late Pleistocene ESR Chronology of core QC_2 in the Southern Huanghai Sea. Chinese Bulltin, 1993a, 38, 1280-1284.
4. Ye Yuguang, He Jie, Diao Shaobo et al., Study on ESR Ages of Late pleistocene Coastal Aeolian Sands. Marine Geology and Quaternary Geology, 1993b, 13, No.3, 85-90.
5. Ye Yuguang, He Jie, Diao Shao bo etal., ESR Dating of fluvial Sediments Using Ge Center in Quartz. Chinese Geographical Science, 1993c, 3, 284-288.
6. Ye Yuguang, Diao Shaobo, He Jie et al., ESR Dating of the coastal wind sand of China. Oceanologia et Limnologia Sinica, 1995, 26, 488-493.

ESR Measurement of Alluvial Zircons as an Aid to Provenance Determination in Geology

Wyss W.-S. Yim

Department of Earth Sciences, The University of Hong Kong, Pokfulam Road, Hong Kong

Abstract. ESR measurement of naturally occurring detrital zircon is useful in geology for provenance determination. The ESR signal is a method of fingerprinting radioactive impurities present within the mineral with the intensity increasing with the geological age. Because the method is highly sensitive, non-destructive, fast and relatively low cost compared to existing methods of dating including fission track, U/Pb, and ion probe, it has potential in geological studies for the identification of zircons with different origin and as a rapid semi-quantitative method of dating. In the present study, alluvial zircons of known ages ranging from Quaternary to the Devonian and unknown ages have been measured. Based on the results obtained, some applications are outlined. In order for ESR to be useful as a quantitative method of dating zircons, it is necessary to determine the uranium content and the total radiation dose.

1 Introduction

Since the pioneering work of Zeller et al. (1967) showing the potential of the ESR method for dating, ESR has been used increasingly in geology for Quaternary dating. Comprehensive reviews of the method in dating are available (e.g. Hennig and Grun, 1983; Ikeya, 1985, and Grun, 1989).

Zircon ($ZrSiO_4$) is a common accessory mineral in siliceous igneous rocks, crystalline limestones, schists, and gneisses, in sedimentary rocks derived therefrom, and in beach and alluvial placer deposits (Bates and Jackson, 1984). Due to its relatively high density in comparison to quartz (SG 4.6-4.7 as compared to 2.65), its hardness and chemical stability, zircon is a detrital heavy mineral which survives multiple cycles of erosion. It is therefore important to study the provenance of zircons to provide clues on landcape evolution to assist geological studies. Furthermore, because zircon contain appreciable quantities of U and Th, age determinations can be carried out either based on the Th/U ratio or the ratios of various Pb isotopes to each other or to isotopes of U or Th (Deer et al., 1992). Widely used methods for dating zircons include fission track, U/Pb, and ion probe. However, these methods involve relatively complicated sample treatment and expensive equipment not available widely.

Zeller et al. (1967) first suggested that the ESR method was of little promise for the age determination of zircon. Because the ESR peak resulting from the trapped charges were found near the surface, the mineral may instead be ideal for geothermometry and climatic investigations (Zeller, 1968). A study of optical absorption and ESR signal of zircon by Samaddar and Ghosh (1969) led to the conclusion that the colouration of zircon is produced by high energy irradiation from radioactive impurities present. This is confirmative that ESR measurements are reliable for the fingerprinting of radioactive impurities present within zircons. In contrast to Zeller et al. (1967), Taguchi et al. (1985) showed that pre-Pleistocene zircons may be datable. Subsequently, Hayashi et al. (1990) and Kasuya and Danhara (1992) confirmed that the intensities of the signal g= 2.0083 in natural zircons was proportional to their fission-track ages.

The ESR signal of alluvial zircons derived from the erosion of zircon-bearing igneous rock was originally formed by the volcanic activity. In the present study, alluvial zircons of both known and unknown ages, the former obtained either directly using fission-track dating or indirectly using potassium-argon dating of the zircon-bearing source rock are measured using ESR to provide information on their provenance. Based on using the zircons with known ages for calibration, the approximate ages of zircons with unknown ages may be determined.

2 Materials and Method

In the present study, alluvial zircons of known and unknown ages from New South Wales and Tasmania, Australia, and Thailand were selected for ESR measurements. Table 1 provides a summary of provenances and ages of zircons at five localities studied. All samples were collected from naturally occurring unconsolidated alluvial deposits with the zircons concentrated through panning or sluicing. Except for the Tolly's Gully sample, all zircons have been fission-track dated by Prof. A.J.W. Gleadow. The age of the Tolly's Gully zircon was inferred from the the K/Ar age of Tertiary basalt (F.L. Sutherland, pers. comm.). In addition to those zircons, a number of zircons of unknown ages from Tasmania were also selected for measurement.

In the laboratory, zircons above and below 2 mm were separated from the mineral concentrates in two ways using the methods described by Yim (1990). About 0.5 g weight of zircon (either single crystal or multiple crystals) were ground into a powder using a percussion steel pestle and mortar followed by hand grinding with an agate pestle and mortar. 0.09 g of the ground sample was placed in an annealed silica tube for measurement using a JEOL model JES-FE3X ESR spectrometer (X-band) with 100 KHz field modulation at room temperature. ESR signals were recorded on graph paper to permit comparison of peak heights between samples.

3 Results

Figure 1 shows the ESR signals obtained from the five alluvial zircons ranging in age from Quaternary to the Devonian presented in Table 1. Signals in decreasing intensity are found for the Weld River, Mutual Tin Mine, Tolly's Gully, Cambodian border, and Chantaburi zircons in that order which is in agreement with their respective ages obtained from either fission-track or K/Ar dating. The sample of Devonian zircons from the Weld River, which was identified by Yim et al. (1985) to have been derived from granite, is found to show appreciably stronger signals than the other samples. This is in spite of a scale difference of just below one-third. The results show that the ESR signals are in agreement with the geological ages obtained by the other methods.

ESR signals obtained for five alluvial zircons of unknown ages suspected to have been derived from Tertiary basalts in Tasmania are shown in Fig. 2. With the possible exception of the Bells Plain and Mussel Roe samples which show slightly weaker signals and consequently younger ages, all other samples show intensities resembling each other. Because of the similarity in ESR signal to the fission-track dated Mutual Tin Mine sample shown in Fig. 1, they are also likely to have been derived from the older basalts in the higher parts of the Blue Tier and ca. 47.1 Ma in age based on the fission-track age of 6 zircon grains (Yim et al., 1985). The Bells Plain and Mussel Roe samples are likely to be under 47.1 Ma and over 47.1 Ma in age respectively based on the difference in ESR signals to the other samples.

4 Discussion

In the case of the Pleistocene-aged Chantaburi zircon, the ESR signal shows that the

Table 1. Provenance and age of zircons at five localities selected for the present study

Locality	Provenance	Age and reference
Weld River, Tasmania, Australia	alluvial zircons derived from Devonian granites	367 ± 15 Ma - fission track (Yim et al., 1985)
Mutual Tin Mine, Tasmania, Australia	alluvial zircons derived from Tertiary basalts	47.1 ± 1.4 Ma - fission track (Yim et al., 1985)
Tolly's Gully, New South Wales, Australia	alluvial zircons derived from Tertiary basalts	ca. 37 Ma - K/Ar dating (F.L. Sutherland, pers. comm.)
Gemstone mine near Cambodian border, Thailand	alluvial zircons derived from Tertiary basalts	12.1 ± 1.1 Ma - fission track (A.J.W. Gleadow, pers. comm.)
Chantaburi Gemstone Mine, Thailand	alluvial zircons derived from Pleistocene basalts	1.3 ± 0.1 Ma - fission track (A.J.W. Gleadow, pers. comm.)

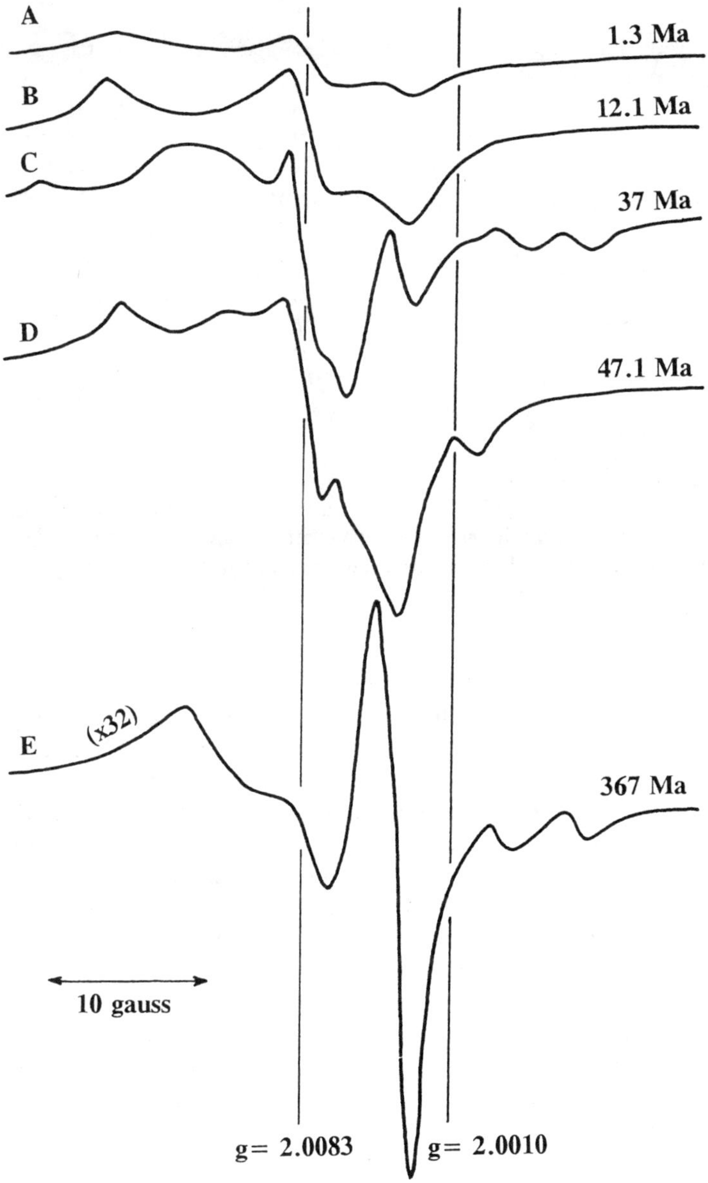

Fig. 1. ESR signals of the five dated alluvial zircons. These zircons range in age from Quaternary to the Devonian. Instrumental conditions used include microwave frequency 9.26 GHz, magnetic field 3305.5 ± 25G, sweep time 4 minutes, modulation frequency 100 KHz, modulation amplitude 2G, and time constant 0.3 ms. The gain used is one hundred unless indicated. Sampling localities: A- Chantaburi Gemstone Mine; B- gemstone mine near Cambodian border; C- Tolly's Gully; D- Mutual Tin Mine, and E- Weld River.

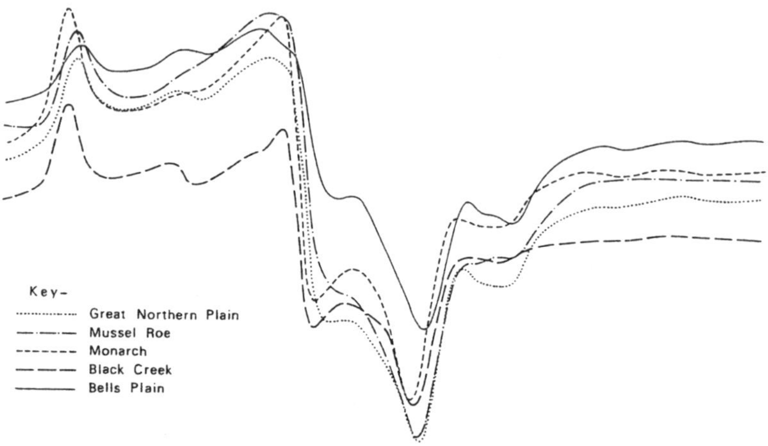

Fig. 2. ESR signals of alluvial zircons of unknown ages from five localities in Tasmania. Except for the Bells Plain sample which shows a slightly weaker intensity and consequently a younger age, the similarity in intensity in the other zircons are suggestive of the same age and provenance. Instrumental conditions are the same as in **Fig. 1** and the gain used is one hundred. Redrawn from Yim (1991).

method used was sufficiently sensitive for detection. Unlike the study of Taguchi et al. (1985), ESR signals are found to be sufficiently sensitive to permit the identification of Pleistocene zircons. This shows that the ESR method is more sensitive than previously realised. Furthermore, it may be possible to increase the sample size used in the ESR measurement to lower the minimum age of detection.

The ESR method used possesses a number of advantages over existing methods of zircon dating including fission track, U/Pb, and ion probe. Most notable is the speed of measurements. A scan typically can be made within 5 minutes. The latter has the potential of giving a superior mean age for single zircon crystals which is complementary to the ion probe method in which only a small part of the zircon crystal is measured. Besides powdered samples, single grain crystals can also be measured. The method is found to be suitable for the investigation of the provenance of alluvial zircons where rapid initial concentration of the mineral can be made by panning followed by hand picking. This is further aided by the occurrence of large zircon grains exceeding 2 mm which may be derived from megacrysts and xenocrysts from basalts. For example, such zircons are abundant in the alluvial gemfields of eastern Australia including northeastern Tasmania, Yunnan Province and Hainan Island in China, and Indochina. If the zircon occurring is below 2 mm, composite samples made of multiple grains may be homogenised by grinding prior to measurement. Follow-up work is however necessary to determine the ideal sample weight best suited for ESR measurements. Since stronger ESR signals can be expected for older zircons, the sample weight used for measurement may be decreased below the 0.09 g used in the present study.

For polycyclic landscapes of antiquity associated with stable continental areas, the provenance of zircons identified through dating has proven to be of great importance in geology. In northeastern Tasmania, Yim et al. (1985) and Yim (1991) demonstrated that the identification of at least two sources of alluvial zircons distinguishable visually through particle size and shape, one from Devonian granite and the other from Eocene Tertiary basalts, is crucial to the understanding of the long geological history involved in the formation of stanniferous placers. The ESR method is therefore useful for heavy mineral provenance studies including palaeoenvironmental reconstruction, basin analysis, and mineral exploration.

For more precise determination of geological ages, the ESR method also shows potential. In order to do this, it is necessary to carry out further studies including the determination of the uranium content of the zircons in addition to carrying out irradiation studies to obtain the total radiation dose. However, if the extra costs are taken into account, the total cost for ESR dating of zircons would be comparable to the existing methods of zircon dating. The ESR method is therefore particularly advantageous when semi-quantitative age information is required.

5 Conclusions

The ESR method is found to be a useful geological tool for fingerprinting the provenance of zircons. Since the ESR signals obtained are based on the amount radioactive impurities present which decays as a function of time, this also provides a rapid and non-destructive semi-quantitative means of dating zircons. The major advantages of the method are the speed and relatively low cost compared to the established geological methods. It is also non-destructive. For more precise age determination, it is necessary to determine the uranium content and the total radiation dose of the zircons. Because the method permits the recognition of recylced zircons of different ages, major areas of geological application include landscape evolution, basin analysis, and mineral exploration.

Acknowledgements

I am grateful to Prof. A.J.W. Gleadow, Ms W. Zhu, Drs M. Kasuya and J.C. van Moort for their assistance. I would also like to thank R. Hansawek for the provision of zircons from Thailand. This work was supported by research grants from the Hui Oi Chow Trust and the University of Hong Kong.

References

1. E.J. Zeller, P.W. Levy, and P.L. Mattern, Geologic dating by electron spin resonance, in Proceedings Symposium on Radioactive Dating and Low level Counting, International Atomic Energy Authority, Vienna (1967) 531-540.

2. G.J. Hennig, and R. Grun, ESR dating in Quaternary geology, *Quat. Sci. Rev.* **2** (1983) 157-238.
3. M. Ikeya, Dating methods of Pleistocene deposits and their problems: IX. Electron spin resonance, *Geoscience Canada* **2** (1985) 73-87.
4. R. Grun, Electron spin resonance (ESR) dating, *Quat. Int.* **1** (1989) 65-109.
5. R.L. Bates, and J.A. Jackson, *Dictionary of Geological Terms*, 3rd ed., Anchor Press, Doubleday (1984).
6. W.A. Deer, R.A. Howie, and J. Zussman, *An Introduction to Rock-forming Minerals*, 2nd ed., Longman, Hong Kong (1992).
7. E.J. Zeller, Chap. 5.1 Use of electron spin resonance for measurement of natural radiation damage, in *Thermoluminescence of Geological Materials*, edited by D.J. MacDougall, Academic Press, London (1968) 271-279.
8. M. Samaddar, and P.K. Ghosh, On the origin of coloration in zircon crystals, *Technology* **6/1** (1969) 42-44.
9. S. Taguchi, M. Harayama, and M. Hayashi, ESR signal of zircon and geologic age, in *ESR Dating and Dosimetry*, edited by M. Ikeya and T. Miki, Ionics, Tokyo (1985) 191-196.
10. M. Hayashi, I. Shinno, S. Taguchi, and S. Sugihara, ESR signals of zircon irradiated with thermal neutrons and γ-rays, *J. Min. Petr. Econ. Geol.* **85** (1990) 27-33.
11. M. Kasuya, and T. Danhara, Single grain ESR analysis of zircon: comparison with fission-tracks, *Quat. Int.* **13/14** (1992) 131-134.
12. W.W.-S. Yim, *Heavy Mineral Provenance and the Genesis of Stanniferous Placers in Northeastern Tasmania*, unpublished Ph.D. thesis, University of Tasmania (1990).
13. W.W.-S. Yim, A.J.W. Gleadow, and J.C. van Moort, Fission track dating of alluvial zircons and heavy mineral provenance in northeast Tasmania, Australia, *J. Geol. Soc. London* **142** (1985) 351-356.
14. W.W.-S. Yim, Tin placer genesis in northeastern Tasmania, in *The Cainozoic in Australia: A Re-appraisal of the Evidence*, edited by M.A. Williams, P. de Deckker and A.P. Kershaw, Special Publication No. 18 (1991) 235-257.

ESR Study of Two Paleoanthropological Sites on Lower Reaches of Yangtze River

Renyou LIANG, Renhua XU, Zicheng PENG, Peihua HUANG

Structure Research Laboratory, University of Science and Technology of China, Academia Sinica, Hefei, P.R.China, 230026

Abstract. We have used ESR method for dating Hexian and Yinshan sites, Anhui province, P. R. China. The age of Hexian site is between 171-359 ka, the age of Yinshan site is between 123-212 ka. Anthropologists suggest that the Hexian Man belongs to late Homo erectus while the Yinshan Man to early Homo sapiens. Considering the ESR results, we think that it is possible for two races of early man being contemporaneous in south China.

1 Introduction

At the beginning of the 1980's, two paleoanthropological sites have been found one by one along the Yangtze River in Anhui province. They are Hexian site and Yinshan site (see Fig.1). A complete skull and tooth fossils of Homo Erectus were excavated at Longtan Cave, Taodian village, Hexian county in 1980 (Huang et al. 1982, Huang et al. 1981). This is the first one of Homo Erectus discovered in South China. In 1981 and 1982, a maxilla and an occipital bone of Homo Sapiens were found in layer 2 of

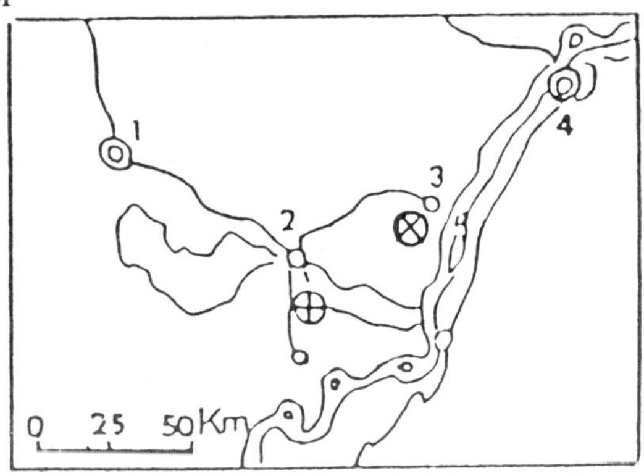

Fig. 1. The location of the Hexian site and Yinshan site, P. R. China 1.Hefei, 2.Chaoxian, 3.Hexian, 4.Nanjing, ⊗ Hexian site, ⊕ Yinshan site

another limestone cave at Yinshan village, Chaohu city (Xu et al. 1984). The latter is 50km from Hexian Man site. Only during two years, discovery of both sites in the lower Yangtze River is doubtless important in Paleoanthropology. It is significant to study of geography distribution of Homo Erectus and Homo sapiens , especially their evolution in South China.

By analysis of discovered fossils, anthropologists suggest that the Hexian Man belongs to late Homo erectus(Wu et al. 1982), while the Yinshan Man to early Homo Sapiens(Xu et al. 1986, Zhang et al. 1986). However, it is possible that they lived in the same time according to many dating results. This problem carried perplexity to anthropologists. They paid attention to these discoveries and made much research using different methods. According recent new findings, it is possible that the H. erectus overlapped in time with H. sapiens (C. C. Swisher et al. 1996). We used ESR method for further dating both sites .

2 Samples and Measurements

Longtan cave of Hexian site is developed in Cambrian dolomite. It extends along the strike of strata. The excavated cave is 9 m long with west-east direction and 3-4 m wide in north-south, 5 m high. The sediments in the cave could be divided into four layers from lower to upper (Huan et al. 1982). The fossil skull was discovered in the second layer, with yellow brownish silty sand. Deer and rhinoceros tooth enamels, which associated with fossil skull in Layer 2 was selected as dating samples (see Fig.2).

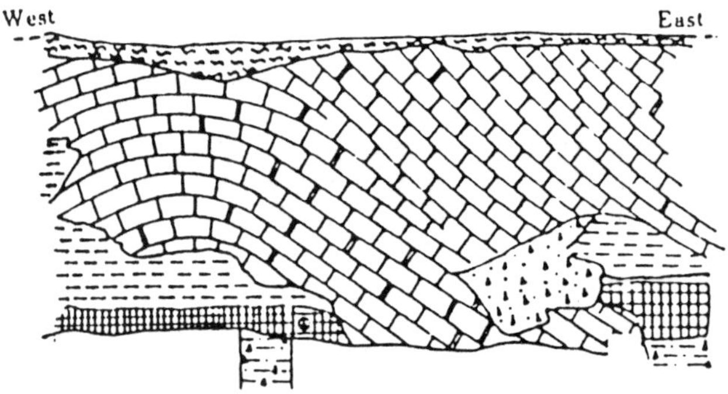

Brownish red and brownish black clay
Brownish red clay
Yellowish green silt
Yellowish hrownish silty clay
Yellowish grey silty clay
Breccia
Skull of Hexian Man

Fig. 2. Profile section of Longtan cave, Hexian site

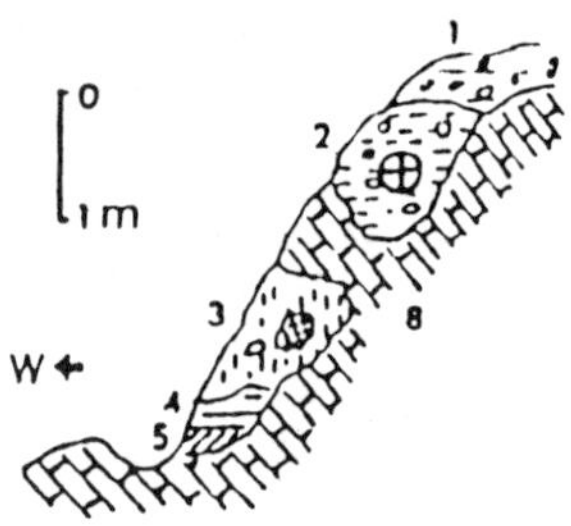

1. Grey deposit
2. Brown conglomerate
3. Yellow
4. Bron deposit
5. pale yellow deposit
6. Limestone ⊕: fossils

Fig. 3. Profile section of Yinshan site

The cave top of Yinshan site was collapsed. The caves extend due the north-south direction. The area of the cave sediments is about 200 m^2, but not continuous. The sediments could be divided into five layers from upper to lower (Xu et al. 1984). The hominid remainings has been excavated from Layer 2 (see Fig.3). Three mammal fossils as dating samples include the deer teeth collected from Layer 2, horse and bovidae teeth from Layer 3, while tapir teeth from Layer 4.

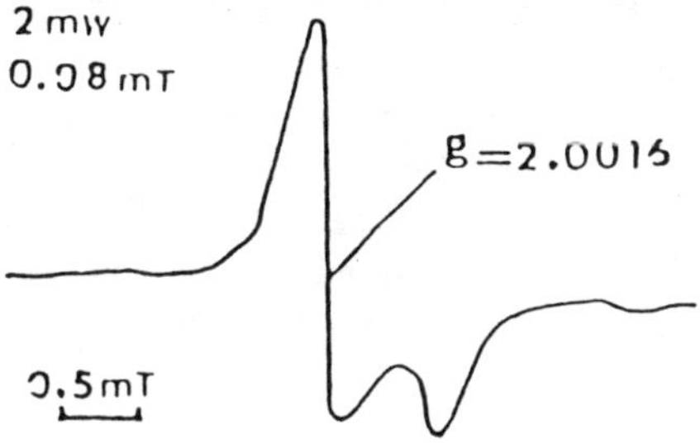

Fig. 4. The ESR signal at g=2.0016 of teeth enamal

Six fossils samples collected from two sites were dated by tooth ESR methods. After treating, the samples were irradiated with different irradiation doses using a ^{60}Co γ-source. The ESR signal at g=2.0016 was measured with a Bruker ER-200D ESR X-band spectrometer with the following measurement conditions: 2 mW microwave power, $0.08mT_{pp}$ modulation amplitude (see Fig.4). The equivalent dose (D_E) was determined from back-extrapolation of the measured intensity values using the computer program EIT-SIM (Grün, pers. comm.). The uranium contents in the enamels and the dentine (Ue,Ud) and the uranium, thorium and potassium contents (Use,Thse,Kse) in the sediments are gotten by neutron activation analysis. The alpha efficiency was estimated by irradiation with an ^{241}Am source. The $^{234}U/^{238}U$ ratio was estimated by U-series analysis. The analytical results were shown in Table 1.

Table 1. The experiment data of fossil enamel samples in Hexian and Yinshan site

Site	Sample	Layer No.	D_E (Gy)	Ue (μg/g)	Ud (μg/g)	Use (μg/g)	Thse (μg/g)	Kse (1%)
Hexian	Rhino. tooth	2	431.7 ± 42.8	1.25 ± 0.06	29.10 ±0.90	5.06 ± 0.06	9.09 ± 0.04	1.59 ± 0.01
Hexian	Deer tooth	2	909.8 ± 69.1	2.75 ± 0.08	29.20 ±0.90	5.06 ± 0.06	9.09 ± 0.04	1.59 ± 0.01
Yinshan	Deer tooth	2	1601 ± 188	20.1 ± 0.8	86.2 ± 2.6	14.1 ± 0.42	7.81 ± 0.23	0.80 ± 0.02
Yinshan	Horse tooth	3	1282 ± 171	5.70 ± 0.17	41.4 ± 1.2	9.58 ± 0.29	4.30 ± 0.13	0.83 ± 0.02
Yinshan	Bovidae tooth	3	1563 ± 141	5.65 ± 0.17	63.4 ± 1.9	13.8 ± 0.41	6.50 ± 0.20	1.40 ± 0.04
Yinshan	Tapiruse tooth	4	1134 ± 114	2.41 ± 0.07	41.4 ± 1.2	10.31 ±0.31	3.70 ± 0.11	0.55 ± 0.02

When determining the ESR ages of teeth, two effects must be taken into account: Uranium accumulation and Rn escape. In order to account for this uranium uptake process two models have been postulated by Ikeya (Ikeya 1982): (i) early U-uptake (EU) and (ii) linear, continuous U-uptake (LU). Without further information about the possible mechanism of the uranium uptake, the correct age of a sample lies within the range presented by the EU and LU estimates as usual.(Gr et al. 1991). It has been suggested that about 15-20% of the Rn may escape from the teeth (Ikeya 1982). ESR age estimates ESR age estimates of six samples were calculated using the computed program DATD(Grün, pers. comm.) and under the consideration of 20% Rn escape, EU and LU model. (see table 2)

On the basis of ESR results, the age of the fossil samples are between 171-359ka for Layer 2 0f Hexian site, while that between 123-212 ka for Layer 2 of Chaoxian Yinshan site. The hominids are included in the both layers.

Table 2. The ESR age results of teeth fossil samples from Hexian and Yinshan site (for 20% Rn escape)

Site	Sample	Layer	EU(ka)	LU(ka)
Hexian	Rhino tooth	2	171±18	228±23
Hexian	Deer tooth	2	264±23	359±29
Yinshan	Deer tooth	2	123±19	212±30
Yinshan	Horse tooth	3	228±31	361±49
Yinshan	Bovidae tooth	3	238±23	363±34
Yinshan	Tapiruse tooth	4	319±34	456±47

3 Discussion

Li Huhou et al.(1983) measured the sediment age of second layer at Hexian site by TL dating method and suggested the age of Hexian-men is not over 200ka. Wan Jianke et al. (1986) determined animal fossil age of Hexian site with amino acid method. The results are 200-300ka. Eight fossil bones samples from layer 2 were analyzed by U-series dating by Chen Tiemei (Chen et al. 1987). Four samples indicated that there was no or only little uranium mobilization in the $^{230}Th/^{234}U$ and $^{231}Pa/^{235}U$ systems occurred in the past. The age estimates were in the range of 150 to 190 ka. Samples suffered by uranium mobilization showed a larger scattering in the age estimates of 150 to 270 ka. The ESR age results in this study are 171 to 359 ka, which correspond to the U-series age and amino acid age results, but older than the TL age results. The layers from 1 to 3 of Longtan cave were formed through scouring process. The grit and soil were brought into cave by running water. In the deposited process, the fossils of ape-man and animal were buried, then transported and buried again. Considering this situation, it is possible that the fossils ages are older than the sediment ages. As usual the TL age is a reflection of the time for the latest deposition. The symbiotical mammal with ape-man could live from 359ka to 171ka up to now.

Table 3. The age results of Layer 2 at Hexian and Yinshan site by different dating methods

Site	ESR(ka)	U-series(ka)	TL(ka)	Amino acid(ka)
Hexian	171-359	150-270	<200	200-300
Yinshan	123-212	140-310	—	—

Seven fossil bones samples from layer 2 of Yinshan site were analyzed by U-series dating (Chen et al. 1987). Three samples showing uranium in close system had the age estimates between 160 and 200 ka, while four samples with uranium loss had ages of ranged 140 to 310 ka. The ESR ages of Layer 2 are in the range of about 123 to 212 ka, that is consistent with the U-series ages.

Based on the ESR dating results, the ages of two sites are overlapped in some range. The age of Hexian site is between 171-359 ka, the age of Yinshan site is between 123-212 ka. The shape and properties of Hexian man skull and other fossils were studied by anthropologists Huan et al. (1982) and Wu et al (1982), who thought that the Hexian man is Homo erectus. Xu et al. (1986) and Zhan et al (1986) suggested that the Yinshan Man belongs to early Homo sapiens. They are different races and Hexian man must be older than Yinshan man. The analysis of the animal fauna and ancient climate led to the same consideration. However, the geochronology and the geomorphology show that it is possible for two races of early Man living in the closed times, even in the same time (Xu et al. 1991, Chen et al. 1987). C. C. Swisher at al. (1996) dated the fossil bovid teeth from Ngadong and Sambungmacan, Central Java, using ESR and U-series methods. They thought that the age results raise the possibility that H. erectus overlapped in time with anatomically modem humans (H. sapiens) in southeast Asia. The ESR results here also gave the possibility that H. erectus is contemporaneous with H. sapiens in south China.

References

1. Huan Wangpo, Fang Duosheng and Ye Yongxiang, Preliminary study of Human fossils from Hexian, Anhui and relative problems, *Vertederata Tilisiatica*, **3** (1982) 248-256.
2. Huan Wangpo, Fang Duosheng and Ye Yongxiang, Observation of skull fossil of Homo Erectus discivered at Longtan Cave of Hexian, Anhui, *Chinese Science Bulletin*, **26** (1981) 1508-1510.
3. Xu Chunhua, Zhang Yinyun, Chen Caidi and Fang Duosheng, Human occipital bone and mammalian fossils from Chaoxian, Anhui, *Acta Anthropologica Sinica*, **3** (1984) 202-209.
4. Wu Rukang, Dong Xingren, Preliminary study of Human fossils at Hexian, Anhui, *Acta Anthropologica Sinica*, **1** (1982) 2-13.
5. Xu Chunhua, Zhang Yinyun and Fang Dusheng, Human fossil newly discovered at Chaoxian, Anhui, *Acta Anthropologica Sinica*, **4**(1986) 305-310.
6. Zhang Yinyun, The dental remaing of early Homo Sapiens found in China, Acta Anthropologica Sinica, **2** (1986) 103-113.

7. Ikeya, M., A model of a linear Uranium accumulation for ESR age of Hetdelberg (Mauer) and Tautavel bones, *Japanese Journal of Applied Physics Letters*, **21**(1982) L690-L692.
8. Grün, R. and Stringer, C. B. Electron spin resonance dating and the evolution of modern Humans, *Archaeometry*, **2** (1991) 153-199.
9. Li Huhou and Mei Yi, Superior limit of age of Hexian man, *Chinese Scince Bulletin*, **11** (1983) 703.
10. Wan Jiangke, Hong Huahua and Zhong Yueming, Amino acid dating and its application in study of archaeology and paleoanthropology, *The Study of Prehistory*, **1-2** (1986) 161-171.
11. Chen Tiemei, Yuan Sixun, Gao Shijiun and Hu Yanqui, Uranium series dating of fossil bones from Hexian and Chaoxian fossil human sites, *Acta Anthropologica Sinica*, **3** (1987) 249-254.
12. Xu Chunhua, Zhang Yinyun and Fang Duosheng, Relation of human fossils from Hexian and Chaohu, *Cultural Relics Research*, **7** (1991) 64-77.
13. C. C. Swisher, W. J. Rink, S.C. Anton, H. P. Schwarcz, G. H. Curtis, A. Suprijo and Widiasmoro, Latest Homo erectus of Java: Potential Contemporaneity with Homo sapiens in Southeast Asia, *Science*, **274** (1996) 1870-1874.

EPR Characterisation of Gold Mineralisation in Weathered Terrain, Jim's Find South, Tanami Desert , NT, Australia

L. Xu and J.C. van Moort

Geology Department, University of Tasmania, GPO Box 252-79, Hobart 7001, Tasmania, Australia

Abstract. Electron Paramagnetic Resonance (EPR) spectroscopy can be used as a measurement of lattice defects of quartz and other silicates. EPR's association with quartz in the proximity of ore is because such quartz has a considerable amount of lattice defects. This study investigates the EPR characterisation of gold mineralisation in regolith samples from the Tanami Desert. Samples were treated with hot $HCl+HNO_3+H_2SO_4$ acid to remove weathering products. The residue consists essentially of quartz. EPR results show that the acid insoluble residues associated with gold mineralisation are markedly paramagnetic. The EPR 326.5±5 mT and the 326.5 ±5 mT × ±100mT sweeps correlate well with the primary gold mineralisation and the EPR anomaly is larger than the ore body.

1. Introduction

Electron paramagnetic resonance (EPR) spectroscopy, which measures microwave absorption by unpaired electrons in a magnetic field, is often used to characterise materials. The method has been used since the seventies as exploration tool for gold and other metals in quartz veins, eg. Scherbakova et al. (1976). Van Moort (1986, 1989), van Moort and Barth (1987), van Moort and Russell (1987), and van Moort and Brathwaite (1988), used simple room temperature EPR powder spectra to characterise vein sequences and to discriminate between barren and auriferous vein systems.

In order to make the method more generally applicable, EPR was subsequently used to characterise rock pulps, in particular at Carlin, Nevada (van Moort et al. 1995) and at Rosebery, Tasmania (Pwa, 1996)

This article represents a further extension of the use of EPR, viz. in the assessment in highly weathered terrain of original wall rock alteration patterns. The present study concentrates on the use of EPR to characterise rock pulps derived from percussion drill hole chips. The aim is to delineate primary gold mineralisation, and its surface expression. In order to remove interfering impurities the rock pulps were leached with hot acids.

Three areas of increasing complexity are of interest in connection with the EPR spectra of the weathered regolith samples: (1) the shape of the EPR spectra, (2) greater peak height of mineralised samples, (3) an additional peak associated with the main signal.

2. Regional setting

The Tanami Desert is a relatively new prospecting area approximately 600 km NW of Alice Springs in the Northern Territory, Australia, with incipient mining activity. The type of mineralisation is gold in quartz-carbonate veining and associated envelopes of sericite-pyrite alteration. Ore reserves are about 600,000 tones with 3.7 g/t Au. The present study is on Jim's Find South Prospect, some 23 km south-southwest of the Tanami Mine, which is within the Granites-Tanami Complex (Fig. 1). The area is underlain by a sequence of interbedded pillow basalt, mudstone, greywacke and chert, cut by two major north-south faults dividing the stratigraphy into three lithologically similar sequences The oldest rocks belong to the Early Proterozoic Tanami Complex

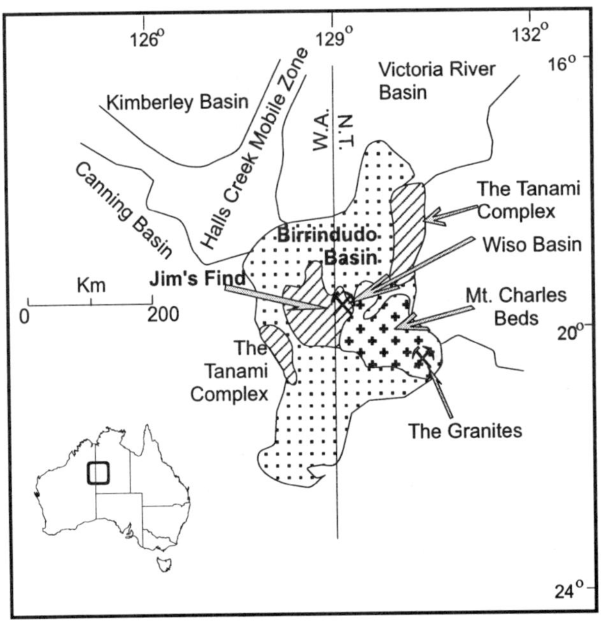

Fig. 1. The Granites - Tanami complex (After Ireland, 1989)

with stratigraphic equivalents, the Halls Creek Group and the Arunta complex occurring to the northwest and south respectively (Ireland, 1989 and Eupene, 1989). The regolith comprises the entire low-temperature altered, unconsolidated or secondarily recemented cover that is found overlying more coherent bedrock.

Rusulting from surficial or near surficial process, regolith materials may form by in situ weathering, erosion, transport and/or deposition of older materials. The weathering extends to 90m from the surface.

Primary gold mineralisation occurs in veins containing 2 to 20% quartz-carbonate and surrounding altered wall rock. The veins range from 0.5 mm to 0.5 m in thickness and are preferentially oriented parallel to the mineralised body. Mineralised zones within fresh bedrock are characterised by chlorite-sericite-pyrite alteration. Within the regolith, mineralisation characteristically occurs in zones of brecciated and sheared quartz sericite-clay chlorite altered rocks with limonite after pyrite (Stott, 1994).

3. Sample selection and preparation

One hundred forty samples were taken from twenty-three drill holes from three section lines of the area. The depth of samples taken from the drill holes in these lines are from a few meters up to 110 meters.

All the samples were pulverised in a manganese steel mill to less than 120 μm and subsequently treated with hot HCl, HNO_3 and H_2SO_4 for 12 hours at 80°C in order to destroy carbonates, sulphides, oxides and weathering products inclusive kaolinite. After that, the samples were washed three times with distilled water. The residue consists essentially of quartz, some sericite, feldspar and some resistate minerals like rutile.

4. EPR spectroscopy

Perfect quartz neglecting surfaces, does not of course, contain unpaired electrons and thus shows no EPR signals. Actual quartz however contains numerous atoms out of place and various substitutional and interstitial impurities. Some of these are inherently paramagnetic (e.g., Fe^{3+}, Mn^{2+} etc.) In other cases spin unpairing by irradiation creates EPR detectable species.

The effects of damage produced by cosmic radiation or from radioisotopes in the rock itself are studied by EPR (ESR) for dating purposes. Experience has shown in exploration geology that the intensity of various components of the EPR signal correlate with trace elemental abundance in quartz (Russell et al 1996, Pwa 1995) and studies of EPR for exploration purposes appear to be related to the presence of lattice substitution.

In this study, a measure of paramagnetism of all samples was obtained by analogue electron paramagnetic resonance powder spectroscopy at the Central Science Laboratory of the University of Tasmania in Hobart.

5. EPR signals observed

Two type of EPR X-band first derivative signals were found to be of practical use: the large V-shaped signal at g≈2 with a width of about 2 mT, often superimposed on a signal about 50 mT wide (Fig. 2). The first signal is best observed in the ±5 mT sweep and the second signal often not more than an off-set, is best observed in the ±100 mT sweep.

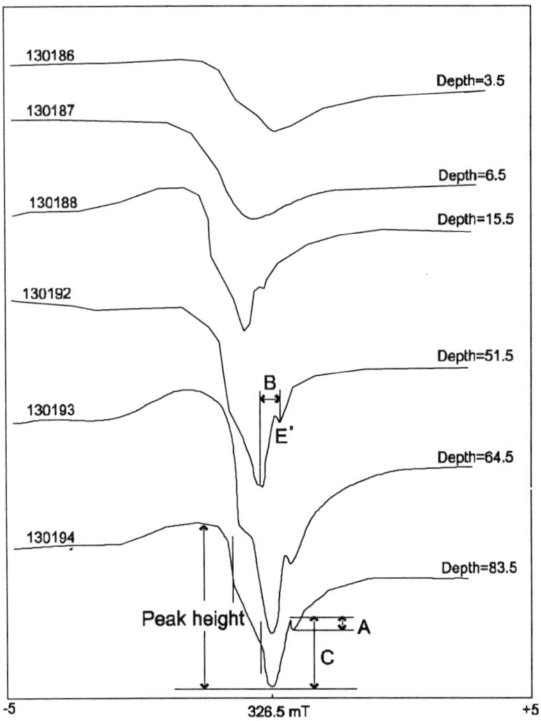

Fig. 2. First derivative EPR X- band 326.5±5 mT spectrum at room temperature of powdered samples from drill hole JRC 031

The major centre is likely to be the $[AlO_4]^\circ$ centre (Petrov et al. 1989, Russell 1995). The nature of the broad centre, observable in the larger sweep is unknown. In some samples, the $[GeO_4/M^+]$, Fe^{3+}, E′ or oxygen hole centres can be observed (van Moort and Barth, 1989). These additional centres do not correlate in anyway with the gold content.

The major signal in the samples from near the surface exhibits a flat and wide signal. The spectra show a signal of greater peak to peak height for samples from greater depth (Fig. 2). This central EPR line can be observed in all samples, but the intensity changes with depth and also with increased mineralisation. The individual signal width of 2mT does not change and the individual peak-to-peak widths, B of the central signal to E′, does not change from a value of about 0.2 mT. The peak heights of E′ to C shown as A/C in fig. 2, which changes with the depth and mineralisation C/A from 1:1 to 8:1. For the deeper mineralised samples the ratio of two peaks C/A are larger, and for the barren and near surface samples, the ratios are smaller. This additional peak is likely to be the E′ defect structure in the SiO_2 lattice, which is the result of an electron trapped in the vacancy when a bridging oxygen is removed from the lattice (Yip and Fowler, 1975).

6. EPR signals and occurrence of Au

To compare the trend of the distribution of the EPR intensity and the gold mineralisation, the distribution of Au in the regolith was determined by neutron activation analyses of the 140 samples from depths down to 110 m on three regolith sections (Stott, 1995). Where the complete regolith profile is preserved, Au is present as laterally extensive sub-horizontal zones of secondary Au enrichment (100ppb to 2g/t) in lateritic gravels and within the saprolite above the hypogene ore zone (Stott, et al. 1996). According to the above, the gold mineralisation zone has been outlined. The washed quartz samples from the hypogene mineralised zone show approximately three times stronger EPR ±5 mT and ±100 mT signals, in contrast to the samples from near the surface and the barren zone (Fig. 2).
The heights of the major peak in the ±5 mT signal and the broad signal in the ±100 mT sweep of the samples correlate significantly with the Au content.

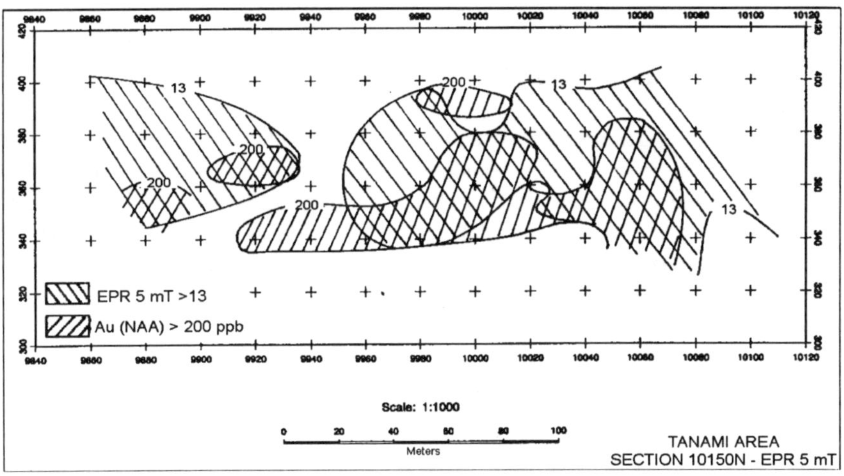

Fig. 3. The distribution map of EPR ±5 mT and gold along section 10150N with drill holes JRC 029, JRC 031, JRC 032, JRC 033, JRC 101, JRC 204, JRC 166 and JRC 203

The distribution of EPR ±5mT intensity shows there are two halos separately overlain on the gold anomaly, which extend upward to the surface (Fig. 3). The gold mineralisation zone was wrapped by a big EPR ± 100 mT halo (Fig. 4). The product of EPR ±5 mT × ±100 mT signal intensities is found to agree more accurately with the primary gold mineralisation. All of the EPR ±5 mT, EPR±100 mT and ±5mT × ±100 mT open to the east along the gold anomaly. This evidence suggests that such a horizontal anomaly may have existed toward the east as an extension of the gold mineralisation, and provides a potential target for further attention. The EPR and the gold anomaly both end at the same bottom, except from holes 10040N to 10080N where the distribution of EPR and gold goes down to the bedrock.

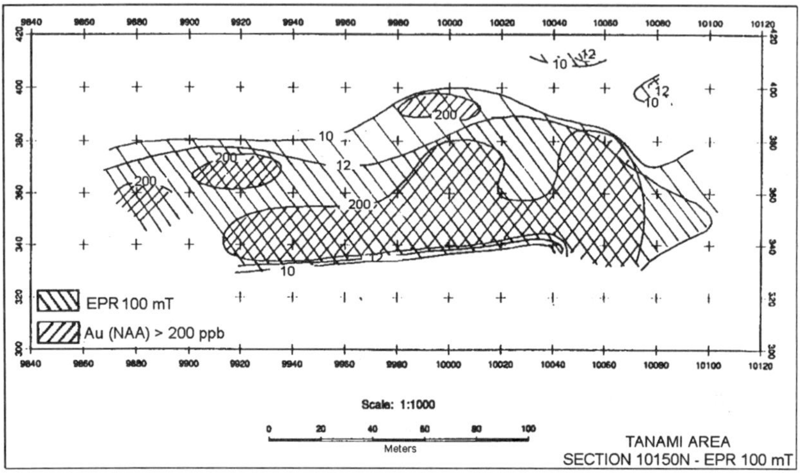

Fig. 4. The distribution of EPR ±100 mT and gold along section 10150N with drill holes JRC 029, JRC 031, JRC 032, JRC 033, JRC 101, JRC 204, JRC 166 and JRC 203

EPR spectra of non-acid treatment samples were also determined, but the spectra are quite different to those of the cleaned samples because of unremoved impurities inside the sample. A strong Mn^{2+} resonance line may be observed in absorption signal existing of the untreated samples and no positive correlation Au content. We interpret the change of line shape as being due to a scatter of magnetic fields arising from magnetic order among the Mn^{2+} ions smearing out the Al^{3+} narrow line spectrum.

7. Conclusion

Acid treatment of the samples is essential. EPR spectra of non-acid treatment samples do not clearly demonstrate the presence or absence of gold mineralisation. The quartz associated with gold mineralisation show typical EPR spectra. This study demonstrates that the application of EPR analysis to monitor possible additions and depletions to the mineralisation. The results indicated that EPR ± 5mT and their indices EPR ±5mT × ±100mT are correlate well with the primary gold mineralisation and that the paramagnetic halos are larger than the ore bodies.

The EPR spectra of the sample from near the surface show low paramagnetism compared with the samples from deeper places. The ±5 mT spectra from deeper samples have two EPR peaks, where the apparent separation between the two peaks is smaller in barren samples and larger in mineralised samples. The spectra of the mineralised samples have a greater peak height for the main signal. These EPR features and intensity outline the primary gold mineralisation

Acknowledgment

This study would not have been possible without the sample collection and research undertaken by C.L. Stott as a honours student at the Geology Department of the University of Tasmania. The authors wish to thanks Dr. D.R. Hutton of the Physics Department of Monash University for valuable comments.

References

1. Aung Pwa, 1995. Application of rock geochemistry and electron paramagnetic resonance of rock in mineral exploration, Rosebery mine area, western Tasmania. Unpublished Ph.D. thesis, University of Tasmania 303 pp.
2. Eupene, G.S., Nicholson, P.M. and Hickey, S.H. 1989. The Tanami gold project, In Jones, D.G. (Editor), Northern Territory. Northern Territory gold deposits, Publication of the Geology Department and University extension, the University of Western Australia. 16. P. 59
3. Ireland. T.J. 1989. *The geology of the Granites gold deposits,* Northern Territory. In Jones, D.E. (Editor), Northern Territory gold deposits, Publication of the Geology Department and University extension, University of Western Australia. 16. p. 59-63
4. Petrov, I., Agel, A. Bershov, L.V. and Hofner, S.S. 1990. Aluminium traces in quartz as indicator for temperature of formation. In R. Emmerman and P. Giese (Editors). KTB Report 90-3. Niedersächsisches Landesaint for Bodenforschung, Hannover, 561
5. Russell D.W., 1995. *The characterisation of quartz and gold in the Beaconsfield and Lefroy areas, Northern Tasmania,* Ph.D. thesis, 230pp
6. Russell, D.W., Bailey, G.M., van Moort, J.C. and Cohen, D.D. 1996. Chemical composition and paramagnetism of vein quartz from the Tasmania gold mine, Beaconsfield, Northern Tasmania, *Nucl. Instr. and Meth. in Phys. Res.*, B49
7. Stott, C., 1994. A *regolith study of the Jim's South gold anomaly, Tanami Desert, Northern Territory*. Honours thesis, Codes Key Centre, University of Tasmania, PP137.
8. Stott, C., Xu, L., Butt, C.R.M., Bailey, G.M. and van Moort, J.C. 1996. Gold and associated elements in the lateritic regolith at Jim's Find Tanami Desert, Northern Territory, Australia, *30th International Geological Congress*, Beijing 1996, Vol 3, p64
9. Scherbakova, M.Y., Sotinikov, V.I., Proskuryakov, A.A., Mashkavisev, R.I. and Soltntsev, V.P. 1976. Use of quartz EPR spectra in the analysis of ore content (as examplified by gold-ore and coper molybdenum mineralisation). Geologiya Rudnykh Mestorozhdenii, 18: 63-69 (in Russian)
10. van Moort, J.C., 1986. Electron spin resonance spectra of quartz as guide to mineralisation. In R. Large (Editor), The Mount Read Volcanics and associated Ore Deposits, *Geol. Soc. Aust. Tasmania Div.,* 85 pp.

11. van Moort, J.C. and Russell, D., 1987. Electron spin resonance of auriferous and barren quartz at Beaconsfield, Northern Tasmania. *J. Geochem. Explor.*, 27:227-237
12. van Moort, J.C. and Brathwaite, R.L., 1988: Electron paramagnetic resonance powder spectra of epithermal quartz from the Martha Hill gold-silver deposit, Waihi, New Zealand. Bicentennial gold 88. Geol. Soc. Australia Extended Abstracts 2: 575-578
13. van Moort, J. C. and Barth, W.H., 1989: Defects in crytalline quartz: use of EPR for the characterisation of quartz from gold bearing quartz vein in Australia. Abstract I: Symp. on Petrogenesis and Mineralisation of Granitoids. Chinese Soc. Mineralogy, Petrology and Geochemistry: 220-221
14. van Moort, J.C., Hotchkis M.A.C. and Aung Pwa 1995. EPR spectra and lithogeochemistry of jasperoids at Carlin, Nevada: distinction between auriferous and barren rocks, *J. of Geochem. Explor.,* 55:283-299
15. Yip, K.L. and Fowler, W.B., 1975. Electronic structure of E' centers in SiO_2. *Phys. Rev.* B. 11:2337-2338

EPR in the Research of the Formation of Highly Dispersed Powders of Refractory Compounds

Marina Vlasova, Mykola Kakazey

Institute for Materials Sciences Problems NASU, Kiev 252180, Ukraine

Abstract. The method of "paramagnetic probe" was used to analyze carbothermal reduction of a number of oxides, carbide- and nitride - formation, to reveal complex structure of the ultrafine particles.

1 Introduction

Development of the high-tech ceramics requires increased properties of the powders. In its turn, these properties are specified by the synthesis conditions. Transition from the starting components to the final synthesis products undergoes a number of subsequent and parallel special-phase-structural-kinetic process which are a necessary prerequisite for designing powder particles with needed properties. To obtain such information it is necessary to extend the research techniques available. Regarding the paramagnetic properties of different intermediate synthesis products the EPR method can be applied to analyze such processes. Due to the differences in such products paramagnetism the phase-structural as well as spatial properties are identified using EPR.

2 Specimens Preparation

To study carbothermal reduction of the silicon, chromium and boron oxides, carbide- and nitride -formation, the powder mixtures with carbon-containing component or mixtures prepared by the sol-gel method were used. The mixtures were heat treated in argon or nitrogen flow in the tubular furnaces.

EPR spectra were registered using radiospectrometer SE/x 2547 in 3-cm range at T~ 300-500 K.The measurements were carried out both in air or in vacuum, which made it possible to as certain "oxygen-sensible" formation. In the last case the pressure was equal 0,01 Pa, the time of evacuation was up to 10 minutes.

3 Experimental Results and Discussion

We present only the EPR results don't going into details of complex processes which accompany the carbothermal reduction of oxides and formation of highly dispersed nonmetal particles. Namely they have allowed for the first to trace the diversity of carbon system rearrangement in the presence of oxide's dispersed component and to detect the reactive forms of carbon, and consequently to foresee the methods of control of synthesis process.

For carbothermal reduction of oxides different forms of carbon are used involving crushed and activated graphite, black, products of thermodestruction (TDS), etc. So far, it was considered that efficiency of reduction, carbide- and nitride-formation is only specified by carbon particles size [1]. However, our studies indicated the presence of defects in the carbon lattice to be the second significant factor in these processes.

Let us consider reaction $SiO_2 + C \rightarrow SiC + 2CO$ where TDS products of saccharose, black serve as the carbon component, and aerosil and silicon acid - as the silicon one.

During heat treatment of numerous organic compounds in vacuum or inert atmosphere as a result of its TDS a porous carbon material with the developed surface is formed. It is known, that due to the formation of the small sized aromatic regions of polyconjugation (ARP) with subsequent increase in their content, their coarsening and polycondensation [2-4]. The process is accompanied with the appearance of paramagnetic centers (PC), which are unpaired π-electrons delocalized along the polyconjugation regions [5]. Depending on ARP the parameters and properties of EPR signals are somewhat changed. With subsequent increase in the initial mixture treatment temperature first (up to $T_{tr} \sim 673$ K) a signal with the width of $\Delta B \sim (5\text{-}7).10^{-4}$ T and g~2,0025 is observed followed by a narrow signal with $\Delta B \sim (2\text{-}4)\cdot 10^{-4}$ T (at $T_{tr} \geq 873$ K). The latter is characterized by "an oxygen effect" when the EPR signal widens or even disappears on oxygen adsorption. It is caused by the exchange between PC and adsorbed oxygen via the system of moving carriers. On lamella formation and microcrystal growth a widening of "narrow" signal and exchange line are observed instead of it [6]. It is known that during material graphitization owing to relaxation of paramagnetic centers through conduction electrons the EPR signal is unobserved even during specimens evacuation [7].

At present we consider our results. Figure 1 shows change in the amplitude and width of the carbon EPR signal typical of saccharose after heat treatment [8]. They reflect formation of small-sized ARP, their content increase (up to $T_{tr} \sim 673$ K) with subsequent coarsening of the ARP with appreciable "oxygen effect" (see curves 1 and 2). Decrea

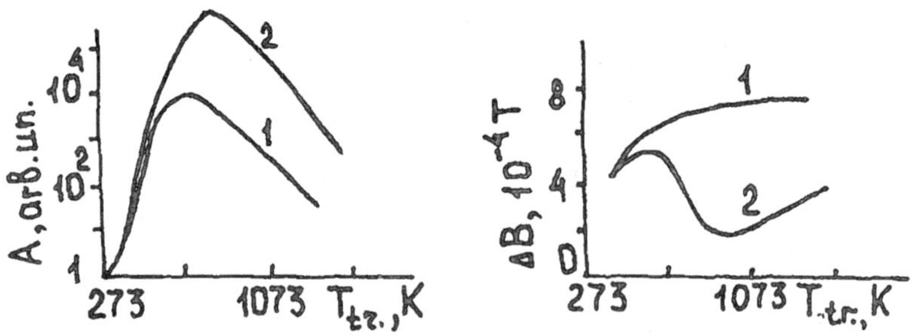

a b

Fig. 1. Amplitude (a) and width (b) of carbon EPR signals versus treatment temperature of saccharose in an argon flow. 1- measurement in air. 2 -in vacuum.

se in EPR signal amplitude and some widening of the line at $T_{tr} > 673$ K corresponds to the elements of three-dimensional ordering, though saccharose TDS products are considered to be nongraphitized materials.

During heat treatment of both the aerosil-saccharose mixture and "pure" saccharose formation of the carbon substance starts from ARP nucleating and their coarsening [9] (Fig.2).At $T_{tr} \sim$ 373-873 K EPR signal is observed with g =2,0025±0,0003, ΔB ~ (10-8).10^{-4} T. After T_{tr}= 473-2073 K during specimens evacuation a narrower signal with ΔB ~ (3-1,5).10^{-4} T is observed (see curve 2). It should be noted that with the rise in T_{tr} from 1073 to 1673 K a wide resonance signal with g ~ 2,008 is observed (curve 3) with some widening of the narrow signal. Its g-factor shifts to the value of 2,0027. Such effects indicate three-dimensional ordering in the carbon material, formation of microcrystals, appearance of the charge carriers. On the other hand, decrease in the wide signal amplitude and its narrowing in the specimens after T_{tr}> 1673 K indicate destruction of these ordered carbon regions and their participation in carbide formation.

During heat treatment of the aerosil-black mixture [10] EPR signal with g ~ 2,0026 only appears after 1273 K. With the rise in T_{tr} it narrows (Fig.3). Decrease in ΔB corresponds to destruction of the more ordered carbon forms and formation of the two-dimensional carbon structures. This process is reflected by the increase in EPR signal amplitude up to T_{tr}= 1973 K. Subsequent decrease in the signal amplitude is caused by participation of the two-dimensional carbon particles in carbide formation. No "oxygen effect" is observed in case of two-dimensional carbon particles - large sized ARP (see Fig.2,3). It means that carbon PC are isolated from the outer gaseous atmosphere. It suggests formation of some "screening shirt" from the interaction products on the carbon particles. The results of the IR absorption spectra permitted to conclude that at the C- SiO_2 interface the microstrains arise causing Si-O-Si bonds destruction and formation of Si-O-C and Si-C bonds. The gaseous products evolved (CO, CO_2) destroy-"activate" black particles or carbon microcrystals, homogenize Si-O-C mixture thus, intensifying reduction-carbide formation. Therefore a rise in EPR signal amplitude from ARP with its subsequent sharp decrease is observed (see Fig.2,3). Such a process predetermines process predetermines complex structure of the carbide particles under the reaction being not completed. There is a carbon nucleus in the form of ARP in the center of the carbide particles, and nonstoichiometric silicon oxide - in the outer layer.

During heat treatment of H_2SiO_3-saccharose in argon and nitrogen flow [9,11] TDS of saccharose and carbon component formation are characterized by specific behavior (Fig.4). Namely, in case of PC formed at T_{tr}> 1073 K the "oxygen effect" disappears or weakens; during measurements in vacuum no increase is observed in the amplitude from large-sized ARP for which ΔB ~ 2.10^{-4} T. The thermal treatment in air at 873-973 K (this is the temperature of oxidation of free carbon) didn't lead to disappearance of the carbon EPR signal. At the same time the "oxygen effect" for PC took place after careful rubbing of the specimen. It suggests that such PC are "screened" from the outer gaseous atmosphere by the products of components reduction and interaction. According to IR spectroscopy this is SiO_x, there $x<2$. Appearance of the second

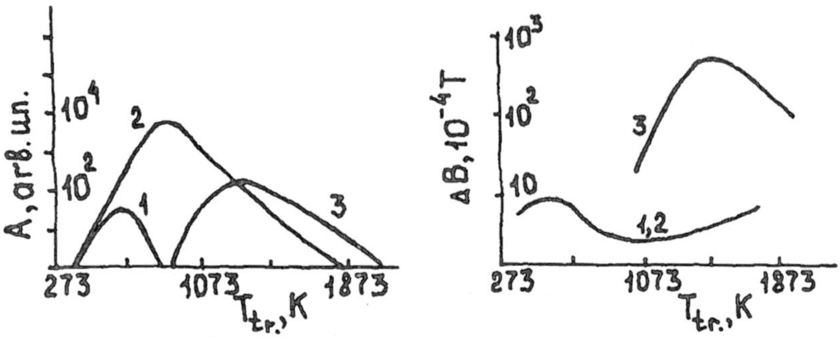

a b

Fig. 2. Amplitude (a) and width (b) of carbon EPR signals versus treatment temperature of aerosil-saccharose mixture in an argon flow. 1,3- measurement in air, 2,3- in vacuum.

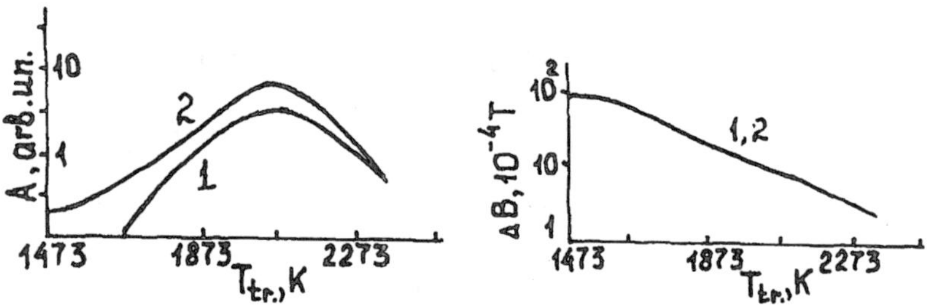

a b

Fig. 3. Amplitude(a) and width (b) of carbon EPR signals versus treatment temperature of aerosil-black mixture in an argon flow. 1-measurement in air, 2- in vacuum.

maximum on $A = f(T_{tr})$ curves indicates destruction of carbon formations with some elements of three-dimensional ordering on ARP. The carbon signal from ARP in nitride particles available up to high synthesis temperatures indicates similarity of the processes accompanying carbide- and nitride formation. It was found out that with the carbon nucleus available α -Si_3N_4 is formed. In case of its complete disappearance (due to complete participation in a reaction) actually the whole nitride formed is β -phase.

While analyzing carbothermal reduction of Cr_2O_3 it is possible to trace changes in the oxide phase. Cr_2O_3 is an antiferromagnetic with $T_N = (306 + 0{,}5)$ K. In the paramagnetic region ($T_{mes} > T_N$) a symmetrical signal of the Lorentz form is observed with g =1,985 ± 0,001, $\Delta B \sim 540.10^{-4}$ T [12,13]. At $T_{mes} = 290$ K asymmetrical signals ob served with $\Delta B \sim 2000.10^{-4}$ T, g-factor of which is shifted to 2.9>g>2,0 region. Such signals are usually associated with the existence of the ferro- and paramagnetic phases (inclusions) in the starting specimen or their arising at the heat treatment. They may be products of a varying composition from CrO_2 to $CrO_{1,56}$ arising due to the re-arrangement while annealing the starting nonstoichiometric and nonhomogeneous specimen [14]. It may suppose that gradual transition of highvalent chromium oxides

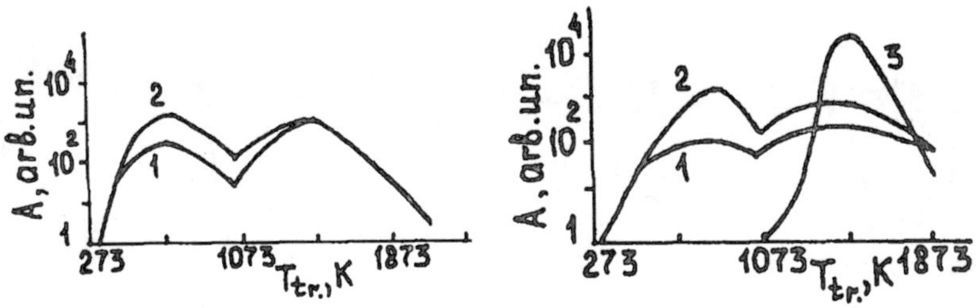

a b

Fig. 4. Amplitude of carbon EPR signals versus treatment temperature of H_2SiO_3-saccharose mixture in flows an argon (a) and a nitrogen (b). 1- measurement in air, 2- in vacuum. 3-carbon signal with $\Delta B \sim (10^2\text{-}10^3).10^{-4}$ T, measurements in air and in vacuum.

to Cr_2O_3 while annealing in argon explains nonmonotonic character of curve 1 in Fig.5. In the presence of saccharose TDS products rearrangement - reduction of the starting chromium oxide is intensified (see Fig.5, curves 2-4). With the increase in saccharose content in the mixture the signal from Cr_2O_3 phase ($T_{mes} > T_N$) attenuates at lower temperatures of treatment. Accordingly, the temperature of the carbide formation initiating decreases. The use of readily thermally destroyed compounds, e.g. acetic chromium, results in low temperature chromium carbide "resonance" synthesis when reactive two-dimensional forms of carbon and Cr_2O_3 are formed practically simultaneously. Chromium oxide reduction is intensified (see Fig.5,curves 5) and according to X-ray data [12,13] the carbide formation is also intensified.

During heat treatment of B_2O_3- saccharose mixture [15] the TDS products are found to be isolated from the outer gaseous atmosphere due to low melting point of boron anhydride (p.m. ~ 700 K). Therefore, the "oxygen effect" may be only observed prior to the melt formation (Fig.6a). Moreover, favorable conditions are created for lamellas and microcrystals formation indicated by a sharp widening in the "narrow" carbon signal (Fig.6b). However, starting already from T_{tr} ~ 1073 K, destruction-activation of the ordered carbon regions is observed, viz., the signal narrows and the

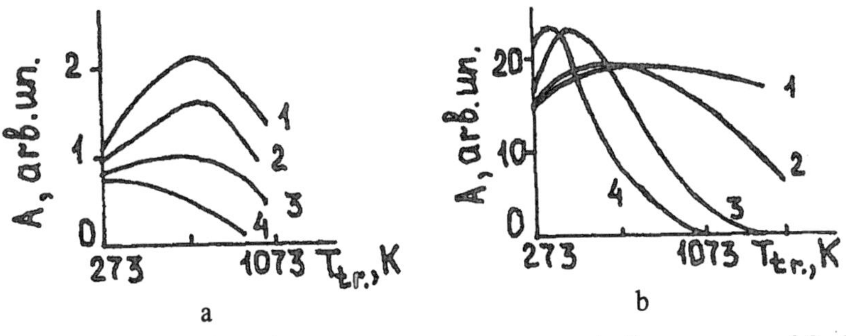

a b

Fig. 5. Amplitude of EPR signals versus treatment temperature in flow on argon of Cr_2O_3 (1) and mixtures 40% mass. saccharose + 60% mass. Cr_2O_3 (2), 67% mass. saccharose + 33% mass. Cr_2O_3 (3), 21% mass. saccharose + 79% mass. $Cr(Ac)_3$ (4). a- measurement at 290 K, b- 360 K.

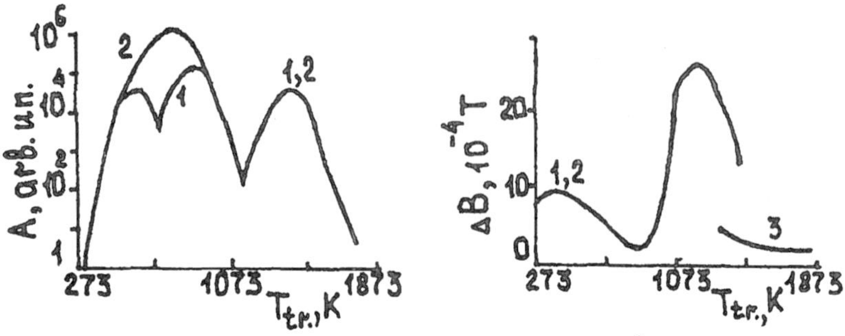

a b

Fig. 6. Amplitude (a) and width (b) of carbon EPR signals versus treatment temperature of B_2O_3 + saccharose mixture in an argon flow. 1,3- measurement in air, 2,3 - in vacuum.

second maximum appears on the curves A = f(T_{tr}). Formation of B_6O, B_4C or BN causes decrease in the amplitude and ΔB of EPR wide signal. In the case of uncompleted reaction of reduction-carbide(nitride) - formation a narrow EPR signal is registered from the carbon residues in the particles core. Using additives, e.g., NH_4Cl, CH_3COOH, LiOH, etc., it is possible to control saccharose TDS [16,17], thus creating favorable conditions for either the two-dimensional carbon structures stabilization, or three-dimensional ordering and hence, to control oxides reduction. These processes are easily observed regarding the amplitude change, width and g-factor of the EPR line (see Fig. 6,7).

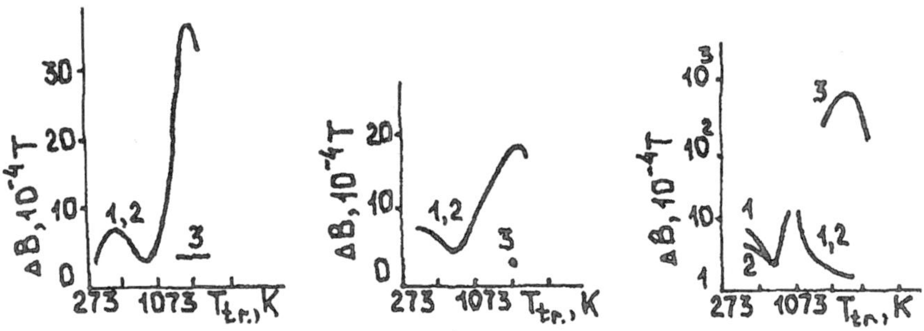

a b c

Fig. 7. Width of carbon EPR signals versus treatment temperature of mixtures B_2O_3 + saccharose with additives of NH_4Cl (a), CH_3COOH (b), LiOH (c). 1,3 -measurement in air, 2,3 -in vacuum.

4 Conclusion

The use of the EPR method and its information content while studying formation of the ultrafine nonmetallic powders is based on the two conditions, viz., availability of paramagnetic centers in the products or their arising during technological treatment; and a feasibility to carry out analysis of PC rearrangement while varying one of the technological parameters with subsequent complulsory agreement with the results of other research methods. Such an approach made it possible to ascertain that in the carbothermal oxide reduction with subsequent carbide- and nitride-formation the reactive paramagnetic forms of carbon, i.e. two-dimensional carbon particles take part. During heat treatment any carbon component of the mixture is continuously rearranging, i.e. is ordered, destroyed, etc. PC participation in phase formation of a number of carbides and nitrides at C-oxide interface permitted monitoring of complex structure of the particles formed.

References

1. T.Ya.Kosolapova, T.V.Andreeva, T.S.Bartnitskaya et al, *Non-Metallic Refractory Systems*, Metallurgiya, Moscow (1985).
2. L.S.Singer. Electron paramagnetic resonance (EPR) in carbons, in *Proceedings of the Workshop on Electrochemistry of Carbon* (Pentagon, New-Yourk 1984) 26-39.
3. V.I.Kasatochkin (ed.), *Structural chemistry of carbon and coals*, Nauka, Moscow (1969).
4. A.S.Fialkov, *The carbon-graphite materials*, Energiya, Moscow (1979).
5. V.M.Samoylov, C.M.Rumyancev, B.G.Ostronov, The investigation by method EPR of the mesophase transformation kinetics in oil pitch, *Chemistry of solid fuel* N2 (1983) 35-38.
6. A.A.Shklyaev, T.P.Miloshenko, A.F.Lukovnikov, The EPR spectra and orientation ordering in the structure of anisotropic semicoke, *Chemistry of solid fuel* N1 (1988) 70-75.
7. A.S.Kotosonov, Electron paramagnetic resonance in the graphitized and nongraphitized carbon materials, *DAN USSR* **196** (1971) 637-640.
8. M.V.Vlasova, N.G.Kakazei, V.N.Minakov, G.A.Puchkovskaya, V.S.Sinel'nikova, T.V.Tomila, V.I.Scherbina, Silicon carbide formation in the reaction of silicon reduction by carbon, *Powder Metallurgy* N9 (1989) 57-62.
9. M.V.Vlasova, L.T.Domasevich, N.G.Kakazei, N.P.Pavlenko, T.V.Tomila, S.Yu.Artyukh, V.I.Ivchenko, A.A.Kilimnik, Yu.I.Lezhnenko, Effect of initial state of the Synthesis component in the reaction of carbothermal reduction of silica on structure of silicon carbide. I. Aerosil-saccharose system, *Powder Metallurgy* N4 (1993) 64-74.
10. M.V.Vlasova, L.T.Domasevich, N.G.Kakazei, N.P.Pavlenko, T.V.Tomila, S.Yu.Artyukh, V.I.Ivchenko, A.A.Kilimnik, Yu.I.Lezhnenko, Effect of initial state of the synthesis component in the reaction of carbothermal reduction of silica on structure of silicon carbide. II. Aerosil-soot system, *Powder Metallurgy* N7 (1993) 54-61.
11. M.V.Vlasova, T.S.Bartnitskaya, L.L.Sukhikh, L.A.Krushinskaya, T.V.Tomila, S.Yu.Artyukh, Mechanism of Si_3N_4 nucleation during carbothermal reduction of silica, *J.Material Science* **30** (1995) 5263-5271.
12. M.V.Vlasova, N.G.Kakazei, V.N.Minakov, V.P.Sergeev, V.S.Sinel'nikova, A.I.Kharlamov, O.T.Khorpyakov, Carbide formation in the systems: chromium oxide-carbon-containing component, *Inorganic Materials* **24** (1988) 1998-2003.
13. M.V.Vlasova, V.S.Sinelnicova, Carbon-thermal reduction of high-dispersed mixture of chromium and nickel oxides, *Powder Metallurgy* N1 (1991) 59-64.
14. T.V.Rode, *The oxygenous compounds of chromium and chromium catalyzers*, Isdatelstvo AN USSR, Moscow (1962).
15. T.S.Bartnitskaya, M.V.Vlasova, T.Ya.Kosolapova, N.V.Kostyuk, I.I.Timofeeva, Boron nitride formation in the process of carbothermal reduction, *Powder Metallurgy* N12 (1990) 55-60.

16. T.S.Bartnitskaya, M.V.Vlasova, T.Ya.Kosolapova, N.V.Kostyuk, I.I.Timofeeva, T.V.Tomila, Role of structural ordering of initial components in formation of boron nitride, *Powder Metallurgy* N6 (1991) 54-61.
17. T.S.Bartnitskaya, M.V.Vlasova, V.I.Lyachenko, T.I.Serebryakova, I.I.Timofeeva, T.V.Tomila, Formation of high-dispersed boron nitride under carbothermal reduction on presence of lithium additions, *Powder Metallurgy* N1 (1993) 64-73.

Hyperfine Structure of EPR Spectra of Distorted Antiferromagnetic Trinuclear Clusters

Alexander I. Shames [1], *Moisey I. Belinsky* [2]

[1)] Department of Physics, Ben-Gurion University of the Negev, P.O.Box 653, 84 105 Be'er-Sheva, Israel
[2)] School of Chemistry, Sackler Faculty of Exact Sciences, Tel-Aviv University, Ramat Aviv, 69 978 Tel-Aviv, Israel

Abstract. The hyperfine electron-nuclear interaction in distorted antiferromagnetic trinuclear metal clusters is considered. The joint effect of antisymmetric exchange, Heisenberg interactions and external magnetic field on the hyperfine structure of the EPR spectrum of trinuclear clusters is taken into account. It is shown that the type of hyperfine structure observed (equilateral, isosceles, scalene) depends upon the ratio between the antisymmetric exchange parameter on the one hand and both the effective parameters of distortions and resonance field on the other hand. Strong C_2 type distortions result in a hyperfine spectrum characterized by intermediate spin. Strong antisymmetric exchange transforms the hyperfine spectrum to the trigonal type.

1 Introduction

The determination of real geometry of trinuclear exchange clusters still remains the subject under study [1]. Space groups of crystals of homonuclear $(M^{III})_3$ complexes determined by X-ray diffraction (at room temperature) usually do not show C_3 axes, but the deviations from trigonal symmetry are generally within the experimental error [2]. However, accurate low-temperature susceptibility and specific heat measurements favored a lowering of symmetry of the triangular cluster so that at least two values of coupling constant J were required [2]. The picture becomes more complicated taking into account the results of Mössbauer study of Fe^{III} carboxylate crystals [3,4] : a single asymmetric quadrupole doublet is observed in zero field but a complicated structure of Mössbauer spectra occurs in a magnetic field. The former also contradicts the simple suggestion of geometrical distortions of metal ion cluster, e.g., inequality of relevant metal-ligand distances and angles.

Thus, different experiments, especially spectroscopic measurements, can show different types of symmetry of the same exchange cluster. Of course, it does not mean the changes of real geometry during the experiment. These statements do not seem such ambiguous in case the notion of *physical symmetry* will be introduced. In any experiment the physical symmetry reveals just the correlation between real geometry of the cluster, set of intra- and inter-cluster interactions and external experimental conditions. For instance, in cited experiments [3,4] the symmetry observed depends on the applied external magnetic field. The peculiarities of these Mössbauer spectra were qualitatively explained for isosceles [5] and scalene [6] clusters. The parameters of the quadrupole Hamiltonian depend upon both exchange parameters and external magnetic field. The antisymmetric (AS) exchange interaction leads to the flattening of the electric field gradient at the nuclei, located in geometrically non-equivalent positions due to distortions. The magnetic field lying in the plane of the cluster reduces the antisymmetric exchange interaction and re-establishes the geometric non-equivalence of the ions.

Hyperfine structures of EPR spectra are also very sensitive to the correlation between the set of exchange parameters and resonance conditions for given EPR transition. The effect of the antisymmetric exchange interaction to the EPR spectra of isosceles cluster was studied in [7]. The main goal of the present paper is to consider hyperfine structure of EPR spectra for the most interesting case of scalene homonuclear antiferromagnetic (AF) cluster as well as to discuss dependence of physical symmetry observed upon experimental conditions.

2 The Antisymmetric Exchange Interaction and the Ground State of a Scalene Trinuclear Cluster

The isotropic metal-metal exchange coupling in a homonuclear trigonal cluster is described by the Heisenberg-Dirac-Van Vleck (HDVV) Hamiltonian of the type

$$\hat{H} = -2J\sum_{l,j=1}^{3} \hat{\mathbf{s}}_l \hat{\mathbf{s}}_j \tag{1}$$

where s_j are the spin operators of the interacting ions. The eigenvalues of Hamiltonian (1) are :

$$E_o = -J[S(S+1) - 3s(s+1)] , \tag{2}$$

where S is the total spin, $\mathbf{S} = \mathbf{S}_{12} + \mathbf{s}_3$, $\mathbf{S}_{12} = \mathbf{s}_1 + \mathbf{s}_2$. The characteristic J values obtained from magnetic measurements for homonuclear clusters are: $|J| > 200\ \mathrm{cm}^{-1}$ for Cu_3 clusters, $|J| \cong 40\ \mathrm{cm}^{-1}$ $\{60\ \mathrm{cm}^{-1}\}$ for Cr_3 {Fe3} carboxylates and $|J| > 100\ \mathrm{cm}^{-1}$ for $[Fe_3O_4]$ clusters.. The states with S = 1/2 degenerated on the intermediate spin S_{12} are the GS of an AF system with half-integer ionic spins. Antisymmetric exchange splits this ground state. The AS Hamiltonian for trimeric cluster has the form [8-10]

$$\hat{H}_{AS} = i\sum_{l,j=1}^{3} \{\mathbf{G}_{lj}^{(1)} \otimes \{\hat{\mathbf{s}}_l^{(1)} \otimes \hat{\mathbf{s}}_j^{(1)}\}^{(1)}\}^{(0)} = \frac{1}{\sqrt{6}}\sum_{l,j=1}^{3} \mathbf{G}_{lj}[\hat{\mathbf{s}}_l \times \hat{\mathbf{s}}_j], \tag{3}$$

where $\mathbf{G}_{jl} = -\mathbf{G}_{lj}$ is the real antisymmetric vector constant. The AS Hamiltonian in the matrix form:

$$\hat{H}_{AS} = \sum_{\nu} D^{\nu}\left\{\hat{\sigma}_{y} \otimes \hat{\sigma}_{\nu}\right\};\ \nu = X, Y, Z;\ D^{\nu} = \gamma\left(s_1, s_2, s_3, S_0\right)\sum_{l,j=1}^{3} G_{lj}^{\nu}\ . \quad (4)$$

In the case of the homonuclear system (S_0 = s -1/2), the expression for γ has the form:

$$\gamma(s) = \left(S_0 + 1\right)^2 / 6\sqrt{2}\ . \quad (5)$$

The following order of the basic functions has been chosen $\left|\frac{1}{2}(S_0)\frac{1}{2}\right\rangle$, $\left|\frac{1}{2}(S_0) - \frac{1}{2}\right\rangle$, $\left|\frac{1}{2}(S_0+1)\frac{1}{2}\right\rangle$, $\left|\frac{1}{2}(S_0+1) - \frac{1}{2}\right\rangle$.

Considering distorted systems, let us introduce, besides the AS exchange, small (δ_{12}, $\delta_{13} << |J|$) rhombic distortions of the cluster :

$$\hat{V}_r = -2\delta_{12}\hat{\mathbf{s}}_1\hat{\mathbf{s}}_2 - 2\delta_{13}\hat{\mathbf{s}}_1\hat{\mathbf{s}}_3 + \frac{1}{4}Tr\left(2\delta_{12}\hat{\mathbf{s}}_1\hat{\mathbf{s}}_2 + 2\delta_{13}\hat{\mathbf{s}}_1\hat{\mathbf{s}}_3\right),\ Tr(\hat{V}_r) = 0\ . \quad (6)$$

The matrix representation of the Hamiltonian (6) in the basis $\left|S(S_{lj})M\right\rangle$ has the following structure:

$$\hat{V}_r = \delta_1(\hat{\sigma}_Z \otimes \hat{1}) + \delta_2(\hat{\sigma}_X \otimes \hat{1}), \quad (7)$$

where δ_1 and δ_2 are effective parameters of distortions:

$$\delta_1 = \left(\delta_{12} - \frac{\delta_{13}}{2}\right)\left(S_0 + 1\right),\ \delta_2 = \delta_{13}\frac{\sqrt{3}}{2}\left(S_0 + 1\right). \quad (8)$$

In the case of small distortions (δ_1, $\delta_2 << |J|$), it follows from the geometry of the system that the D^X, $D^Y << D^Z$ relation holds for the AS vector parameter. Therefore we will consider further a model in which $D^X = D^Y = 0$, $D^Z = D$.

Diagonalizing the Hamiltonian $\hat{H} = \hat{H}_{AS} + \hat{V}_r$, we obtain the splitting of the S=1/2 in two Kramers doublets with energy levels (Fig.1)

$$E^{\pm} = \pm\Delta;\quad \Delta^2 = D^2 + \delta^2;\quad \delta^2 = \delta_1^2 + \delta_2^2\ . \quad (9)$$

The "distortion" parameters δ_1 and δ_2 were obtained by magnetic and spectroscopic studies for different clusters and may be values of the order 1-10 cm^{-1}. The AS exchange parameter, for example, for Cu_3 cluster with a strong HDVV exchange, is $D \sim 10\ cm^{-1}$ and for Fe_3 clusters in carboxylates $D \sim 1\ cm^{-1}$. For weak distortions and small AS exchange the 2Δ splitting of the ground state may be of the order of Zeeman splittings.

3 Zeeman Splittings and the Line Intensities in the EPR Spectrum

The interaction of the system with an external magnetic field is described by the isotropic Zeeman Hamiltonian

$$\hat{H}^Z = \hat{H}_e^Z + \hat{H}_n^Z = g_e\mu_B\mathbf{B}\sum_{j=1}^{3}\mathbf{s}_j + \gamma_n\mathbf{B}\sum_{\alpha=1}^{3}\mathbf{I}_{\alpha}. \quad (10)$$

We will assume that the electronic splittings in the magnetic field considerably exceed both the HF effects and nuclear Zeeman splittings. A magnetic field removes the Kramers degeneracy. Diagonalizing the electronic part of the Hamiltonian (10) together with the Hamiltonian $\hat{H} = \hat{H}_{AS} + \hat{V}_r$, one can obtain the following expression for the electronic energy of GS sublevels for the scalene cluster in the magnetic field

$$E^{\pm}_{1(2)}(\mathbf{B}) = \pm\left[\Delta^2 + H^2 \underset{(-)}{+} 2H\left(\Delta^2 - D^2\sin^2\theta\right)^{1/2}\right]^{1/2}, \quad \mathrm{H} = \frac{g_e m \mid\ \mid ö}{2} \tag{11}$$

where θ is the angle between the direction of the external field and the normal $\vec{n}$ to the plane of the triangle. The AS interaction leads to an anisotropy of the splittings of the cluster electronic sublevels even for the simple model of isotropic Zeeman interaction.

The intensities $W_k = W_{ij}$ of different RF transitions (Fig. 2) between Zeeman sublevels i and j are determined by the matrix element of the form

$$(S = S_X \quad iS_Y), \qquad W_{ij} = \left|\left\langle \Psi_i \middle| \hat{S} \middle| \Psi_j \right\rangle\right|^2. \tag{12}$$

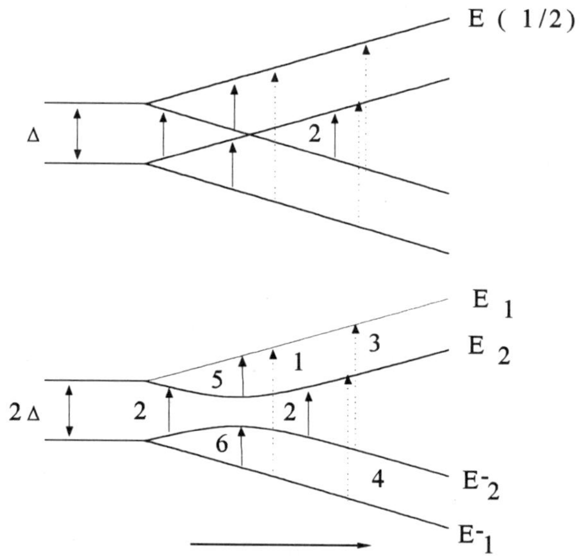

Electronic Zeeman energy sublevels of the trinuclear system. The direction of increase in the magnetic field is shown by horizontal arrows. The EPR transitions are shown by vertical arrows : a) magnetic field parallel to the C_3 axis of the cluster (q = 0); b) magnetic field perpendicular to the C_3 axis of the cluster (q = p/2).

Before the analysis of both EPR transition probabilities and hyperfine splittings, let us note that in the previous papers such an analysis was done by comparison of the exchange parameters (D, δ etc.) with the energy of RF quantum hν (so called low- and high-frequency transitions). When the eigenvalues and eigenvectors are dependent on the direction and value of magnetic field, this approach seems to be insufficiently correct. It is obvious, that in general case the ratio between exchange parameters and the value of magnetic field describes the chosen transition more correct. Indeed, transitions 2 and 2' (Fig.1) are observed at the same RF quantum but at different values of the external magnetic field. As it was shown in [5,7], strong external magnetic field reduces the Dzialoshinskii interaction and reveals all the effects of distortions. It means, for the high-field transition 2' one will observe another probabilities and hyperfine splittings than for the low-field transition 2 in spite of the both EPR transitions are observed at the same RF quantum!

The transition probabilities in the parallel field are the following

$$W_1^{\|} = W_2^{\|} = W_{2'}^{\|} = \frac{D^2}{2\Delta^2}; \qquad W_3^{\|} = W_4^{\|} = W_5^{\|} = W_6^{\|} = \frac{\delta^2}{2\Delta^2}. \quad (13)$$

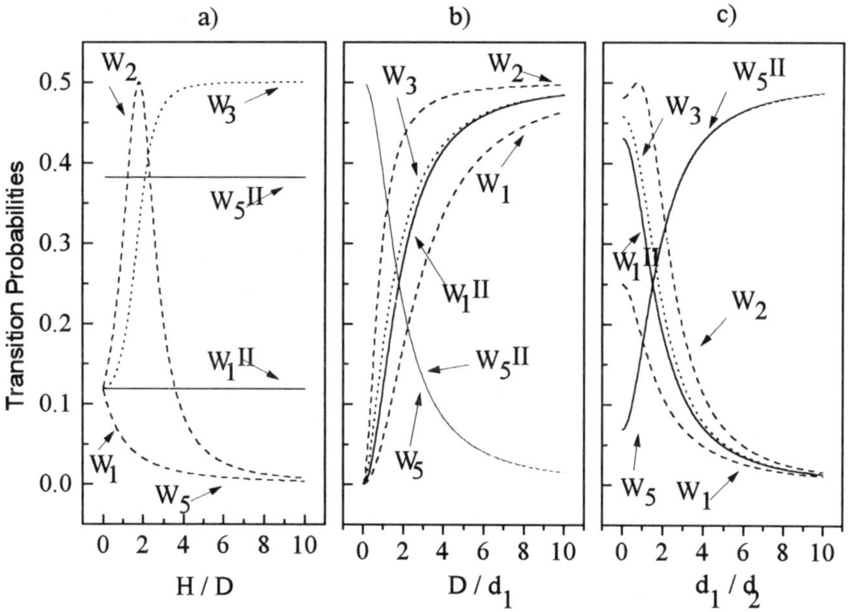

Fig. 2. The dependencies of the probabilities of EPR transitions for parallel and perpendicular orientations of magnetic field on : a) external magnetic field value, $d_1 = 0.5$, $d_2 = 0.2$, D = 0.3; b) Dzialoshinskii interaction' parameter, $d_1 = 0.5$, $d_2 = 0.2$, H = 0.3; c) ratio between parameters of distortions, $d_2 = 0.2$, D = 0.3, H = 0.3.

Figs. 2a-c show the dependencies of the EPR transition probabilities on the ratios between exchange parameters and the magnetic field. For the chosen model of axial

AS exchange, the probabilities in parallel field do not depend on the magnetic field value (Fig. 2a). AS exchange leads to the burning of the transitions 1, 2 and 2′ forbidden in the HDVV model and at the same time to the reduction of the allowed transitions 3, 4, 5 and 6 (Fig. 2b). In the limiting case of strong AS exchange, which occurs in an equilateral system, the transitions 3-6 are quenched and only the transitions 1, 2 and 2′ remain. In a scalene system ($\delta_2 \neq 0$), the transition probablities depend on the ratio between the effective parameters of the distortions. For the case $\delta_1/\delta_2 = 0$, the probabilities of the transitions 1, 2 and 2′ are stronger than the probabilities of the transitions 3-6. For the isosceles system ($\delta_{12} >> \delta_{13}$) only the allowed transitions will be observed (Fig. 2c).

In the perpendicular field, the transition probabilities essentially depend on the value of the magnetic field. The transition probabilities have the form

$$W_1^{\perp} \quad \frac{D^2}{2E_1^2} \qquad W_2^{\perp} \quad W_{2'}^{\perp} \quad \frac{D^2}{2E_2^2}$$

$$W_3 = W_4 = \frac{\{D^2 + [E_1 - (\ + H)][E_2 + (\ - H)]\}^2}{8E_1E_2[E_1 - (\ + H)][E_2 + (\ - H)]};$$

$$W_5 = W_6 = \frac{\{D^2 - [E_1 + (\ + H)][E_2 + (\ - H)]\}^2}{8E_1E_2[E_1 + (\ + H)][E_2 + (\ - H)]}. \tag{14}$$

Fig. 2a shows the field dependence of the “perpendicular transitions” W_i , $i = 1-6$ at fixed D value. When the magnetic field increases, the transitions 5 and 6, as well as the transition 1, are quenched and the transitions 3 and 4 are burned. This corresponds to the change from the allowed low-field lines to the same high-field one and to the quenching of all lines in the HDVV model. The most interesting one represents the region of low and intermediate magnetic field values. Here all six transitions co-exist with non-zero probabilities. Moreover, the probabilities of the transitions 2 and 2′ reach their maximal value (~ 0.5) only for a low-field region. The dependencies of the transition probabilities on the ratio between the exchange parameters are similar to these for parallel field (Figs. 2b, c) except the abnormal intensities of the transitions 2 and 2′ at the ratio $\delta_1/\delta_2 < 3$.

4 Hyperfine Structure of the EPR Transitions

The hyperfine electron-nuclear spin-spin interaction is described by the Hamiltonian [11] :

$$H_{hf} = \sum_{a=1}^{3} a_a s_a I_a \cdot \tag{15}$$

In the case of strong electronic Zeeman splittings $(g_e\mu_B B >> a\ )$, one can obtain the Hamiltonian (15) in the form :

$$\hat{H} \ = \ _{=}\gamma \quad (E\ )\hat{}\ , \tag{16}$$

where $\mathbf{B}_\alpha(E_I)$ is the effective magnetic field on the α-th nucleus, induced by the polarization of the electronic spin of the cluster in the E_I-th state.

The field shift due to the HF interaction from the resonance position H_k of the line corresponding to the allowed $(\Delta m_\alpha = 0)$ hyperfine EPR transition number k between the electronic sublevels i and j, can be written as

$$\Delta H \ = -\frac{\gamma}{g\,\mu_B} \sum^{3} [B\ (E\ ) - B\ (E\ )]m\ , \tag{17}$$

where $B_\alpha^x(E_i)$ is the projection of the effective HF field on the direction of the external magnetic field. In the case of a homonuclear cluster $(a_1 = a_2 = a_3 = a)$ in the parallel field, the resonance field shifts are

$$\mathrm{D}H^{hf\|}_{1,2,2'} \quad \frac{}{3\,g_e \quad B}(m_1 \quad m_2 \quad m_3) \tag{18}$$

$$\Delta H^{hf\|}_{3(4)} = \Delta H^{hf\|}_{5(6)} = -\frac{a}{3g_e\mu_B}\left\{(m_1+m_2+m_3) \underset{(-)}{+} \underset{(-)}{+} \frac{2s+1}{2\Delta}[(\delta_1+\sqrt{3}\delta_2)m_1+(\delta_1-\sqrt{3}\delta_2)m_2-2\delta_1 m_3]\right\} \tag{19}$$

Assuming, for simplicity, that the nuclear spins are $I_1 = I_2 = I_3 = 1/2$ (the nuclear spins of the ^{57}Fe isotope), in the limit of large rhombic distortions (D << δ) which suppresses the AS exchange, we find the result of ref. [12]. (Fig. 3a) for allowed transitions 3 - 6. At intermediate values of the AS exchange (D ~ δ) peculiarities appear in the HF splittings of the allowed transitions due to the scalene geometry of the cluster. They are displayed in additional splittings of the HF lines which are degenerate for the equilateral and isosceles models. The HF splittings for forbidden transitions 1,2 and 2′ do not depend on the ratio between the exchange parameters (Fig. 3b). It is necessary to note that the HF structure coincides with the HF structure of the EPR signal of the ferromagnetic magnetic ground state with S_{max}. The length of the HF splitting is a in this case and does not depend on the exchange parameter. the intervals between these HF lines are $a/3$. As we see, the HF structure of the EPR spectra for the AF ground state depends essentially on the exchange parameters and magnetic field. The maximal length of the AF HF spectra is $5a/3$ for transitions 3-6.

In the limiting case of a strong AS interaction (D >> δ) all the aforementioned peculiarities disappear as well as any manifestations of non-equivalence of the cluster's ions. The HF splittings for all allowed and forbidden transitions are the same (Fig. 3b) and defined by expression (18).

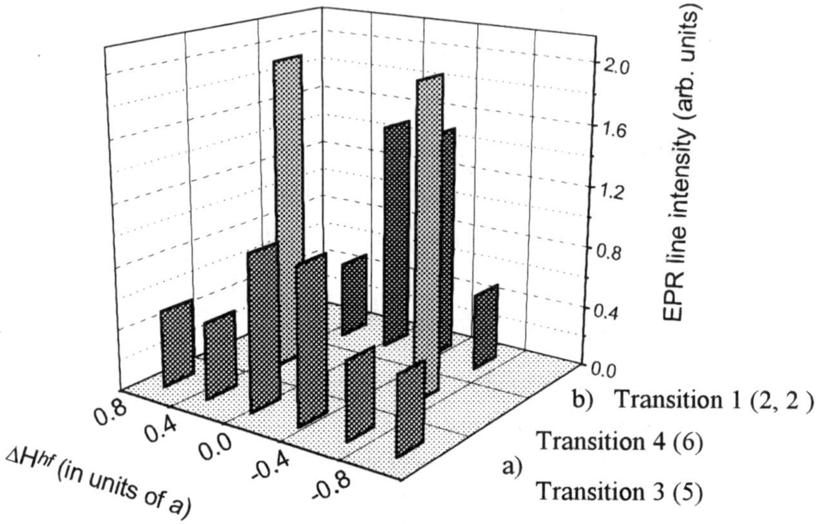

Fig. 3. Hyperfine structure of distorted homonuclear ($I_1 = I_2 = I_3 = 1/2$) cluster in parallel field : a) allowed transitions, $D/d_1 = 0.0006$, $d_2 = 0.002$; b) allowed and forbidden ransitions, $D/d_1 = 600$, $d_2 = 0.002$.

For a perpendicular orientation of the magnetic field, the non-linear behavior of all levels leads to a field dependence of the HF splittings. The resonance field shifts have the form

$$DH^{hf\perp}_{1(2,2')} = \frac{}{(+)\,3\,g_e\ \ _B}\;\frac{\overset{+}{(\ \ )}\,H}{E_{1(2)}}(m_1 + m_2 + m_3) \tag{20}$$

$$\Delta H^{hf}_{3(4)} = -\frac{a}{(\ )\,6g_e\mu_B}\left\{\left[\frac{\delta(-)H}{E_{1(2)}} - \frac{\delta(+)H}{E_{2(1)}}\right](m_1 + m_2 + m_3) - \right.$$

$$\left. - \frac{2s+1}{\delta}\left[(\delta_1 + \sqrt{3}\delta_2)m_1 + (\delta_1 - \sqrt{3}\delta_2)m_2 - 2\delta_1 m_3\right]\right\} \tag{21}$$

$$\Delta H^{hf\perp}_{5(6)} = -\frac{a}{6g_e\mu_B}\left\{\left[\frac{\delta(\overset{+}{-})H}{E_{1(2)}} + \frac{\delta(\overset{-}{+})H}{E_{2(1)}}\right](m_1+m_2+m_3) \underset{(-)}{+} \right.$$
$$\left. \underset{(-)}{+} \frac{2s+1}{\delta}\left[(\delta_1+\sqrt{3}\delta_2)m_1 + (\delta_1-\sqrt{3}\delta_2)m_2 - 2\delta_1 m_3\right]\right\} \qquad (22)$$

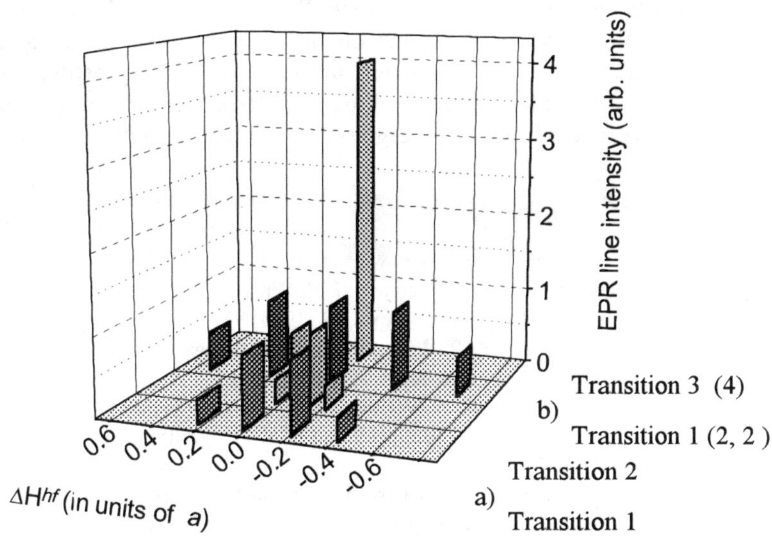

Fig. 4. **Hyperfine structure of distorted homonuclear ($I_1 = I_2 = I_3 = 1/2$) cluster in perpendicular field : a) D/H = 3.33, transitions 1 and 2; b) D/H = 1000, transitions 1- 4.**

For the cases of weak AS interaction (or strong magnetic field, D << H), the HF splitings for all transitions look like the corresponding ones at the parallel field (see Fig. 3). For the transitions, which occur at magnetic field values of order D, the HF structure of the EPR spectra displays a large variety of HF splittings (Fig. 4 a). This case is the most complicated for the analysis of the real geometry of the cluster. In contrast, at strong AS exchange (D >> H) a very simple HF structure reveals for all transitons (Fig. 4b). The appearance of an intensive singlet line in the spectrum of the forbidden transition is the best evidence of strong AS exchange. In spite of the aforementioned reduction of the effects of geometrical non-equivalence by a strong AS interaction in the HF structure in the parallel field, at a perpendicular field for the same D value the HF splittings of the allowed transitons 3 and 4 depend only on the ratio between the parameters of the distortions.

5 Conclusions

The hyperfine structure of the EPR spectra of distorted trinuclear metal clusters reveals the correlation between the real geometry of the cluster, the set of intracluster parameters and experimental conditons. For the parallel field, all peculiarities of the scalene cluster can be observed when the AS exchange is smaller than, or of the same order as, the distortions. Strong antisymmetric interaction in a distorted system smoothes the manifestation of geometrical non-equivalence of the cluster's ion. Strong AS exchange reduces the action of the exchange "distortions" δ_1 and δ_2, and results in an equilateral type of HF structure of the EPR spectrum. In contrast, for the perpendicular field, even with strong AS exchange, the hyperfine splittings of the allowed transitions still depend on the ratio between the parameters of the distortions and reveal the real geometry of the cluster. The hyperfine structure of the EPR lines of the antiferromagnetic ground state of the trimeric cluster essentially depend on the ratio of the Heisenberg and antisymmetric exchange parameters.

References

1. U. A. Jayasooriya, R. D. Cannon , R.P. White, J. A. Stride, R. Grinter and G. J. Kearley, J. Chem. Phys. 98 (1993) 9303.
2. R. D. Cannon and R. P. White, Progr. Inorg. Chem. 36 (1988) 195.
3. M. Takano, J. Phys. Soc. Japan 33 (1972) 1312.
4. B. D. Rumbold and G. V. Wilson, J. Phys. Chem. Solids 34, 1887 (1973)
5. V. E. Fainzil'berg, M. I. Belinsky and B. S. Tsukerblat, Molec. Phys. 44(5) (1981) 1195.
6. A. I. Shames, V. E. Fainzil'berg, M. I. Belinsky and B. S. Tsukerblat, Zhurn. Khim. Phys. 2(3) (1983) 316; ibid 2(6) (1983) 723.
7. M. I. Belinsky, B. S. Tsukerblat and A. V. Ablov, Sov. Phys. Solid State 16(4) (1974) 639.
8. B. S. Tsukerblat, M. I. Belinsky and A. V. Ablov, Phys. Stat. Sol. (b) 51 (1972) K71.
9. T. Moriya, Phys. Rev. 117 (1960) 635; ibid 120 (1960) 91.
10. I. E. Dzyaloshinskii, J. Exper. Teor. Phys. 32 (1957) 1547; ibid 33 (1957) 1454.
11. S. Geschwind, in : Hyperfine Interaction, ed. By A. J. Freeman and R. B. Frankel, Academic Press (1967).
12. T. Shimizu, J. Phys. Soc. Japan 25(4) (1968) 1021.

Correlation between Photoluminescence and P_b-like Center of Porous Silicon: ESR Image Analysis for Porous Silicon

Hiroshi Furuta,[1] *Chihiro Yamanaka,*[1] *Mitsuru Okigawa,*[2] *Hiroyasu Nakata*[3] *and Motoji Ikeya*[1]

[1] Department of Earth and Space Science, Graduate School of Science, Osaka University, 1-1 Machikaneyama, Toyonaka, Osaka 560, Japan
[2] Semiconductor Business Headquarters LSI Division Sanyo Electric Co., Ltd., 180, Ohmori, Anpachi-Cho, Anpachi-Gun, Gifu, 503-01, Japan
[3] Department of Physics, Graduate School of Science, Osaka University, 1-1 Machikaneyama, Toyonaka, Osaka 560, Japan

Abstract. The spatial distribution of P_b-like center in porous silicon was compared with that of photoluminescence (PL) intensity. They were measured with a microwave scanning ESR microscope and CCD camera respectively. The PL intensity had a negative correlation with the high density area of P_b-like centers. On the other hand, the PL intensity was positively correlated with the low density area. The production process of P_b-like defects might accompany with the production of another radiative PL-related mechanism. The PL efficiency may decrease PL quenching by P_b-like defects as a recombination center.

1 Introduction

1.1 Porous Silicon (PS)

Porous silicon (PS) has a nano-size skeletal structure which is made by anodic etching of crystalline silicon in hydrofluoric acid. Since strong visible photo luminescence (PL) was reported by Canham (1990), PS has attracted a great deal of attention as a material of silicon based photonic devices. P_b-like centers in porous silicon can be measured with ESR. The spectrum of P_b-like center is similar to that of P_b-center, which is a dangling bond located at Si/SiO_2 interface.

Tischlere et al.(1991) reported that photo-oxidation increases paramagnetic defects in PS and decreases PL intensity, and ascribed the negative correlation to dangling bond formation. However, Ookubo et al. (1992) proposed another mechanism that the PL degradate is caused by size changing of nano-structure of PS not by the presence of dangling bonds. Their discussion is based on their results that PS loses its PL even when the dangling bonds produced by annealing in vacuum are terminated by hydrogen by annealing in H_2 gas. On the other hand, Mochizuki et al. (1995) obtained ODMR results which support the former mechanism.

In this paper, we studied the relation between visible red PL intensity and the density of P_b-like defects using a scanning ESR microscope.

1.2 A Microwave Scanning ESR Microscope

A microwave scanning ESR microscope has been developed to study the spatial distribution of spins in a sample (Furusawa et al. 1988). Their subsequent papers reported ESR images reflecting a preferential inclusion of nitrogen at (111) sectors in synthetic diamond (Furusawa et al. 1990). Recent studies of ESR imaging are also summarized by Ikeya (1993). A piece of flat sample is set on an aperture on a wall of TE_{111} mode cylindrical cavity and is moved by X-Y stage as shown in Fig. 1. ESR signals are measured at each position. The spatial resolution of an ESR image is determined by the size of the aperture and by the scanning step which can be smaller than the diameter of the aperture.

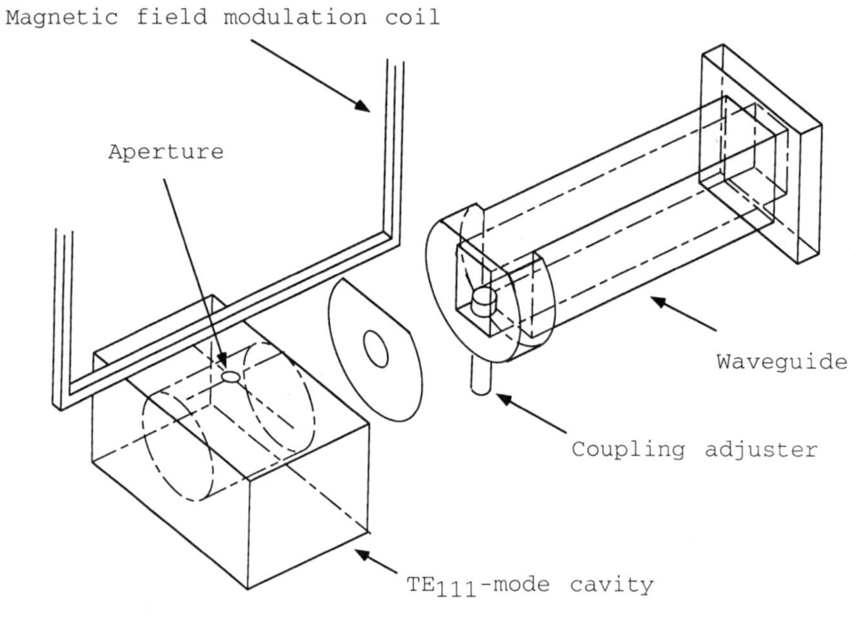

Fig. 1. A cylindrical TE_{111}-mode cavity with an aperture. A sample placed over an aperture is moved by an X-Y stage and stepping motors.

3. Sample Preparation

Disks of porous silicon were prepared from a B-doped p-type (100) crystalline silicon wafer (ρ = 8-20 Ω cm) in HF-based ethanoic electrolytes ($HF:C_2H_5OH:H_2O$ = 1:2:1) in anodizing container (Fig. 2). Silicon wafers were provided by Sanyo Co., Ltd. A platinum net was used as a cathode. A teflon tube with a diameter of 20 mm was used for container resistant to HF. PS was formed within a circle region with 20 mm diameter of the Si plate, where the Si plate was etched by the solution. The silicon wafers were exposed to light from a Tungsten lamp during this etching process. This enhances the anodizing by producing holes on the surface of silicon. Anodizing conditions of samples are summarized in Table 1. Sample A1 and A2 are prepared in the same condition. All as-prepared porous silicon were rinsed by ethanol, dried in air.

Table 1. Anodizing condition

Sample	Current	Time
A1	30mA/3cm^2	20min
A2	30mA/3cm^2	20min
A3	30mA/3cm^2	10min
B	10mA/3cm^2	40min

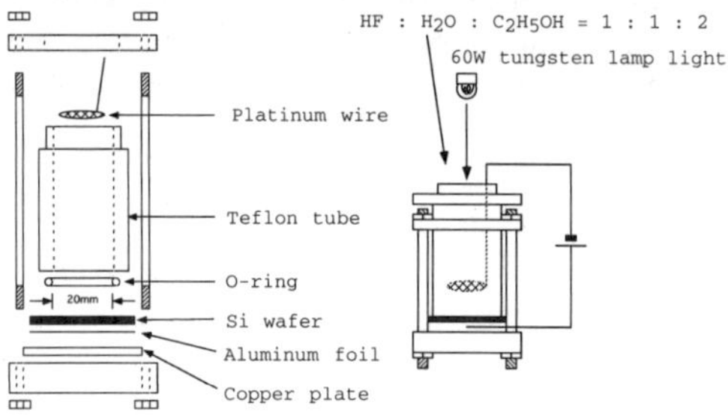

Fig. 2. Anodizing container

4 Experimental

4.1 As-anodized Porous Silicon

As-anodized PS showed strong visible red photo luminescence under 254 nm UV light exposure. A mercury lamp was used as a UV light source with a filter (Melles Griot 03 FIM 016, wavelength = 256.14 nm, bandwidth = 8.68 nm), which cut a visible light. Distributions of PL intensity from PS were obtained by a CCD video camera (Vixen, B05-3M) with a video image analyzing system (HAMAMATSU, DVS-3000).

A signal of P_b-like center was detected in all samples at room temperature by an X-band ESR spectrometer (RE-2X, JEOL) operating at 100 kHz modulation of magnetic field with a cavity with an aperture of 3 mm at the orientation of (011)//H as shown Fig 3a. Distributions of P_b-like center was measured by a scanning ESR microscope. Figure 4a shows the spatial distribution of P_b-like center in sample A1. Step size of sample scan is 1 mm. Figure 4b shows the spatial distribution of the visible red PL intensity from sample A1. These images show the following two features.

a) The area of about 1mm in width from the edge has no PL (non-PL area) in its etched area of about 18 mm.

b) P_b-like center concentrated at the edge of the etched area with no PL (non-PL area) or of the area with PL (PL area).

The one dimensional distributions of PL and P_b-like centers in Sample A2 were obtained as shown in Fig 5. A scan step of sample A2 was 0.5 mm. This figure shows the following two features that

b2) P_b-like centers concentrated at the inside edge of the PL area, not of the non-PL area and

c) The distribution of PL intensity had negative correlation with P_b-like center.

The one dimensional distribution of PL and P_b-like centers in Sample A3 and Sample B were measured as shown in Fig 6 and Fig 7 respectively. They shows that

d) The distribution of PL intensity had a positive correlation with that of P_b-like center.

The feature of d) seems opposite to that of c). The phenomena of b), b2) were probably caused by a concentration of an anodizing current to the edge of etching area. Such an edge effect is much smaller in Sample B. The following interpretation might explain our present results consistently. The process of producing P_b-like defects accompanies the formation of radiative defects which cause visible PL. The apparent intensity of P_b-like center has a positive correlation with the PL intensity. However, when the concentration of P_b-like center is high, the intensity has a negative correlation as described c). If the concentration of P_b-like center still increases with anodization while PL-related radiative center is saturated. PL efficiency of radiative defects is decreased by PL quenching caused by P_b-like defects as a recombination center. And it is accelerated by high concentration of P_b-like center.

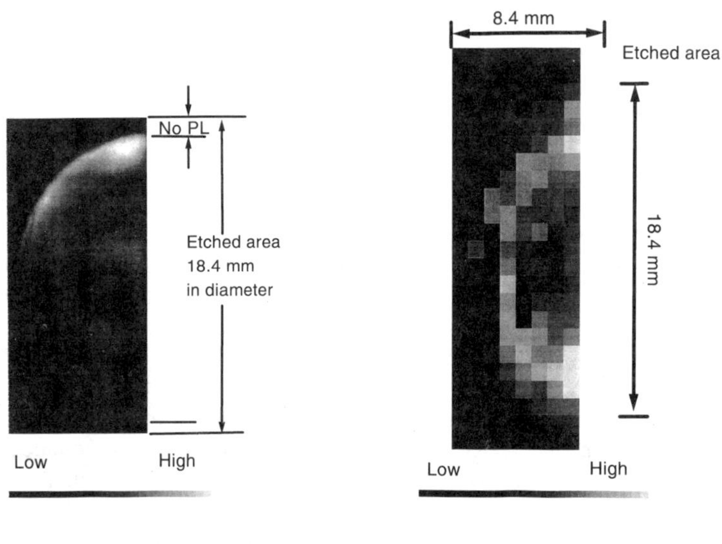

Fig. 3a. Distribution of visible red PL intensity from Sample A1. **b.** Distribution of P_b-like center of Sample A1.

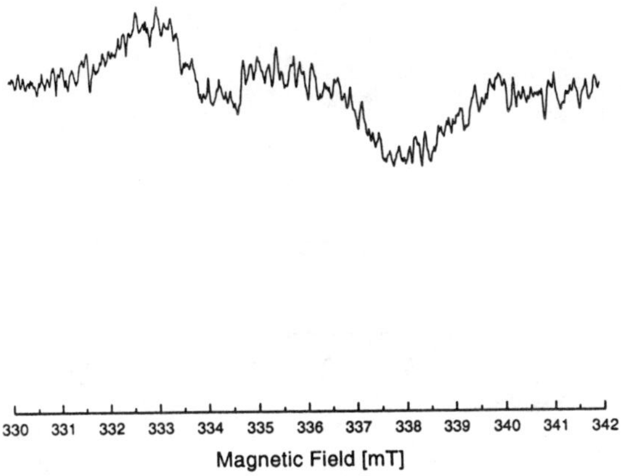

Fig. 4. ESR spectrum of as-anodized porous silicon (sample B) at 300K measured with a microwave scanning ESR microscope for the direction of the magnetic field parallel to [011].

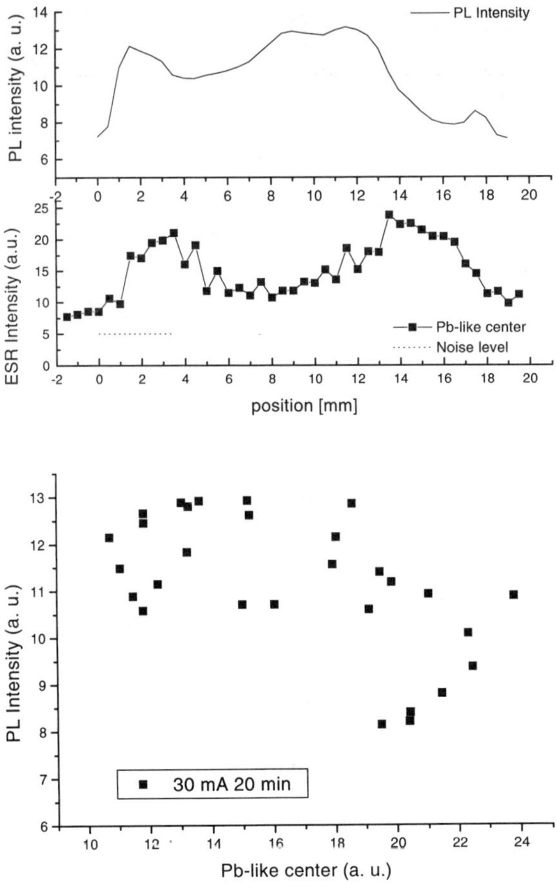

Fig. 5. A relation between PL intensity and P_b-like density in Sample A2

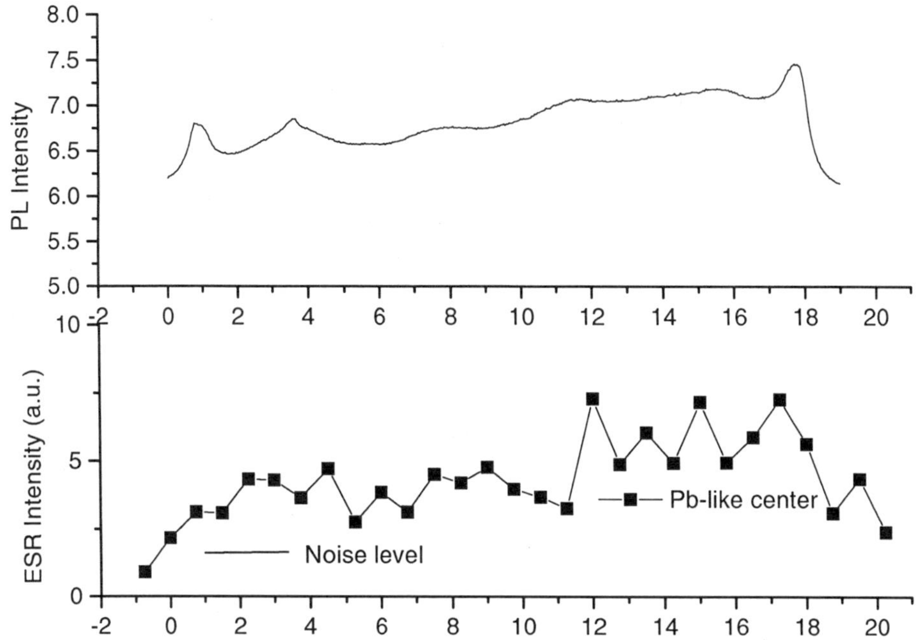

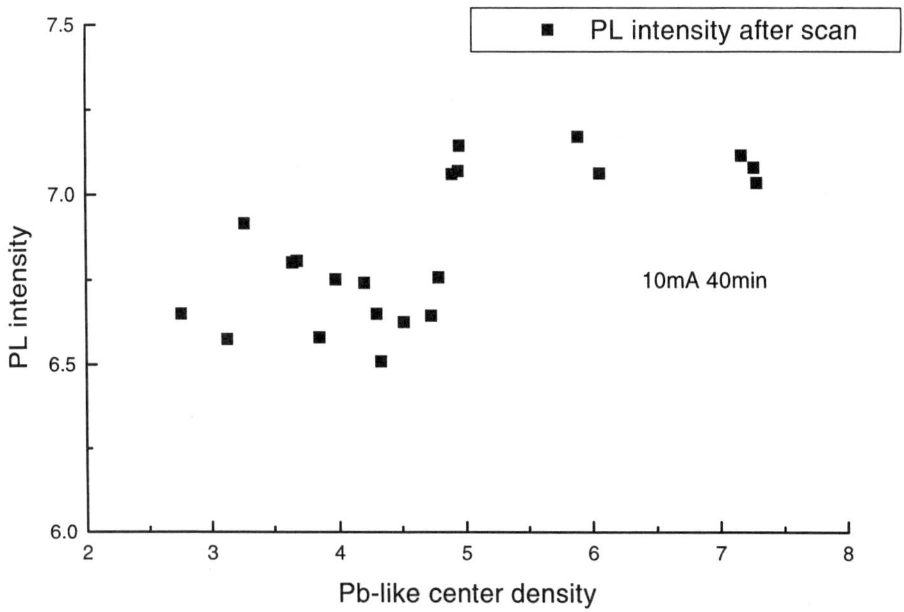

Fig. 6. A relation between PL intensity and the P_b-like density in Sample B

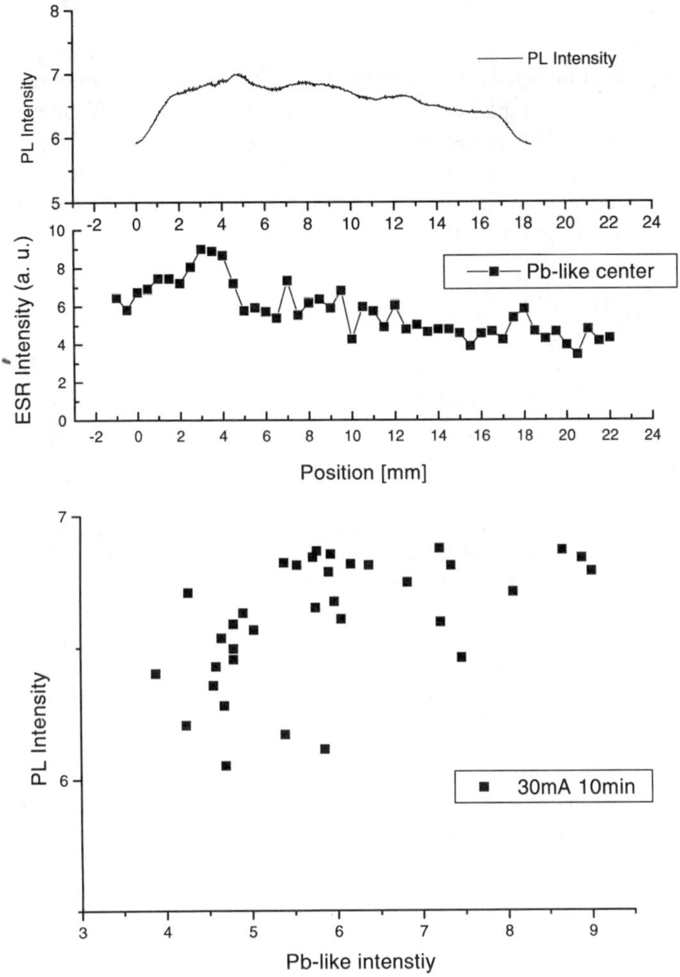

Fig. 7. A relation between PL intensity and the P_b-like density in Sample A3

4.2 Oxygen Ion Implantation Experiments

Annealing in air are known to decrease the PL intensity and increase numbers of both dangling bonds (g = 2.0055) and P_b-like center(B. K. Meyer. et al. 1993). It is due to thermal oxidization process.

1E15 Oxygen ions with energy of 2 MeV were implanted into PS (30 mA, 20 min) with the currency of 0.4 μA in order to produce dangling bonds. Oxygen related defects (g_1= 2.0036, g_2 = 2.0064, g_3 = 2.061) were produced in PS. ESR signal of dangling bonds (g = 2.0056) was overlapped on this signal. PL from PS was decreased with oxygen implantation. PL intensity had a negative correlation with the density of dangling bonds. The existence of dangling bonds or destruction of radiative mechanism caused the PL quench.

5 Conclusion

Paramagnetic dangling bonds in electrochemically etched porous silicon was studied with a microwave scanning ESR microscope. PL intensity had a negative correlation with the density of P_b-like center in as-anodized PS for the density area, while it had a positive correlation for the low density area. Extremely high concentration of P_b-like center may accelerate the PL quenching. Oxygen ion implantation produced a high density of dangling bonds in PS and caused PL quenching.

Acknowledgement

We appreciate discussion with Dr. Kohno, Toyoda, and various helps by Mr. Sasaoka, Hirai, Dr. Ogoh, Prof Katakuse and, Ichhara, Matsumoto.

References

1. L. T. Canham, Silicon quantum wire array fabrication by electrochemical and chemical dissolution of wafers *Appl. Phys. Lett.* 57 (1990) 1046
2. M. Furusawa and M. Ikeya. Electron spin resonance microscopic imaging of fossil crinoid utilizing localized microwave field. *Anal. Sci.* 4 (1988) 649-651
3. M. Furusawa and M. Ikeya. Distribution of nitrogen and nickel in a syntehtic diamond crystal observed with scanning ESR imaging. *J. Phys. Soc. Jpn.* 59, 2340-2343.
4. M. Ikeya. New Application of Electron Spin Resonance -Dating, Dosimetry and Microscopy. World Scientific (1993)
5. M. A. Tischler, R. T. Collins, J. H. Stathis, and J. C. Tsang: Luminescence degradationin porous silicon. *Appl. Phys. Lett.* 60 (1992) 639
6. N. Ookubo, H. Ono, Y. Ochiai,Y. Mochizuki,and S. Matsui. Effects of thermal annealing on porous silicon photoluminescence dynamics. *Appl. Phys. Lett.* 61, (1992) 940
7. B. K. Meyer, D. M. Hofmann, W. Stadler, V. Petrova-Koch, F. Koch, P. Omiling and P. Emanuelsson. Defects in porous silicon investigated by optically detected and by electron paramagnetic resonance techniques *Appl. Phys. Lett.* 63 (1993) 2120
8. Y. Mochizuki and M. Mizuta. Role of dangling bond centers on radiative recombination processes in porous silicon, *Appl. Phys. Lett.* 67 (1995) 1396
9. B. K. Meyer, V. petrova-Koch, Muschik, H, Linke, P. Omling, V. Lehmann. Electron spin resonance investigation of oxidized porous silicon, *Appl. Phys. Lett.* 63 (1993) 1930-1932

Intercomparison With the Alanine Dosimeters From IAEA

M. Lin, Y. D. Chen, H. S. Ye and G. Q. Zhang

Radiometrology Centre, China Institute of Atomic Energy, P.O.Box 275 (20), 102413 / Beijing, P.R.China

Abstract. An intercomparison between Radiometrology Centre of CIAE and IAEA was made in September 1994. The alanine dosimeters were sent from IAEA, compared with four types of liquid chemical dosimetric systems, such as Fricke, silver dichromate, potassium (silver) dichromate and ceric-cerous sulfate dosimeters made at CIAE. Since these dosimeters have different dose ranges, different ways were used to make the experiments of ^{60}Co γ ray irradiations. The agreement was excellent. The minimum relative difference was 0.19 % and the maximum 2.1 %.

1 Introduction

The four kinds of liquid chemical dosimeters prepared at CIAE, all based on UV-visible spectrophtometric analysis of radiolytic oxidization or reduction of ions in an aqueous acidic solution, have been studied as precise high-dose dosimeters. The Fricke dosimetric system is used as one of four classical methods to measure the absorbed dose of ionizing radiation (Matthews, 1982; McLaughlin et al., 1989). The two dichromate dosimeters with different initial concentrations of silver dichromate ions and the ceric-cerous sulfate dosimeters are now being used as reference dosimeters for radiation processings (McLaughlin et al., 1989; Sharpe et al., 1985). The dose response characteristics for high-dose ^{60}Co γ ray reference dosimetry have been investigated (Chen et al., 1993; Ye et al., 1994 (a,b); Zhang et al., 1994).

The alanine dosimeters sent from IAEA are based on electron spin resonance (ESR) spectrometry of radiation-induced free radicals in alanine (Regulla et al., 1982). This dosimetric system has been developed by several researchers and widely used for reference and transfering dosimetry for high-dose ^{60}Co γ ray radiation processings for IDAS (International Dose Assurance Service). (ASTM Standards Practice E1607-94, 1994).

The aim of this study is to check the aqueous chemical dosimeters made at CIAE, which had been used as the second-standard reference dosimetric systems.

2 Experimental Procedure

2.1 Preparation of the Dosimeters

The liquid chemical dosimeters were prepared by using the third redistillated water. The composition of each kind of dosimeters is shown in Table 1. The dosimeter solution was sealed in the standard glass ampoules (12 mm in o.d.).

Table 1. Composition, relative dose converting coefficients K_D and temperature-reponse coefficients K_T of the aqueous chemical dosimeters

Dmtr.*	Composition of the dosimeter	K_D (Gy)	K_T (% / ℃)
F	1.0 mM Fe^{2+} and 1.0 mM NaCl in 0.4 M H_2SO_4	208.7	0.17
DA	0.35 mM $Ag_2Cr_2O_7$ in 0.1 M $HClO_4$	8403	-0.22
DK	2.0 mM $K_2Cr_2O_7$ and 0.5 mM $Ag_2Cr_2O_7$ in 0.1 M $HClO_4$	5.74×10^4	-0.22
CC	0.01 M Ce^{4+} and 4.0 mM Ce^{3+} in 0.4 M H_2SO_4	724.8	-0.25

*F — Fricke dosimeters; DA — Silver dichromate dosimeters; DK — Potassium (silver) dichromate dosimeters; CC — Ceric-cerous sulfate dosimeters

The alanine dosimeters were prepared 3 mm in diameter and 30 mm in length. The dosimeter package contained 4 dosimeter capsules (about 12 mm in o.d., 40 mm in length and 4 mm in wall thickness) accommodated in 2 sealed compartments, in which three capsules were used to be irradiated and the rest one used for control purposes.

2.2 Irradiation and Calibration

The irradiations were performed at CIAE by using the squirrel-cage ^{60}Co sources of 0.74 PBq in September 1994. The reproducibility of 0.1 % to the source was measured. A water phantom was put into the cylindrical hole of the stainless steel shell. The central hole of the phantom was used as the intercomparison spot. All the dosimeters were irradiated at this position one by one. It's dose rate was calibrated by the air-saturated Fricke dosimeters. And the side-dose values of the irradiaiton position for every lift and drop of the sources were taken into account.

The alanine dosimeters were irradiated at two dose levels of about 2.5 kGy and 10 kGy, respectively. The liquid chemical dosimeters were irradiated by two means due to their different dose ranges. One was linear irradiations for Fircke and silver dichromate dosimeters and the other one was accumulative irradiations for potassium (silver) dichromate and ceric-cerous sulfate dosimeters as well as alanine dosimeters.

2.3 Analysis of dosimeters

The four kinds of aqueous chemical dosimeters were analyzed by using a double-beam UV-visible spectrophotometer (Cary-3E) at CIAE. Four different wavelengths were 303 nm for Fricke dosimeters, 350 nm for silver dichromate dosimeters, 440 nm for potassium (silver) dichromate dosimeters and 319 nm for ceric-cerous sulfate dosimeters, respectively. The measuring temperature was controlled at 25.0 ± 0.1 ℃ by the temperature controlling systems connected with thc spectrophotometer. The quoted nominal dose values for the Fricke and silver dichromate dosimeters were calculated by equation (1), while for the potassium (silver) dichromate and ceric-cerous sulfate dosimeters by equation (2), in which D is absorbed dose, Gy; $\Delta\dot{A}$ is difference of absorbance per minute, min^{-1}; t is irradiated time, min and K_D is dose converting coefficient calibrated by Fricke dosimeters, Gy, shown in Table 1.

$$D = \Delta\dot{A} \cdot t \cdot K_D \tag{1}$$

$$D = \Delta A \cdot K_D \tag{2}$$

The irradiated alanine dosimeters and unirradiated ones were posted back to IAEA about one week later after the irradiations with the results of our liquid chemical dosimeters.

Table 2. Estimated dose of alanine dosimeters from IAEA reports

Seriel №	#161	#162	#163	#165	#166	#167
Estimated dose (kGy)	10.13	9.890	9.864	2.577	2.563	2.561
Average dose (kGy)	9.961			2.567		

3 Results and Discussion

The results from IAEA reports are shown in Table 2 and the final results of this intercomparison between CIAE and IAEA are shown in Table 3. In this intercomparison, two different irradiaiton and dose calculation meathods were used. Linear irradiaitons were used for Fricke and silver dichromate dosimeters, which the quoted nominal dose values were calculated by equation (1). While accumulative irradiations were performed the same as alanine dosimeters for the potassium (silver) dichromate and ceric-cerous sulfate dosimeters, which the quoted nominal dose values were workcd out by equation (2). Since the K_T values of aqueous chemical dosimeters (shown in table 1) are very small, the temperature corrections were ignored for this intercomparison.

From the results shown in Table 3, the minimum relative difference is 0.19 %, and the maximum one is 2.1 %, all within the relative uncertainties (at 95 % confidence level) of aqueous chemical dosimeters.

Table 3. Dose intercomparison results

	Lower dose point			Higher dose point			
CIAE Dmtr.	D_{CIAE} (kGy)	D_{IAEA} (kGy)	R.D.* (%)	D_{CIAE} (kGy)	D_{IAEA} (kGy)	R.D.* (%)	Uncertaity of Dmtr. (%) (K=2)
F	2.60	2.567	1.3	9.98	9.961	0.19	2
DA	2.58		0.51	9.92		-0.41	4
DK	2.62		2.1	10.05		0.89	4
CC	2.58		0.51	9.81		-1.5	4

* Relative difference between CIAE and IAEA.

4 Conclusions

The results of the four kinds of aqueous chemical dosimeters prepared at CIAE are in good agreements with the alanine dosimeters from IAEA, all within the uncertainty of each kind of dosimeters at two dose levels of 2.5 and 10 kGy, respectively. By using the two different ways of irradiations and dose calculations, it is a way for checking the different reference dosimeters between the second-standard laboratories and IAEA by IDAS.

Acknowledgements

The authors wish to appreciate Ms. Lanzhen Liu for her assistance with ^{60}Co γ ray irradiations and Dr. Kishor Mehta for his supplying alanine dosimeters and sending reports to us.

References

1. R. W. Matthews, Aqueous chemical dosimetry, *Appl. Radiat. Isot.* 33 (1982) 1159-1165.
2. W. L. McLaughlin, A. W. Boyd, K. H. Chadwick, J. C. McDonald, and A. Miller, *Dosimetry for radiation processing*, Taylor & Francis, London (1989)

81-87, 148-149, 173-175.

3. P. H. G. Sharpe, J. H. Barret, and A. M. Berkley, Acidic aqueous dichromate solution as reference dosimeters in the 10-40 kGy range, *Appl. Radiat. Isot.* 36 (1985) 647-652.
4. Y. D. Chen, H. F. Long, and G. Q. Zhang, Ferrous sulfate dosimeter, *Atomic Energy Science and Technology*, 27 (1993) 138-144.
5. H. S. Ye, Y. D. Chen, G. Q. Zhang, and M. Lin, Potassium dichromate dosimeter, *Atomic Energy Science and Technology*, 28 (1994a) 284-288.
6. H. S. Ye, Y. D. Chen, G. Q. Zhang, and M. Lin, Development of a silver dichromate dosimeter, *IAE-0132* (1994b) 1-6.
7. G. Q. Zhang, Y. D. Chen, H. S. Ye, and M. Lin, A ceirc-cerous sulfate dosimeter, *IAE-0131* (1994) 1-8.
8. D. F. Regulla, and U. Deffer, Dosimetry by ESR spectroscopy of alanine, *Appl. Radiat. Isot.* 33 (1982) 1101-1114.
9. ASTM Standards Pracitice E1607-94, Standard practise for use of the alanine/ESR dosimetry system, *Annual Book of ASTM Standards,* 12.02 Nuclear (Ⅱ), solar, and Geothermal Energy (1994) 855-860.

Studies of SEESR and Its Data-System

Yayan Liu *Donghong Zheng*

Changchun Institute of Applied Chemistry, Chinese Academy of Sciences, Changchun,130022, P. R. China

Abstract. Previously designed cells for ESR-electrochmical measurements have been limited in their range of applicability because of mutually exclusive constraints places by the ESR cavity quality factor and electrochemical iR drop problems. In this paper a series of easily constructed ESR-electrochemical cells have been designed. The advantages of simultaneous electron chemical ESR (SEESR) experiments using these two kinds of cells for the studies of methylene green (MG) and methyl viologen(MV^{2+}) are discussed, and the results indicated that these cells were suitable for the studies of SEESR. In addition, the application of using signal-averaging technique for the detection of short-lived radials are discussed. Finally, an ESR-data system for ESR spectrum analyzer was studied.

1 Introduction

Simultaneous electron chemical ESR (SEESR) has been used extensively in electrochemical studies to detect and identify paramagnetic intermediates[1-2], because the combining electrochemistry with ESR spectroscopy allows more confident identification of intermediates and products and enables the elucidation of mechanisms and the evaluation of kinetic parameters. But the number of recent applications does not reveal the full capabilities of the technique. The main problem is the difficulty in the designing a very good SEESR cell. In order to solve the difficulty in cell design, a set of cells fitting to different systems have been designed in order to find some suitable cells for SEESR.

2 Experimental Section

2.1 Reagents

Methyl viologen(MV^{2+}) and methylene green(MG) were used as received from the Aldrich Chemicals Company as reagent grade. All other chemicals were also reagent grade. All solutions were prepared with double distilled water and deoxygenated with passing high pure nitrogen. All experiments were carried out at room temperature. The concentration of Methyl Viologen is 10^{-2}mol/L in 0.1mol/L kCl solution.

2.2 Instrumentation

Electron spin resonance measurements were carried out with JES-FE-3AX spectrometer equipped with TE_{011} cylindal cavity, 100-kHz field modulation, and a low noise-high power . Electrochemical experiments were done with Modle 175 Universal programmer, Model 173 potentiostat (Princeton Applied Research). The reference electrode(RE) was a conventional Ag/AgCl electrode saturated with KCl.

2.3 The Design of Electrochemical -ESR Cells

The design of the cells for simultaneous ESR and electrochemical measurements is extremely critical. A SEESR cell should have:

1. A counter electrode(CE) is as far away as possible from the working electrode(WE) in order to avoid some restriction its use in kinetic studies. Because products from the CE may react with those at the WE after a relatively short period of time and there may be interference in the ESR response from products generated at the CE.
2. The close proximity of WE and reference electrode(RE) satisfies many of the conditions required for a good electrochemical response.

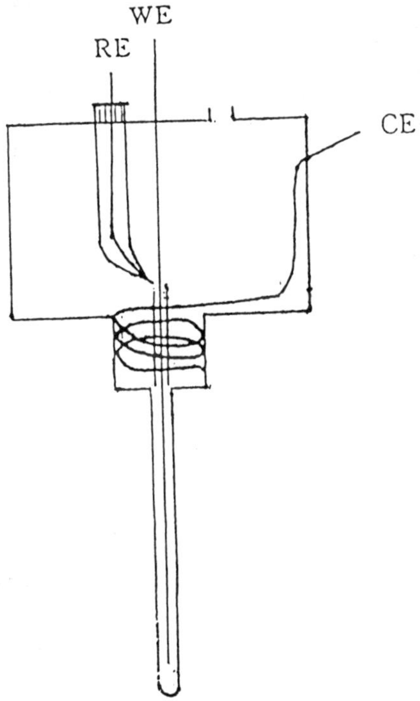

Fig.1. Electrochemical-ESR cell for use with sationary solutions (CE:counter electrode RE:reference electrode, WE:working electrode)

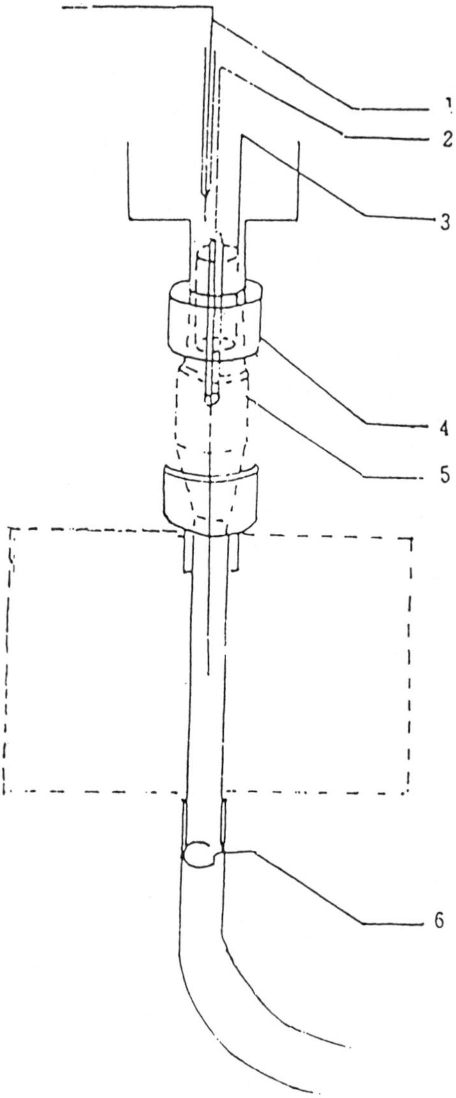

Fig.2. Electrochemical-ESR cell for use with flowing solutions (1.RE 2.WE 3.CE 4.external fastener 5. Teflon internal connecting tube 6.CE)

Many of these requirements are mutually conflicting. For example, the effective surface area of the WE is required to be large for an adequate rate of electrolysis when short-lived species need to be detected. However, the larger the electrode area, the more distorted is the electrochemical response because of the increased iR drop. Additionally, large amounts of metal in the ESR cavity reduce the Q (quality factor) of the cavity which leads to reduction of the ESR detection sensitivity. Different sample tubes such as thin-layer tube can be assembled by external fastener in order to design different cells.

3 Results and Discussion

First all, two easily constructed SEESR cells were designed, as shown in Fig.1. for stationary mode and Fig.2. for flow-through mode. The cell in Fig.1 is constructed from a pyrex glass tube. The working electrode was used a piece of platinum wire.The reference electrode was Ag/Agcl electrode. The working electrode and the tip of the reference electrode are positioned very close to each other to minimize the iR drop. In the flow-though mode(Fig.2), the glass tube house the reference electrode. The counter electrode (CE) is normally a platinum wire, the end of which is positioned above the bottom of the cell. The designs shown in Fig.1 and 2 can minimize the possibility of electrolysis products generated at the counter electrode (CE) reaching the center of highest ESR sensitivity where the working electrode is place.

The quality of the ESR performance of the present cell is presented in Fig.3.. The substance chosen for electrochemical reduction in these experiment was Methyl Viologen as the dication(MV^{2+}) which is easily reduced to a stable cation radical($MV^{+\cdot}$). Fig.3. gives the ESR spectrum of MV^{+}generated by 5 min of reduction controlled potential elecrolysis of MV^{2+} in 0.1mol/L Kcl solution. The ESR-voltammetric behaviour for MV^{2+} in water resembles that reported for reduction of Methyl Viologen[3]. The above examples reported demonstrate that solvents with high dielectric constants can readily be used for electrochemical-ESR experiments with the newly designed cell.

Moreover, cyclic voltammegram of Methylene green(MG) was also studied with ESR technique by these cells. MG is a kind of dye. ESR spectra of MG is shown in Fig.4.. Working conditions are as follow: constant potentials 0.4V vs Ag/AgCl (saturated KCl), supporting electrolyte: phosphate buffer (pH 7.5)+0.1M KCl. The electrochemical behavior of MG has been investigated by using spectroelectrochemistry in situ. The result demonstrated that ESR signals vary with different pH values.

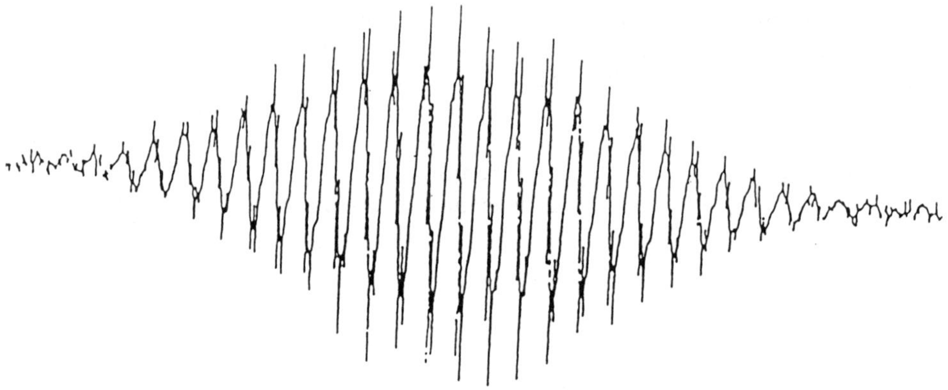

Fig 3. ESR spectrum of 10^{-2} mol/L Methyl viologen in 0.1mol/L KCl after five minutes electrolysis at -0.75V(vs.SCE)

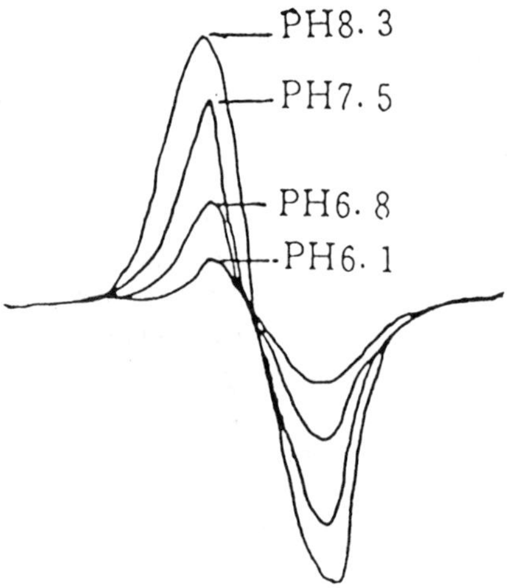

Fig.4. ESR spectrum of MG obtained with flow-through cell

A computer system combined with a JES-FE series ESR spectrometer was studied; the computer system can carry out acquisition, processing and simulation of measured data. The system consists of an IBM-PC microcomputer, an interface consisting of A/D and D/A converters and application software. The sampling and signal average can be made by the system for a very wide rage of sweep time. It improved greatly the ESR spectrometer in measuring the short life time of free radicals and the weak signal. The system has good functions and specifications, but it is inexpansive.

4 *Conclusion*

The above examples reported in water solvent demonstrate that solvents with high dielectric constants can readily be used for electrochemical-ESR experiments with the newly designed cell. The cells are suitable for virtually all solvents used in electrochemical investigations, giving excellent electrochemical and ESR performance for solvents with low and high dielectric constant. The cells are easy to manufacture and handle. The ESR data system for acquisition and analysis data of ESR spectra is successful.

References

1. I.B. Goldberg and T.M. McKinney in P.T. Kissinger and W.R. Heineman. *Labroratory Techniques in eletroanalytical Chemistry.* Marcel Dekker. New York, (1984) pp.675-728. and references cited therein.
2. R. N. Bagchi, A.M. Bond and R.Colton. *J.Electroanal.Chem..* **199**(1986)297-302
3. R.N.Bagchi, A.M.Bond and F.Scholz. *J.Electroanal.Chem..* **252**(1988)259-267

The Nature of Amino Radicals in Frozen Ammonia-Water for Future ESR Dating of Comets

Hirayuki Toda, Hideki Sasaoka, Hiroshi Kanamori, Kazunori Kanosue, Makoto Hirai and Motoji Ikeya

Department of Earth and Space Science, Graduate School of Science, Osaka University, 1-1 machikaneyama, Toyonaka, Osaka 560, Japan

Abstract. Radiation effects of frozen ammonia-water ($H_2O:NH_3$) were studied with ESR for future dating of comets and outer planets in future. Signals of OH, NH_2, and HO_2 in irradiated ice of $H_2O:NH_3$ were analyzed and confirmed using the semi-molecular orbital calculation (MOPAC, PM-3). The spectrum of NH_2 shows as a triplet line with a g-factor of g = 2.005 and a hyperfine splitting constant of A_H = 2.6 mT; the hf structures due to N atoms (I = 1) were smeared out. The lifetime of NH_2 radicals were obtained from the isothermal annealing experiment to be 10^6 year at about 65K, i.e., at ambient temperature of satellites around Saturn. The activation energy of 0.29 ± 0.02 eV and the frequency factor of 5.7×10^8 s^{-1} were calculated from the Arrhenius plot.

1 Introduction

Some satellites of outer planet world far away from the Sun are covered with icy materials. The ambient temperature is extremely low in the outer planet world. Radiation effects caused by solar wind protons and cosmic rays have been accumulated due to the long lifetime of radical species. Therefore, ESR dating of impact craters, volcanic activities, and so on in outer planet world would also be made with ESR dating based on radiation effects as applied to terrestrial minerals and fossils such as SiO_2, $CaCO_3$ and $CaSO_4$ (Ikeya 1993). Solid CO_2 were studied to date the polar caps of Mars, comets and icy satellites by Hirai et al. (1994).

A considerable amount of impurities seems to be involved in comet ice (H_2O) from several studies. It may be considered that NH_3, SO_2, CH_4, CO, CO_2, etc. are dissolved in water and solidified. Amino radicals (NH_2) produced from NH_3 among these impurities are of special interests considering the origin of life. The spectrum of NH_2 was observed as a triplet line due to the large anisotropic hyperfine (hf) splitting due to nitrogen nucleus in NH_2 in solid NH_3

(D. R. Smith et al. (1970)) or strong hydrogen bonds in H_2O doped with NH_3 (B. S. Al-Naimy et al. (1966)). We have studied the nature of NH_2 to use the signal for future ESR dating.

2 Experimental

Aqueous solution of NH_3 (5%) was solidified by cooling to liquid nitrogen temperature and was irradiated by γ -rays up to dose of 18 kGy (1Gy = 1 J/kg) from a source of ^{60}Co. UV-irradiation was made at energy deposit of 100 μ W at 77K, using a deuterium lump (Hamamatu-L2196). The sample was powdered and then transferred into ESR standard sample tubes.

ESR measurement were performed using an X-band spectrometer (JEOL 1-X, 2-X) with a cold gaseous nitrogen flow system. The sample temperature was controlled by changing the temperature of the gaseous nitrogen. ESR spectra were taken at a microwave power of 5 mW. Semi-molecular orbital calculation of NH_2 radical in the crystal of ice (H_2O) were made using MOPAC (PM-3) in CAChe worksystem programmed by J. J. P. Stewart.

3 Result and discussion

3.1 Hyperfine splitting constant of N (A_N) broadening of NH_2 in H_2O

A spectrum of a triplet line having the hf splitting constant of A = 2.6 mT and the g-factor of g = 2.005 was observed in the UV-irradiated sample at 110K as shown in Fig.1 (a). These parameters agree well with the values reported for NH_2 radicals by D.R. Smith et al. (1970) and S. N. Foner et al. (1958). Fig.1 (b) and (c) show ESR spectra of frozen ammonia-water γ -rayed at 77K and measured at 110K and 125K, respectively. The spectra was separated into the signals of OH, NH_2, and HO_2. The spectra and g-factor of OH and HO_2 corresponded to those obtained by Janusz Bednarek et al. (1987). The signal of OH was observed at 110K, but HO_2 has not yet been formed at this temperature. On the other hand, HO_2 was generated at 125K, where OH was annealed out. The signal of NH_2 was present at both temperatures. The ESR spectrum of NH_2 consists of three lines, though the spectrum should theoretically have nine lines: one line is separated into three lines with the intensity ratio of 1:1:1 by the hf splitting of the nuclear spin of N (I_N = $\pm$1,0) and further splitted into three lines with the intensity ratio of 1:2:1 by two H (I_H = $\pm$1/2). The measured hf splitting with the intensity ratio of 1:2:1 are due to H nuclei and not to an N nucleus as in other studies.

There are two reasons for the absence of the hf splitting constant by a nitrogen nucleus: a) the large anisotropy of an N atom in NH_2 and b) the very slow rotation of NH_2. These may be confirmed with the semi-molecular orbital calculation (MOPAC).

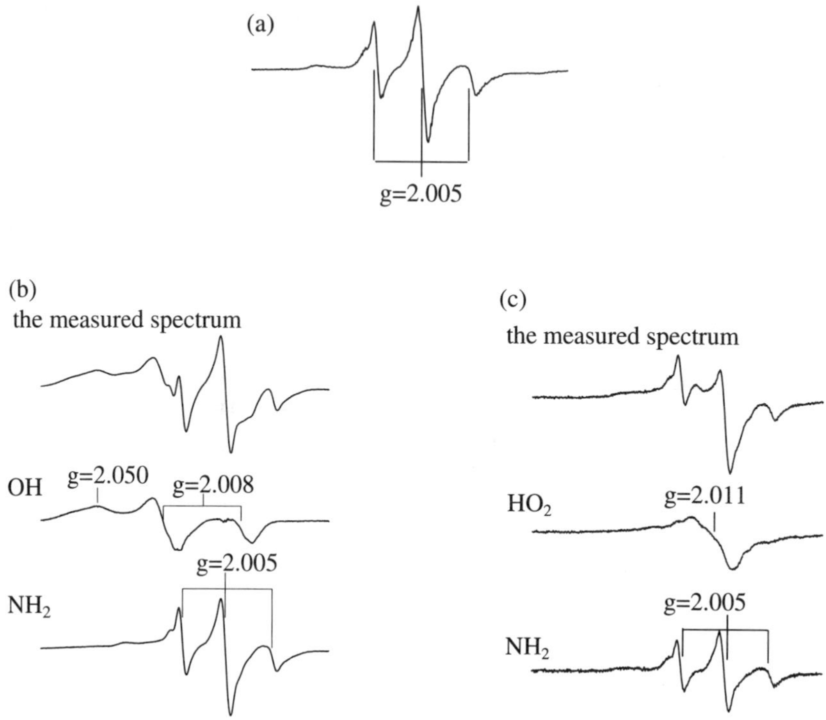

Fig. 1. ESR spectra of frozen ammonia-water irradiated by UV-light at 77K and measured at 110K (a), γ-rayed at 77K and measured at 110K (b) and at 125K (c). The spectrum at 110K or 125K irradiated by γ-ray was separated with OH and NH_2, or H_2O and NH_2, respectively, based on the the spectrum produced by UV-irradiation, measured at 110K, and recognized as NH_2 radical.

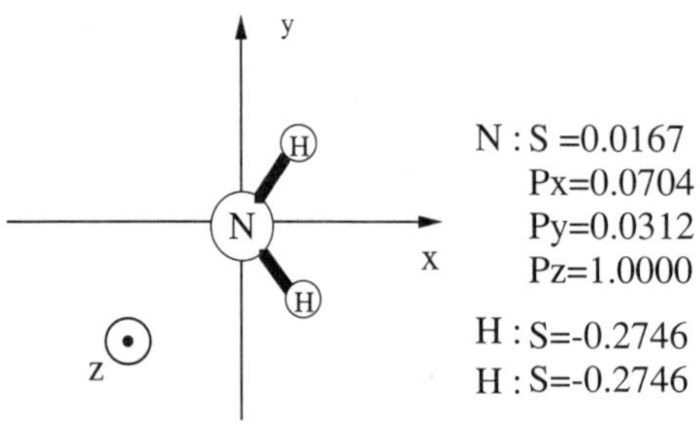

Fig. 2. The spin population of NH_2 radical was calculated using MOPAC (PM3, UHF). It is related to hf splitting constant.

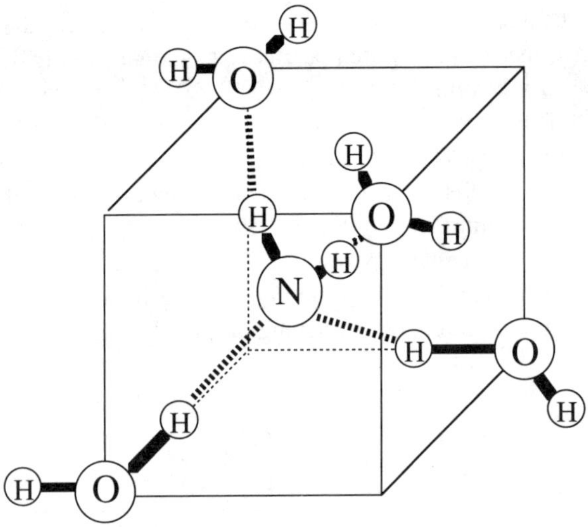

Fig. 3. The model based on the MOPAC result. NH_2 surrounded by four H_2O molecules obtained by MOPAC calculation with locked N and O. H_2O was replaced by NH_2 in a crystal of ice. Hydrogen bondings between N and H and O and H stabilized NH_2 in H_2O.

The spin population of NH_2 calculated with MOPAC is related to the hf splitting as shown in Fig. 2. The value of the spin population along z-axis (P_z) is much larger than that along other axises (P_x, P_y), i.e., the value of the hf splitting constant along z-axis is much larger. The large differences between the values of P_x, P_y, and P_z cause the broadening of $I_N = \pm 1$ lines because each radical has anisotropic hf splitting constant , A, as

$$A = A_s + A_p(3\cos^2\theta - 1)$$

where A_s and A_p are isotropic and anisotropic hyperfine splitting constants, respectively, and θ is the angle between the direction of the magnetic field and the axis of p-orbital.

The hf splitting constant A was averaged if radicals are in rotation. Then, the hf splitting of the nitrogen in NH_2 was observed. There are four hydrogen bonds around NH_2 as shown in Fig 3 as supported by MOPAC calculation. Hence, NH_2 doesn't rotate and the hyperfine splitting of I_H around $I_N = 0$ only appeares.

3.2 The isothermal annealing experiments and the possibility of ESR dating

Isothermal annealing studies of NH_2 radical produced by γ-rays and UV were performed. An isothermal annealing curve of NH_2 follows the first order decay with the lifetime τ at the temperature T expressed as $dI/dt = (-1/\tau)I$, $1/\tau = \nu_0\exp[-E/kT]$, where I is the relative ESR signal intensity, E is the activation energy, ν_0 is the frequency factor.

The Arrhenius plot of the lifetimes, i.e., semilogarithmic plot of the intensity against the reciplocal temperature is shown in Fig.4 giving as E = 0.28±0.04 eV (6.5 kcal/mol) and ν_0 = 5.6×10^8 s^{-1} for the sample irradiated by γ-ray, and E = 0.29±0.02 eV (6.7 kcal/mol) and ν_0 = 5.7×10^8 s^{-1} for the sample produced by UV-irradiation.

The lifetime of NH_2 is about 10^6 year at about 65K, that is, the ambient temperature of a satellite around Saturn, Rhea, as shown in Fig. 4. The ESR dating of NH_2 radicals can be performed for a period of 10^6 years ago, if the radicals are not saturated. The ESR dating may be done in the next century by a sample-return program or by launching a portable ESR for remote sensing (Ikeya 1993).

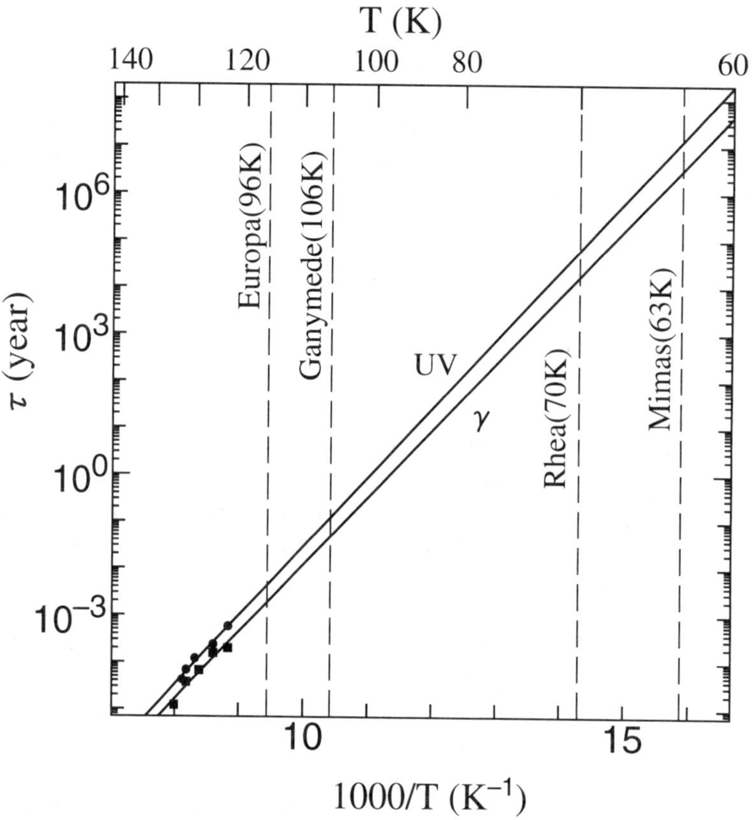

Fig. 4. The Arrhenius plot of the lifetime of NH_2 at various temperature. The liner lines give E = 0.28±0.04 eV and ν_0 = 5.6 ×10^8 s^{-1} for the sample irradiated by γ-ray, and E = 0.29±0.02 eV and ν_0 = 5.7×10^8 s^{-1} for the sample produced by UV-irradiation. The ambient temperature at satellites were indicatced in the field.

Acknowledgement

We are indebted to Mr. Ikeda for ^{60}Co irradiation performed at the Institute of Industry Science of Osaka University. We thank members of this laboratory for helpful discussion.

References

1. M. Ikeya, New Application of Electron Spin Resonance-Dating, Domistry and Microscopy (World Scientific, Singapore, 1993)
2. M. Hirai, M. Ikeya, Y. Tukamoto and C. Yamanaka, Radiolysis of solid CO_2 and future electron spin resonance dating on outer planets, *Jpn. Appl. Phys. Vol. 33* (1994) pp. **L**1453-**L**1455.
3. D. R. Smith and W. A. Seddon, electron spin resonance spectra of ^{15}N labelled amino radicals, *Canadian Lournal of Chemistry,* 48, (1970) 1938-1942.
4. B. S. Al-Naimy, P. N. Moorthy, and J. J. Weiss, An electron spin resonance study of the γ radiolysis and the Photolysis of frozen ammonia-water systems, *J. Phys. Chem.* 70, (1966) 3654-3660.
5. J.J.P. Stewart, *J. Comput. Chem.* **10** (1993) 221.
6. S.N.Foner, E. L. Cocheran, V. A. Bowers, and C. K. Jen, *electron spin resonance spectra of the NH_2 and ND_2 free radical at 4.2 K, Phys. Rev. Letters,* 1, (1958) 91-92.
7. Janusz Bednarek and Andrzej Plonka, single-crystal electron spin resonance studies on radiation-produced species in ice I_h, *J. Chem. Soc.,Faraday Trans.1,* **83** (12), (1987) 3725-3735.
8. Janusz Bednarek and Andrej Plonka, single-crystal electron spin resonance studies on radiation-produced species in ice I_h, *J. Chem. Soc.,Faraday Trans. 1,* **83** (12), (1987) 3737-3747.

Investigation on Phosphorus Porphyrin Radical by Electron Spin Echo Envelope Modulation (ESEEM) Method

Jun Yamauchi,[a] Katsuichi Kanemoto,[a] Yong Li,[a,b] and Jin L. Guo[b]

[a] Graduate School of Human and Environmental Studies, Kyoto University, Kyoto 606, Japan
[b] Department of Chemistry, Tsinghua University, Beijing 100084, China

Abstract. Electron spin echo envelope modulation (ESEEM) measurements were carried out for a dichloro(tetraphenylporphyrinato)phosphorus(V) neutral radical in toluene. The radical is assigned to be a π radical and the unpaired electron is not located on the center phosphorus ion because no modulation frequency related to the phosphorus nucleus was detected in the FT-ESEEM spectra. The modulation frequencies due to proton, nitrogen and chlorine nuclei were observed. Based on the intense peak at 2.5MHz, which is attributed to the modulation frequency of the chlorine nuclei, the isotropic hyperfine constant and quadrupole parameters of the axially coordinated chloride ions were estimated.

1 Introduction

Porphyrin complexes of almost all metals have been prepared and extensively studied. For complexes with non-metallic elements, on the other hand, the studies have been focused on phosphorus porphyrins.[1-7] In the series of phosphorus porphyrin complexes, the dichloro(5,10,15,20-tetraphenylporphinato)phosphorus(V) chloride ($[PCl_2(tpp)]Cl$) is one of the most important compounds, because $[PCl_2(tpp)]Cl$ can be converted to other P-tpp complexes with different axial ligands by substitution of the axial chloro ligands.[1-7] Therefore, the detailed investigation of $[PCl_2(tpp)]Cl$ and its related species would be useful to further studies on the structures and properties of phosphorus porphyrin complexes.

In general, radicals of metal porphyrin complexes are not stable and their life time are only several minutes or hours in air. However, Marrese and Carrano have found that by means of electrochemical reduction of $[PCl_2(tpp)]Cl$, one can easily obtain a very stable radical, $[PCl_2(tpp)]\bullet$. However, this species provides only a single-line ESR spectrum and little information can be obtained.[8] We have tried, but failed in carrying out ENDOR measurements for $[PCl_2(tpp)]\bullet$ because the ESR signal of $[PCl_2(tpp)]\bullet$ is difficult to saturate even at low temperature. In the present research, electron spin echo envelope modulation (ESEEM) measurements were carried out for $[PCl_2(tpp)]\bullet$ in toluene to obtain in more detail information on the radical.

2 Experimental

$[PCl_2(tpp)]Cl$ was prepared according to the procedure of Marrese and Carrano.[8] $[PCl_2(tpp)]$• powder was obtained from $[PCl_2(tpp)]Cl$ in acetonitrile by electrochemical reduction at a fixed potential of -0.50V (it was reported that the first reduction potential is -0.30V and the second -0.68V in acetonitrile *vs.* SCE[8]). The elemental analysis indicates $[PCl_2(tpp)]$• containing $4H_2O$, Found: P, 4.44; C, 67.22; H, 4.24; N, 7.23; Cl, 9.45. Calc.: P, 3.94; C, 67.18; H, 4.61; N, 7.12; Cl, 9.01%. $[PCl_2(tpp)]$• powder was dissolved in toluene, and the solution was degassed for ESR and ESEEM measurements.

All ESR and ESEEM measurements were performed on a JEOL PX1050 FT-ESR spectrometer at 10K. Fig.1 shows the ESR spectra of $[PCl_2(tpp)]$• in toluene. The ESEEM spectra were measured using the two-pulse echo sequence (90º(20ns)- τ - 180º(40ns)- τ -echo) or the three-pulse one (90º(30ns)- τ -90º(30ns)-T-90º(30ns)- τ - echo, τ =300ns) at the magnetic field position (314.3mT) marked with an arrow in Fig. 1 and under the microwave frequency of 8.820GHz. In order to eliminate unwanted two-pulse echoes in the three-pulse ESEEM, phase-cycling[9] was used in all the three-pulse ESEEM measurements. The data processing and the Fourier transformation (FT) were performed using ESPRIT-425 PULSE ESR SYSTEM PROGRAM of the FT-ESR instrument.

3 Results and Discussion

3.1 ESR Spectra

As previously reported,[8] $[PCl_2(tpp)]$• only showed one ESR line with g = 2.0054 and the linewidth of about 0.7mT (Fig. 1). Although all phosphorus, nitrogen, chlorine and hydrogen atoms in the radical have nuclear spins, no hyperfine structure was detected. This finding suggests that the unpaired electron is not located on the center phosphorus ion and the radical may be a π one. As a result, a hyperfine structure due to the phosphorus nucleus was not observed, whereas a hyperfine constant would become large for a phosphorus centered radical. Since the unpaired electron is delocalized on the porphyrin ring which leads to the small spin density on each nucleus, the hyperfine constants due to other nuclei would be small. In addition, their hyperfine structures are overlapped for each other, thus $[PCl_2(tpp)]$• only showed one ESR peak, and little information can obtained from ESR measurements.

3.2 FT-ESEEM Spectra

The two-pulse and three-pulse FT-ESEEM spectra are shown in Fig. 2 and 3, respectively. The peak at 13.4MHz (Fig. 2 and 3) can clearly account for hyperfine modulation frequency of protons in the porphyrin. The sum proton peak at about

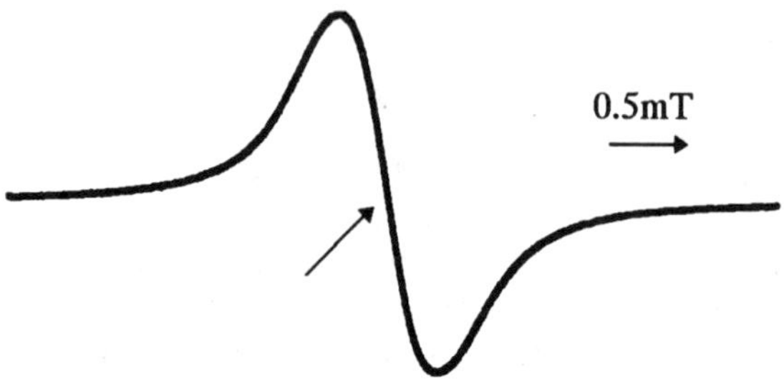

Fig. 1. The ESR spectrum observed for $[PCl_2(tpp)]^{\bullet}$ in toluene at 10K. The ESEEM measurements were carried out at the magnetic field marked by an arrow.

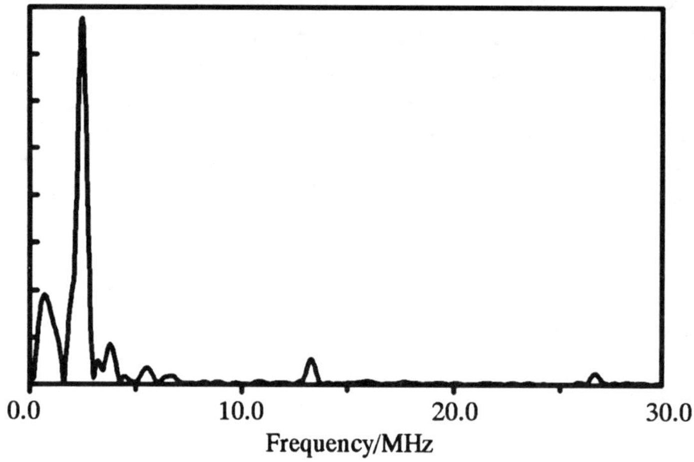

Fig. 2. The two-pulse FT-ESEEM spectrum observed for $[PCl_2(tpp)]^{\bullet}$ in toluene at 10K.

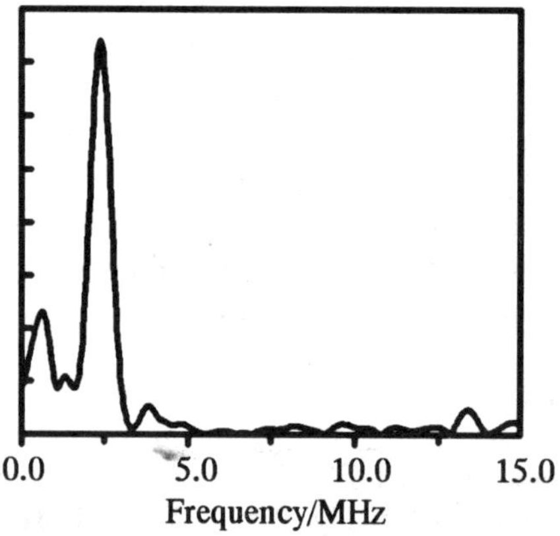

Fig. 3. The three-pulse FT-ESEEM spectrum observed for $[PCl_2(tpp)]^{\bullet}$ in toluene at 10K.

26.8MHz was also observed (Fig. 2). The Larmor frequency of ^{31}P ($I = 1/2$) at the magnetic field applied is about 5.4 MHz.[10] However, it is interesting that no frequency peak due to ^{31}P was detected at this region in the three-pulse spectrum (Fig. 3) and in the higher frequency domain in the two-pulse spectrum (Fig. 2), except for the peaks due to protons. Considering the ESR linewidth of about 0.7mT (*ca.* 19.6MHz), this observation would be explained in terms of no important hyperfine interaction between the phosphorus and the electron spin. It also means no change in the valence state of the phosphorus ion by the reduction and the added unpaired electron entered into π orbital of tpp to produce a π radical. This fact is excellent agreement with the above-mentioned assumption deduced from the ESR measurements.

In the previous work, the coupling constants for porphyrin N nuclei (I=1) in π radicals have been reported in the range from 0.07 to 0.26mT.[11-13] Within this range of the coupling, FT-ESEEM spectra calculated by E. J. Reijerse et al. for the case of S=1/2 and I=1[14] showed the complicated patterns with several peaks. However, none of the spectra exhibited such a strong one-line peak as our case. Thus, in the present study, an intense peak at 2.5MHz in Fig. 2 and 3 may be attributed to the modulation frequency of chlorine nuclei ($I = 3/2$). The Larmor frequencies (ν_n) of chlorine nuclei at the magnetic field applied are about 1MHz.[10] According to calculations for the case of $S = 1/2$ and $I = 3/2$ made by E. J. Reijerse et al.,[14] a sharp peak is observed at a quadrupole frequency

$$\nu = 3Q_{zz}(1+\eta^2/3)^{1/2} \approx 2.5\text{MHz} \tag{1}$$

when $\nu_n \approx a_{iso}/2 \approx 1.0$ MHz in the particular case that anisotropic term T_{zz} is almost equal to zero, where the quadrupole parameters Q_{zz} and η ($= |(Q_{xx}-Q_{yy})/Q_{zz}|$) have their usual meaning,[15] and it is supposed that $Q_{zz} \approx 0.8$MHz and $\eta \approx 0.5$ in this case. Furthermore, Q_{xx} and Q_{yy} values can be estimated to be -0.2MHz and -0.6MHz, respectively, according to the above values and the relation $Q_{xx} + Q_{yy} + Q_{zz} = 0$. The fact that the η value is not zero means $[PCl_2(tpp)]^\bullet$ does not have an axial symmetry. The intense frequency peak of the chlorine nuclei and the rather large isotropic hyperfine constant ($a_{iso} \approx 2.0$ MHz), compared with the ESR linewidth, indicates the chloride ions are also the axial ligands in the radical. In our previous work, the X-ray analysis of $[PCl_2(tpp)]Cl$ revealed that $[PCl_2(tpp)]^+$ does not have an axial symmetry and the two chloride ions are axially coordinated to the center phosphorus ion.[16] That is to say, although $[PCl_2(tpp)]^+$ is reduced to $[PCl_2(tpp)]^\bullet$, the structures between the two species may be not essentially different.

For the present radical, the coupling constant of the Cl atom is larger than that of the centered atom P. This fact can be explained as follows. The axially coordinated Cl atom interacts with the π orbital of tpp through not only p_z orbital but also p_x and p_y orbitals. Therefore, the spin density on the Cl atom may be larger than that on the P atom. The other peaks below 7MHz in Fig. 2 and below 4MHz in Fig. 3 are assigned to be mainly due to nitrogen nuclei of the porphyrin. Usually, patterns of FT-ESEEM spectra due to ^{14}N nuclei ($I = 1$) are different with other nuclei and very complicated.[14,17] In the present work, the peaks due to ^{14}N nuclei are also quite complicated, and it is extremely difficult to extract direct information from them about the magnitude of hyperfine and quadrupole interaction.

References

1. C. J. Carrano and M. Tsutsui, Unusual metalloporphyrine. Phosphorus complexes of tetraphenylporphine, *J. Coord. Chem.*, **7**, (1977) 79-83.
2. P. Sayer, M. Gouterman, and C. R. Connell, Porphyrins. 34. Phosphorus complexes of octaethylporphyrin, *J. Am. Chem. Soc.*, **99**, (1977) 1082-1087.
3. M. Gouterman, P. Sayer, E. Shankland, and J. P. Smith, Porphyrins. 41. Phosphorus mesoporphyrin and phthalocyanine, *Inorg. Chem.*, **20**, (1981) 87-92.
4. S. Mangani, E. F. Meyer, Jr., D. L. Cullen, M. Tsutsui, and C. J. Carrano, Crystal and molecular structure of dihydroxo(5, 10, 15, 20-trtraphenylporphinato)phosphorus(V) hydroxide dihydrate, *Inorg. Chem.*, **22**, (1983) 400-404.
5. C. A. Marrese and C. J. Carrano, Mechanism of electrochemical reduction of the (5, 10, 15, 20-tetraphenylporphinato)dihydroxophosphorus(V) cation, *Inorg. Chem.*, **23**, (1984) 3961-3968.
6. H. Segawa, K. Kunimoto, A. Nakamoto, and T. Shimidzu, Synthesis of axial dialkoxy phosphorus(V) porphyrin derivatives — novel hypervalent phosphorus compounds, *J. Chem. Soc., Perkin Trans. I*, (1992) 939-940.
7. H. Segawa, N. Nakayama, and T. Shimidzu, Electrochemical synthesis of one-dimensional donor-acceptor polymers containing oligothiophenes and phosphorus porphyrins, *J. Chem. Soc., Chem. Commun.*, (1992) 784-786.
8. C. A. Marrese and C. J. Carrano, Synthesis, characterization, and electrochemistry of (5, 10, 15, 20-trtraphenylporphinato)dichlorophosphorus(V) chloride, *Inorg. Chem.*, **22**, (1983) 1858-1862.
9. J.-M. Fauth, A. Schweiger, L. Braunschweiler, J. Forrer, and R. R. Ernst, Elimination of unwanted echoes and reduction of dead time in three-pulse electron spin-echo spectroscopy, *J. Magn. Reson.*, **66**, (1986) 74-85.
10. R. C. Weart and M. J. Astle (eds), *CRC Handbook of Chemistry and Physics*, 62nd edition, CRC Press, Florida (1981) E67-69.
11. R. H. Felton and H. Linschitz,, Polarographic Reduction of Porphyrins and Electron Spin Resonance of Porphyrin Anions, *J. Am. Chem. Soc.*, **88**, (1966) 1113-1116.
12. N. S. Hush and J. R. Rowlands, Hyperfine Structure in the Electron Spin Resonance Spectra of Reduced Porphyrins, *J. Am. Chem. Soc.*, **89**, (1967) 2976-2979.
13. J. Fajer, D. Borg, A. Forman, D. Dolphin, and R. Felton, π-Cation Radicals and Dications of Metalloporphyrins, *J. Am. Chem. Soc.*, **92**, (1970) 3451-3459.
14. E. J. Reijerse and C. P. Keijzers, Model calculations of frequency-domain ESEEM spectra of disordered systems, *J. Magn. Reson.*, **71**, (1987) 83-96.
15. C. P. Poole, Jr. and H. A. Farach, *Theory of Magnetic Resonance*, 2nd edition, John Wiley & Sons, New York (1987) pp103-105.
16. J. L. Guo, F. Sun, Y. Li, and N. Azuma, Molecular and crystal structure of dichloro(5, 10, 15, 20-trtraphenylporphinato)phosphorus(V) chloride dichloromethane solvate, *Polyhedron*, **14**, (1995) 1471-1476.
17. S. A. Dikanov and Yu. D. Tsvetkov, *Electron Spin Echo Envelope modulation (ESEEM) Spectroscopy*, CRC Press, Florida (1992) pp187-208.

Mn^{+2} ESR Signals of Coral Reefs in the South China Sea: Implcations for Paleoclimate Indication

Yu-Guang Ye, Shi-Guang Zhou and Xin-Bo Liu

Institute of Marine Geology, Ministry of Geology and Mineral Resources, Qingdao, 266071, China

Abstract. This paper focuses on ESR signals of Mn^{+2} in the Holocene coral reef sequence from Well Sanya-3 drilled on Hainan Island. The normalized ESR signal based on 100% aragonite shows a very similar changing pattern to the Holocene paleoclimatic fluctuations of China revealed previously by other scholars. The reflection of ESR signals of Mn^{+2} in coral reefs to paleoclimatic changes is preliminarily interpreted as climatic control over the input of runoff-carried terrigenous materials into seawater in which coral reefs developed. Our research suggests that ESR signals of Mn^{+2} in coral reefs promise to be a potential proxy index of paleoclimate.

1 Introduction

Among various proxy indexes of paleoclimate, trace elements and oxygen isotope are most commonly used. Since Emiliani (1955) initiated the oxygen-isotope stratigraphy, research works on the relationship between $\delta^{18}O$ and paleoclimate and on its mechanism have been done quantitatively, and some theories have been established and widely applied. The oxygen-isotope stratigraphy, however, is limited to be applied to coral reef sequences, because the changing proportions of various biospecies in the analyzed whole-rock samples and post-sedimentation diageneses taking place in coral reefs greatly affect indication of oxygen isotope to paleoclimate. Is it likely to find out a new proxy paleoclimatic index with higher sensitivity? This paper aims to evaluate a potential paleoclimatic indicator: ESR signals of Mn^{+2} in coral reefs.

2 Mn and Paleoclimate

Mn in seawater occurs dominantly in the form of ions. It is the only trace element which is never effectively removed from solutions by precipitation, so it moves extensively. A large part of Mn in seawater derives from terrigenous materials carried into the sea by various agents whose change in energy is closely related to paleoclimate. In addition, such parameters as salinity and volume of seawater as well as the growth and development of oceanic organisms, which have a great deal to do with climate, bear a tie with Mn content in coral reefs.

Recently several researchers have revealed the change of Mn content during geological times. In studies on paleooceanography of specific areas in west Pacific Ocean since 150 ka BP, Cang et al. (1993) found Mn abnormality in sediments at the boundary between the oxygen isotope stage 1 and stage 2 (Holocene/ Late Pleistocene) and at the boundary between the oxygen isotope stage 3 and stage 4 (Wurm II/ Wurm III), that is, Mn content jumps when it come to the junction from cold periods to warm periods. Therefore the authors suggested that the Mn abnormality resulted from the rapid change of climate from cold to warm and may be regarded as an indicator of dramatic change in climate over a small region. Guo et al. (1993) also reported fluctuations of Mn content in Core Nanyong-1 (152 m long). They found the high-low alternation of Mn content in the core is correlated with the paleoclimatic cycles from Luochuan loess sequence, deep-sea oxygen isotope stages and organic-carbon stages in Well CK-6 from Caidamu Basin. Besides, Glen et al. (1991) studied the changes of Mn content in corals during different seasons.

3 ESR Signals of Mn^{+2} in Coral Reefs

Mn^{+2} ions in carbonate crystals are paramagnetic and thus among the most suitable matters to be tested by ESR technique under which the six spectrum lines with superhyperfine structure of Mn^{+2} ions are very easy to identify. Mn^{+2} content in coral reefs is relatively low, generally tens of ppm, but well falls in the testing range (0.1-0.0001%) of the ESR spectrometer. The determined concentration of Mn^{+2} in coral reefs by chemical analyses is the total content, including not only the Mn^{+2} ions from carbonate crystals but also those in sulphate, organic complex and suspended load with manganiferous fragments, while the Mn^{+2} content obtained by ESR method only includes Mn^{+2} ions in carbonate crystals. Moreover, ESR signals of Mn^{+2} in carbonates are very stable and independent of chronology. Based on above-noted reasons, it appears to us that the intensity of ESR signals of Mn^{+2} in coral reefs reflects the nature of Mn^{+2} ions in carbonates.

4 Experiments and Results

Fifty-eight samples were taken from Well Sanya-3 , which was drilled on Hainan Island. The well is about 15 m deep, the part from 0 to 13.9 m is coral reefs and below 13.9 m terrigenous materials and basement rock. Experiments involved in this paper were carried out in the Experiment-Test Center of Marine Geology, MGMR, Qingdao.

ESR spectra were measured with ECS-106 ESR spectrometer made by BRUKER Company, Germany. The measurement was under the following conditions: room temperature, microwave power 2 mW, magnetical field 348±10 mT, modulation amplitude 0.1 mT, transform time 5.12 ms, and time constant 327.68 ms. For comparison, grain sizes of all samples were selected in the range between 0.15 and 0.2 mm. All samples were measured three times consecutively, and the average of three data was taken as the final measuring value. Typical ESR spectrum of samples is shown in Fig. 1. with clear peaks of Mn 3 and Mn 4 located respectively on the left and right of central magnetical field. The ESR result is given in Fig. 2a.

Mn and Fe content of samples, measured with JY38S ICP-AES made by JION YVON Company of France, is shown in Fig. 2b. and Fig. 2c. Mineral components of samples were determined with DMax-RA X-ray diffractometer made by RIGAKU Company of Japan. Concentrations of aragonite and low-Mg calcite are shown in Fig. 2d. and Fig. 2e. respectively.

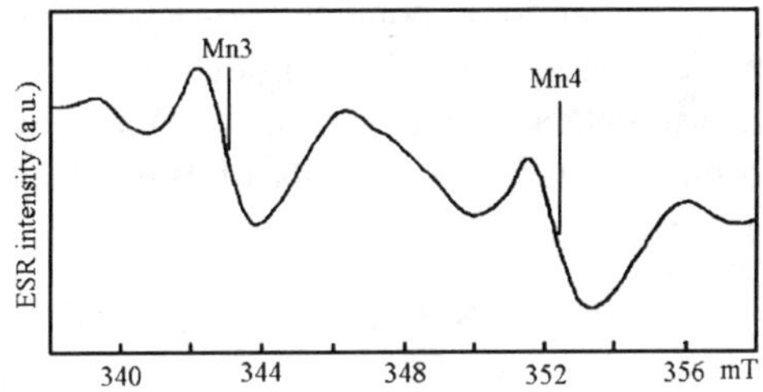

Fig. 1. A Typical ESR Spectrum of the Sample

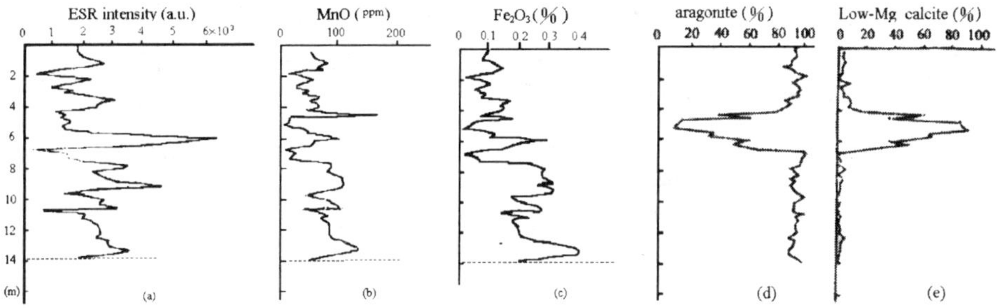

Fig. 2. ESR intensity of Mn^{+2}(a), MnO(b), Fe_2O_3(c), Aragonite(d), Low-Mg calcite(e) change with the depth

5 Discussion and Conclusion

Fig. 2. indicates that in Well Sanya-3 the variation of ESR signals of Mn^{+2} is in rough agreement with that of Mn and Fe content, except at 4.5 m depth where chemical Mn content is much higher than the ESR signal. This suggests that as a whole, Mn determined by chemical analysis is chiefly from Mn^{+2} incorporated into crystals of carbonate and far less from other components.

Fig. 2d. and Fig. 2e. show that the coral reef sequence is composed largely of aragonite. However, an obvious increase in low-Mg calcite from 4 to 7 m suggests a meteoric diagnesis which altered the chemically unstable aragonite into stable low-Mg calcite.

Fig. 2. also demonstrates that the ESR signal of Mn^{+2} is positively correlated to the calcite content but inversely to aragonite content. This can be explained by the fact that the radius of Ca^{+2} ions in aragonite is different from that in calcite. Aragonite is of rhombic system with nine coordination numbers, and the Ca^{+2} ion (0.126 mm) is bigger than that (0.106 mm) in calcite, which is of trigonal system with six coordination numbers. Mn^{+2} ions, 0.091 mm in radius and approximately equal to Ca^{+2} in calcite, are easily replaced with Ca^{+2} in calcite and to enter into the lattice of calcite.

Because of a change in mineral components between 4 and 7 m in the well, we can't judge whether the ESR signal of Mn^{+2} in the same section is attributable to the mineral changes or to paleoclimatic change. To eliminate the effect of mineral components, we normalized the ESR signals of Well Sanya-3 based on 100% aragonite (Fig. 3.).

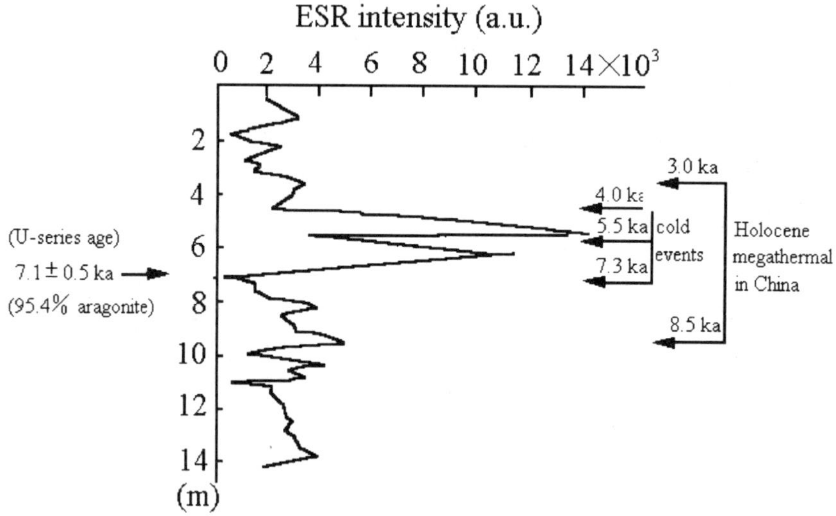

Fig.3 ESR intensity of Mn^{+2} change with depth on 100% aragonite

Research on the U-series age of Well Sanya-3 (Ye et al., 1991) indicated that the reef sequence was formed primarily in the Holocene, and a sample at 7.1 m depth, which was formed in a chemically closed system (containing 95.47% aragonite) yielded an U-series age of 7.1±0.5 ka. We may roughly consider every meter of the reef sequence as to span of 1 ka.

According to abundant data of palynology, ice cores, paleolimnology, paleopedology, archaeology and coastal changes, Shi and Kong (1992) argued that the warmest period in Holocene in China was between 8.5 and 3.0 ka BP, in which Fig.3 shows the highest ESR signals. After studying fossil assemblage of Holocene mammals in China, Ji (1996) pointed out an evident southward shift of mammals during 7.6 to 6.0 ka, suggesting a warmer climate at that time. He also found a complex assemblage mixed with mammal types living in the south and north during 5.6 to 3.0 ka, indicating a period characteristic of violent climatic fluctuations. Additionally, a short-term drop in temperature around 5.5 ka BP was revealed too by Ji (1996), after which time the temperature rose rapidly. This climatic fluctuations reported by Ji are consistent with the change of ESR signals in Fig.3. Furthermore, the cold events at 7.3 ka and 4.0 ka clearly correspond to the decrease of ESR signals in Fig.3. Therefore we think that the ESR signals revealed from Well Sanya-3 reliably reflect the Holocene climatic changes.

The mechanism that ESR signals of Mn^{+2} or Mn content in coral reefs recorded paleoclimatic changes may be tentatively interpreted as climatic control over the input of runoff-carried terrigenous materials into the sea. During warm periods, more terrigenous matter was carried into the sea by runoffs , increasing the concentration of Mn^{+2} in seawater and thus more Mn^{+2} being incorporated into lattices of coral reefs. But during cold times, the case is reversed.

Though ESR signals of Mn^{+2} and Mn content in coral reefs possibly both reflect paleoclimatic changes, the former is more sensitive and reliable than the latter. For example, the short-term lowering of temperature around 5.5 ka can't be seen in Fig.2a but is clear in Fig. 3. The normalized ESR signals of Mn^{+2} shows a changing pattern consistent with that of the previously-reported paleoclimatic change during the Holocene, and can thus be regarded as a new proxy paleoclimatic index.

Acknowledgments

We express thanks to professor Ming-shu Zhang for providing us the samples of Well Sanya-3 and are grateful for the financial support provided by the Chinese Foundation for Development of Geological Sciences and Techniques (No.9507).

References

1. Cang shuxi, Yan Jun, Li Tiegang and Xue Shenggu, Paleo-Oceanographic Researches for a Certain Area of the West Pacific Ocean since 0.15 Ma .B.P. Marine Geology & Quaternar Geology (1993), 13(3), 97-101.
2. Emiliani .C, Pleistocene Temperatures, J .Geol. (1955), 63, 538-578.
3. Glen T.S., Todd M .C., Robert B.D., Gerard M.W., Mittcell W.C.and Peter W.G., Paleochemistry of Manganese in Corals from the Galapagos Islands. Coral Reefs (1991), 10, 91-100.
4. Guo Lifen, Chen Wanyan and Chen Lihong, Palaoclimatic Variation of Yongshu Reef in Nansha Island over the past Million Years. Tripic Oceanology. (1993), 12 (4), 39-46.
5. Ji Hongxiang , Paleoclimatic Fluctuations and Mammals in Holocene Megathermal in China, Marine Geology & Quaternary Geology, (1996),16(1),5-16.
6. Shi Yafen and Kong Zhaochen (eds), The Climatic and Environments of Holocene Megathermal in China, China Ocean Press, Beijing (1992).
7. Ye Yuguang, He Jie, Diao Shaobo et al., ESR and Uranium Series Ages of Holocene Coral Reefs in the South China Sea, Geological Review, (1991),37,165-171.

Determination of a Weak Exchange Interaction in $[KAs_4W_{40}Cu_2(H_2O)_2O_{140}]^{23-}$ by Single Crystal EPR Spectroscopy

Young Hwan Cho, Hyunsoo So,* and Michael T. Pope†*

*Department of Chemistry, Sogang University, Seoul 121-742, Korea
† Department of Chemistry, Georgetown University, Washington, DC 20057, U.S.A.

Abstract. Single crystal EPR spectra of $(NH_4)_{23}[KAs_4W_{40}Cu_2(H_2O)_2O_{140}]\cdot nH_2O$ exhibit an orientation-dependent fine structure which may be attributed to the exchange and magnetic dipole interactions between the two Cu^{2+} ions. Single crystal and polycrystalline EPR spectra were simulated by solving 64 x 64 energy matrices. The resulting parameters are $g_{\parallel} = 2.437$, $g_{\perp} = 2.082$, $A_{\parallel} = 0.0084$ cm^{-1}, $A_{\perp} = 0.0023$ cm^{-1}, $Q' = 0.0009$ cm^{-1}, $J = 0.0060$ cm^{-1}, and the Cu...Cu distance = 6.7 Å. The metal separation based on the point-dipole model is 0.3 Å larger than that in the dicobaltate analogue. The magnitude of J suggests that the unpaired electrons on two Cu^{2+} ions interact directly and/or through a sequence of ten bonds involving four tungsten atoms and five oxygen atoms.

1 Introduction

Dimeric transition metal ion complexes have attracted the continuing attention of chemists. For some systems electron paramagnetic resonance (EPR) studies have provided remarkable insight into the nature of the interaction between the two metal ions. The EPR spectrum of the binuclear copper(II) acetate, first studied by Bleaney and Bowers in 1952,[1] has revealed that both exchange and dipolar interactions are important for this complex, in which the Cu...Cu separation is 2.6 Å.

The dipolar interaction is determined by the interspin distance and the angles between the metal-metal vector and the g axes of the metal ions. It would be very useful to deduce structural information from EPR spectra of dimeric systems which have unknown structure. Smith *et al.* derived perturbation expressions for the spin Hamiltonian of a dimeric system, and used them to deduce the metal separations for a number of dimers of copper(II), oxovanadium(IV), and titanium(III).[2, 3] However, general perturbation expressions could not be derived for the system in which the principal g axes of the two metal ions are not parallel. It was found later that approximate perturbation expressions could be used for such systems when the exchange interaction was sufficiently small.[4]

More recently Eaton *et al.* have used these perturbation expressions to analyze frozen solution and single crystal EPR spectra of spin-labeled transition metal complexes, in which one unpaired electron is on a transition metal ion and the second unpaired electron is on a nitroxyl radical.[5,6] They determined the isotropic exchange coupling constants, J, ranging from -35 to +10 x 10^{-4} cm^{-1} and the interspin distances ranging from 9.2 to 15.5 Å for a series of spin labeled copper(II), vanadyl(IV), and silver(II) porphyrins.[7-9] However, the interspin distances could not be determined by X-ray crystallography, because these compounds were doped into single crystals of zinc tetraphenylporphyrin.

So we have tried to find a binuclear metal system which exhibits good single crystal EPR spectra and structural information of which can be obtained from X-ray crystallography. In order to get well-resolved EPR spectra, one has to dilute the paramagnetic species in a diamagnetic host crystal. It is not so easy to find isomorphic host crystals for binuclear metal complexes. Fortunately, paramagnetic polyoxometalates exhibit EPR spectra with hyperfine structures even in the concentrated single crystals, for the large size of the polyanions weakens the intermolecular dipolar interaction.

Binuclear compounds can be prepared using the large heteropolytungstate anion $[As_4W_{40}O_{140}]^{28-}$ as a ligand. This anion, consisting of four AsW_9O_{33} subunits and joined by four extra WO_6 octahedra, provides two coordination sites for various transition metal ions.[10] The structure of the dicobaltate, $(NH_4)_{23}[(NH_4)As_4W_{40}Co_2(H_2O)_2O_{140}]\cdot nH_2O$ (**1**) has been determined;[11,12] see Figure 1. Each Co^{2+} ion in this compound is coordinated by four oxygen atoms of the heteropolyanion forming a plane, and has an aqua ligand in an axial position. The anion belongs to the C_{2v} point group, and the two CoO_5 groups are related by a mirror plane. The Co...Co vector and the Co-OH_2 vectors lie in the second mirror plane, with the angle between the vectors being 33°. The Co...Co distance is 6.4 Å.

In this paper we report the single crystal EPR spectra of the dicuprate(II), $(NH_4)_{23}[KAs_4W_{40}Cu_2(H_2O)_2O_{140}]\cdot nH_2O$ (**2**). The powder spectrum of $(NH_4)_{23}[(NH_4)As_4W_{40}Cu_2(H_2O)_2O_{140}]$ was reported previously by Kokoszka *et al.* without detailed analysis.[13]

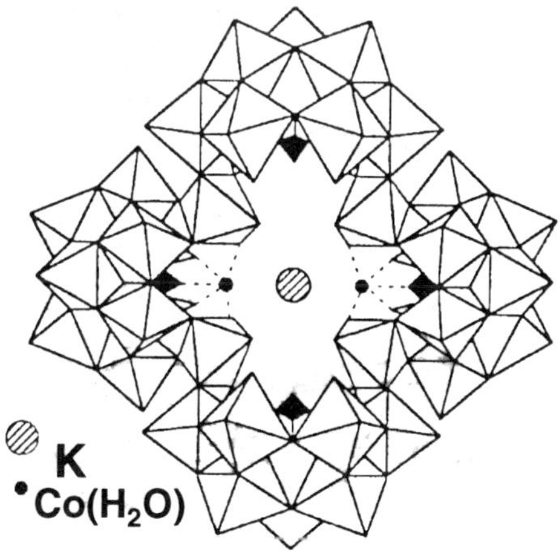

Fig. 1. Polyhedral representation of $[(NH_4)As_4W_{40}O_{140}Co_2(H_2O)_2]^{23-}$. After [12]

2 Experimental Section

$(NH_4)_{23}[KAs_4W_{40}Cu_2(H_2O)_2O_{140}] \cdot nH_2O$ was synthesized by the following method. Aqueous 0.5 M $CuSO_4$ (10.2 mL, 5.1 mmol) was added with stirring to a solution of 30 g (2.5 mmol) of the potassium salt of 40-tungsto-4-arsenate dissolved in 200 mL water. After 10 m., 10 g of solid ammonium nitrate was added. The precipitate was collected on a filter paper, redissolved in 300 mL water and reprecipitated with 20 g NH_4NO_3. This operation was performed five times in order to obtain a salt free of potassium ions except the encrypted one. Single crystals were grown by slowly evaporating an aqueous solution of **2** in a refrigerator. The powder EPR spectrum agreed with the reported one.

A plate-shaped crystal of **2** was mounted on a quartz rod attached to a goniometer, and spectra were recorded every 5° with the magnetic field in three orthogonal planes. The microwave frequency was measured by an Anritsu frequency counter and DPPH was used as a *g* marker. The magnetic susceptibility measurements at 5 - 300 K were performed using a SQUID magnetometer.

3 Results and Discussion

3.1 Single Crystal Spectra

When the magnetic field was in the plane (Plane 1) perpendicular to the long axis of the crystal, the spectrum exhibited one pair of four lines partially overlapped. This indicates that the molecular C_2 axis is parallel to the long axis of the crystal. When the rotation axis of the goniometer was perpendicular to the long axis of the crystal and parallel to the largest face of the crystal, two pairs of four lines were observed indicating that orientations of the two copper ions are not equivalent with respect to the magnetic field.

There are two copper ions, Cu(1) and Cu(2), the principal axes of which do not coincide (Figure 2). Each copper (I = 3/2) ion has eight electron-nuclear spin states, and it is necessary to consider 64 spin states to interpret EPR spectra of a binuclear copper system. Since perturbation theory cannot be used in general for this kind of system, 64 x 64 matrices were numerically solved to get accurate energy levels and transition probabilities.

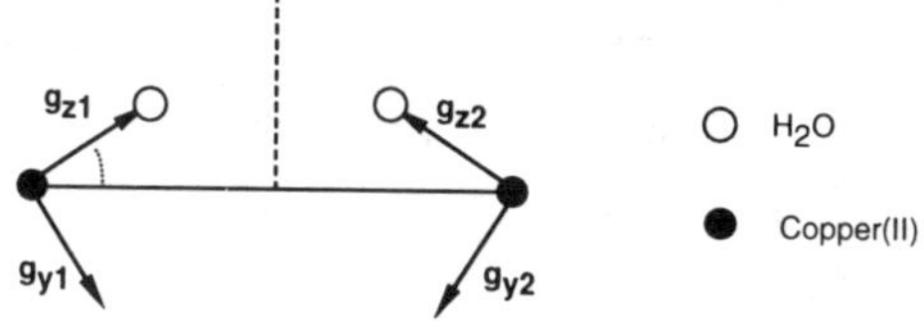

Fig. 2. Orientations of principal *g* axes in two $CuO_4(H_2O)$ groups. Two g_x axes are perpendicular to the yz plane.

The system can be described by the following spin Hamiltonian:

$$\hat{H} = \sum_{i=1}^{2} (\mu_B \mathbf{B} \bullet \mathbf{g}_i \bullet \mathbf{S}_i + \mathbf{S}_i \bullet \mathbf{A}_i \bullet \mathbf{I}_i + \mathbf{I}_i \bullet \mathbf{P}_i \bullet \mathbf{I}_i - g_N \mu_N \mathbf{B} \bullet \mathbf{I}_i) - J\mathrm{S}_1 \bullet \mathrm{S}_2 + \sum_{i,j = x_1,y_1,z_1} J_{ij}\, \mathrm{S}_{1i}\, \mathrm{S}_{2j}$$

where

$$J_{ij} = g_{1i} \{\sum_{k = x_2,y_2,z_2} g_{2k}\, d_{jk}\, (d_{ik} - 3\sigma_i \sum_{l = x_1,y_1,z_1} d_{lk}\, \sigma_l)\} \mu_B^{\,2} / r^3$$

$\mathbf{g}_i$, $\mathbf{A}_i$, and $\mathbf{P}_i$ represent g, hyperfine coupling, and nuclear quadrupole coupling tensors, respectively. J is the isotropic exchange coupling constant, and J_{ij} are components of the dipolar coupling tensor. The expression for J_{ij} is based on the point-dipole model.[2] Here r is the Cu...Cu distance, d_{jk} etc. are direction cosines connecting the 1 and 2 coordinate frames, and σ_i's are direction cosines of the Cu...Cu vector with respect to x_1, y_1, and z_1. Since preliminary calculations had revealed axial symmetry for the g and A tensors for each electron, the axial quadrupole coupling term $Q'[I_z^2 - I(I+1)/3]$ was used. Other symbols have their usual meanings.

It was assumed that for each electron the g and A tensors were collinear. Since the two copper ions are equivalent, their EPR parameters are the same; only the orientations of their principal axes are different. The principal axes of Cu(2) were expressed in terms of those of Cu(1), and a 64 x 64 energy matrix was set up. The matrix was diagonalized numerically for every 0.5 mT in the range 260 - 360 mT, and energy levels and transition probabilities were calculated. Then the positions of the EPR lines in the interval of 0.5 mT were determined by using $h\Delta\nu = g\mu_B\Delta B$.

By simulating the spectra with the magnetic field in three mutually orthogonal planes, the following parameters were determined: $g_x = g_y = 2.082$, $g_z = 2.437$, $A_x = A_y = (-)0.0023\ \mathrm{cm}^{-1}$, $A_z = (-)\ 0.0084\ \mathrm{cm}^{-1}$, $Q' = 0.0009\ \mathrm{cm}^{-1}$, $J = 0.0060\ \mathrm{cm}^{-1}$, $r = 6.7$ Å, and $\alpha = 33°$. (α is the angle between the Cu...Cu vector and a g_z axis.) Copper consists of two isotopes (^{63}Cu 69.17 %, ^{65}Cu 30.83 %), which have slightly different nuclear g factors and nuclear quadrupole moments. Since the hyperfine lines from the two isotopes were not resolved, the simulation did not include both isotopes. Thus the resulting hyperfine coupling and quadrupole coupling constants may be the weighted mean values due to the two isotopes.

The largest g value in Plane 1 is expected when the magnetic field is along the Cu...Cu vector. The EPR spectrum for this orientation consists of two sets of four hyperfine lines separated by exchange and dipolar interactions; see Figure 3a. The positions of the outer lines can be simulated by using different combinations of J and r. But a correct intensity pattern is obtained only when $J = 0.0060\ \mathrm{cm}^{-1}$ (Figure 3b). Figures 3c and 3d show that intensity patterns are quite different especially in the center of the spectrum when J = 0 or $-0.0060\ \mathrm{cm}^{-1}$. The sensitivity of the intensity pattern to J has allowed us to determine an accurate J value.

The smallest g value in Plane 1 is expected when the magnetic field is perpendicular to the Cu...Cu vector or parallel to the g_x axis in Figure 2. Measured and simulated spectra for this perpendicular orientation are shown in Figure 4.

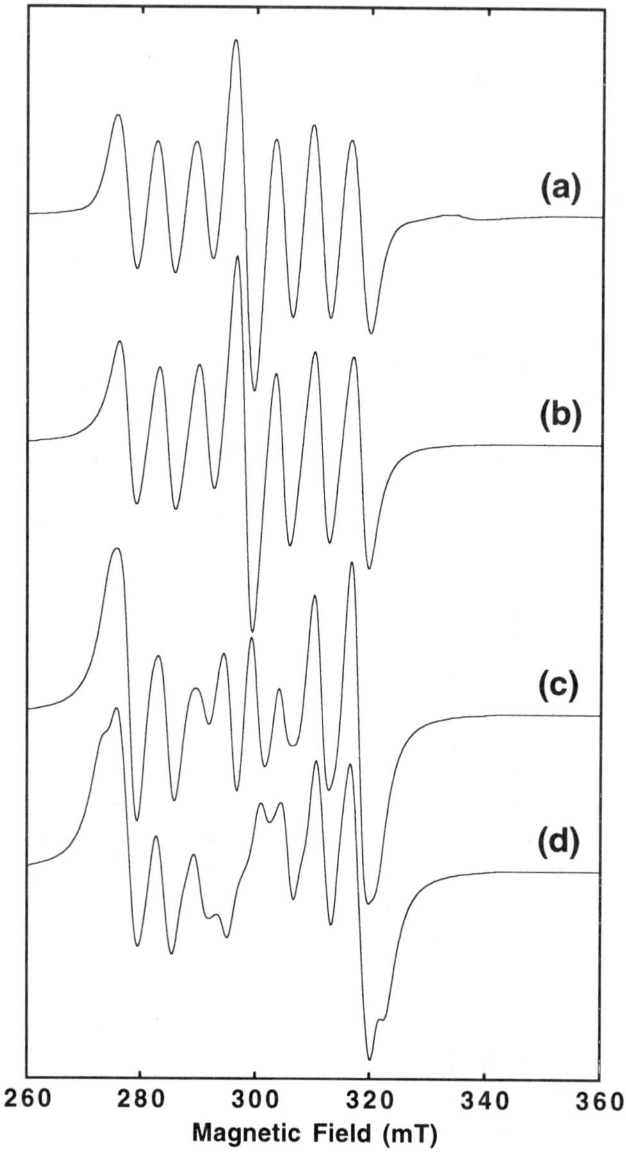

Fig. 3. Single crystal EPR spectra with the magnetic field parallel to the Cu...Cu vector. (a) Measured at room temperature; simulated using (b) $J = 0.0060$ cm^{-1} and $r = 6.7$ Å, (c) $J = 0$ and $r = 6.3$ Å, and (d) $J = -0.0060$ cm^{-1} and $r = 6.07$ Å. See the text for other parameters used.

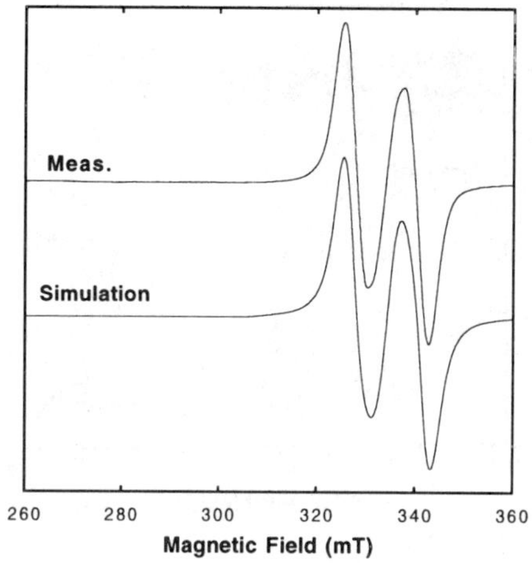

Fig. 4. Measured and simulated EPR spectra with the magnetic field parallel to the g_x axis in Plane 1.

3.2 The Polycrystalline Spectrum

The powder spectrum measured at room temperature is shown in Figure 5a. The simulated spectrum using the parameters determined from the single crystal spectra agrees satisfactorily with the experimental spectrum (Figure 5b). The six parallel lines (actually two overlapping sets of four lines) correspond to the spectrum with the magnetic field parallel to a g_z axis. However, the parallel portion of the spectrum could be simulated reasonably by using various combinations of J, r, and α. So one might find it difficult to determine an accurate J value when only polycrystalline spectrum is available without any structural information.

3.3 Perturbation Calculations

We have examined the reliability of perturbation calculations in simulating the EPR spectra for this system. The simulated spectrum using perturbation expressions agrees satisfactorily with the measured one when the magnetic field is not far from the g_z axis. Thus the parallel portion of the powder spectrum and the single crystal spectrum with the magnetic field parallel to the Cu...Cu vector (Figure 3) could be simulated using perturbation expressions. However, the fit deteriorated considerably when the magnetic field was near the $g_\perp$ direction.

It is of interest to find out how small J should be in order that the approximate perturbation expressions can be used. It has been found that the simulated X-band spectrum using perturbation expressions deviates significantly from the exact calculation even when the magnetic field is parallel to the Cu...Cu vector, when $|J|$ is larger than 0.2 cm^{-1}.

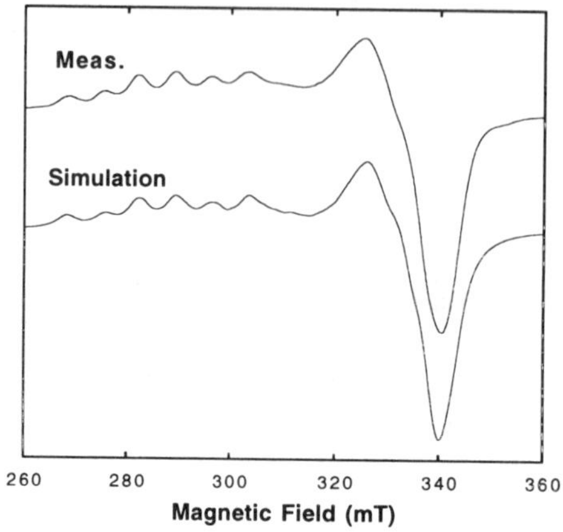

Fig. 5. (a) Measured and (b) simulated powder spectra at room temperature.

3.4 The Exchange Interaction

The most interesting result of this study is determination of a weak ferromagnetic interaction between the two copper ions separated by *ca.* 6 Å. The J value (0.0060 cm^{-1}) is too small to be detected by magnetic susceptibility measurements. Measured magnetic susceptibilities for **2** in the 5 - 300 K temperature range could be interpreted to follow the Curie law expected for two noncoupled copper(II) ions.

Recently Hoffmann *et al.* reviewed weak exchange interactions ($|J| < 0.1$ cm^{-1}) determined by single crystal EPR spectroscopy for various copper compounds.[14] Since hyperfine lines were rarely observed in concentrated single crystals, most J values were determined by using the merging effect. They analyzed the measured J value in terms of two competing contributions, a ferromagnetic contribution, J_F, and an antiferromagnetic one, J_{AF}. It was suggested that J_F was temperature-independent and that $|J_{AF}|$ increased on cooling.

In order to determine the temperature-dependence of J, we have analyzed the single crystal EPR spectra with the magnetic field parallel to the Cu...Cu vector at 300 - 95 K. Some EPR parameters for the 95 K spectrum ($J = 0.0043$ cm^{-1}, $r = 6.63$ Å, and $A_z = 0.0093$ cm^{-1}) are quite different from those of the 298 K spectrum. The J value is found to decrease on cooling, indicating that $|J_{AF}|$ indeed increases on cooling.

The unpaired electron occupies the $3d_{x2-y2}$ orbital in the Cu^{2+} ion, and the two unpaired electrons can interact via three different pathways: (1) a direct overlap between the two $3d_{x2-y2}$ orbitals, (2) through ten bonds between the two Cu^{2+} ions, and/or (3) through a more direct route (four bonds) involving the encrypted potassium cation. A J value (0.005 cm^{-1}) comparable to that of **2** was reported for two copper(II) atoms separated by 6.6 Å and having no atomic bridge between them.[15] So the direct interaction cannot be ruled out. The long pathway consists of ten bonds involving four tungsten atoms and five oxygen atoms. For a long-range exchange interaction, the number (and types) of bonds between paramagnetic centers should be important factors in determining the magnitude of J. Long-range intramolecular exchange interactions have been observed in spin-labeled transition metal complexes.[7-9] Some copper and oxovanadium compounds exhibit such interactions through ten bonds with $J = 0.004\ cm^{-1}$, which is comparable to that of **2**. In order to find out the effect of the encrypted cation, we have measured the EPR spectra of the heteropolyanion having NH_4^+ or Ba^{2+} as the encrypted cation. No significant change has been found in their EPR spectra. Therefore the short superexchange pathway may be ruled out. So probably both the direct overlap of the two d_{x2-y2} orbitals and the long pathway involving ten bonds contribute to the exchange interaction in **2**.

We have also found that the exchange interaction for the oxovanadium analogue of **2**, $[KAs_4W_{40}(VO)_2O_{140}]^{23-}$ (**3**), is antiferromagnetic and its J value ($-0.0730\ cm^{-1}$) is much larger than that of **2**. Since the direct interactions in **2** and **3** should be similar, the superexchange pathways must be more important in **3** than in **2**. More work is needed to elucidate the role of the superexchange pathways in the exchange interactions for these polyoxometalates.

Acknowledgment

This research has been supported by the Korean Ministry of Education (BSRI-95-3411). We thank Prof. Y. Jeannin for allowing us to reproduce one of his drawings.

References

1. A. Bleaney and K. D. Bowers, *Proc. Roy. Soc. (London)* **A214** (1952) 451.
2. T. D. Smith and J. R. Pilbrow, *Coord. Chem. Rev.* **13** (1974) 173.
3. S. G. Carr, T. D. Smith, and J. R. Pilbrow, *J. Chem. Soc. Faraday II* (1974) 497.
4. J. F. Boas, P. R. Hicks, J. R. Pilbrow, and T. D. Smith, T. D. *J. Chem. Soc. Faraday Trans.* **74** (1978) 417.
5. G. R. Eaton, and S. S. Eaton, *Accts. Chem. Res.* **21** (1988) 107.
6. S. S. Eaton, K. M. More, B. M. Sawant, P. M. Boymel, and G. R. Eaton, *J. Magn. Reson.* **52** (1983) 435.
7. R. Damoder, K. M. More, G. R. Eaton, and S. S. Eaton, *J. Am. Chem. Soc.* **105** (1983) 2147.
8. R. Damoder, K. M. More, G. R. Eaton, and S. S. Eaton, *Inorg. Chem.* **22** (1983) 2836.
9. R. Damoder, K. M. More, G. R. Eaton, and S. S. Eaton, *Inorg. Chem.* **22** (1983) 3738.
10. M. Leyrie and G. Hervé, *Nouv. J. Chim.* **2** (1978) 233.
11. F. Robert, M. Leyrie, G. Hervé, A. Tézé, and Y. Jeannin, *Inorg. Chem.* **19** (1980) 1746.
12. Y. Jeannin, G. Hervé, and A. Proust, *Inorg. Chim. Acta* **198-200** (1992) 319.
13. G. Kokoszka, F. Padula, and A. R. Siedle in *Biological & Inorganic Copper Chemistry,* edited by K. D. Karlin and J. Zubieta, Adenine Press (1985) p. 209.
14. K. Hoffmann, W. Hilczer, and J. Goslar, *Appl. Magn. Reson.* **7** (1994) 289.
15. C. Balagopalakrishna, M. V. Rajasekharan, S. Bott, J. L. Atwood, and B. L. Ramakrishna, *Inorg. Chem.* **31** (1992) 2843.

EPR Study of Weak Exchange in Some Copper(II) Metallomesogens

Venkatraman Sujatha[a], *B.K.Sadashiva*[b], *Kazuchika Ohta*[c], *Monisha Bose*[d] *and Sankaran Subramanian*[a*]

a. Regional Sophisticated Instrumentation Centre, Indian Institute of Technology, Madras - 600 036, INDIA.
b. Raman Research Institute, Bangalore -560 080, INDIA.
c. Department of Functional Polymer Science, Shinshu University, Ueda, Nagano, 386 JAPAN.
d. S.N.Bose National Centre For Basic Sciences, Calcutta, INDIA.
*. author for correspondence.

Abstract. Polycrystalline samples of the disc-like complexes namely, bis[(1-(p-n-octylphenyl)3-(p-n-octyloxyphenyl)propane-1,3-dionato]copper(II), bis[1,3-di(p-n-heptylphenyl)propane-1,3-dionato]copper(II), abbreviated as C_8OCuC_8 and C_7CuC_7 respectively have been studied by EPR. These compounds yield a rich variety of multiline spectra in the low field parallel region. The observation of singlet-triplet forbidden EPR transitions in these complexes allow one to determine exchange parameters. The spectra have been simulated on the basis of a weak isotropic exchange interaction. The observed EPR spectra are discussed on the basis of crystal structure packing.

1 Introduction

There has been extensive research on the synthesis and characterization of metallomesogens (Ohta et al. 1986; Sadashiva et al. 1991) However, EPR studies of only a few compounds are reported so far (Eastman et al. 1987; Martinez et al. 1995). These compounds are important for the EPR study on the basis of their one and two dimensional stacking which leads to exchange interaction, with the dimensionality depending on the lattice structure.

The appearance of an EPR spectrum of undiluted compounds are modulated by (i) nuclear hyperfine interaction, (Abragam and Bleaney, 1970), (ii) formation of polymeric units (typical example of this, is the dimer $[\{Cu(O_2CMe)_2(H_2O)_2\}]$ (Abe et al, 1953), where seven hyperfine lines from two equivalent nuclei are resolved, (iii) the electron-electron dipolar interaction which dominates the hyperfine term in the usual case of 3-5 Å separation between paramagnetic sites (Abragam and Bleaney, 1970), (iv) strong electronic exchange over a series of paramagnetic centres (Soos et al. 1977) (v) weak electronic exchange over many centres (Simpson et al. 1978) and (vi) quadrupolar effect which is generally seen only in dilute systems.

More commonly, the EPR spectrum of a strong exchange coupled system gives single narrowed line with no hyperfine structure. Single crystals of C_6CuC_6 (Sujatha et al. 1997) and C_8CuC_8 (Eastman et al. 1987) gives a single exchange narrowed EPR line which exhibits the properties of one dimensional exchange systems. A rare situation is the one in which the electronic exchange interaction between adjacent paramagnetic centres are comparable to or less than the metal hyperfine interaction. This has been observed in the one dimensional exchange system $[NBu_4]_2[Cu(mnt)_2]$ (Plumlee et al. 1975). In this paper, from the analysis of the multiline EPR spectrum for C_8OCuC_8 and C_7CuC_7, we postulate a weak exchange interaction. The molecular structure of these compounds are given in Figure 1.

2 Experimental Section

The Compounds under study were prepared according to the procedures described in the literature (Ohta et al. 1986; Sadashiva et al. 1991). X- and Q- Band EPR spectra were recorded with a Varian E-112 spectrometer. Ec-268 /E-257 Variable temperature accessory was used for variable temperature EPR measurements. All simulations were carried out using a DEC 3000 Model 300 series AXP computer.

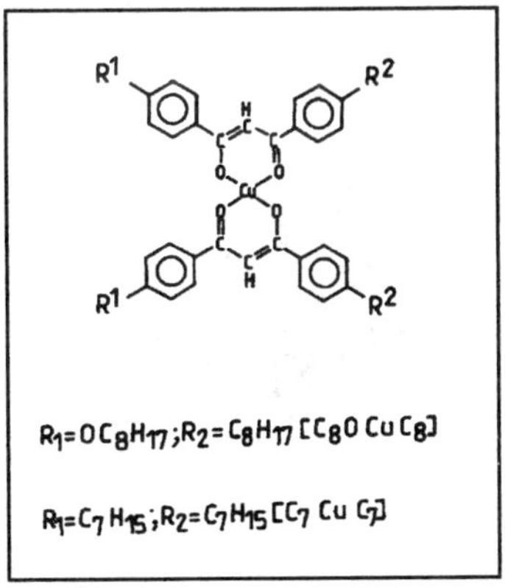

Fig.1. Basic structural formula

3 Results and Discussion

3.1 Crystal Structure and molecular Packing

The compound C_8OCuC_8 crystallises in the space group $P\bar{1}$ in the triclinic symmetry with two molecules per unit cell (Z = 2). The unit cell parameters are a =11.754(10)Å, b =15.479(10)Å, c =17.977(2)Å, α =69.92(10)°, β = 86.66(10)°, γ =68.69(10)°. The complete refinement of the X-ray data is underway. The Cu-Cu distance in this lattice is 6.52Å

The compound C_7CuC_7 also crystallises in the space group $P\bar{1}$ in the triclinic symmetry with one molecule per unit cell (Sujatha et al. 1997). The unit cell parameters are a =10.065(9)Å, b =10.297(3)Å, c =13.300(4)Å, α =71.32(3)°, β =89.64(5)°, γ =87.08(6)°. This consists of disc-like molecules which are arranged in a two dimensional array parallel to the crystallographic ab plane (Figure 2.). The Cu-Cu distance along the stack is 10.1Å.

3.2 Interpretation of EPR Results of C_8OCuC_8

The room temperature EPR spectrum of polycrystalline C_8OCuC_8 (Fig.3) shows well resolved multilines in the parallel region. Careful inspection shows that the four lines (indicated by *) in Figure 3 form a quartet centered at $g_{\parallel}$ with a spacing of A = 170 G. Spaced roughly between the quartet lines are the intense lines which are also centered at $g_{\parallel}$ with a spacing of 170 G. These lines appears to be the result of overlapping absorptions. Additional resonances are also visible on the outer edges of the m = ±3/2 peaks.

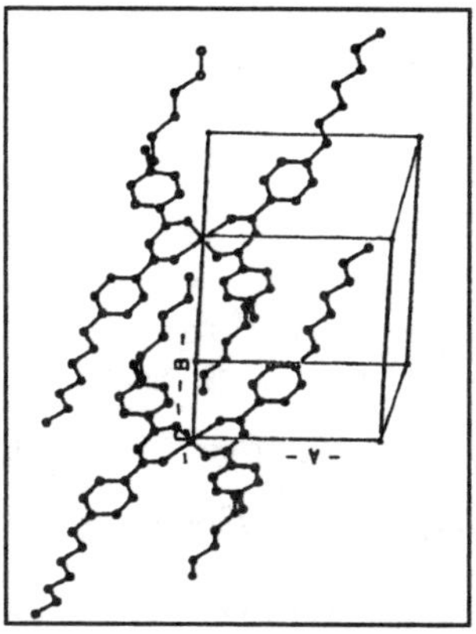

Fig. 2. Orientation of the molecule C_7CuC_7 w.r.t the unit cell axes of the crystal

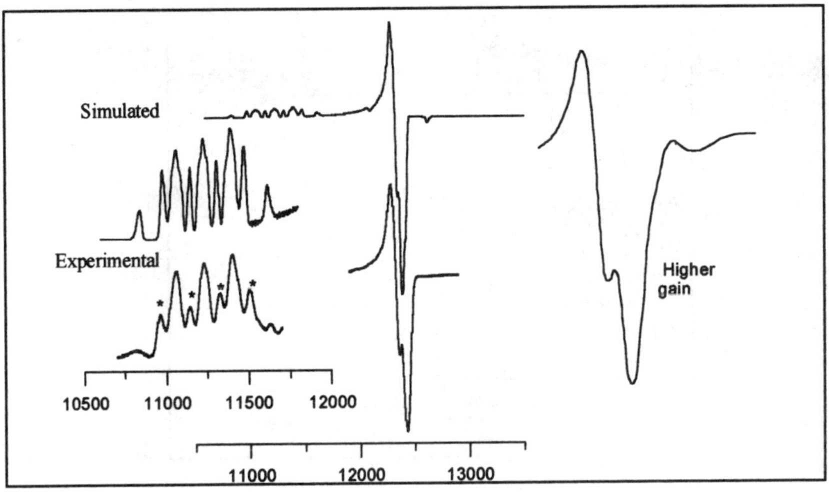

Fig. 3. EPR Spectrum of C_8OCuC_8 . Parallel region is shown in the insert. The lines (indicated by *) are the normal Cu(II) hyperfine lines.

The middle intense lines have resemblance to "forbidden" hyperfine lines $\Delta m_I = \pm 1$ and $\Delta m_I = \pm 2$, but could not be interpreted due to quadrupolar effects since these give minimal effects in the 'parallel' region. Moreover, the room temperature EPR spectrum of the compound under study in the corresponding palladium host lattice shows no evidence of quadrupolar interactions.

Also, when it is taken into account that interdimeric (pseudo dimer) Cu-Cu distance is 6.5Å, a direct dipolar interaction which is known to vary propotional to $< r^{-3} >$, r being interspin distance, can be assumed to be very small ($< 10^{-2}$ cm^{-1}). This is also supported by the appearance of weak $\Delta Ms = \pm 2$ half-field transition at g = 4.1. Thus the observed EPR spectrum can be explained on the basis of weak exchange interaction between Cu(II) ions.

In the EPR spectrum of C_8OCuC_8 the two weak signals located at lower and higher field values of the normal $\Delta Ms = \pm 1$ transitions can be assigned as forbidden transitions between the singlet state and triplet state of exchange coupled Cu(II) ions. Furthermore, Itoh et al. (1969) demonstrated the anomalous feature of the hyperfine structure in the weakly coupled cases where J is comparable to the electronic Zeeman energy. In C_8OCuC_8, the behavior of the hyperfine structure of two Cu(II) ions (with same $g_{\|}$ & $A_{\|}$ values) which are weakly coupled is shown in Figure 4 which makes the evaluation of |J| possible. In this case the energies of the singlet and triplet states are close and they are mixed with each other by the Zeeman and hyperfine interactions. Here the outermost weak lines are located at ±(A+J) from the centre of the spectrum in the parallel region. The value of |J| calculated directly from the

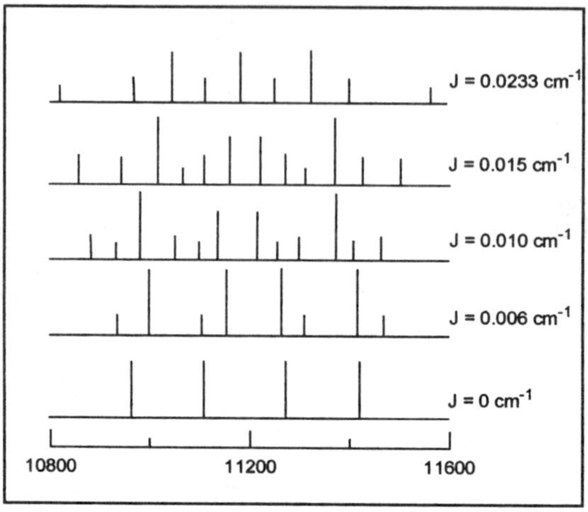

Fig. 4. The calculated EPR line spectra (parallel region) of a weakly interacting Cu(II) pair with change in the magnitude of |J| and A = 170 G at Q-band frequency.

spectrum is 0.023 cm^{-1}. We observed that while the signals associated with triplet $\Delta Ms = \pm 1$ transition remains nearly unchanged in the range of temperature 298K-343K, a shift with temperature is detected for the two singlet-triplet transitions. The temperature dependence of |J| is given in Figure 5. Beyond 343K the system will enter the discotic liquid crystalline phase. It has been suggested that changes in |J| value with temperature arise for weakly coupled cases from lattice shrinkage effect (Michael Duggan et al. 1974). Then, it is not surprising, owing to the two copper(II) ions coming close to each other to have distinct thermal dependence of |J|.

3.2.1. EPR Spectral Simulation

In discussing the interpretation of the EPR spectra of C_8OCuC_8, we have considered "magnetic interactions" such as exchange or pseudo-dipolar interactions. For a system in which pairs of copper (II) ions are interacting, the EPR spectrum can be described by the spin Hamiltonian

$$\hat{H} = \sum_{i=1,2} \{\beta\left(g_{\|}H_zS_{iz} + g_{\perp}H_xS_{ix} + g_{\perp}H_yS_{iy}\right) + AS_{iz} + B\left(S_{ix}I_{ix} + S_{iy}I_{iy}\right)\} + \hat{H}_{dip} - JS_1.S_2 \quad (1)$$

When J < 30 cm^{-1} no " pseudo dipolar" contributions to Eq(1) is needed. In the weak exchange limit (J<A), S is not a good quantum number and there is mixing up of singlet and triplet states. This results in the appearance of J flips at the outer edges

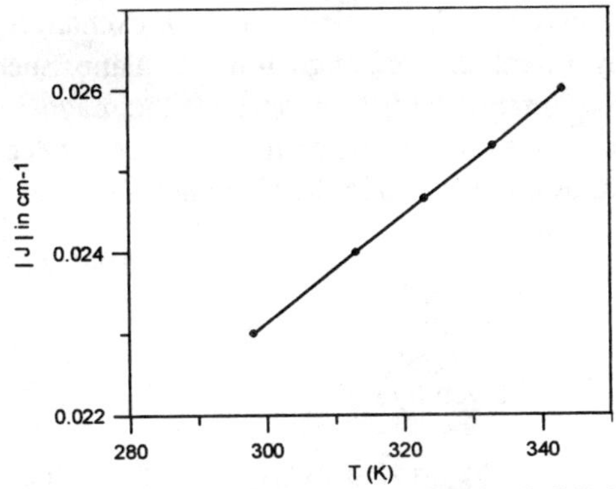

Fig. 5. Temperature dependence of |J| for C_8OCuC_8

of $\Delta M_I = \pm 3/2$ transitions. We have simulated the EPR spectrum using the program GNDIMER (Smith et al. 1974) where the Eq(1) is solved using the perturbation theory in the Zeeman representation. The simulated spectrum matches well with the experimental one (see Figure 3). The parameters used for simulation are given in Table 1.

Table 1. EPR parameters obtained from simulation.

System	$g_\parallel$	$g_\perp$	$A\times10^{-4}$ cm^{-1}	$B\times10^{-4}$ cm^{-1}	$\lvert J\rvert$ cm^{-1}
C_8OCuC_8	2.256	2.054	170	41	0.0233
C_7CuC_7	2.214	2.053	165	39	0.0103

3.3 EPR Results of C_7CuC_7

We report here a detailed study of EPR of C_7CuC_7 polycrystalline compound. X-ray crystal structure indicates that the compound forms crystallographically regular stacks. The central feature of the analysis is that the exchange coupling J along the chain is weak (J<A). So there exists stacking exchange in the presence of nuclear hyperfine interaction of comparable magnitude. The EPR spectrum (Fig .6) shows resolved copper hyperfine quartet centered at $g_\parallel$ with A= 165 G. Spaced midway are the three less intense lines centered at $g_\parallel$. Additional shoulders centered at $\pm(3A+2J)/2$ in the parallel region are also seen. The value of |J| calculated from the spectrum is c.a. 0.01 cm^{-1}. The large separation between paramagnetic sites (~10.1Å) in C_7CuC_7 leads to dipolar interactions which are very much smaller than hyperfine interaction and hence neglected. Thus the smallness of |J| leads to unusual features in the EPR spectrum.

From the crystal structure it can be seen that the Cu(II) ions are more-or-less uniformly distributed at 10.1Å apart, along both the a and b axes and the EPR feature unequivocally indicates the presence of weak exchange (J<A) and therefore should be discribed by a weak 2D exchange in the ab plane. Such an interaction also should also lead to very interesting EPR spectra for the magnetic field exploring the ab plane, as compared to any other plane such as ac or bc. Indeed single crystal EPR spectra show very interesting patterns in the ab plane as expected. These results will be published later in detail.

3.3.1 EPR Spin Hamiltonian

The exchange interaction is given by

$$V_{ex} = J\Sigma S_i . S_{i+1} \tag{2}$$

Neglecting exchange in zeroth order, the spin Hamiltonian for each C_7CuC_7 stack contains Zeeman and hyperfine interactions

$$\hat{H}_0 = \sum_{i=1}^{\infty} \hbar(\omega_m)_i S_i^z \tag{3}$$

The transition between m (m=3/2,1/2,-1/2,-3/2) levels are slow on the EPR time scale. Then the Zeroth order Hamiltonian H_0 is based on the distribution of m values with all 2I+1 values for nuclear spin I occurring with equal probability of $(2I+1)^{-1}$. .The sites are magnetically equivalent when $m_i = m_i+1$ and even a weak exchange mixes the degenerate spin states. Thus we are dealing with degenerate perturbation theory (H_0+V_{ex}) . The chain can be assumed to be formed of segments within which adjacent spin have same m value and bordered by spins with different m values.

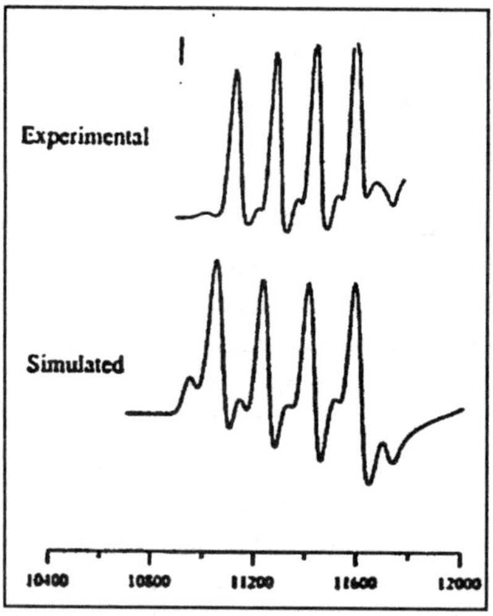

Fig. 6. EPR spectrum of C_7CuC_7 (parallel region)

Each segment will provide a different kind of a spectrum which can be calculated assuming that the population of different segments are statistically determined. The probability of having a segment of length N is given by

$$P_N = \left[\frac{2I}{2I+1}\right]^2 \times \frac{1}{(2I+1)^{N-1}} \tag{4}$$

Where N = 1,2,...∞. The condition that the total number of sites is conserved imposes

$$\sum_{N=1}^{\infty} NP_N = 1 \tag{5}$$

The general problem of $H_0 + V_{ex}$ reduces to computing the effects of the perturbation

$$V_N = J[S_0.S_1 + S_N.S_{N+1}] \tag{6}$$

for interaction between an N-mer and the spins on either side. V_N introduces splittings as the first order corrections to (m,m,m) sequence which is given by

$$E^{(1)} = \left(\frac{J}{4}\right) S_1{}^z\left(S_0{}^z + S_2{}^z\right) \tag{7}$$

and depends on the electron spin projections $S_i^z = 1/2$ of the sites. The statistical probabilities of other sequences are as given by Plumlee et al. (1975).

3.3.2. EPR Spectral Simulation

A computer program has been written for powder simulation correct to second order based on the Hamiltonian H_0+V_{ex} and various possible N-mer sequences. The resulting properly weighted spectrum is calculated with cylindrically symmetric g and A tensors and the powder spectrum (θ = 0 to 90°) is generated using θ dependent line intensities. The simulated spectrum is shown in Figure 6. The parameters obtained for the best fit of the experimental and the simulated spectrum are given in Table 1. The degenerate perturbation theory is valid only when $J < A$.

3.4 EPR at Discotic Temperatures

At the discotic temperature region (346-416K) of C_8OCuC_8, the weak signals corresponding to the singlet -triplet transition in the solid state and the multiplet in the parallel region move inward with an overall reduction in hyperfine as well as decrease in intensity of lines. Finally, at the isotropic temperature around 416K, all lines merge giving an EPR spectrum corresponding to monomeric copper (II) species. Due to the random motion of the molecules in the liquid crystalline state with respect to each other there seems an overall loss in the spectral resolution. The exchange pathway that exist in the uniform crystal packing in the solid state is lost.

On the other hand, in C_7CuC_7 at discotic temperature the hyperfine feature is completely lost. The EPR spectrum corresponds to an increased coupling where the Cu-Cu distance is presumably small. So it is assumed that the molecules which are arranged in discs come closer to each other, in the mesophase which are otherwise stacked uniformly in the crystal lattice showing weak exchange properties. For both these compounds under study the room temperature EPR spectra obtained after a thermal cycling are never equal to those of the freshly prepared compounds revealing a hysteresis.

4. Conclusions.

The agreement between experimental and simulated EPR spectra confirms the postulated weak exchange interaction. We have dealt with two regimes of weak exchange interaction. In C_8OCuC_8, |J| is marginally greater than A and in C_7CuC_7, |J| is less than A which accounts for the difference in the appearance of the spectrum.

It is very pleasing that the nature of exchange (pseudo dimer in C_8OCuC_8, and 2D exchange in C_7CuC_7) can be clearly discerned by the arrangement of the molecules in the lattice. The above two systems have a rare combination of weak exchange and well resolved spectra which makes it more challenging to arrive at the correct model. We are in the process of getting more information from the single crystal EPR studies of these systems.

Acknowledgment

Thanks are due to Dr.G.V.R.Chandramouli for helping us in the computer simulation.

References

1. K. Ohta, H. M. A. Takagi, I. Yamamoto and Keimatszaki, Discotic liquid crystals of transition metal complexes, *Mol. Cryst. Liq. Cryst.* **135** (1986) 247-264.
2. B.K.Sadashiva, Archana Ghode and P.Rani Rao, Synthesis and mesogenic properties of some β-diketones and their Copper (II) and Palladium(II) chelates, Mol. Cryst., *Liq. Cryst.* **200** (1991) 187-196.
3. P.Eastman, Miin-Liang Horng, B.Freina, and K.W.Sheu, ESR of a one dimensional Heisenberg antiferromagnet which undergoes a crystalline to discotic phase transition, *Liq. Cryst.* **2** (1987) 223-228.
4. J.I.Martnez, D.W.Bruce, D.J.Price and P.J.Alomo, Mesophase order of a new smectic paramagnetic copper complex detected by EPR, *Liq. Cryst.* **19** (1995) 127-132.
5. A.Abragam and B.Bleaney, Electron paramagnetic Resonance of Transition ions, Clarendon press, Oxford, 1970, Ch.9. 491-535.

6. H.Abe and J.Schimada, Anomalous magnetic resonance absorption of Copper acetate, *Phys. Rev.* **90** (1953) 316.
7. Z.G.Soos, R.T.McGregor, T.T.P. Cheung, and A.J.Silverstein, Antisymmetric and anisotropic exchange in ferromagnetic Copper (II) layers, *Phys. Rev. B* **16** (1977) 3036-3018.
8. G.D.Simpson, R.L.Belford and R.Biagioni, EPR study of internuclear exchange coupling in an undiluted Vanadyl complex, *Inorg.Chem.* **17** (1978) 2424-2428.
9. Venkataraman Sujatha, Kazhuchika Ohta, Monisha Bose and S.Subramanian (unpublished results).
10. K.W.Plumlee, B.M.Hoffmann, J.A.Ibers, and Z.G.Soos, Weak exchange in the Heisenberg linear chain : Structure and EPR of $|N(n\text{-}Bu)_4|2|Cu(mnt)_2|$, *J. Chem. Phys.* **63** (1975) 1926-1942.
11. K.Itoh, H.Hayashi, and S.Nagakura, Determination of the singlet-triplet separation of a weakly interacting radical pair from the ESR spectrum, *Mol. Phy*, **17** (1969) 561-577.
12. D.Michael Duggan, and N.Henderckson, Magnetic Exchange Interaction in Transition Metal Dimers. *Inorg. Chem.*, **13** (1974) 2929-2940.
13. T.D.Smith,and J.R.Pilbrow, The determination of structural properties of dimeric transition metal ion complexes from EPR spectra, *Coord. Chem. Rev.* **13** (1974) 173-278.

Impact of Spectral Diffusion on Apparent Relaxation Times for the Stable Radical in Irradiated Glycylglycine

James R. Harbridge, Gareth R. Eaton, and Sandra S. Eaton

Department of Chemistry and Biochemistry, University of Denver, Denver, Colorado 80208 USA

Abstract. When continuous-wave power saturation is used to measure electron spin relaxation times for inhomogeneously broadened EPR spectra, spectral diffusion can cause the apparent value of $(T_1T_2)^{1/2}$ to be shorter than the true value. Spectral diffusion also can impact time-domain measurements of relaxation times. Spectral diffusion was detected in samples of γ-irradiated glycylglycine by comparison of continuous wave power saturation, electron spin echo, inversion recovery, saturation recovery, and electron-electron double resonance data.

1 Introduction

For a homogenously broadened single-line EPR spectrum the lineshape of the first-derivative EPR signal, including the effect of power saturation, is given by equation (1).

$$\chi''(\omega) = \frac{1}{2}\chi_0\omega_0 T_2 \frac{1}{1+(\omega-\omega_0)^2 T_2^2 + \gamma^2 B_1^2 T_1 T_2} \tag{1}$$

where χ'' is the amplitude of the EPR signal, χ_0 is the magnetic susceptibility, ω and ω_0 are the observing frequency and the resonant frequency, respectively, γ is the magnetogyric ratio, B_1 is the microwave magnetic field, and T_1, T_2 are the longitudinal and transverse electron spin relaxation times, respectively. Continuous wave (CW) power saturation can be used to determine the product $(T_1T_2)^{1/2}$ based on equation (1) [1].

Homogeneous broadening implies that the lineshape is Lorentzian and that the linewidth is determined by the relaxation time T_2.

Interpretation of CW power saturation data is more complicated for multiline spectra and inhomogenously broadened lines than for single-line homogenously broadened spectra. Inhomogeneous broadening may arise from unresolved hyperfine splitting, magnetic field inhomogeneity, distributions in Hamiltonian parameters, and/or anisotropy in Hamiltonian parameters. In such samples processes that move saturation through the spectrum and shift spins out of the resonator bandwidth contribute to the relief of saturation and to the apparent value of $(T_1T_2)^{1/2}$ obtained by CW power saturation. These are collectively known as spectral diffusion, which includes the following processes. 1) Cross relaxation involves mutual spin flips as described by the operators $S_{1+}S_{2-}$ or $S_{1-}S_{2+}$. If we are observing S_1 this process can provide relief of saturation although the total set of spins has not returned to equilibrium. In the most general model of spectral diffusion S_1 and S_2 can be either electron or nuclear spins. Cross relaxation is called spin diffusion by some authors. 2) Another type of process involves nuclei to which the unpaired electron is coupled. Motion between magnetically inequivalent environments or spin flips of these nuclei can shift the resonant frequency for the unpaired electron. 3) Motion of a molecule with anisotropic Hamiltonian paramcters can shift the resonance of saturated spins. This is exploited in saturation transfer EPR.

When spectral diffusion processes occur with time constants that are shorter than T_1 or T_2, the apparent value of $(T_1T_2)^{1/2}$ obtained by CW power saturation is decreased. Typical X-band CW resonators have quality factor, Q, of about 3000, which corresponds to a bandwidth at half-height of about 3 MHz, or roughly 1 G. Thus a shift of resonant frequency by about a gauss is sufficient to move a saturated spin off-resonance, thereby contributing to relief of saturation. Spectral diffusion processes also may impact time-domain measurements of relaxation times.

Although the potential importance of spectral diffusion was recognized in the early literature [2], there are few cases in which the processes have been well characterized. In this report we examine the impact of spectral diffusion on apparent relaxation times for the stable radical in γ-irradiated glycylglycine. The radical in this sample has been characterized by single-crystal EPR [3] as

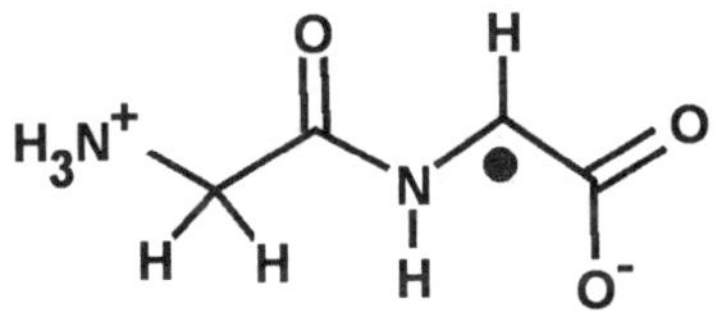

A CW power saturation study at room temperature found $(T_1T_2)^{1/2} = 4.0\text{x}10^{-7}$ s [4]. In the work reported here, values of T_1 from inversion recovery and saturation recovery experiments and of T_2 from electron spin echo experiments were compared with $(T_1T_2)^{1/2}$ obtained by CW saturation and the importance of spectral diffusion in this system was confirmed by electron-electron double resonance (ELDOR).

2 Experimental

Polycrystalline glycylglycine was irradiated at room temperature to a total dose of 9.8 Mrad with ^{60}Co γ-rays. The approximately 17-G doublet splitting of the EPR spectrum in the irradiated sample was in good agreement with the literature [3].

CW, 3-pulse inversion recovery, and electron spin echo (ESE) data were recorded on a Bruker ESP380E spectrometer with a split-ring resonator. The microwave B_1 in the resonator was calibrated by comparison of the power saturation curve for a strong pitch sample with a curve that had been obtained previously on a Varian E9 with a TE_{102} rectangular resonator. The B_1 in the TE_{102} resonator had been measured by the method of perturbing spheres [5]. Our ESP380E has a low-power (2 W) microwave amplifier and additional detection path for long-pulse saturation recovery (SR) experiments. SR data were recorded on the ESP380E or on a locally constructed spectrometer [6]. Spin-echo-detected ELDOR experiments were performed by modification of a locally constructed spin-echo spectrometer [7]. The additional microwave frequency was generated with a Wavetek MicroSweep Model 965.

3 Results and Discussion

CW power saturation of the irradiated glycylglycine sample at room temperature gave an apparent $(T_1T_2)^{1/2}$ = 4.0x10^{-7} s, which is in good agreement with the literature value of 4.0x10^{-7} s [4]. In both cases the samples extended the full length of the resonator, and the effective value of B_1 reflected the distribution over the resonator.

Spin echo measurements were performed as a function of B_1 and extrapolated to infinitely small turning angle to correct for the effects of instantaneous diffusion [8]. The resulting value of the phase memory relaxation time, T_m, was 0.32 μs. For the current calculations this was used as an estimate of T_2. Spectral diffusion may contribute so the dephasing, so 0.32 μs is a lower limit on the value of T_2. A more precise value will be determined by a Carr-Purcell-Meiboom-Gill sequence.

Inversion recovery experiments were performed with a 24 ns inverting pulse followed by a 2-pulse spin-echo detection sequence. The inversion recovery curves were not single exponentials. Fits of the data to the sum of three exponentials gave time constants of about 3, 20, and 70 μs. The presence of multiple exponentials indicates the presence of multiple processes that contribute to the return to equilibrium.

The saturation recovery curves also indicated the presence of multiple exponentials. For a pump pulse of 0.4 μs a fit of the data to the sum of three exponentials gave time constants of 3, 20, and 60 μs. These values are similar to the ones obtained in the inversion recovery experiment. As the length of the pump pulse was increased from 0.4 to 240 μs, the time constants increased to 14, 47, and 90 μs. In the presence of spectral diffusion the long pump time permits diffusion of saturated spins into other portions of the spectrum, and the subsequent recovery curve has decreased contributions from the spectral diffusion process(es). Of this set of time constants, the longest value is the best approximation to T_1, although this time constant may still include some contributions from spectral diffusion processes.

Based on the estimates of $T_2 = 0.32$ μs from the spin echo data and $T_1 = 90$ μs from the saturation recovery data, $(T_1T_2)^{1/2} = 5.3\text{x}10^{-6}$ s, which is more than an order of magnitude longer than the value obtained by CW saturation. The longer value of $(T_1T_2)^{1/2}$ obtained by the time-domain techniques than by CW saturation, the presence of multiple exponentials in the inversion recovery and saturation recovery curves, and the pump-time dependence of the saturation recovery data suggest that spectral diffusion is a major factor in this spin system. To observe the spectral diffusion more directly an ELDOR experiment was performed.

In the ELDOR experiment a long saturating pulse at frequency ν_1 was applied to one line of the doublet signal and a two-pulse spin echo sequence at frequency ν_2 was used to monitor the signal at the other line of the doublet. The time between the pulse at ν_1 and the beginning of the two-pulse spin echo sequence at ν_2 was incremented. In the absence of spectral diffusion, the echo intensity at ν_2 would be independent of the pulse at ν_1. However, in the presence of processes that move saturation between lines of the spectrum, the intensity of the echo at ν_2 is dependent upon the time constant of the spectral diffusion process and on the time between the pulses at ν_1 and ν_2.

An ELDOR curve for the irradiated glycylglycine sample is shown in Figure 1. In this experiment the frequency for the first pulse, ν_1, was set to the resonance of the low-field line and the frequency for the spin echo sequence was set to observe the high-field line. The resonator was strongly overcoupled to yield a bandwidth that encompassed both frequencies. The separation between ν_1 and ν_2 corresponded to an offset in resonant field of about 19 G. The long low-power pulse at ν_1 was highly selective and only a narrow band of frequencies was excited. The pulses at ν_2 for the spin-echo detection had a bandwidth of about 5 G. Thus there was negligible overlap between the set of excited spins and the set of detected spins. As the time between the pulses increased, the intensity of the spin-echo signal decreased and then increased, finally returning to the value observed in the absence of the pulse at ν_1 (Figure 1). The initial decrease reflects transfer of spins saturated at ν_1 to spins observable at ν_2, which is unambiguous evidence for a spectral diffusion process. The system then returns to equilibrium by a slower process. Curves with similar shapes were obtained when the roles of the two frequencies were reversed, i.e. ν_1 was set to saturate the high-field line and ν_2 was set to observe the low-field line. Note that the pulsed ELDOR experiment yields information complementary to the more common CW ELDOR experiment. Unique initial conditions can be created via the preparation pulse, and no microwaves are incident on the sample during signal detection. Work is currently underway to characterize the spectral diffusion processes in this system and to determine the characteristic time constants.

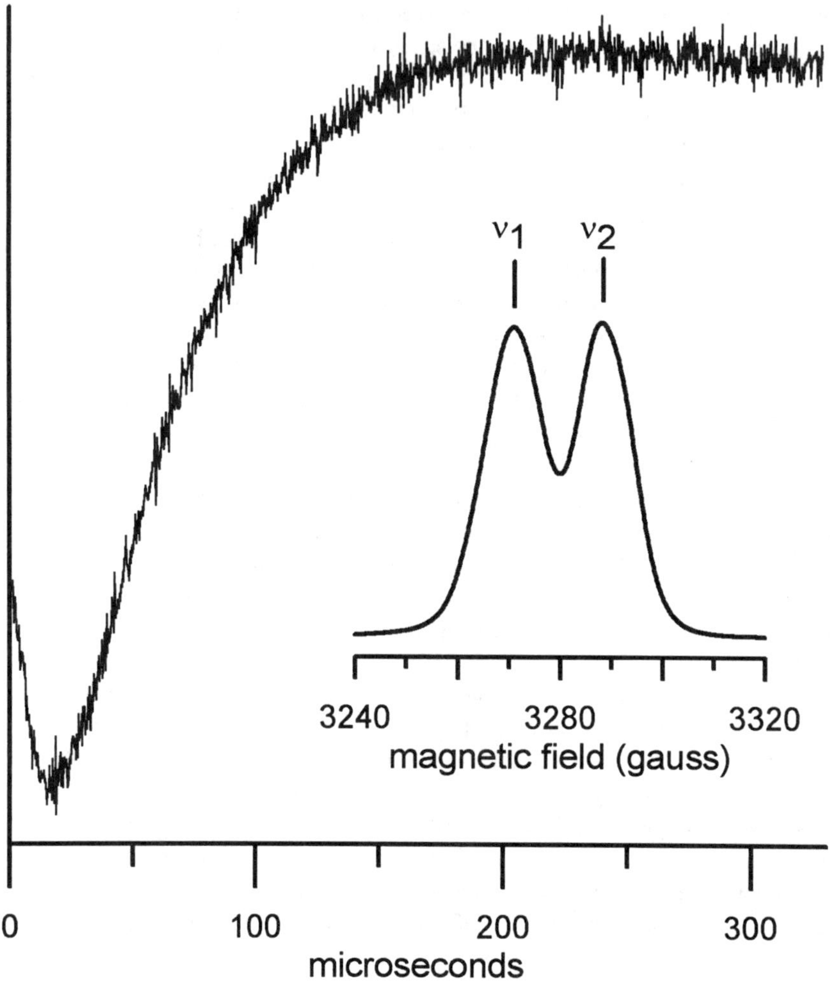

Fig. 1. Saturation ELDOR data for irradiated glycylglycine at room temperature obtained with magnetic field = 3288 G, ν_1 = 9.2259 GHz and ν_2 = 9.1697 GHz. The length of the first pulse was 6144 ns. The time axis is measured from the end of the first pulse. The inset displays the CW absorption spectrum of the radical obtained with 0.065 mW microwave power and modulation amplitude of 1.0 G at 100 kHz. The positions in the CW spectrum that were in resonance at the two microwave frequencies and constant magnetic field of the ELDOR experiment are shown in the inset.

Acknowledgments

The financial support of this work by National Institutes of Health Grant GM21156 and National Science Foundation shared instrumentation grant CHE-9318714 for purchase of the Bruker ESP 380E is gratefully acknowledged.

References

1. C. P. Poole, Jr. and H. A. Farach, *Relaxation in Magnetic Resonance*, Academic Press, New York (1971), ch. 3.
2. Ref. 1, ch. 15.
3. M. Katayama and W. Gordy, Electron spin resonance of γ-irradiated glycylglycine, *J. Chem. Phys.* **35** (1961) 117-122.
4. E. S. Copeland, A simple method for estimating H_1 in ESR experiments - the microwave power saturation of γ-irradiation induced glycylglycine radicals, *Rev. Sci. Instrum.* **44** (1973) 437-442.
5. D. R. Thomas, L. R. Dalton, and J. S. Hyde, Rotational diffusion studied by passage saturation transfer electron paramagnetic resonance, *J. Chem. Phys.* **65** (1976) 3006-3024 and references therein.
6. R. W. Quine, S. S. Eaton, and G. R. Eaton, Saturation recovery electron paramagnetic resonance spectrometer, *Rev. Sci. Instrum.* **63** (1992) 4251-4262.
7. R. W. Quine, G. R. Eaton, and S. S. Eaton, Pulsed EPR Spectrometer, *Rev. Sci. Instrum.* **58** (1987) 1709-1723.
8. S. S. Eaton and G. R. Eaton, Irradiated fused-quartz standard sample for time-domain EPR, *J. Magn. Reson.* **A102** (1993) 354-356.

Pulsed Electron-Electron Double Resonance to Determine Distances Between Paramagnetic Species in Photosystems

Asako Kawamori[1], *Hideyuki Hara*[1], *Kazuhito Shigemori*[1], *and Andrei V. Astashkin*[2].

[1]Faculty of Science, Kwansei Gakuin University, Nishinomiya 662, Japan.
[2]Institute of Chemical Kinetics and Combustion, Russian Academy of Sciences, 630090 Novosibirsk, Russia.

Abstract. A pulsed electron-electron double resonance (ELDOR) method was applied to measure the dipole interactions between paramagnetic species in plant Photosystem II (PS II). '2+1' pulse sequence, a special case of ELDOR, was applied to the overlapped signals of the primary electron acceptor quinone Q_A^- and Y_D^+ radical pairs in cyanide-treated PS II. The distance between them has been determined to be 38.5 ± 0.7 Å. The distance between Q_A^- and Y_Z^+ was determined to be 38 ± 1 Å. These distances and the previously determined distance between Y_D and the Mn-cluster, 27 ± 0.5 Å in the S_2 state of the oxygen-evolving PS II, are shown in the membrane scheme.

1 Introduction

Photosystem II (PS II) in higher plants is composed of several intrinsic and extrinsic polypeptides with unknown structures. Among them, the heterodimer of D1 and D2 proteins is believed to bind almost all electron-transfer components of PS II, the Mn-cluster in oxygen-evolving complex (OEC), the tyrosine donors Y_Z and Y_D, the primary electron donor P680, pheophytins, and the primary (Q_A) and secondary (Q_B) electron acceptor quinones [1,2]. In Fig. 1 a supposed model structure based on a bacterial reaction center is shown. On the donor side of PS II, a photooxidized P680 is reduced by the electron donated from Y_Z. Y_Z^+ is, then, reduced by the electron from the Mn-cluster that is oxidized stepwise through five catalytic states S_0-S_4 and finally produces an oxygen molecule [3]. Ca^{2+} and Cl^- are other indispensable cofactors for the oxygen evolution.

To elucidate the location of the electron carriers in PS II, the EPR measurements of the magnetic relaxation times [4-8] and a selective hole burning method have been applied [9]. In the previous work, the dipole interactions between the paramagnetic species in PS II were probed by a pulsed ELDOR (ESE ELDOR) method [10]. The pulse sequence employed was 3-pulse ESE method, but the microwave (m.w.) pulses with two different carrier frequencies were used. This method was applied particularly to estimate the distance, 27 Å, between Y_D^+ and the Mn-cluster in PS II [10].

The distance between Y_D^+ and Y_Z^+ was estimated to be 29-30 Å using the '2+1' ESE method [11]. The '2+1' electron spin echo (ESE) technique used in [11] is a special case of general electron-electron double resonance (ELDOR) methods. It employs a sequence of three m.w. pulses with the same carrier frequency and is applicable when the EPR transitions of the studied paramagnetic centers can be efficiently excited by the pulses used. This implies that the EPR spectra of the studied species have to be reasonably narrow (of the order of one-two mT) and overlapping. The Y_D^+ and Y_Z^+ pair studied in [10,11] presents an example of this kind. In this work a '2+1' ESE technique was applied to different radical pairs, consisting of a acceptor quinone Q_A^- and Y_D^+ or Y_Z^+. The determined distances between paramagnetic species are presented in the PS II membrane scheme.

2 Experimental

The oxygen-evolving PS II membranes (500 μM O_2/mg Chl per h) were isolated from market spinach by the method of Kuwabara and Murata [12] Cyanide (CN^-)-treated PS II was prepared according to Sanakis et al. [13] after Tris-treatment of the oxygen-evolving PS II. CN^--treated PSII was suspended in a buffer (340 mM KCN, 0.34 M sorbitol, 15 mM NaCl, 5 mM $MgCl_2$ and Tricine/NaOH (pH 8.0 at 20 °C)) containing 50 % glycerol (w/v). These preparations eliminate three extrinsic proteins and four manganese atoms, and convert the paramagnetic non-hem ion on the acceptor side into a diamagnetic state. The CN^--treated PS II membranes with concentration of 12 mg Chl/ml were transferred into Suprasil quartz tubes with the inner diameter of 4 mm and stored in liq. N_2 until use.

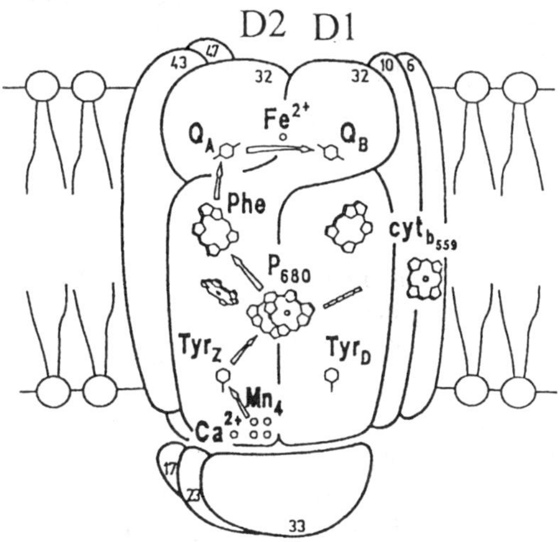

Fig. 1. A model structure of photosystem II. (a) Primary donor P680 is supposed to be a chlorophyll dimer located at the symmetric center of D1 and D2 proteins. Electron transfer occurs mainly on the D1 side from water to the acceptor quinones along the arrows.

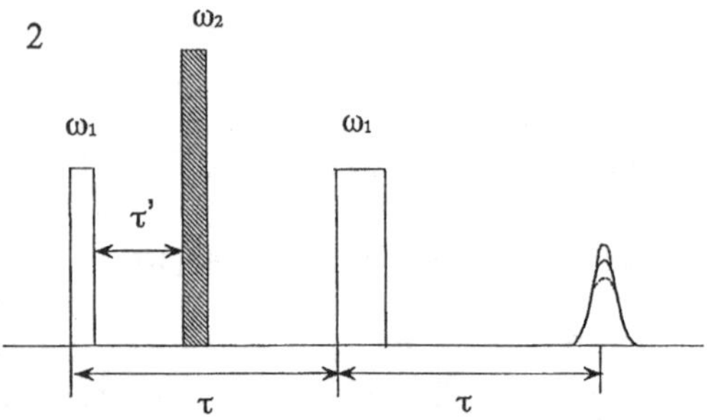

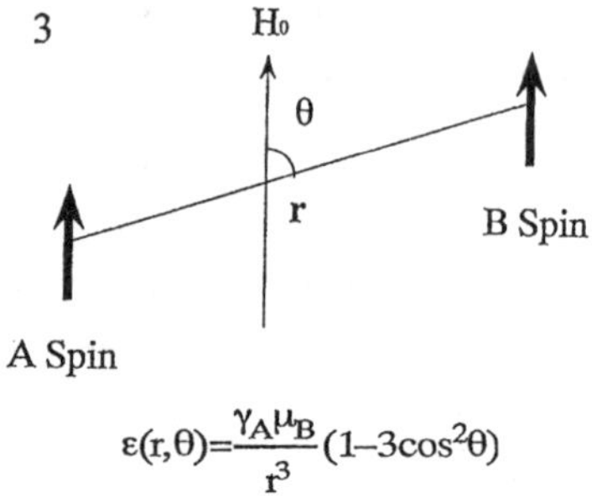

Fig. 2 The pulse sequence of the ESE ELDOR method. The primary ESE signal is formed by the first and third m.w. pulses with the carrier frequency ω_1, separated by the time interval τ. The amplitude of the ESE signal is observed as a function of the position τ' of the second m.w. pulse with the carrier frequency ω_2.

Fig. 3 Dipolar interaction between two spins A and B. The energy $\varepsilon(r,\theta)$ is a function of the distance r and the orientation of r relative to the applied magnetic filed $\gamma_A = 2\pi g_A \beta / h$.

The ESE ELDOR measurements have been performed on a pulsed EPR spectrometer ESP-380 (Bruker) using a pulse sequence shown in Fig. 2. Another m.w. generator was used as a second m.w. frequency source. The output of this generator was fed into the resonator through the m.w. pulse former unit of the pulsed EPR spectrometer giving the second pulse shown in Fig. 2.

The pulsed EPR spectrometer was equipped with a cylindrical dielectric cavity (ER4117DHQ-H, Bruker), and a helium gas flow system (CF935, Oxford Instruments). The measurement temperature was about 80 K and m.w. pulses of 16, 24 and 16 ns duration were used. In '2+1' sequence used to study radical pairs with overalapping EPR spectra the frequencies of ω_1 and ω_2 were same [11]. The pulse amplitudes for this sequence are adjusted to give spin rotation angles 30°, 60° and 30° degrees, respectively.

3 Theory

Consider a spin system composed of pairwise-distributed radicals, with every pair consisting of two types of spins, A and B. In the pulsed ELDOR method used in this work, the spin system is excited by three m.w. pulses (Fig. 2). The first and third pulses, separated by the time interval τ, have a carrier frequency ω_1 resonant with the EPR transitions of spins A and form the primary ESE signal of these spins. The second pulse, separated from the first one by the time interval τ' ($\tau' \leq \tau$), has a carrier frequency ω_2. This pulse is resonant with spins B and changes their projections from $|\alpha\rangle$ to $|\beta\rangle$ and vice versa. If the magnetic dipolar interaction between the pairwise-distributed spins is appreciable, the flip of a B spin changes the local magnetic field for its partner in the pair (A spin). As a result, the magnetization of the A spins after the third pulse cannot be completely refocused at the time 2τ and the amplitude of the primary ESE signal exhibits an oscillating dependence on the second pulse position (i.e., on τ') [10]:

$$V(\tau,\tau') \propto 1 - p\,[1-\cos(D\tau')] \tag{1}$$

where

$$D = 2\pi D_0(1-3\cos^2\theta) \tag{2}$$

is the secular component of the dipole interaction between spins A and B. The value p ($0 \leq p \leq 1$) is a fraction of spins excited by the pumping pulse and depends on the EPR spectrum width of spins B and on the pumping pulse intensity and duration [10].

From Eq. (1) it follows that the frequency of the oscillations in the dependence of the primary ESE signal on τ' (briefly, ELDOR trace) is determined by the dipole interaction and their amplitude is determined by the excitation factor p. The typical value $p \approx 0.3$ found in our experiments described below indicate an incomplete excitation of the Y_D^+ signal by the pumping pulse. This result is not surprising as the Y_D^+ spectrum width ($\approx$2 mT) is larger than the excitation width of the 60° pumping pulse of 24 ns duration ($\leq$1.4 mT). Besides, the non-oscillating background due to cytochrome (Cyt) b559 also contributes to all ELDOR traces recorded in this work, thus decreasing the effective p values. As the p values are of no special importance in this study, they will not be discussed further.

In a nonoriented system Eq. (1) is to be averaged over the angle θ between the radius-vector **r** connecting the radicals in pair, and the direction of the external magnetic field $\mathbf{H}_0$:

$$\langle V(\tau,\tau')\rangle \propto \int_0^{\pi} V(\tau,\tau')\sin\theta d\theta \tag{3}$$

From the value of the dipole interaction D_0 determined in the ELDOR experiment, the distance between the radicals can be found using a point-dipole approximation :

$$D_0 = (g\beta)^2/hr^3 \tag{4}$$

In case of '2+1' sequence method with same second frequency $\omega_1 = \omega_2$, being applied, instead of Eq. (1), a more complicated expression is used as described in [11]

$$V(\tau,\tau') \propto 1-2\langle S_3\rangle\sin^2(D\tau/2) - 2\langle S_2\rangle\sin^2(D\tau'/2) + \langle S_2S_3\rangle[\sin^2(D\tau'/2) - \sin^2(D(\tau-\tau')/2 + \sin^2(D\tau/2)] \tag{5}$$

where $S_{2,3} = [\omega_1^2/\omega_n^2]\sin^2(\omega_n t_p/2)$, t_p is the duration of the corresponding m.w. pulse, $\omega_1 = 2\pi g\beta H_1$ is its amplitude, $\omega_n = (\omega_1^2+\Delta\omega^2)^{1/2}$ is the nutation frequency of the electron spin during the pulse, and $\Delta\omega = 2\pi g\beta\Delta H_0$ is a difference between the m.w. frequency and the resonance freqyuency of the spin. Among these terms, the third term is most efficeiently observed.

4 Results and Discussion

Before '2+1' ESE experiment, the radical pairs Y_D^+ and Q_A^- were trapped by illumination of the CN^--treated PS II at 273 K for 8 min followed by dark adaptation for 3 min and then freezing at 77 K. The formation of the radicals was examined by continuous wave (CW) EPR as shown in Fig. 4a.

The '2+1' ESE experiment was performed at a fixed τ = 1080 ns, with τ' varying from 40 to 1000 ns. The magnetic field was fixed at the peak position of the field sweep ESE signal that shows an overlapped signal of Y_D^+ and Q_A^- similar to the spectra in Fig. 4a. The dependence of the primary ESE amplitude on τ', measured in

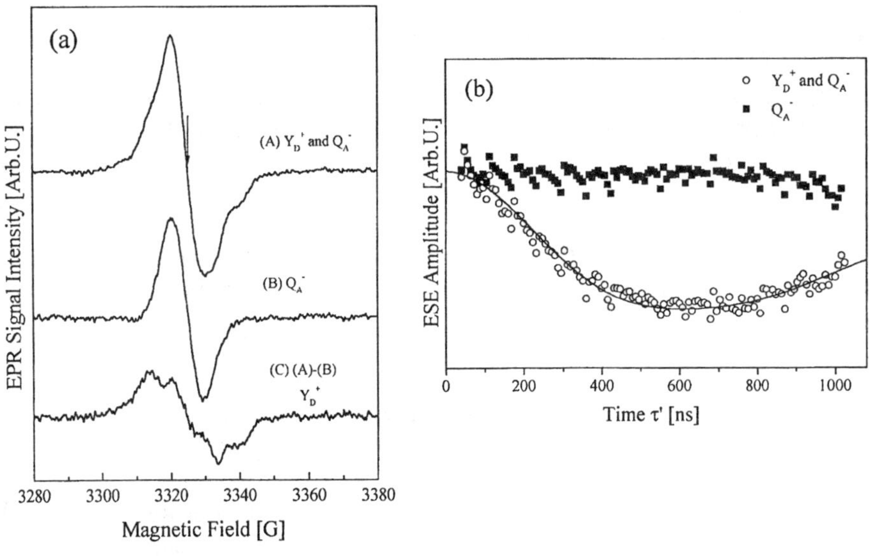

Fig. 4(a) CW EPR of Y_D^+ and Q_A^- in the CN^--treated PS II observed at 77K. (A) Overlapped EPR spectra of Y_D^+ and Q_A^-. Y_D^+ and Q_A^- were trapped by illumination for 8 min followed by dark adaptation for 3 min at 273 K. (B) EPR of Q_A^- (reduced with DTN($Na_2S_2O_4$)). (C) EPR spectrum of Y_D^+ obtained by a subtraction (A)-(B). EPR conditdions: Microwave frequency 9.3 MHz microwave power 0.25 mW; field modulation frequency, 100 kHz; modulation amplitude 3.2 G

Fig. 4(b) The '2+1' ESE traces of Y_D^+ and Q_A^- in CN^--treatment PS II measured at 80K. Open circles: ESE amplitudes for Y_D^+ and Q_A^-, trapped by illumination. The magnetic field was fixed at the peak of the field sweep ESE signal. Filled squares: the trace of Q_A^- (reduced with DTN). Solid line: calculated using Eqs. (2-5) for the dipole interaction constant D_0= 0.89 MHz. All traces are normalized to unity at τ'= 0 .

the CN$^-$-treated PS II with trapped $Y_D^+Q_A^-$ radical pair, is shown in Fig. 4b by open circles. This time profile reveals about one period of low frequency oscillation and differs remarkably from that obtained in the CN$^-$-treated PS II with chemically reduced Q_A^- (filled squares), with its CW spectrum shown in Fig. 4a. Therefore the difference between these two traces shows an oscillation that can be ascribed to the dipole interaction between Y_D^+ and Q_A^-. To determine the value of the dipole interaction constant D_0 for the $Y_D^+Q_A^-$ radical pair, simulations were carried out using Eqs. (2-5) to fit the experimental curve.

As a result of these simulations, the dipole interaction constant D_0 was found to be 0.89 MHz. From the value of D_0, the distance between Y_D^+ and Q_A^- is estimated to be 38.8 ± 0.7 Å using Eq.(5). The error of 0.7 Å has no physical meaning, since it is much smaller than the molecule size of Y_D and Q_A. Therefore, the round value r ≅ 38.5 Å can be considered as a good estimate for the distance between Y_D^+ and Q_A^-. As '2+1' ESE method can detect selectively the magnetic dipole interaction, the obtained value is more accurate than those estimated with other various methods [7,8,9].

The radical pairs $Y_Z^+Q_A^-$ were trapped by illumination of a CN$^-$-treated PS II sample for 20 sec at 253 K and freezing at 77 K during illumination. Then the distance between this radical pair was determined to be 38.3 ± 1.0 A using '2+1' ESE experiment.

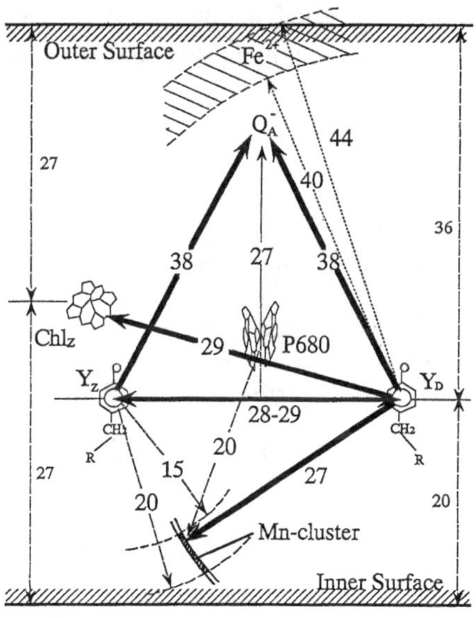

Fig. 5. Summary of the distances between paramagnetic species in PS II. The distances obtained by ELDOR and '2+1' methods are indicated by bold lines. The distance between Y_D and Y_Z has been recalculated from [10]. The position of P680 is assumed to be in an equal distance both from Y_D and Y_Z. The distances of Y_D and Y_Z from the membrane surfaces are assumed to be same. The distance between Y_Z and the Mn cluster has been obtained by the measurement of the spin-lattice relaxation time T_1 of Y_Z^+ in the S_1 state of Ca^{2+}-depleted PS II [14]. The sizes of each area show error ranges for the center of gravity of spin distributions. The distance between P680 and Q_A has been obtained by ESEEM of spin polarized radical pair induced by a laser pulse [15].

Considering Q_A to be hydrogen-bonded with a histidine coordinated to the non-heme iron the Fe^{2+} is located at the outer side to the membranes as in a purple bacterial reaction center, the distance between Y_D^+ and the non-heme iron may be estimated to be about 40 Å. The distance between Y_D and the non-heme iron was estimated to be 42 ± 2 Å by using the pulsed EPR measurement of a selective hole burning method [9] and 37 ± 5 Å from the spin-lattice relaxation time [8]. Therefore, their results seem to be in reasonable agreement within the experimental error.

5 Conclusion

The information on the distances between the electron transfer paramagnetic species in PS II obtained in this work and in others [8,9,10,11,14, 15] is collected in Fig. 5. Compared to other methods such as spin-lattice relaxation time measurement [7,8] and selective hole burning methods [9] the accuracy has been improved considerably. The ESEEM of a spin polarized radical pair $P680^+Q_A^-$ [15] in the CN^--treated PS II has also a similar accuracy. The distances from the Fe^{2+} to both tyrosines are approximately same, which suggest that the plane span by Q_A-Fe^{2+}-Q_B in Fig. 1 for a model structure is perpendicular to the plane of Y_D-Fe^{2+}-Y_Z. The Mn-cluster on the donor side resides closer to Y_Z. Thus the distance determination by ESE method has proved to be useful in complicated protein structures for which the X-ray data are not available.

References

1. A.W. Rutherford, A. Boussac and J-L. Zimmeremann, *New J. Chem.* **15** (1991) 491-500.
2. B. Svensson, I. Vass, E. Cedergren, E. and S. Styring, *EMBO J.* **7** (1990) 2051-2059.
3. B. Kok, B. Forbush, and M. McGloin, *Photochem. Photobiol.*, **11** (1970) 457-475
4. R.G. Evelo, S. Styring, A.W. Rutherford, and A.J. Hoff, *Biochim. Biophys. Acta* **973** (1989) 428-442.
5. J.B. Innes, and G.W. Brudvig, *Biochemistry* **28** (1989) 1116-1125.
6. Y.Isogai, S. Itoh, and M.Nishimura, *Biochim. Biophys. Acta* **1017** (1990), 204-208.
7. Y.Kodera, K.Takura, and A. Kawamori, *Biochim. Biophys. Acta* **1101** (1992) 23-32.
8. D.J. Hirsh, W.F. Beck, J.B. Innes,. and G.W.Brudvig, *Biochemistry* **31** (1992), 532-541.
9. H. Hara, and A. Kawamori, *Applied Magn. Reson.*, accepted
10. H. Hara, A. Kawamori, A.V. Astashkin, and T. Ono, *Biochim. Biophys. Acta* **1276** (1996) 140-146,
11. A.V. Astashkin, Y. Kodera, and A. Kawamori, *Biochim. Biophys. Acta* **1187** (1994) 89-93.
12. T. Kuwabara, and N. Murata, *Plant Cell Physiol.* **23** (1982) 533
13. Y. Sanakis, V. Petrouleas, and B.A. Diner, *Biochemistry* **33**, (1994) 9922-9928
14. Y. Kodera, H. Hara, A.V. Astashkin, A.Kawamori, and T. Ono, *Biochim. Biophys. Acta* **1232** (1995) 43-51.
15. *Biochim. Biophys. Acta*, submitted

The Electron Spin Relaxation Analysis of Conducting Polypyrrole Films under Oxygen

K. Kanemoto,[a] *A. Adachi,*[b] *and J. Yamauchi* [a]

[a] Graduate School of Human and Environmental Studies, Kyoto University, Kyoto 606, Japan
[b] Advanced Material Laboratory, Japan High Polymer Center, Ibaraki 305, Japan

Abstract. The electron spin dynamics of conducting polypyrrole heavily doped with benzensulfonate ions has been studied through measurements of spin-spin and spin-lattice relaxation times with CW ESR or pulsed one. The result that two relaxation times are equal indicates that electron spins in polaron are involved in the electron transport. Under oxygen, the relaxation rate increases with the rise of partial pressures of oxygen. The variation in the relaxation rate is explained on the basis of Langmuir' adsorption mechanism. Then, it is shown that the adsorbed oxygen molecules on polypyrrole give an additional dipole field, which leads to the increase of relaxation rate.

1 Introduction

Polypyrrole (PPy) is well known as one of the best candidates for technological application because of its stability upon exposure to the atmosphere [1]. Nevertheless, it was proved that PPy heavily doped with benzensulfonate ions exhibits a slight conductivity decrease depending on the partial pressure of oxygen molecule [2]. In order to investigate paramagnetic oxygen effects on PPy, ESR measurements are considered to be effective, where the interaction between oxygen and polaron (PPy^+) plays an important role. However, with the use of ESR parameters such as hyperfine interaction and g value, this effect cannot be elucidated because of the apparent absence of hyperfine splitting and the negligible variation in g value [3].

In this paper, we present approaches to examine oxygen effects on PPy^+ using the data of relaxation times derived from pulsed ESR as well as CW one. These electron spin relaxation experiments will lead to a quantitative analysis of its spin dynamics under oxygen.

2 Experimental

ESR measurements have been carried out on electrochemically grown, conductive PPy films heavily doped with benzensulfonate ions, which are provided by BASF (Lutamer Elektr Leifahigkeit Co., Germany). Those thickness and density are 35μm and 1.4 g/cm^3, respectively. The conductivity at room temperature is about 56 S/cm. Spin-lattice relaxation time T_1 was measured with a JEOL PX1050 FT-ESR spectrometer at X band (9.1GHz) using an inversion recovery pulse sequence (180°-τ-90°), while spin-spin relaxation time T_2 measured with the FT-ESR spectrometer using an 1 pulse free induction decay (FID) signal or with a JEOL FE1XG CW ESR spectrometer at X band using peak-to-peak line width (ΔH_{pp}). A Mn^{2+}-MgO solid solution was used as a reference for the g value and magnetic field calibration.

3 Results and Discussion

3.1 Relaxation Time Analysis Under Vacuum

The CW ESR spectra at 50-300K of PPy under vacuum (<10^{-3} Torr, 1Torr = 133.3Pa) have exhibited no hyperfine splitting and the negligible variation in g value (g = 2.0023-24). Despite a solid state, the line widths are very narrow (ΔH_{pp} at 293K is 0.021mT), which is compatible with the previous data [3, 4]. This narrow line width indicates the presence of highly mobile spins in the polaron.

A FID signal and the inversion recovery observed at 293K are shown in Figs. 1(a) and 1(b), respectively. The decay profile of 1(a) is expressed by a single exponential as well as the recovery one of 1(b), indicating that only one type of spins is relevant in these experimental observations.

Figure 2 shows the temperature dependence of T_1 and T_2 from 50 to 300K measured with a pulsed ESR under vacuum. This result indicates that two relaxation times are nearly equal, $T_1 = T_2$, whose relation can be realized under the condition $\omega_0\tau_c \ll 1$ [5], that is, a correlation time τ_c is sufficiently short compared to the inverse of a

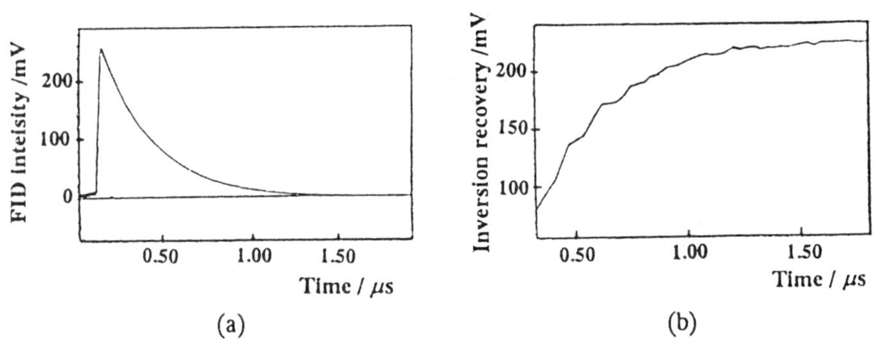

Fig. 1. Typical FID signals (a) and the inversion recovery (b) at 293K.

resonance frequency ω_o. This relation is often recognized in such very rapid motional electrons as in semiconductors. While it was thought that electron spins in polarons observed in ESR measurements and the carriers giving rise to electrical transport were different [6], Chauvet et al. [7] came to the conclusion, by measuring magnetoresistance, that both polarons and bipolarons participate in the transport. Consequently, the result observed here will be one of the proofs that polarons are involved in the transport.

3.2 Relaxation Time Analysis Under Oxygen

In Fig. 3, CW ESR spectra at 293K under vacuum and 152 Torr of oxygen molecule are shown. The linewidth (ΔH_{pp}) dramatically changes from 0.024mT (vacuum) to 0.208mT (152Torr), which indicates the presence of a strong magnetic interaction between electron spins in polaron and oxygen molecules. Then, Fig. 4 gives the dependence on the partial pressure of oxygen, of spin-spin relaxation rate (T_2^{-1}) derived from pulsed and CW ESR measurement. Properly, T_2^{-1} should be determined by dynamic method, pulsed ESR as in Fig. 2, yet the results by CW ESR below 20Torr have exhibited a similar tendency to those by pulsed one. Therefore, over 20Torr, where pulsed measurement cannot be made due to dead time, T_2^{-1} determined by CW method will be adequate over the whole range of pressure. Then it is confirmed that T_2^{-1} decreases with the rise of partial pressure of oxygen.

As under vacuum, it was identified under oxygen that $T_1 = T_2$, τ_c is sufficiently short. Under this condition, two relaxation rates are given by [5]

$$\frac{1}{T_1} = \frac{1}{T_2} = 2\gamma_e^2 H^2 \tau_c \,. \tag{1}$$

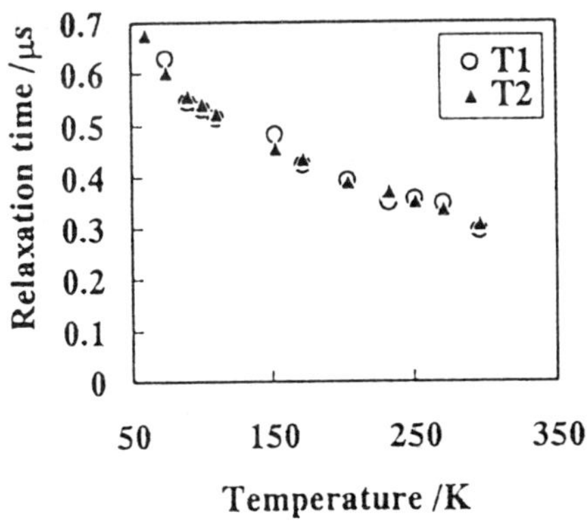

Fig. 2. Temperature dependence of T_1 and T_2 measured with a pulsed ESR under vacuum.

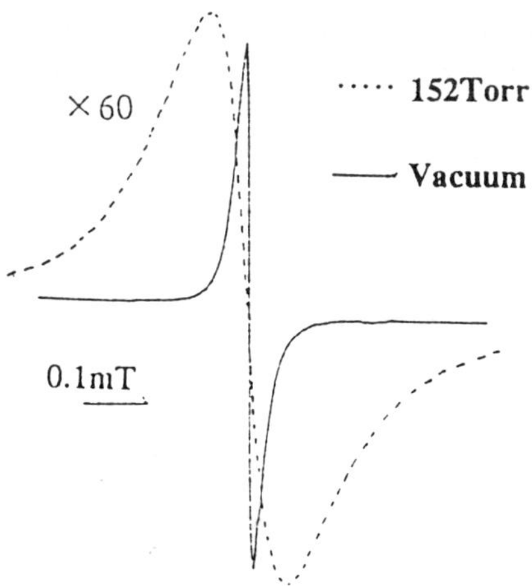

Fig. 3. Typical CW ESR spectra under vacuum and 152 Torr of oxygen at 293K.

Here γ_e is the gyromagnetic ratio of the electron spin, and H is an isotropic fluctuating field in the limit of very rapid motion. Since τ_c modulated by the motion of electron spins, that is, concerned with the conductivity, exhibits only a slight variation in conductivity depending on the rise of oxygen pressure [2], we are able to assume that τ_c is nearly constant under oxygen. Then, it is concluded that the increase of T_2^{-1} in Fig. 4 is caused by a enhanced fluctuating field.

As shown in Fig. 3, the CW ESR spectrum under 152 Torr seems to consist of one type of electron spins, which indicates that oxygen molecules penetrate into PPy films. Furthermore, the spectral variation in Fig. 3 is reversible within about half a day. By these facts, oxygen molecules are considered physically adsorbed on PPy films. Therefore, the increase of T_2^{-1} in Fig. 4 is attributed to an additional dipole field due to electron spins of adsorbed oxygen molecules. Then T_2^{-1} can be divided into two parts :

$$\frac{1}{T_2} = \frac{1}{T_{2,v}} + 2\,\gamma_e^2\,\Delta H^2\,\tau_c \tag{2}$$

where $T_{2,v}$ is the spin-spin relaxation time under vacuum, and ΔH is the additional dipole field due to electron spins of adsorbed oxygen molecules.

In order to examine the increase of T_2^{-1} due to adsorbed oxygen molecules in more detail, we analyze this data on the basis of the relationship of the Langmuir adsorption as follows :

$$\theta = \frac{KP}{1+KP} \tag{3}$$

where θ represents the proportion of the adsorption amount, and K, the equilibrium constant of adsorption, is given by $K = k_a / k_d$ with rate constants k_a for adsorption and k_d for desorption. Then, on the presumption that ΔH is proportional to the amount of adsorbed oxygen molecules around electron spins in polarons, θ can be expressed by

$$\theta = \frac{\Delta H}{\Delta H_\infty} \tag{4}$$

where ΔH_∞ is the additional dipole field in the maximally adsorptive state. Thus, utilizing (2), (3) and (4), we get

$$P = a\,P\left(\frac{1}{T_2} - \frac{1}{T_{2,v}}\right)^{-1/2} - \frac{1}{K} \tag{5}$$

$$\text{with}\quad a = \left(2\,\gamma_e^2\,\Delta H_\infty^2\,\tau_c\right)^{1/2} = \left(\frac{1}{T_{2,\infty}} - \frac{1}{T_{2,v}}\right)^{1/2} \tag{6}$$

where $T_{2,\infty}$ is the spin-spin relaxation time in the maximally adsorptive state. From equation (5) the graph, plotted as P vs. $P(1/T_2 - 1/T_{2,v})^{-1/2}$, is expected to become a straight line. Fig. 5 shows the relationship between P and $P(1/T_2 - 1/T_{2,v})^{-1/2}$, indicating that our view is valid except at low partial pressure of oxygen, under which alternative mechanisms may be involved. The slope of this straight line gives $T_{2,\infty}$ = 0.0182μs and moreover, its intercept determines $K = k_a / k_d$ = 0.0212 Torr^{-1}. Then, with K^{-1} = 47.2 Torr, we get at the equilibrium state the relation $\theta = P / (P + 47.2)$.

In Fig.6, the time dependence of T_2^{-1} determined by CW ESR for the variation in partial pressure of oxygen from 0.02 to 152 Torr is shown. Here, T_2^{-1} goes up gradually and after once reaching 38.5 μs^{-1}, it returns to the equilibrium value (T_2^{-1} =32.1μs^{-1}) half a day later. Such a long time may be taken to average out the adsorbed sites on PPy to the equilibrium state. Then, on the rising step, the above mentioned mechanism of the adsorbed oxygen will be applicable. According to Langmuir's relationship, the time variation of θ is expressed by the following relation :

$$\frac{d\theta}{dt} = k_a P(1-\theta) - k_d \theta \, . \tag{7}$$

This relation, when solved on the initial condition $T_2 = T_{2,0}$ at t = 0, can be changed from (4)(6) to the next expression :

$$\ln\left|\left(\frac{1/T_2 - 1/T_{2,\mathrm{v}}}{1/T_{2,\infty} - 1/T_{2,\mathrm{v}}}\right)^{1/2} - \frac{KP}{KP+1}\right| = b - (k_a P + k_d)t \tag{8}$$

with $$b = \ln\left|\left(\frac{1/T_{2,0} - 1/T_{2,\mathrm{v}}}{1/T_{2,\infty} - 1/T_{2,\mathrm{v}}}\right)^{1/2} - \frac{KP}{KP+1}\right| .$$

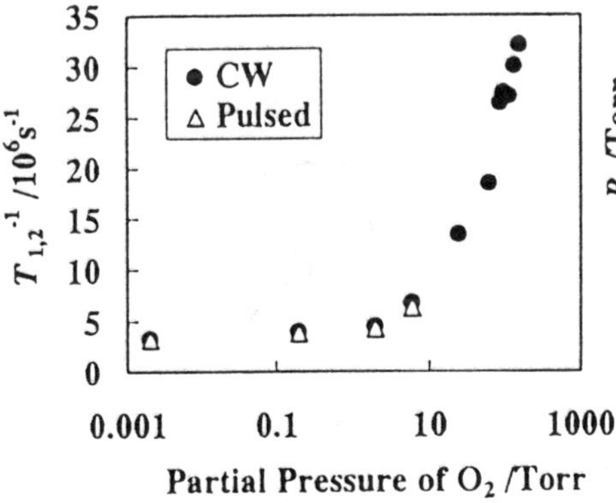

Fig. 4. T_2^{-1} vs. partial pressure of oxygen at 293K.

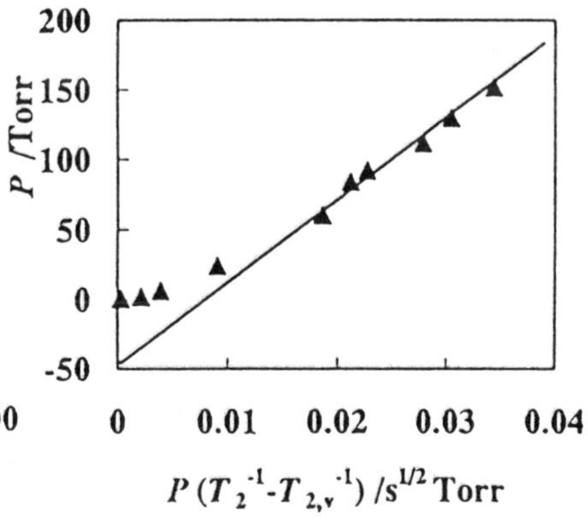

Fig. 5. Partial pressure of oxygen (P) vs. $P(1/T_2 - 1/T_{2,v})^{-1/2}$.

Using this equation, the result in Fig. 6 gives a straight line (Fig. 7) on the rising step of T_2^{-1}. Therefore our mechanism is proved to be valid in this range similarly as in Fig. 5. Performing the analysis of the slope $(k_a P + k_d)$, we can determine from k_a / k_d = 0.0212 Torr^{-1} that $k_a = 1.14 \times 10^{-5}$ ($\mathrm{s}^{-1}\ \mathrm{Torr}^{-1}$) and $k_d = 5.38 \times 10^{-4}$($\mathrm{s}^{-1}$).

4 Conclusion

Through the electron spin dynamic study of polypyrrole (PPy) heavily doped with benzensulfonate ions, we have been able to show :

1) In spite of the presence or the absence of adsorbed oxygen molecules, spin-spin and spin-lattice relaxation times are equal ($T_1 = T_2$) ; the correlation time is very short. This property, often exhibited by conduction electrons in semiconductors, will be one of the proof that electron spins in polaron are also involved in the electron transport.

2) Relaxation rate (T_2^{-1}) has increased remarkably with the rise of the partial pressure of oxygen. This variation is proved to be explained in terms of the dipole field due to the adsorbed oxygen, not by the correlation time. Provided that the increase of T_2^{-1} is dependent on the number of adsorbed oxygen molecules around electron spin in polaron, the variation in T_2^{-1} under oxygen can be fitted *via* an adsorptive model of Langmuir's type.

3) The model also gives a description for the time dependence of T_2^{-1} under oxygen, but the equilibrium value of T_2^{-1} is obtained after reaching to the maximum peak value. This may be because it takes somewhat long time to average out the adsorbed sites on PPy.

References

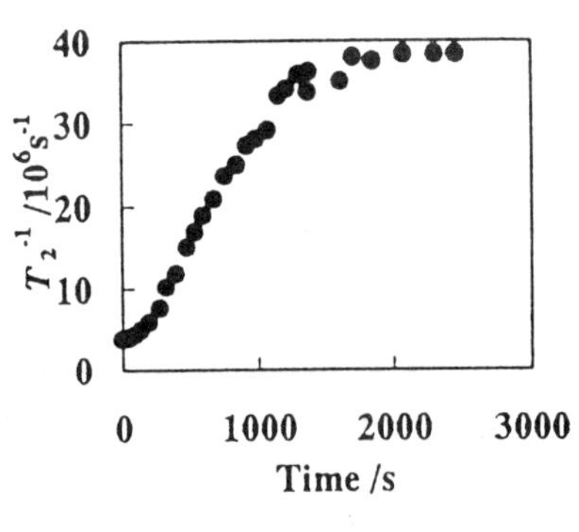

Fig. 6. Time dependence of T_2^{-1} for the variation in partial pressure of oxygen from 0.02 to 152 Torr at 293K.

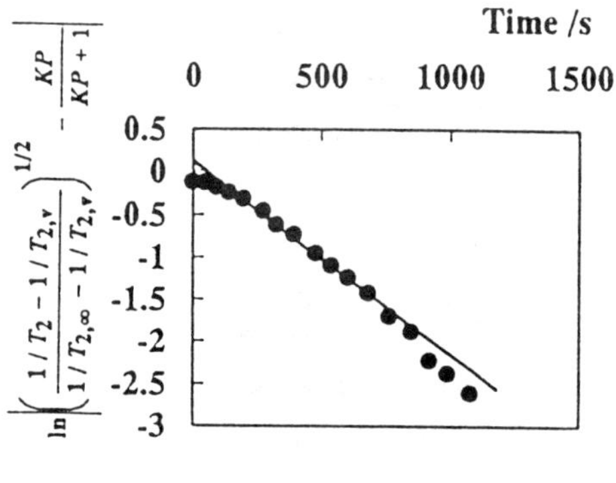

Fig. 7. $\ln|\{(T_2^{-1} - T_{2,v}^{-1}) / (T_{2,\infty}^{-1} - T_{2,v}^{-1})\}^{-1/2} - KP / (KP+1)|$ vs. time.

1. A. Adachi and J. Yamauchi, *Synth. Met.* **73** (1995) 101-105
2. A. Adachi and J. Yamauchi, *Bull. Chem. Soc. Jpn.* **69** (1996) 811-814
3. G. B. Street, T. C. Clarke, M. Krounbi, K. K. Kanazawa, V. Lee, P. Pfluger, J. C. Scott, and G. Weiser, *Mol. Cryst. Liq. Cryst.* **83** (1982) 253-264
4. H. S. Nalwa, *Phys. Rev. B,* **39** (1989) 5964-5974
5. C. P. Slichter, *Principles of Magnetic Resonance, 3rd ed.*, Springer-Verlag, Berlin (1990)
6. P. Pfluger, G. Weiser, J. Campbell Scott and B. Street, in T.A. Skotheim (ed), *Handbook of Conducting Polymers*, Marcel Dekker, New York (1986)
7. O. Chauvet, S. Paschen, L. Forro, L. Zuppiroli, P. Bujard, K. Kai, and W. Wernet, *Synth. Met.* **63** (1994) 115-119

Photo-induced Hydrogen and Electron Transfer in Solution as Studied with FT-EPR

R. Groß, M. Plüschau*, K.-P. Dinse*, and D. Beckert***

*) Physikalische Chemie III, Technische Hochschule Darmstadt
**) Chemisches Institut der Universität Leipzig

Abstract. Application of FT-EPR with its inherent time and frequency resolution to the study of photo-induced reactions in solution leads to an unambiguous identification of radicals on a sub-microsecond time scale. The technique was used to study the isotope effect in a OH-bond breaking reaction and also to explore the hydrogen addition reaction to [60]-fullerene. In the former case we found that the mass transfer rate was decreased by a factor of 2 at room temperature, an isotope effect much less than anticipated. In the latter case we found that the hydrogen addition reaction in which various intermediate radicals are involved proceeds via a correlated triplet/doublet pair leading to strong electron spin polarisation of the final reaction product.

1 Introduction

The inherent time and frequency resolution of Fourier-transform EPR (FT-EPR) render this experimental method an ideal tool for the study of light-induced hydrogen and electron transfer reactions in solution, because unambiguous identification of paramagnetic reaction products can be performed on a nanosecond time scale. This feature is particularly important if parallel reactions, for instance with solvent molecules, or "branching" in the reaction pathways can occur. In a previous study /1/ we have shown that in the photo-induced hydrogen abstraction reaction in the photosystem anthraquinone (AQ)/4-methyl-2,6-di-*tert*.-butylphenol (DTBP) the hydrogen transfer can be described by a 2-step process in which the primary electron transfer first constitutes a correlated radical ion pair. This radical pair has a finite probability to separate in free radical ions before the final proton transfer completes the reaction, yielding the neutral 9-hydroxy-anthroxyl ($AQH^{\cdot}$) radical and the corresponding phenoxyl radical. Under the experimental conditions studied, the "escape" probability amounted to 7(2) % in 2-propanol at room temperature.

The aim of the present investigation was to determine if the rate constant of the mass transfer step shows an isotope effect. In our experimental set-up, such an isotope-dependence could be detected conveniently by measuring the escape probability of the primary reaction product, the anion radical $AQ^{-\cdot}$. Any decrease of the rate constant of the mass transfer step would lead to an increase of the production yield of $AQ^{-\cdot}$. By measuring the relative yield of $AQ^{-\cdot}$ vs. $AQH^{\cdot}$ (or $AQD^{\cdot}$), the difficult problem of an absolute yield determination is bypassed.

2 Results and Discussion

As was shown in ref. 1, EPR lines of AQH$^\cdot$ and AQ$^{-\cdot}$ can be selectively detected, and analysis of signal build-up and decay as a function of delay time to the laser pulse can be used to determine the relative concentration of both radicals. If AQD$^\cdot$ instead of AQH$^\cdot$ is generated, the resulting EPR spectrum is considerably more crowded and difficulties arise to determine the signal intensities of AQD$^\cdot$ and AQ$^{-\cdot}$ lines independently from each other. This problem was solved by utilising the small g-factor difference of both radicals, by which small changes of the microwave frequency of the spectrometer can be converted into a differential shift of both spectra. As a result, even for the congested case AQD$^\cdot$ / AQ$^{-\cdot}$ unperturbed lines of each radical could be identified, as is shown in Fig. 1.

In Fig. 2, the EPR signal intensity in the protonated and the deuterated system is shown for both radicals as a function of delay time. Using the same kinetic scheme as described in ref. 1, the time dependence of the experimental data can be well described over 4 decades in time. From the ratio of signal intensities observed

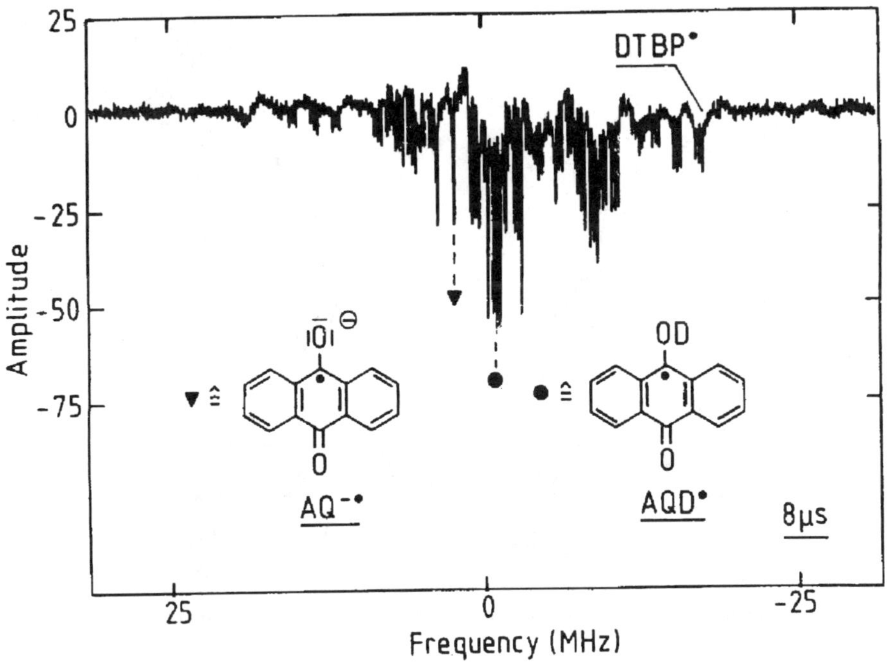

Fig. 1. FT-EPR spectra of the photoreaction products of AQ/DTBP(deut.) in 2-propanol taken at a delay time of 8 μs with respect to the laser pulse. All observed lines can be attributed to hfs lines of AQ$^{-\cdot}$, AQD$^\cdot$, and the neutral 4-methyl-2,6-di-*tert.*-butylphenoxyl radical. Representative lines used for the evaluation of the radical polarisation are indicated.

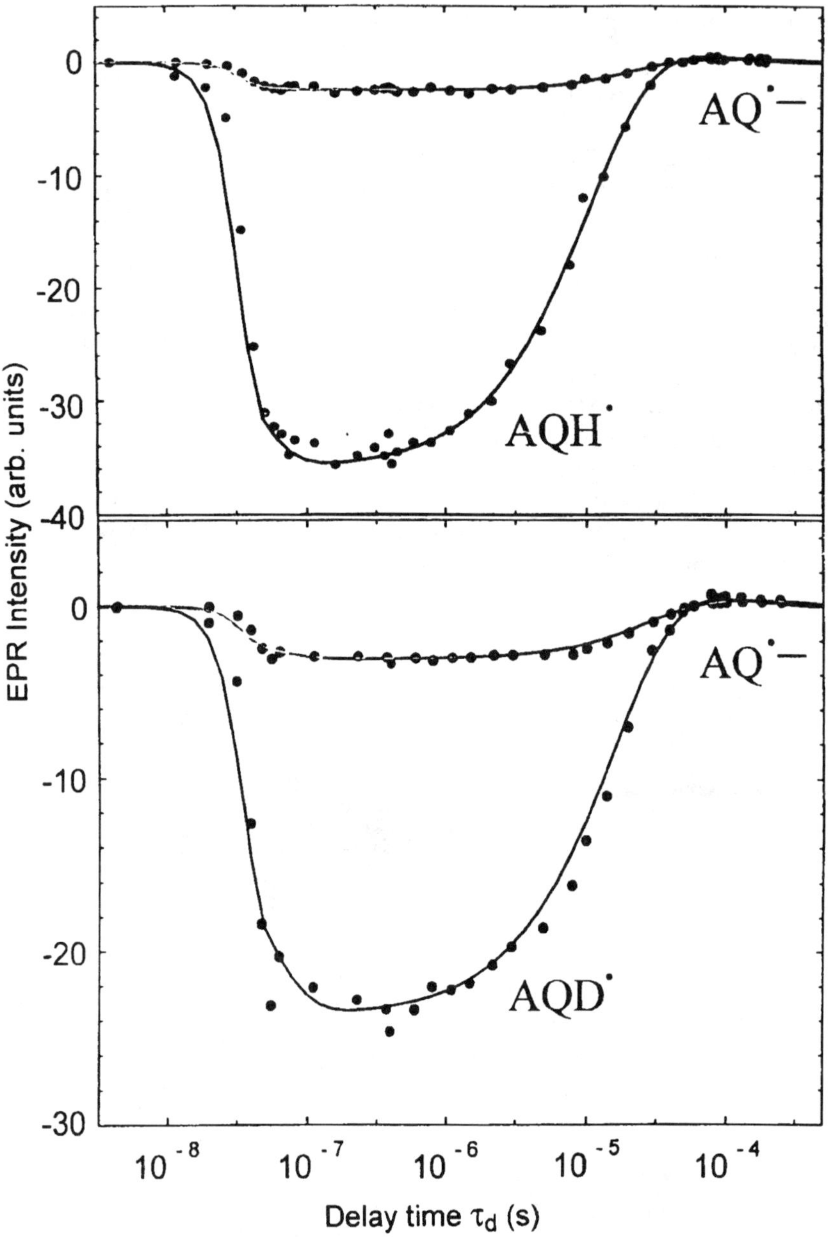

Fig. 2. EPR signal intensity of the protonated and deuterated system as a function of delay time

between 10^{-7} and 10^{-6} s, the escape probability of the anion radical can be deduced with good accuracy even without invoking details of the kinetic scheme. This results from the fact that for these early times, spin lattice relaxation is not effective, and therefore one can directly convert the (emissive) EPR signal intensity into a ratio of radical concentrations, because both radicals originate from the same triplet precursor.

From the simulation a ratio $k(H^+)/k(D^+) = 2.0(2)$ is obtained, much smaller than would be expected for a simple bond-breaking mechanism. It should be noted, however, that this value is similar to that deduced by van Willigen and Hirota /2/ from a study of the isotope effect in the hydrogen abstraction reaction in the system acetone/2-propanol. We tentatively explain the relatively small isotope effect by considering sterical arguments, which implies that matching of relative orientation of reaction partners is more important than bond breaking. Further experiments to explore the influence of solvent molecules are in progress.

In a further study we investigated the light-induced hydrogen addition reaction to [60]-fullerene. Experiments with c.w. light irradiation have demonstrated that a highly spin polarised spectrum of $^2HC_{60}^{\cdot}$ can be observed in solution /3/ and a detailed reaction scheme was proposed /4/. In order to verify the complicated reaction mechanism which invoked various intermediate radicals and to determine the hitherto unexplained spin polarisation mechanism, time resolved experiments were necessary.

According to ref. 4, the system C_{60}/benzophenone/2-propanol in decalin led to the most intense $^2HC_{60}^{\cdot}$ signals under steady-state conditions. First we verified the existence of $^2(C_6H_5)_2COH^{\cdot}$ and $^2(CH_3)_2COH^{\cdot}$ transient radicals when exciting a solution of $2 \cdot 10^{-3}$ M benzophenone and 1 M 2-propanol with 308 nm laser pulses. When adding 10^{-4} M [60]-fullerene, the otherwise dominating emission/absorption-polarised $^2(C_6H_5)_2COH^{\cdot}$ signal disappeared, and only triplet [60]-fullerene and $^2HC_{60}^{\cdot}$ signals were observed. As evident from Fig. 3, the $^2HC_{60}^{\cdot}$ signal is polarised in an emission/absorption pattern. This polarisation pattern is generated on a sub-µs time scale. This short-time polarisation cannot result from a $^2HC_{60}^{\cdot}/\ ^3C_{60}$ "doublet-triplet" encounter, as was proposed previously, because such a bi-molecular reaction would lead to an estimated rise time of the polarised signal of more than 30 µs. (For this estimate a diffusion-controlled rate constant of $3 \cdot 10^9\ M^{-1}s^{-1}$ and a concentration of $^3C_{60}$ of 10^{-5} M was used.) Instead, the observed polarisation in the S=1/2, I=1/2 spin system of the $^2HC_{60}^{\cdot}$ radical, which can also be termed "dipolar order" because of the missing electron and nuclear Zeeman polarisation, must originate from the spin polarisation of the precursor $^2(C_6H_5)_2COH^{\cdot}$ radical. This polarisation, however, can only be transferred to $^2HC_{60}^{\cdot}$ invoking hyperfine interaction and level-anti-crossing (LAC) in the reacting $^3C_{60}/\ ^2(C_6H_5)_2COH^{\cdot}$ triplet/doublet pair.

Qualitatively, this mechanism can be described as follows: In the incipient encounter phase, triplet C_{60} and the doublet precursor radical can be described approximately in the product representation $|S=1, M_S\rangle|S'=1/2, M_S', M_I\rangle$ leading to nearly degenerate electron spin levels except for $|M_S + M_S'| = 3/2$. Because of the spin-polarised doublet component, these close lying levels will also exhibit different populations. Hyperfine mixing of electron and nuclear spin states can then lead to a partial transfer of this population difference into nuclear spin levels. On closer contact of the reaction partners, the increasing exchange interaction necessitates description of the pair in a coupled basis. Hydrogen transfer can only occur within the doublet component, whereas pairs in the quartet spin state will re-dissociate.

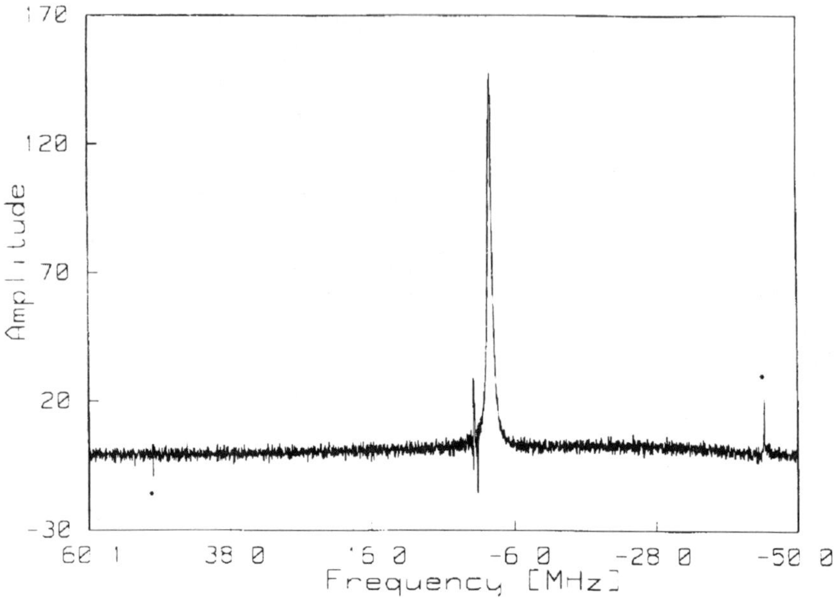

Fig. 3. EPR signal of triplet state [60]-fullerene (centre) and of the $^2HC_{60}^{\cdot}$ radical (*) as observed in the system [60]-fullerene/benzophenone/2-propanol/decalin at 8 μs after the laser pulse (308 nm).

"Dipolar order" in the reaction product $^2HC_{60}^{\cdot}$ is generated because one of the M_S sublevels of $^2HC_{60}^{\cdot}$ originates exclusively from the formerly degenerate electronic spin levels which were subject to hyperfine mixing, whereas the other M_S sublevel is formed from the "isolated" $|M_S + M_S'| = 3/2$ level. For that reason this particular electron spin sublevel carries no nuclear spin polarisation, whereas the other one is nuclear spin polarised. As a consequence, EPR transitions within this 4-level system are oppositely polarised, which is the signature of dipolar order.

From the instant when the final product $^2HC_{60}^{\cdot}$ is formed, no further spin mixing can occur because the hydrogen donor has become diamagnetic. Nevertheless, from the observed polarisation in the $^2HC_{60}^{\cdot}$ ("emissive/absorptive") the sign of the proton hyperfine interaction constant can be derived as positive. For this conclusion, only the precursor radical polarisation (also "emissive/absorptive") is relevant, because in our case the g values of $^3C_{60}$ and $^2(C_6H_5)_2COH^{\cdot}$ are quite similar. At first sight the predicted positive sign of the proton hyperfine interaction constant seems surprising because large proton coupling constants in aromatic hydrocarbons are all negative. Here, however, hydrogen addition provokes rehybridization of the attachment position and subsequent localisation of spin density at next neighbour carbon position, as was already shown by Negri et al. /5/. Quantum chemical calculations using UHF 3/31G with subsequent LDA correction confirmed sign and size of the experimentally observed coupling constant, thus confirming that the proposed spin polarisation scheme is qualitatively correct. Further work has to be performed in order to verify the reaction and polarisation scheme also quantitatively.

Acknowledgement

Financial support of the Deutsche Forschungsgemeinschaft (Di 182/12) and of the Volkswagenstiftung is gratefully acknowledged. The quantum chemical calculations were performed by Dr. J. Schulte.

References

1. M. Plüschau, G. Kroll, K.-P. Dinse, and D. Beckert, *J. Phys. Chem.* **96** (1992) 8820-8827.
2. K. Tominaga, S. Yamauchi, and N. Hirota, *J. Phys. Chem.* **95** (1991) 3671-3676; P. R. Levstein and H. van Willigen, *J. Chem. Phys.* **95** (1991) 900-908.
3. J. R. Morton, K. F. Preston, P. J. Krusic, and L. B. Knight Jr., *Chem. Phys. Lett.* **204** (1993) 481-485.
4. R. Klemt, E. Roduner, and H. Fischer, *Chem. Phys. Lett.* **229** (1994) 524-530.
5. J. R. Morton, F. Negri, and K. F. Preston, *Can. J. Chem.* **72** (1994) 776-782.

Superhyperfine Interaction of ^{127}I with Cu^{2+} probe in $Na_4KCu(IO_5OH)_2 \cdot 12H_2O$ crystal

Ruiyun Zhan, Mingmei Wu, Jie Xu, Guizhen Liu, Yayan Liu and Qiang Su

Changchun Institute of Applied Chemistry, Chinese Academy of Sciences, Changchun 130022, P. R. China

Abstract. ESR method has been used to study the superhyperfine interaction of ^{127}I with a Cu^{2+} probe in $Na_4KCu(IO_5OH)_2 \cdot 12H_2O$ complex. The main purpose of this paper is to confirm the existence of unpaired electron spin on the iodine atom and to find a reasonable explanation for the spin distribution from the Cu^{2+} ion. The calculation of ESR parameters shows that about 0.77% of the unpaired electron spin is located on each iodine atom.

Recently bis (hydrogen periodato) cuprate complex has been synthesized. Its structure study indicated that the complex contained a CuO_4 plane configuration and four chains of Cu-O-I in each repeated lattice unit. [1]. This configuration is analogous to those in cupric oxide and some copper-containing ceramic compounds. Generally, the latter has ferromagnetic or antiferromagnetic properties due to spin superexchange interaction [2,3]. Therefore, any information that can be obtained from magnetic resonance about its magnetic properties is very important. In this paper, we first report the ESR studies on the superhyperfine of ^{127}I with Cu^{2+} probe, which confirm the electronic transfer from copper along the Cu-O-I chains..

1 Experimental

1.1 Preparation of the Sample

The synthesis of the complex was described in ref [1]. The chemical and instrumental analysis indicated that the complex was a coordination complex. Most of Cu ion had a trivalent state. But about 0.5% of Cu ions had been found to be divalent. The ESR silence of Cu^{3+} at room temperature is probably due to large zero field splitting in the S=1 unit. But the Cu^{2+} ions can be used as a spin probe for ESR study.

1.2 ESR Experiment

ESR spectra were recorded on a JEOL JES-FE3AX spectrometer operating at X-band, with the microwave frequency 9.42GHz. A JEC-UCR-ZX angular rotator was used to obtain the angular dependent ESR spectra.

2 Results and Discussion

2.1 Paramagnetic Center

The X-ray diffraction study indicated[1] that the single crystal of the $Na_4KCu(IO_5OH)_2 \cdot 12H_2O$ complex belongs to the monoclinic system with a space group $P2_1/c$, the parameters of cell are a=0.6108nm, b=2.5055nm, c=1.1716nm, $\beta = 97.77°$, V=2.231 nm^3, Dc=2.555g/cm^3, Z=4.

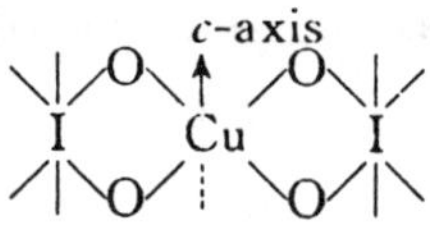

Fig.1 Structure diagram of paramagnetic center ion.

The copper atom is chelated by two groups of distorted octahedral $IO_5(OH)$ to form a CuO_4 square plane with average Cu-O distance of 0.183nm. The coordination number of iodine is 6 and each $IO_5(OH)$ has a distorted octahedral structure. The nearest neighbour of the copper atom is an oxygen atom of a water molecule at 0.2688nm over the plane. Therefore, the copper atom and five oxygen atoms form a tetragonal monopyramidal configuration. The Cu^{2+} with nearby compensating charge is assumed to has an approximate C_{2v} symmetry, see Fig.1.

2.2 Analysis of ESR Spectra

Assuming uniaxial symmetry of the paramagnetic center for the c-axis, the spin Hamiltonian is written as

$$H = \beta\left[g_{//}H_zS_z + g_{\perp}(H_xS_x + H_yS_y)\right] + A_0S_zI_{0z} + B_0(S_xI_{0x} + S_yI_{0y}) + \sum_{i=1}^{2}\left[A_iS_zI_{iz} + B_i(S_xI_{ix} + S_yI_{iy})\right] \quad (1)$$

where the first term is the Zeeman energy of the electron spin S(=1/2).The second term and third term show the hyperfine interaction between S and nuclear spin

I_0(=3/2) of Cu^{2+}. A_0 and B_0 are the parallel and perpendicular components of the hyperfine coupling tensor, respectively. The last term is the superhyperfine interaction between S and iodine nuclear spin I_i (=5/2, i=1,2). A_i and B_i are their parallel and perpendicular components of superhyperfine coupling tensors, respectively. Considering the assumed symmetric properties of paramagnetic Cu^{2+} center, we define a resultant nuclear spin $I=I_1+I_2$.

The allowed transitions of ESR are those in which $\Delta M=\pm 1$, $\Delta m_0=0$, $\Delta m=0$, where M is the electron magnetic quantum number, m_0 and m are the nuclear magnetic quantum numbers of $^{63,65}Cu$ and ^{127}I, respectively. For a single crystal sample, the ESR spectra are of angular dependence. The ESR parameters can be determined by a tensor elements measurement and a tensor diagonalization procedure. But, in our case, due to the overlap of some superhyperfine lines, it is difficult to obtain spectral parameters by a tensor measurement method. However, when the external field is parallel or perpendicular to the c-axis the spectra recorded will be more simple. Their resonance transition positions can be written as

$$H(m_0,m,\theta=0^\circ)=\frac{g_D}{2g_{//}}(H_0-A_0m_0-Am)+\frac{g_D}{2g_{//}}\{(H_0-A_0m_0-Am)^2 -2B_0^2(\frac{15}{4}-m_0^2)-2B^2(30-m^2)\}^{1/2} \quad (2)$$

$$H(m_0,m,\theta=90^\circ)=\frac{g_D}{2g_\perp}(H_0-B_0m_0-Bm)+\frac{g_D}{2g_\perp}\{(H_0-B_0m_0-Bm)^2 -(A_0^2+B_0^2)(\frac{15}{4}-m_0^2)-(A^2+B^2)(30-m^2)\}^{1/2} \quad (3)$$

Where $H_0=h\nu/g_D\beta$, g_D is a g-factor of DPPH.

A typical experimental result is shown in Fig. 2. When the external field is along the c-axis , there are four main peaks in the spectrum (see Fig.2a), which are attributed to the hyperfine splitting of $^{63,65}Cu$. Some weak lines overlap on the main peaks. They arise from the superhyperfine interaction of the iodine nuclei. When the external field is perpendicular to c-axis, the ESR spectrum is shown in Fig.2b. In this spectrum, no main peaks of Cu^{2+} ions can be observed directly. But then, many superhyperfine lines had been detected. Some of them overlapped each other and were scattered over a wide region.This result indicated that the hyperfine and the superhyperfine interaction are of the same order in this direction.On the other hand,we had seen an obvious flotation of base-line in the spectra. This effect may arise from the high-spin Cu^{3+} ions. The ESR powder spectrum of the complex is shown in Fig.3. It is a typical spectrum of Cu^{2+}-containing powder sample. The fine lines on the enveloped curve of the Cu^{2+} spectrum also revealed that the effect of the superhyperfine interaction of iodine is remarkable.

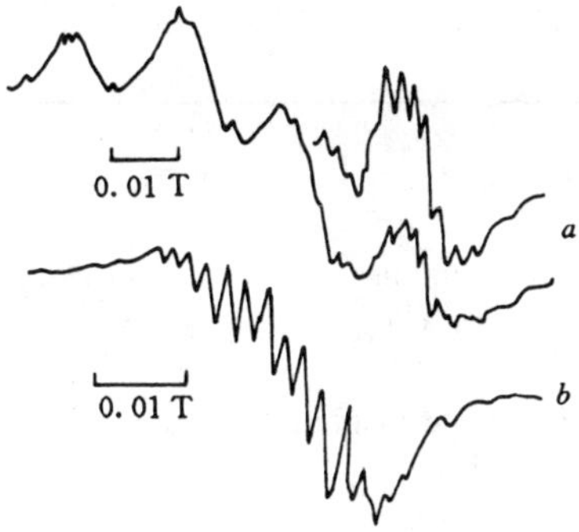

Fig 2. ESR spectra of $Na_4KCu(IO_5OH)_2 \cdot 12H_2O$ complex. The external field is (a) along and (b) perpendicular to c-axis, respectively.

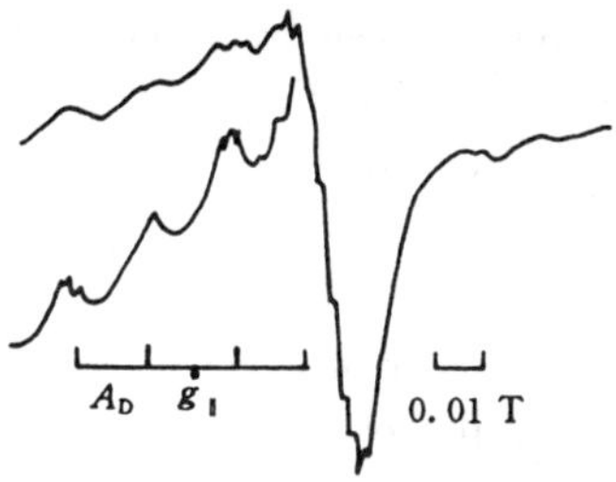

Fig 3. ESR powder spectrum of $Na_4KCu(IO_5OH)_2 \cdot 12H_2O$ complex.

In order to obtain all parameters, following method had been used [5]. In the first place, we estimated approximate values of $g_{\parallel}$, $g_{\perp}$, A_0, B_0, A, B and put these initial values into eq.(2) or (3), and calculated all resonance lines of positions. The calculated resonance pósitions H^c are compared with the experimental data H^e. We can obtain a deviation parameter Δ H, which is defined by

$$\Delta H = \left[\sum_i (H_i^c - H_i^c)^2 \right]^{1/2} \tag{4}$$

The calculation of eq.(4) is replaced by slowly changing the resonance parameters untill the minimum Δ H is obtained. All parameters are listed in Table 1.

Table 1. ESR parameters of $Na_4KCu(IO_5OH)_2\cdot 12H_2O$ complex

g factor	$g_{//}$=2.236	$g_{\perp}$=2.031
Hyperfine coupling constant of $^{63,65}Cu$	$\lvert A_0 \rvert$=165.43	$\lvert B_0 \rvert$=27.81
Superhyperfine coupling constant of ^{127}I	$\lvert A \rvert$=16.66	$\lvert B \rvert$=21.48

The unit of coupling constants is 10^{-4} T

2.3 Unpaired Electron Density on Iodine Atoms

In CuO_4 plane, the covalency bonds consist of $3dx^2$-y^2 orbital of Cu ion and 2s-2p hybrid orbital of oxygen ions. The out-of-plane π bond and in plane π bond are not important. Therefore, the unpaired electrons of Cu^{2+} ion are mainly located around the Cu-O bonds. However, the superhyperfine interaction of ^{127}I in ESR spectra indicates that a part of unpaired electrons are located on the iodine atoms. In other words, some unpaired electrons of Cu^{2+} ion had migrated to iodine atoms along the Cu-O-I chains. The transfer mechanism would be interpreted by a spin polarization effect, in which α spin on Cu^{2+} ion results in spin polarization of filled oxygen 2p orbital encouraging β spin on iodine ion .This explanation is analogous to those in CuO[4] and some copper-containing ceramic materials[2,3], which produce a ferromagnetic or antiferromagnetic superexchange interaction on Cu-O-Cu chain. According to this mechanism, a negative spin density would be generated on the iodine atom . Based on ESR theory, the superhyperfine interaction constants A and B can be written as follows,

$$A = A_s + A_p + A_d \tag{5}$$

$$B = A_s - (A_p - A_d)/2 \tag{6}$$

where A_s and A_p are isotropic and anisotropic superhyperfine interaction terms of iodine nuclei, respectively. A_d is the dipolar contribution.These terms are given by

$$A_s = (8\pi /3) g\beta g_n\beta_n \lvert \psi(0) \rvert^2 f_s \tag{7}$$

$$A_p = (4/5) g\beta g_n\beta_n \langle r^{-3} \rangle_p f_p \tag{8}$$

$$A_d = g\beta g_n\beta_n R^{-3} \tag{9}$$

where g_n is g-factor of iodine nucleus and β_n is nuclear Bohr magneton.$|\psi(0)|^2$ is the density of 5s electrons on the iodine nucleus place. r in $<r^{-3}>_p$ is the distance of 5p electron from the iodine nucleus and R is the atomic distance between copper and iodine atom. f_s and f_p are the unpaired electron density on the 5s and 5p orbital of iodine atom, respectively. The f_s+f_p is the resultant electronic density on iodine atom. Considering a larger R value of 0.2879nm, the dipolar interaction term A_d is negligible. Therefore, the isotropic term can be calculated to be -1.987mT and anisotropic coupling constant to be A_p=3.21, -1.61 and -1.61 mT, respectively. When the unpaired electrons are entirely located in 5s or 5p orbitals , the value of A_s and $(A_p)_z$ are 41600 MHz and 1624.8MHz, respectively[6]. Therefore the f_s+f_p can be determined by using eq.(7) and eq.(8). The results show that about 0.77% of unpaired electron are on the each iodine atom.

References

1. Mingmei Wu, Qiang Su, Ninghai Hu, et al., *J. Inorg. Chem.*(Chinese), **19**(1993)48.
2. R. T. Cava, *Scientific American,* **263(8)**(1990)24.
3. J. R. Morton, *J. Chem. Soc., Faraday Trans. I,* **85**(1989)4091.
4. M. O' Keeffe and F. S. Stone, *J. Phys. Chem. Solids*, **23**(1962)261.
5. S. Twads, K. Okuda and M. Date *J. Phys. Soc. Jap.*, **47**(1979)512.
6. W. Jr. Weltner, *Magnetic Atoms and Molecules*, New York: Scientific and Academic Edition, (1983), pp.343.

EPR Investigation of MoO_3 Doped with Vanadium Catalysts

Nguyen Tien Tai

Institute of Chemistry - National Centre for Natural Science & Technology of Vietnam.
Nghia do- Tuliem - Hanoi - Vietnam.

Abstract. EPR was applied in a study of the phase transition of V / MoO_3 catalyst. The samples were prepared by thermal decomposition of the mixtures: NH_4VO_3 and $(NH_4)_6$ $Mo_7O_{24} \cdot 4H_2O$. The decomposition temperatures varied within the range of 473 - 723 K. The observed EPR signals consist of 24 lines, which correspond to anisotropic hyperfine structure (hfs) components. The hfs arises from interaction of the unpaired electron with the nuclear spin of ^{51}V. Each of the 24-hfs lines, registered at 77 K, exhibits further structure: extrahyperfine. Two phases of MoO_3, hexagonal and rhombic, formed at low and high decomposed temperatures respectively, may be distinguished by EPR spectra. The phase transition temperature was determined and compared with XRD data.

Keywords: EPR, hyperfine structure, extrahyperfine structure, phase transition.

1. Introduction

The presence of impurity in oxide systems influences their electromagnetic and other properties. The behaviour of such systems have been effectively studied by means of EPR techniques. One of the most interesting is a MoO_3 doped with vanadium.

Electromagnetic behaviour of monocrystalline V/MoO_3 has been investigated by a number of investigators, for example [1,2]. The powder samples have been studied by others [3,4] in a concern with the selective catalytic oxidation of hydrocarbons. The observation of a hexagonal-rhombic phase transition of pure MoO_3 by means of EPR method was shown in [5], where the observed paramagnetic species were Mo^{5+}. The aim of the present work was to apply the EPR method to study the hexagonal-rhombic phase transition of V/MoO_3, where ions V^{4+} play a role of spin probe.

2. Experiments

2.1 Materials

The starting materials are $NH_4 VO_3$ and $(NH_4)_6 Mo_7 O_{24} \cdot 4H_2O$. In order to obtain the mixture of vanadium and molybdenum salts, the weighed portion of the starting materials were disolved in a diluted aqueous solution of ammonia. After filtering the two solutions were mixed and dried on a water bath. The obtained samples with an atomic ratio 1:50 of V:Mo were passed through a 0.3 - 0.4 mm sieve. The quartz ampoules containing 150 - 200 mg of the powdered samples were heated up to expected temperatures in the range from 473 - 723 K, maintained for 4 h, then cooled to the room temperature, at which temperature the EPR spectra were recorded.

2.2 EPR and XRD Spectra

The EPR measurements were carried out using a EPR-B Spectrometer, X-band, (Russian). MgO (Mn^{2+}) monocrystalline was used as a standard for g-factor and spin concentration. XRD analysis was performed by means of an X-Ray Diffractometer model DRON-2 (USSR), using Cu-Kα radiation.

3. Results and Discussion

In MoO_3 doped with vanadium we did not observe a signal of Mo^{5+} with $4d^1$ configuration. The EPR spectra of V/MoO_3 consist of two different signals, designated as A and B. Both consist of 24 lines, which correspond to their anisotropic hfs components and arise from interaction of unpaired electrons with the nuclear spin V^{51} (I=7/2). It is clear that the localization of d-electrons occurs preferentially in vanadium 3d-orbitals and not in molybdenum 4d-states.

The hfs parameters of signals A and B are presented in the table 1. Three features are apparent from the results:

1. The g-values of signals A and B are significantly different.
2. X- and Y- hfs components of signal A are unresolved and considered as perpendicular hfs components while those of signal B are clearly resolved.
3. Each of the 24-hfs lines of signals A and B, recorded at 77 K, exhibits further structure, which is different from signal A to B (fig.1). The lines of signal A are tentatively attributed by us to the extrahyperfine structures arising from the interaction of the unpaired electron with the nitrogen-ligand nuclei, while in the case of signal B: molybdenum-ligand nuclei. The nitrogen and molybdenum extrahyperfine splitting constants are: 2.1 gauss and 3.2 gauss, respectively.

Signal A has been found in the samples, heated up to 573K, while signal B: over 543K. The signal B is shown in Fig. 2. In the temperature region from 543 to 573K,

Table 1. Hfs parameters of EPR signals on V/MoO_3 .

EPR signals	Heated temperature (K)	Ax (gauss)	Ay (gauss)	Az (gauss)	Gx	Gy	Gz
A	473-543	5 6		1 8 6	1.9 6 5		1.908
A+B	543-573						
B	573-723	5 6	5 7	180	1.976	1.974	1.921

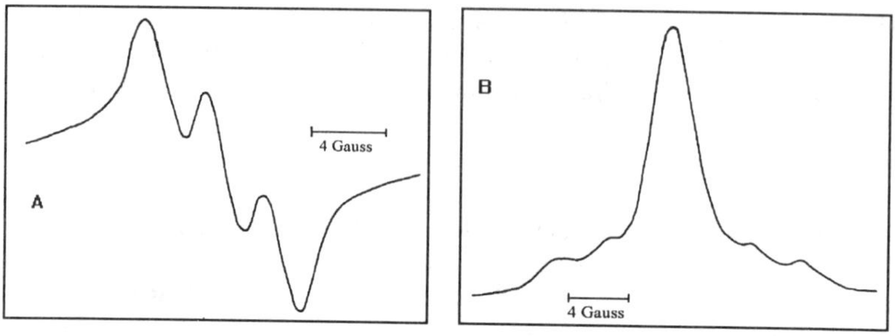

Fig. 1. Extrahyperfine structures of EPR spectra of V/Mo oxide catalyst; (Second Y-Components, $T_{rec.} = 77K$);

a) Signal A, second derivative of the absorption, decompositon temperature of sample was up to 543K.

b) Signal B, the first derivative of the absorption, decomposition temperature of sample was over 573K.

where signals A and B were accumulated, their relative amount changes reversibly: by increasing of decomposed temperature, signal A was decreased and almost vanished at decomposed temperatures over 573K while signal B was increased. The presence of a rhombic phase in the samples heated over 573K was clear from the X-ray diffraction data.

Combining data from the extrahyperfine structure and the decomposed condition of the sample, it seems reasonable to assume that in the low temperature phase (hexagonal), a little amount of nitrogen was not decomposed. The observed extrahyperfine structures of signal A (fig.1.a.) arise from the interaction of those nitrogen-ligand nuclei with the unpaired electron. In a case of high temperature phase (rhombic), nitrogen was almost decomposed and the observed extrahyperfine structures of signal B are logically attributed to molybdenum-ligand nuclei of the oxide lattice.

The structural identification of the paramagnetic centre, corresponding to signals A and B, however, is not possible on the base of the present results.

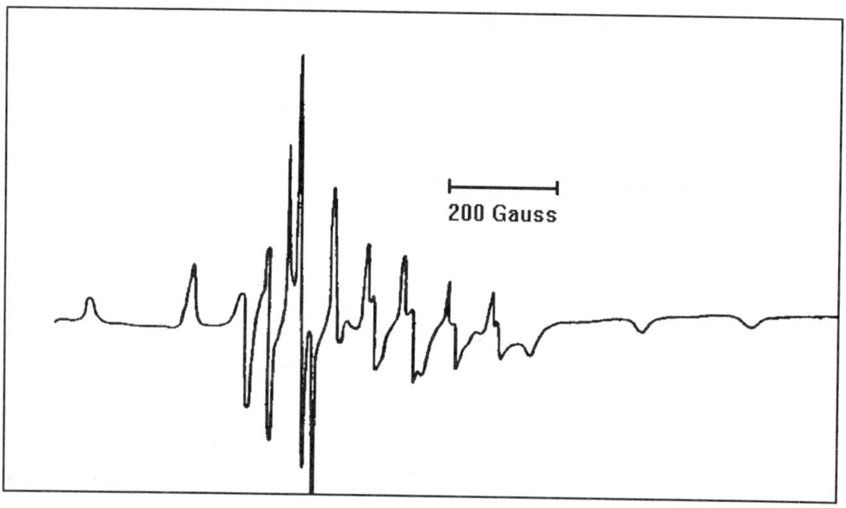

Fig. 2. EPR spectra of rhombic V/Mo oxide catalyst- Signal B; $T_{rec.} = 273$ K.

Thus, the hexagonal-rhombic phase transition taking place in the temperature region 543-573K was firmly established by EPR and XRD data. The low and high temperature phases of the catalyst are characterized by their respective EPR signals.

The temperature of phase transition, observed in V/MoO_3, was only slightly different from that of pure MoO_3, reported in [5] (573K). It may therefore be concluded that doping of vanadium impurity with an atomic ratio of V:Mo=1:50 results in litle change in the MoO_3 lattice.

References

1. A. Ioffe, I. B. Patrina and E. V. Zeleneskaya. EPR Spectra of MoO_3 Single Crystals Doped with Vanadium and Niobium; Phys. Stat. Sol. 37, (1970), 425-432.
2. Sperlich, P. H. Zimmermann and G. Keller. Magnetic Susceptibility and EPR of the Semi-conducting Mixed Crystal V_2MoO_8. Z. Physik, 270, (1974), 267-275.
3. M. A. Makarova, B.V. Rozentuller, M. IU. Kutyrov, O. V. Crylov. Proc. IV USSR Conf. on Mechanism of Catalytic Reactions, Moscow, (1986), 37-41.
4. Nguyen Tien Tai. IVth European Conference on Solid State Chemistry, Dresden, (September 1992).
5. E. G. Ixmailov, G. A. Craves, N. X. Oxmainov, A. X. Cotelnicova. DAN USSR, 283, (1985), 151-155.

EPR Spectra of Some Gadolinium(III) Complexes

Ryo MIYAMOTO and Susumu SUDOH

Department of Chemistry, Faculty of Science, Hirosaki University, Bunkyocho-3, Hirosaki, Aomori, 036, JAPAN

Abstract. EPR spectra of some coordination complexes of gadolinium(III) ion with various ligands were observed as disordered samples. Those spectra have different line-shapes with each other, indicating 4f-electrons are affected by strength and structure of ligand field of coordinating ligands. Zero-field splitting parameters were estimated by spectral simulations.

1 Introduction

Since the rare-earth elements are characterized by their 4f-electrons, how the 4f-electrons contribute to the chemical bonds is very important problem for understanding the properties of rare-earth ion complexes. If the behavior of the 4f-electrons can be observed directly, it is expected that fruitful information about them would be obtained. Electron paramagnetic resonance (EPR) is one of the powerful methods used to directly observe the paramagnetic properties of the 4f-electrons, that are owing to their electron spin and orbital angular momentum.

Though rare-earth ions in metal oxides and ionic crystals of alkali or alkali-earth elements have measured by EPR in order to investigate the effects of crystal-fields or lattice constants, few coordination complexes of rare-earth ion have been studied. One of the advantages of the use of coordination complexes for studying the nature of 4f-electrons is that the local environment of the rare-earth ion can be controlled and modified by the coordinating ligands. It should lead us the new aspects of the nature of 4f-electrons.

In the previous study,[1] we reported the EPR spectra of some Gd(III)–β-diketonate complexes, and discussed the effects of the coordinating ligands. Here, the EPR spectra of Gd(III) complexes having macrocyclic ligand in which nitrogen atoms are coordinating to Gd(III) ion were observed, and the effects of axial and equatorial ligands were discussed.

2 Experimental

All of the chemicals used here were commercially obtained. Gd(III) complexes were synthesized in the usual methods with minor modification.[2, 3]

The EPR spectra were observed using a JEOL RE3X X-band EPR spectrometer at 77–120K and room temperature. Frozen solution and/or magnetically diluted powder, where corresponding Y(III) complexes were used as matrices, were used to observe the EPR spectra. In order to investigate the effects of axial ligands, Gd(III) complexes were dissolved in various coordinative solvents, such as acetonitrile, ethanol, dimethylformamide and dimethylsulfoxide. Furthermore, anions, such as CN^-, SCN^-, Br^-, I^-, Cl^- and NO_3^-, were added to the solutions of Gd(III) complex.

In order to estimate zero-field splitting (ZFS) parameters, EPR spectral simulations for disordered systems were performed using original software based on the spin Hamiltonian for $^8S_{\frac{7}{2}}$ at Hirosaki University Center for Computer and Communications. The eigenfield method was used to obtain resonance fields.[4)]

3 Results and Discussion

Selected X-band EPR spectra of some Gd(III) complexes are shown in figure 1.[a] These spectra have resonance signals in wide range of the magnetic fields indicating that these are coming from 4f-electrons of Gd(III) ion, whose electronic state is $^8S_{\frac{7}{2}}$. It is also shown that the observed spectra were different from each other upon changing the ligand. These results indicate that 4f-electrons, which are situated in the inner shell and shielded by the outer 5s and 5p electrons, are affected by the strength and/or the coordination structure of the ligand fields.

EPR spectral simulations were attempted to obtain ZFS parameters of these Gd(III) complexes. Although these simulation spectra don't show perfect agreement with the observed ones, qualitative information about the 4f-electrons can be obtained. Typical $|D|$ values are shown in table 1. These values are comparable with Gd(III) ions in metal oxides and metal halides.[5, 6)] This table shows that the ZFS parameter ($|D|$ value) of $Gd(acac)_3$ complex is larger than that of the other complexes. The donating oxygens of the $acac^-$ ligand have negative charge and basicity of that is strong, then it is considered to give a strong ligand field. So, the remarkable result is that it seems correct for rare-earth ion complexes too that the stronger ligand field, the larger $|D|$ value.

ZFS parameter of Gd(III) complex with hexaazamacrocyclic ligand (ham1, ham2) was not as large as expected to the covalent property of coordinating nitrogens. This is because the cavity of the macrocyclic ligand is slightly large, hence the bond length of Gd–N is rather long in comparison with Gd–O in $Gd(acac)_3$ complex, and because the macrocyclic ligand has a neutral charge, while the $acac^-$ ligands have negative charges which means a strong ligand field. Differences in the effect of coordination between Gd–N and Gd–O are not so clear, and study is in progress.

---

[a]Hacac = acetylacetone; Hthd = tetramethylheptanedione; Hhfac = hexafluoroacetylacetone; 18c6 = 18-crown-6; ham1 = 2, 7, 13, 18-tetramethyl-3, 6, 14, 17, 23, 24-hexaazatricyclo[17. 3. 1. $1^{8,12}$]tetracosa-1(23), 2, 6, 8, 10, 12(24), 13, 17, 19, 21-decaene; ham2 = 2, 11, 17, 26-tetramethyl-3, 10, 18, 25, 31, 32-hexaazapentacyclo[24. 3. 1. $1^{12,16}.0^{4,9}.\ 0^{19,24}$]ditriaconta-1(31), 2, 4, 6, 8, 10, 12, 14, 16(32), 17, 19, 21, 23, 25, 27, 29-hexadecaene.

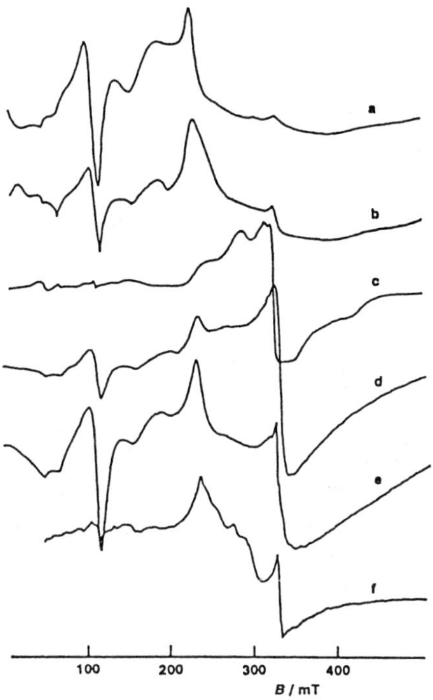

Fig. 1. X-band EPR spectra of some Gd(III) complexes. a) $Gd(acac)_3$ frozen solution of toluene–ethanol (1:1 v/v), b) $Gd(thd)_3$ frozen solution of toluene–ethanol, c) $Gd(hfac)_3$ frozen solution of toluene–ethanol, d) $Gd/Y(18c6)(PF_6)_3$, e) $Gd/Y(ham1)(PF_6)_3$, f) $Gd(ham2)Cl_3$ frozen solution of methanol.

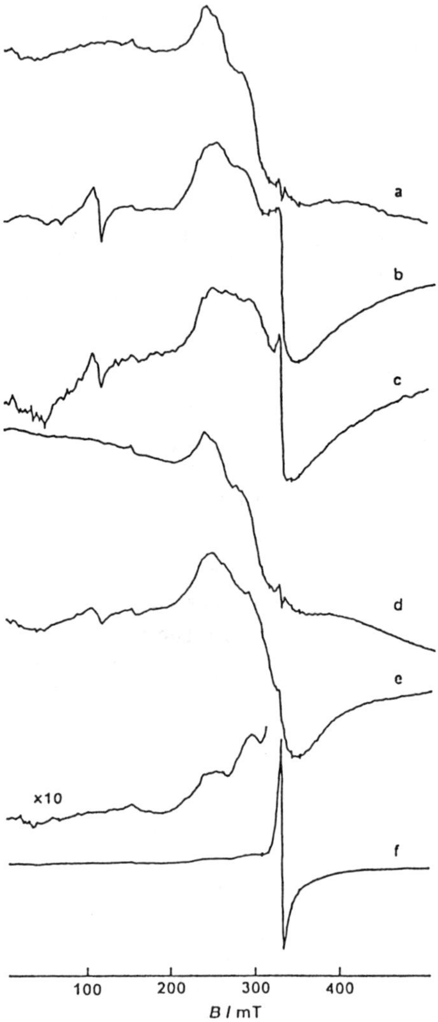

Fig. 2. X-band EPR spectra of $Gd(ham2)Cl_3$ in CH_3CN frozen solution. Various anion of tetrabutylammonium salt were added as axial ligands. Measurement conditions, temperature: 77K, microwave power: 5mW, field modulation, 100kHz, $1mT_{p-p}$. a) CN^-, b) SCN^-, c) Br^-, d) I^-, e) none, f) NO_3^-.

Table 1. ZFS parameters of some Gd(III) complexes.

Complexes	$(\|D\|/hc)/10^{-4}\text{cm}^{-1}$ †
$Gd(acac)_3$	730 (2.2)
$Gd(hfac)_3$	200 (0.7)
$Gd/Y(ham1)(PF_6)_3$	530 (1.6)
$Gd/Y(18c6)(PF_6)_3$	330 (1.0)

† Numbers in parenthesis are $(|D|/h)$/GHz.

Addition of some anions modifies in EPR spectra of the Gd(III) complex, $Gd(ham2)^{3+}$, which are shown in figure 2. Furthermore, EPR spectra were different one another when the Gd(III) complex was dissolved in various solvents. These results indicate that the solvent molecules were coordinated to the $Gd(ham2)^{3+}$ complex in axial position. These axial ligation means that the changes of ligand fields are only made in those directions. In that case, the only f-orbitals whose lobes are spread to the axial directions are affected by those ligands. These results indicate that axial ligation to the Gd(III) ion with the macrocyclic planar ligand may be affected to 4f-electrons in f_{z^3}, f_{xz^2}, or f_{yz^2} orbitals (in general set). These results are very interesting with regard to the properties of 4f-electrons in chemical bonds.

4 Conclusion

In this work, the EPR spectra of some Gd(III) complexes were observed. These spectra are different each other upon changing ligand, indicating that the ligand field and/or coordination structure affected inner 4f-electrons. EPR is a powerful method to study the behavior of 4f-electrons of rare-earth complexes.

For more precise discussion, single crystal study would be required.

Acknowledgment

This research was supported by a Grant-in-Aid for Scientific Research on Priority Areas "New Development of Rare Earth Complexes" Nos. 06241202, 07230202 and 08220204 from The Ministry of Education, Science and Culture.

References

1) R. Miyamoto, S. Sudoh, *Bull. Chem. Soc. Jpn.* **68** (1995) 3439–3443.
2) R. B. King, P. R. Heckley, *J. Am. Chem. Soc.* **96** (1974) 3118.
3) L. De Cola, D. L. Smailes, L. M. Vallarino, *Inorg. Chem.* **25** (1986) 1729.
4) G. G. Belford, R. L. Belford, J. F. Burkhalter, *J. Magn. Reson.* **11** (1973) 251.
5) R. C. Nicklin, J. K. Johnstone, R. G. Barnes, D. R. Wilder, *J. Chem. Phys.* **59** (1973) 1652.
6) A. Abragam, B. Bleaney, *Electron Paramagnetic Resonance of Transition Ions*, Dover (1986), Oxford(1970), Chap. 5.

EPR in situ Investigation of Reduction of Rhombic V-Mo Oxide by Allyl Alcohol

Nguyen Tien Tai *, *B.V. Rozentuller* **, *O.V. Crylov* **.

* Institute of Chemistry - National Centre for Natural Science & Technology of Vietnam. Nghia do- Tuliem - Hanoi - Vietnam.
** Institute of Chemical Physics, Russian Sci. Acad. Koshigin 4 - Moscow - Russian.

Abstract. EPR in situ, combined with Gas Chromatography Analysis (GCA), was employed in a kinetic study of the reduction of rhombic V-Mo oxide by allyl alcohol. The samples with an atomic ratio of V : Mo = 1 : 50 were obtained by thermal decomposition of mixtures of molybdenum and vanadium salts. The kinetic parameters of the reduction process were obtained by GCA and EPR in situ and compared with those of hexagonal V-Mo oxide catalyst.

Keywords: EPR in situ, kinetics, reduction.

A knowledge of the reduction processes occurring in the catalyst upon adsorption of the reactants is essential for the proper understanding of the mechanism of catalytic oxidation of hydrocarbons. For this purpose, studies of the reduction of hexagonal V-Mo oxide catalyst by allyl alcohol were undertaken [1,2], where EPR combined with GCA was applied in a deeper study of the reduction processes. The main features concluded in [1,2] are summarized as follow: The active centers on the reduction of hexagonal V-Mo oxide catalyst are the reduced vanadium ions V^{4+}. During reduction, acrolein was created on the surface by the interaction of allyl alcohol with oxygen of the oxide lattice and in the bulk of the catalysts the oxidation degree of vanadium was changed: $V^{5+} \rightarrow V^{4+}$. Two processes occur synchronously. The equation selected for the description of the processes was:

$$V^{5+}O^{-} + C_3H_5OH \rightarrow V^{4+}\square^{o} + C_3H_4O + H_2O \qquad (1)$$

where $\square^{o}$ is a vacancy.

The total kinetic equation was:

$$W_{C_3H_4O} = 3.6 \times 10^{-13} \exp\left(-\frac{1600}{RT}\right)[C_3H_5OH]\,(Nv - [V^{4+}]) \qquad (2)$$

Where:

$N^* = 5.6 \times 10^{19}$ ions/g, approximately 50 % of total number of doped vanadium ions in catalyst;
N^{v}: number of V^{4+} in the sample.

EPR characterization of V-Mo oxide was studied and reported in [3,4]. Two paramagnetic V^{4+} centers of different symetry and extrahyperfine structure of the EPR spectra were observed. One of them characterizes for hexagonal, and the other the rhombic phase of V-Mo oxide catalyst.

To continue the study on V-Mo oxide system we carried out the experimental study focused not on the hexagonal but on the rhombic phase of the oxide system. The results are briefly presented in this communication.

The catalysts were prepared by thermal decomposition of a mixture of NH_4VO_3 (pure, Reachim USSR) and $(NH_4)_6 Mo_7 O_{24} . 4H_2O$ (pure, Reachim USSR). Catalysts with an atomic ratio of V:Mo = 1:50 were obtained. The BET specific surface area of the preparation was 10 ± 1 m^2/g. In all experiments 20 mg samples were used. The system of kinetic measurement with special quartz ampoules permits determination of reaction products and the content of V^{4+} during reduction: the reaction mixture was analysed by CGA and the number of V^{4+} was determined by EPR-B Spectrometer (Russian). MgO (Mn^{2+}) monocrystalline was used as a standard for g-factor and spin concentration. A description of the catalyst and EPR in situ experimental technique was detailed in [1].

The reduction of rhombic V- Mo oxide catalyst by allyl alcohol was studied, similarly as with the hexagonal phase of the catalyst, and may be also interpreted as the result of two synchronous processes: surface- and bulk-, represented respectively by equations:

$$\mathrm{Ln}\left\{ W_{C_3H_4O} / (W_{C_3H_4O})^{o} \right\} = K_{\mathrm{eff.}} .t \qquad (3)$$

and

$$\mathrm{Ln}\left\{ 1 - \frac{[V^{4+}]}{[V^{4+}]^{\infty}} \right\} = K_{\mathrm{eff.}} .t \qquad (4)$$

The data concerning the reduction for T= 114 °C are given as an example in Fig.1.

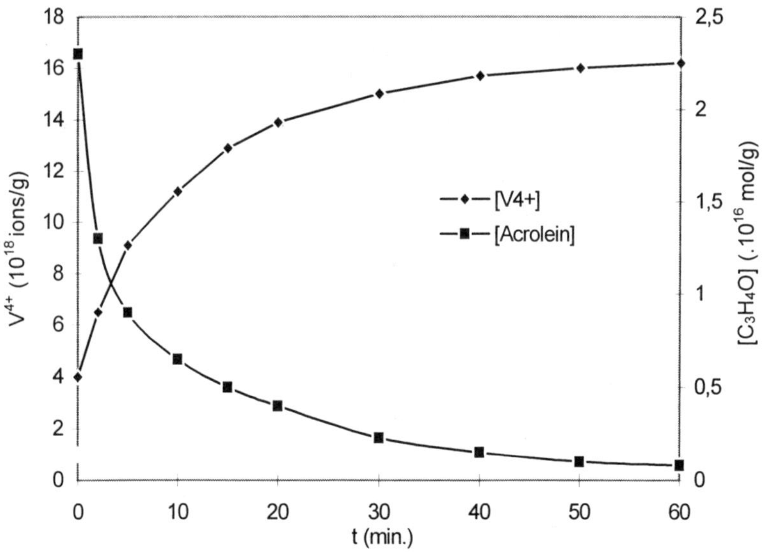

Fig.1: a. Kinetic of reduction of V-Mo oxide catalys by allyl alcohol. $[C_3H_5OH] = 7.4 \times 10^{16}$ mol/g.

The kinetic and EPR data shown in Fig.1 were treated on the basis of Eqs. (3) and (4). The plots of Ln{$W_{C_3H_4O}$ / $(W_{C_3H_4O})^o$} and Ln{1-([V^{4+}] / $(V^{4+})^\infty$)} versus the time must be straight lines with a slope of $K_{eff.}$ (Fig. 2.) The effective activation energy was determined from an Arrhenian temperature dependence of the $K_{eff.}$ and is given in Table.1.

Tab. 1. The temperature dependence of the effective rate constant of reduction of V-Mo oxide by allyl acohol; [C_3H_5OH] = 6.7 x 10^{17} mol./cm^3.

	$T_{red.}$ 114 °C	$T_{red.}$ 124 °C	$T_{red.}$ 144 °C	$T_{red.}$ 164 °C	E_a (Kcal/Mol.K)
Hexagonal [1] (10^{-3}/s)	0.3	0.4	1.0	2.4	16
Rhombic / Eq. (3) (10^{-3}/s)	0.5	0.6	1.1	1.9	11
Rhombic / Eq. (4) (10^{-3}/s)	0.6	0.8	1.4	2.2	11

The values of effective activation energy obtained with EPR in situ data are virtually identical with the kinetic values. The observed kinetic parameters of rhombic V/Mo oxide catalyst reveal small changes in comparison to those of hexagonal [1]. The kinetics of the reduction is therefore quite analogous to that of the hexagonal phase which has been presented in [1,2].

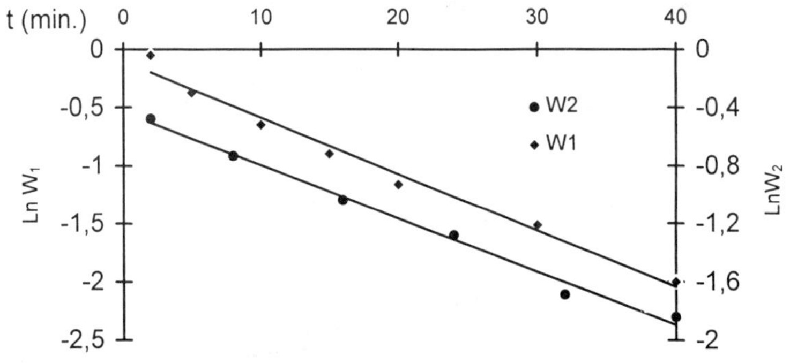

Fig. 2. Plots of Eqs. (3) and (4).

References

1. M. A. Makarova, Thesis. Moscow, 1987.
2. Makarova M. A., Rozentuller B. V., Crylov O.V. Kinetic i Kataliz, 28,1987, 1143-1147.
3. Nguyen Tien Tai. IVth European Conference on Solid State Chemistry, Dresden, (September 1992).
4. Nguyen Tien Tai. APS'97, Hongkong(1997).

Slow Molecular Motion of Spin Probe in the Glassy Polymer Matrix Detected by Double Modulation ESR

B. Rakvin and Z. Veksli

Ruđer Bošković Institute, P.O. Box 1016, HR-10000, Zagreb, Croatia

Abstract. The slow molecular motion of perdeuterated nitroxyl spin probe in amorphous polymer matrix of polybutadiene (PB), polyisobutylene (PIB), poly(methyl methacrylate) (PMMA) and paraffin have been studied by double modulation electron spin resonance (DMESR). The interrelationship between the spin-lattice relaxation rates obtained by DMESR and the local molecular dynamics of side chain groups was observed. An electron proton dipolar interaction as a possible mechanism which can affect detected relaxation rates was discussed.

1 Introduction

Among other techniques, the ESR spectroscopy of nitroxyl spin probes, has been proved to provide information about the motional dynamics in the slow motional region ($< 10^{-5}$ s) in various solid state matrices. Usually, a nitroxyl probe can be employed for testing an internal molecular motion in the solid host by measuring slow relaxation processes and magnetization transfer between different spectral sites. In these processes one measures experimental parameters such as spin-spin relaxation time (T_2), spin-lattice relaxation time (T_1) or nuclear spin-lattice relaxation time (T_{n1}) of a probe. One can expect, as discussed by Schwartz et al. (1986), that the measurement of T_1 would give an ability to reach the smallest detectable local dynamical effects (approximately the same order of magnitude as $1/T_1$ which can be detected by the ESR probe method in the vicinity of a probe), since for a typical spin probe in the slow motional regime the homogeneous linewidth $1/T_2$ is larger than $1/T_1$. An extensive study of nitroxyl T_1 in glassy solvents in the sub-T_g region by employing saturation recovery technique, SR-ESR, provided by Du et al. (1995) shows well defined behavior of T_1 as a function of temperature ($1/T_1 \propto T^k$, k=2.34±015) regardless of the solvent. From these results one expects a small impact of solvent vibration on nitroxyl T_1. Indeed, these expectations are supported by isotope experiments (^1H was replaced by ^2H in solvent matrix) where only 10% of measured T_1 in the sub-T_g region can be attributed to intermolecular relaxation processes. However, in some cases in the sub-T_g region one expects that the effect of intermolecular and intramolecular dynamics takes an important role in the process. Therefore, to gain more information from intramolecular and intermolecular relaxation processes one can also employ different types of alternative techniques, such as simultaneous combination of pulsed SR-ESR and electron-electron double resonance, ELDOR, as shown by Robinson et al. (1994), or two-dimensional ELDOR, as suggested by Dubinskii et al. (1994), which may better couple these effects. It is shown that the probe motions in glassy polymer matrices are restricted but, little progress is made in the complete understanding of polymer dynamics as well as the coupling of these motions to the intramolecular spin probe motion or the packing of polymer around the probe.

Feasibility to detect the experimental evidence for molecular dynamics of different side groups, which are attached to the polymer backbone in the glassy phase by employing simple spin probe, will be the focus of the present work. It is expected that even in the sub-T_g region the polymers containing side groups should exhibit faster relaxation time of a probe than the polymer chains without side groups.

In order to determine whether local dynamical changes in polybutadiene (PB), polyisobutylene (PIB), poly(methyl methacrylate) (PMMA) and paraffin matrices could be detected by a spin probe as a function of temperature in the sub-T_g region (between 100K and 190K) a double modulation electron spin resonance method, DMESR, has been employed. As shown earlier by Belov et al. (1988), Alekseev et al. (1990) and Rakvin (1994), DMESR technique can be employed as an alternative method to detect T_1 for an inhomogeneously broadened line under the condition that $T_1 >> T_2$. Recently it was also demonstrated by Rakvin et al. (1991) and Valić et al. (1992) that a modulation spectrum can be used for monitoring the slow motional dynamics of spin probe and internal free rotation of CH_3 groups attached to the piperidine ring of a probe. It should be mentioned that the pulse ESR and DMESR measurements on model systems, tempone and PD-tempone in glycerol-water mixture, have been discussed in detail in our previous work, Rakvin et al. (1991), showing that the T_1^{DM} (T_1 detected by DMESR) can be approximated to the T_1* values obtained by the stimulated echo sequence. Since in the present study the contribution of the proton electron-nuclear dipolar relaxation rate, END, will be discussed, it is interesting to test a sensitivity of T_1^{DM} to this particular relaxation mechanism. Following the results of Du et al. (1995) on T_1, obtained by saturation recovery, one finds 10% difference between the T_1 of deuterated and nondeuterated spin probe in glycerol-water matrix at 100K. The similar experiment performed earlier by Rakvin et al. (1991) shows that T_1^{DM} is roughly of the same order of magnitude but the difference between the deuterated and nondeuterated probe in the same matrix is about 90%. Thus, one expects that T_1^{DM} can detect an END type relaxation at a large scale.

2 Experimental

The perdeuterated spin probe, 4-oxo-2,2,6,6-tetramethyl-1-piperidinyl-oxy (PD-tempone) was dispersed in polymer matrices at low concentration not exceeding 0.02 mass %. Four polymers differing in chain structure and side chain groups and (corresponding) relaxation modes were selected. Atactic poly(methyl methacrylate) (PMMA), $M_w=4.4\times10^5$ and polyisobutylene (PIB), $M_w=8.4\times10^5$ with two side chain groups per molecular segment and two polymers without side chain groups; polybutadiene (PB) $M_w=3\times10^3$ having random distribution of cis-1,4 and trans-1,4 structures and paraffin 52/54.

DMESR measurements were performed on a Varian E-109 X-band spectrometer by using 100 kHz modulation of Zeeman magnetic field as the first modulation. The second modulation (sweeping in the frequency interval from 250 kHz to 350 kHz) was performed by an additional modulation coil and a Wavetek radiofrequency source. The modulation peak at 300 kHz was detected and accumulated in the standard PC computer. The linewidth at half height of the DMESR peak, Δ_{obs}, obtained by fitting Lorentzian lineshape to the DMESR spectrum. Δ_{obs}, is related to the spin-lattice relaxation time under the condition that $T_1 >> T_2$:

$$\frac{1}{\Delta_{obs}\pi} = T_1^{DM} \sim T_1 \qquad (1)$$

Below et al. (1988) have tested Equation 1 by simultaneously detected T_1 for paramagnetic centers in the irradiated quartz with pulse ESR and DMESR techniques.

3 Results and Discussion

The temperature dependence of PD-tempone dispersed in the matrices of four different polymers is shown in Fig.1. The expected relaxation rate for PD-tempone in 1:1 glycerol-water mixture is also calculated ($1/T_1 = 0.2T^{2.34}$) according Du et. al. (1995) and shown in Fig.1. As the temperature is decreased in the sub-T_g region the slope of relaxation rates becomes similar to the slope of glassy phase measured by SR-ESR method. In this temperature region one expects that $1/T_1^{DM}$ is affected only by the local dynamics in the amorphous polymer matrix. The shorter relaxation time for PIB than the one for PB should be ascribed to the influence of the side chain methyl groups on the PIB backbone. The temperature dependence of $1/T_1^{DM}$ in the PMMA polymer matrix indicates almost the same behavior as in the PIB matrix with a slightly smaller value of $1/T_1^{DM}$. In order to confirm that the amount of relaxation time in the sub-T_g region, closely associated with polymer side chain groups, we also measured a few points of $1/T_1^{DM}$ for the paraffin matrix having polymer chains of a simple structure. The result is shown in Fig.1 as the longest relaxation times detected for PD-tempone in the monitored temperature region. Present experimental results clearly demonstrate that in the sub-T_g region contribution of the motional dynamics of polymer side chain groups to the spin-lattice relaxation time of spin probe can be detected by the perdeuterated spin probe and DMESR technique.

It is obvious that the proton dynamics of the side chain groups in polymer matrices modulate electron spin-lattice relaxation rate and in the further discussion the possible underlying relaxation mechanisms will be discussed. It was shown by Du et al. (1995) that the relaxation rate of the nitroxyl radical trapped in the glassy matrices exhibits $T^{2.34}$ dependence. It was also suggested that this relaxation rate has an intramoleclar origin. The spin probe relaxation rates in the glassy PMMA exhibit similar behavior and approximately the same value as in the PIB matrix (Fig. 1). In addition the relaxation rate of PMMA also exhibits a small extremal behavior in the sub-T_g region (already found by Valić et al. (1990)). The data for PMMA as in Fig. 1 given in Fig. 2 in the enlarged scale

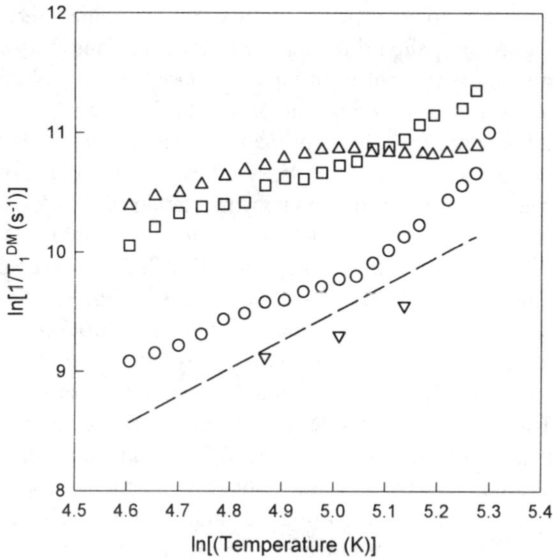

Fig. 1. DMESR relaxation rates vs temperature represented in ln-ln plot.for PD-tempone dispersed in different polymer matrices: triangle for PMMA, square for PIB, circle for PB and reverse triangle for paraffin. Dashed line represent expected $1/T_1$ for PD-tempone in water-glycerol mixture obtained by SR- ESR (Du et al. (1995)).

reveal a clear maximum of the relaxation rate around 160K. The extremal behavior of the relaxation rate can be tentatively examined within the relaxation mechanism which consists of two parts. In the first part one can expect a contribution from the polymer backbone without side chain groups, such as PB, which shows nearly the same slope as a probe in glassy phase measured by SR-ESR. The second part is related to protons of polymer side chain groups which can couple electron spin by an END mechanism. One expects an appearance of the extremal behavior in the relaxation rate only in the case when rotational correlation times of methyl protons reach the values around $1/\omega_n$ or $1/\omega_e$ (where ω_n represent nuclear Larmor frequency and ω_e electron Larmor frequency). Therefore, the total relaxation rate for PMMA can be described within the END mechanism combining glassy-type modes and classical orientational correlation times. This leads to the first part of a contribution which can be described with temperature dependence T^k and two other therms which describe extremal behavior,

$$\frac{1}{T_1^{DM}} = a\,T^{k} + b\left[\frac{2\omega_n\tau}{1+(\omega_n\tau)^2}\right] + c\left[\frac{2\omega_e\tau}{1+(\omega_e\tau)^2}\right] \qquad (2)$$

where *a, b* and *c* represent the constants depending on an amount of END interaction. Rotational correlation time is represented by the Arrhenius equation, $\tau=\tau_0\exp(E/kT)$, with temperature independent activation energy E. It can be noted that the Equation 2 contains too many parameters for an efficient fitting of the experimental results. However, by assuming that parameter k should be changed between 2 and 3 as expected in the glassy phase the minimum of χ^2 error in fitting procedure had been obtained for k=2.4. The continuous line in the Fig. 3 represents the best fit to Equation 2. Contribution from the first term (dashed line 1 in Fig. 3) leads to the linear coefficient a=0.09$K^{-2.4}s^{-1}$ giving an expectable good description of the glassy phase which coincides, within the experimental error, with an exponent for the glassy relaxation rate, as found earlier by Du et al. (1995). The larger extremal behavior (dashed-dotted line 2 in Fig. 3) is obtained for a fast proton motion with b=25.1kHz, τ_0=0.82ps, and E=2.2kJ/mol by employing electron Larmor frequency. The third contribution (dashed line 3 in Fig. 3) is described by the nuclear Larmor frequency and slower proton dynamics c=15.7kHz, τ_0=1.2ps and E=12.5kJ/mol. This model description leads to the two types of methyl groups dynamics. One type of methyl groups exhibits fast motional dynamic comparable to the electron Larmor frequency $1/\omega_e$=2.7×10^{-11}s) while the second group performs slower motion comparable to the nuclear Larmor frequency $1/\omega_n$=2.7×10^{8}s in the observed temperature interval. Earlier NMR study by Powels et al. (1964) and recent neutron scattering study by Floudas at al. (1992) of the proton correlation time in PMMA also clearly show the two well resolved motional dynamics corresponding to the faster ester methyl groups

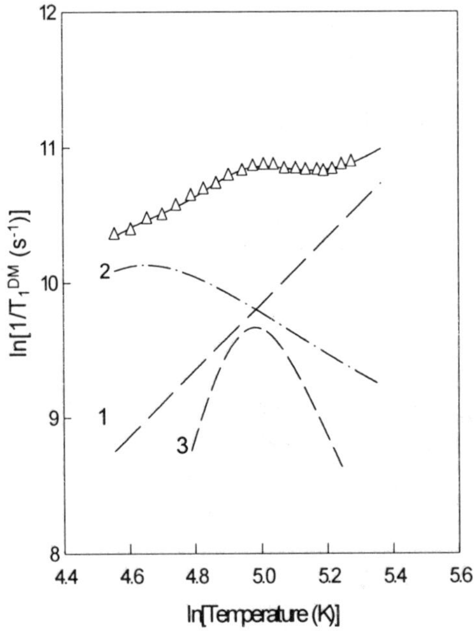

Fig. 2. Ln-ln plot of $1/T_1^{DM}$ vs temperature of PD-tempone dispersed in PMMA matrix (triangles). The dashed line 1 represents the contribution from the first term of Equation 2, the dashed-dotted line 2 corresponds to the contribution of the electron Larmor frequency term and the dashed line 3 represents the nuclear Larmor frequency term. The solid line is the sum of all these contributions in Equation 2.

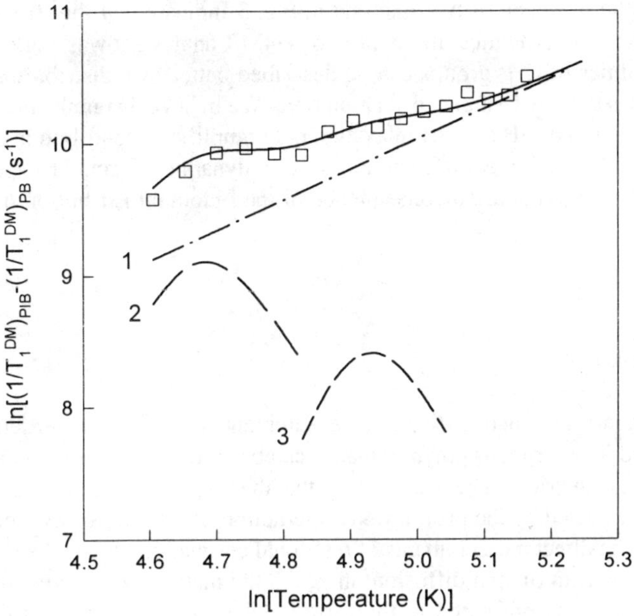

Fig. 3. Ln-ln plot of $1/T_1^{DM}$ vs temperature for subtracted relaxation rates PIB-PB (squares). The dashed-dotted line 1 represents the contribution from the first term in Equation 2, the dashed lines 2 and 3 correspond to the contribution of the nuclear Larmor frequency term for the two different CH_3 groups. The solid line is the sum of all three contributions in Equation 2.

motion with a smaller activation energy (E~0.7kJ/mol) and the slower main chain methyl group motion with a larger activation energy (E~12kJ/mol).

In the case of PIB, a contribution of the side chain groups motion can be separated from the backbone dynamics by subtracting relaxation rate of PB from the relaxation rate of PIB. The results of the substraction of relaxation rates of this two polymers and the corresponding fitting curves are shown in Fig. 3. By assuming the similar relaxation mechanism as in the case of PMMA the contribution of the first glassy term (k=2.3 amd a=0.23K$^{-2.3}$s^{-1}) exhibits slightly larger contribution than in the case of PB. The first extremal behavior at lower temperature with a maximum contribution constat b_1=4.5kHz, (τ_0=0.1ps, E=13.3kJ/mol) and the second at higher temperature b_2=9.0kHz, (τ_0=0.3ps, E=9.3kJ/mol) are obtained with the nuclear Larmor frequencies. This model description leads to the two types of methyl groups dynamics. Nolle et al. (1959) have studied motional dynamics of the methyl groups in PIB polymer matrix by the NMR spectroscopy. It was found that near 140K a small fraction of the methyl groups rotate about C_3 axis, while the others are interlocked. The second maximum in Fig. 3 about 140K coincide with the NMR results.

However, the presence of two maxima in Fig. 3 indicate that the PIB exhibits two different CH_3 groups dynamics. In the previous NMR analysis it was suggested that the motional dynamics of CH_3 groups can be described better by a distribution of rotation correlation times than by a single correlation time. We believe that in the present analysis the difference between PIB and PB relaxation rates amplifies a contribution of side chain groups thus enabling to better differentiate two CH_3 dynamics. It could be speculated that the two DMESR extrema are a consequence of the binomal distribution of CH_3 group motion.

4 Conclusions

These experiments demonstrate that the measurements of T_1^{DM} for a perdeuterated spin probe dispersed in amorphous polymer matrix can be employed to study local molecular dynamics of these matrices. The data indicate that the relaxation rate detected by DMESR is significantly affected by the proton END interaction. It was shown by Robinson et al. (1994), that in the direct measurements of rotational correlation time in the slow motional region the mechanism of spin diffusion, in which the matrix protons are coupled to the electron of a probe is presented in the form similar to the second and third term in Equation 2. The possible maximum contribution of Robinson et al. (1994) terms can be compared with the maximum contribution of terms in Equation 2. It can be noted that the obtained maximum value of constant *b* and *c*, respectively, are well below of the expected maximum diffusion rate (~130kHz) for the PD-tempone probe estimated by Robinson et al. (1994).

It is shown that a local dynamic of side chain groups contributes to an increase in the effective relaxation rate of the probe. In addition, a temperature dependence of the effective relaxation rate can be used to elucidate the local molecular dynamics of the side chain groups and the molecular packing of amorphous polymer chains due to the sequence structure. The obtained experimental data suggest that this method could offer valuable data on local dynamics and polymer chain packing of more complicated polymer systems such as block copolymers and polymer mixtures.

References

1. B. F. Alekseev, A. B. Tikhonov, Yu. V. Bogachev, O. G. Gasimov, and V. A. Yanchurov, Nature of very narrow ESR lines with two frequency longitudinal modulation of the magnetic field, *Soviet Phys. Solid State* **32** (1990) 764-767.
2. P. G. Below and A. D. Milov, Double-modulation ESR and Electron spin echo study of paramagnetic centers in fused quartz: Comparison of techniques, *Chem. Phys. Lett.* **151** (1988) 79-82.
3. J-L. Du, G. R. Eaton and S. Eaton, Temperature, Orientation, and Solvent Dependence of Electron Spin-Lattice Relaxation Rates for Nitroxyl Radicals in Glassy Solvents and Doped Solid, *J. Magn. Reson.* **A115** (1995) 213-221.

4. A. A. Dubinskii, G. G. Marech and H. W. Spiess, Two-dimensional EPR spectroscopy of nitroxides: Elucidation of restricted molecular motions in glassy solid, *J. Chem. Phys.* **100** (1994) 2437-2448.

5. G. Floudas and J. S. Higgins, Ester methyl group rotation in poly(methyl methacrylate) and in the blend solution chlorinated polyethylene/poly(methyl methacrylate). A quasielastic neutron scattering study, *Polymer* **33** (1992) 4121-4128.

6. A. W. Nolle and J. J. Billings, Proton Magnetic Resonance in Polyisobutylene at Various Temperatures, Pressures and Frequencies, *J. Chem. Phys*. **30** (1959) 84-90.

7. J. G. Powles, B. I. Hunt and D. J. H. Sandiford, Proton spin lattice relaxation and mechanical loss in a series of acrylic polymers, *Polymer* **5** (1964) 505-515.

8. B. Rakvin, S. Valić and Z. Veksli, Slow molecular motion and intramolecular dynamics of nitroxyde spin probe detected by the double-modulation ESR technique, *Chem. Phys. Lett.* **182** (1991) 171-174.

9. S. Valić, B. Rakvin, Z. Veksli and S. Pečar, Slow molecular motion of different spin probes in model glycerol-water matrix studied by double modulation ESR, *Chem. Phys. Lett.* **199** (1992) 609-614.

10. B. Rakvin, Sensitivity of the Method of Double Modulation ESR Detection of Electron Spin-Lattice Relaxation Time, *J. Magn. Reson.* **A106** (1994) 245 .

11. B. H. Robinson, D. A. Haas and C. Mailer, Molecular Dynamic in Liquids: Spin-Lattice Relaxation of Nitroxide Spin Labels, *Science* **263** (1994) 490-494.

12. L. J. Schwartz, G. L. Millhauser and J. H. Freed, Two-Dimensional Electron Spin Echoes Magnetization Transfer and Molecular Dynamics, *Chem. Phys. Lett.* **127** (1986) 60-66.

13. S. Valić, B. Rakvin, Z. Veksli and Z. Grubišić-Gallot, Study of slow molecular motion of spin probes in polymers by the double modulation ESR method, *Macromolecules* **23** (1990) 5182-5186.

Studies of Piezoelectric Poly(vinylidene fluoride) Film

Yayan Liu *Geixiang Wang* *Jiayan Yu*

Changchun Institute of Applied Chemistry, Chinese Academy of Sciences, Changchun, 130022, P.R.China

Abstract. Poly(vinylidene fluoride)(PVDF) films from the melt were drawn at different temperatures. The structure and properties of the drawn PVDF films were characterized by Wide Angle X-ray Diffraction(WAXD), Infrared(IR) spectroscopy and Electron Spin Resonance(ESR). The properties of the drawn films strongly depend on the drawing conditions. At a given draw ratio, however, the crystal forms changed remarkably with drawing tempertures. The structural changes of the drawn PVDF films were discussed .

KEY WORDS : Poly(vinylidene fluoride), Wide Angle X-ray Diffraction, Infrared, Electron Spin Resonance

1 Introduction

Piezoelectric polymer PVDF has become available by a series of treatments .The structure and properties of this polymer have received considerable attention[1-3] .It has been shown that this polymer takes three stable crystal modifications , referred to as the α , β and γ phases[4]. The most common polymorph of PVDF is the α form , which is produced by cooling from the melt. The crystallite is nonpolar, which has no piezoelectricity. Mechanical deformation of α phase PVDF at 80°C yields the β phase. In β phase[4], the molecules are in a planar zigzag conformation and the unit cell is orthorhombic, space group Cm2m, with lattice constants a=8.58, b=4.91, and c=(chain axis)=2.56Å. There are two polymer chains per unit cell. In this crystal structure the dipole moments of the monomer units are all parallel to the axis. The third crystal form, called the γ phase, is produced by crystallization in certain solvent systems[5] or by high-temperature annealing[6]. Among these crystal forms, the β phase crystal contains a large amount of spontaneous polarization in its unit cell[7,8] , and when it is oriented and then poled under a high d.c electric field, it will exhibit high piezoelectric properties. In previous years, the studies of the methods of preparing oriented β phase .crystal with a high degree of crystal perfection have attracted the interest of many researchers[9-11]. The oriented β phase crystal is usually obtained by the drawing of α phase crystal at atmospheric pressure or by low-temperature annealing crystallization .

This paper presents our work on one aspect of the effect of drawing temperature on crystal transformation, i.e., the dependence of the relative contents of α phase and β phase crystal in a drawn PVDF sample on its drawing temperature.

2 Experimental

2.1 Materials and Film Preparation

The starting material was a commercial PVDF resin obtained from Russia. The samples were in pellet form. Sheets of a certain thickness (0.2-0.3mm) were compression molded in a press at 210°C and quenched to room temperature. The sheet (60mmx30mm in area) were uniaxialy drawn in an Instron 1121 tensile machine at different drawing temperatures (22°C-130°C) at a constant strain rate of 5mm /min to about four times their original length, then allowed to relax in the drawn state for 10 min at 120°C.

2.2 WAXD

Wide Angle X-ray Diffraction (WAXD) scans of the specimens were taken on a D/MAX- II B diffractometer, using Ni-filtered Cuk_α radiation. The films were scanned from 5 °to 50° at a rate of 2 °/min. Graphite crystal was utilized as monochromator and analyzer. The currency and voltage in X-ray tube are 20mA and 40kv, respectively.

2.3 IR

Infrared (IR) spectra of the drawn films at different drawing temperatures were recorded in the infrared region from 400-860cm $^{-1}$ on the BIO-RAD IR spectrometer.

2.4 ESR

For Electron Spin Resonance(ESR), weighted specimens (50G) were put in glass tubes of 2mm in diameter, and connected with a vacuum system. At a pressure of 0.1 Pa, the tubes were sealed. Under ^{60}Co X-rays, they were irradiated for 24h. Using a JES-FE3AX ESR spectrometer, the activities of the trapped radicals were investigated.

3 Results and Discussion

In the first group of experiments, the phase transition from phase α to β occurring during the drawing of PVDF film has been followed at various temperatures, ranging from 22 ° C to 130 °C. A set of typical IR spectra of the samples under investigation is shown in Fig.1.

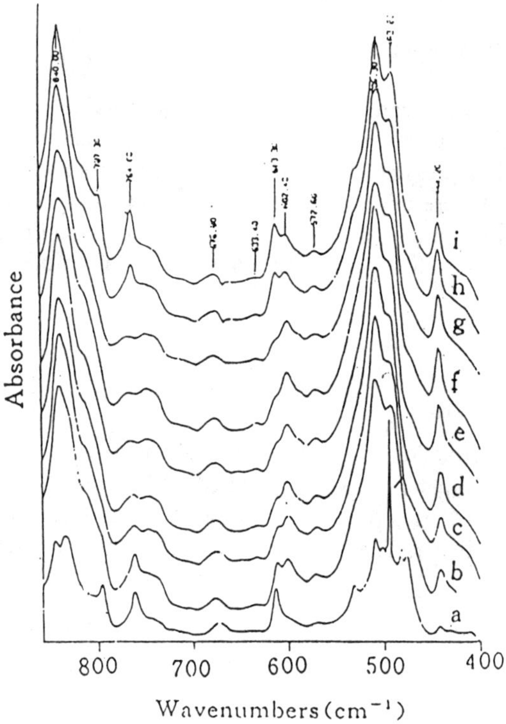

Fig.1. Infrared spectra of the drawn PVDF films (draw ratio: 4) drawing temperatures a)undrawn ; b)22°C; c)40°C; d)60°C; e)80°C; f)90°C; g)100°C; h)120°C; i)130°C

According to previous band assignments[12], some bands clearly associated with both phases can be studied in the spectral range between 400 and 860cm $^{-1}$, Thus the absorption band at 766cm $^{-1}$ is characteristic of the α phase, and the band at 840cm $^{-1}$ (symmetric CF_2 stretching and C-C symmetric stretching) is characteristic of the β phase. The absorption band near 766cm^{-1} increases with increasing temperature. At the highest experimental temperature(130°C), the intensity at 764cm^{-1} become stronger. In order to quantitatively study the drawing temperature effect on the crystal conversion ratio, the PVDF films were examined with IR spectroscopy and the crystal contents of α phase and β phase were estimated from the observed spectra. The plot of intensity ratio $I_{840}/I_{766}+I_{840}$ versus temperature is given in Fig.2.. The intensity ratio represents the evolution of both phases during the deformation process at different temperatures . Judging the curve in Fig.2. , the relative content of β phase increases from 71%~77% when temperature is raised to 80°C. The relative content of β phase reaches its maximum value near 80°C, then it decreases with temperature above 80°C. NO further significant phase transformation takes place at higher drawing temperature above 130 °C.

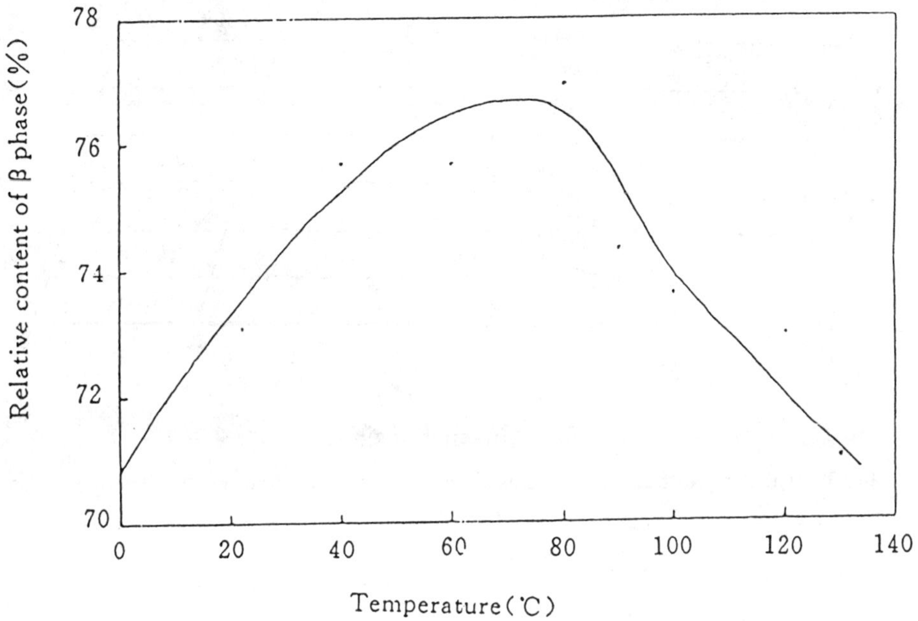

Fig. 2 . The relative content of β phase versus drawing temperature for PVDF films

Moreover, X-ray diffraction measurements were made to examine structural changes induced by drawing. Fig. 3 .shows a series of diffractograms of PVDF films drawn at different temperatures. In the 5°-50° range of diffraction angle 2θ, there exists a sharp peak near 20° and a broad shoulder at its lower angle side for all the samples. The peak near 2θ =20° is assigned to (110) (200) reflection from β phase. With further increase in temperature, the (020) reflections around 18° from α phase crystal strengthen in their intensities, the other α phase (200) reflections around 36° show an increase. It reveals that the degree of crystal conversion from α phase to β phase depends on the drawing temperature. At 80°C, the conversion of α → β reaches its maximum, however, when temperature reaches 90°C, the conversion of α → β reverses and α phase appears again.

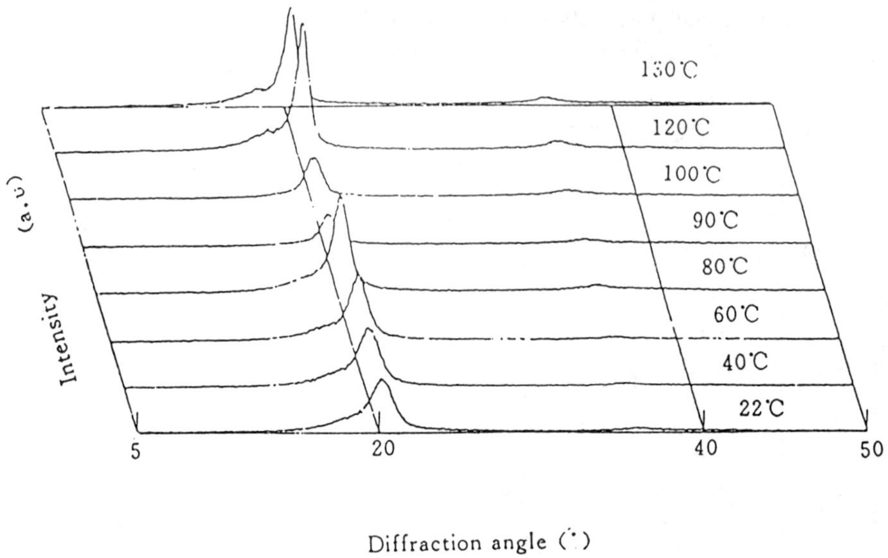

Fig.3. The X-ray diffraction patterns observed at various drawing temperâtures (draw ratio : 4)

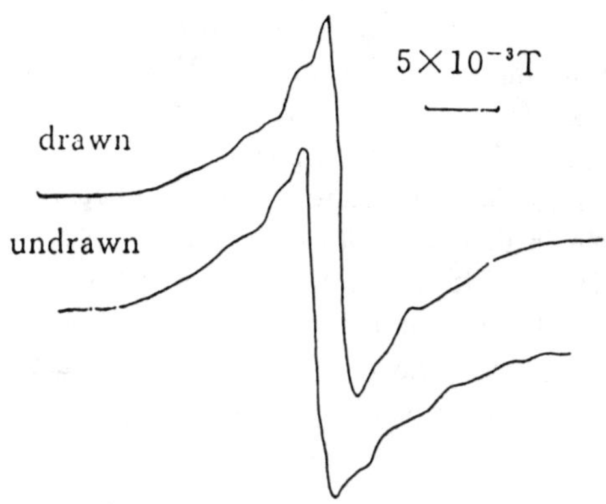

Fig.4 . First-derivative ESR spectra of drawn and undrawn PVDF films at room temperature

Fig. 4 . shows the ESR spectra of a drawn and an undrawn PVDF films at room temperature. It shows that at least two radicals species are brought about when PVDF is subjected to γ-irradiation , which differ in line-width. From the spectra of the irradiated specimens mentioned above, there is a strong absorption centered in the spectrum.

Comparing the spectra of the irradiated specimens drawn and undrawn, it is noted that superfine structure is observed in the irradiated undrawn specimen . According to previous ESR research on irradiated polyenthyl[13], PVC[14]. The peak centered in spectrum is attributed to $\text{-}CH_2-\dot{C}H\text{-}(CH=CH)_n-$. Since the unpaired electron is sited in delocated carbons, consequently, spectrum of drawn and undrawn PVDF films is similar. Another stable radical is $-CH_2-\dot{C}H-CH_2-$. According to theoretical calculation, the number of spectra of isonuclei is 2nI+1 (wherein I is nuclear spin, n is the number of proton).The number of spectrum line is 10. However, due to that some spectrum lines overlap each other , six lines appears.

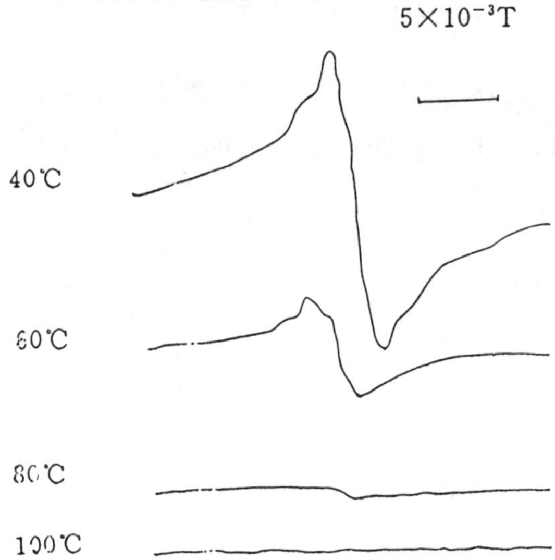

Fig.5 . Some ESR spectra for drawn PVDF film at different temperatures

Fig. 5 . shows the ESR spectra at different temperatures, which suggest that the stable radicals are attributed to $\text{-}CHF\text{-}CH_2-\dot{C}H\text{-}(CH=CH)_n-CHF\text{-}$ and $\text{-}CHF\text{-}CH_2-\dot{C}H\text{-}CH_2\text{-}CHF\text{-}$. It can be seen from the data in Table 1, which are the percentage of radicals at different temperatures . Moreover, it is the case that free radical reaction with each other takes place, so the concentration of the radical decrease when temperature rises. Consequently, ESR signal strength decreases with increasing in temperature. When temperature rises to 100 ° C, ESR signal strength is zero.

Table 1. The decay rates of the two stable radicals

Temperature	strong peak	weak peak
25°C	100%	100%
40°C	49%	47.4%
60°C	15.1%	12.3%
80°C	1.3%	1.8%
100°C	0	0

Conclusions. The various methods used in this work offer the possibility to obtain more detailed information on the drawing-induced $\alpha \rightarrow \beta$ phase transition of PVDF .From our results, it can be concluded that this transition is highly dependent upon temperature, i.e., as the temperature increases , the extent of transformation decreases. Transformation of α phase to β phase though drawing is more efficient when the process is carried out at temperatures between 70 and 80 °C. Since the β phase is the more stable phase below 80°C, which gives a higher efficiency of the $\alpha \rightarrow \beta$ transformation. Above 100°C, the consequent increase in chain mobility allows crystals orientation in the draw direction, without considerable deformation of the crystals. Moreover, we have demonstrated the advantage of using ESR spectroscopy to study the free radicals of γ-irradiated PVDF films.

References

1. Kawai , H. , *Jpn. J. Appl. Phys.* ,**8**, (1969)975-976
2. D. Naegele, D.Y.Yoon and M. G. Broadhurst, *Macromolecules*, **11**,(1978)1297-1298.
3. G.R.Davies, H.Singh, *Polmer* ,**20**,(1979)772-774.
4. R.Hasegawa, Y. Takahashi, Y.Chatani and H. Tadokoro, *Polym. J.*,**3**,(1972)600-610
5. M.Kobayashi, K.Tashiro and H. Tadokoro, *Macromolecules*, **8** , (1975)158-171
6. W, M.Prest., Jr., and D.J.Luca, *J. Appl. Phys.*,**46**,(1975)4136-4138
7. Doll.W.W and Lando, J.B.,*J .Macromol. Sci.(B)*, **2** ,219-221(1968), ibid, ,**4** ,889-893(1970)
8. R. Hasegawa, M. Kobayashi and H. Tadokoro ,*Polym. J.* , **3**, (1972)591-599
9. K.Nakamura and Y.Wada, *J.Polym.Sci., A-2*,**9**,(1971)161-173
10. R.Hayakawa and Y.Wada, *Adv.Poly.Sci.*,**11**,(1973)1-4
11. G.w.Day,C.A.Hamilton,R.L.Peterson,R.T.Phelan,andL.O.Mullen, *Appl.Phys.Letters*,**24**,(1974)456-458
12. S.Osaki ,T.Kotaka, *Ferroelect.*,**32**, (1981)1-3
13. Ohnishi, S., et al, *J .Polym. Sci*,**149**,(1960)503-506
14. E.J,Lawton , J.S. Balwit , Phys Chem,33,(1960)395-397

Paramagnetic Rhodium Species in Silicoaluminophosphate Type 11 Molecular Sieves

Tomasz Wasowicz and Larry Kevan

University of Houston, Department of Chemistry, Houston, TX 77204-5641

Abstract. Electron spin resonance spectroscopy is used to monitor the behavior of paramagnetic rhodium species in Rh-SAPO-11, Rh-AlPO4-11 and RhAPSO-11 molecular sieve materials. Because of the low ion exchange capacity of SAPO-11 molecular sieve, liquid phase ion exchange does not incorporate enough Rh(III) ions to subsequently stabilize paramagnetic rhodium species in SAPO-11. However, solid state ion exchange at 400 ^{o}C gives higher rhodium ion incorporation into SAPO-11 to form Rh-SAPO-11. Paramagnetic rhodium species are detected in solid state ion-exchanged SAPO-11. The initially observed species B seems analogous to a weak rhodium species B formed in X zeolite and is tentatively assigned to Rh(0). Subsequent evacuation at higher temperature or H_2 reduction leads to less intense A and C paramagnetic rhodium species which are also analogous to species observed in X zeolite and which are assigned to Rh(II). The introduction of Rh(III) into the SAPO-11 synthesis mixture gives RhAPSO-11 in which the stabilized Rh(0) species B is more stable than in solid state ion exchanged Rh-SAPO-11.

1 Introduction

The aluminophosphate (AlPO4-n) and silicoaluminophosphate (SAPO-n) molecular sieves belong to a new class of microporous materials[1],[2]. The modification of these materials by isomorphous replacement of framework cations by transition metal ions or by incorporation of such ions into extraframework positions can play a significant role in catalytic reactions. Since the location and oxidation state of transition metal ions can affect the catalytic activity, it is important to study the chemical environment and location of such transition metal ions. The SAPO-11 molecular sieve is composed of 10-ring straight channels surrounded by four 4-ring and six 6-ring channels. This structure has the AlPO4-11 structure[3],[4] except that some of the framework tetrahedral phosphorus (P) and aluminum (Al) sites are partially substituted by silicon (Si). This substitution produces a net negative framework charge which is balanced by H+ ions which can be ion-exchanged to some extent by transition-metal cations.

Rhodium species in molecular sieves and other microporous materials have been reported to be effective catalyst sites for processes such as hydrogenation and dimerization of ethylene[5], methanol carbonylation[6] and propylene hydro-formylation[7]. Depending on the specific catalytic process, paramagnetic Rh(0) and nonparamagnetic Rh(I) were proposed as active species. There are no electron spin resonance (ESR) studies on rhodium ion-exchanged SAPO-11 molecular sieves or on rhodium ion incorporation into the SAPO-11 framework to our knowledge.

2 Experimental Section

SAPO-11 was synthesized according to the procedure described by Azuma and Kevan.[8] To a mixture of 13 g distilled water and 9.19 g aluminum isopropoxide 5.13 g of an 85 wt % H_3PO_4 was added and stirred 1 h. A solution of 1.00 g of 30 wt % SiO_2 in 1 g distilled water was added. Finally 2.56 g of diisopropylamine was added and the mixture was covered and stirred overnight at room temperature. The final gel composition was:

$$1.0\ {}^{i}Pr_2NH : 0.2\ SiO_2 : 0.9\ Al_2O_3 : 0.9\ P_2O_5 : 50\ H_2O.$$

The homogeneous gel was sealed in a teflon-lined autoclave and heated for six days at 220 °C under autogenous pressure. The product was separated from the gelatinous gel by filtering. The identity and crystallinity of this *as-synthesized SAPO-11* material was verified by powder X-ray diffraction (XRD). The material was calcined at 550 °C under flowing oxygen for 18 hours. After calcination the XRD pattern was also verified. $AlPO_4$-11 was synthesized according to a similar procedure except that no silica source was added.

Solid state ion exchange was accomplished by mixing 1 g of calcined SAPO-11 with 0.002 g, 0.004 g and 0.008 g of powdered $RhCl_3 \cdot 3H_2O$ with a few drops of distilled water. The mixture was dried briefly at 90 °C, placed in a tube oven and heated slowly up to 400 °C under oxygen flow at a rate of about 1.5 °C/min. After reaching 400 °C the temperature was maintained for about 16 h. Electron microprobe analysis showed that the actual Rh content in the SAPO-11 phase was 0.1, 0.3 and 0.65 wt % respectively. Rhodium (III) was introduced into calcined $AlPO_4$-11 using solid state ion exchange in an analogous way. Liquid phase ion-exchange does not incorporate enough Rh(III) ions to subsequently stabilize paramagnetic rhodium species. The ion-exchanged material is designated Rh-SAPO-11.

Samples were also prepared with rhodium (III) introduced during synthesis of SAPO-11. These samples are designated RhAPSO-11. Rhodium was added during the synthesis in two ways (a) as a solution of 0.025 g or 0.05 g Rh_2O_3 in 5.13 g of 85 wt % H_3PO_4 and (b) as a solution of 0.1 g or 0.05 g $RhCl_3 \cdot 3H_2O$ in 6 g H_2O. In both cases after the synthesis the XRD pattern was in agreement with the pattern for SAPO-11. As-synthesized RhAPSO-11 material was calcined in the same way as the SAPO-11 material. Electron microprobe analysis showed that the actual rhodium concentration was 0.1 and 0.3 wt % respectively in calcined RhAPSO-11.

All ESR spectra were recorded with a Bruker ESP 300 X-band spectrometer at 77 K. The magnetic field was calibrated with a Varian E-500 gaussmeter. The microwave frequency was measured by a Hewlett Packard HP 5342A frequency counter.

3. Results

3.1 General Remarks

For all Rh-SAPO-11 samples a prominent narrow ESR singlet with $\Delta H_{pp} \approx 5$ G at $g = 2.002$ was observed which is due to matrix defects formed in all samples during the calcination process. This signal appears also in SAPO-11 samples after calcination. Most probably the defects are formed as a result of dehydration during calcination of the SAPO-11 material. In figures the top and the bottom part of this signal was cut off to allow a better presentation of the spectra. The other ESR signals are assigned to rhodium species formed by thermal reduction during the solid state ion-exchange heating. The rhodium signals are very weak which indicates that only a very small part of the Rh(III) can be reduced and stabilized in paramagnetic form. In contrast, in Ni-SAPO-11 the EPR intensity of paramagnetic Ni(I) species is about two orders of magnitude larger and the intensity of the defect signal is not dominant.[9,10,11]

3.2 Solid State Ion Exchange

After short evacuation (5 min) ion-exchanged samples reveal an ESR signal B0 at $g_\perp$=2.056 and $g_\parallel$=2.250. The intensity of the signal correlates with the amount of Rh(III) used for the solid state ion exchange, which shows that the signal is related to rhodium. The shape of the signal is slightly perturbed by the framework defect signal. As the evacuation temperature is raised to 100 °C a superhyperfine multiplet separated by 20.9 x 10^{-4} cm^{-1} appears on the perpendicular component. The ESR parameters of this signal B are $g_\parallel \approx 2.253$ and the $g_\perp \approx 2.043$ (Fig. 1a). A similar spectrum was reported by Goldfarb and Kevan[12,13] in Rh-exchanged Na-X zeolite with ESR parameters of $g_\perp$=2.060 and $g_\parallel > 2$ with a sextet hyperfine structure on the perpendicular component of about 20 x 10^{-4} cm^{-1}. The hyperfine structure does not appear on the parallel component or is hidden under the linewidth. As suggested by Goldfarb and Kevan, the hyperfine structure can be assigned to an interaction with ^{27}Al (I=5/2).

As the heating temperature reaches 300 °C the overall spin concentration decreases, but after continued heating to 400 °C the intensity of the ESR signal increases again and a weak new signal A with $g_\perp = 2.530$ and $g_\parallel \approx 1.90$ (Fig. 1c.) becomes visible. A similar signal was reported by Goldfarb and Kevan in RhCa-X zeolite[14] with $g_\perp$= 2.532 and $g_\parallel$=1.891. But in contrast to RhCa-X zeolite, in SAPO-11 the intensity of signal A is very weak and signal A exists together with signal B. After exposing the sample to oxygen and heating at 400 °C only signal B is observed and signal A disappears.

For comparison, solid state ion-exchanged Rh-$AlPO_4$-11 was also prepared even though $AlPO_4$-11 nominally does not have any ion-exchange capacity. In those samples after 5 min evacuation at room temperature a signal B0 with $g_\perp$ = 2.057 and $g_\parallel$ =2.289 is observed. After overnight evacuation at room temperature superhyperfine structure becomes visible with a line separation of about 11.8 x 10^{-4} cm^{-1} (Fig. 1b).

If heated and then oxidized Rh-SAPO-11 is exposed to 100 torr hydrogen at room temperature, after 0.5 h there appear three signals: A at g=2.508, C1 at g=2.421 and

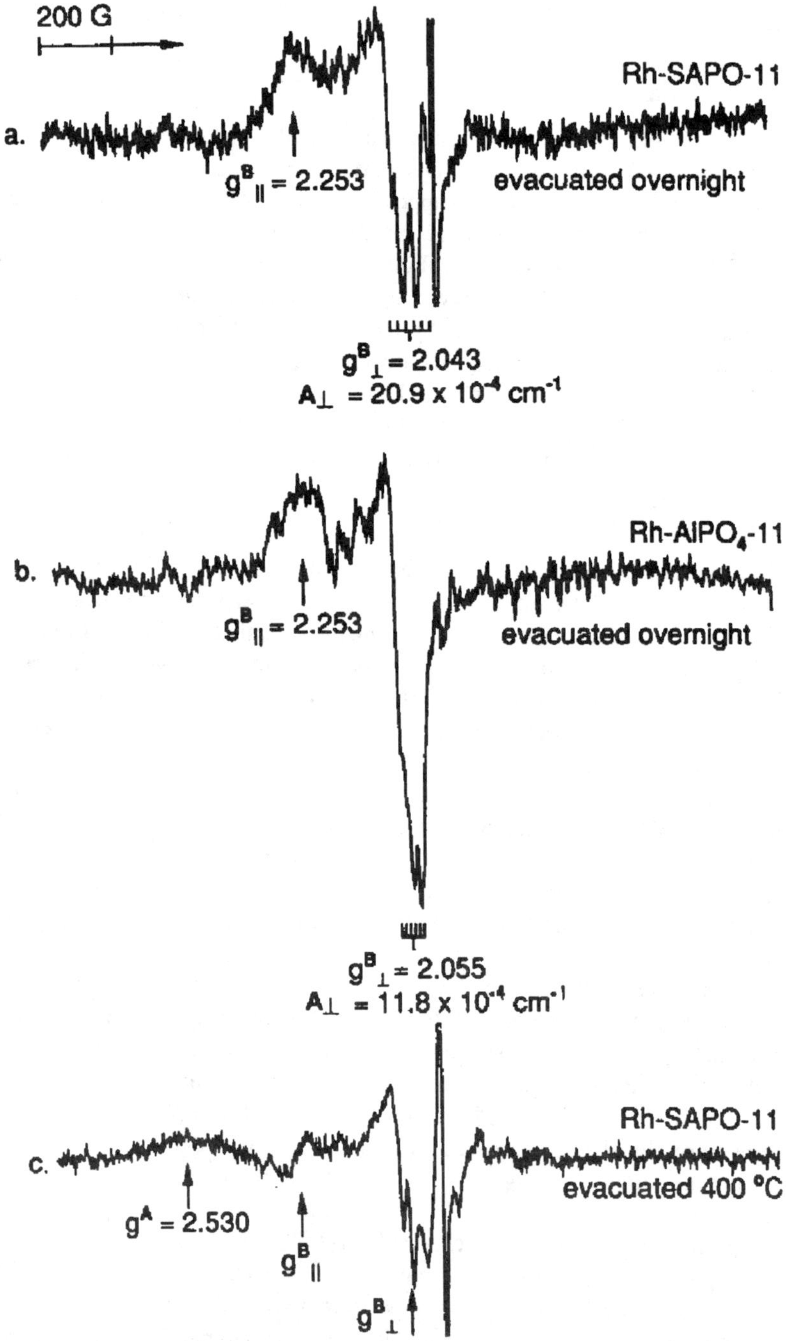

Fig. 1. ESR spectra of solid state ion-exchanged (a) Rh-SAPO-11 evacuated at RT 16 h, (b) Rh-$AlPO_4$-11evacuated at RT 16 h, (c) Rh-SAPO-11 evacuated 16 h at 400 °C.

C2 at g=2.330 (Fig. 2a). If Rh-SAPO-11 containing 100 torr hydrogen is heated to 100 °C these three signal intensities increase. At 200 °C signal A moves to g=2.430 (signal A1) with a decrease of intensity and signals C1 and C2 transform to one signal C at g=2.334 (Fig. 2c). A signal similar to signal C was found during hydrogen reduction of RhNa-Y zeolite.[15]

If Rh-SAPO-11 is heated and oxidized at 500 °C and then is exposed to 50 torr ethylene, after 3 minutes at room temperature there appear weak signals Et0 at g= 2.525 and Et1 at g=2.368. After four days the signals at g=2.368 and g=2.525 decrease and form a signal Et2 at g = 2.25 which is visible as a shoulder at g=2.181. After evacuation of the ethylene from Rh-SAPO-11 and heating in vacuum at 100 °C the intensity of signal Et2 (which overlaps with the parallel component of signal B) decreases but remains present in the spectrum. Similar signals were reported by Bass and Kevan at g=2.25 after exposure of RhCa-X zeolite to ethylene[16], at g = 2.26 after exposure of RhCa-L zeolite to ethylene[17] and at g=2.26 after exposure of Rh/SiO_2 to ethylene.[18] Oxidation of Rh-SAPO-11 at 500 °C restores the initial spectrum observed before ethylene adsorption.

3.3 RhAPSO-11

RhAPSO-11 samples after 10 min evacuation at room temperature reveal an ESR spectrum which consists mainly of an anisotropic signal S0 with $g_\perp$=2.040 and $g_{||}$=2.174 . After evacuation of the sample at 100 °C for 1 h the signal changes its parameters to $g_{||}$=2.271 and $g_\perp$=2.068 (S1). This spectrum remains unchanged during heating to 500 °C. After dehydration at 500 °C, RhAPSO-11 exposed to 500 torr oxygen at room temperature and heated for 1 h at 500 °C shows ESR spectral changes (Before each ESR measurement the oxygen was evacuated at room temperature for 10 min). Signal S1 decreases and a new signal S2 appears with g_{xx} = 2.107 g_{yy} = 1.943 and g_{zz} = 2.268. RhAPSO-11 containing signal S2 and exposed to 100 torr hydrogen at room temperature for 1 h, shows signal S1 with a new signal SH at g=2.010. Further increase of the temperature up to 500 °C does not cause any change in the observed signals.

4 Discussion

SAPO-11 has the same crystalline structure as $AlPO_4$-11 which is neutral. The introduction of a small amount of silicon to replace more phosphorus than aluminum in the framework of $AlPO_4$-11 generates a net framework negative charge which is balanced by cations in ion-exchange positions. This generates some ion-exchange capacity. In zeolites where alumina units are surrounded by silica units, theframework is also negative but the framework charge is several times greater than in SAPO materials. Thus, zeolites have a greater ion-exchange capacity.

In this work the main Rh-related ESR signal which dominates all other signals for various pretreatments is an anisotropic signal with an axially symmetric g tensor: $g_\perp$ = 2.055; $g_{||}$ = 2.250. An ESR signal with $g_\perp$ = 2.060 related to paramagnetic rhodium species in Na-X zeolites was reported by Goldfarb and Kevan.[12,13] In those papers the relative amount of this species was small and the possible origin of that signal was

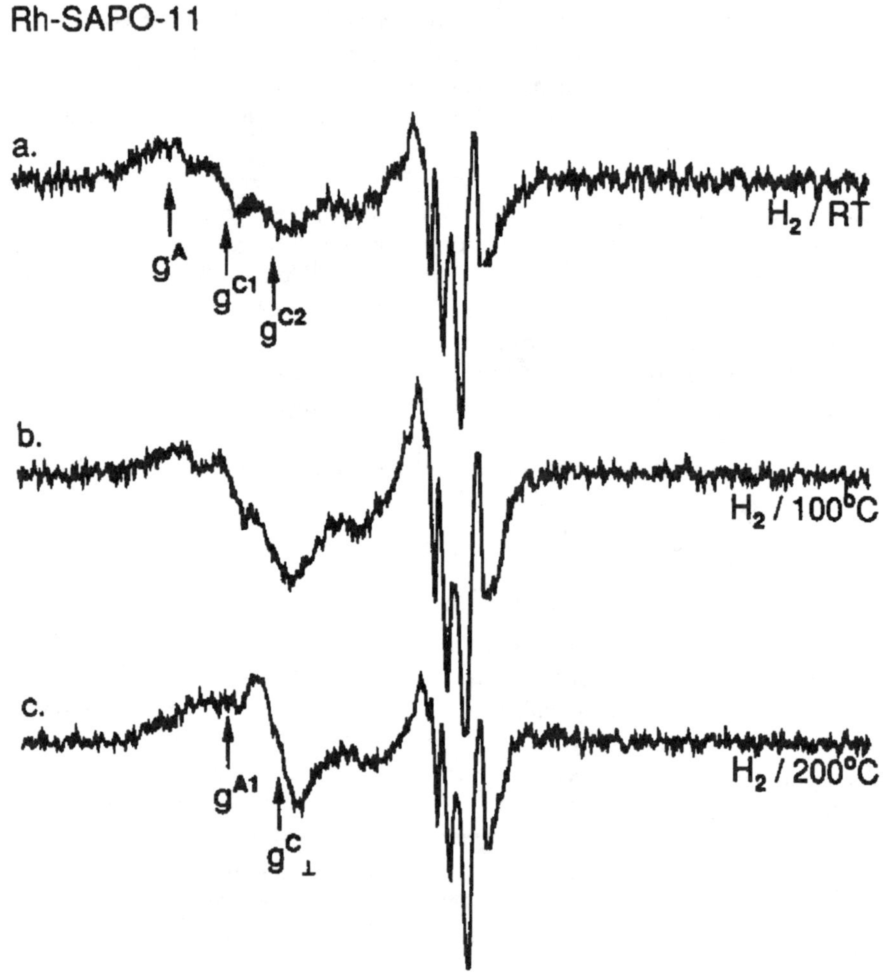

Fig. 2. ESR spectra at 77 K of solid state ion-exchanged Rh-SAPO-11 after H_2 reduction at (a) room temperature, (b) 100 °C and (c) 200 °C.

not discussed. In SAPO-11 the $g_\perp = 2.055$ signal dominates other weaker signals which are assigned to Rh(II) species. It is not surprising that the same paramagnetic Rh species are observed in X zeolite and in SAPO-11, but it is surprising that the relative species abundances are so different in these two materials. In a relatively highly charged zeolite framework Rh(II) species are most efficiently stabilized. In contrast, in SAPO-11 with a lower framework charge the stabilization of divalent, charged species seems to be less effective. It seems plausible that the $g_\perp = 2.055$ signal in SAPO-11 may represent a paramagnetic rhodium species with a lower oxidation state. Rh(0) has a $4d^9$ electron configuration. Analogous $4d^9$ electron configurations and ESR signals with similarly located g components occur for Pd(I)[19,20], Ag(II)[21,22,23] and for $3d^9$: Cu(II)[24,25,26,27]. This assignment to Rh(0) is supported by results obtained for solid state ion-exchanged Rh-$AlPO_4$-11. In this molecular sieve with a neutral framework, it is probable that stabilization of neutral Rh(0) will occur. In Fig. 1 the spectrum recorded for Rh-SAPO-11 with the B0 signal is compared with a spectrum recorded for Rh-$AlPO_4$-11. The ESR parameters of both spectra strongly suggest that it is the same paramagnetic species, namely Rh(0). The transformation of the B0 signal during heating under vacuum to signal B may be caused by decreasing the number of water molecules coordinated around the rhodium atom and increasing the interaction with the SAPO-11 framework. The hyperfine structure on the perpendicular component of signal B may arise with decreased water content and a consequent decrease of the interaction with water protons. This allows resolution of a hyperfine interaction of the unpaired electron with possibly ^{27}Al nuclei (I = 5/2) in the framework.

The formation of signal A1 during heating above 400 °C is attributed to autoreduction of the trivalent rhodium ions to Rh(II) with a d^7 electron configuration and $g_{\parallel} < g_{\perp}$ assuming a d_{z^2} or d_{xy} ground state depending on the strength of the tetragonal ligand field.[25] After oxidation signal A1 disappears which suggests the oxidation of Rh(II) to Rh(III).

After hydrogen reduction there arise signals A, A1, C , C1 and C2 attributed to Rh(II) species in slightly different environments. Based on the shape of these signals, they are assigned to $g_{\parallel}$ with the $g_{\perp}$ hidden in the g = 2.00 region.

Different behavior exists for Rh(0) species in RhAPSO-11. The S1 signal has a higher $g_{\parallel}$ than the B0 signal observed in ion-exchanged Rh-SAPO-11, and the perpendicular component of S1 does not reveal any hyperfine structure. The change of the signal during dehydration at 100°C and the stability of the spectrum up to 500 °C suggests that the Rh(0) species in RhAPSO-11 may be localized in a different site than in the solid state ion-exchanged Rh-SAPO-11 material. The lack of hyperfine structure also supports a different location in RhAPSO-11. The lower Rh concentration in RhAPSO-11 means that the Rh atoms are more distant from each other and are less mobile even at elevated temperature. This supports that the hyperfine structure on the B0 signal in the solid state ion exchanged Rh-SAPO-11 may came from ^{103}Rh interactions rather than from ^{27}Al. The stability of the signal during activation also suggests that Rh(0) in RhAPSO-11 is in less accessible sites. This is still not enough evidence to state that the Rh is incorporated into the framework, but the rhodium introduced into SAPO-11 material during synthesis is more stable chemically.

5 Conclusions

ESR spectroscopy was used to trace the redox behavior of Rh-ions introduced into SAPO-11 material by solid state ion exchange. During such exchange some of the Rh ions are thermally reduced in the presence of a small amount of water to Rh(0) atoms. In the synthesized RhAPSO-11 material Rh(0) seems to be more chemically and thermally stable.

Acknowledgment. This research was supported by the Robert A. Welch Foundation and the National Science Foundation. The authors thank M. Hartmann for helpful discussions.

References

1. Wilson, S. T.; Lok, B. M.; Flanigen, E. M. U.S. Patent 4 310 440, 1982.
2. Wilson, S. T.; Lok, B. M.; Messina, C. A.; Cannan, T. R.; Flanigen, E. M. *J. Am. Chem. Soc.* **1982**, *104*, 1146.
3. Bennett, J. M.; Richardson, J. W.; Pluth, J. J.; Smith, J. V. *Zeolites* **1987**, *7*, 160.
4. Richardson, J. W.; Pluth, J. J.; Smith, J. V. *Acta Crystallogr.* **1988**, *B44*, 367.
5. Okamoto, Y.; Ishida, N.; Imanaka, T.; Teranishi, S. *J. Catal.* **1979,** *58*, 82
6. Lars, S.; Andersson, T; Scurrell, M. S. *J. Catal.* **1981**, *71*, 233.
7. Davis, M. E.; Rode, E.; Taylor, D.; Hanson, B. E. *J. Catal.* **1984**, *86*, 67.
8. Azuma N.; Lee C. W.; Kevan L. *J. Phys. Chem.* **1994**, *98*,1217.
9. Hartmann, M. private communication
10. Hartmann, M; Kevan, L. *J. Chem. Soc. Faraday Trans.,* **1996**, *92*, 1429.
11. Azuma, N.; Hartmann, M.; Kevan, L. J. Phys. Chem. **1995**, *99*, 6670.
12. Goldfarb D.; Kevan L. *J. Phys. Chem.* **1986**, *90*, 264.
13. Goldfarb D.; Kevan L. *J. Phys. Chem.* **1986**, *90*, 2137
14. Goldfarb D.; Kevan L. *J. Phys. Chem.* **1986**, *90*, 5787.
15. Goldfarb D.; Kevan L. *J. Am. Chem.Soc.* **1987**, *109*, 2303.
16. Bass S. J.; Kevan L. *J. Phys. Chem.* **1990**, *94*, 1483.
17. Bass S. J.; Kevan L. *J. Phys. Chem.* **1990**, *94*, 4640.
18. Bass S. J.; Kevan L. *J. Chem. Soc. Faraday Trans.* **1990**, *86*, 3015.
19. Taarit Y. B.; Vedrine J. C.; Dutel J. F.; Naccache C. *J. Magn. Res.*, **1978**, *31*, 251.
20. Yu, J.-S.; Lee, C. W.; Kevan, L. *J. Phys Chem.* **1994**, *98*, 5736.
21. Michalik, J.; Kevan, L. *J. Am. Chem. Soc.* **1986**, *108*, 4247.
22. Morton, J. R.; Preston, K. F. *J. Magn. Res.* **1986**, *68*, 121.
23. Wasowicz, T.; Michalik, J. *Radiat. Phys. Chem.* **1991**, *37*, 427.
24. Flentge, D. R.; Lunsford, J. H.; Jacobs, P. A.; Uytterhoeven, J. B. *J. Phys. Chem.* **1975**, *79*, 354.
25. Pöppl, A; Newhouse, M; Kevan, L *J. Phys. Chem.* **1995**, *99*, 10019.
26. Zamadics, M; Chen, X; Kevan, L *J. Phys. Chem.* **1992**, *96*, 5488.
27. Zamadics, M; Chen, X; Kevan, L *J. Phys. Chem.* **1992**, *96*, 2652.

Magneto-structural Correlations of $[Cu\ (dpt)\ OAc]_n\ (ClO_4)_n$: A μ - Acetato Bridged Quasi One-Dimensional Complex

*Rajendra Bose Muthukumaran, Babu Varghese, and Periakaruppan T. Manoharan**

Regional Sophisticated Instrumentation Centre and Department of Chemistry, Indian Institute of Technology, Madras – 600 036, India.

Abstract. A complex of formula $[Cu(dipropylenetriamine)(CH_3COO)](ClO_4)$ has been investigated by two techniques to elucidate its one-dimensional behavior. X-ray diffraction study confirmed structural monodimensionality that consists of chains of Cu(II) ions with zigzag arrangement of these ions running parallel to crystallographic *a*-axis. Then EPR data are presented to show that the magnetic system is also one-dimensional. The single crystal EPR spectral data show the line shapes are Lorentzian in all orientations of the crystal, which suggests significant interchain spin diffusion, while line width has an angular dependence of the type $(3\cos^2\theta-1)^{4/3}$, showing the characteristics of one-dimensional systems.

Keywords: acetate, EPR, low-dimensionality, and lineshape

1 Introduction

Electron paramagnetic resonance (EPR) spectroscopy is a powerful technique for the investigation of low-dimensional magnetic materials. In a sense the decisive evidence for the low-dimensional magnetic properties rests on the detailed analysis of spin dynamics as manifested in EPR studies.

Because of the low-dimensional structural properties of $[Cu\ (dpt)\ OAc]_n\ (ClO_4)_n$, an investigation to refine the structural data and to elucidate the magnetic properties was undertaken. Structural determination show that this material consist of a series of linear chains of Cu(II) ions. The present work gives more detailed information than the one appeared in an earlier account of this crystal structure (Skeleton, et. al., 1972). Though this earlier work referred to the ploymeric nature of this complex, it was not very clearly brought out as we would need it for a study on one dimensional interaction. Hence in the present work, the structural data are presented first and used as an evidence to point out the one dimensional character of this material.

Then the EPR data which shows the quasi one-dimensionality of the spin dynamics will be treated in detail.

2 Structural Properties

At room temperature [Cu(dpt)OAc](ClO_4) crystallizes in non-centrosymmetric space group Pna2_1. The unit cell constants a = 8.575(3)Å, b = 18.511(3)Å, c = 8.9021(4)Å were determined from automatic centering of 25 high angle reflections and refined by least-square methods using an Enraf-Nonius CAD4 computer controlled diffractometer. The salient feature of the structure is that copper(II) ions are bridged in a zigzag manner by *syn-anti* conformation of acetate group to form the polymeric chains of $[Cu(dpt)OAc]^+$ cations running parallel to the crystallographic *a*-axis (Fig.1).

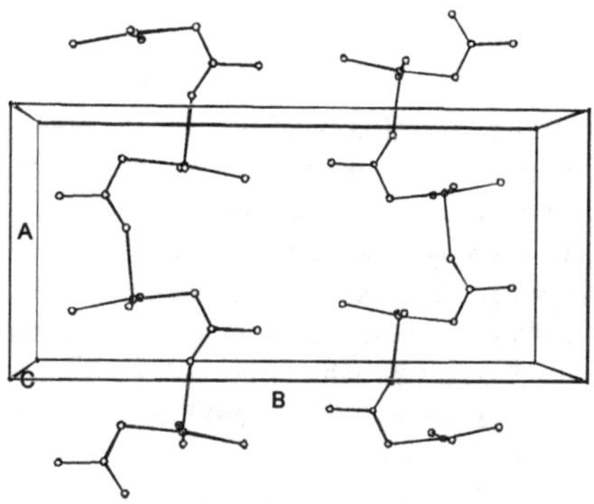

Figure.1 . View of the Chain $[Cu(dpt)(OAc)]^+$ running parallel to *a*-axis. Each monomeric unit in the chain is related to the neighbor through an *a*-glide in the *ac* plane.

Each unit in the chain is related to the next through an *a*-glide in the *ac* plane. The Cu-Cu distance is 4.581Å along the chain direction. The present refinement of 1070 unique reflections with $I \geq 3.0\sigma(I)$ observed with Mo-Kα intensity data, using full-matrix least-square methods with VAX version of the SHELX-93 (Sheldrick) computer program is in agreement with the previous study (Skeleton, 1972). With these data, the non-hydrogen atoms are shown to undergo anisotropic thermal vibrations.

3 EPR Spectral Data

The polycrystalline and single crystal EPR measurements were made on a Varian E112 EPR spectrometer using 100kHz modulation at X-band (9.5GHz) and Q-band (35GHz) frequencies. Magnetic susceptibility measurements which were made on a powder sample at a field of 0.1 T using SQUID susceptometer (model MPMS, Quantum Design, inc.) in the temperature range 4.2-300 K support the one-dimensionality of the magnetic system with the exchange coupling constant along the chain, 2J = 5.01 ± 0.04 cm^{-1}, with a interchain interaction of zJ′ = 0.0027 ± 0.0005 cm^{-1} that is ferromagnetic. Padè expansion series fro a ferromagnetic S = 1/2 chain was used to fit the susceptibility data.

The polycrystalline EPR spectra of $[Cu(dpt)OAc]^{+}$, which were recorded at X-band at room temperature and at liquid nitrogen temperatures show an axial symmetry. The general features of the spectrum are similar to those for copper ions in a $d_{x^2-y^2}$ ground state (Hathaway, 1987). There is no appreciable change in the lineshape and there is no increase in resolution with the temperature change. However, at Q-band the polycrystalline spectrum resolved to give a spectrum with rhombic symmetry.

3.1 g-factor

The single crystal EPR spectra at X-band and Q-band at room temperature consist of a single resonance line in all crystal orientations. The angular dependence of g^2 is shown in Fig.2 for rotations about the crystallographic orthogonal axes *a*, *b*, and *c* respectively.

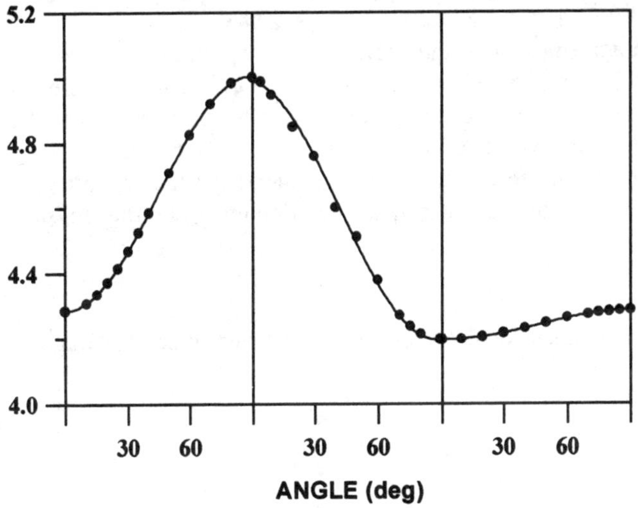

Fig. 2. Plot of the angular dependence of g^2 of the single crystal EPR spectra of $[Cu(dpt)(OAc)](ClO_4)$. The solid lines represents the result of the equation.

An angular g-factor dependence in the ith crystal rotation can be described by the general g^2 tensor expression (Waller, et. al., 1973),

$$g^2(i) = \alpha_i + \beta_i \cos 2\theta + \gamma_i \sin 2\theta \tag{1}$$

where θ is the azimuthal angle of the applied magnetic field. The anisotropic parameters α,β, and γ can be calculated by a least squares method from equidistant (in angle θ) experimental points from the expressions

$$\alpha_i = n^{-1} \sum_{\theta=0}^{m} g^2(\theta)$$
$$\beta_i = 2n^{-1} \sum_{\theta=0}^{m} g^2(\theta)\cos 2\theta \qquad (2)$$
$$\gamma_i = 2n^{-1} \sum_{\theta=0}^{m} g^2(\theta)\sin 2\theta$$

where n = 36, m = 175° for data points collected in 5° intervals and n = 18, m = 170° for 10° intervals. The g^2 plots with parameters calculated from equations are presented as solid lines in Fig.2, where a good fit to the experimental points may be seen.

A collection of α, β and γ parameters for three orthogonal rotations enable the determination of the g^2 tensor components. The following g factors were obtained from the rotational data for [Cu(dpt)OAc] (ClO_4),

$$g_a = 2.238 \; ; \; g_b = 2.07 \; ; \; g_c = 2.043.$$

in extremely good agreement with the polycrystalline spectrum (at Q-band) results of $g_1 = 2.235$, $g_2 = 2.07$, and $g_3 = 2.048$ indicating that the tensor axes are along the crystallographic axes. Though the the g-values refer to the crystal axes they also happens to be molecular g-values of the monomeric unit for two reasons: () all four molecules of the unit cell are magnetically equivalent; (ii) the crystallographic a-axis is the chain axis with the largest g-value, defining the g_{zz} for the molecule.

3.2 Linewidths

The linewidth of one-dimensional complexes is predicted to be (McGregor, et. al., 1976)

$$\Delta H_{pp} \cong \left(\frac{4}{3}\right)^{2/3} \frac{M_2^{2/3}}{J^{1/3}} \qquad (3)$$

where M_2 is the second moment of the EPR absorption and J is the intrachain exchange coupling frequency. The second moments depend on a geometrical factor which depends on the angle between the static magnetic field and the vector connecting the two interacting spins, on the sixth power of the spin-spin distance,

$$M_2 = \left\{ \begin{array}{c} \frac{3}{4} S(S+1)\mu^2 g^2 \sum_j r_{ij}^{-6} \{(3\cos^2\theta_{ij} - 1) + \\ \sin^4\theta_{ij} \exp[-(1/2)(2\omega_0/\omega_e)^2] + \\ 10\cos^2\theta_{ij} \sin^2\theta_{ij} \exp[-(1/2)(\omega_0/\omega_e)^2]\} \end{array} \right\} \tag{4}$$

where ω_0 and ω_e are the Zeeman and exchange frequencies respectively and θ_{ij} and r_{ij} are respectively the angle formed by static magnetic field with the vector connecting the two dipolar interacting centers i and j and the distance between them. The first term in the braces is the secular component, while the other two are non-secular components.

In one dimensional complexes, the dipolar coupling is expected to be the largest between the nearest neighbors in the chain due to their close proximity. As a consequence the linewidth should be greatest along the chain axis due to the secular part of the second moment. If the interchain distances are too long compared to the intrachain distances, the angular dependence of the linewidth is approximated by

$$\Delta H_{pp} \propto \left|3\cos^2\theta - 1\right|^{4/3} \tag{5}$$

where θ is the angle between the chain axis and the applied field. This expression predicts a linewidth of zero for θ = 54.7°, 'the magic angle'. However, at angles close to magic angle, the non-secular contributions to the second moment dominates and give linewidth larger than predicted..

The X-band spectra of a single crystal at room temperature show a strong angular dependence of the peak-to-peak linewidth in the static magnetic field (Fig.3). The linewidth has its maximum of 95 G along the *a*-axis, which corresponds to the chain direction, goes through a minima of 23 and 41 G at 55° from *a*-axis and reaches a second maximum of 28 and 49 G perpendicular to *a* in the ab and ac planes respectively. In the limit of accuracy of the experiment, the linewidth is isotropic in the *bc* plane, and consequently we report the angular dependence of linewidth in the *ab* and *ac* plane rotations.

The angular dependence of the linewidth of the EPR spectra at room temperature conforms nicely to that expected for a one-dimensional ferromagnet.The solid line shown in the angular dependence of linewidth plot (Fig.3) corresponds to the best fitting values calculated with the expression

$$\Delta H_{pp} = 52.7\left|3\cos^2 - 1\right|^{4/3} \tag{6}$$

which is similar to the expression used to describe the linewidth behavior in TMMC (Dietz, et. al., 1971). The experimental linewidths show relatively small differences from the calculated values.

There are two major reasons why these differences occur. first, is the presence of interchain interactions, that is predicted by the magnetic susceptibility measurements and the EPR spectral lineshape analysis. However, the small (<10 G)

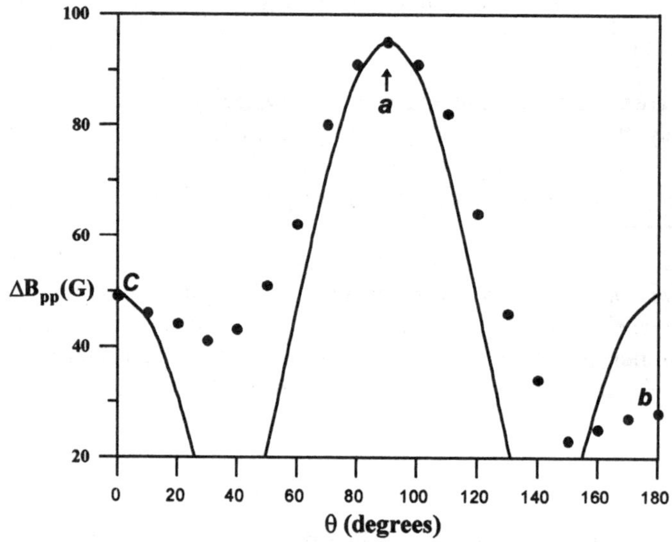

Fig. 3. Angular dependence of the X-band peak-to-peak derivative linewidth of [Cu(dpt)(OAc)](ClO_4) at room temperature. θ is the angle of applied field with the *a*-axis and lines correspond to the best fitting values calculated with the expression $\Delta H_{pp}=A|3\cos^2\theta-1|^{4/3}$

variation in the angular dependence of the linewidth in the *ac* plane (where the chain propagates through *a* glide) from the expected dependence suggests that the interchain dipolar contributions to the second moment are much less than those due to intrachain dipolar interactions.

The second and more important factor in the angular dependence of the linewidth in [Cu(dpt)OAc] (ClO_4), is that the angle between the nearest neighbor Cu–Cu vector in the chain, and the crystallographic *a* axis is 20.8°. This means that the angular dependence of the linewidth depends greatly on θ′, the angle between the Cu–Cu vector the magnetic field direction ($\theta \neq \theta'$). The difference between the linewidths at H∥*c*, and at H∥*b* can be explained by this fact as can the shift in the minima from the expected value of $\theta = 54.7°$ to $\theta \approx 60°$.

3.3 Lineshapes

In one dimensional systems the movement of the spin deviation is very slow due to the restricted alternative paths for the disturbance to move from one site to another site. Therefore the spectra of the one dimensional system are mostly affected by the long time part of the spin correlation function, which is dominated by the spin-diffusion effects (Bencini et. al., 1990). This long time spin-diffusion effects along the chain results in a lineshape described by the Fourier transform of an exponential relaxation function with $\exp(-t^{3/2})$ time dependence (Richards, 1976). Because of this, EPR theory for one-dimensional complexes predicts that the spectral lineshape be non-Lorentzian when the applied magnetic field is parallel to the chain axis (Dietz, et. al., 1971).

The single crystal EPR spectra of [Cu(dpt)OAc](ClO_4) contain Lorentzian lineshape in all crystal orientations. This is due to the fact that the inter chain interactions are significantly strong enough to suppress the dominance of the long-time diffusive tail in the spin correlation function, which leads to the Lorentzian lineshape. It has been shown by Hennessy, et. al.,(1973) that the lineshapes are more sensitive to the inter chain couplings than the linewidths. Thus in [Cu(dpt)OAc](ClO_4), it is possible to revert to a Lorentzian lineshape, while retaining the characteristic $(3\cos^2\theta-1)^{4/3}$ angular dependence.

The analysis of the EPR spectra provides much additional information in the magnetic structure of [Cu(dpt)OAc](ClO_4). In fact, in one-dimensional systems, the competition between the dipolar interactions which tend to give broad Gaussian lines with angular dependence of the type $(3\cos^2\theta-1)^2$, and the exchange interaction, which tend to give Lorentzian lines with an angular dependence of the type $(1+\cos^2\theta)$, yields an intermediate situation in which the angular dependence is of the type $(3\cos^2\theta-1)^{4/3}$ and the lineshape is Lorentzian only at magic angle, while it is intermediate between Gaussian and Lorentzian at other angular settings (Richards, 1976). In the present case, we observed the characteristic $(3\cos^2\theta-1)^{4/3}$ type angular dependence of linewidths, while the lineshapes are Lorentzian due to the inter chain spin diffusion, which dominates the short-time contributions to the spin correlation function.

4 Conclusion

The single crystal EPR spectra of [Cu(dpt)OAc](ClO_4) give interesting results. The analysis of the anisotropy of linewidth at room temperature conforms nicely to that expected for a one-dimensional ferromagnet, while the lineshapes indicate the significance of the interchain interactions that lead to Lorentzian line shape.

Acknowlegement

One of the us (RM) is thankful to CSIR, New Delhi, for a fellowship and CSIR, New Delhi and DST, New Delhi for the travel assistance. We thank the Department of Science and Technology, New Delhi, for a grant (SP/S1/F-47-90) to PTM and CSIR for another grant to PTM.

References

1. B. W. Skeleton, T. N. Waters, and N. F. Curtis, Crystal and molecular structure of catena-μ-acetato[di-(3-aminopropyl) amine]copper(II) perchlorate, *J. Chem. Soc.,(Dalton Trans.)* (1972) 2133-2138.
2. G. M. Sheldrick, *Acta Crystallogr.* Manuscript in preparation.
3. B. J. Hathaway, Copper, Comprehensive Coordination Chemistry, **5** (1987) 533-774.
4. W. G. Waller, M. T. Rogers, a generalization of methods for determining g tensors, *J. Magn. Resonance* **9** (1973) 92-107.
5. P. M. Richards, Magnetic Resonance in One-, and Two-Dimensional Systems, In Local Properties at Phase Transitions, Muller, K. A., and Rigamonti, A. (eds.), North Holland: Newyork (1976) p. 539–605.
6. R. E. Dietz, F. R. Merritt, R. Dingle, D. Hone, B. G. Silbernagel, and P. M. Richards, Exchange Narrowing in One-dimensional Systems, *Phys. Rev. Lett.* **26** (1971) 1186–88.
7. A. Bencini, D. Gatteschi, Electron paramagnetic resonance of exchange coupled systems, Springer-Verlag, Berlin (1990).
8. K. T. McGregor, and Z. G. Soos, Anisotropic Exchange in Linear Chain Complexes of Copper(II), *J. Chem. Phys.* **64** (1976) 2506–2517.
9. M. J. Hennessy, C. D. McElwee, and P. M. Richards, Effect of Interchain Coupling on Electron-Spin Resonance in Nearly One-Dimensional Systems, *Phys. Rev.* **B7** (1973) 930-947.

EPR as an exploration tool to assess quartz reefs, Beaconsfield Gold Mine, Tasmania, Australia

D.W. Russell [1,2] *and J.C. van Moort* [1]

1 Geology Department, University of Tasmania, GPO Box 252-79, Hobart, Australia
2 Don College, Devonport

ABSTRACT: A study of three hundred and seven quartz samples from the Beaconsfield goldfield indicates that the intensity of the $[AlO_4]^{\circ}$ EPR centre, determined at room temperature, can be used to separate reef quartz from barren quartz.

Introduction

In a set situation miners have always been able to distinguish visually between barren and auriferous quartz veins. A variety of methods has been used to quantify the differences between these types of quartz, including fluid inclusion studies, decripitometry, thermal release gas evolution mass spectrometry, oxygen isotope and trace element studies, thermoluminescence and cathodoluminescence. Scherbakova et al. (1976) used powder EPR at the temperature of liquid nitrogen "to analyse the gold, copper and molybdenum content of quartz veins".

Since 1985 van Moort and co-workers have investigated the use of room temperature EPR to characterise generations of quartz veins and to define wall rock alteration patterns. The following article outlines the use of electron paramagnetic resonance in the study of auriferous quartz reefs at Beaconsfield.

Geology

The 10 km long Beaconsfield/Salisbury goldfield is situated in Ordovician sediments west of the Tamar Fracture System in central northern Tasmania. The main production came from the Tasmania Reef, a Devonian fissure reef with quartz emplaced on a pre-existing fault zone. The reef is 395 m long and varies from 20 mm to 8 m, averaging 2.5 m wide. There are many smaller vein systems in the area, both auriferous and mined or barren and explored only. The veins consist of microcrystalline quartz, sometimes with a core of ankerite. The drill holes from which material was used for the present study intersect the reef below the deepest level of the old mine workings. Gold values in the Tasmania Reef vary with depth. Over the first 120 m an average grade of 38 g/t was maintained, dropping gradually to 4 g/t at 415 m. However at 450 m the average grades had risen to 20 g/t. The geology and mineralogy of the goldfield was discussed in detail by Russell and van Moort (1992) and Russell (1995).

Sample selection and sample preparation

The samples represent quartz samples of several grams each from core of diamond-drill hole B4, off-shoots B4A and B4B, three wedges of diamond-drill hole B11 (all with high Au, base metals and moderate As at reef intersection), DDHs B10, B12, B15A, B16 (low Au, high As at intersection), DDHs B9, B17 (low mineralisation) and DDH B13 (no Au, no As). All three hundred and seven samples were finely ground (approximately 10 μm) in an agate mortar, leached overnight in hot fuming nitric acid, rinsed three times with distilled water and dried at 90° C. This process removed all carbonate and sulphide impurities intergrown with the quartz.

Standardisation

All spectra were standardised against Mn^{2+} diffused through MgO powder, inserted in a double cavity for all measurements. All spectra simultaneously recorded the Mn^{2+} lines number 3 and 4 and the quartz EPR spectra. The E' centre is relatively more pronounced at low microwave power. At higher microwave power the E′centre diminishes through saturation, whilst the central $[AlO_4]^{\circ}$ centre grows proportionally up to microwave power 100 mW. See Figure 1. The nature of this major signal will be discussed below.

The relation between the peak height and the absorption of the microwave power as expressed by double integral of the signal is proportional at high microwave power. See Figure 2. Several early Beaconsfield samples have provided reproducible spectra over a period of eight years and have been used as a measure of the reproducibility of the EPR results (i.e. 3-4%). Standard instrument settings have been observed throughout the project.

Generalised description of the EPR spectra.

The commonly occurring EPR spectra of vein quartz were summarised by van Moort and Barth (1992). The first derivatives of natural quartz EPR spectra are often complex. Workers in this area have identified a range of "centres" in spectra of single crystals. In an early paper, Weil (1984) describes several forms of lattice defects in a single quartz crystal, as observed in EPR spectra, viz. electron donors replacing Si (e.g. Al or rarely Ga), electron acceptors (e.g. Ge, Ti, Fe), compensating ions (e.g. H+, $Li,^{+}$ Na^{+}) and defects caused by irradiation (e.g. broken bonds).

Of practical importance for the interpretation of quartz powder spectra are Friebele and Griscom's (1979) and Griscom's (1980) papers on electron paramagnetic resonance of amorphous materials as glasses, discussing the spectra in terms of impurities, the oxygen vacancy E' and non linking oxygen bonds ("oxygen hole centres"). A general discussion of metastable electron-hole centres and other defects is beyond the scope of this practical study. Attention, however, will be given to the most prominent EPR signal in the sample.

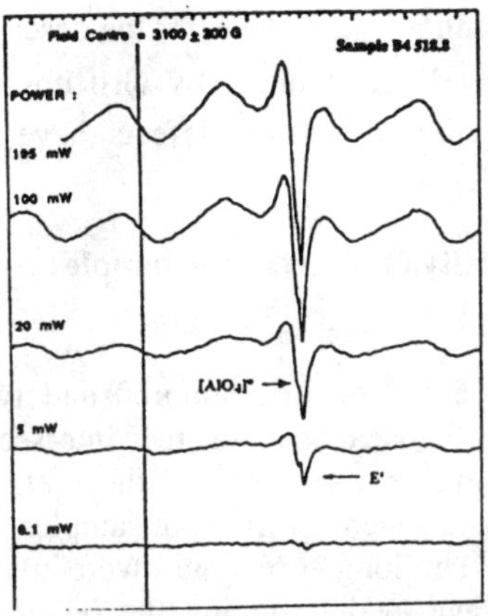

Fig.1: Different spectral traces of the same sample, as a function of microwave power

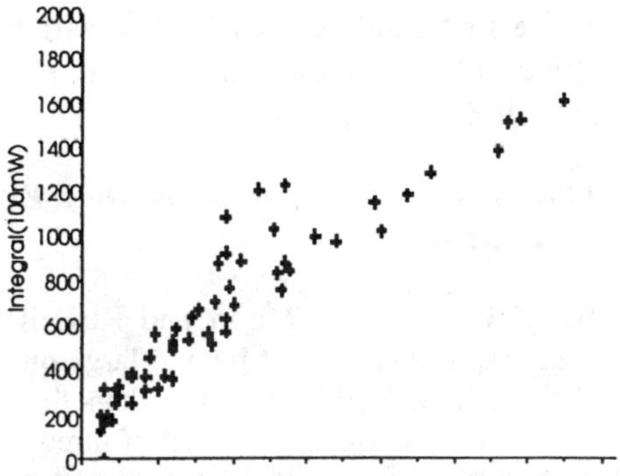

Figure2:- Graphical presentation of the strong relationship between Peak Height and Integral for 100mW conditions. (N=95, R_S=1.000

The major signal located between g=2.0127 and g=1.9969

McMorris (1970) was one of the early workers to observe a pronounced EPR powder signal at g≈2 for "a dozen extracts of igneous, sedimentary and metamorphic quartz, several pegmatitic quartzes and one alpha active obsidian". He attributed the signal to α-radiation damage, but did not substantiate this claim. The same centre can also be observed in some silica glasses (Schnadt and Räuber, 1971), flint (Garrison et al., 1981), volcanic rocks (Shimakowa, et al., 1984) and in vein quartz as described by van Moort and co-workers (1987, 1988, 1991, 1995). After irradiation and immediate EPR analysis at <100 K, the signal frequently shows hyperfine splitting due to the presence of aluminium. This observation indicates that there is a relationship between the signal and the presence of aluminium in the sample. Presumably, this centre represents what Scherbakova et al. (1976) and Matyash et al. (1982) observed in vein quartz and called the "Al-O" centre which, they claim, can only be observed at cryogenic temperatures. There is confusion in the literature on this issue, as can be demonstrated by the following descriptions of this EPR centre in quartz. Shimakowa et al. (1984) report that the centre (with a minimum at g=2.0025-2.0030) is visible at room temperature and cannot be enhanced by x-rays. Imai et al. (1985) report that it is visible at 77 K only and can be enhanced by x-rays. Then Shimakowa and Imai (1985) reported that there are two superimposed centres. An earlier study of glasses led Friebele and Griscom (1979) and Griscom (1980) to the belief that there is a double oxygen link in the silica lattice, which they called "peroxy centre" or "peroxy radical centre". This suggested centre occurs at the g value of 2.0025-2.0030, but their explanation is not universally accepted. The major signal is well developed in

Beaconsfield (Figure 3) and in those few samples subjected to x-ray irradiation, it does not significantly change its intensity. On close examination, the signal consists of two adjacent first derivatives, one with a crossover at g=2.007 and the other with a crossover at g=2.003. These crossovers are reminiscent of the g_{yy} and g_{xx} of the $[AlO_4]^\circ$ centre as described by Petrov et al. (1990). They identified the following $[AlO_4]^\circ$ centres in quartz, g_{zz}=2.059, g_{yy}=2.009 and g_{xx}=2.003. After the description of the first single crystal EPR spectrum of $[AlO_4]^\circ$ in α-quartz by Griffiths et al. (1954, 1955), subsequent descriptions of a family of related defects have been published.

Pilot studies on the relation between EPR intensity of quartz spot samples core Au content

Early work by van Moort and Russell (1987), based on one hundred and twelve samples, indicated that by and large quartz from the Beaconsfield reef intersections showed higher EPR intensity than those from quartz veins outside the reef. See Figure 4. Subsequently a total of three hundred and seven quartz spot samples from core (with known gold content as determined on 1 m long core splits) were analysed by EPR as a further pilot project. The first stage in determining the relationship between quartz and gold mineralisation was to separate the samples in two subsets. All mineralised samples with a $Au_{(fire)}$ content ≥0.02 ppm (the cut-off used at Beaconsfield by the mining companies) were extracted from the initial data pool. The result of this subdivision is illustrated in the table 1, where the EPR peak height represents the amplitude of the first derivative signal as indicated in Figure 3. Both the total data pool and the non-mineralised samples displayed a very strong correlation between the lack of gold and peak height (R_S=0.469 and R_S=0.500 for 307 and 204 samples respectively, at a 99.999% confidence level, p = 0.001).

Table 1: $Au_{(fire)}$ vs EPR Peak Height (cm), with the subdivision of the initial population into mineralised (N=103) and non-mineralised samples (N=204).

Population	Number	Gold Mean (ppm)	Peak Height Mean (cm)	Sp. Rank Coeff.	@ p=0.001
All Samples	307	14.5	7.4	0.469	0.190
Au<0.02 ppm	204	0	5.7	0.500	0.190
Au≥0.02 ppm	103	43.1	10.9	-0.084	0.200

The data appear to provide a process whereby the mean peak height of the EPR signal can be used to identify whether a sample is potentially auriferous or not. Both the mean peak height of the non-mineralised group (5.7 cm) and the total data pool (7.4 cm), suggest that a value between 5 cm and 7 cm could be used as a peak height cut-off point (i.e. for standard instrument settings). To test this the samples were split into two groups of respectively a peak height of less than 5 cm and more than or equal to 5 cm. This process demonstrated that there is a strong relationship between peak heights less than 5 cm and the absence of gold. The group with the higher peak height, however, contained too many samples without gold to give a significant correlation. The EPR method, as used, provided consequently only a cheap and efficient technique to indicate where there is *no* gold.

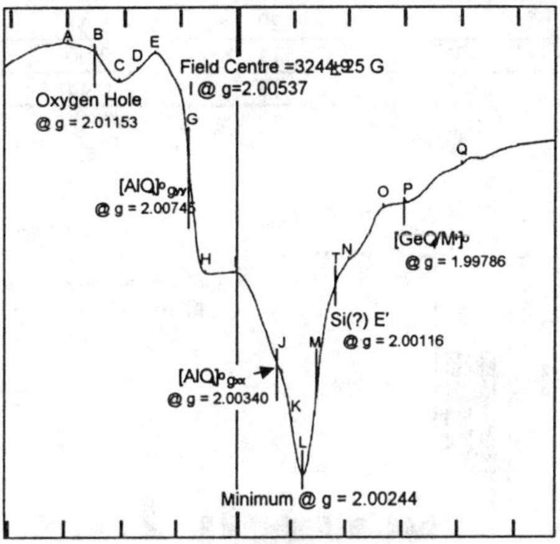

Figure 3: The calibration plot of sample, M22 (B_{11} 403).

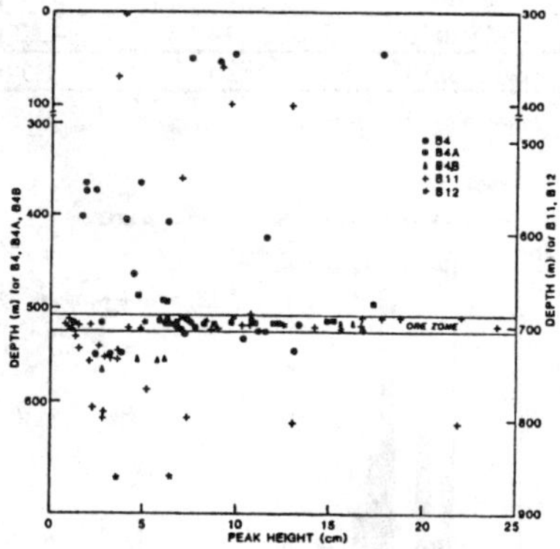

Figure 4: A plot of EPR peak heights as a function of depth for quartz powders from diamond drill holes intersecting mineralisation at Beaconsfield. (after van Moort and Russell, 1987).

Results of the final study

Out of the three hundred and seven samples one hundred and twenty were re-analysed by EPR at microwave power 0.5 and 100 mW. When the population (N=120) was subdivided into reef samples and barren samples, there were several correlations between the gold fire assay values and the EPR results. In the auriferous subset, the normalised peak height (100mW), integral (100mW) and integral (0.05mW) demonstrated significant correlations of R_S=0.200 (N=63), R_S=0.400 (N=47) and R_S=0.300 (N=51) respectively with $Au_{(fire)}$ values.

After removal of surface samples, tiny veins with inadequate information on gold content and duplicates a final seventy auriferous and barren samples remained in a final pool. Statistical analysis of this group found significant correlations between $Au_{(fire)}$ and the normalised peak height (100mW) and standardised integral (100mW). These correlations were R_S=0.410 (N=70, r=0.380 at p=0.001) and R_S=0.380 (N=57, r=0.330 at p=0.01) respectively with $Au_{(fire)}$ values. This indicated that the EPR signals at this *higher* microwave setting could be used as an indicator of gold mineralisation. Frequency histograms were plotted for the reef and barren populations, to determine the most suitable range of EPR values to predict mineralisation. These plots indicated a "cut-off point" of 25 units. Hence, the use of this "cut-off point" (25 units), will select 33 (80%) of the auriferous samples and at the same time will discard 18 (62%) of the barren samples. See Figures 5 and 6. Further statistics highlighting the use of EPR as a discriminator for gold mineralisation are listed below in table 2.

Table 2:- Demonstrating the use of the discriminatory Peak Height value of 25 units. N_R = number of reef samples, N_B = number of barren samples.

Population	N_t	Au Mean	N_R	N_B	$N_R:N_B$
Total	70	31.6 ppm	41	29	1.41
x >25 units	44	49.3 ppm	33	11	3.00
x<25 units	26	1.5 ppm	8	18	0.44

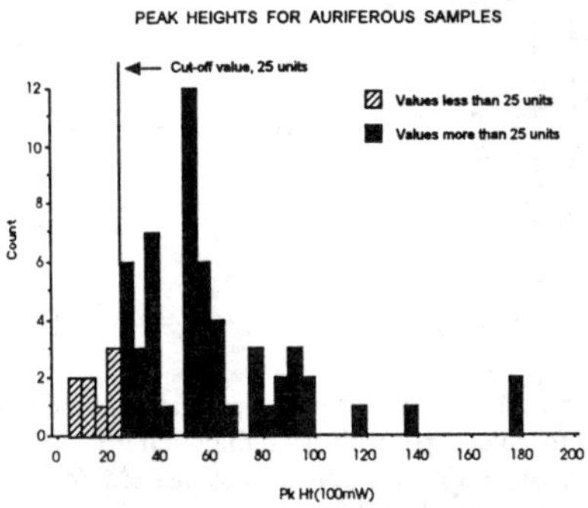

Figure 5: Frequency histograms plotted for mineralised samples, N=44. These EPR values are relative Peak Height units measured at a power of 100 mW. These plots demonstrate the "cut-off" at 25 cm, for standard instrument settings

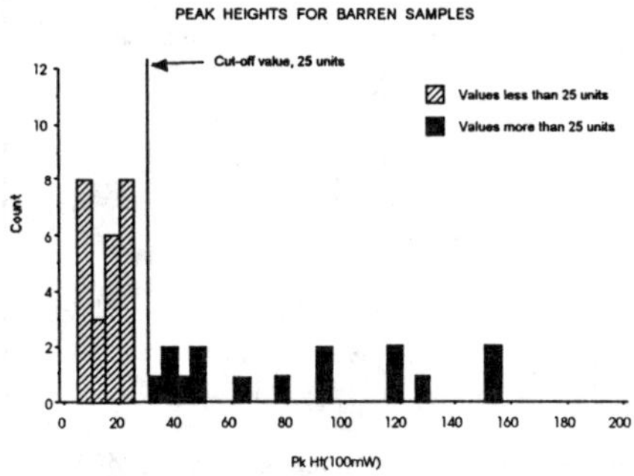

Figure 6: Frequency histograms plotted for barren samples, N=26 . These EPR values are relative Peak Height units measured at a power of 100 mW. These plots demonstrate the "cut-off" at 25 cm, for standard instrument settings

This table demonstrates the concentration of gold samples above the cut-off (mean 53.8 ppm) v's below the cut-off (mean 3.6 ppm) . The ratio of reef samples v's barren samples (i.e. $N_R:N_B$) further demonstrates the screening power of this cut-off value (4.36 above the cut-off value v's 0.40 below this value.)

EPR and trace elements in quartz

Of the three hundred and seven quartz samples analysed by EPR the subset of one hundred and twenty samples was chemically analysed by PIXE/PIGME spectroscopy (Russell at al 1996).

When the total population is statistically analysed by Spearman Ranking (Table 3), there is a strong primary relationship between EPR and K (R_S=0.660), Ti (R_S=0.653), Fe (R_S=0.583) and Al (R_S=0.546). There is also a group of elements with a weaker, but significant secondary(?) relationship with EPR. These are $Au_{(fire)}$ (R_S=0.354), Li (R_S=0.343), and As (R_S=0.338). Five elements (i.e. Ti, Cl, Mn, Ge, Zr) have two distinct populations: a subset of high chemical analyses for that element, and a subset of chemical analyses which were below the MDL for that element.

Table 3 : Correlation between EPR Intensity and Trace Elements in quartz powders. (ns = not significant)

EPR INTENSITY v's ELEMENT	K	Ti	Fe	Al	Li	Au(f)	Zr	As	Cl	Ge	r@ p=0.001
All; N=120	0.660	0.653	0.583	0.546	0.343	0.354	ns	0.338	ns	ns	0.32
Cl<20ppm N=51	0.575	0.634	0.485	0.423	ns	ns	ns	ns	ns	ns	0.443
Cl≥20ppm N=69	0.771	0.704	0.672	0.731	0.328	0.325	ns	ns	0.375	ns	0.38
Ge<1ppm N=65	0.629	0.701	0.543	0.44	ns	ns	0.471	ns	ns	ns	0.395
Ge≥1ppm N=55	0.741	0.608	0.638	0.757	0.425	0.432	ns	0.418	ns	0.526	0.425
Ti<4ppm N=54	0.779	0.568	0.401	0.629	0.543	0.690	ns	ns	ns	0.512	0.519
Ti≥4ppm N=66	ns	0.404	0.433	ns	ns	ns	ns	ns	0.487	ns	0.426

EPR and atomic ratios

The relationship between the presence of substitutional Al and compensating H, Li and Na in selected vein quartz crystals was originally investigated by Bambauer (1961). In a study on chalcedony, Flörke et al. (1982) extended this substitutional balance to K, Li, Na plus Ca and Mg on the one hand and Al plus Fe on the other hand. For Beaconsfield the compensating ions of Na, Li and K were assessed in the light of the strength of their correlation with Al and Fe. See table 4. Ca and Mg are not significant in the Beaconsfield samples.

Table 4: A matrix of Rs values for N=120 and r=0.320 demonstrating the importance of K^+ as a compensating ion.

mol (Element)	Fe	Al	Al+Fe
Na	0.241	0.492	0.485
Li	0.228	0.242	0.241
Na+Li	0.331	0.564	0.557
K	0.744	0.850	0.855

While there are only weak correlations between Li and Fe, Al or Al+Fe, there are significant correlations between K and each of Fe(Rs=0.744Rs), Al (Rs=0.492) and also Al+Fe (Rs=0.485) and highly significant. This suggests that K^+, not withstanding its large ionic radius, is the most prominent compensating ion in the Beaconsfield quartz.

Electron microprobe analysis did not reveal any potash feldspar or muscovite in the quartz and the (mol K)/(mol K + mol Al) and (mol K)/(mol K + 3 mol Al) ratio in the quartz exceeds the values for these minerals (Russell 1995). This indicates that the elemental impurities represent lattice substitutions rather than mineral inclusions.

The intensity of the EPR signal has a significant but scattered correlation with $Au_{(fire)}$, Li, K, Ti and Al. Fig.7 shows the Al versus 100 mW EPR signal.

Discussion

The study of the auriferous Beaconsfield quartz reef forms part of a project to assess the practical use of EPR as an exploration tool. studies were undertaken on quartz from gold mines throughout Western Australia, Victoria and Queensland. The most detailed of the early studies is a study of the Fosterville goldfield near Bendigo, Victoria (Nand,1989; van Moort et al 1991). Van Moort and Brathwaite (1988) used EPR spectra in a more general way to describe sequences of quartz veins at Waihi, New Zealand. Reef studies have now been carried out throughout the world, e.g. at Rodalquilar, Spain, Homestake, South Dakota, Huangshan, China. Recent work involves studies on auriferous Carlin jasperoids (van Moort et al, 1995).

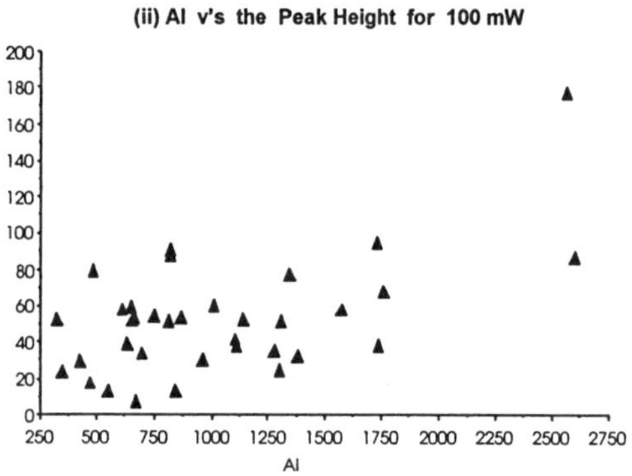

Figure 7: Correlation between EPR signals and Al or K in Reef Quartz.Al (ppm) v's Peak Height for 100mW; N=35,Rs=0.345, p=0.05 for r=0.3246.

EPR spectroscopy was used in all these studies as a precursor to slower and more expensive chemical analysis of which the results have been published in most cases. In contrast EPR data, so much easier to collect, have rarely been published, mainly because of restrictions imposed by sponsoring mining companies.

Conclusions

The results of this work are that mineralised quartz of reefs at Beaconfield has on average a more pronounced $[AlO_4]^{\circ}$ signal and that barren quartz shows only infrequently a strong EPR signal. The relations are more of a statistical nature, are more useful to discriminate between auriferous veins and barren veins and should not be used as a measure of gold content.

References

1. Bambauer, H.U., 1961, Spurenelementgehalte und γ-farbenzentren in Quarzen aus Zerrkluften der Schweizer Alpen. Schweiz. Min. und Petr. Mitt. 4:361-369

2. Flörke, O.W., Köhler-Herbertz, B., Langer, K., and Tönges, I., 1982, Water in microcrystalline quartz of volcanic origin: Agates. Contrib. Mineral. Petro., 80: 324-333

3.Friebele, E.J., and Griscom, D.L., 1979. Radiation effects in glass. In Tomoizawa, M. and Doremus, R.H. (editors), Treatise on Material Science. New York, Academic Press, 17:257-351.
4.Garrison, E.G., Rowlett, R.M., Cowan, D.L. and Holroyd, L.V., 1981. ESR dating of ancient flint. *Nature* (London), 290:44-45.
5.Griffiths, J.H.E., Owen, J. and Ward, I.M., 1954. Paramagnetic resonance in neutron irradiated diamond and smoky quartz. *Nature*, 173:439-442.
6.Griffiths, J.H.E., Owen, J. and Ward, I.M., 1954. Magnetic resonance in irradiated diamond and quartz. Report of the Bristol Conference - Defects in Crystallline Solids. Phys. Soc. London, 81-07.
7.Griscom, D.L., 1980. Electron spin resonance in glasses. *J. Non-Crystalline Solids*, 40:211-272.
8.Imai, N., Shimakowa, K. and Hirota, M., 1985. ESR dating of volcanic ash. *Nature*, 314:81-83.
9.Matyash, I.V., Brik, A.B., Monakov, V.I. and Derskii, L.I., 1982. Formation of native gold in quartz according to EPR data. Geokhimya, pp1048- 51.
10.McMorris, D.W., 1970. ESR detection of fossil alpha damage in quartz. *Nature*, 22:146-148.
11.Nand, A.S., 1989. The geochemistry of the Fosterville Goldfield, Victoria. Unpub. Hons. Thesis, Geology Department, University of Tasmania. 114p.
12.Petrov, I., Agel, A., Bershov, L.V. and Hafner, S.S., 1990. Aluminium traces in quartz as indicator for temperature of formation. In R. Emmerman and P. Giese (Editors). KTB report 90-3. *Niedersächsisches Landesamt fur Bodenforschung, Hannover,* 561.
13.Russell D.W., 1995. The characterisation of quartz and gold in the Beaconsfield and Lefroy areas, northern Tasmania. Unpub. Ph.D. thesis, Geology Department, University of Tasmania. 230pp.
14.Russell, D.W. and van Moort, J. C., 1992. Mineralogy and stable isotope geochemistry of the Beaconsfield, Salisbury and Lefroy goldfields. In: Tasmania: An Island of Potential. Tasmanian Department of Mines, *Geological Survey, Bulletin* 70:208-225.
15. Russell, D.W., Bailey, G.M., van Moort, J.C. and Cohen, D.D., 1996, Chemical composition and paramagnetism of vein quartz from the Tasmanian gold mine, Beaconsfield, northern Tasmania, Nuclear Instruments and Methods in Physics Research B 109/110, 598-600
16.Scherbakova, M.Y., Sotnikov, V.I., Proskuryakov, A.A., Mashkavtsev, R.I. and Soltntsev, V.P., 1976. Use of quartz EPR spectra in the analysis of ore content (as examplified by gold-ore and copper molybdenum mineralisation). *Geologiya Rudnykh Mestorozhdenii,* 18:63-69 (in Russian).
17.Schnadt, R. and Räuber, A., 1971. Motional effects in the trapped-hole center in smoky quartz. *Solid State Commun.*, 9:159-161.
18.Shimakowa, K., Imai, N. and Hirota, M., 1984. Dating of volcanic rock by electron spin resonance. *Isotope Geoscience*, 2:365-373.
19.Shimakowa, K. and Imai, N., 1985. ESR dating of quartz in tuff and tephra. In M. Ikeya and T. Miki (Editors), ESR Dating and Dosimetry. Ionics, Tokyo, pp 181-185.
20.Van Moort, J.C. and Barth , W.H., 1992. The application of electron paramagnetic resonance to the characterisation of vein quartz. In: Tu Guanzhi, Xu Kequin and Qui Yazhao (Editors), *Petrogenesis and Mineralisation of Granitoids. Proc. International Symposium on Petrogenesis and Mineralisation of Granitoids.* Science Press, Beijing, pp 1072-1086.
21.Van Moort, J.C. and Brathwaite, R.L., 1988. Electron paramagnetic resonance powder spectra of epithermal quartz from the Martha Hill gold-silver deposit, Waihi, New Zealand. *Bicentennial Gold 88, Geol. Soc., Australia, Extended Abstracts.* **23**:575-578.
22. Van Moort, J.C. Hotchkis, M.A.C. and Aung Pwa, 1995. EPR spectra and lithogeochemistry of jasperoids at Carlin, Nevada: distinction between auriferous and barren rocks. *J. Geochem. Expl.* 55:283-299.
23.Van Moort, J.C., and Russsell, D.W. 1987. Electron spin resonance of auriferous and barren quartz at Beaconsfield, northern Tasmania. *J. Geochem. Expl.* 27:227-237.
24.Van Moort, J.C., Nand, A.S., Cohen, D.D., Newman, S. and Pwa, A. 1991. The use of electron paramagnetic resonance spectroscopy and trace element content of vein quartz in mineral exploration. In Mrna, F. (Ed.) Exploration Geochemistry 1990. *Proc. Geol Survey and Czech Academy of Sciences, Prague,* pp221-229.
25.Weil, J.A., 1984. A review of electron spin spectoscopy and applocation to the sturdy of paramagnetic defects in quartz. *Phys. Chem. Miner.*, 10:149-165.

Electron Spin Echo Envelope Modulation (ESEEM) Study of Cu(II) Complexes of 15-Crown-5 and Benzo-15-Crown-5

Yong Li and Jun Yamauchi*

Graduate School of Human and Environmental Studies, Kyoto University, Kyoto 606, Japan

Abstract. Electron spin echo envelope modulation (ESEEM) measurements were carried out for $3d_{z^2}$ ground-state Cu(II) complexes of 15-crown-5 and benzo-15-crown-5 in BF_3-ether-water system and in 60% $HClO_4$. Two kinds of proton modulations were detected and are ascribed to the protons of axially coordinated water molecules and of the crown ethers respectively. In BF_3-ether-water system, the FT-ESEEM spectra are complicated because BF_3 is coordinated to oxygen atoms of the crown ethers. In 60% $HClO_4$ solution, chlorine nuclear modulations of ClO_4^- anions, which are in the axial directions through the hydrogen bonds with the axially coordinated water, were also observed.

1 Introduction

Cu(II) complexes of 15-crown-5 (15C5) and benzo-15-crown-5 (B15C5) have been shown to take a seven-coordinate pentagonal bipyramidal geometry with a distinctive electronic configuration of $3d_{z^2}$ ground-state doublet.[1,2] The ESR spectra revealed unusual g-anisotropy ($g_z < g_x, g_y$). In BF_3-ether-water system, the presence of water molecules coordinated at axial positions have been confirmed based on the detection of ^{17}O (I=5/2) superhyperfine splittings, derived from $H_2{}^{17}O$, in the g_z region of the ESR spectra.[3] It is also interesting that the complexes in BF_3-ether-water system, compared to those in other solutions, are very stable with a very large complex formation constant.[4] However, the role of BF_3-ether in this case was not clarified. Although X-ray analyses of the Cu(15C5) and Cu(B15C5) complex single crystals obtained from $HClO_4$ solution demonstrated that the Cu(II) ion is situated almost in the center of the crown ether ring and is axially coordinated to two water molecules,[5] no direct proof was given in the previous work that the axial ligands are water molecules in $HClO_4$ solution.

In order to answer the above-mentioned questions and to investigate in more detail the coordination and solvent behavior in solution, electron spin echo envelope modulation (ESEEM) measurements were carried out for the Cu(15C5) and Cu(B15C5) complexes in BF_3-ether-water system and in 60% $HClO_4$ solution and for

* Visiting researcher from Department of Chemistry, Tsinghua University, Beijing 100084, China.

a powder sample of 0.5% Cu(II) doped in Mg(B15C5) crystal, to detect the weak interaction between the unpaired electron of Cu(II) ion and nuclear spins of the ligands and/or solvent, which can not be observed in the usual ESR measurements.

2 Experimental

The Cu(15C5) and Cu(B15C5) complexes in BF_3-ether-water system and in 60% $HClO_4$ solution, and a powder sample of 0.5% Cu(II) doped in Mg(B15C5) crystal were prepared according to the previous methods.[4,5,6] All ESR and ESEEM measurements were performed using a JEOL PX1050 FT-ESR spectrometer at 10K and under the microwave frequency of 8.815GHz. Fig.1 shows the ESR spectra of the Cu(15C5) and Cu(B15C5) complexes in BF_3-ether-water system, in which a glassy state can form at low temperature, and also of a powder sample of 0.5% Cu(II) doped in Mg(B15C5) crystal. The ESR spectra of the complexes in 60% $HClO_4$ solution are similar to those in BF_3-ether-water system. The ESEEM spectra were measured using two- and three-pulse echo sequences at the magnetic field positions marked with alphabet as shown in Fig. 1. The pulse widths are 20, 40ns for two-pulse ESEEM measurements, and 30, 30, 30ns for three-pulse ones. In order to eliminate the influence of unwanted two-pulse echo on the three-pulse ESEEM, the phase-cycling technique[7] was used in all the three-pulse ESEEM measurements. The data processing and the Fourier transformation to obtain power spectra were performed using ESPRIT-425 PULSE ESR SYSTEM PROGRAM of the FT-ESR instrument.

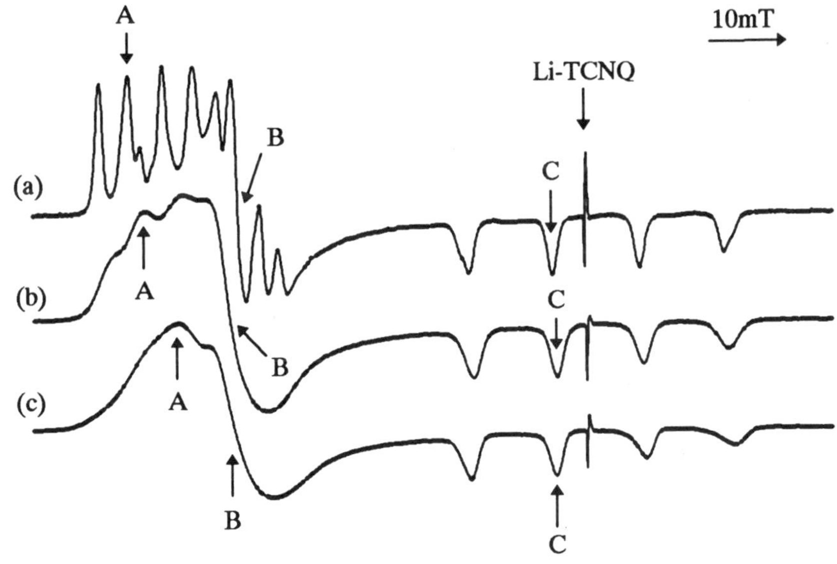

Fig. 1. The ESR spectra observed for the Cu(15C5) and Cu(B15C5) complexes in BF_3-ether-water system ((b) and (c)) and the powder sample of Cu(II) doped in Mg(B15C5) crystal (a). The ESEEM measurements were carried out at A, B and C points. The g-value of Li-TCNQ is 2.0025.

3 Results and Discussion

3.1 ESEEM Spectra of the Complexes in BF_3-ether-water System and BF_3 Coordinated to the Crown Ring

Fig. 2 shows some typical FT-ESEEM spectra observed for the Cu(15C5) complex in BF_3-ether-water system. The FT-ESEEM spectra observed at A and B points are similar, and no important difference was detected in the FT-ESEEM spectra between Cu(15C5) and Cu(B15C5) complexes. As shown in Fig. 2, the spectra are complicated and a lot of peaks are observed in the low frequency domain, even for three-pulse FT-ESEEM spectra (Fig. 2(c)). It is clear that only proton can not give such complicated spectra and there must be nuclear modulations from other nuclei. In BF_3-ether-water system, the nuclei having nuclear spin are hydrogen, boron and fluorine, the Larmor frequency of which at the magnetic field of 300mT are $\nu(^1H$, I=1/2)=12.8MHz, $\nu(^{19}F$, I=1/2)=12.0MHz、$\nu(^{10}B$, I=3, natural abundance=19.58%) =1.4MHz and $\nu(^{11}B$, I=3/2, natural abundance=80.42%)=4.1MHz, respectively.[8] Thus, the complicated peaks in the low frequency domain are ascribed to be mainly due to boron nuclei. BF_3 is a strong Lewis acid and is easily coordinated to oxygen atoms of the crown ether. Then boron and fluorine nuclei can interact with the unpaired electron at the Cu(II) ion. In fact, it has been reported that three BF_3 molecules are coordinated to one free 15C5 on the average.[9] The present ESEEM investigation indicates that BF_3 is also coordinated to the Cu(II) complexes of 15C5 and related O_5-crown ethers. Since boron has two isotopes which can cause nuclear quadrupole interaction, and since the number of BF_3 molecules coordinated to the crown ring of one complex molecule is not fixed, the observed FT-ESEEM spectra became complicated.

The fact that BF_3 molecules coordinated to the crown ring can well explain why $3d_{z^2}$ ground-state Cu(II) complexes of 15C5 and related O_5-crown ethers are easily formed in BF_3-ether-water system. Usually, it is difficult to form Cu(II) complexes of 15C5 and related O_5-crown ethers in the presence of water because water is a much stronger ligand than the crown ethers.[1,2] In BF_3-ether-water system, although water exists in the solution, Cu(II) complexes of 15C5 and related O_5-crown ethers can be easily formed because BF_3 molecules coordinated to the crown ring can protect the Cu(II) ion, which is located in the center of the crown ether cavity, from the coordinations of solvent water molecules. This behavior means that the thermodynamic equilibrium is largely shifted to the complexation of the Cu(II) ion and the crown ether than to the hydration of Cu(II) ion.

3.2 Explanation of ESEEM spectra observed for the complexes in 60% $HClO_4$

Fig. 3 shows some typical FT-ESEEM spectra observed for the Cu(B15C5) complex in 60% $HClO_4$. Like in BF_3-ether-water system, the FT-ESEEM spectra observed at A and B points are essentially the same, and the corresponding FT-ESEEM spectra of Cu(B15C5) and Cu(15C5) complexes are similar. However, the spectra observed for the complexes in 60% $HClO_4$ are rather simpler than in BF_3-ether-water system. The

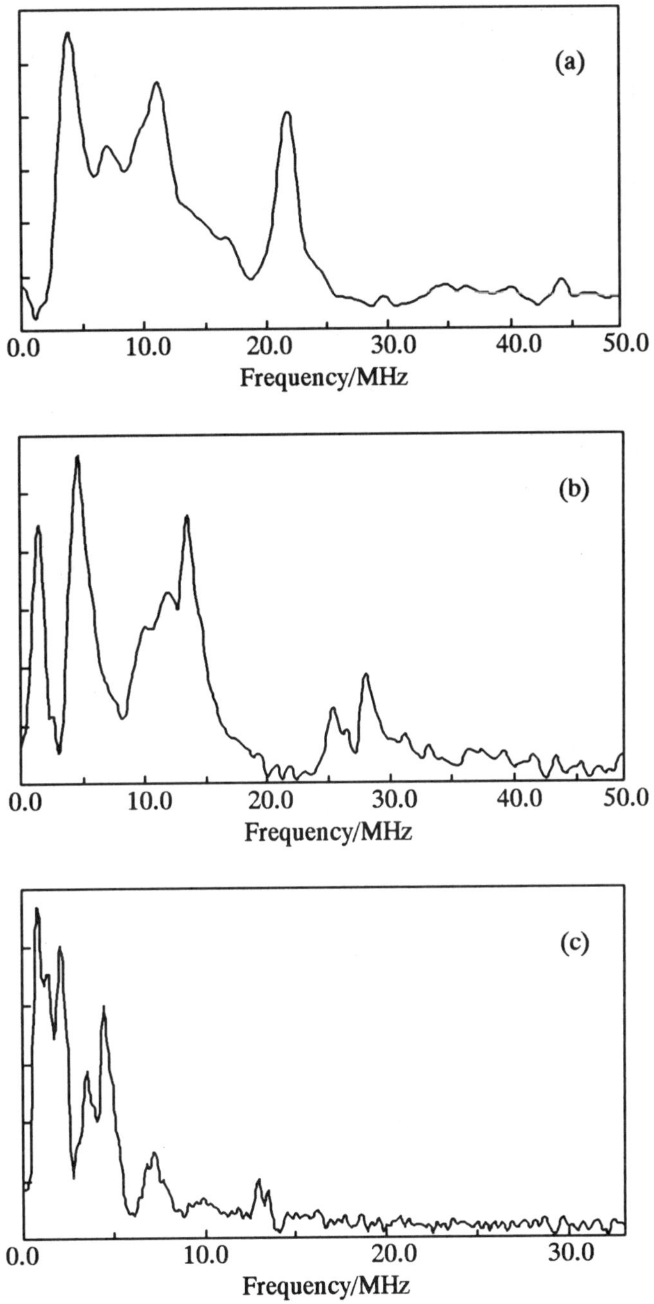

Fig. 2. The FT-ESEEM spectra observed for the Cu(15C5) complex in BF_3-ether-water system. (a) and (b): The two-pulse spectra observed at A and C points respectively; (c): The three-pulse spectrum observed at C point.

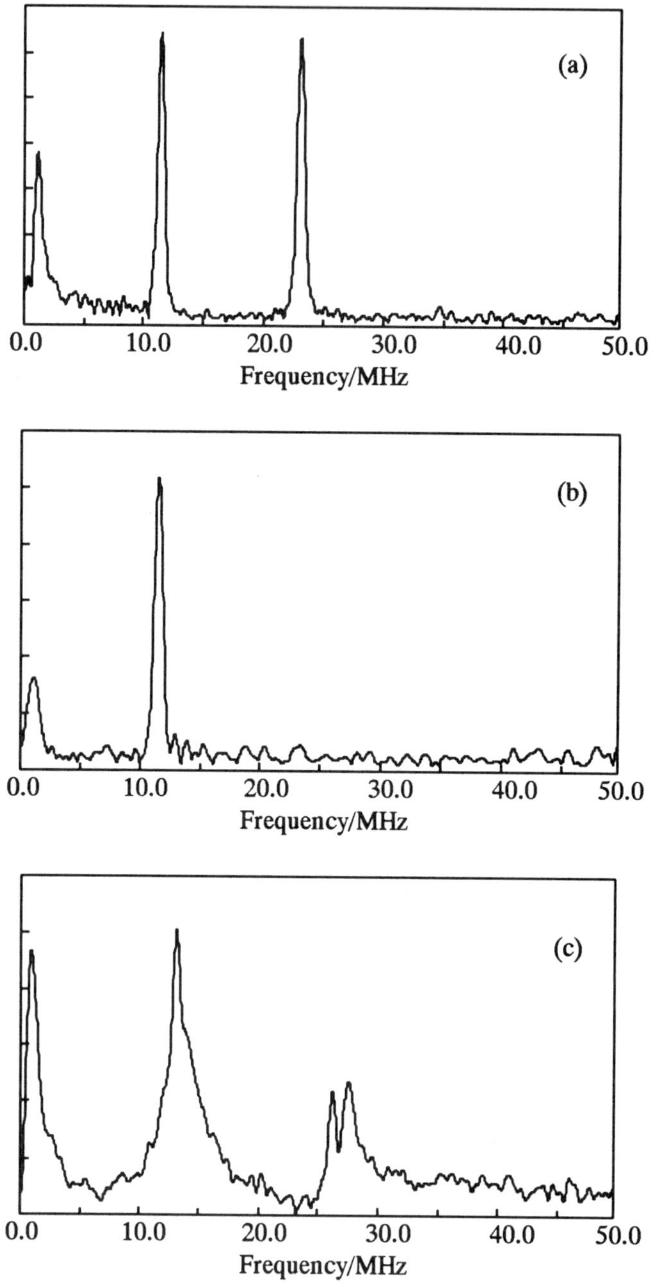

Fig. 3. The FT-ESEEM spectra observed for the Cu(B15C5) complex in 60% $HClO_4$. (a) and (b): The two-pulse and three-pulse spectra observed at B point respectively; (c): The two-pulse spectrum observed at C point.

peaks at 11.4, 11.6 and 13.6MHz in the FT-ESEEM spectra observed at A, B (Fig. 3(a) and (b)) and C (Fig. 3(c)) points respectively are obviously due to the proton Larmor frequency. It is interesting that two proton sum peaks (26.4 and 27.7 MHz) are observed at C point (Fig. 3(c)), whereas only one sum peak (Fig. 3(a)) for A and B points. Of course, this phenomenon is also detected in BF_3-ether-water system (Fig. 2(b)). The fact means that two types of protons with different coupling parameters exist at least. In fact, the peak at 13.6MHz due to protons observed at C point (Fig. 3(c)) is also broader.

The contribution to the proton modulation from free water in the present case can be first excluded, because, similar to the complexes in 60% $HClO_4$, two sum proton peaks are also detected in the two-pulse FT-ESEEM spectrum observed at C point for the powder sample of Cu(II) doped in Mg(B15C5) crystal, which does not contain any crystal water as judged from the X-ray analysis.[10] Therefore, the proton frequency peaks are ascribed to be due to the hydrogen nuclei of the crown ring and the axially coordinated water molecules. Since a strong axial coordination is necessary for the formation of $3d_{z^2}$ ground-state Cu(II) complexes, the distance from the hydrogen atom of the axial coordinated water molecule to the center Cu(II) ion(ca. 0.26nm) is much shorter than that from the hydrogen atom of the crown ring to the center ion(ca. 0.40nm). It is to say that there are two kinds of the hydrogen atoms in the different environment and it is proper to reveal the two proton sum peaks. This observation provides a positive proof for the water molecules axially coordinated to the Cu(II) ions of the present complexes in 60% $HClO_4$ solution. In addition, two sum proton peaks from equatorial and axial ligands in frozen solutions of Cu(II) complexes had been observed by using appropriate field positions in the ESR spectrum.[11,12] This result also indicates that the above-mentioned assignment of the proton peaks is reasonable. In ESEEM measurements sum peaks are particularly useful to discriminate different protons, which produce broad peaks and usually cannot be estimated in the proton Larmor frequency domain.

All FT-ESEEM spectra observed for the Cu(II) complexes in 60% $HClO_4$ exhibited a signal at about 1MHz. This peak can be evidently assigned to chlorine Larmor frequency from solvent ClO_4^-, because the same peak was also detected in the three-pulse ESEEM spectra (Fig. 3(b)) and only chlorine atoms have a nuclear spin except for hydrogen atoms in the 60% $HClO_4$ solution. Considering that the similar peak was also observed for the powder sample of Cu(II) doped in Mg(B15C5) crystal, and furthermore that the X-ray analysis indicated the counter ion ClO_4^- is bound to the axially coordinated water molecule through a hydrogen bond, then it is suggested that the observed chlorine Larmor frequency in the 60% $HClO_4$ solution is mainly due to the ClO_4^- ions at axial directions. In addition, since the present Cu(II) complexes take a $3d_{z^2}$ ground-state doublet, the population of the unpaired electron in the $3d_{z^2}$ orbital is larger. Therefore, the real distance between the nucleus at the axial position and the unpaired electron is shorter than the apparent one. As a result, a nuclear modulation from the nucleus at the axial position is easily observed in ESEEM measurements for $3d_{z^2}$ ground-state Cu(II) complexes.[13]

3.3 ESEEM Spectra of the Powder Sample of Cu(II) Doped in Mg(B15C5) Crystal

The FT-ESEEM spectra of the powder sample of Cu(II) doped in Mg(B15C5) crystal are similar to those observed for the Cu(B15C5) complex in 60% $HClO_4$. Then the assignment and explanation of the FT-ESEEM spectra are in agreement with those mentioned above for the Cu(II) complexes in 60% $HClO_4$. However, the relaxation times are different between the complex in the frozen solution and in the solid. The Cu(B15C5) complex doped in Mg(B15C5) crystal has the shorter relaxation time. Thus the modulation intensities were weaker and resulted in the FT-ESEEM spectra with the lower signal-to-noise ratio.

References

1. K. Ishizu, T. Haruta, K. Nakai, K. Miyoshi, and Y. Sugiura, ESR g-anisotropy inversion observed for Cu(II)-crown ether complexes in solution, *Chem. Lett.*, (1978) 579-582.
2. K. Ishizu, T. Haruta, Y. Kohno, K. Mukai, K. Miyoshi, and Y. Sugiura, Electron spin resonance study of Cu(II)-crown ether complexes with $3d_z2$ ground state doublet in solution, *Bull. Chem. Soc. Jpn.*, **53**, (1980) 3513-3516.
3. Y. Li, H. Tomoda, K. Tajima, K. Ishizu, and N. Azuma, ESR studies on a new $3d_z2$ type copper(II)-crown ether complex with axially coordinated water molecules, *Chem. Lett.*, (1985) 961-964.
4. Y. Li, K. Tajima, K. Ishizu, and N. Azuma, ESR studies of a new $3d_z2$-type copper(II)-crown ether complexes with pentagonal bipyramidal geometry, *Bull. Chem. Soc. Jpn.*, **60**, (1987) 557-562.
5. Y. Li, X-ray structures and spectroscopic studies of diaqua- and dichlorocopper(II) complexes of 15-crown-5 and 4'-substituted benzo-15-crown-5 with a $3d_z2$ ground-state doublet, *Bull. Chem. Soc. Jpn.*, **69**, (1996) 2513-2523.
6. Y. Li, K. Tajima, K. Ishizu, and N. Azuma, ESR studies of copper(II) ion doped in single crystal of benzo-15-crown-5 complex, *Bull. Chem. Soc. Jpn.*, **61**, (1988) 4067-4070.
7. J.-M. Fauth, A. Schweiger, L. Braunschweiler, J. Forrer, and R. R. Ernst, Elimination of unwanted echoes and reduction of dead time in three-pulse electron spin-echo spectroscopy, *J. Magn. Reson.*, **66**, (1986) 74-85.
8. R. C. Weast and M. J. Astle (eds), *CRC Handbook of Chemistry and Physics*, 62nd edition, CRC Press Inc., Florida (1981) E67-69.
9. R. A. Bartsch, P. N. Juri, and J. L. Mills, Interaction of boron Lewis acids with polyethers, *Inorg. Chim. Acta*, **45**, (1980) L41-L42.
10. Y. Li, J. Yamauchi, and N. Azuma, Explanation of differences in ESR spectra and structures between $[Cu(B15C5)(H_2O)_2](ClO_4)_2$ in $[Mg(B15C5)(H_2O)_2](ClO_4)_2$ crystal and in frozen 60% $HClO_4$ by means of vibronic interaction theory, *J. Coord. Chem.*, (submitted).
11. A. Schweiger, New trends in pulsed electron spin-resonance methodology, in *Modern Pulsed and Continuous-Wave Electron Spin Resonance*, edited by L. Kevan and M. K. Bowman, John Wiley & Sons, New York (1990) pp43-118.
12. M. Mohl, A. Schweiger, and H. Motschi, Models of phosphate binding to copper(II): investigation of the electtron spin echo envelope modulation of complexes on surfaces and

in solutions, *Inorg. Chem.*, **29**, (1990) 1536-1543.

13. F. Jiang, K. D. Karlin, and J. Peisach, An electron spin envelope modulation (ESEEM) study of electron-nuclear hyperfine and nuclear quadrupole interactions of dz2 ground state copper(II) complexes with substituted imidazoles, *Inorg. Chem.*, **32**, (1993) 2576-2582.

Single Crystal EPR Studies on Exchange Coupled Copper(II) Dipropylene Triamine Thiocyanate Complex

*Murugesan VELAYUTHAM, K. R. Justin THOMAS and Sankaran SUBRAMANIAN**

Department of Chemistry & Regional Sophisticated Instrumentation Centre, Indian Institute of Technology, Madras 600 036, India

Abstract. Metal polyamine complexes are a unique class of interesting systems known to exhibit low dimensional cooperative magnetic phenomena associated with the columnar crystallographic packing in the solid state. Angular, frequency dependent EPR line widths and line positions have been studied in detail to estimate the intrastack (J) as well as the interstack (J') exchange interactions. The calculated correlation time t_c for interchain spin diffusion is 3.4 x 10^{-9} s using the dependence of line width at X- and Q-band frequencies. The $|3\cos^2\theta - 1|^{4/3}$ dependence of line width variation and the ratio $J' / J = 2.87 \times 10^{-2}$ at 298K are signs of 1-D behaviour. The exchange interaction pathway is most probably through the thiocyanate sulphur atom.

1 Introduction

Copper complexes with nitrogen co-ordination, mimic the active sites in biological enzyme systems, are used for cleaving DNA and also used as magnetic materials.(1-3) The magnetic properties of the paramagnetic systems can be studied by magnetic susceptibility measurement and EPR. The columnar arrangement of the molecules in the crystal with low dimensional exchange coupling can lead to anisotropic magnetic properties. In most of the systems the intrachain exchange coupling is greater than the interchain interactions.(4) Systems with strong exchange coupling will give exchange narrowed lines in EPR. Studies of temperature and frequency dependent angular variation g-value and line shape analysis, will give information about the local magnetic interactions. Magnetic interactions of value less than 1cm^{-1} cannot be studied by susceptibility measurements. The ratio of interstack / intrastack (J'/J) will give the dimensionality of the magnetic interactions. Kuppusamy et al.(4) have estimated the intersite exchange coupling J' in $[NMe_4]_2[Cu(mnt)_2]$ (mnt = maleonitriledithiolato) using frequency dependent line width and temperature variation studies. The important structural features, angular variation of g-values as well as line width, and the exchange interaction of the title complex are highlighted in this article.

2 Experimental

Dipropylene triamine(dpt), copper chloride and imidazole in the ratio of 1:1:1 were mixed in water at 60°C. The blue colour solution was filtered and crystals were obtained after few days. Crystals formed were Cu(H2Im)(HIm)Cl polymer.[5] The title complex, however, was prepared by mixing dpt, copper chloride imidazole and potassium thiocyanate in the molar ratio of 2:2:1:1. X- and Q- band EPR spectra were recorded on a Varian E112 spectrometer. UV-Vis spectra were recorded on a Shimadzu UV-3100 spectrometer.

3 Results and Discussion

3.1 Crystal Structure

Crystals of the title compound belong to monoclinic, space group $P2_1/n$ with $Z = 4$. For the complex $\tau = 0.28$, $T^5 = 0.95$.[6] Of the four molecules per unit cell, two are magnetically inequivalent and are related by a 2-fold rotation about the *b* axis. The other two are related to the first two by an inversion symmetry and hence are magnetically equivalent to their inversion counterparts. The structure of the molecule is given in Fig.1. Projection of the crystal structure along the c axis is given in Fig.2. Molecules at (0,0,0) and (1/2,1/2,1/2) face each other through the basal plane of the pyramid with a Cu····Cu distance of 4.5 Å. The copper chains run parallel to the *c* axis. The distance between the axial thiocyanate sulphur atom of one molecule and the equatorial thiocyanate sulphur atom of the neighbouring molecule is 4.46 Å.

3.2 Electron Paramagnetic Resonance Spectroscopy

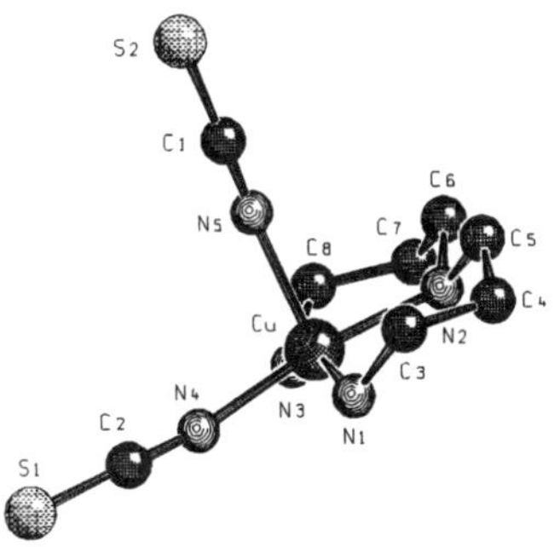

Fig. 1. Molecular structure of $Cu(dpt)(NCS)_2$

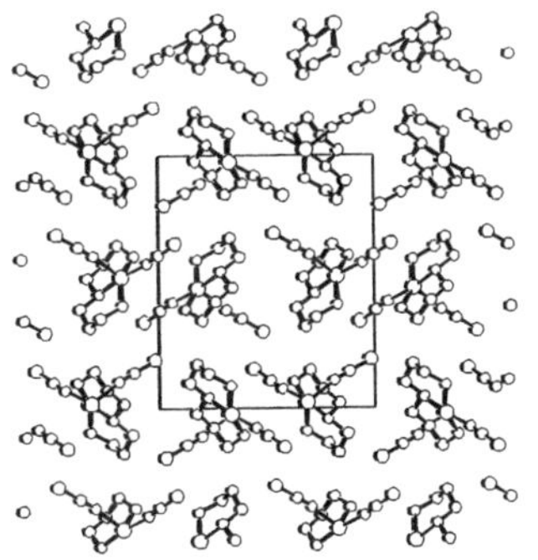

Fig. 2. Projection of the crystal structure along the *c* axis

[Reproduced by Permission of The Royal Society of Chemistry]

3.2.1 Polycrystalline EPR

X-band spectrum at 298K corresponds to orthorhombic g values (Table.1) without any resolved nuclear hyperfine structure. The G value $[(g_{\parallel}-2)/(g_{\perp}-2)]$ is 2.09, which is less than 4, and hence corresponds to a reasonable magnitude of exchange coupling (*J*). Since the highfield g value is greater than 2.05 the SOMO should be $d_{x^2-y^2}$. In X-band the four molecules appear magnetically equivalent and the exchange coupling dominates the hyperfine value. At 77K, X-band powder spectrum shows considerable broadening and a possible decrease in exchange. The Q-band powder spectrum at 298K shows well resolved orthorhombic g-values (Table 1) without resolved nuclear hyperfine structure. Hence the exchange coupling value is greater than copper hyperfine coupling and is of the order of or less than Δg at Q-band.

3.2.2 Solution EPR

The g and A values for the solution spectrum of the title compound in different solvents are given in the Table.1. The g_{iso} values are different from the average g_{ave}-value of powder due to change in symmetry of the complex in solution. In all the solvents $g_{\parallel} > g_{\perp}$ corresponding to the $d_{x^2-y^2}$ ground state. The fact that $g_{\parallel} < 2.3$ indicates appreciable covalency in the metal ligand bonds. At 77K, frozen aqueous solution spectrum gives a broad peak due to aggregation. The g-value of the broad peak is different from powder spectrum presumably due to coordination of water molecule at the sixth site. A frozen mixture of pyridine, water solution of the

Table 1. EPR parameters

sample	Temp	g_{iso}	g_1	g_2	g_3	A^*_{iso}	A^*_1	A^*_2	A^*_3
Q-band									
powder	298K	2.1279	2.2025	2.1238	2.0574	-	-	-	-
X-band									
powder	298K	2.1266	2.1942	2.1256	2.0602	-	-	-	-
	77K	-	Broad	line	2.0549	-	-	-	-
In aceto-nitrile	298K	2.1355	-	-	-	6.17	-	-	-
In water	298K	2.1384	-	-	-	6.33	-	-	-
	77K	-	2.1126	-	-	-	-	-	-
In methanol	298K	2.1148	-	-	-	7.00	-	-	-
	77K	2.0978	2.2338	2.0580	2.0017	6.85	16.75	1.90	1.90
In water + pyridine	298K	2.1291	-	-	-	7.33	-	-	-
	77K	2.1291	2.2393	2.0740	2.0740	7.33	18.00	2.00	2.00

* values are expressed in mT.

complex exhibits EPR spectra with superhyperfine structure from the surrounding nitrogen atoms. From the A_{iso} values the strength of the ligands are in the order pyridine > methanol > water > acetonitile. The resolution of the isotropic spectra at 298K also follows the same order.

3.2.3 Single Crystal EPR

g-Tensor. EPR measurements were made with the magnetic field vector scanning the three orthogonal planes at both X- and Q- band frequencies. The three planes are represented as *ac**, *bc** and *ab*. From Fig. 1. the g_{zz} direction is along the CuN_5 axis, the second g direction is the bisector of the CuN_3 and CuN_4 axes and the third g direction is perpendicular to the above. The angular variation of g values in the three planes are given in the Fig. 3,4 & 5. At X-band only a single line was observed at all orientations, indicating that the exchange between the two magnetically inequivalent sites is stronger than half the separation between them, i.e., $J' > (\Delta g\beta B) / 2$. However, at Q-band in *bc** plane two resonances could be seen, well resolved for most of the orientations, with the maximum separation between the two lines being 950G. In fig. 4, the average of the Q-band resonances at various orientations as well as X-band g-value variation for the single site observed are also shown. The angular variation in the other two planes gives only one peak at all directions. The theoretically calculated g values agree very well with the experimental values. A simple diagonalization procedure yielded the principal values of the g tensor (Q-band) to be $g_1 = 2.2563$ $g_2 = 2.0540$ $g_3 = 2.0605$, for both the spatially distinct sites. The direction cosines calculated theoretically agree within $\pm 2°$ to the direction cosines calculated from the angular variation of g-values.

Line width and line shape. The angular variation of the line width at X- and Q-band for *ac** and *ab* planes are given in the Fig. 6 & 7 respectively. From the crystal structure (and EPR) the two sites are magnetically equivalent in *ac** plane. The line

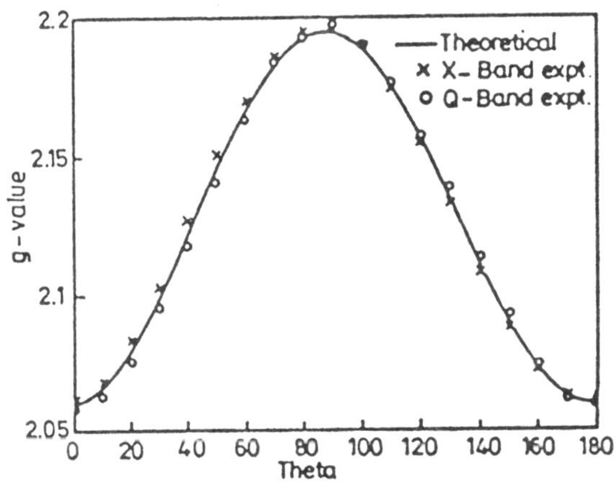

Fig. 3. Experimental and calculated angular variation of g in the *ac** plane

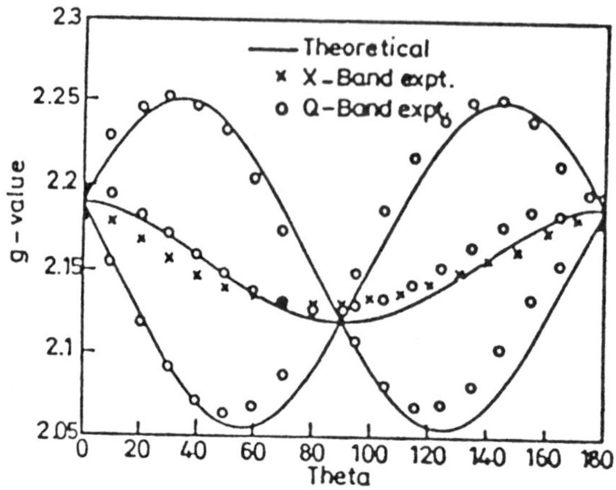

Fig. 4. Experimental and calculated angular variation of g in the *bc** plane

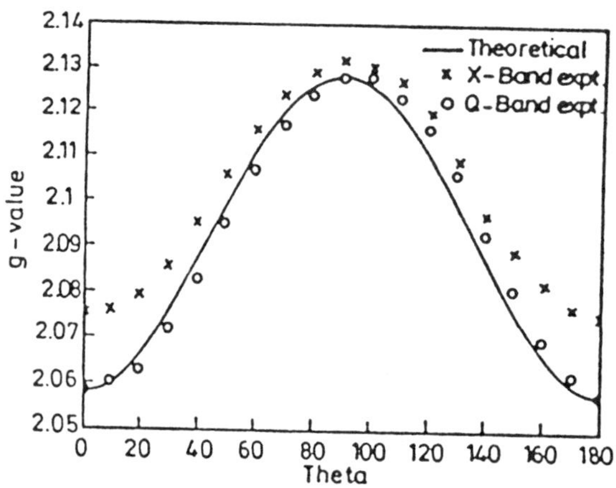

Fig. 5. Experimental and calculated angular variation of g in the *ab* plane

width variation follows a $|3\cos^2\theta - 1|^{4/3}$ dependence and it is evident that the dominating interaction contributing to the line width anisotropy is dipolar in nature. Significant line width variation was observed in *ab* and *ac** planes between X- and Q- band measurements. This is expected when the magnitude of exchange coupling lies in between the X- and Q- band g anisotropies, i.e.

$$\Delta g\, \beta\, B_o \text{ (X-band)} < J < \Delta g\, \beta\, B_o \text{ (Q-band)}$$

In this intermediate case contribution of nonsecular components to the Q-band line widths are only marginal. Generally line widths are larger in Q- band than in the X- band (Fig.7). At some orientations line width variation in the X- band is greater

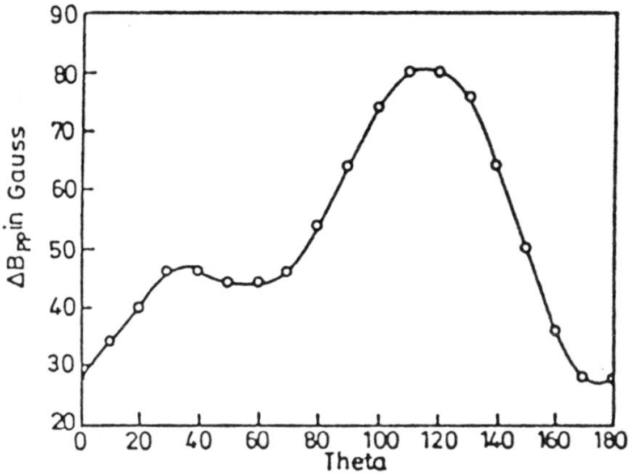

Fig. 6. Angular variation of line width in the *ac** plane

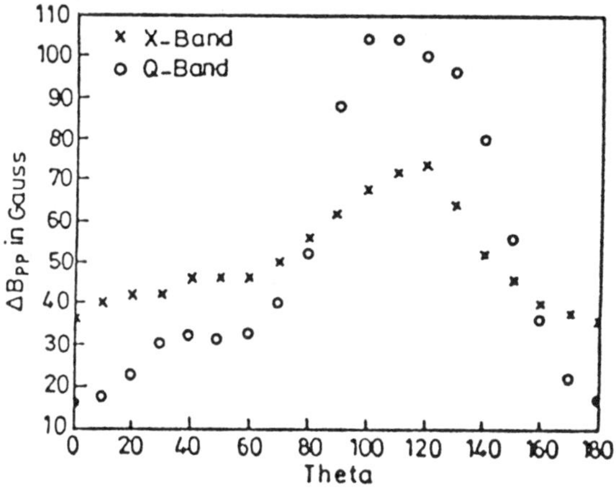

Fig. 7. Angular variation of line width in the *ab* plane

than the Q- band, indicating that the secular and nonsecular Fourier components are sensitive to the frequency of measurement.[7]

Interstack Exchange, J'. The line width data at X-and Q-band frequencies in the *ab* plane were used to estimate the effective intersite exchange field, H_e , as well as J'. For $J' > (\Delta g)\beta B/2$, we would observe a single line and the theories of exchange averaging predict that the width of the line varies as the square of the splitting ΔB (in gauss) between the sites[7]

$$\Delta B_{pp} = (\Delta B)^2 / H_e \tag{1}$$

where H_e is defined as the effective intersite exchange field. If the measurements are done at two different frequencies, e.g, X- and Q- band, then[7]

$$\Delta B_{pp}^{Q} - \Delta B_{pp}^{X} = [(\Delta B_Q)^2 - (\Delta B_X)^2]/H_e \qquad (2)$$

where ΔB_X and ΔB_Q are the differences in the expected resonance fields of the two sites at X- and Q- band, respectively, in the absence of J'. The quantity H_e depends on the rate of exchange between inequivalent sites and is given by[7]

$$(g\beta/\hbar)\, H_e = 4\, t_c^{-1}\, (3^{1/2}/2) \qquad (3)$$

where $t_c = 0.16\,\hbar\,|J|^{1/3}/\,|J'|^{4/3}$ is the correlation time for interchain spin diffusion. The calculated H_e of 172 G from the *bc* plane gives a value of 3.38×10^{-9} s for t_c.

Intrastack Exchange, J. The intrastack (isotropic) exchange parameter J can be estimated from the line width variation in the ac^* plane. In this plane the two sites are equivalent and the effective field is given by [8]

$$B_e = B \,.\, \Delta g \qquad (4)$$

The effective field B_e is seen perpendicular to the ac^* plane where the axis of quantization defined by B . g lies and will give rise to nonsecular broadening effects. The above effect is shown to be negligible when $\omega_o t_c >> 1$.[7] In our case $t_c = 3.38 \times 10^{-9}$ s (*vide supra*) and hence $\omega_o t_c = 30$ at X- band and 118 at Q- band. Therefore, the nonsecular contribution to line width can be neglected. In the strong exchange narrowing case the line width ΔB_{pp} is given by[9]

$$(3^{1/2}/2)\, \Delta B_{pp} = M_2^{t}/J \qquad (5)$$

where M_2^t is the sum of dipolar and hyperfine second moments. Secular part of the second moment is calculated as described elsewhere [10] and its variation is shown in Fig.8. It follows the line width variation in this plane. Since the interchain Cu····Cu

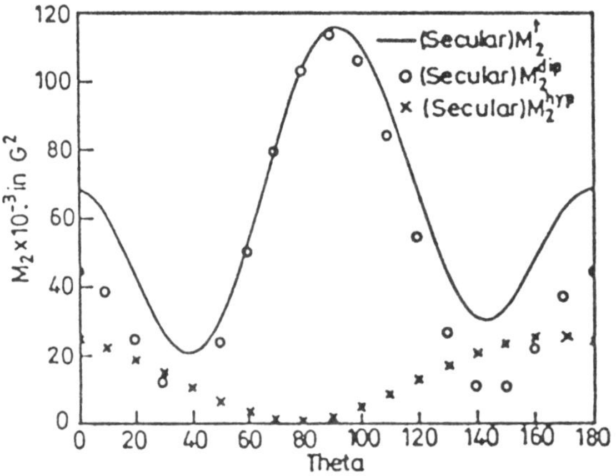

Fig. 8. Angular variation of local fields in the *ac** plane

distance is 4.5 Å the dipolar second moment contribution is higher than that from the hyperfine.[11] One orientation value is sufficient to calculate the J using equation (5) and the value is 612 ± 5 G. From the values of J and t_c the calculated J' is 17 ± 2 G. Plot between ΔB_{pp} and M_2^t shows that there is a linear relation.

4 Summary and Conclusions

The ratio of $J' / J = 2.87 \times 10^{-2}$ is low enough to make the interactions almost one-dimensional. The exchange coupling is not along the copper chain. Owing to the specific diffuseness of their valence orbitals, the sulphur atoms could be involved in Cu–S···S–Cu exchange pathways.[12] The orbitals of axial thiocyanate sulphur atom of one molecule overlap with those of the equatorial thiocyanate sulphur of the neighbouring molecule. Such a ligand S···S overlap provides a super exchange pathway leading to a 1-D Heisenberg chain along –Cu–S····S–Cu–S····S–Cu– parallel to b axis. Hence the chain is moving in a zig-zag manner. The interchain Cu····Cu distance of 4.5Å leads to a sizeable contribution to the dipolar second moment to the linewidth as well as coupling constant J'.

References

1. W. Kaim and J. Rall, Copper - A "Modern " Bioelement, *Angew. Chem., Int. Ed. Engl.* **35** (1996) 43 - 60.
2. G. Pratviel, J. Bernadou, and B. Meunier, Carbon - Hydrogen Bonds of DNA Sugar Units as Targets for Chemical Nuclease and Drugs, *Angew. Chem., Int. Ed. Engl.* **34** (1995) 746 - 769.
3. H. M. Buschbaum, *Angew. Chem., Int. Ed. Engl.* **30** (1991) 723 - 744
4. P. Kuppusamy, and P.T. Manoharan, Exchange Interactions in Bis(tetramethylammonium) Bis(maleonitriledithiolato)cuprate(II), $[NMe_4]_2Cu(mnt)_2]$: A quasi-1-D Weak-Exchange system, *Inorg. Chem.* **24** (1985) 3053 - 3057.
5. A. Bencini, C. Benelli, D. Gatteschi, and C. Zanchini, Magnetic Coupling in a Linear-Chain Copper(II)-Imidazolate Compound. Mechanism of the Exchange Interaction through Bridging Imidazolate Ligands, *Inorg. Chem.* **25** (1986) 398 - 400.
6. M. Cannas, G. Carta, A. Cristini, and G.Marongiu, Crystal structures of thiocyanate polyamine copper(II). Part (VI) Di -(3-aminopropyl)amine bis(isothiocyanato) copper(II), *J.C.S Dalton Transaction* (1974) 1278 - 1281.
7. B. L. Ramakrishna, and P.T. Manoharan, Exchange Interactions in Transistion metal complexes and in some ion-radical salts *Mol. Phy.* **52** (1984) 65 - 80.
8. M. Yokota, and S. Koide *J. Phys. Soc. Jpn.* **9** (1954) 953 -
9. J. H. Van Vleck, *Phys. Rev.* **74** (1948) 1168 -
10. K. T. McGregor, and Z. G. Soos *Chem. Phys.* **64** (1976) 2506 - 2517
11. A. Bencini, and D. Gatteschi, EPR of Exchange Coupled Systems, Springer-Verlag, (1990) chapter 10.
12. Y. Journaux, F. Lloret, and O. Kahn, Single-Crystal EPR study of copper(II) Trinuclear Compounds: Exchange-Averaging effects, *Inorg. Chem.* **29** (1990) 3048 - 3051.

Chapter 3

Physics and Materials Science

Submillimeter Wave ESR of Magnetic Materials in High Magnetic Fields

Mitsuhiro Motokawa

Institute for Materials Research, Tohoku University, Katahira, Sendai, 980-77 Japan

Abstract. Techniques, advantages and some experimental results on magnetic materials of submillimeter wave ESR are described.

1 Introduction

In order to investigate the magnetic properties of strongly correlated magnetic materials, ESR in submillimeter wave region with use of high magnetic fields is considered to be one of the most powerful tools, because the frequency of the collective modes of the spin systems are often above 10 cm^{-1}. From the technical point of view, however, the experiment in this frequency region, which is just an in between of microwave and light, is not easy due to weak radiation sources and low sensitivity detectors. The present author has developed ESR equipment using FIR lasers, Backward Traveling Wave Tubes and Gunn diodes for the radiation sources and pulsed high magnetic fields up to 40 T. Even for a metal sample, a pulsed field is useful because the penetration depth of the magnetic field is much deeper than the skin depth of the electromagnetic radiation. The advantages of this kind of experiment are observations of the following effects:

(1) Short relaxation time (broad width): strong interaction between spins produces widths 1 T.

(2) Powder samples: Even if the g-values are distributed, absorption peaks from $g_{//}$ and $g_{\perp}$ factors can be separated and their values and widths can be obtained from the line shape simulation.

(3) Exchange splitting: Two absorption lines due to two sites in a crystal amalgamated at low frequencies by strong exchange interaction can be separated. This experiment yields microscopic exchange interactions rather than other statistical measurements like susceptibility or specific heat.

(4) Spin systems with large energy gaps: (a) spin wave gap of strongly correlated materials like copper oxides, (b) impurity spin state in ferro- or antiferromagnetic systems, (c) paramagnetic resonance with large D parameters from DS_z^2 in a spin Hamiltonian, (d) paramagnetic resonance with large g-values.

(5) Nonlinear effects: generally S^nH^m (n+m=even) terms are included in a spin Hamiltoinan but are usually neglected because they are too small. But this gives nonlinear g-values in this wave length region. In a spin-pair system, the g-value which is supposed to be same as that of a single spin changes due to strong exchange interaction. This is also observable in this frequency region.
(6) Magnetic phase transitions in high magnetic fields: magnetic field plays the most important role for this phenomena and consequently high frequency is necessary. (a) Antiferromagnetic resonance in high field phase. (b) Resonance above the meta-magnetic transitions. (c) Quantum effects of low dimensional magnetic materials. (d) Nonmagnetic-magnetic transition (singlet-triplet transition) due to quantum effect.

2 Methods of Submillimeter Wave ESR

ESR equipment is usually consists of an oscillator, a transmission line, a cavity, a detector and a magnet. The most important parts of the equipment are the oscillator and the magnet. Figure1(a) shows the possible combination of available oscillators and magnetic fields. As a source of millimeter waves, klystrons were used formerly but are now replaced by Gunn oscillator. Gyrotron, a very strong millimeter wave osc-illtor, is not listed here because it is too special. The thick lines are the combinations already used and the thin lines are possible options but not used yet. In the frequency-field diagram of Fig.1(b), the directions to scan the field or frequency are shown by arrows and the alphabet "A~F" correspond to those of Fig.1(a).

The oldest method of submillimeter wave ESR is to use FIR spectroscopy with a thermal light source. This can be done even at zero field for the magnetic resonance which has a energy gap at zero field like antiferromagnetic resonance. This is indicated by "A" in Fig.1(a),(b). It was 1960 that the first experiment was performed

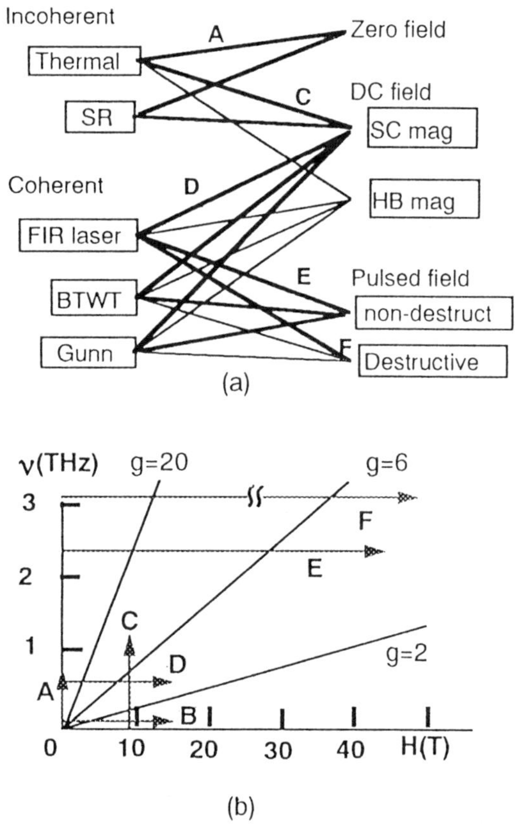

Fig. 1. (a)Combination of light sources and magnetic fields and (b)indications of scanning directions. "B" is the case of microwave and pulsed field which was done in early stage of this kind of experiments [8].

on antiferromagnetic resonance of NiO by Kondoh[1] and immediately followed by Ohlman and Tinkham on FeF_2 in 1961[2]. This method, however, is low in sensitivity and observation of magnetic dipole transition is difficult as in ESR. The old fashioned type spectrometer was of grating rotation type but now a Fourier transformation method of interferometer is usually used due to the development of computerized commercially available FTIR spectrometers. For FIR spectroscopy, the magnetic field must be static. Superconducting magnets are favored but the field intensity is limited to 20 T or less depending on budget. This kind of measurement indicated by "C" was carried out by Torrance and Tinkham(3). Use of a hybrid magnet, which is a combination of a wide-bore superconducting magnet and a water cooled magnet installed inside. In this way one can achieve over 30 T. This method is being tried in our laboratory. The water cooled magnet consumes approximately 10 MW electric power at 30 T and the running cost is quite expensive. It is available at only six places in the world and we have one in our Institute. It should be noted that the thermal light source does not work near the big magnet because of strong leakage flux. In the case of FIR spectroscopy, use of pulsed field is difficult.

Synchrotron radiation(SR) may also be used for FIR spectroscopy but it requires a superconducting magnet, but not pulsed fields. So the available field is limited to 20 T but so far only up to 10T has been used. The hybrid magnet is difficult to use with SR because both facilities are too big. When a coherent light source is used, the frequency is kept constant during the measurement and the magnetic field must be scanned. In this case both static and pulsed fields can be used.

The advantage of the coherent light source method is high resolution. The resolution of FIR spectroscopy is approximately 1 cm^{-1}, while 0.01 cm^{-1} is possible in the case of scanning the field at a constant frequency. To obtain coherent radiation in the submillimeter wave region, the so called FIR laser is convenient. At the beginning of the development of this kind of laser, only pulsed lasers were available. The first trial was made in 1968 with a superconducting magnet[4]. This method corresponds to "D" and has been a standard method[5] but with the CW laser mentioned later. The present author first used pulsed HCN and H_2O gas lasers with pulsed fields[6] and observed the impurity resonance of Mn in FeF_2 in 1985 at 2.53 THz and at 36.3 T[7]. This is indicated by "E" in Fig.1.

The gas lasers optically pumped by strong CO_2 laser are commercially available and now widely used as the CW coherent light sources. The stability of this type of laser is not good without a stabilizer but there is no problem when it is used with pulsed fields as we are doing. The frequencies can be changed by changing CO_2 laser modes and by changing gas but not continuously. As will be shown later many lines are available in our FIR laser. Recently EPR of $CuGeO_3$ was observed at 3.1 THz using a destructive pulsed field over 100 T, which is indicated by "F".

Backward Traveling Wave Tubes (BTWT) are also convenient. The frequency can be changed continuously within a band. Now this equipment is available as a (BWO) from Russia. The most useful oscillator is the Gunn oscillator whose highest frequency is limited to 135 GHz but two or three times higher frequencies are available by using multipliers. This works with a low voltage power supply and the frequency can be changed slightly but not as much as with a BTWT.

Recently a sophisticated "Vector Network Analyzer" in the milli- and submillimeter wave regions has become commercially available. The electromagnetic wave is supplied by the higher harmonic generator. This should be used with a static magnetic field.

3 Our Experimental Technique

We have an optical pumping laser which covers the frequency band shown in Fig.2 where only very strong lines are shown. We have two BTWT which oscillate at 220 and 370 GHz bands and some Gunn oscillators which cover frequencies between 90 and 180 GHz. We now use only pulsed fields up to 40 T and the low temperature is available down to 0.4 K using a ^{3}He refrigerator. As shown in Fig.3, The electromagnetic wave from an oscillator is guided by a light pipe and goes through a sample which is located at the center of the magnet and then arrives at a detector of In:Sb or impurity doped Ge in liquid He. We do not use a cavity because the wavelength is too short. Pulsed fields are produced by discharging a capacitor bank of 100 kJ to a magnet coil, and a half-sine shape field as a function of time is obtained. As shown in Fig.4(a), a resonance absorption is observed both in increasing and decreasing fields and this figure is easily converted to an absorption signal as a function of magnetic field by computer as shown in Fig.4(b).

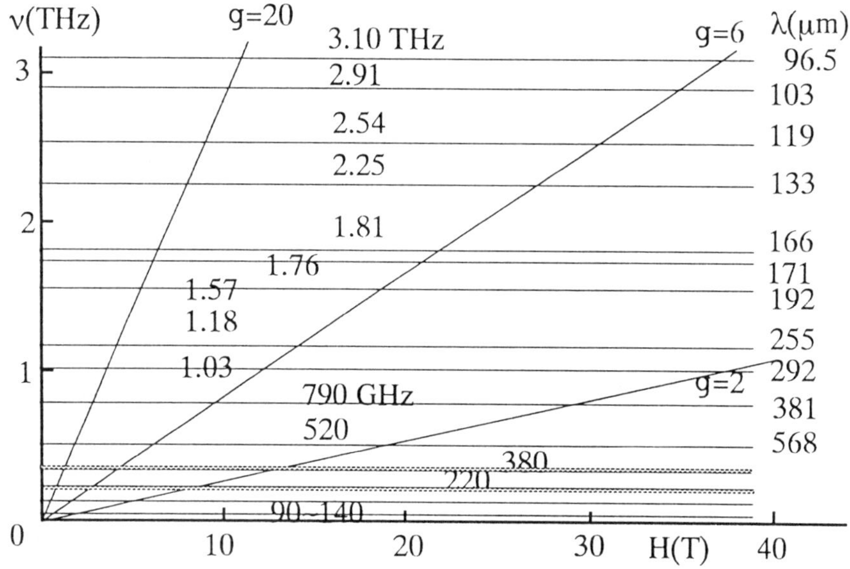

Fig. 2. Available oscillation lines and pulsed magnetic fields

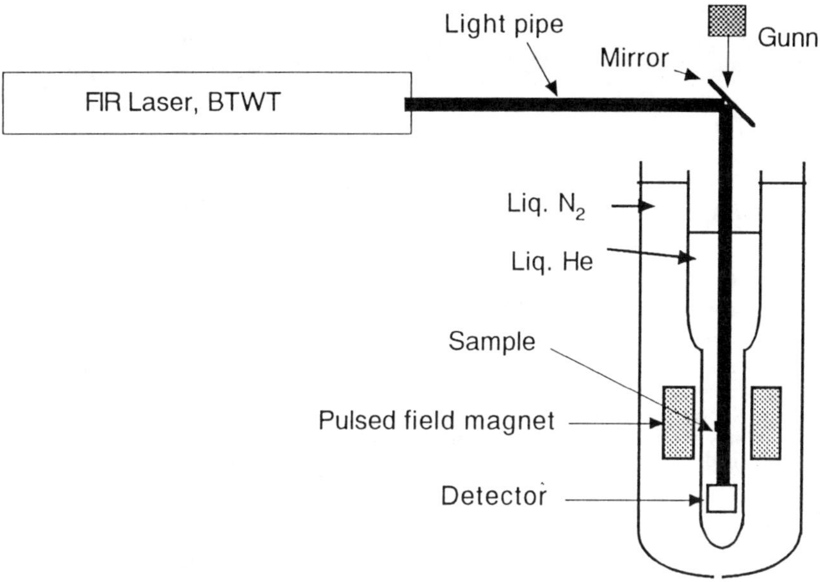

Fig. 3. Alignment of equipment. Gunn is set at the mirror position

4 Experiments

4.1 Short Relaxation Time (broad width)

Due to a strong non secular interaction between spins, the resonance width is sometimes broader than 1 T. In this case a high magnetic field is necessary to observe whole line shape.

4.2 Powder Sample

As shown in Fig.5, two broad absorption lines with different g-values can be separated in high fields. Even if the g-values are distributed, as a powder sample, absorption peaks of $g_{//}$ and $g_{\perp}$ can be distinguished and their values and widths can be obtained from the line shape simulation. We obtained information about paramagnetic and antiferromagnetic resonance of powder Li_2CuO_2 with almost the same accuracy for a single crystal [9].

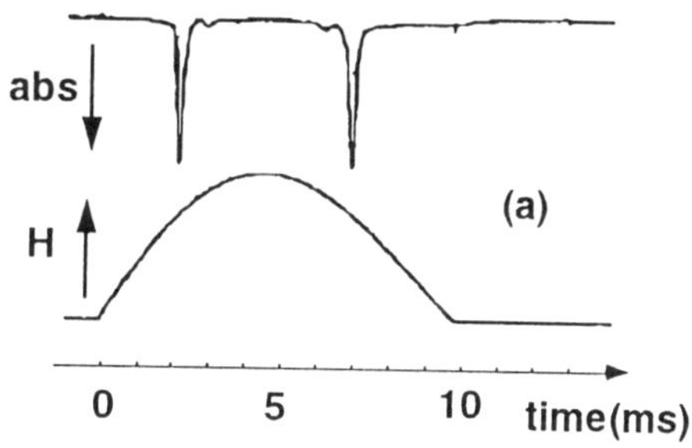

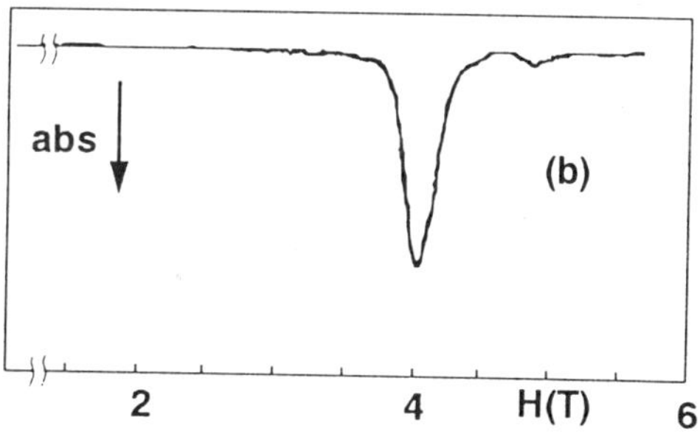

Fig. 4. (a) Example of an absorption line as a function of time in a pulsed field at 373.5 GHz and at 4.2K. (b) The signal is converted to an absorption line vs magnetic field

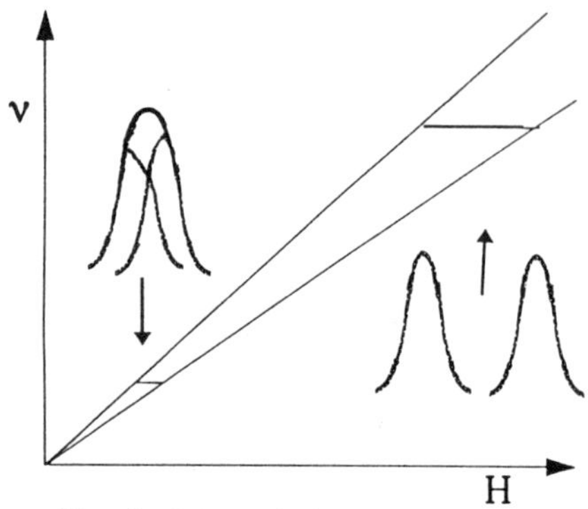

Fig. 5. Schematic figure to explain how two lines overlap at low field and split at high field.

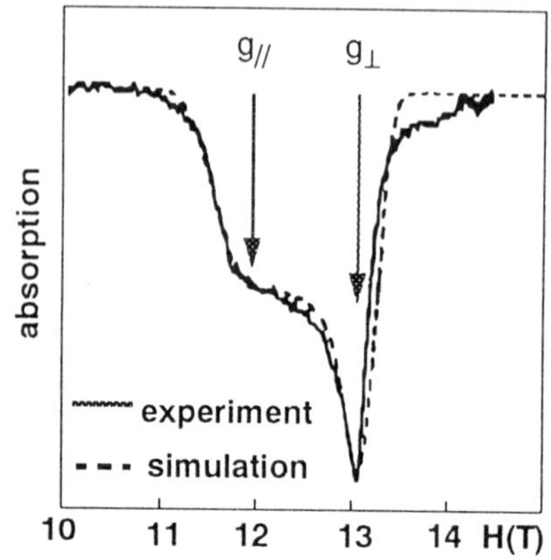

Fig. 6 Line shape of EPR of powder Li_2CuO_2 with simulation.

An example of paramagnetic resonance observed at 370.4 GHz and at 90 K is shown in Fig.6 with a correspondence simulation. From this simulation, g-values were obtained as $g_{//}$=2.30±0.01 and $g_{\perp}$=2.01±0.02.

4.3 Exchange Splitting

If there are two different crystalline field sites in a unit cell, two absorption lines supposed to be observed when an external field is applied to an arbitrary direction. But they are amalgamated to one absorption line at low frequencies as shown in Fig.7 due to an exchange interaction J between spins at the different sites. This is a sort of the exchange narrowing effect. At a sufficiently high field satisfying $\Delta g\mu_B H>J$, they split again as shown in Fig.7. This effect is called as the "exchange splitting". This experiment gives microscopic local exchange interactions, while other measurements such as susceptibility or specific heat give statistically averaged values. We have performed experiments on $(NH_4)_2CuCl_4 2H2O$, $Cu(NH3)_4SO_4H_2O$ and $CuCl_2 2H_2O$[10].

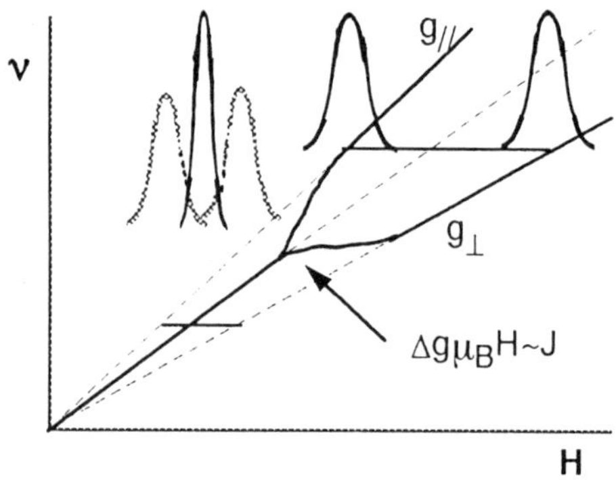

Fig. 7 Schematic figure of exchage splitting

4.4 Spin Systems with Large Energy Gaps

4.4.1 Spin Wave Gap of Strongly Correlated Materials

This is so called antiferromagnetic resonance. The energy gap is given by

$$\frac{\omega}{\gamma} = \sqrt{2H_E H_A} \pm H_0 \tag{1}$$

where ω and γ indicate resonance frequency and gyromagnetic ratio, respectively. H_E, H_A, and H_0 denote the exchange field which reflect exchange interaction between antiferromagnetically ordered spins, anisotropy field and external field, respectively. Recently many oxide compounds were found to be interesting and they usually have strong exchange interactions. Equation(1) corresponds to the spin wave gap at q=0 and it is important to know this energy in these materials as complementary information compared with neutron scattering results which give q≠0 spin wave energy. We did AFMR experiments in high magnetic field on, for example, Bi_2CuO_4[11].

4.4.2 Impurity Spin State in Ferro- or Antiferromagnetic Systems

An isolated magnetic moment caused by an impurity spin in an ordered magnet experiences a strong exchange field from surrounding host spins. Then it has large energy gap and by determining this energy we can estimate the mixing of wave functions between impurity and host spins. We performed this experiment on Mn^{2+} impurity in FeF_2[7] for which experimental results are shown in Fig.8 and 9.

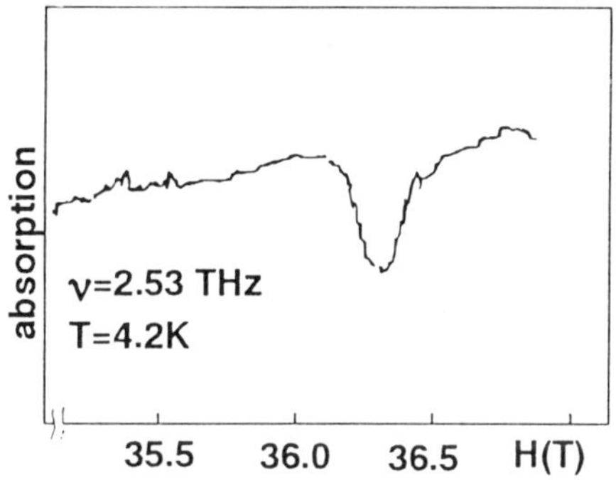

Fig. 8 An example of impurity resonance of Mn^{2+} spin in FeF_2

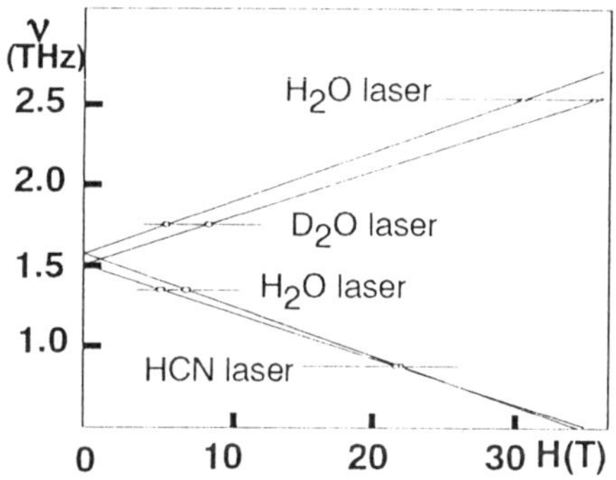

Fig. 9 Frequency-field diagram of host AFMR and impurity resonance

4.4.3 Paramagnetic Resonance with Large D in a Spin Hamiltonian

The usual spin Hamiltonian of paramagnetic materials is written as

$$H = g\mu_B SH + DS_z^2 \tag{2}$$

where μ_B is the Bohr magneton. In the case of S=1, D<0 and H//z, the energy levels split as Fig.10. If D is very large, we need a high frequency comparable to it

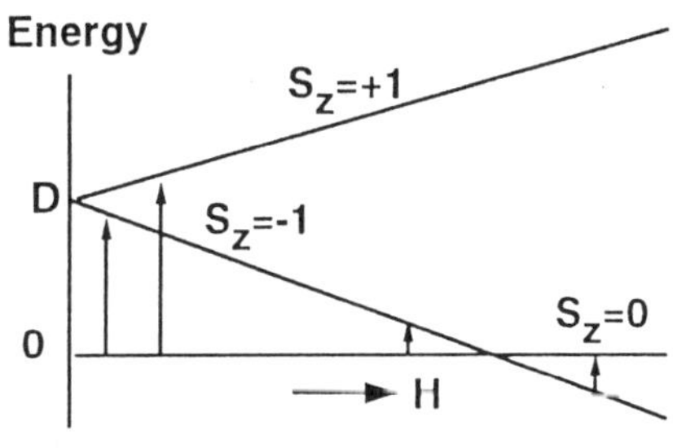

Fig. 10 Energy levels of eq.(2)

at low field as indicated by long arrows. But at high field, the resonance is observable even at a low frequency as indicated by short arrows. Transitions within the doublet are forbidden. We did this experiment on $CdPS_3$ and determined D as -24 cm^{-1} [12].

4.4.4 Paramagnetic Resonance with Large g-values.

Some rare earth ions have very large g-values and complicated energy levels due to large S values. In this case submillimeter wave ESR is helpful to determine the crystalline field parameters[4]. By this method the complicated energy levels were identified in DyAl Garnet[13].

4.5 Nonlinear Effects

Generally S^nH^m terms, where n+m must be even, are included in a spin Hamiltoinan but usually they are neglected when n+m>2 because they are too small. But in high magnetic fields, these terms contribute to the resonance condition and effectively give nonlinear g-values. We observed this effect in Co^{2+} spin systems[14]. As is well known, The energy state of Co^{2+} spin is given by six Kramers doublets. Usually only the ground doublet contributes to ESR because the energy separation between the ground doublet and next excited doublet is approximately 100 cm^{-1}. In high magnetic field, however, where the Zeeman energy is greater than 10 cm^{-1}, the down-going branch of next excited doublet decreases and has cross over effect with the up-going branch of the ground doublet because of mixing of wave functions of both doublets as shown in Fig.11. This is due to the following reason. According to the old theory by Abragam and Pryce[15], the six Kramers doublets are diagonalized at zero field but when the Zeeman term is introduced in the Hamiltonian, they are not completely diagonalized and the g-values are given by the first order perturbation. In the case of large Zeeman energy, the Hamiltonian must be rediagonalized and it gives the cross over effect near the crossing point of the down-going branch of the next excited doublet and the up-going branch of the ground doublet.

Another example of this kind of mixing effect is ESR observed in paired

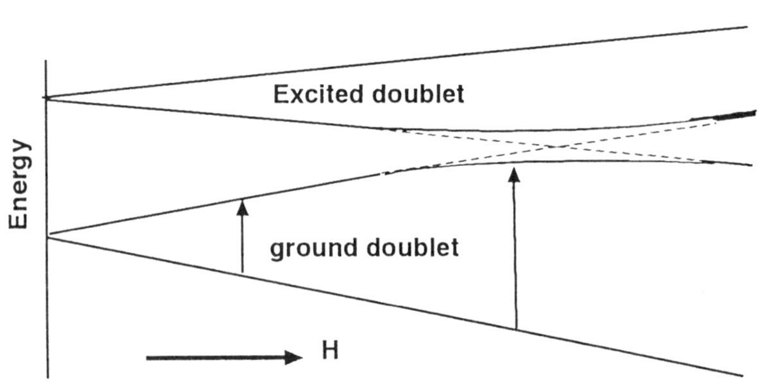

Fig. 11. Cross over effct between the graound and next excited doublets in high field

spin systems of Co^{2+} ions in Rb_2MgF_4[16]. This crystal has a two-dimensional square lattice. In the case of high density, ~30%, of Co^{2+} ions substituted by Mg^{2+} ions, there are isolated single spins and also statistically some paired spins connected by an antiferromagnetic exchange interaction J. The energy levels of paired spins are illustrated by Fig.12. The ground state is a singlet and an excited triplet state is separated by J. The ESR is observed by intra-triplet transitions as indicated by arrows in Fig.12 and the transition between the ground singlet and excited triplet is forbidden.Then the resonance absorption can be observed at finite temperatures. Even in such cases, the g-values are usually supposed to be same as those of a single spin. We observed, however, g=6.63 for isolated single spin but g=5.88 for paired spin. This is due to an effect that the strong exchange antiferromagnetic interaction causes mixing of the wave functions of the ground doublet and the excited doublets and the g-values of the ground doublet are reduced by the mixing of small g-values of the next excited doublet. There have been some ESR experiments on paired spin systems, but such effects were too small because only the cases of small exchange interaction were studied. By this high field ESR method, such a strong exchange case is observable.

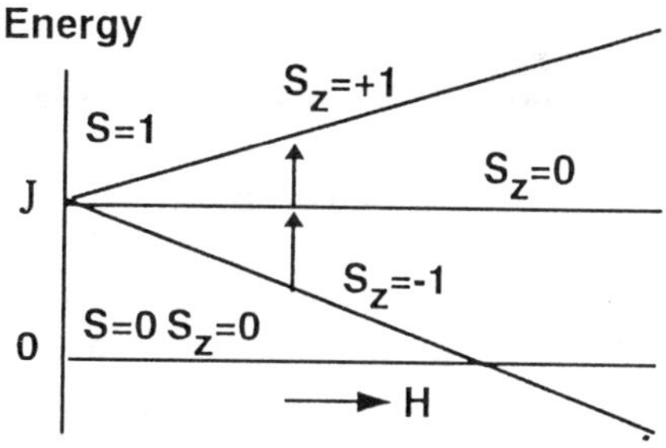

Fig. 12 Ground singlet and excited triplet

4.6 Magnetic Phase Transitions in High Magnetic Fields

Some ordered magnetic materials show phase transitions at high magnetic fields, and the spin structure usually changes. Then the resonance conditions completely change in other phases. To observe this kind of effect, we need a wide frequency range and high magnetic fields.

4.6.1 Antiferromagnetic Resonance in High Field Phase

When an external field is applied to the easy axis of an antiferromagnetic material, the spin structure suddenly changes to "spin-flop" phase and the magnetization abruptly increases at

$$H_C = \sqrt{2H_E H_A} \tag{3}$$

as shown in Fig.13, and finally saturates. The resonance condition

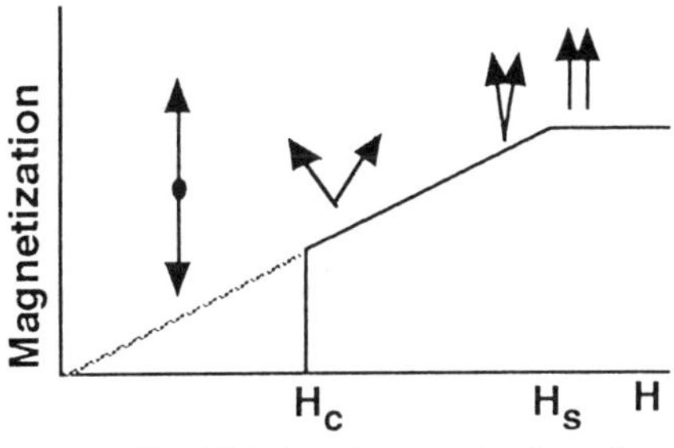

Fig. 13 Induced magnetization of an antiferromagnet in a field

completely changes, but many studies have been done in this field so far.

4.6.2 Resonance Above the Metamagnetic Transitions

Metamagnetic transitions occur when the anisotropy energy of an antiferromagnet is stronger than the exchange energy and an applied field exceeds the exchange field. Above the transition field, the induced magnetization is constant. $FeCl_2$ and $CoCl_2 2H_2O$ are well known examples and the magnetic resonance experiments in high field phase were done many years ago. Recently we found an interesting metamagnetic material $Y_2Cu_2O_5$[17]. In the case of Cu^{2+} spins, the anisotropy energy K is considered to be of the order of

$$K \sim \left(\frac{\Delta g}{2}\right)^2 J \tag{4}$$

where Δg is the anisotropy difference of g-values and the coefficient of J should be less than unity. Then the metamagnetic transition is not expected. Figures 14 and 15 are the magnetization curves and frequency-field diagrams for magnetic resonance, respectively. The mystery of metamagnetic transitions and resonance modes in metamagnetic phases have not been solved yet.

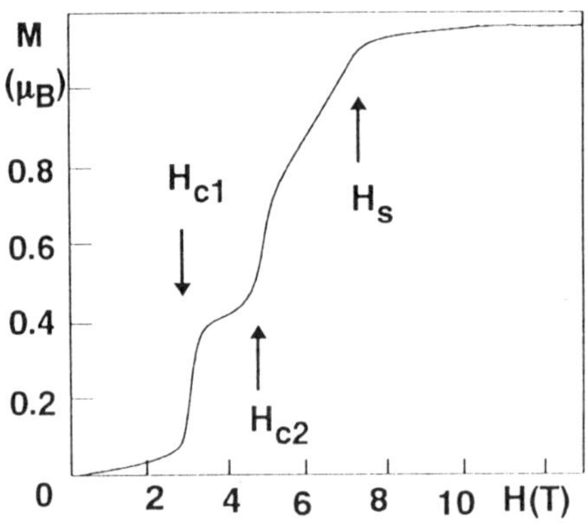

Fig. 14 Magnetization of $Y_2Cu_2O_5$

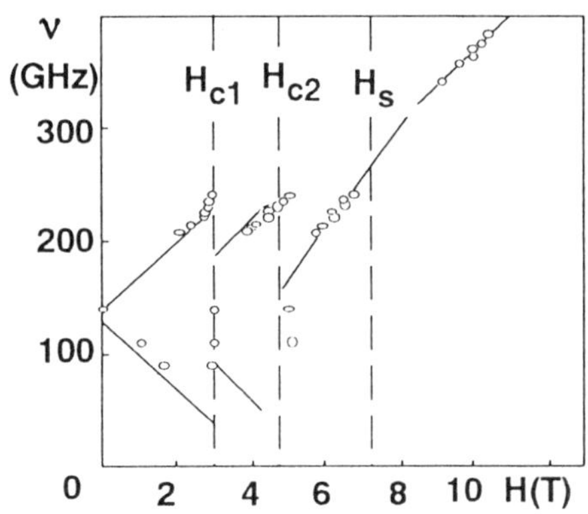

Fig. 15 Frequency-field relation of $Y_2Cu_2O_5$

4.6.3 Quantum Effects of Low Dimensional Magnetic Materials

In quantum spin systems, i.e. low dimensional and small spin systems, sometimes strange magnetic phase transitions occur at high magnetic fields. We found a strange magnetic transition at 12.5 T in $CsCuCl_3$ in 1978 but the strange behavior was not explained until a few years ago. It turned out by theorists that this transition is due to quantum effect[18]. In general, in the spin system like triangular lattice magnets, the energy state are degenerated and the ground state is caused by quantum fluctuation. In this case the anisotropy energy suppresses the fluctuation and the 120°-structure is stable at zero field. In high field, however, the magnetic field acts to destroy the anisotropy effect and the quantum fluctuation revives. Then the magnetic structure changes from 120°-structure to co-linear structure. The resonance condition is also influenced by this transition. Details are described in the reference [20].

4.6.4 Singlet-triplet Transition Due to Quantum Effect

Quantum spin systems often fall into a singlet state due to large fluctuations of spins caused by antiferromagnetic interactions in a chain. The important and interesting examples are Haldane gap and spin-Peierls phenomena.

In a spin system with S=even, the ground state is a spin singlet and the magnetic state has a gap. This is called the Haldane gap. The transition between the singlet and the excited triplet is forbidden but the absorption caused by this transition was observed[19]. This is due to the staggered field acting on spins at the inequivalent Ni^{2+} spin sites[21] and is not directly connected with the Haldane problem.

The study of the spin-Peierls transition is old. It was found more than 20 years ago in organic one-dimensional magnetic compounds. The fundamental behavior has been investigated and almost explained. This phase transition also creates an energy gap between ground singlet and excited triplet by the dimerization of spins in a chain but accompanied by lattice distortion. Recently a new type of inorganic spin-Peierls material was found[22] and has been intensively studied. From the point of view of magnetic resonance, much information can be obtained from g-shift and resonance widths, while in the case of organic compounds the ESR shows always very sharp absorption with g=2.00. In this case, the forbidden transition between singlet and excited triplet was observed[23]. We studied ESR experiments on this material[24] and found that in the so called M-phase where magnetic moment is induced, the dimerized state is destroyed and a new type of three-dimensional order arises.

References

1. H.Kondoh, Antiferromagnetic Resonance in NiO in Far-Infrared Region, *J.Phys.Soc.Jpn.*, **15** (1960) 1970-1975
2. R.C.Ohlmann and M.Tinkham, Antiferromagnetic Resonance in FeF2 at Far-Infrared frequencies, *Phys.Rev.*, **123** (1961)425-434
3. J.B.Torrance,Jr and M.Tinkham, Magnon Bound States in Anisotropic Linear Chains, *Phys.Rev.*, **187** (1969) 587-594
4. J.Boettcher, K.Dransferd and K.F.Reuk, Far-Infrared High Resolution Spectroscopy on EPR Lines with a Laser, *Phys.Letters*, **26A** (1968)146-148
5. R.W.Sanders, R.M.Belanger M.Motokawa, V.Jaccarino and S.M.Rezende, Far-infrared Laser Study of Magnetic Polaritons in FeF_2 and Mn Impurity Mode in FeF_2:Mn, *Phys.Rev.*, **B23** (1981)1190-1204
6. M.Date, M.Motokawa, A.Seki, S.Kuroda, K.Matsui, H.Nakazato and H. Mollymoto, Submillimeter Electron Spin Resonance II,Megagauss Magnets and ESR in $CuCl_22H_2O$, *J.Phys.Soc.Jpn.*, **39** (1975)898-904

7. M.Motokawa, C.Uyeda, A.Otsuka, M.Date and R.M.Toussaint, Submillimeter Wave ESR of Mn^{2+} Impurity in FeF_2 in Pulsed High Magnetic Field, *J. Phys. Soc. Jpn.* **54** (1985) 2317-2322
8. S.Foner, high Field Antiferro-, Ferri- and Paramagnetic Resonance at Millimeter Wavelengths, *J.Phys.radium*, **20** (1959)336-340
9. H.Ohta, N.Yamauchi, T.Nanba, M.Motokawa, S.Kawamata and K.Okuda, EPR and AFMR of Li_2CuO_2 in Submilimeter Wave Region, *J. Phys. Soc. Jpn.* **62** (1993) 785-792
10. M.Motokawa, S.Kuroda and M.DateSubmillimeter ESR Experiments Using Pulsed High Magnetic Field, *J. Appl. Phys.* **50** (1979) 7762-7767
11. H.Ohta, K.Yoshida,T.Matsuya, T.Nanba, M.Motokawa, S.Hosoya,K.Yamada, and Y.Endoh, EPR and AFMR of Bi_2CuO_4 in Submilimeter Wave Region, *J. Phys. Soc. Jpn.* **61** (1992) 2921-2929
12. M.Motokawa, High Field Electron Spin Resonance in the Submillimeter Region, *High Field Magnetism ed. M.Date, North Holland* (1983) 219-227
13. P.Jansen and W.P.Wolf, Far infrared parmagnetic resonance in DyAl Garnet, *Physica* **B155**(1989)328-331
14. S.Kuroda, M.Motokawa and M.Date, SH^3 Term in the Submillimeter Electron Spin Resonance of Cobalt-Tutton Salt, *J. Phys. Soc. Jpn.* **44** (1978) 1797-1803
15. A.Abragam and M.H.L.Pryce, The Theory of Paramagnetic Resonance in Hydrated Cobalt Salts, *Proc.Roy.Soc.* **A206** (1951)173-191
16. M.Motokawa, H.Ohta, N.Makita and H.Ikeda, Electron Paramagnetic Resonance of Co:Rb_2MgF_4 and Anomalous g-value of Paired Spin System, *J. Phys. Soc. Jpn.* **61** (1992) 322-328
17. S.Kimura, H.Ohta, M.Motokawa, S.Mitsudo, W,J,Jang, M.Hasegawa and H.Takei, Submillimeter Wave ESR Measurements of Metamagnetic $Y_2Cu_2O_5$, *Int.J. Infrared and Millimeter Waves* **17** (1996)833-841
18. T.Nikuni ;and H.Shiba, *J.Phys.Soc.Japan* **69** (1993) 3268
19. M.Motokawa, M.Arai, H.Ohta, M.Mino, H.Tanaka and K.Ubukata, Quantum Efffect in the High Field Phase of $CsCuCl_3$, *Physica* **B211** (1995) 199-204
20. W.Lu, J.Tuchendler, M von Ortenberg and J.P.Renard, Direct Observation of the Haldane Gap in NENP by Far-Infrared Spectroscopy in High Magnetic fields, *Phys.Rev.Lett.*, **67** (1991) 3716-3719
21. H.Shiba, T.Sakai, B.Lutthi, W.Palme and M.Sieling,ESR absorption from the Ground State in Haldane Systems:Theory and Experiment, *J. Magn. & Magn. Mater.* **140-144** (1995) 1590-1594
22. M.Hase, I.Terasaki,and K.Uchinokura, Observation of the Spin-Peierls Transition in Linear Cu^{2+}(Spin-1/2) Chains in an Inorganic Compound $CuGeO_3$, Phys. Rev .Lett.**70** (1993) 3651-3654
23. T.M.Brill, J.P.Boucher, J.Voiron, G.Dhalenne, a.Revcolevschi andJ.P.Renard, High-Field Electron Spin Resonance and Magnetization in the Dimerized Phase of $CuGeO_3$, *Phys. Rev.Letters* **73** (1994) 1545-1548
24. Y.Yamamoto, H.Ohta, M.Motokawa, O.Fujita and J.Akimitsu, The Observation of g-shift in Spin-Peierls Material $CuGeO_3$ by submillimeter Wave ESR, to be published in *J.Phys. Soc. Jpn.* **66**(1997)

EPR Study of Paramagnetic Defects in $LiNbO_3$ Crystals

S. H. Choh, I. -W. Park[†] *and S. S. Kim*[‡]

Department of Physics, Korea University, Seoul 136-701, Korea
[†] Seoul Branch KBSI, Korea University Campus, Seoul 136-701, Korea
[‡] Department of Physics, Changwon National University, Changwon 641-773, Korea

Abstract. EPR results on intrinsic defects and iron group transition metal impurities in $LiNbO_3$ crystals are summarized. EPR signals of the transition metal ions in congruent crystals were recorded as a main and additional weak ones, having similar spin Hamiltonian parameters. From these results, the site symmetry of centers and the charge compensation mechanism of the intrinsic and transition ion centers in $LiNbO_3$ and Mg-doped $LiNbO_3$ crystals are discussed. New interesting experimental findings are also included.

1 Introduction

Since the successful growth of $LiNbO_3$(LN) by the Czochralski method,[1] LN has been one of the most important electro-optical materials.[2-4] The optical and electrical properties of LN have been strongly influenced by the presence of iron group impurities and other defects. The photorefractive damage, which strongly limits its application as the laser host, was reduced drastically after adding more than a few mol% of MgO or ZnO to the melt.[5,6] The crystal structure was found to be hexagonal as shown in Fig. 1, and the composition of congruent $LiNbO_3$ (CLN) is $[Li_{1-5x}Nb_{5x}]Nb_{1-4x}O_3$ with x=0.0118, indicating 5.9 % of the niobium antisite (Nb_{Li})[7] is present in CLN. At the congruent composition with [Li]/[Nb]=0.942,[8] the crystal has the same [Li]/[Nb] ratio as the melt. Consequently, such crystals may contain a very high concentration of intrinsic defects like lithium vacancy(V_{Li}), niobium vacancy(V_{Nb}) and niobium at lithium site(Nb_{Li}). One of widely accepted intrinsic defects is the Nb_{Li}-V_{Nb} complex in the CLN crystal, as shown in Fig. 2(c).[8] However, the controversy concerning the actual sites and their charge compensators for the dopant ions in LN crystals still remains because of the similar environments for the Li and Nb sites in the crystal.

In this study we summarize our previous reports[9-12] on intrinsic defects and TM ions in LN and some new findings in Mg-doped LN(MLN) crystals.[13] The site symmetry of centers and the charge compensation mechanism of the intrinsic and dopant centers are compared and discussed.

2 Experiment

The $LiNbO_3$ crystals employed in this study were grown by the Czochralski method from the congruent melt. Typical dimensions of measured samples are 2, 3, and 4 mm along the a-, b-, and c-axis, respectively. All EPR data were measured with Bruker ESP300 X- and Q-band spectrometers at KBSI, and with an Oxford ESR900 cryostat together with an ITC4 temperature controller for low temperature measurements.

3 Experimental Results and Discussions

3.1 Thermal Reduction

It has been known that the reduction of LN causes the material to become black.[14] We were able to make nominally pure LN crystals into black after heating it at 900℃ in CO atmosphere. However, it recovered its transparency by re-heating in air at 900℃ for 2 hours.

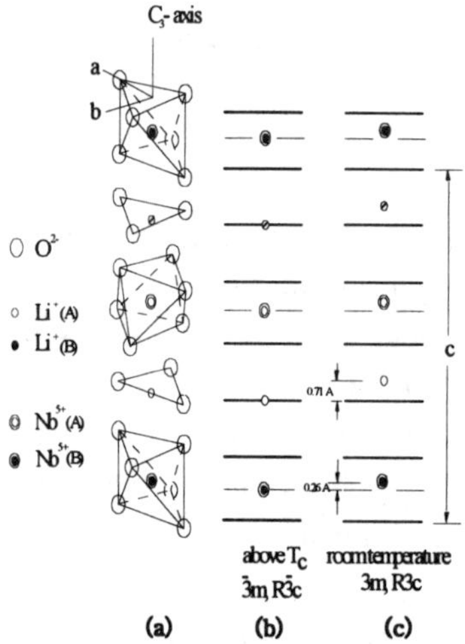

Fig. 1 Crystal structure of $LiNbO_3$. (a) The atomic array along the c-axis. (b) The schematic diagram of cations above T_c, and (c) that at room temperature. Thick solid lines in (b) and (c) mean the oxygen planes.

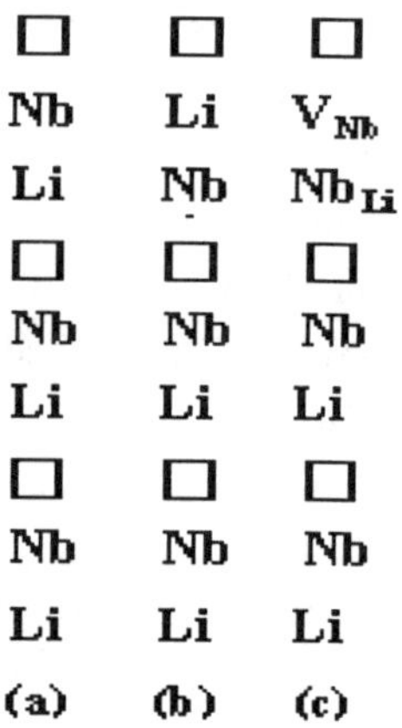

Fig. 2 Cations along the c-axis for (a) the stoichiometric and (b) local ilmenite structure of LN. (c) The Nb_{Li}-V_{Nb} complex defect in CLN.

Nominally pure LN crystals with different ratios of [Li]/[Nb] concentration were prepared, but they turned out to contain ppm levels of Fe^{3+} as a trace impurity. From Fe^{3+} EPR spectra shown in Fig. 3 for B ∥ c, the ratio of [Li]/[Nb] was roughly estimated by referring to the result of Malovichico et al.[15] The sharper linewidth and more symmetrical line shape in Fig. 3 means the closer composition to the stoichiometry compared to that of CLN. Therefore, according to the results of their analysis,[15] the compositions of [Li]/[Nb] for the samples are $0.942 = A < B < C < D < 1$, respectively.

Fig. 4 shows $Nb^{4+}(4d^1)$ and $Ti^{3+}(3d^1)$ signals for samples B and C after the thermal reduction. EPR spectra for both samples were obtained with the same experimental settings including sample dimensions. Before the reduction Nb^{5+} and Ti^{4+} were diamagnetic. Appearance of Nb^{4+} and Ti^{3+} centers were only after the thermal reduction, therefore, these centers originate from Nb^{5+} and Ti^{4+}, respectively. From these results, it may be inferred that these ions take an electron from neighboring oxygen ions since these centers are observed only in the reduced black samples and the black coloration is due to oxygen vacancies.[16]

The linewidth of the sample C in Fig. 3 is plainly sharper than that of the sample B. Therefore, the concentration of intrinsic defect in C is less than that in B because C is closer to stoichiometry than B. Consequently, it may be reasonably argued that

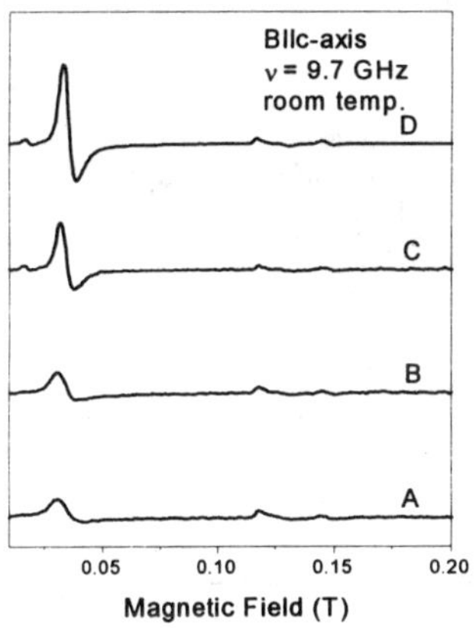

Fig. 3 Fe^{3+} EPR spectra measured with LN samples having various [Li]/[Nb] ratios. See the text for A, B, C, and D.

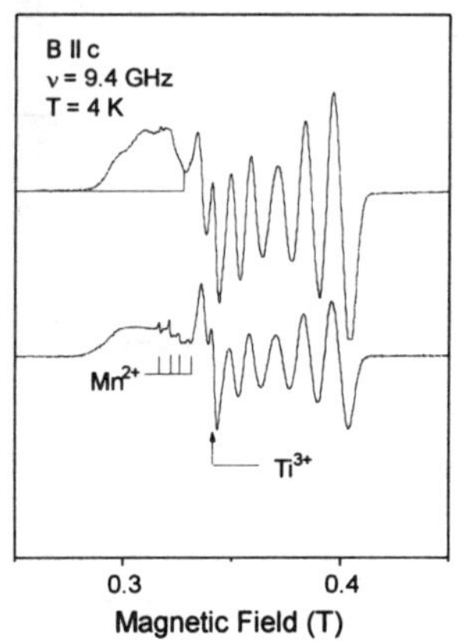

Fig. 4 Nb^{4+} and Ti^{3+} EPR spectra, after thermal reduction for B and C. Signals are not fully resolved due to overlaps near 0.31 T.

the appearance of Nb^{4+} center is related with the concentration of intrinsic defect and is probably generated at the complex defect, Nb_{Li}-V_{Nb}, shown in Fig. 2(c).

3.2 Divalent Transition Metal Ions in Congruent $LiNbO_3$

In this section we discuss the charge compensation mechanism and the relation between the divalent (Cu^{2+} and Mn^{2+}) impurities and intrinsic defects in CLN. The Cu^{2+} ion has the $3d^9$ (S=1/2) electronic configuration and its spin Hamiltonian(SH) can be described as follows due to the nuclear spin I=3/2 of ^{63}Cu and ^{65}Cu:

$$H_{1/2} = \mu_B \boldsymbol{B\, g\, S} + \boldsymbol{I\, A\, S} - \beta_n \boldsymbol{B\, g_n\, I} + \boldsymbol{I\, Q\, I} \tag{1}$$

where the first term is the electronic Zeeman, the second the hyperfine, the third the nuclear Zeeman and the last the nuclear quadrupole term. All symbols are conventional ones. However, because $^{55}Mn^{2+}$ has the $3d^5$ (S=5/2) electronic configuration with I=5/2, the SH of Mn^{2+} with the three-fold symmetry should include the zero-field splitting (ZFS) terms like eq. (2).

$$H_{5/2} = H_{1/2} + b_2^0\, O_2^0 / 3 + b_4^0\, O_4^0 + b_4^3\, O_4^3 + b_4^{-3}\, O_4^{-3} / 60 \tag{2}$$

Therefore, the difference between the Cu^{2+} and Mn^{2+} centers in LN is the inclusion of ZFS terms.

Fig. 5 shows Cu^{2+} EPR spectrum at liquid helium temperature for the magnetic field parallel to the c-axis. Four signal lines arise from the hyperfine interaction with both nuclei of ^{63}Cu and ^{65}Cu with I=3/2. Considering negligible contribution of the nuclear

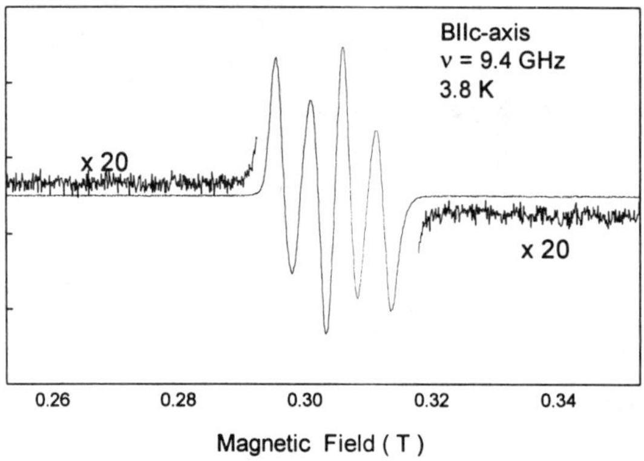

Fig. 5 Cu^{2+} EPR spectrum for B || c at liquid helium temperature. x 20 means the magnification factor of sensitivity.

Zeeman and nuclear quadrupole terms in eq. (1), we are able to analyze Cu^{2+} EPR spectra in terms of only the electronic Zeeman and hyperfine interaction as before.[17]

The Mn^{2+} EPR spectrum for B || c is shown in Fig. 6 for comparison with that of Cu^{2+}. In Fig. 6 the linewidths of hyperfine signals in the satellite lines ($\pm$1/2> $\pm$3/2>), marked by arrows closest to the central transition in the small boxes are broader than others. Consequently, the hyperfine signals in satellite transitions of Mn^{2+} were recorded relatively less symmetric than those of Cu^{2+}.

Previous reports suggest that Cu^{2+} and Mn^{2+} ions substitute for the Li^+ site in CLN with a charge compensator of V_{Li}.[18] Accepting these reasonable proposals of divalent ions (DI) to be located at the Li^+ site, we can understand that Nb_{Li}-V_{Nb} as well as V_{Li} can be charge compensators, because the complex defect is also deficient of a hole for the charge neutrality. If this is the case, there exist two DI centers, DI-V_{Li} and DI-(Nb_{Li}-V_{Nb}). Local environment surrounding these two DIs especially along the c-axis is not the same. Nevertheless it is believed that Cu^{2+} ions associated with V_{Li} and Nb_{Li}-V_{Nb}, respectively, are not distinguished, since only one set of four hyperfine signals of Cu^{2+} are recorded in Fig. 5. However, asymmetric signals of Mn^{2+} may arise from the presence of two different Mn^{2+} centers in CLN. The stacking orders of Mn^{2+} along the c-axis are V_{Li}-Nb-□-Mn^{2+}-Nb with the charge compensator of V_{Li}, and Nb_{Li}-V_{Nb}-□-Mn^{2+}-Nb with the charge compensator of the complex, as shown in Fig. 7(b) and (d), respectively. Since Mn^{2+} (S=5/2) should include in its SH the ZFS

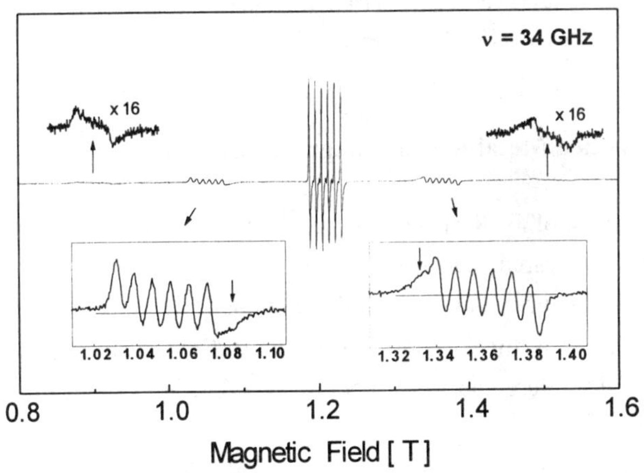

Fig. 6 Mn^{2+} EPR spectrum in CLN for B || c with the Q-band at room temperature. The small boxes are with the increased vertical sensitivity and horizontal scale.

terms, which are more sensitive than the g-value and hyperfine interaction under the locally perturbed environment, thus, slightly different ZFS parameters for two centers may cause deviated positions of Mn^{2+} signals. This arises Mn^{2+} signals asymmetric, even though not fully resolved. If these cases are actually realized, Mn^{2+} can be electrically neutralized with the charge compensator of V_{Li} or Nb_{Li}-V_{Nb}. Moreover, these charge compensators can not change the C_3 symmetry of Mn^{2+} because they lie along the c-axis.

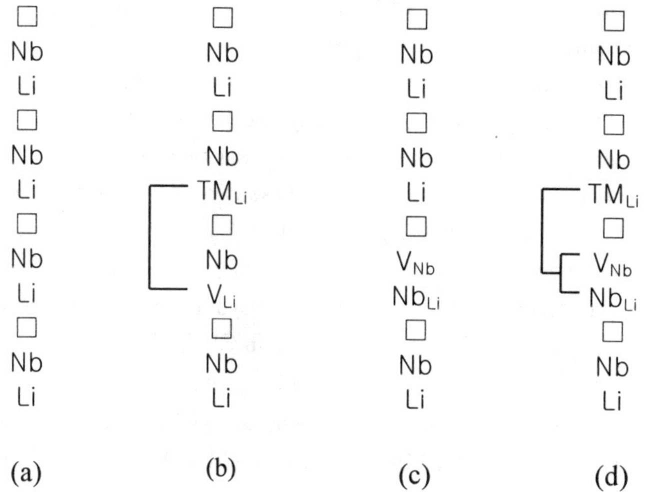

Fig. 7 Schematic diagram for two TM^{2+} centers in CLN. (a) Normal LN stacking, (b) TM^{2+}-V_{Li} in CLN, (c) Nb_{Li}-V_{Nb} defect, and (d) TM^{2+} associated with the Nb_{Li}-V_{Nb} defect.

3.3 Trivalent Transition Metal Ions in Congruent $LiNbO_3$

Fig. 8 displays an array of EPR spectra for Cr^{3+} in CLN, the so-called *stack-plot*, measured in the crystallographic ca-plane. The marked signals with Fe^{3+} are identified as those from the iron impurity unintentionally contained. All transitions except those from Fe^{3+} appear to be from the Cr^{3+} ion. Fig. 9 shows Cr^{3+} signals for B ∥ c+20° at an increased sensitivity. Here the Cr^{3+} spectra are recorded as one main and two additional weak lines.

Similar structures were more evidently observed with Fe^{3+} for B ∥ c as shown with arrows in Fig. 10. The appearance of weak additional signals for TM^{3+} centers can be understood with ZFS terms similar to that of Mn^{2+}. However, the splittings for TM^{3+} centers are far clearer than the case of Mn^{2+} in CLN. This means that the difference of deviated ZFS terms for the TM^{3+} center is larger than that of the Mn^{2+}.

Futhermore, there are at least three distinctive signals of Cr^{3+} as shown in Fig. 9 and of Fe^{3+} as displayed for B ∥ c in Fig. 10, respectively. This fact is different from the case of TM^{2+} in CLN. Consequently, there are at least three different TM^{3+} centers along the c-axis. The strongest one may be originated from the structure of $TM^{3+}{}_{Li}$ and two neighboring Li vacancies. This center is properly charge compensated. The other two may arise from TM^{3+} associated with the Nb_{Li} + V_{Nb} complex, which is approximately 6% content in CLN. Consequently these structures are less populous in the crystal than $TM^{3+}{}_{Li}$ with two Li vacancies.

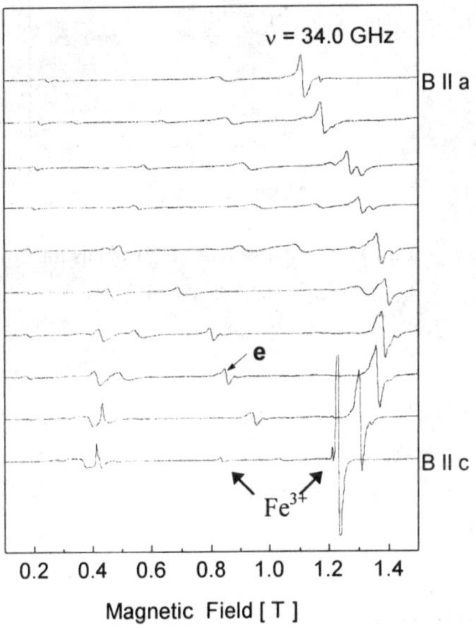

Fig. 8. Cr^{3+} EPR spectra in the ca-plane measured with a Q-band.

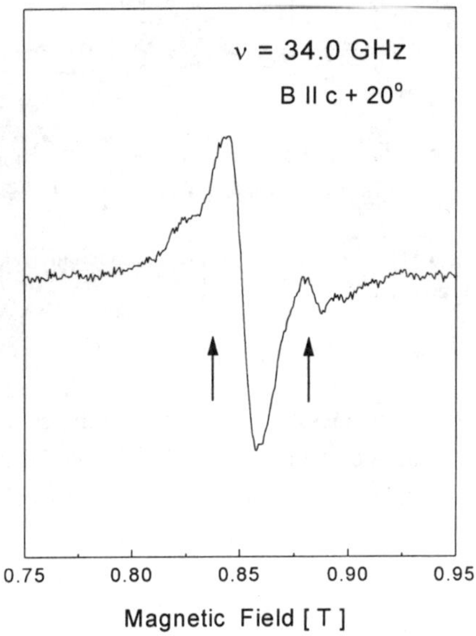

Fig. 9. An expanded signal, marked with e in Fig. 8 for B ∥ c+20^0.

3.4 Trivalent Transition Metal Ions in Mg-doped $LiNbO_3$

Since the report that the optical damage could be drastically reduced in MLN,[5] many papers on defects in MLN have been reported.

Fig. 11 is a comparison of Mn^{2+} in CLN and MLN. As can be seen, there is no remarkable difference except the broader linewidth in MLN. Mg ions in MLN occupy Li sites, forming the Mg_{Li} defect, and reduce the complex defect. Therefore, the broader linewidth of Mn^{2+} signals in MLN can be attributed to the relatively remote Mg_{Li}^{2+} ions.

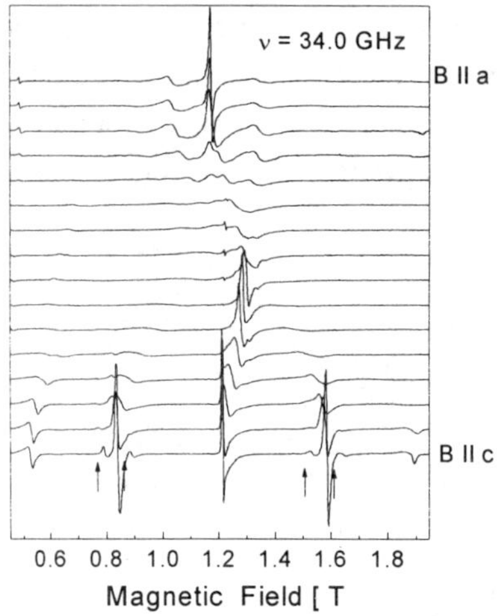

Fig. 10 Fe^{3+} EPR spectra in CLN measured in the ca-plane.

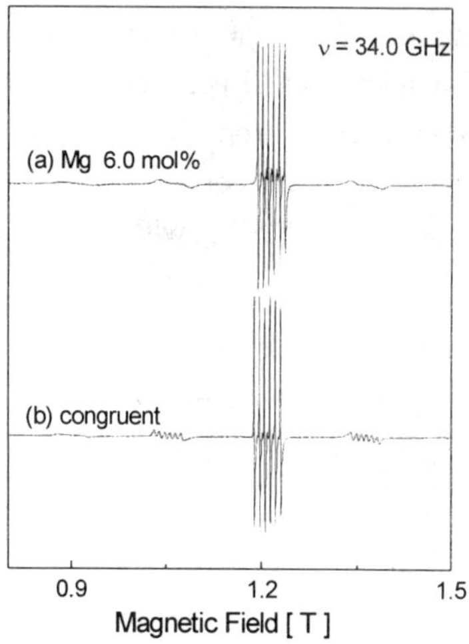

Fig. 11 Comparison of Mn^{2+} signals for B ∥ c in CLN and 6 mol% MLN.

However, there is a remarkable difference for Fe^{3+} EPR spectra between in CLN and in MLN, as shown in Fig. 12. Up to 1 mol% of Mg-adding, the similar discussion for the Fe^{3+} center holds as Mn^{2+} in MLN. But totally different Fe^{3+} EPR spectra were obtained in the 6.3 mol% of Mg-doped LN compared with that in CLN.

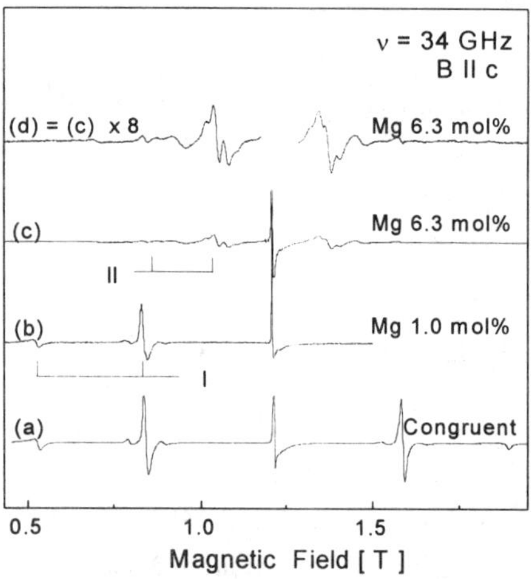

Fig. 12 Comparison of Fe^{3+} signals for B ∥ c in CLN and MLN. (d) is with the increased sensitivity by 8 times for (c).

Mg ions replace Li ions, as discussed in Section 3.2. Then, the local charge state becomes different, because the valence of the Mg ion is +2 while that of Li is +1. Consequently, the occupation of Fe^{3+} at the Nb site is more favorable to reduce the excess charge due to the Mg_{Li} defect. The report, the D value for Fe^{3+} at the Nb site is reduced to a half of the case for Fe^{3+} at the Li site, based on the calculation with the superposition model,[19] is consistent with our study. As can be seen in Fig. 12(c), the fine structure splittings are about a half compared with that of CLN.

The superposed signals for Fe^{3+} in MLN (Fig. 12(d)) can be understood by the same analogy with that in CLN. Because when a few mol% of Mg is added into the crystal, a few percentage of Mg_{Li} defects are created, also causing a small difference of ZFS of Fe^{3+} at the Nb site, similar to that of CLN. Böker et al[20] proposed that Fe^{3+} ions substituted for Nb^{5+} instead of Li^{+} ions in MLN from their EPR results, and their proposal was confirmed in terms of the charge state of dopant ions in our discussion.

3.5 Three-fold Symmetry

To investigate the effects of three-fold symmetry on EPR, we adopted Cr^{3+}(S=3/2) and Fe^{3+}(S=5/2) paramagnetic centers in LN. As can be seen in Fig. 1, Li^{+} ions in the crystal occupy two different sites with chemically equivalent but physically inequivalent 3-fold symmetry. These paramagnetic ions locate at the three-fold symmetric site, preferably Li^{+}.

The SH for the S=3/2 center with three-fold symmetry may be described by eq. (3), without the nuclear terms for I = 0 (approximately 90 % of the Cr isotopes).

$$H_{3\,2} = \mu_B\, \mathbf{B}\, \mathbf{g}\, \mathbf{S} + b_2^0\, O_2^0 /3 \tag{3}$$

For the S=5/2 center with the same symmetry, the fourth order ZFS terms should be added to eq. (3):

$$H_{5\,2} = H_{3\,2} + b_4^0\, O_4^0 + b_4^3\, O_4^3 + b_4^{-3}\, O_4^{-3} /60 \tag{4}$$

Cr^{3+} and Fe^{3+} EPR spectra are shown in Figs. 8 and 10, respectively. The central transition of Cr^{3+} signals was recorded as one line for each center no matter where the magnetic field was applied. Even though Cr^{3+} ions may occupy two different Li^{+} sites in the crystal, they can not be distinguished with EPR, because the ZFS term for S=3/2 must be limited up to the second order. However, when a $S \geq 2$ ion substitutes for the 3-fold symmetric site, the fourth order terms of ZFS must be included in its SH.

EPR signals of Fe^{3+} ions at two different Li^{+} sites in a LN crystal were actually splitted into a pair when the magnetic field was applied off the c- and a-axis, respectively. This splitting for the central transition (|1/2> ↔| -1/2>) was observed for B ∥ c+66° and c+72° in Fig. 10. However, the splitted Fe^{3+} signals collapsed into one, when the magnetic field was applied along the c- or a-axis. This can be explained with the O_4^3 term which becomes zero when the applied magnetic field is parallel to the c- or a-axis. Unfortunately the line splittings for many orientations are overlapped due to the broad linewidth for the Fe^{3+} center in CLN. However, if we choose the principal X-axis of ZFS as the a-axis for one Fe^{3+} center (I), that of the other Fe^{3+} center (II) becomes the a+π/3 axis due to the crystal structure (see Fig. 1(a)). When the applied magnetic field is parallel to either the c- or a-axis, the $b_4^3O_4^3$ term of two Fe^{3+} centers are simultaneously zero. But when the magnetic field is applied off the c- and a-axis, the term of both Fe^{3+} centers have different signs with equal magnitude. However, the $b_4^{-3}O_4^{-3}$ term does not contribute in the ZX-plane of ZFS because $O_4^{-3}|_{\varphi=0} = 0$ in the conventional spherical coordinates.

Therefore, the splitting of the central transitions of Fe^{3+} EPR signals in Fig. 10 can be explained in terms of the ions at two physically inequivalent sites in LN and the nonzero b_4^3 term, which determines the magnitude of the splitting.[12]

4 Summary

EPR results of intrinsic defects, divalent and trivalent TM ions in LN crystals are discussed. The Nb^{4+} center appears to be generated at the complex defect of Nb_{Li}-V_{Nb} after the thermal reduction of CLN. A few TM centers are related with the complex in CLN crystals, while other TM centers take V_{Li} as a charge compensator. The line splitting of the central transition of Fe^{3+} centers in LN is explained by Fe^{3+} ions occupying two physically inequivalent Li sites in the crystal. Fe^{3+} ions in MLN seem to be favoring the Nb site due to the local excess charge from neighboring Mg ions.

Acknowledgements

This work is supported in part by Korea Science and Engineering Foundation through the RCDAMP at Pusan National University(1994-1997), by Korea Basic Science Institute(1996), and in part by the BSRI Program(BSRI-96-2410).

References

1 A. A. Ballman, J. Am. Ceram. Soc. **48**, 112 (1965).

2 A. R uber, in *Current Topics in Materials Science* (Edited by E. Kaldis), North Holland, New York(1978), pp 481-601.

3 A. M. Prokhorov and Yu. S. Kur'minov, *Physics and Chemistry of Crystalline Lithium Niobate(*Adam Hilger, New York, 1990).

4 D. A. Bryan, R. R. Rice, R. Gerson, H. E. Tomaschke, K. L. Sweeney and L. E. Halliburton, Opt. Eng. **24**, 143 (1985).

5 G. G. Zhong, J. Jin and W. Zhong-Kang, 11th International Quantum Electronics Conference, p.631. IEEE Cat. No. 880 CH 1561-0 (1980).

6 T. R. Volk, V. J. Pryalkin and N. M. Rubinina, Opt. Lett. **15**, 997 (1994).

7 S. C. Abrahams and P. Marsh, Acta Crystallogr. **B 42**, 61 (1986).

8 H. J. Donnerberg, S. M. Tomlinson and C. R. A. Catlow, J. Phys. Chem. Solids **52**, 201 (1991).

9 I. -W. Park, S. H. Choh, K. J. Song, T. H. Yeom, and C. Rudowicz, J. Korean Phys. Soc. **26**, 168 (1993).

10 T. H. Yeom, Y. M. Chang, C. Rudowicz, and S. H. Choh, Solid State Commun. **87**, 245 (1993).

11 T. H. Yeom, S. H. Choh, Y. M. Chang, and C. Rudowicz, phys. stat. sol.(b) **185**, 417 (1994).

12 I. -W. Park and S. H. Choh, Bull. Magn. Res. **17**, 276 (1995).

13 I. -W. Park, D. K. Kim, S. H. Choh and S. S. Kim, J. Korean Phys. Soc. **29**, S699 (1996).

14 K. L. Sweeney and L. E. Halliburton, Appl. Phys. Lett. **43**, 336 (1983).

15 G. I. Malovichko, V. G. Grachev, E. P. Kohanyan, O. F. Shirmer, K. Betzler, B. Gather, F. Jermann, S. Klauer, U. Schlarb and M. Wö hlecke, Appl. Phys. **A56**, 103 (1993).

16 I. -W. Park, Ph. D thesis (Korea University, Seoul, 1995).

17 R. M. Krishina and S. K. Gupta, Bull. Mag. Res. **16**, 239 (1994).

18 G. Corradi, H. Sö the, J-M. Spaeth, and K. Polgar, J. Phys: Condens. Matter **2**, 6603 (1990).

19 Q-R. Zhang and X-Q. Feng, Phys. Rev. **B43**, 12019 (1991).

20 A. Bö ker, H. Donnerberg, O. F. Schirmer and F. Xiqi, J. Phys.:Condens. Matter **2**, 6865 (1990).

EPR Studies of Spin Transitions: A Review

Y. Servant[1)], *C. Cantin*[1)], *O. Kahn*[2)] *and J. Kliava*[1)]

[1)]Centre de Physique Moléculaire Optique et Hertzienne, Université Bordeaux-I, Talence, France
[2)]Laboratoire des Sciences Moléculaires, Institut de Chimie de la Matière Condensée de Bordeaux, Pessac, France

Abstract. The electron paramagnetic resonance has been widely used to evidence and investigate spin transitions in d^5 to d^7 ion compounds. Direct studies of Fe^{3+}, Ni^{3+}, Co^{2+} ions and indirect studies of Fe^{2+} are reviewed and shown to provide valuable data on structural transformations accompanying the spin transitions and mechanisms thereof.

1 Introduction

Several molecular materials containing transition group ion complexes which exhibit thermally-induced spin transitions can potentially be used as molecular electronics devices, namely, for the purpose of information recording and processing. In this event, a number of questions must be answered, concerning the existence and the characteristics of hysteresis associated to the spin transition. It is evident that, in order to present a sufficiently large hysteresis loop, a spin transition must go with a structural phase transition. Thus, structural analysis of spin transition materials becomes of a primary importance.

Experimental techniques used to study the spin transitions fall into two groups. The first one is concerned with the electronic characteristics of the material, *viz.*, magnetic susceptibility, optical response and yields no or little information on the crystallographic structure. The second group comprises "direct" methods of structural analysis, such as X-rays diffraction, EXAFS, *etc.*, which, in turn, are insensitive to the "magnetic" structure. Moreover, in the case of polycrystalline or non-crystalline materials, these methods are only of a limited sensitivity, as far as some finer details of the short-range ordering are concerned. Namely, they yield only radial distribution functions of atomic density, that is, binary correlation functions and give no direct information on the local symmetry which require higher than first order correlation functions.

In this context, it should be noted that EPR, which is a *magnetic spectroscopy* technique and at the same time a *structure-sensitive* method of analysis, is particularly appropriate for studies of this kind. EPR is by far more informative than any static magnetic experiment with regard to modifications of magnetic interactions during the spin transition. As a method of structural research, EPR can provide data on subtle structural transformations in the environment of transition ions.

2 Direct EPR Studies

Because of the possibility to obtain two distinct signals for the high-spin (HS) and low-spin (LS) ions, and important information about the electronic states involved in the transition, EPR was widely used in *direct* studies of spin transition Fe^{3+} compounds. EPR can be potentially of use in *direct* studies of spin transition Ni^{3+} compounds but these studies are still rare because of the unusual oxidation state of this ion and its preferential stabilisation in the LS state. For Co^{2+} compounds, the EPR studies are more delicate because the HS spectrum is only observable at low temperatures.

2.1 Fe^{3+} ($3d^5$) Compounds

The HS $S = 5/2$ EPR spectrum is described by a spin hamiltonian which comprises axial and rhombic zero-field quadrupole terms, the corresponding parameters being D and E. In a strong crystal field case, for axial symmetry, $E = 0$, the effective g-factor has the principal values $g_{//} = 2$ and $g_{\perp} = 6$, whereas for rhombic symmetry, $E/D = 1/3$, $g = 4.29$. Thus the observation of an apparent g-factor greater than or equal to 4.29 is the manifestation of the HS state. In the LS state, $S = 1/2$, the EPR spectrum reduces to a single pattern at $g \approx 2$. It should be noted however that a $g \approx 2$ signal can also be due to HS Fe^{3+} in low or intermediate crystal fields.

An abrupt and almost complete spin transition of Fe^{3+} at 159 K in [Fe(3-OCH_3-SalEen)$_2$]PF_6 was studied by EPR from 290 to 77 K [1]. At room temperature, the EPR spectrum is characteristic of the HS ions in a nearly axial symmetry (a major line at low magnetic field and a minor line at $g \approx 2$). The most dramatic change in the EPR spectrum occurs within a range of a few degrees around $T_c = 159\,K$, which is the critical temperature of the spin transition. Below this temperature, a three-component rhombic LS signal ($g_x = 2.220$, $g_y = 2.187$ and $g_z = 1.946$) becomes dominant. A discussion of the principal g-values shows that in the $^2T_{2g}$ ground state of Fe^{3+}, the unpaired electron resides in the d_{xy} orbital. It is interesting to note that at low temperatures a residual, very weak HS signal has $g \approx 4.3$, showing that the microscopic environment of the HS ferric centres also becomes rhombic. Other compounds of the same family exhibit a gradual and incomplete spin transition and similar information about the $^2T_{2g}$ ground electronic state can be deduced from EPR.

Grinding and doping this system result in a more gradual and incomplete spin transition [1]. The grinding leads to the appearance of a new EPR signal at $g \approx 4.3$ at 290 K which does not disappear at 77 K. For the ground cobalt-doped compounds, a new HS signal at $g \approx 5.5$ occurs at room temperature, its intensity increasing with the concentration of dopant.

EPR spectra at various temperatures are given for polycrystalline samples of [Fe(acen)dpp](BPh_4), [Fe(acen)(pic)$_2$](BPh_4) and [Fe(acen)(lut)$_2$](BPh_4) exhibiting a gradual spin transition between 100 and 300 K [2]. Two broad EPR signals are observed at room temperature, one at $g \approx 4.1$ and the other at $g \approx 2.1$. The presence of the $g \approx 4.1$ signal for HS Fe^{3+} indicates that the molecular symmetry is rhombic.

Below the transition temperature, a three-component LS Fe^{3+} signal at $g \approx 2.1$ is observed, analogous to that described in [1].

Samples of $[Fe(SalAPA)_2]ClO_4$ (identified as FeP) and $[Fe(SalAEA)_2]ClO_4$ (identified as FeE) were investigated both in solid state and as frozen solutions [3]. In an unperturbed FeP, the EPR spectrum between 301 and 108 K exhibits two signals: a broad isotropic $g \approx 4.3$ HS signal and a rhombic LS pattern in the $g \approx 2$ region. The evolution of relative intensities for the LS and HS species evidences a gradual but incomplete spin transition. At 301 K, the sample contains approximately equal numbers of LS and HS species, whereas at 108 K only traces of HS species remain. Similar results are obtained for a ^{57}Fe-enriched FeP. In the case of FeE, the magnetic susceptibility data alone do not provide unequivocal proof of a spin-transition, so, spectroscopic techniques are necessary to demonstrate the presence of LS complexes. In the LS state of Fe^{3+}, FeP complexes have a rhombic **g**-tensor, while FeE complexes have an axial **g**-tensor. For all complexes, in the $^2T_{2g}$ ground state the unpaired electron is almost totally in a d_{xy} orbital. The most notable difference between the two categories of complexes is the energy spacing of the three Kramers doublets: the first excited-state doublet is somewhat higher in energy in the axial complexes, but the second excited-state doublet is much higher in the rhombic complexes.

A solvated sample of $FeP.CH_2Cl_2$ was studied by EPR from 301 to 77 K [3]. Solvation dramatically lowers the critical temperature, to 152 K from 295 K for the unsolvated FeP. In addition to the $g \approx 4.3$ HS signal, a low-field shoulder corresponding to $g \approx 6$ is visible below 174 K, which may indicate the presence of HS sites. The LS EPR spectrum of $FeP.CH_2Cl_2$ covers a significantly larger range of g-values ($g_x = 2.355$, $g_y = 2.092$, $g_z = 1.966$) than that for FeP ($g_x = 2.314$, $g_y = 2.129$, $g_z = 1.998$). The authors suggest that there is little change in the crystal field felt by Fe^{3+} ions upon solvation, but just a change in the nature of low-symmetry distortion of the LS ground state. Grinding increases the relative intensity of the HS signal as also the linewidths of the LS signal at all temperatures. This broadening suggests that the residual HS complexes are randomly distributed within the same crystallites as the majority HS complexes and are not in separate domains of any appreciable size.

In $[Fe(acpa)_2]Bph_4.H_2O$ EPR signals at $g \approx 4$ and at $g \approx 2$ are observed at room temperature [4]. At 78 K, the $g \approx 2$ signal is well resolved, increases in intensity relative to the other signal, and is characterised by axial symmetry in contrast with rhombic symmetry in HS state. The d_{xy}-character of the lowest energy Kramers doublet of $^2T_{2g}$ origin is also established.

Gradual but nearly complete spin transitions of Fe^{3+} in $[Fe(3\text{-}OEt\text{-}SalBzen)_2]BPH_4.CH_3CN$ (I) and $[Fe(3\text{-}OEt\text{-}SalBzen)_2]BPH_4$ (III) were studied by EPR from 300 to 132 K [5]. For both samples, the axial LS signal ($g_{//} = 2.18$, $g_{\perp} = 1.94$) increases in intensity at low temperatures at the expense of the HS signal. Although the intensity of the HS signal is strongly temperature-dependent, no temperature dependence of the effective g-values ($g = 8.36,\ 5.25$) has been detected. So, the authors infer that the ratio E/D remains constant throughout the spin transition. This constancy of E/D may be explained by an insignificant role of the intermolecular interactions in these BPh_4^- complexes, to which the identical HS g-values of I and III seem to give some credence. It is also consistent with a domain mechanism of the spin transition.

A gradual and nearly complete spin transition of Fe^{3+} in [Fe(3-OEt-SalAPA)$_2$]ClO_4.C_6H_6 was studied by EPR from 298 to 77 K [6]. The LS ($g \approx 2$) and HS ($g \approx 4$) signals are very similar to those found in [3]. The three g-tensor components ($g_x = 2.379,\ g_y = 2.077,\ g_z = 1.974$) indicate that the first excited electronic state is at least 5000 cm^{-1} higher in energy than the ground state. Thus, in the LS state, a temperature-independent magnetic moment μ_{eff} is expected, so that any temperature-dependent changes of μ_{eff} can be attributed to the spin transition. The details of the g-tensor analysis presented in a subsequent report [7] indicate that the unpaired electron in the ground Kramers state is essentially in a d_{xy} orbital.

A gradual but complete spin transition of Fe^{3+} in [Fa(acen)(N-Me-imd)$_2$]BPh_4 was studied by EPR [8]. At 295 K, the spectrum shows a complex pattern with $g \approx 8.4$ and 4, indicative of a distorted HS Fe^{3+} complex, as well as a $g \approx 2$ signal. Cooling to 77 K results in a much simpler spectrum with lines at $g_x = 2.32,\ g_y = 2.14,\ g_z = 1.93$ typical of LS $S = 1/2$ Schiff-base systems. The authors claim that relaxation and/or exchange interactions are responsible for the unusual aspect of the room-temperature spectrum or that the relaxation between the HS and LS states may be extremely rapid, ~10^{-10} s, resulting in an averaged spectrum. It should be noted however that the presence of a $g \approx 2$ signal is quite typical of the HS Fe^{3+} EPR spectra in the case of intermediate crystal fields.

EPR spectra from 298 to 78 K are given for [Fe(3-CH_3OSPH)$_2$]PF_6 and [Fe(3-CH_3OSPH)$_2$]BPh_4, exhibiting a gradual but complete spin transitions [9]. At 298 K, two EPR lines are observed; one at $g \approx 4.2$ and the other at $g \approx 2.0$. At low temperature, the EPR signal near 2.0 is well resolved, grows in intensity at the expense of the other signal $g \approx 4.2$ as the temperature is decreased, and is characterised by rhombic symmetry.

In the case of a gradual but nearly complete transition in [Fe(AE)$_2$]BPh_4 [10], the spectrum at room temperature comprises a broad symmetric signal at $g \approx 4$ assignable to the HS species and a very weak feature at $g \approx 2$. At 90 K, both HS and LS signals are observed. At liquid helium temperatures, the resolution is still improved, allowing the observation of a prominent $g_x = 2.42,\ g_y = 2.18,\ g_z = 1.97$ LS pattern and much weaker HS signals at $g = 4.38$, $g = 4.08$ and $g = 1.46$. For $E/D = 0.25$, the three effective g-values of the three Kramers doublets arising from the $^6A_{1g}$ state would be $g_x = 0.49,\ g_y = 0.38,\ g_z = 9.84;\ g_x = 4.08,\ g_y = 3.76,\ g_z = 4.76$ and $g_x = 1.45,\ g_y = 9.38,\ g_z = 0.92$. Thus, the most intense HS signal, with $g = 1.46$, must be attributed to the lower doublet, whereas the $g = 4.38$ and $g = 4.08$ features may result from the overlap of the three lines of the middle doublet. In the 90 K spectrum, some lines may arise from a second HS species with $E/D \approx 0.12$. The LS spectra provide similar information to that of [3].

EPR has also been used to study the spin transitions in [Fe(3-OEt-SalAPA)$_2$](ClO_4).C_6H_5Cl [11], [Fe(5MeO-salen)(NiL^1)Cl] [12] and Fe(amben)Br.H_2O [13].

2.2 $Ni^{3+}(3d^7)$ Compounds

The HS (S=3/2) EPR spectrum reduces to an anisotropic line, with $g=\frac{5}{3}g_s-\frac{2}{3}g_\ell$, where $g_s=2$ and $g_\ell=-3/2$, whence $g=13/3=4.33$. The LS (S = 1/2) EPR spectrum consists of an isotropic or anisotropic line with $g\approx 2$. Thus, as in the previous case (§ 2.1), an apparent g-factor close to 4 is a manifestation of HS (S = 3/2), whereas a g-factor close to 2 is a manifestation of the LS (S = 1/2).

A very gradual spin transition of Ni^{3+} in $BaLaNiO_4$ (K_2NiF_4-type structure) was studied by EPR from 293 to 6 K [14]. At room temperature, the EPR spectrum is mainly composed of an anisotropic HS line characteristic of an elongated octahedral surroundings, with an apparent g-factor close to 4.3 and a minor LS line at $g\approx 2$. As the temperature is lowered, the LS line increases in intensity at the expense of the HS line. At 6 K, only an anisotropic signal ($g_{//}=2.04$, $g_\perp=2.15$) is observed, which can be attributed to the Ni^{3+} ions in the LS state. The slight excess of $g_\perp$ over $g_{//}$ characterises a tetragonally elongated octahedral surroundings.

2.3 $Co^{2+}(3d^7)$ Compounds

The LS (S = 1/2) EPR spectrum usually consists of an anisotropic line (*e.g.*, $g_x=2.289, g_y=2.206, g_z=2.046$) whose average g-value is not too different from 2 [15]. The HS (S = 3/2) EPR spectrum is only observable at low temperatures, because of short spin-lattice relaxation times for HS Co^{2+} ions. The EPR spectrum clearly shows a rhombic distortion (*e.g.*, $g_x=5.75, g_y=4.54, g_z=2.22$) [15]. The average HS g-value of (4.17) is close to 4.3, as in the previous cases (§ 2.1 and 2.2).

A $Co(H_2fsa_2phn)(H_2O)_2$ compound exhibiting a spin transition with a hysteresis loop with $T_c\downarrow$~101 K and $T_c\uparrow$~105 K, was studied by EPR from 230 to 4 K [16]. Towards ~230 K, a broad signal appears at $g=2.3$. At lower temperatures, this signal splits into two (at ~130 K) then into three (at ~80 K) components. The g-value pattern ($g_x=2.44, g_y=2.32, g_z=2.06$) is characteristic of Co^{2+} in LS state. It is typical of a spin doublet in a quasi-axially elongated environment, suggesting a mainly z^2-type orbital for the unpaired electron. Moreover, a broad signal at $g\approx 5.11$, characteristic of Co^{2+} in HS state, can be observed below ~70 K. Although the HS state is predominant from room temperature to ~100 K, the corresponding EPR signal only appears at lower temperatures because of the short spin-lattice relaxation time in the $S=3/2$ state. The plot of $I.T$ vs. T, I being the intensity of the LS EPR line for decreasing temperatures, confirms the spin transition in an interval of a few degrees around 101 K.

EPR spectra of $Co(H_2fsa_2en)(H_2O)_2$ exhibiting a spin transition with a hysteresis loop at $T_c\downarrow$~81.5 K and $T_c\uparrow$~84.6 K, were measured as a function of temperature by cooling the sample from 300 to 4 K [17]. The results are quite similar to the previous ones [16]. The LS pattern ($g_x=2.661$, $g_y=2.318$, $g_z=2.031$) obtained at 20 K is indicative of a pronounced rhombic distortion in the Co^{2+} environment. The sharp decrease in $I.T$ values vs. T, observed between ~80 and 87 K, gives further evidence of the spin transition.

$Co(H_2(fsa)_2en)(4\text{-}Mepy)_2$ (G) and $Co(H_2(fsa)_2en)(4\text{-}t\text{-}Bupy)_2$ (H) complexes exhibit, respectively, a very smooth and incomplete spin transition and a discontinuous spin transition with the critical temperatures $T_c\downarrow$~138 K and $T_c\uparrow$~154 K [18]. The EPR spectrum of G at 4.2 K provides a clear evidence for the coexistence of two spin states, characterised by a signal at $g = 2.25$ for the LS complexes and signals at $g = 5.63$ and 4.30 for the HS complexes. The EPR spectrum of H at 6 K shows two distinct features. The predominant LS species was identified by its stronger component at $g_\perp = 2.23$, which partially hinders the weaker one, $g_{//}$, located in the 2.0-2.1 range. This pattern is indicative of a nearly tetragonal axially elongated symmetry. The $g = 4.35$ small feature was associated with the residual HS species.

The spin transition of Co^{2+} in $[Co(dmvi)(phen)_2]ClO_4.3H_2O$ was studied by EPR from 293 to 5 K [19]. At room temperature, the EPR spectrum is reduced to a weak and nearly symmetric LS signal at $g = 2.15$. Its intensity increases significantly with decreasing the temperature, and it reveals progressively an axial symmetry. Another weak line, assignable to a residual HS fraction, appears at 5 K, showing that the transition is not complete.

3 Indirect EPR Studies: $Fe^{2+}(3d^6)$ Compounds

A *direct* observation of spin transitions in ferrous compounds by EPR of Fe^{2+} is difficult, if not impossible, since the Fe^{2+} ion is 'EPR silent' not only in its LS ($S = 0$) *diamagnetic* state, but, in practice, also in its HS ($S = 2$) *paramagnetic* state due to a very short spin-lattice relaxation time and/or a very large zero-field splitting. The spin transition experienced by the Fe^{2+} ions can however be detected by EPR in an *indirect* way, from an analysis of modifications of EPR spectra of some *paramagnetic probes* - foreign paramagnetic ions incorporated in the matrix. Note that there can be two different causes of such modifications: first, *spin-spin interactions* between the foreign ions and Fe^{2+} in its HS state, second, a *structural change* occurring in the matrix in the course of the spin transition. In the case of foreign paramagnetic ions, the modifications caused by this structural change are of a particular interest, since they yield data not on (more or less trivial) *short-range* transformations in the environment of the Fe^{2+} ions but rather on *long-range* ones involving the whole matrix. Note also that doping spin-transition compounds with foreign species, besides introducing paramagnetic probes, can modify their characteristics, namely, the transition temperatures and parameters of the hysteresis loop, which can be very interesting from the viewpoint of their technical applications.

Up to now, only two foreign paramagnetic ions have been used to dope ferrous spin transition compounds, viz., Cu^{2+} ($3d^9$) and Mn^{2+} ($3d^5$).

3.1 Cu^{2+}-doped Compounds

In the two-dimensional ferrous complex $[Fe(NCS)_2(btr)_2]\cdot H_2O$ doped with Cu^{2+} ($S = 1/2, I = 3/2$), the HS$\leftrightarrow$LS transition of Fe^{2+} ions shows a hysteresis loop with

$T_c\downarrow$~123.5 K and $T_c\uparrow$~144.5 K [20, 21]. The EPR spectrum of Cu^{2+} in the LS state of Fe^{2+} (diamagnetic phase) is characteristic of an elongated tetragonal environment ($g_{//} = 2.29$, $g_{\perp} = 2.05$), and shows a hyperfine structure and a nine-component superhyperfine structure due to *four* nitrogen atoms ($I_N = 1$) from the triazole ring located in the basal plane. In the HS state of Fe^{2+} (paramagnetic phase), the resolution of EPR spectra is limited by the dipolar and exchange broadening, because of small minimal distances between paramagnetic ions.

The EPR studies on single crystals of Cu^{2+}-doped $[Fe(NCS)_2(btr)_2]\cdot H_2O$ and $[Fe(NCSe)_2(btr)_2]\cdot H_2O$ (for the latter $T_c\downarrow$~215 K, $T_c\uparrow$~220 K) [22] suggest that the diamagnetic phase of these compounds has a crystalline structure different from that of the paramagnetic phase. Two reasons are advanced to corroborate this viewpoint: one, the elongation of the coordination polyhedron of Cu^{2+} in the diamagnetic phase, as compared with the compression of the one of Fe^{2+} in the paramagnetic phase inferred from X-ray studies, two, an absence of magnetically non-equivalent Cu^{2+} sites in the diamagnetic phase, in disagreement with the X-ray data for the paramagnetic phase. (The first argument is not convincing since the elongation of the coordination polyhedron of Cu^{2+} is doubtless caused by the static Jahn-Teller effect.)

The Cu^{2+} spectrum in the paramagnetic phase of single crystals of $[Fe(ptz)_6](BF_4)_2$ complex [23] ($T_c\downarrow$~128 K, $T_c\uparrow$~134 K) is averaged by the dynamic Jahn-Teller effect ($g = 2.16$), however it still shows a resolved hyperfine structure, because of longer distances between paramagnetic ions (~ 1.086 nm) as compared to other copper-doped spin transition complexes. In the diamagnetic phase, the Jahn-Teller effect is frozen and Cu^{2+} gives rise to a typical anisotropic spectrum ($g_{//} = 2.303$, $g_{\perp} = 2.076$), with well-resolved hyperfine and superhyperfine structures, similar to those reported in [20, 21]. Two different diamagnetic phases, a quickly cooled phase I and a slowly cooled phase II, could be detected by using Cu^{2+} as a probe.

In Cu^{2+}-doped $[Fe(trz)(Htrz)_2](BF_4)$ compound [24] ($T_c\downarrow$~345 K, $T_c\uparrow$~374 K), the EPR spectrum of Cu^{2+} in the LS state of Fe^{2+} is anisotropic, with $g_x = 2.04$, $g_y = 2.12$, $g_z = 2.28$. The hyperfine structure could only be resolved well below the spin transition temperatures. The spin transition is indicated by an abrupt broadening of the Cu^{2+} EPR spectrum, the linewidth in the paramagnetic phase being determined by dipolar and exchange broadening. It is interesting to note that the g-value of Cu^{2+} in the HS state of Fe^{2+}, $g \approx 2.05$, is very different from the mean g-value in the LS state of Fe^{2+}, $g \approx 2.15$, which suggests the existence of a structural transformation related to the HS↔LS transition.

3.2 Mn^{2+}-doped Compounds

Mn^{2+} ($S = 5/2, I = 5/2$) was used for the first time as a paramagnetic probe of spin transitions in $[Fe(phen)_2(NCS)_2]$ and $[Fe(2\text{-pic})_3]Cl_2\cdot C_2H_5OH$ powders ($T_c = 182$ K and 121 K, respectively) [25]. In $[Fe(phen)_2(NCS)_2]$, the EPR spectra at all temperatures are broad and hyperfine splitting can hardly be resolved. In $[Fe(2\text{-pic})_3]Cl_2\cdot C_2H_5OH$ below the transition temperature, a typical powder spectrum is observed, with the zero-field splitting less than the Zeeman splitting and linewidths narrow enough to give well-resolved hyperfine patterns. In the course of the spin transition no change occurs in the zero-field splitting parameters D and E, other than a usual gradual temperature dependence typical of systems which do not undergo such a transition. Thus, the authors conclude that any structural changes accompanying the spin transition are short-range ones (inside the Fe^{2+} complex) and not long-range ones (involving the whole lattice). On the other hand, the spin transitions manifest themselves by considerable modifications of the EPR spectra shapes, interpreted in [25] as an overall EPR line broadening in the temperature range of the spin transition. This broadening was ascribed to a dramatic increase of the lifetime of individual M_S states of HS Fe^{2+} ions, which produces magnetic fields resulting in a dipolar broadening of the Mn^{2+} EPR spectra. Above the transition temperature the lifetime is shortened and these magnetic fields are averaged, so that the EPR lines are narrowed again.

However, in the subsequent EPR study of Mn^{2+}-doped single crystals of $[Fe(2\text{-}pic)_3]Cl_2{\cdot}C_2H_5OH$ [26], the latter interpretation has been disapproved by the finding that the EPR line broadening in the spin transition range was increasing with applied magnetic field and thus could *not* be due to magnetic dipolar mechanism. The observed transformations of the EPR spectra in the course of the spin transition have been re-interpreted as being due to a co-existence in the host matrix of at least three domains, presumably: (i) all HS, (ii) half HS and half LS, (iii) all LS, each of these domains characterised by a particular set of zero-field splitting parameters of Mn^{2+}.

In *one* of the two crystalline forms of $[Fe(C_6H_8N_2S_2)_2(NCS)_2]$, which undergoes a spin transition with hysteresis between 171.2 K and 180.9 K, a single-crystal EPR study of Mn^{2+} [27] reveals well-resolved features, typical of zero-field splitting comparable to the Zeeman splitting. Very sharp hyperfine components are observed in the diamagnetic phase, while in the paramagnetic phase they are much broader and tend to broaden as the temperature is lowered. This broadening is tentatively ascribed to magnetic fields generated by HS Fe^{2+} ions or, perhaps, to strain effects causing a distribution throughout the crystal of zero-field splitting parameters of Mn^{2+}. The small linewidths in the diamagnetic phase are also observed in the case of co-existence of LS and HS domains. The Mn^{2+} EPR linewidth in the LS Fe^{2+} domains formed in the single crystals in the course of the spin transition indicates that these domains are of much larger sizes than 50 to 55 molecules, as proposed in earlier studies on powders. This distinction can also account for the considerable slowing down of the spin transition rate in the large single crystals with respect to the powder samples, detected by EPR of Mn^{2+}. The Mn^{2+} zero-field splitting parameters observed for the LS and HS phases in the spin transition range within experimental error are the same, indicating that the change in crystal geometry must be very minor. A pronounced difference is observed, however, at liquid nitrogen temperature: in the LS state of Fe^{2+} $D = 495 \cdot 10^{-4}$ cm^{-1}, while in the HS state of Fe^{2+} $D = 380 \cdot 10^{-4}$ cm^{-1} (a complete disappearance of the HS state at this temperature takes about 1 hour).

Mn^{2+} has been used as a dopant in single crystals of $[Fe(PTZ)_6](BF_4)_2$ [23] (see the previous section for the results concerning Cu^{2+}). The HS↔LS transition is indicated by a sudden reduction of Mn^{2+} EPR linewidth as well as by a more rapid decrease of the D-value with decreasing temperature. The existence of two distinct diamagnetic phases, the quickly cooled phase I and the slowly cooled phase II, *vide ultra*, was also observed in the EPR spectra of Mn^{2+}. Moreover, the same orientation of the zero-field splitting tensor is observed in the paramagnetic phase and in the diamagnetic phase I and its parameters in the paramagnetic phase, $D = 184.8 \cdot 10^{-4}$ cm^{-1}, $E = 0$, $F - a = -18.8 \cdot 10^{-4}$ cm^{-1}, and in the diamagnetic phase I, $D = 153.7 \cdot 10^{-4}$ cm^{-1}, $E = 0$, $F - a = -24.2 \cdot 10^{-4}$ cm^{-1}, ınfer one and the same local symmetry, that of a trigonally distorted octahedron. As far as the diamagnetic phase II is concerned, only the $D = 240 \cdot 10^{-4}$ cm^{-1} value could be determined. Nevertheless, the fact that the orientation of the zero-field splitting tensor in this phase is quite different from that in the paramagnetic phase, indicates a symmetry change. On the basis of these results, the authors argue that the HS↔LS transition in one and the same system may occur both with a crystalline phase transformation (if the diamagnetic phase II is formed), or without such a transformation (if the diamagnetic phase I is formed) and that, generally, the crystalline phase transformation is driven by the spin transition rather than being the cause of the spin transition.

In $[Fe(NCS)_2(btr)_2]\cdot H_2O$ and $[Fe(NCSe)_2(btr)_2]\cdot H_2O$ compounds [22] (see the previous section for the results concerning Cu^{2+}), the resolution of Mn^{2+} EPR spectra is low. Relatively high D-value have been found in powder $[Fe(NCSe)_2(btr)_2]\cdot H_2O$ below the transition temperature: $D = 0.12$ cm^{-1}. In single crystals of $[Fe(NCS)_2(btr)_2]\cdot H_2O$ above and below the transition temperature $D = 0.066$ cm^{-1}, $a = 0$ and $D = 0.11$ cm^{-1}, $a = -0.0018$ cm^{-1}, respectively. These data indicate the existence of a structural transformation accompanying the spin transition.

3.3 Non-doped Compounds

Finally, we indicate a possibility of indirect EPR studies of the spin transition of Fe^{2+} ions which has been overlooked in some previous works [20, 21]. Ferrous compounds often contain a certain proportion of Fe^{3+} ions which in this case always are in the HS state. These ions serve as good spin probes, since, even at elevated temperatures, Fe^{3+}, an S-ion, provides EPR spectra exhibiting a well-resolved zero-field splitting. By analysing the latter, one can get information on the sites of Fe^{3+} ions which can be used with a certain caution to characterise the structure of environment of Fe^{2+} ions, as well.

In the $[Fe(trz)(Htrz)_2](BF_4)$ compound about 0.01 of the total number of iron atoms, occur in the samples in Fe^{3+} state [24]. In the LS state of Fe^{2+}, the EPR spectrum consists of a relatively narrow line at $g_{ef} = 4.3$, accompanied by a plateau extending to $g_{ef} = 9.7$, and of a broad absorption at $g_{ef} = 2.0$, which is quite typical of Fe^{3+} ions in a relatively strong low-symmetry crystal field. This spectrum changes in a drastic manner in the course of the spin transition experienced by Fe^{2+} ions, reducing itself to a very broad absorption at $g_{ef} = 2.0$. It is interesting to note that the evolution of this Fe^{3+} line in the course of the HS↔LS transition of Fe^{2+} ions is exactly the opposite of that observed for the HS↔LS transition experienced by Fe^{3+} (and Ni^{3+}) ions themselves. The mean D value is estimated as 0.3 cm^{-1} in the diamagnetic phase of Fe^{2+} and lower than 0.15 cm^{-1} in the paramagnetic phase of Fe^{2+}. One can conclude that the LS to HS transition of Fe^{2+} ions is accompanied by a weakening of low-symmetry crystal field distortion at the Fe^{3+} sites. This finding is corroborated by EXAFS studies indicating a 0.18 Å Fe^{2+}–N bond lengthening in the LS to HS transition. One can see that similar structural transformations are observed in the environment of ions undergoing and not undergoing the spin transition. This suggests that the structural changes accompanying the spin transition in this case are not local but involve the whole crystalline structure.

4 Conclusion

EPR is certainly too 'heavy' a technique to be employed just to detect a spin transition. Its application can only be justified if it is able to detect structural changes accompanying this transition and thus probe into the mechanism responsible for it.
In the case of octahedral Fe^{3+}-, Ni^{3+}- and Co^{2+}-compounds, variable temperature EPR study *directly* evidences the spin transition by modifications of magnetic interactions. For octahedral ferrous compounds doped with Mn^{2+} or Cu^{2+}, variable temperature EPR study *indirectly* evidences the spin transition and simultaneously provides data on subtle structural transformations which occur in the course of the spin transition.

At present, only a few indirect EPR studies of spin transitions exist, but this technique can certainly be applied to a number of other systems, in particular, involving other spin transition ions than Fe^{2+}.
As concerns the choice of a paramagnetic probe for EPR studies of the spin transitions, we would like to emphasize that ions with zero-field splitting, such as Mn^{2+}, Fe^{3+}, etc. should certainly be preferred, the corresponding spin Hamiltonian parameters being very sensitive to the structure and symmetry of environment.

References

1. M.S. Haddad, M.W. Lynch, W.D. Federer and D.N. Hendrickson, Spin-crossover ferric complexes: curiosities observed for unperturbed solids, *Inorg. Chem.* **20** (1981) 123-131; Spin-crossover ferric complexes: unusual effects of grinding and doping solids, *Ibid* **20** (1981) 131-139.
2. H. Ohshio, Y. Maeda and Y. Takashima, Rapid electronic relaxation phenomenon in a $^2T \leftrightarrow {}^6A$ spin-equilibrium system, *Inorg. Chem.* **22** (1983) 2684-2689.
3. W.D. Federer and D.N. Hendrickson, Dynamics of spin-state interconversion and cooperativity for ferric spin-crossover complexes in the solid state. 1. The unperturbed fast spin-flipping N_4O_2 complex [Fe(SalAPA)$_2$]ClO$_4$, *Inorg. Chem.* **23** (1984) 3861-3870; 2. Perturbation of the fast spin-flipping N_4O_2 complex [Fe(SalAPA)$_2$]ClO$_4$, *ibid.* 3870-3877.
4. Y. Maeda, N. Tsutsumi and Y. Takashima, Examples of fast and slow electronic relaxation between ^2T and ^6A, *Inorg. Chem.* **23** (1984) 2440-2447.
5. M.D. Timken, D.N. Hendrickson and E. Sinn, Dynamics of spin-state interconversion and cooperativity for ferric spin-crossover complexes in the solid state. 3. Bis[*N*-(2-(benzylamino)ethyl)salicylaldiminato]iron(III) complexes, *Inorg. Chem.* **24** (1985) 3947-3955.

6. M.D. Timken, A.M. Abdel-Mawgoud and D.N. Hendrickson, Dynamics of spin-state interconversion and cooperativity for ferric spin-crossover complexes in the solid state. 6. Magnetic and spectroscopic characterizations of $[Fe(3\text{-}OEt\text{-}SalAPA)_2]X$ ($X=ClO_4^-$,BPh_4^-), *Inorg. Chem.* **25** (1986) 160-164.
7. M.D. Timken, C.E. Strouse, S.M. Soltis, S.A. Daverio, D.N. Hendrickson, A.M. Abdel-Mawgoud and S.R. Wilson, Dynamics of spin-state interconversion and cooperativity for ferric spin-crossover complexes in the solid state. 5. Variable-temperature spectroscopic, magnetic, and single-crystal X-ray structural characterizations of the spin-state and order-disorder transformations of a Schiff base complexe, *J. Am. Chem. Soc.* **108** (1986) 395-402.
8. B.J. Kennedy, A.C. McGrath, K.S. Murray, B.W. Skelton and A.H. White, Variable temperature magnetic, spectral, and X-ray crystallographic studies of "spin-crossover" iron(III) Schiff-base–Lewis-base adducts. Influence of noncoordinated anions on spin-state interconversion dynamics in $[Fe(salen)(imd)_2]Y$ species ($Y=ClO_4^-$, BF_4^-, PF_6^-; imd=imidazole), *Inorg. Chem.* **26** (1987) 483-495.
9. M. Mohan, N.S. Gupta, L. Chandra, N.K. Jha and R.S. Prasad, On the nature of the high-spin ($^6A_1$) $\leftrightarrow$ low-spin ($^2T_2$) transition in $[Fe(3\text{-}CH_3OSPH)_2]Y$ complexes, *Inorg. Chim. Act.* **141** (1988) 185-192.
10. J.P. Costes, F. Dahan and J.P. Laurent, Structural characterization and magnetic study (EPR and static susceptibility measurements) of a novel ferric spin-crossover complex: Bis(7-amino-4-methyl-5-aza-3-hepten-2-onato(1^-)) iron(III) Tetralphenylborate, *Inorg. Chem.* **29** (1990) 2448-2452.
11. A.J. Conti, R.K. Chadha, K.M. Sena, A.L. Rheingold and D.N. Hendrickson, Dynamics and phase transitions in spin-crossover complexes: X-ray structures and basic crossover phenomena in the solvate series $[Fe(3\text{-}OEt\text{-}SalAPA)_2](ClO_4).S$, *Inorg. Chem.* **32** (1993) 2670-2680.
12. C.T. Brewer, G. Brewer, G.B. Jameson, P. Kamaras, L. May and M. Rapta, Structure and magnetism of electronically distorted iron(III) Schiff-base complexes, *J. Chem. Soc. Dalton Trans.*, **1** (1995) 37-43.
13. C.T. Brewer, J. Jasinski, W. Mahany, L. May and S. Prytkov, Spin crossover complexes of iron(III) with an N_4 tetradentate Schiff base ligand, *Inorg. Chim. Acta*, **232** (1995) 183-187.
14. G. Demazeau, J.-M. Marty, B. Buffat, J-M. Dance, M. Pouchard, P. Dordor et B. Chevalier, Sur la stabilisation dans un réseau oxygéné du nickel(+III) à spin fort en champ octaédrique, *Mat. Res. Bull.* **17** (1982) 37-45.
15. J. Zarembowitch and O. Kahn, Magnetic properties of some spin-crossover, high-spin, and low-spin cobalt(II) complexes with Schiff bases derived from 3-formylsalicylic acid, *Inorg. Chem.* **23** (1984) 589-593.
16. J. Zarembowitch, R. Claude and P. Thuéry, a novel example of spin transition in a six-coordinate cobalt(II) complex: magnetic properties of bisaqua-N,N'-O-phenylenebis(3-carboxysalicylaldiminato)cobalt(II), *Nouv. J. Chim.* **9** (1985) 467-471.
17. J. Zarembowitch, R. Claude and O. Kahn, High-spin$\leftrightarrow$low-spin transition in solid $Co(H_2fsa_2en)(H_2O)_2$ [H_4fsa_2en = *N,N'*-Ethylenebis(3-carboxysalicylaldimine)]: a magnetic investigation, *Inorg. Chem.* **24** (1985) 1576-1580.

18. P. Thuéry and J. Zarembowitch, Spin state of cobalt(II) in five- and six-coordinate Lewis base adducts of (*N,N'*-Ethylenebis(3-carboxysalicylaldiminato))cobalt(II). New spin-crossover complexes, *Inorg. Chem.* **25** (1986) 2001-2008.
19. J. Faus, M. Julve, F. Lloret, J.A. Real and J. Sletten, (dimethylviolurato)bis (phenanthroline)cobalt(II), a novel spin-crossover octahedral Co(II) complex. Synthesis, crystal structure, and magnetic properties of $[Co(dmvi)(phen)_2]ClO_4.3H_2O$, *Inorg. Chem.* **33** (1994) 5535-5540.
20. W. Vreugdenhil, J.G. Haasnoot, O. Kahn, P. Thuéry and J. Reedijk, A copper(II) dope as a detector for the high-spin↔low-spin transition in the two-dimensional compound [trans-bis(thiocyanato)bis(4,4'-bi-1,2,4-triazole)iron] hydrate, *J. Am. Chem. Soc.* **109** (1987) 5272-5273.
21. W. Vreugdenhil, J.H. Van Diemen, R.A.G. De Graaff, J.G. Haasnoot, J. Reedijk, A.M. Van Der Kraan, O. Kahn and J. Zarembowitch, High-spin↔Low-spin transition in $[Fe(NCS)_2(4,4'\text{-bis-1,2,4-triazole})_2](H_2O)$, X-ray crystal structure and magnetic, Mossbauer and EPR properties, *Polyhedron* **9** (1990) 2971-2979.
22. A. Ozarowski, Y. Shunzhong, B.R. McGarvey, A. Mislankar and J.E. Drake, EPR and NMR study of the spin-crossover transition in $Fe(4,4'\text{-bi-1,2,4-triazole})_2(SCN)_2.H_2O$ and $Fe(4,4'\text{-bi-1,2,4-triazole})_2(SeCN)_2.H_2O$. X-ray structure determination of $Fe(4,4'\text{-bi-1,2,4-triazole})_2(SeCN)_2.H_2O$, *Inorg. Chem.* **30** (1991) 3167-3174.
23. A. Ozarowki and B.R. McGarvey, EPR study of manganese(II) and copper(II) in single crystals of the spin-crossover complex $Fe(PTZ)_6(BF_4)_2$, *Inorg. Chem.* **28** (1989) 2262-2266.
24. C. Cantin, J. Kliava, Y. Servant, L. Sommier and O. Kahn, Fe^{2+} spin transition in $[Fe(trz)(Htrz)_2](BF_4)$ as evidenced by a variable temperature EPR study, *Applied Magn. Reson.* **12** (1997) 81-86; Variable temperature EPR study of Fe^{2+} spin transition in Cu^{2+}-doped $[Fe(trz)(Htrz)_2](BF_4)$, *ibid.* **12** (1997) 87-93.
25. P.S. Rao, A. Reuveni, B.R. McGarvey, P. Ganguli and P. Gütlich, Electron spin resonance and Mössbauer study of the high-spin–low-spin transition in $[Fe(phen)_2(NCS)_2]$ and $[Fe(pic)_3]Cl_2.EtOH$, *Inorg. Chem.* **20** (1981) 204-207.
26. P.E. Doan and B.R. McGarvey, EPR study of manganese(II) in single crystals of the spin-crossover complex $Fe(2\text{-pic})_3]Cl_2.C_2H_5OH$. Evidence for domains in transition, *Inorg. Chem.* **29** (1990) 874-876.
27. A. Ozarowki, B.R. McGarvey, A.B. Sarkar and J.E. Drake, EPR study of manganese(II) in two crystalline forms of $Fe(C_6H_8N_2S_2)_2(NCS)_2$ and the high-spin–low-spin transition that occurs in only one form. X-ray structure determination of both forms, *Inorg. Chem.* **27** (1988) 628-635.

Submillimeter EPR of Non-Kramers Ions

V. F. Tarasov[1], *G. S. Shakurov*[1], *B. Z. Malkin*[2], *V. A. Ulanov*[1] *and M. M. Zaripov*[1]

[1] Kazan Physical-Technical Institute of the Russian Academy of Sciences, Kazan, 420029, Russia
[2] Kazan State University, Kazan, 420008, Russia

Abstract. The paper concerns some peculiarities of EPR spectroscopy of non-Kramers transition ions in solids. Wide band EPR spectrometer destined for this purpose and operating in the frequency band 65 - 535 GHz is described. Results of investigations of Cr^{2+} in Mg_2SiO_4, CdF_2, CaF_2, SrF_2 and BaF_2 as well as Ho^{3+} in KY_3F_{10} are presented.

1 Introduction

EPR is well known as a very informative physical method for study the crystal structure and chemical bonds in paramagnets. It is really true for the Kramers ions with an odd number of electrons. In the case of non-Kramers ions with integer spin energy levels can be split into singlets if the ion is located in a low symmetry crystal field. In this case information about crystal field parameters and g-factor can be obtained by studying of magnetic and/or electric dipole transitions between these singlet states. As far as zero field splitting (ZFS) is very much different for various ions and crystal environment, the EPR spectrometer destined for this purpose should operate in a wide frequency band. It is not simple to construct such spectrometer. For this reason a number of experimental works, devoted to an investigation of the non-Kramers ions in solids, is much less than it is for the Kramers ions. At that time a theory of the non-Kramers ions in solids is in progress [1-3] and new experimental data are necessary to prove it.

2 Spectrometer

The main problem which should be resolved under the wide band spectrometer creation is a choice of microwave generator. In several EPR spectrometers [4,5] optically pumped far infrared gas lasers have been used. These spectrometers can operate over wide frequency band with wavelengths from a few millimeters to dozens of micrometers. The frequency of generation is determined by active medium and pumping frequency and can not be easily changed. The second possible generators are backward wave oscillators (BWO). They are relatively simple vacuum tubes which can operate at frequencies up to 1 THz. The frequency can be changed easily during an experiment. These tubes also have been successfully used in the EPR spectrometers [6-8].

We constructed wide band EPR spectrometer operating at the frequencies between 65 and 535 GHz on the basis of BWO. The whole frequency band is covered by a set of 5 tubes with the individual span ratio about 1.4. Output power of each BWO depends strongly on frequency and varies between 2.8 and 55 mW. A quasi optic guide is used to transfer microwave power from the exit slit of the tube onto the sample. We don't use a cavity in the spectrometer, because we change the microwave frequency during the experiment. An n-type indium antimonide crystal at 4.2 K is used to detect the microwave power passed through the sample. Magnetic field is produced by a conventional electromagnet. The field can be varied from –0.05 to 1 T. Therefore resonance transitions can be observed exactly at zero magnetic field. A polarization attenuator with a polarization plane rotator may be installed in the quasi optic guide. It enables us to align direction of microwave field (B_1) along or transverse the direction of constant magnetic field (B_0). It is important, because this way we have a possibility to separate resonance transitions with different selection rules.

A resolution of the spectrometer depends on frequency stability and radiation band width. We have relative frequency stability near $5 \cdot 10^{-5}$ per 10 minutes and band width less than $1.5 \cdot 10^{-5}$. The frequency is measured by a Fabry-Perot interferometer with an accuracy ±0.5 GHz. The detailed description of the spectrometer is published in [9].

3 Results

We have used this spectrometer to investigate different transition metal and rare-earth non-Kramers ions in solids. All the objects can be separated in three groups.

3.1 Transition Metal Ions in Low Symmetry Crystal Field

Excellent example of this situation presents chromium doped synthetic forsterite (Mg_2SiO_4:Cr). It is promising material for near-IR tuneable lasers and Q-switches for solid femtoseconds lasers. In this connection considerable attention was attracted to investigation of paramagnetic centers formed by the impurity Cr ions. It was known by EPR that Cr^{3+} substitutes Mg^{2+} in the two different crystallographic sites with pseudo-octahedral coordination[10]. Several works were devoted to investigation of lasing Cr^{4+} substituting the Si^{4+} in the pseudo-tetrahedral site [11-13]. As far as it concerns Cr^{2+} there was the only work (as we know), where the Cr^{2+} centers were presumably detected by EPR [14]. The authors of this work observed two different non-Kramers Cr centers in forsterite. They didn't establish certainly the origin of these centers and theoretical treatment was made for both possible cases, if these centers were due to the Cr^{4+} with S = 1 or Cr^{2+} with S = 2.

Our study of Cr-doped forsterite in the submillimeter band revealed three Cr non-Kramers centers. Investigation in the wide frequency band enables us to identify reliably the origin of these centers. The first center was found to be analogous to the Cr^{4+} in the tetrahedral position, described earlier [11]. For each of the other two centers seven resonance transitions were observed at frequencies between 70 and 350 GHz. The experimental results are presented by circles in Fig.1. Dependencies of resonance frequencies on magnetic field are described very well by the classic scheme of energy levels for the S = 2 ion in the orthorhombic field [16]. Values of the spin Hamiltonian parameter E were determined directly as 1/6 of the ZFS between |+1> and |−1> levels, D values were fitted for best description of the experimental results. Calculated resonance frequencies versus magnetic field are presented in Fig.1 by the solid lines. Obtained values of D end E together with the data from [14] are presented in Table 1. We observed 4 magnetically non-equivalent positions for the first Cr^{2+} center and 2 positions for the second center. Accordingly we attribute these centers to the Cr^{2+} in the M1 and M2 octahedral positions.

Table 1. Comparison of obtained D and E values (in cm^{-1}) for the Cr^{2+} centers in forsterite with the literature data.

	D	D [14].	E	E [14]
M1 center	−2.49	−2.1	0.15	−0.10
M2 center	2.23	2.1	0.4	0.4

We observed and studied also Ni^{2+} non-Kramers ions in forsterite [17] and gave a new description of this center, which is different from the previously known [18].

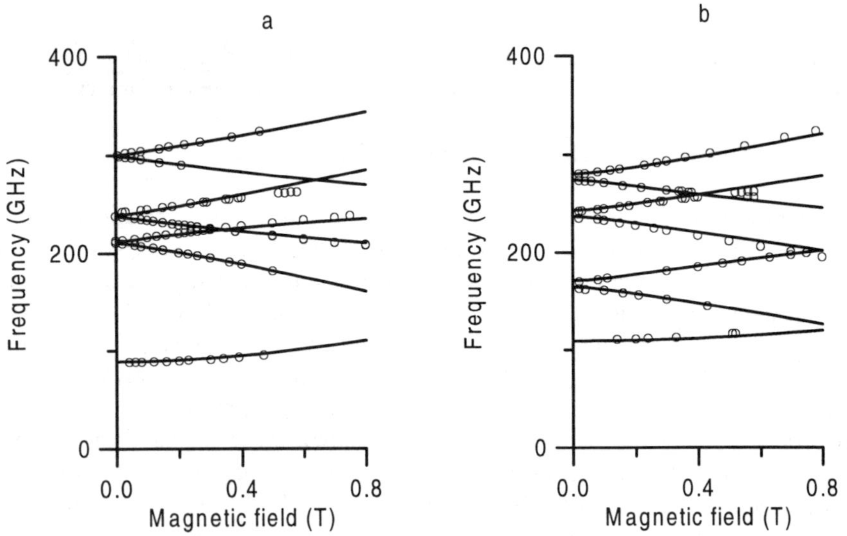

Fig. 1. Cr^{2+} resonance frequencies versus magnetic field at $B_0 \parallel b$ crystal axis: M1 center (a) and M2 center (b).

3.2 Jahn-Teller Transition Metal Ions in Cubic Hosts

Cubic crystals containing non-Kramers transition metal ions with a degenerate orbital ground state are the second class of materials which can be studied effectively by means of the wide band EPR spectroscopy. It is well known that in the cubic crystal field the ground energy level of transition metal ion can be doublet or triplet [16] and one would think that conventional EPR could be always applied. Really it is not so. In many cases high symmetry of local crystal field in the ion position is reduced due to the Jahn-Teller effect and the orbital ground state again is found to be a singlet. Therefore wide band spectroscopy again becomes more effective than the conventional EPR.

It is known [19] that character of the crystal field distortion depends on a mode of lattice vibrations that most effectively interacts with the impurity ion. Predominant interaction with the e_g - type vibrations forms centers with tetragonal symmetry. Trigonal centers appear as a result of the more efficient interaction with the t_{2g} - vibrations. However, under certain conditions interactions with both vibration modes are found to be important. In this case the centers with orthorhombic symmetry should be formed. It was established earlier [20,21] that for the Cu^{2+} and Ag^{2+} ions in crystals of fluorite structure the symmetry of the Jahn-Teller distortion depends on relation between the size of impurity ion and parameters of crystal lattice. In this connection it was interesting to check this thesis for the non-Kramers Cr^{2+} ion.

Before our work it was known [22] that Cr^{2+} ions in CdF_2 formed the orthorhombic centers. We investigated symmetry of Cr^{2+} impurity centers in CdF_2, CaF_2, SrF_2 and BaF_2. Rotational diagrams as well as resonance frequencies versus magnetic field were obtained in the submillimeter band. When it was possible orientation diagrams at X and Q bands were obtained and analysed too.

As a result the orthorhombic centers were found in CdF_2, CaF_2, and SrF_2 and tetragonal center in BaF_2 [23,24]. For all samples parameters of the spin Hamiltonian were found by exact diagonalization of the corresponding matrices. The results are presented in Table 2.

Table 2. Parameters of the spin Hamiltonian (in cm^{-1}) for Cr^{2+} in crystals of fluorite structure.

	b_2^0	b_2^2	b_4^0	b_4^2	b_4^4
CdF_2	918.67	30.17	1.33	0	1.47
CaF_2	949.33	59.67	0.17	0	1.37
SrF_2	930	28.33	0.40	0	1.53
BaF_2	703.33		0		0

We should note that parameters for CdF_2 are much different from the results of [22]. As far as they are obtained from direct measurements of the ZFS we suppose that there is a mistake in [22].

Analysis of these date confirmed the connection between the size of impurity ion and the character of Jahn-Teller distortion.

3.3 Non-Kramers Rare-Earth Ions in Low Symmetry Crystal Field

The ground state of the rare-earth non-Kramers ion also can be a singlet if the crystal field has low symmetry. Often ZFS between the ground singlet and the first excited level is too large to be bridged by the X - or Q - band spectrometers.

This situation is realized in cubic double fluorides KY_3F_{10} doped with Ho^{3+} ions. The Ho^{3+} replace Y^{3+} in three magnetically non-equivalent positions, which differ only by the orientation of the fourfold local symmetry axes along the crystallographic directions [100], [101], and [110]. The nearest-neighbour surrounding of the Ho^{3+} presents a square antiprism with slightly different dimensions of the bases. In this tetragonal crystal field the ground and the first excited states of the Ho^{3+} multiplet $^5I_8$ are A_1 and A_2 singlets:

$$A_1\rangle = \alpha\,(|{+8}\rangle + |{-8}\rangle) + \beta(|{+4}\rangle + |{-4}\rangle) + \gamma|0\rangle$$
$$|A_2\rangle = \alpha(|{+8}\rangle - |{-8}\rangle) + \beta(|{+4}\rangle - |{-4}\rangle),$$

where $\alpha \ll \beta$, $\beta \approx \sqrt{1/2}$. These states don't have magnetic moments at zero magnetic field. Longitudinal component of magnetic field mixes A_1 and A_2 states and magnetic moment appears parallel to the symmetry axis of the crystal field. This results in the two consequences:

1. Probability of the magnetic dipole transitions between these levels should be maximum when B_1 is parallel to the z-axis of the paramagnetic center.
2. Effectiveness of magnetic field modulation also increases when B_0 is directed along the z-axis. Consequently optimal conditions for observing magnetic dipole transitions between the A_1 and A_2 states are realized when B_0 and B_1 both are parallel to the C_4 crystal axis.

The Ho nucleus has spin I = 7/2. Hyperfine interaction leads to the formation of mixed electronic - nuclear states so that additional hyperfine structure of the Stark levels appears due to the second order terms even in the absence of magnetic field. Numerical calculations produced in the full basis of 136 electronic - nuclear states of the ground multiplet $^5I_8$ showed [25] that for an arbitrary orientation of the external magnetic field the probabilities of all $(2I+1)^2 = 64$ transitions between the hyperfine components of the two electronic singlets are non-zero. These probabilities depend strongly on magnetic field orientation, resulting in consequent variation of observed spectra.

Figure 2(a) presents results of these calculations for collinear directions of B_0 and B_1. When magnetic field is parallel to the C_4 crystallographic axis only 8 hyperfine transitions with $\Delta m_I = 0$ are allowed. Therefore spectrum consists from 8 lines, which are due to the one of the three magnetically non-equivalent centers. For the other two centers $B_1 \perp C_4$ and transition probabilities equal to zero. Under the magnetic field rotation the hyperfine transitions with $\Delta m_I = \pm 1$ become allowed.

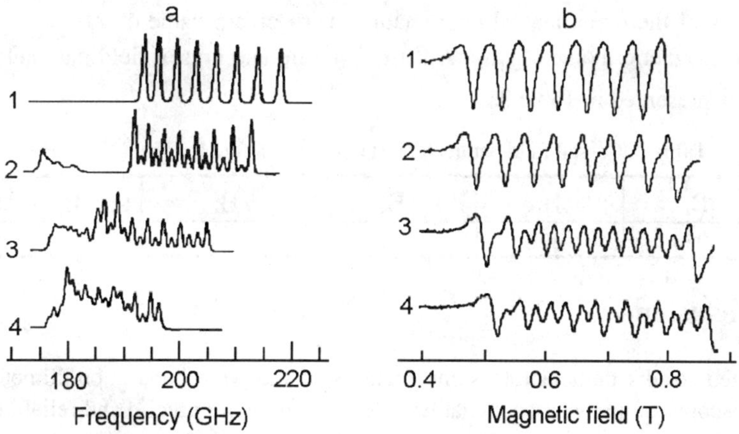

Fig. 2. Spectra of Ho^{3+} for different orientations of collinear B_0 and B_1 relative to the C_4 crystal axes; a) - theory at B_0 = 0.8 T: 1 - 0°, 2 - 20°, 3 - 40°, 4 - 60°; b) experiment at frequency of 194.5 GHz: 1 - 0°, 2 - 10°, 3 - 20°, 4 - 25°.

As a result the first spectrum complicates and additional resonance lines appear due to the other two centers. Further rotation complicates the common spectrum dramatically.

The experimental spectra, measured at different angles between B_0 and C_4 crystal axis [25] are presented in Fig.2(b). They demonstrate good agreement with the theoretical predictions and confirm once more validity of the original exchange charge model for the crystal field calculations [26]. Dependence of resonance frequencies on B_0 was also measured experimentally at $B_0 \parallel C_4$. Results are presented in Fig. 3.

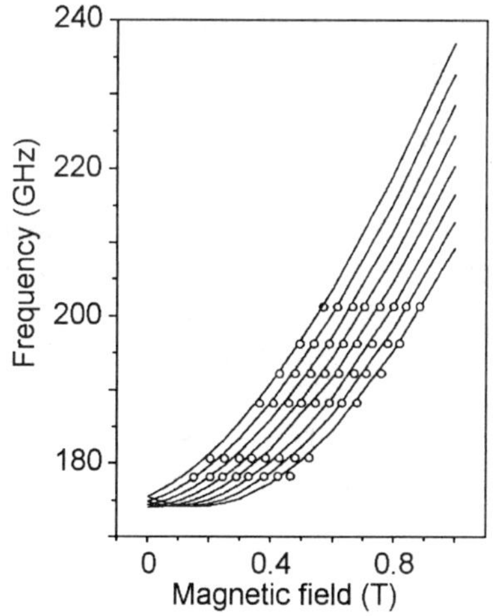

Fig. 3. Resonance frequencies versus magnetic field for Ho^{3+} in KY_3F_{10}:Ho, circles - experiment, lines - theory.

Analysis of the experimental data enables us to obtain value of ZFS = 5.8 cm^{-1}, induced magnetic moment $g_j\mu_B <A_1|J_z|A_2> = 5\ \mu_B$ and crystal field parameters B_p^q. They are presented in Table 3.

Table 3. Crystal field parameters B_p^q (in cm^{-1}) for Ho^{3+} in KY_3F_{10}:Ho.

B_2^0	B_4^0	B_6^0	B_4^4	B_6^4
−276	−172	42	394	−28.5

4 Conclusion

Presented results demonstrate some features of the wide band submillimeter EPR spectroscopy. In many cases it enables one to obtain more precise and reliable results relative to conventional EPR spectroscopy.

Acknowledgements

The work was supported by the RFBR grants No 95-02-05838 and 96-02-18263 and the by the Academy of Sciences of Tatarstan.

References

1. C. Rudowich, On the EPR of $3d^4$ and $3d^6$ Ions in Magnetic Fields, *Physica* **B155** (1989) 336-339.
2. C. Rudowicz, Y-Y. Zhou, W-L. Yu, Crystal Field Analysis for $3d^4$ and $3d^6$ Ions with an Orbital Singlet Ground State at Orthorhombic and Tetragonal Symmetry Sites, *J. Chem. Solids,* **53** (1992) 1227-1236.
3. C.-Y. Jiang, M.-L. Du, Y-Y. Zhou, Contributions to Zero-Field Splittings from Spin Triplets of $3d^4$ and $3d^6$ Ions in Tetragonal Symmetry, *Phys. Rev.*, **B50** (1994) 949-954.
4. P. Janssen, M. Mahy, W. P. Wolf, EPR in Dysprosium Aluminium Garnet at Far-Infrared Frequencies, *Phys. Rev.*, **B37** (1988) 4851-4863.
5. F. Muller, M. A. Hopkins, N. Coron, M. Grynberg, L. C. Brunel, G. Martinez, A High Magnetic Field EPR Spectrometer, *Rev. Scient. Instrum.* **60**, (1989) 3681-3684.
6. E. A. Vinogradov, N. A. Irisova, T. S. Mandelstam, T. A. Shmaonov, Wide Band Submillimeter Radiospectrometer for Studying of Microwave Absorption in Solids, *Prib. Techn. Exper. (Rus)*, No 5 (1967) 192-195.
7. J. Magarino, J. Tuchendler, J. P. D'Haenens, A. Linz, Submillimeter Resonance Spectroscopy of Ho^{3+} in Lithium Yttrium Fluoride, *Phys. Rev.* **B13**, (1976) 2805-2808.
8. K. N. Kocharyan, A. A. Mirzakhanyan, Sensitive EPR spectrometer of millimeter band, *Izvestiya AN Armyanskoy SSR, fisika, (Rus)*, **11**, (1976) 484-488.
9. V. F. Tarasov, G. S. Shakurov, Submillimetre EPR spectrometer, *Appl. Magn. Res.* **2**, (1991) 571-576.

10. H. Rager, Electron Spin Resonance of Transient Chromium in Forsterite, Mg_2SiO_4, *Phys. Chem. Mineral.*, **1** (1977) 371-378.
11. K.R. Hoffman, J. Casas-Gonsalez, S.M. Jacobsen and W.M.Yen, Electron-Paramagnetic-Resonance and Fluorescence-Line-Narrowing Measurements of the Lasing Center in Cr-doped Forsterite, *Phys. Rev.* **B44**, (1991) 12589-12592.
12. D. E. Budil, D. G. Park, J. M. Burlitch, R. F. Geray, R. Dieckmann, J. H. Freed, 9.6 GHz and 34 GHz Electron Paramagnetic Resonance Studies of Chromium-Doped Forsterite, *J. Chem. Phys.*, **101** (1994) 3538-3548.
13. A. Regev, J. Freed, Site Selective Electron Paramagnetic Resonance Study of Photoexcited Chromium Doped Forsterite, *J. Chem. Phys.*, **103** (1995) 5315-5325.
14. J. Casas-Gonsalez, S. M. Jacobsen, K. R. Hoffman, and W. M. Yen, Electron Paramagnetic Resonance Spectroscopy of Chromium-Doped Forsterite (Mg_2SiO_4), *OSA Proceedings on Advanced Solid-State Lasers*, **10**, (1991) 64-68.
15. V. F. Tarasov, G. S. Shakurov, A. N. Gavrilenko, Submillimeter EPR of Chromium Ions in Synthetic Forsterite, *Phys. Solid State (translation of Fiz. Tverd. Tela)* **37** (1995) 270-272.
16. A. Abragam, B. Bleaney, *Electron Paramagnetic Resonance of Transition Ions,* Clarendon Press, Oxford, 1970.
17. V. F. Tarasov, G. S. Shakurov, EPR Spectroscopy of Transition Non-Kramers Ions in Synthetic Forsterite in Submillimetre Range, *Optics and Spectroscopy* **81** (1996), in press.
18. H. Rager, S. Hosoya, G. Weiser, Electron Paramagnetic Resonance and Polarized Optical Absorption Spectra of Ni^{2+} in Synthetic Forsterite, *Phys. Chem. Mineral.* **15** (1988) 383-389.
19. C. A. Bates, Jahn-Teller Effects in Paramagmetic Crystals, *Phys. Rep.*, **35** (1978) 187-304.
20. M. M. Zaripov, V. A. Ulanov, M. L. Falin, Investigation of Silver Impunity Centers in Fluoride-Type Crystals, *Sov. Phys.- Solid State (translation of Fiz. Tverd. Tela)* **29** (1987) 1264-1266.
21. M. M. Zaripov, V. A. Ulanov, Hyperfine and Quadrupole Interactions and Model of $[CuF_8]^{-6}$ Complexes in CdF_2 and CaF_2 Crystals, *Sov. Phys. - Solid State (translation of Fiz .Tverd. Tela)* **30** (1988) 896-897.
22. R. Jablonski, M. Domanska, B. Krukowska-Fulde, T. Niemyski, Chromium-Doped CdF_2 Crystals by EPR Spectroscopy, *Mat. Res. Bull.* **8** (1973) 749-755.
23. M. M. Zaripov, V. F. Tarasov, V. A. Ulanov, G. S. Shakurov, M. L. Popov, Jahn-Teller Chromium Ions in SrF_2 Crystals: EPR Study in the Range 9.3 - 300 GHz, *Phys. Solid State (translation of Fiz. Tverd. Tela)* **37** (1995) 437-440.
24. M. M. Zaripov, V. F. Tarasov, V. A. Ulanov, G. S. Shakurov, Paramagnetic Complexes of Divalent Chromium in BaF_2 Crystals: Investigation by EPR in the 9.3 - 300 GHz frequency band, *Fiz. Tverd. Tela (St.Peterburg)*, **38** (1996) 452-459.
25. B. Z. Malkin, V. F. Tarasov, G. S. Shakurov, Submillimeter EPR Spectroscopy of van Vleck Paramagnets in Parallel Fields. *JETP Lett.*, **62** (1995) 811-817.
26. B. Z. Malkin, Crystal Field and Electron-Phonon Interaction in Rare-Earth Ionic Paramagnets, in *Spectroscopy of Solids Containing Rare-Earth Ions*, edited by A. A. Kaplyanskii and R. M. Macfarlane, Elsevier, Amsterdam, (1987) Chap. 2, 13-50.

Submillimeter Wave ESR Measurements of Haldane System Y_2BaNiO_5

Shojiro Kimura[1], Hitoshi Ohta[1], Mitsuhiro Motokawa[2], Tetsuya Yokoo[3], and Jun Akimitsu[3]

[1]Department of Physics, Faculty of Science, Kobe University, 1-1 Rokkodai, Nada-ku, Kobe 657, Japan
[2]Institute for Materials Research, Tohoku University, 2-1-1 Katahira, Aoba-ku, Sendai 980-77, Japan
[3]Department of Physics, Aoyama-Gakuin University, 6-16-1 Chitosedai, Setagayaku, Tokyo 157

Abstract. Y_2BaNiO_5 is known as one of the model substance of the Haldane system. We have performed the submillimeter wave ESR measurements of Y_2BaNiO_5 single crystals for the first time in the frequency region from 50 to 383GHz and in the temperature region 4.2K to 70K using pulsed magnetic field up to 16T. Due to the excess oxygens, a few percent of Ni^{2+} ions of Y_2BaNiO_5 which we used are substituted by Ni^{3+} ion which has S=1/2. The ESR signals coming from the S=1/2 degrees of freedom originated by the Ni^{3+} at the end of the finite chain are observed at 4.2K. The absorption lines which may arise from the transitions between the excited triplet states, are also observed in the temperature region above 15K.

1 Introduction

Since Haldane's prediction that there exists the finite energy gap between the ground state and the first exited state of the linear-chain Heisenberg antiferromagnet (LCHA) with integer spin [1], S=1 LCHA has attracted much attention. The first experimental evidence of the existence of the finite energy gap, so-called Haldane gap, was obtained from the investigations of $Ni(C_2H_8N_2)_2NO_3(ClO_4)$ (NENP) [2]. In NENP, the ESR signals coming from the transition within the first triplet exited states have been observed [3]. Moreover, the ESR signals coming form the direct transition between the ground state and the first exited states, which should be forbidden by the selection rule, are observed in NENP [4-6]. Recently, these ESR signals have been explained explained by the theory based on the staggered field which arises from the staggered inclination of the principal axis of the g-tensor of Ni^{2+} ion in NENP [6, 7]. To confirm this theory, the Haldane system which has no staggered field is desirable. Y_2BaNiO_5 is known as one of the model substances for the Haldane system. From neutron scattering measurement the Haldane gap of this material is estimated to be about 100K [8]. In Y_2BaNiO_5, Ni^{2+} ions are located at the center of the octahedron of six oxygens with orthorhombic symmetry. These NiO_6 octahedra form the linear chain along the a-axis. Since the local symmetry of all Ni^{2+} sites are equivalent, no staggered field is expected in this material in contrast to NENP. Another interesting topic in the ESR measurements on the Haldane system is the study of the S=1/2 degrees of freedom at the end of the finite chain. The existence of the S=1/2 degrees of freedom at the end of the finite chain is confirmed for the first time by the ESR measurements on a single crystal of NENP containing Cu^{2+} impurity by Hagiwara *et al.* [2, 9]. They reported that the experimental results are explained by the VBS model which is considered to describe the ground state of the Haldane system [10]. In the casc of Y_2BaNiO_5, it is reported from the magnetic susceptibility that a small percentage of Ni^{2+} ions are substituted by Ni^{3+} ions which has S=1/2, due to the excess oxygens in the system [8]. We have observed the ESR absorption lines coming from the S=1/2 degrees of freedom resulting from the Ni^{3+} ions at the end of the finite chain by submillimeter wave ESR measurements. The purpose of this paper is to present our results of the submillimeter wave ESR measurements of Y_2BaNiO_5 single crystals.

2 Experimental

Submillimeter wave ESR measurements of Y_2BaNiO_5 single crystals have been performed in the frequency region from 50 to 383GHz and in the temperature region from 4.2K to 70K using pulsed magnetic fields up to 16T. We used Gunn oscillators for the frequency region from 50 to 150GHz and backward traveling wave tubes for 220GHz and 370GHz bands. Figure 1 shows the block diagram of our experimental set up. The light is guided by the light pipe to the sample which is located at the center of the magnet, and the transmitted light through the sample is detected by an InSb detector. A pulsed magnetic field is produced by the discharge of the 23.5kJ capacitor bank (V_{max}=3kV). Two types of cryostats are available. Figure 1 shows the one with the magnet immersed in the liquid He space. This type of cryostat can decrease the temperature down to 1.8K by pumping liquid He and increase the

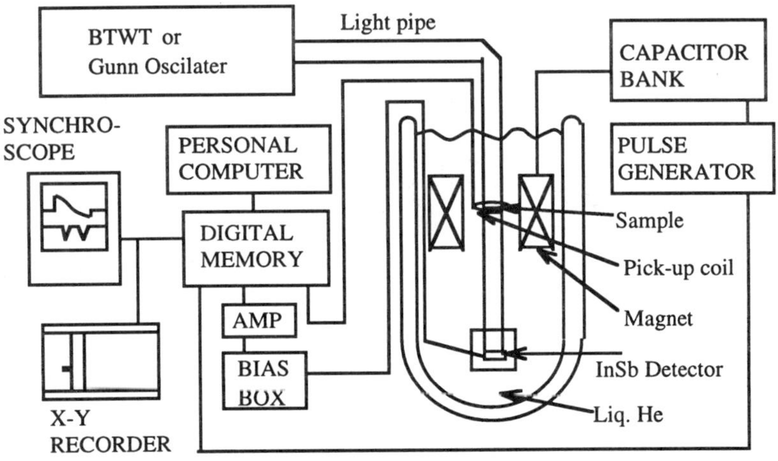

Fig. 1. Block diagram of our experimental set up.

temperature up to 70K by use of a heater. The maximum field is limited to 17T. In another type of cryostat, pulsed magnetic fields of 30T can be produced by the iced magnet [11] and the temperature can be varied from 85K to room temperature. The first cryostat was used in this investigation. Further details of our experimental setups can be found in refs. 12 and 13. Typical size of the single crystals used in our experiments is about 1×2×5mm^2. We used several numbers of single crystals in the measurements.

3 Results and Discussion

Figure 2 shows the temperature dependence of the absorption lines observed at 370.4GHz for H//b. Since the ground state of the Haldane system is the spin singlet state, no ESR absorption line is expected at 4.2K if there is no Ni^{3+} ion. The thermal excitation is not expected at 4.2K because 4.2K is much smaller than the Haldane gap of this material. However, a number of absorption lines are observed at 4.2K as indicated by the arrows in Fig. 2. Even if there is a finite transition probability of the direct transition between the ground state and the first exited states in Y_2BaNiO_5, the direct transition cannot be observed in our measurement because the Haldane gap is much larger than our ESR frequency. Therefore, the absorption lines observed at 4.2K should be considered as the ESR signals coming from the S=1/2 degree of freedom originated by the Ni^{3+} ions at the end of the finite chain. As the temperature is increased, the intensities of the absorption lines, coming from these chain ends decrease while the new absorption lines indicated as I and II in Fig. 2 appear in the temperature region above 15K. The absorption line which shows the same temperature dependence as the absorption line I, is also observed for H//c. The absorption line I may be attributed to the signal coming from the transition between the exited triplet states. The resonance field of the absorption line II shows the shift

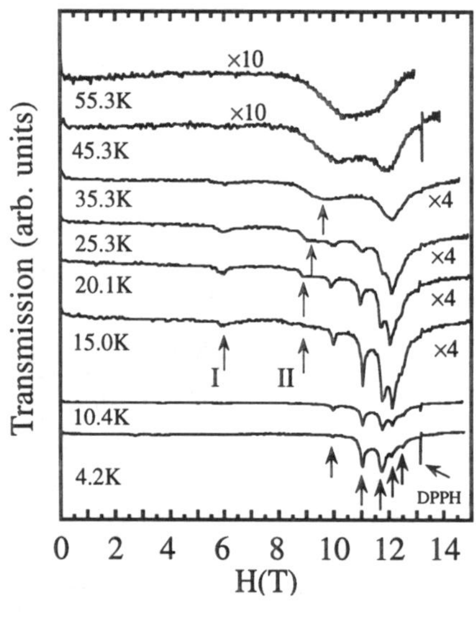

Fig. 2. Temperature dependence of the absorption lines observed at 370.4GHz for H//b.

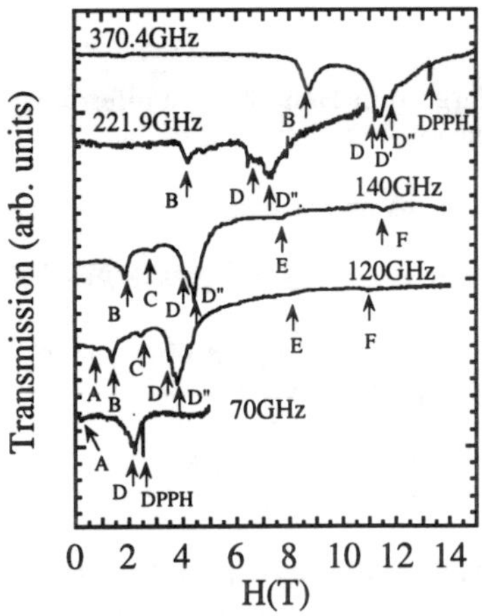

Fig. 3. Frequency dependence of the absorption lines observed for H//a at 4.2K.

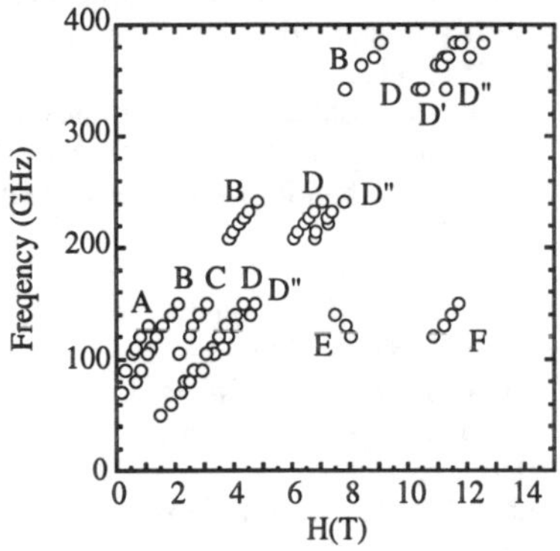

Fig. 4. Frequency field diagram of the absorption lines observed for H//a at 4.2K.

temperature dependence as the absorption line I, is also observed for H//c. The absorption line I may be attributed to the signal coming from the transition between the exited triplet states. The resonance field of the absorption line II shows the shift toward higher field as temperature is increased. The origin of the absorption line II is not clear at this moment because no such absorption line is observed in Cu^{2+} doped NENP which is the Haldane system with the S=1/2 degree of freedom. The frequency dependence of the absorption lines coming from the S=1/2 degree of freedom at the end of the finite chain observed at 4.2K for H//a, is shown in Fig. 3. The intensity of the absorption line B increases as the frequency is increased. The absorption lines C, E and F, which are rather broad compared with other absorption lines, are observed in the frequency region from 105 to 150GHz. Figure 4 shows the frequency-field relations of the absorption lines observed at 4.2K for H//a. The frequency-field relations of the absorption lines D, D' and D" seem to cross the origin. The g-values of the absorption lines D and D" are g=2.39 and g=2.19, respectively. These g-values are anisotropic depending on the field direction. They are g=2.43 and 2.27 for H//b and g=2.16 and 2.11 for H//c, respectively. The extrapolations of the frequency-field relations of the absorption lines A and B have finite frequency about 80GHz at zero magnetic field. The slopes of the frequency-field relations of the broad absorption lines C and F are about twice as large as that of the absorption line D. Observed frequency-field relations cannot be explained by an isolated impurity spin. In Cu^{2+} doped NENP, the absorption lines of which resonance fields are far from g=2 are observed. The absorption lines in Cu^{2+} doped NENP are explained by the model that the Cu^{2+} spin and the S=1/2 degrees of freedom arising at the Ni^{2+} site neighboring the Cu^{2+} site, are coupled by the weak exchange interaction [2, 9]. The analysis of our results using this model, which was used in the case of Cu^{2+} doped NENP, is now in progress. More detailed discussion will be published elsewhere.

Acknowledgments

The authors S K and H O are grateful to Professor T. Nanba of Kobe University for his encouragements and supports. This work was partly supported by a Grant-in- Aid for Scientific Research form the Ministry of Education, Science and Culture.

References

1. F. D. M. Haldane, Nonlinear field theory of large-spin Heisenberg antiferromagnets: Semiclassically quantized solitons of the one-dimensional easy-axis Neel state, *Phys Rev. Lett.* **50** (1983) 1153.

2. K. Katsumata, Experimental studies of one-dimensional quantum spin systems, *J. Magn. Magn. Mater.* **140-141** (1995) 1595-1598.

3. M. Date and K. Kindo, Elementary excitation in the Haldane state, *Phys Rev. Lett.* **65** (1983) 1659.

4. W. Lu, J. Tuchendler, M. von Ortenverg and J. P. Renard, Direct observation of the Haldane gap in NENP by far-Infrared spectroscopy in high magnetic fields, *Phys. Rev. Lett.* **67** (1991) 3716-3919.

5. W. Palme, H. Kriegelstein, B. Luthi, T. M. Brill, T. Yoshida and M. Date, Magnetic resonances in the one-dimensional S=1 spin system NENP, *Int. J. Mod. Phys.* B**7** (1993) 1016-1019.

6. H. Shiba, T. Sakai, B. Luthi, W. Palme and M. Sieling, ESR absorption from the ground state in Haldane systems: Theory and experiment, *J. Magn. Magn. Mater.* **140-141** (1995) 1590-1594.

7. T. Sakai and H. Shiba, Numerical study of a Model for NENP: One-dimensional antiferromagnet in a staggered field, *J. Phys Soc. Jpn.* **63** (1994) 867-871.

8. T. Yokoo, T. Sakaguchi, K. Kakurai and J. Akimitsu, Observation of the Haldane gap in Y_2BaNiO_5 single crystal, *J. Phys Soc. Jpn.* **64** (1995) 3651-3655.

9. M. Hagiwara, K. Katsumata, I. Affleck, B. I. Halperin and J. P. Renard, Observation of S=1/2 degrees of freedom in an S=1 linear-chain Heisenberg antiferromagnet. *Phys. Lev. Rett.* **65** (1990) 3181-3184.

10. I. Affleck, T. Kennedy, E. H. Lieb and H. Tasaki, Rigorous results on valence-bond ground states in antiferromagnets, *Phys Rev. Lett.* **59** (1987) 799-802.

11. M. Motokawa, H. Nojiri and Y. Tokunaga, An idea for the easy construction of a high field magnet; *Physica* B **155** (1989) 96.

12. M. Motokawa, H. Ohta and N. Makita, Submillimeter EPR of Co:Rb_2MgF_4 and anomalous g-values, *Int. J. Infrared MMW* **12** (1991) 149-155.

13. S. Kimura, H. Ohta, M. Motokawa, S. Mitsudo, W.-J. Jang, M. Hasegawa and H. Takei, Submillimeter wave ESR measurements of metamagnetic $Y_2Cu_2O_5$, *Int. J. Infrared MMW* **17** (1996) 833-841.

Simulation of the High Frequency and High Field EPR Transitions for Spin Quintets: Fe^{2+} and Fe^{4+} Ions in YBCO and Related Crystals

Czeslaw RUDOWICZ and Sudhana B. MADHU

Department of Physics and Materials Science, City University of Hong Kong, Kowloon.

Abstract. EPR signals of the spin S=2 $3d^6$ and $3d^4$ ions can hardly be detected in the X- or Q-band due to the large zero-field splitting. Aiming at future high frequency (ν) and high magnetic field (***B***) EPR we have carried out simulations of the energy level and rotation pattern diagrams for $Fe^{2+}(3d^6)$ and $Fe^{4+}(3d^4)$ ions located at the Cu(1)-planar, -pyramidal, and -octahedral sites as well as Cu(2) sites in $YBa_2(Cu_{1-x}Fe_x)_3O_{7-\delta}$. The predicted microscopic spin Hamiltonian parameter sets $\{g_i, b_k^q\}$ were selected for simulations with ***B*** in the range 0-30 T and ν: 35, 100 and 300 [GHz]. The energy level diagrams are plotted with ***B*** along the X, Y, and Z direction. The rotation pattern diagrams are plotted with ***B*** in the XY, ZY and ZX plane. The energy level crossover points, which offer a possibility of EPR studies at lower X- or Q-band frequencies and moderate magnetic fields, are determined. General symmetry of the diagrams with respect to the sign of b_2^0 and b_2^2 as well as the role of the higher order transitions and the effect of the fourth-order zero-field splitting parameters are considered.

1 Introduction

For the spin S = 2 systems only a few cases of the energy level and rotation pattern diagrams exist in the literature and, moreover, only for the low values of the zero-field splitting [*ZFS*] parameters, low magnetic field ***B*** range and X-band frequency [1-3]. The ground state of Fe^{2+} and Fe^{4+} ions in $YBa_2(Cu_{1-x}Fe_x)_3O_{7-\delta}$ [*YBCO*] is found to be a spin quintet well separated from the higher lying states [4]. Using the crystal field [4] and microscopic spin Hamiltonian [*SH*] theory [5,6], the SH parameters b_k^q and g_i were determined [7] for Fe^{2+} and Fe^{4+} (S=2) ions at the Cu(1)-planar, -pyramidal, -octahedral sites and Cu(2) sites in YBCO. The large ZFS, obtained for the pertinent values of the spin-orbit, spin-spin coupling constants and the energy level splittings, hinders the EPR detection in the X- or Q-band.

Simulations [8] of the EPR transitions for selected sets $\{g_i, b_k^q\}$ provide guidelines for high frequency and high field EPR studies of Fe^{2+} and Fe^{4+} (S=2) ions in YBCO and structurally similar crystals. The energy level diagrams are plotted with ***B*** in the range 0-30 T along the X, Y, and Z direction, whereas the rotation pattern diagrams with $\boldsymbol{B}$ in the XY, ZY and ZX plane for the frequency 35 GHz, 100 GHz and 300 GHz. Full sets of the diagrams for are provided in [9].

2 Spin Hamiltonian Parameters for Fe (S=2) Ions in YBCO

Four Cu sites determined from X-ray diffraction studies are considered for $YBa_2(Cu_{1-x}Fe_x)_3O_{7-\delta}$. The local site symmetry is D_{2h} for the Cu(1) fourfold planar and sixfold octahedral sites, whereas C_{2v} for the Cu(1) and Cu(2) fivefold pyramidal sites [4]. For a $3d^6$ or $3d^4$ (S=2) ion in an orthorhombic crystal field [*CF*], with the excited states well separated from the ground state, SH consists of [10, 11]:

$$\tilde{H}_{SH} = \tilde{H}_{Ze} + \tilde{H}_{ZFS} = \mu_B(g_x S_x B_x + g_y S_y B_y + g_z S_z B_z) + f_2(b_2^0 O_2^0 + b_2^2 O_2^2) + f_4(b_4^0 O_4^0 + b_4^2 O_4^2 + b_4^4 O_4^4) \quad (1)$$

where $f_2 = 1/3$, $f_4 = 1/60$ and O_k^q (S_x, S_y, S_z) are the Stevens operators [10, 12].

The superposition model and the CF analysis within the ^5D states [4] provided the energy level values (Δ_i) for the microscopic SH calculations for Fe:YBCO [5, 6]. The SH parameters g_i and b_k^q were calculated for Fe^{2+} and Fe^{4+} ions at the four Cu sites in YBCO for the Fe content x = 0 to 0.18 [7]. Notwithstanding the approximations made [13] it can be expected that the values [7] represent correctly the magnitudes of the parameters b_2^q, whereas the parameters b_4^q tend to be underestimated by the microscopic SH theory within the ^5D approximation.

The representative values of g_i and b_k^q for low doping x = 0.01 in the orthorhombic phase of Fe:YBCO are used for the model calculations. Out of the 64 sets of SH parameters [7], we select as input the sets with the *higher* and *lower* $|b_2^0|$ for each ion/site case (Table 1) to cover the range of magnitudes of the ZFS parameters. Each SHP set is expressed in the axis system which had ensured that most of the originally calculated sets were in the standard range [14], i.e. with the ratio b_2^2/b_2^0 in the range (0 - 1). The corresponding axis systems with respect to the crystallographic axes (a, b, c) are given in Table 1. For all the four Fe^{2+} sites, the higher SHP sets yield on average up to 50% larger energy level splitting than that for the lower ones. Similar behaviour can also be expected for Fe^{4+} ions in YBCO. Hence for Fe^{4+} ions only the higher SHP sets are considered, whereas the parameters b_4^q, which are 100 to 1000 times smaller than for Fe^{2+} ions [7], are neglected.

3 EPR Simulation Program - Results

The program [8] simulates the energy levels of SH (1) for the magnetic field $\boldsymbol{B}$ along a given direction and the allowed EPR transitions ($\Delta M_S = \pm 1$) for a given frequency of the oscillating field $\boldsymbol{B_1}$ ($\boldsymbol{B} \perp \boldsymbol{B_1}$). Using the $\boldsymbol{B}$ range 0-30 T and the SHP sets in Table 1, 33 energy level diagrams for ***B*** || X, ***B*** || Y and ***B*** ||Z are obtained: 24 for Fe^{2+} ion and 9 for Fe^{4+} ion. Similarly, 72 (27) rotation pattern diagrams for ***B*** in the plane XY, ZX and ZY are obtained for Fe^{2+} (Fe^{4+}) ion for the frequency ν = 35, 100 and 300 [GHz]. An illustrative sample of the diagrams [9] is given in Fig. 1 and 2.

Table 1. Input parameters g_i, b_k^q (in cm^{-1}) for Fe^{2+} (upper) and Fe^{4+} (lower) ions in YBCO.

	Cu(1) planar sites (X||b, Y||a, Z||c) SHP set		Cu(1) pyramidal sites (X||a, Y||c, Z||b) SHP set		Cu(1) octahedral sites (X||b, Y||a, Z||c) SHP set		Cu(2) sites (X||c, Y||a, Z||b) SHP set	
	lower	higher	lower	higher	lower	higher	lower	higher
g_X	1.97	1.90	2.08	2.12	2.09	2.13	2.17	2.19
g_Y	2.01	1.94	2.09	2.12	2.07	2.08	2.14	2.17
g_Z	2.37	2.46	2.02	2.02	2.06	2.05	1.89	1.93
b_2^0	-17.1	-30.0	2.8	5.8	1.1	4.9	5.7	8.8
b_2^2	1.4	-0.3	1.6	3.5	-1.33	-2.6	5.6	7.4
b_4^0	0.7	4.2	0.02	0.15	0.04	0.21	3.2	2.5
b_4^2	-0.9	-5.4	-0.002	-0.06	-0.11	-0.15	-3.5	-1.4
b_4^4	0.2	1.1	0.02	0.17	0.16	0.68	0.9	0.2
g_X	2.00	2.00	2.00	2.00	1.97	1.97	2.00	1.99
g_Y	2.00	1.99	1.99	1.99	1.98	1.97	1.99	1.99
g_Z	1.96	1.95	1.97	1.96	2.00	2.00	1.95	1.93
b_2^0	-2.33	-5.15	-2.12	-4.75	1.98	4.64	-2.93	-6.13
b_2^2	0.57	1.48	1.14	2.81	-0.65	-1.48	0.28	0.66

Consideration of the symmetry of the diagrams with respect to the sign of b_2^0 and b_2^2 [9] enables their applications for other S = 2 systems. For the case (a) [$b_2^0 > 0$, $b_2^2 > 0$] the energy level diagrams for ***B*** || Y have similar characteristics as the diagrams for ***B*** || Z, however, with the "inverted" energy level structure, whereas those for ***B*** || X exhibit no energy level crossing. The change of sign of b_2^0 (the case (b): $b_2^0 < 0$, $b_2^2 > 0$) induces an inversion of the energy levels for ***B*** || Z with respect to the case (a) and a swapping of the axes X ⇔ Y together with an inversion, whereas in the rotation pattern diagrams the ZX plane becomes the ZY plane, however, no inversion is observed. The change of sign of b_2^2 (the case (c): $b_2^0 > 0$, $b_2^2 < 0$) induces a swapping of the axes X ⇔ Y without inversion, whereas in the rotation pattern diagrams the ZX plane becomes the ZY plane. For the case (d)

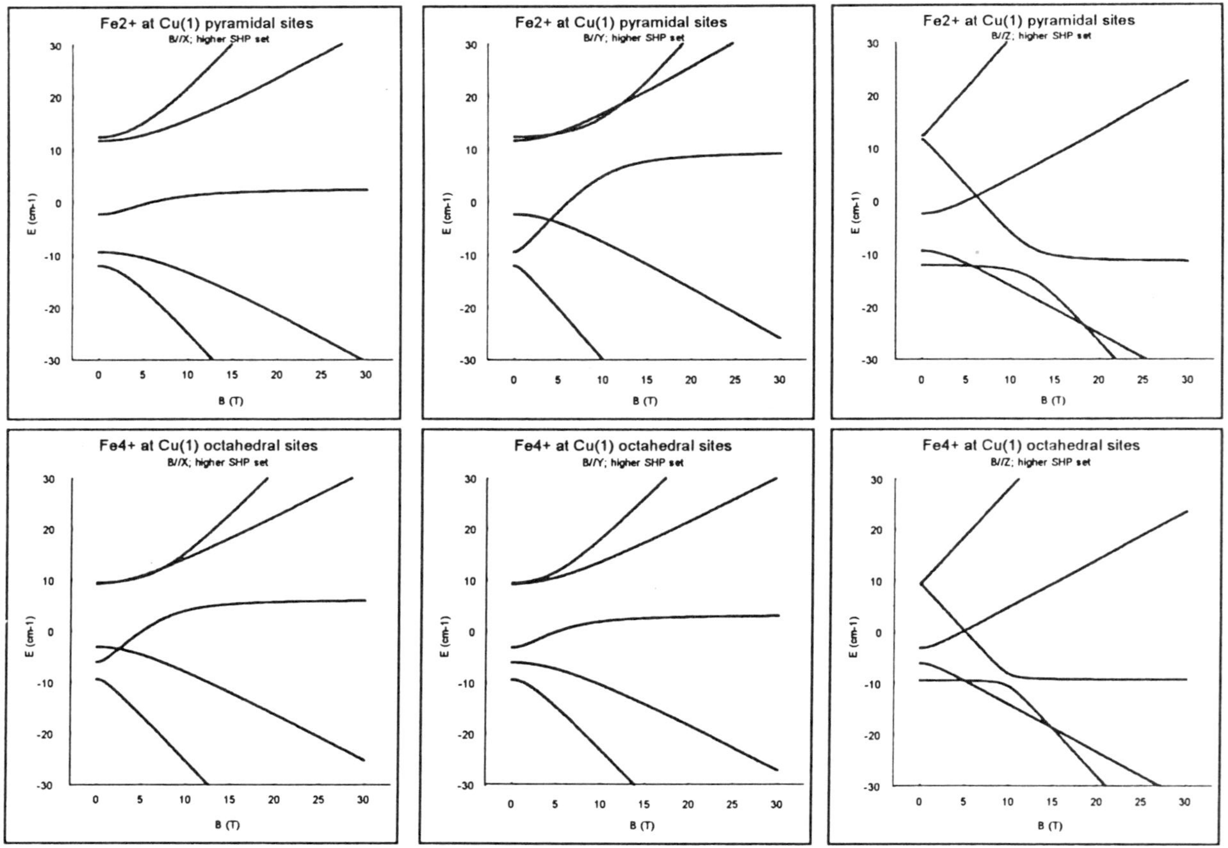

Fig. 1. Energy level diagrams: Fe2+ at Cu(1) pyramidal sites and Fe4+ at Cu(1) octahedral sites in YBCO.

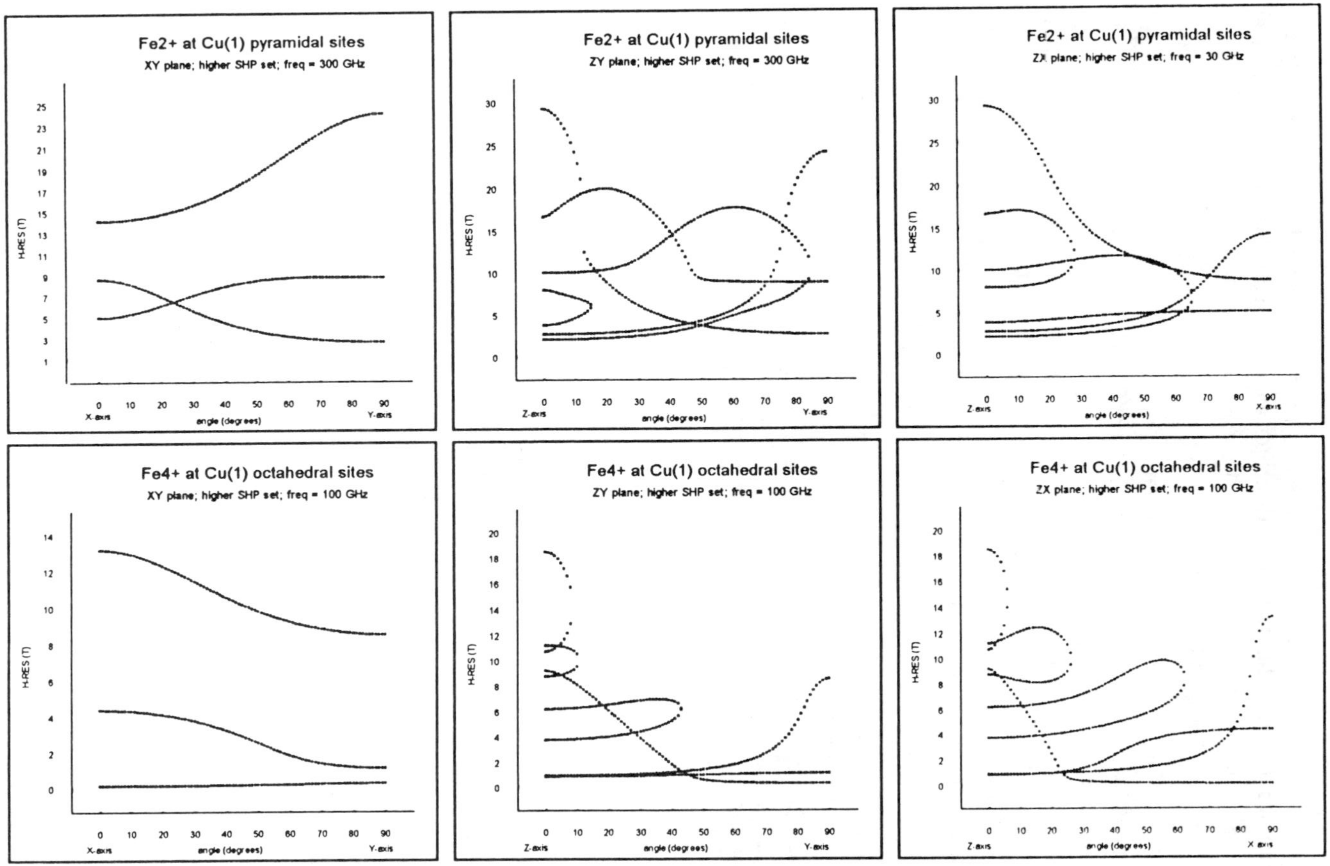

Fig.2. Rotation pattern diagrams: ν= 300 GHz Fe2+ at Cu(1) pyramidal sites and ν= 100 GHz Fe4+ at Cu(1) octahedral sites in YBCO.

[$b_2^0 < 0$, $b_2^2 < 0$] the energy level diagrams for each direction (***B*** || X, ***B*** || Y, ***B*** || Z) correspond to those for the same direction as in the case (a), however, with an inverted structure. Hence no change is observed in the rotation pattern diagrams.

For an orthorhombic S = 2 system with $b_2^0 > 0$ the energy level crossings occur for ***B*** || Z. The predominant ground state component $|M_s\rangle$ varies successively with the increasing ***B*** from $|0\rangle$ to $|-1\rangle$ and then to $|-2\rangle$ [9]. The crossover points B_{c1} and B_{c2} involving the states ($\Phi = a|0\rangle + b|-2\rangle + c|+2\rangle$ and $\Psi = d|-1\rangle + e|+1\rangle$) are listed in Table 2. The characteristic *loops* or *lobes* in the rotation pattern diagrams (Fig. 2) are due to the transitions in the vicinity of B_{c1} and B_{c2}. These transitions offer an interesting opportunity of detection of EPR signals around specified orientations of ***B*** at lower frequencies with moderate or higher fields.

Table 2. The crossover points B_{c1} and B_{c2} for Fe^{2+} (upper) and Fe^{4+} (lower; no crossing is observed for ***B*** || Y) ions in YBCO in the range ***B*** = 0 to 30 T.

	Cu(1) planar		Cu(1) pyramidal		Cu(1) octahedral		Cu(2)	
SHP set:	lower	higher	lower	higher	lower	higher	lower	higher
B || X	5 22.5				2 4	4 9		
B || Y	15	21	2 6	4.5 12.5			6 16	8 22.5
B || Z	17		2.5 9	6 18	1.5 3	4 15.5	23	
B || X				3 11		2.5 7.5		
B || Z				5 16		5 15		7 21

The rotation pattern diagrams for ν=35 GHz show only a few resonance transitions with ***B*** in either of the three planes. This indicates that EPR signals can hardly be detected using the commercial spectrometers. As the frequency increases to 100 GHz and 300 GHz, the rotation patterns are getting richer indicating several more lines. Hence, the possibility of detection of EPR signals increases significantly with increasing frequency. For all SHP sets considered the XY and ZX plane offers a better chance of detection of EPR signals at lower fields with higher frequencies. Regardless of the sign of b_2^0 and b_2^2, the intensity of all transitions in the ZY and XY plane is rather low, except for some data points on the Z- and X-axis, whereas a number of transitions with relatively large intensity occur in the ZX plane.

The rotation pattern diagrams were obtained including the first order transitions [8]. The effect of the higher order (2, 3 or 4) transitions was studied for two SHP sets. In general, we conclude that some additional and weaker resonance lines may be observed at ν = 300 GHz (or higher) due to the higher order transitions [9].

The effect of the fourth-order ZFS parameters b_4^q on the energy levels and rotation patterns was studied for two SHP sets. It is concluded that b_4^q 's have no significant effect on the EPR transitions, except for some isolated data points [9].

4 Conclusions

The detection of EPR signals for Fe^{2+} and Fe^{4+} ions at the four different sites in YBCO and structurally similar crystals is found more feasible in the higher frequencies and/or high magnetic fields. Only a few EPR signals are detectable in the Q-band. The orientations of the magnetic field most favourable for detection of EPR signals are in the ZX plane, especially for ν = 100 GHz and 300 GHz. The energy level crossings B_{c1} and B_{c2} are determined. In the vicinity of B_{c1} and B_{c2} the characteristic *loops* in the rotation patterns are observed, which offer an interesting opportunity of detection of EPR signals. Using the symmetry characteristics of the energy level diagrams for ***B*** along the X-, Y- and Z- direction as well as the rotation pattern diagrams in the three planes XY, ZY and ZX, relevant conclusions may be drawn for other spin quintet systems. The results may provide guidelines for the future high magnetic field and/or high frequency EPR experiments.

Acknowledgements. This work has been supported by the University Grants Committee. Thanks are due to Dr F. Prissok for providing the original simulation program.

References

1. M. Mackowiak, *Il Nuovo Cimento* **29 B** (1975) 207-224.
2. R. D. Dowsing, *J. Mag. Reson.* **2** (1970) 332-337.
3. F. E. Mabbs and D. Collison, *Electron paramagnetic resonance of d transition metal compounds*, Elsevier, Amsterdam (1992).
4. C. Rudowicz and W. L. Yu, *Supercond. Sci. Technol.* **4** (1991) 535-543.
5. C. Rudowicz and Y. Y. Zhou, *J. Phys. Chem. Solids* **55** (1994) 745-757.
6. C. Rudowicz and Y. Y. Zhou, *Comp. & Chem.* **21** (1996) 45-50.
7. C. Rudowicz and W. L. Yu, *Phys. Rev.* **B 47** (1993) 9001-9009.
8. C. Rudowicz and T. H. Yeom, *EPR simulation program manual* - unpublished (1992).
9. C. Rudowicz and S. B. Madhu, *Energy level and rotation pattern diagrams for spin quintets at high frequency and high magnetic fields: Applications to Fe^{2+} and Fe^{4+} ions in YBCO and related crystals,* Research Report AP-97-06, Department of Physics and Materials Science, City University of Hong Kong (1997).
10. A. Abragam and B. Bleaney, *Electron paramagnetic resonance of transition ions*, Dover Publications, New York (1970).
11. J. R. Pilbrow, *Transition ion electron paramagnetic resonance*, Clarendon Press, Oxford (1990).
12. C. Rudowicz, *Magn. Reson. Rev.* **13** (1987) 1-89.
13. C. Rudowicz, M.L. Du, and Y. Y. Yeung, *J. Phys. Chem. Solids* **54** (1993) 733-744.
14. C. Rudowicz and R. Bramley, *J. Chem. Phys.* **83** (1985) 5192-5197.

Variable Temperature EPR Study of Fe^{2+} Spin Transition in an Octahedral Complex $[Fe(trz)(Htrz)_2](BF_4)$

C. Cantin, J. Kliava, Y. Servant[1)], *L. Sommier and O. Kahn*[2)]

[1)]Centre de Physique Moléculaire Optique et Hertzienne, Université Bordeaux-I, F-33405 Talence Cedex, France
[2)]Laboratoire des Sciences Moléculaires, Institut de Chimie de la Matière Condensée de Bordeaux, F-33608 Pessac Cedex, France

Abstract. The electron spin resonance of Fe^{3+} ions has been used for the first time to evidence a low spin ($S = 0$) to high spin ($S = 2$) transition of Fe^{2+} ions in an octahedral ferrous complex $[Fe(trz)(Htrz)_2](BF_4)$. The temperature dependence of the Fe^{3+} resonance shows a spin transition with hysteresis which occurs for the Fe^{2+} ions. The transition temperatures are $T_c\uparrow$ = 374 K in the warming mode and $T_c\downarrow$ = 345 K in the cooling mode. The analysis of the spectral data indicates a structural phase transition accompanying the spin transition.

1 Introduction

In octahedral ligand fields, transition metal ions with d^4 to d^7 configurations generally adopt either a high-spin (HS) or a low-spin (LS) electronic arrangement, according to whether the crystal field strength is lower or higher, respectively, than the mean spin coupling energy. Some compounds, however, exhibit a thermally-induced spin transition between LS and HS states.

In the present paper, we report an EPR study of an octahedral ferrous complex $[Fe(trz)(Htrz)_2](BF_4)$. Iron in this material is mainly in Fe^{2+} (ferrous) state. The Fe^{2+} ions, however, are EPR-silent in the LS *diamagnetic* (S=0) state. Neither do they give rise to any measurable (at elevated temperatures) EPR signal in the HS *paramagnetic* (S=2) state, because of a very short spin-lattice relaxation time.

Yet, a small part of iron, about 0.01 of the total number of iron atoms, occurs in the samples in ferric (Fe^{3+}) state. These ions can serve as excellent spin probes, since even at relatively high temperatures the EPR spectra of Fe^{3+}, an S-ion, exhibit a well-resolved zero-field splitting. By analysing the latter, one can get information on the symmetry and structure of Fe^{3+} sites which can be used (with a certain caution) to characterise the environment of Fe^{2+} ions, as well.

2 Sample, Structure and Optical Response

The preparation of the $[Fe(trz)(Htrz)_2](BF_4)$ complex was described in [1]. EXAFS data strongly suggest that its basic structure is that of a linear chain in which two adjacent iron atoms are triply bridged by triazole ligands through the nitrogen atoms occupying the 1- and 2-positions [2, 3]. See Figure 1.

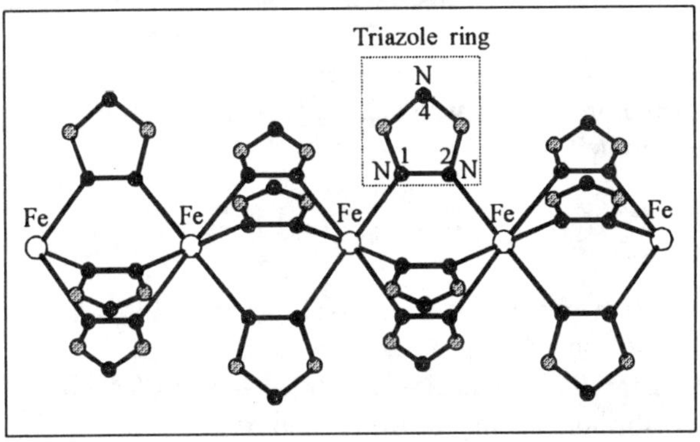

Fig. 1. Structure of the polymeric Fe(II)-1,2,4-triazole compound deduced from EXAFS [3].

The spin transition in the compound is accompanied by a dramatic change of colour from pink (LS) to white (HS). This strong contrast can be observed with an optical device described in a previous work [3]. The result is shown in Figure 2. For the successive thermal cycles, the transition temperature in the cooling mode is constant ($T_c\downarrow$=343 K), whereas the transition temperature in the warming mode is $T_c\uparrow$=380 K for the first cycle and ~372 K for the following cycles.

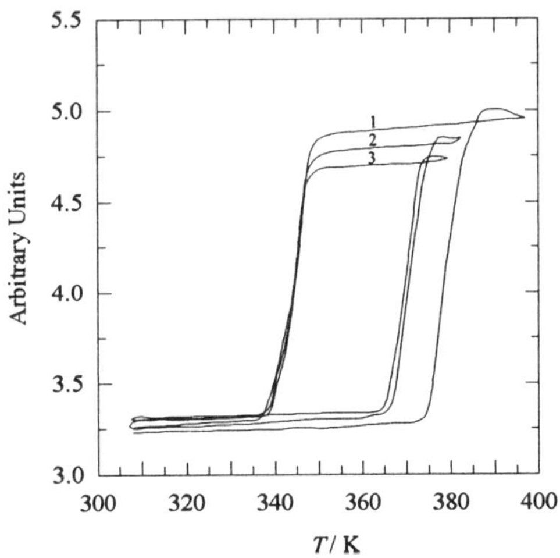

Fig. 2. Optical detection of the spin transition. Successive thermal cycles are marked 1, 2 and 3.

3 EPR Data

At room temperature, 293 K, corresponding to the LS state of Fe^{2+} ions, the EPR spectrum observed is typical of Fe^{3+} ions in a rather strong crystal field [4]. It consists of a relatively narrow line (with the peak-to-peak width $\Delta B_{pp} \approx 23.9$ mT) at $g_{ef} = 4.3$, accompanied by a smooth plateau extending to $g_{ef} = 9.7$, and of a broad absorption at $g_{ef} = 2.0$ which consists of a narrower line superimposed with a broader one. The $g_{ef} = 4.3$ and $g_{ef} = 9.7$ features, as well as the narrow $g_{ef} = 2.0$ component are due to magnetically isolated Fe^{3+} ions which can be described by the "rhombic" spin Hamiltonian, in which all crystal field terms other than the fine structure (FS) are omitted [4, 5]:

$$\mathscr{H} = \beta g \boldsymbol{B} \cdot \boldsymbol{S} + D\left[S_z^2 - \frac{S}{3}(S+1) + \lambda\left(S_x^2 - S_y^2\right)\right]. \tag{1}$$

Here $S = 5/2$ and all symbols have their usual meaning. Instead of the rhombic FS parameter E, we use the rhombic to axial parameter ratio, $\lambda = E/D$, confined, by convention, within the limits $0 \le \lambda \le 1/3$ [6]. The broader $g_{ef} = 2.0$ absorption component can be attributed to magnetically coupled Fe^{3+} ions [7].

The temperature dependence of EPR spectra is shown in Figure 3 in both the warming and cooling modes.

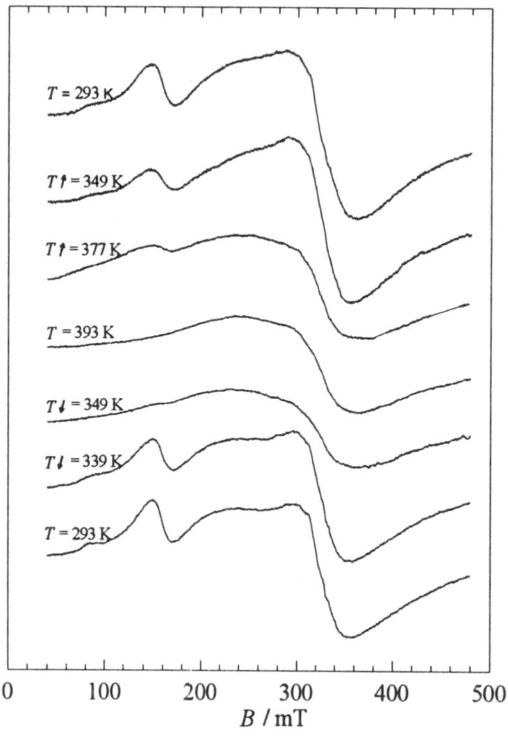

Fig. 3. EPR spectra in the warming ($T_c\uparrow$) and cooling ($T_c\downarrow$) modes.

At the first stage of warming, the intensity of the $g_{ef} = 4.3$ line decreases, while its width remains approximately the same. As the temperature increases above the critical value $T_c\uparrow$, the LS to HS transition takes place for the Fe^{2+} ions. The $g_{ef} = 4.3$ line and the plateau at $g_{ef} = 9.7$ disappear and the EPR spectrum of Fe^{3+} ions reduces to a broad absorption ($\Delta B_{pp} \approx 0.12$ T) at $g_{ef} = 2.0$. A closer inspection of the EPR spectrum of Fe^{3+} in the HS state of Fe^{2+} ions indicates that the absorption at $g_{ef} = 2.0$ is not elementary.

In the cooling mode, as the temperature decreases below the second critical value $T_c\downarrow$, the EPR spectrum approximately recovers its initial shape. However, some discrepancies persist, namely, for the $g_{ef} = 4.3$ line, ΔB_{pp} is now only 22.5 mT and a new ill-resolved feature may be observed at in the range $g_{ef} = 2.2 - 2.5$. These discrepancies disappear within several hours at room temperature.

Figure 4 shows the temperature dependence of the $g_{ef} = 4.3$ EPR line intensity evaluated as:

$$I \propto A\left(\Delta B_{pp}\right)^2, \tag{2}$$

where A is the peak-to-peak amplitude.

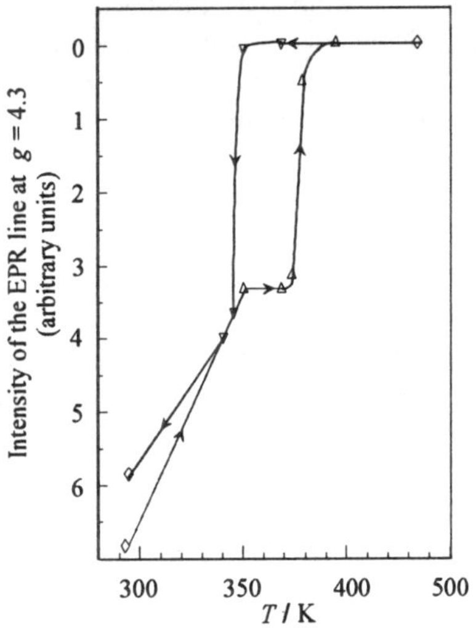

Fig. 4. Hysteresis loop evidenced by the intensity of the EPR line at $g = 4.3$.

One obtains a quite rectangular hysteresis loop with the critical temperatures $T_c\uparrow = 374$ K and $T_c\downarrow = 345$ K, in fairly good agreement with the values obtained from optical measurements, *vide supra*.

4 Discussion

The absence of a detectable EPR signal due to Fe^{2+} ions is confirmed by the fact that there is no temperature dependence of the number of ions responsible for the EPR spectrum, as the overall paramagnetic absorption intensity remains constant.

It is well known that the $g_{ef} = 4\ 3$ line with the concomitant $g_{ef} = 9.7$ plateau is due to magnetically isolated Fe^{3+} ions in a *strong* low-symmetry orthorhombic crystal field [4, 5]. One might argue that as in the HS state the Fe^{2+} ions are paramagnetic, they create magnetic fields at the sites occupied by Fe^{3+} ions, whence a broadening and finally the vanishing of the $g_{ef} = 4.3$ line. However, in this case the $g_{ef} = 9.7$ plateau (arising from the same paramagnetic species) which is already fairly broad in the LS state of Fe^{2+} ions, would also be observable above the transition temperature, which is not the case.

One can conclude that the LS to HS transition is accompanied by a structural phase transition which, on a microscopic level, manifests itself by a weakening of low-symmetry distortions of the crystal field on the Fe^{3+} sites (and consequently, on the Fe^{2+} sites). This conclusion is corroborated by the EXAFS results, namely that of a 0.18 Å Fe–N bond lengthening in the LS to HS transition [3]. The corresponding weakening of the crystal field is consistent with the decrease of the mean axial crystal field parameter D, resulting in a disappearance of the $g_{ef} = 4.3$ line and the $g_{ef} = 9.7$ shoulder in the HS state of Fe^{2+} ions. The reduction of the low-symmetry distortion of the Fe–N octahedron in the LS to HS transition is confirmed by diffuse reflection spectral data obtained on a number of iron - triazole compounds [8].

Note that the Fe^{3+} ions, arising in a relatively small number, certainly have as nearest metallic neighbours the Fe^{2+} ions. One might argue that as the coordination polyhedra of the latter *expand* in the course of the spin transition, they tend to *compress* the coordination polyhedra of Fe^{3+}, in contrast to what is actually observed by EPR. This suggests that in the HS state the iron ions are no more aligned, but rather form a zigzag chain, in which case the distortion of the coordination polyhedra of Fe^{3+} would more or less exactly reproduce those of coordination polyhedra of Fe^{2+}.

Next, we discuss the broadness of the $g_{ef} = 2.0$ Fe^{3+} EPR line observed in the HS state of Fe^{2+} ions. Its broadening may arise from local magnetic fields generated by the paramagnetic Fe^{2+} ions [9] and/or from exchange [10]. In the present case the exchange may occur between Fe^{3+} and paramagnetic (HS) Fe^{2+} ions. The local magnetic fields at the Fe^{3+} sites produced by the HS Fe^{2+} ions are of order of several tens of millitesla. However, their contribution may be strongly reduced because of a very fast spin-lattice relaxation of the HS Fe^{2+} ions. On the other hand, the exchange integrals evaluated from magnetic susceptibility measurements on similar iron-triazole compounds are about several cm^{-1} [3]. Thus, the exchange must significantly contribute in the observed linewidth.

Finally, one can see that the presence (or absence) of the $g_{ef} = 4.3$ line and the $g_{ef} = 9.7$ shoulder in the EPR spectrum of Fe^{3+} ions characterises the low-(or high-) spin state of the Fe^{2+} ions. It is interesting to note that the way in which the Fe^{3+} EPR spectrum evolves in the course of a LS to HS transition experienced by the Fe^{2+} ions ($g_{ef} = 4.3 \rightarrow g_{ef} = 2.0$) is just opposite to the one observed in the case of a "direct" LS to HS transition of Fe^{3+} ions ($g_{ef} = 2.0 \rightarrow g_{ef} = 4.3$). Indeed, as far as the $g_{ef} = 4.3$ line and the $g_{ef} = 9.7$ shoulder are characteristic of the HS state ($S = 5/2$) of Fe^{3+} ions, they disappear in the LS state ($S = 1/2$) of this ion, the latter being characterised by a $g_{ef} = 2.0$ EPR signal.

5 Conclusion

The thermally-induced spin transition of Fe^{2+} ions in $[Fe(trz)(Htrz)_2](BF_4)$, showing a wide hysteresis loop characterised by critical temperatures $T_c\uparrow=374$ K and $T_c\downarrow=345$ K, is studied by Fe^{3+} EPR and optical response. The transformation of the Fe^{3+} EPR spectrum in the course of the spin transition indicates a weakening of the low-symmetry distortion of the coordination polyhedra of iron, suggesting a concomitant structural phase transition. This observation is corroborated by EXAFS and diffuse reflection spectral data.

References

1. J. Kröber, J.-P. Audière, R. Claude, E. Codjovi, O. Kahn, J.G. Haasnoot, F. Groliere, C. Jay, A. Bousseksou, J. Linares, F. Varret, A. Gonthier-Vassal, Spin transitions and thermal hysteresis in the molecular based materials $[Fe(Htrz)_2(trz)](BF_4)$ and $[Fe(Htrz)_3](BF_4)_2.H_2O$ (Htrz=1,2,4-4H-triazole; trz=1,2,4-triazolato), *Chem. Mater.* **6** (1994) 1404-1412.
2. A. Michalowicz, J. Moscovici, B. Ducourant, D. Cracco and O. Kahn, Exafs and X-ray powder diffraction studies of the spin transition molecular material $[Fe(Htrz)_2(trz)](BF_4)$ and $[Fe(Htrz)_3](BF_4)_2.H_2O$ (Htrz=1,2,4-4H-triazole; trz=1,2,4-triazolato), *Chem. Mater.* **7** (1995) 1833-1842.
3. O. Kahn and E. Codjovi, Iron(II)-1,2,4-triazole spin transition molecular materials, *Phil. Trans. R. Soc. Lond.* **A 354** (1996) 359-379.
4. J. Kliava, EPR of impurity ions in disordered solids. Distributions of the spin hamiltonian parameters, *Phys. Stat. Solidi* **(b) 134** (1986) 411-455.
5. E.-M. Yahiaoui, R. Berger, Y. Servant, J. Kliava, L. Cugunov and A. Mednis, Electron paramagnetic resonance of Fe^{3+} ions in borate glass: computer simulations, *J. Phys.: Condens. Matter* **6** (1994) 9415-9428.
6. P.L. Hall, B.R. Angel and J.P.E. Jones, Dependence of spin hamiltonian parameters E and D on labeling of magnetic axes: application to ESR of high-spin Fe^{3+}, *J. Magnetic Resonance* **15** (1974) 64-68.
7. J. Kliava, R. Berger, Y. Servant, J. Emery, J.-M. Grenèche and J. Trokšs, Electron paramagnetic resonance and Mössbauer effect studies in iron-doped ^{57}Fe isotope enriched phosphate glasses, *J. Non - Cryst. Solids* **202** (1996) 205-214.
8. L.G. Lavreneva, V.N. Ikorskii, V.A. Varnek, I.M. Oglezneva and S.V. Larionov, Spin transition in iron (II) coordination compounds with triazoles (in Russian), *Koord. Khim.* **16** (1990) 654-661.
9. P.S. Rao, A. Reuveni, B.R. McGarvey, P. Ganguli and P. Gütlich, ESR and Mössbauer study of high-spin–low-spin transition in $[Fe(Phen)_2(NCS)_2]$ and $[Fe(Pic)_3(trz)]Cl_2EtOH$, *Inorg. Chem.* **20** (1981) 204-207.
10. W. Vreugdenhil, J.G. Haasnoot, O. Kahn, P.Thuéry and J. Reedijk, A copper(II) dope as a detector for the high-spin↔low-spin transition in the two-dimensional compound [trans-Bis(thiocyanato)bis(4,4'-bi-1,2,4-triazole)iron]hydrate, *J. Am. Chem. Soc.* **109** (1987) 5272-5273.

Study of Nonlinear Excitations in Low Dimensional Magnetic Systems by Electron Magnetic Resonance

John E. Drumheller[1], R. S. Rubins[2], K. Subbaraman[1], and C. E. Zaspel[1]

[1] Dept. of Physics, Montana State University, Bozeman, MT, 59717 USA
[2] Dept. of Physics, University of Texas Arlington, Arlington TX, 76019 USA

Abstract. This paper is a short review of the observation of nonlinear excitations in the fluctuations region near the phase transition in low-dimensional magnetic systems as seen by electron magnetic resonance techniques. Specifically reported are the studies of spin-cluster resonance in a one-dimensional Ising ferromagnet and solitons in a two-dimensional nearly classical Heisenberg antiferromagnet.

1. Introduction

Several recent studies have shown that excitations in the fluctuation region near the ordering transition in low-dimensional magnetic systems are not predicted by linear spin-wave theory but nevertheless dominate the thermodynamics of these systems in that temperature range.[1] The one-dimensional (1D) case is particularly illustrative in showing these effects and is fortuitously accessible experimentally. One-dimensional systems do not order except at the 3D crossover temperature, that is, the temperature at which the three-dimensional effects, no matter how small, can be felt. Thus above the 3D crossover there is a broad temperature region in which the system *tries* to order but cannot. That is, there is a range of (sometimes) several Kelvins between the "exchange temperature" and the real 3D ordering temperature over which the fluctuations leading to order can be observed. Similarly the two-dimensional (2D) Heisenberg ferromagnet does not order at finite temperatures but the antiferromagnet shows a broad temperature range over which fluctuations may be observed. What follows is a short review of our studies of the spin-cluster resonances (SCR) in the 1D, Ising ferromagnet $[(CH_3)_3NH]FeCl_3.2H_2O$, known as FeTAC and the observation of two-dimensional solitons in the 2D, Heisenberg nearly classical (S=5/2) antiferromagnet $(C_3H_7NH_3)_2MnCl_4$, known as $(nPA)_2MnCl_4$ or, more simply, PAMC, in which electron magnetic resonance techniques were employed.

2. Spin-cluster Resonace in FeTAC

The compound FeTAC contains well-separated chains of ferrous ions extending in the crystallographic b direction which forms the easy axis of the crystal,[2,3] i.e., the axis along which the spins will tend to align when ordered. In the range 6-12 K, FeTAC is an almost ideal 1D Ising ferromagnet. Neighboring ferrous ions are linked by pairs of chlorines. Weak magnetic coupling between the chains produces a transition to higher dimensional ferromagnetic state below 3.1 K, with a metamagnetic critical field of 90 Gauss.

The low temperature magnetic behavior of FeTAC has been satisfactorily described by an effective spin ½ representation. The elementary excitations of an Ising spin system are localized spin reversals known as spin clusters or, analogous to the Heisenberg case, as bound magnons.[4,5] That is, it costs less energy to reverse two or more spins that are adjacent than to reverse the same number of spins that are separated. For a closed ring of N spins, the first excited set of cluster states lies at an energy of 2J above the ferromagnetic groundstate, where the nearest neighbor exchange has the form

$$\mathcal{H} = 2J\, S_i^z S_j^z \ . \tag{1}$$

A member of this set, known as an n-fold spin cluster, contains n neighboring reversed spins, with $0<n<N$.

Electron magnetic resonance (EMR) in FeTAC is associated with transitions between adjacent spin-cluster levels.[6,7,8] With the external magnetic field H in the plane of the crystal face containing the b axis, the EMR spectrum has a low field extremum when H//b. The angular variation of the spectrum about the b axis is shown in Fig. 1 taken at 5.5 K and 9.3 GHz. At the extremum, the main portion of the spectrum consists of a satellite structure associated with transitions between spin-cluster states of lowest and highest n, with a "main" line at its high field side resulting from transitions of intermediate n. For a long chain (N>>1), most of the spin-cluster transitions would contribute to the main line, which occurs at a field given by

$$h\nu = g_z \mu_B (H - H_{int}) \ , \tag{2}$$

where ν is the microwave frequency and H_{int} is composed of the demagnetization and interchain fields.[9] The values $g_z = 8.2 \pm 0.1$ and $H_{int} = (337 \pm 5)$ Oe were typical of the samples studied. The linear relationship between H and ν is illustrated for one set of data in Fig. 2.

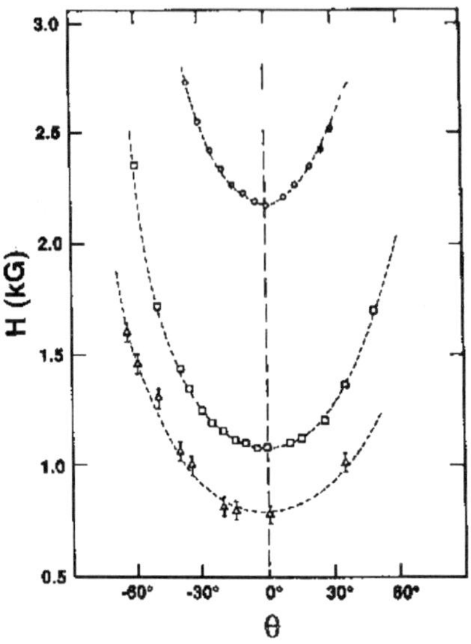

Fig. 1. The angular dependence of the 9.42 GHz EPR spectrum of FeTAC at 5.5 K. The angle $\theta = 90^{\circ}$ corresponds to H//b (the easy axis). The middle line represents the "main" line of the spectrum discussed in the text.

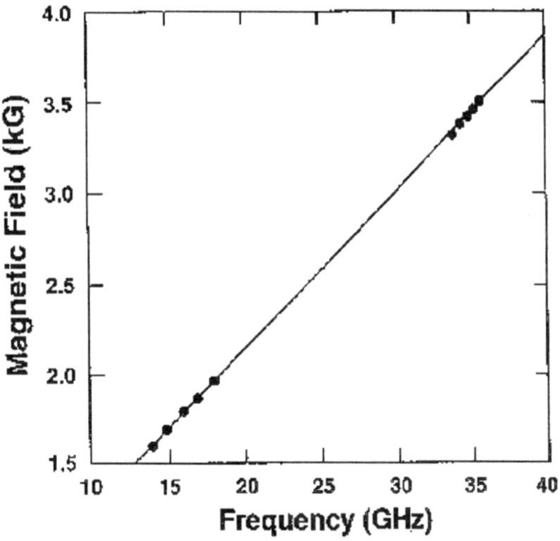

Fig. 2. A plot of the resonance field versus microwave frequency for the main SCR signal. The data are for one particular FeTAC sample at 4.1 K with H//b.

The value of 2J in Eq. (1) was determined from the temperature dependence of the main-line intensity at liquid helium temperatures. This dependence is of the form

$$I \propto \exp\left(-\Delta E/k_B T\right), \tag{3}$$

where ΔE is the excitation at the resonance field of the lowest spin-cluster level contributing to this line. The value of n for this level was obtained by counting the number of satellite lines below the main line, the lowest of which corresponds to the spin-cluster transition ($n = 1 \Leftrightarrow n = 2$). Fig. 3 shows EMR spectra at four different microwave frequencies taken at 3.6 K. As the microwave frequency is increased the number of satellites is seen to decrease while their separations and intensities relative to the main line increase.

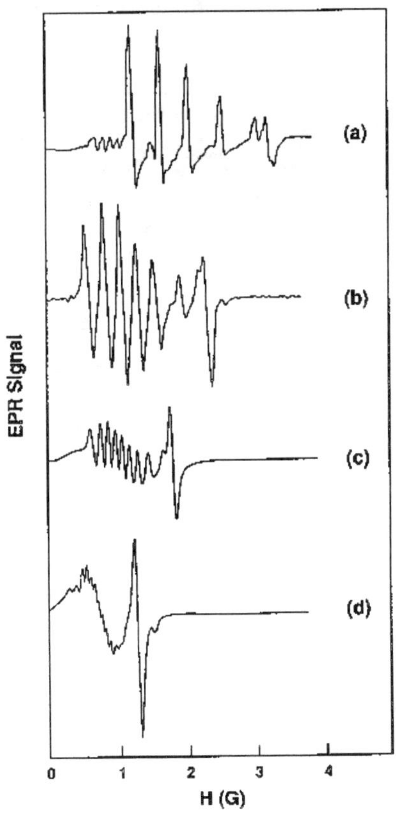

Fig. 3. EPR spectra at 3.6 K with H//b and ν = (a) 35.6 (b) 25.0 (c) 17.6 and (d) 12.4 GHz. The resonance to the right in each case is the "main" line.

An unusual experimental technique allowed these spectra to be taken without moving the sample or allowing it to warm. The sample was placed in liquid helium at the closed end of a Ku-band waveguide the other end of which was connected in turn by tapered sections to several spectrometers working between 10 and 36 GHz. Since no resonant cavity was used, the spectrometer could be operated at any frequency within the range of each klystron—except for the low frequency cut-off near 10 GHz.

Eq. (3) was used to obtain values of ΔE from the low temperature data at several microwave frequencies. Fig. 4 shows a plot of natural logarithm of the intensity versus 1/T for data taken at 14.7 GHz. The straight line fit of the low-temperature data yielded a value 42 K for $\Delta E/k_B$. As spectrum (d) of Fig. 3 shows, the number of satellites present at this frequency could not be counted. However, measurements made between 18 GHz and 36 GHz were used to obtain the value $2J/k_B = (33 \pm 1)$ K, which is close to that of 35 K deduced from susceptibility measurements.[2]

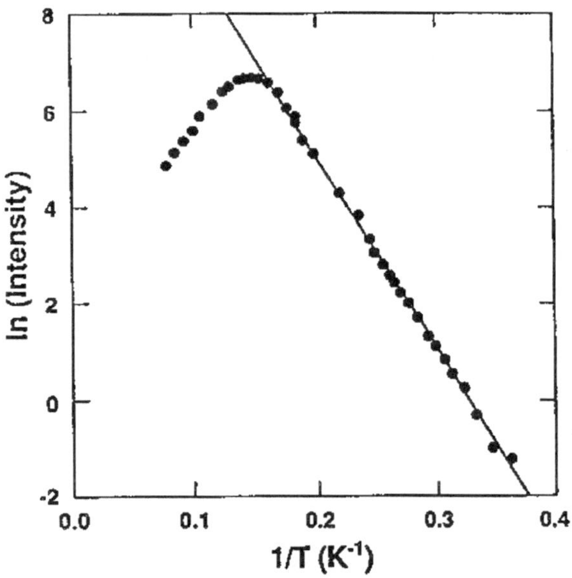

Fig. 4. A plot of the logarithm of the line intensity versus 1/T for data at 14.7 GHz with H//b. The straight line represents a linear least squares fit of the data below 6 K.

The satellite spectra from Fig. 3 were explained by the addition of a term ΔS_i^x to the effective spin-½ Hamiltonian.[8] This type of term is appropriate for describing doublet splittings occurring for a non-Kramers' ion placed in a site of low symmetry. The energy level splittings within a spin-cluster set are described by a sum of single-ion Hamiltonians, each of which has the form

$$\mathcal{H} = g_z \mu_B H_{eff} \sum_{i=1}^{n} S_i^z + \Delta \sum_{i=1}^{n} S_i^x , \quad (4)$$

where H_{eff} = H - H_{int} A very good fit of the satellite data, shown in Fig. 5, was obtained by Sohn for the value $\Delta/k_B = 3.4K$.

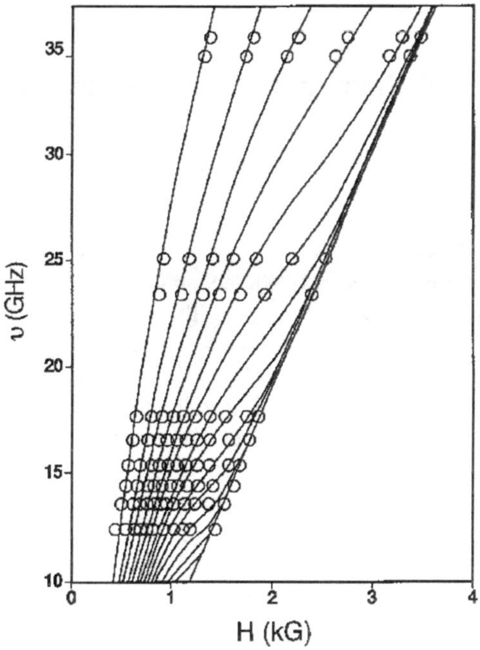

Fig. 5. Microwave frequency versus resonance field. Circles represent the experimental data which were taken at 3.6 K (except for 10 GHz measurement which were made at 4.2 K). The line represents solutions of the Hamiltonian of Eq. (4) with $g_z = 8.2$, $H_{int} = 337$ G, $\Delta/k_B = 3.4$ K, and N = 60.

In recent work, Sohn[9] showed that the relatively large magnetic dipolar interaction between ferrous ions has a relatively minor effect on the satellite structure in the b direction. Rubins et al.[10], investigating the effect of chain length on the temperature dependence of the main-line intensity, found that the longer the chain the lower the was the temperature at which deviations from Eq. (3) became observable. The theory became inapplicable above 6 K because of the non-negligible contributions to the intensity of the higher sets of cluster states. Also, the increasing asymmetry of the line shape with increasing temperature may indicate that the system is entering the "fluctuation" region.

A remaining problem is the apparent discrepancy between the EMR data and those from magnetic and Mössbauer measurements regarding the canting angles of the Fe^{2+} ions.[2] While the latter data were used to deduce canting angles of magnitude 30°, the EMR data are consistent with a much smaller angle. Measurements of the rotational behavior of the 16 GHz EMR spectrum have shown splittings of the main line to be observable at angles of more than 60° from the b direction, with at least six components being present near 80°. These measurements have been used to estimate canting angles with a magnitude of about 5°.

3. EPR Detection of Solitons in the Two-dimensional Heisenberg Antiferromagnet

The two-dimensional classical Heisenberg magnet in the continuum approximation supports solutions which are localized nonlinear excitations called solitons.[11,12] These excitations have an energy $E_S = 4\pi Js^2$ independent of their size which results in thermodynamic quantities having an $\exp(E_S/T)$ temperature dependence due to thermal excitations of the solitons. Waldner[13,14] discussed the possibiltity of 2D solitons dominating the EPR linewidth in 2D antiferromagnetically exchange-coupled compounds; this could be noticed experimentally through the Arrhenius behavior $\exp(b/T)$ of the EPR linewidth above the Neel temperature. For a series of manganese compounds, the measured value of b was within 90% of E_S. However, these experiments do not show conclusively that solitons are observable because spin waves result is a similar Arrhenius behavior.

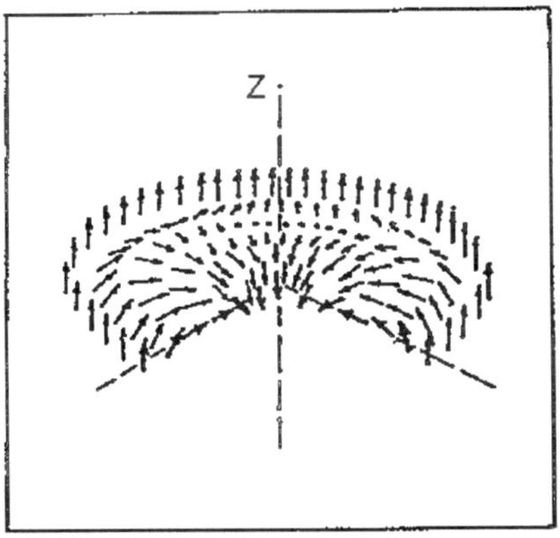

Fig. 6. The spin structure of the 2D magnetic soliton.

These nonlinear excitations were first studied mathematically in the ferromagnet by Belavin and Polyakov[12] as a mapping of the spin space sphere onto the 2D (x,y) lattice plane. Projection of the spins onto the lattice plane results in a vortex-like structure as shown in Fig. 6. The vortex has a large exchange energy at the center as a result of the antiparallel in-plane spin components in this region. However, this large contribution to the energy is minimized by allowing the spins to be approximately perpendicular to the lattice plane (in the -z direction) at the excitation center. Finally it is seen from Fig. 6 that spins far from the vortex center are in the z direction.

Our previous research has focused on the experimental observation of these excitations in 2D magnets and determination of whether solitons of spin waves dominate the dynamics of 2D magnets. To do this we have studied the temperature dependence of the EPR linewidth in the fluctuation region immediately above the ordering temperature. Theoretical calculations of how solitons affect the linewidth are described in detail elsewhere[16,17], but the main result confirms the Arrhenius behavior suspected by Waldner. These calculations have been extended to include the affect of a nonmagnetic impurity at the soliton center on the temperature-dependent EMR linewidth. The main results are summarized: Small solitons (the order of 10 lattice constants) have significant in-plane components at the soliton center resulting in larger energies. However, the presence of a nonmagnetic impurity at the center will eliminate these exchange interactions thereby lowering the energy. These lower energy solitons will tend for form at the impurities. Finally, the lower energy of these solitons will be observed as a different Arrhenius behavior of the EMR linewidth. This calculation was done for very low impurity concentrations (less than 1%) and the excitation energy was determined to decrease lineary as the concentration increased.

First we will review the results of the calculation for no impurities and compare this with the magnon contribution to the linewidth. The magnon contribution has been calculated by extending the result of Chakravarty and Orbach[18] for the quantum Heisenberg antiferromagnet. In Fig. 7 the good qualitative agreement between the observed temperature dependence and the theoretical curves can be seen. The dotted curve represents the contribution to the linewidth from magnons only while the solid curve represents the combined contribution from solitons and magnons. It is also evident that the soliton contribution is the dominant factor in the linewidth temperature dependence in the fluctuation region.

The experiments were done on (n-propylammonium)$_2$Mn$_{1-x}$Cd$_x$Cl$_4$, a quasi two-dimensional antiferromagnet with S = 5/2. The temperature dependent linewidths were obtained with a standard EPR X-band spectrometer with $0.001 < x < 0.01$. At room temperature the compounds exhibited similar angular and temperature dependence with characteristics of spin diffusion.[19] As the temperature is lowered the linewidth decreases until reaching a minimum at about 70 K and then increases sharply as the the Neel point (T_N = 39.2 K) is approached as shown in Fig. 8. The sharp rise in linewidth of course signals the fluctuation region which is dominated by the solitons.

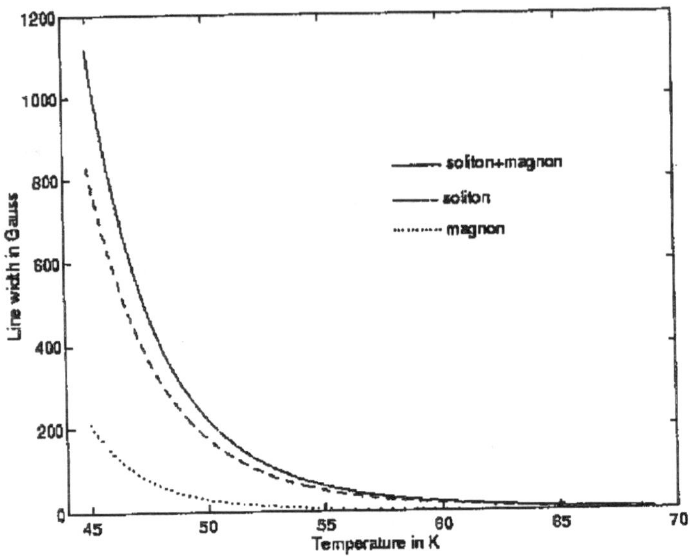

Fig. 7. Predicted linewidth versus temperature depicting the contributions from both solitons and magnons in the critical region.

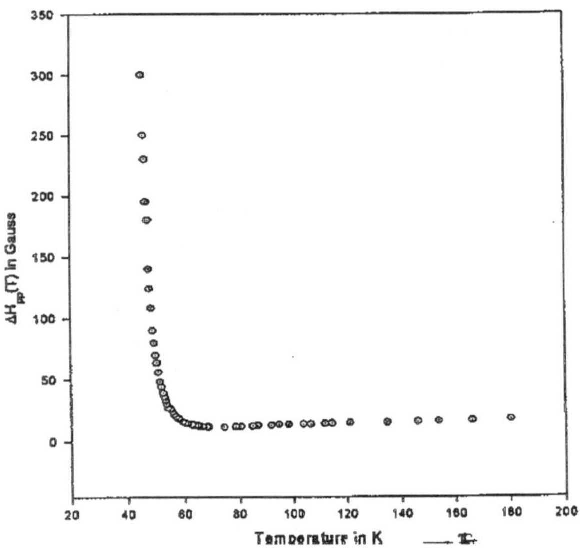

Fig. 8. Linewidth temperature dependence of a 0.15% Cd doped sample.

Fig. 9 shows the logarithm of the linewidth versus inverse temperature in the critical region for four different impurity concentrations. A linear fit to the data yields the soliton excitation energy. A comparison of these energies to those obtained by our calculation is shown in Fig. 10. We find good agreement between theory and experiment. AC susceptibility measurements on compounds with various concentrations of impurities shows little or no change in T_N as a function of impurity concentration. This precludes the possibilty of a variation in T_N causing the observed effect.

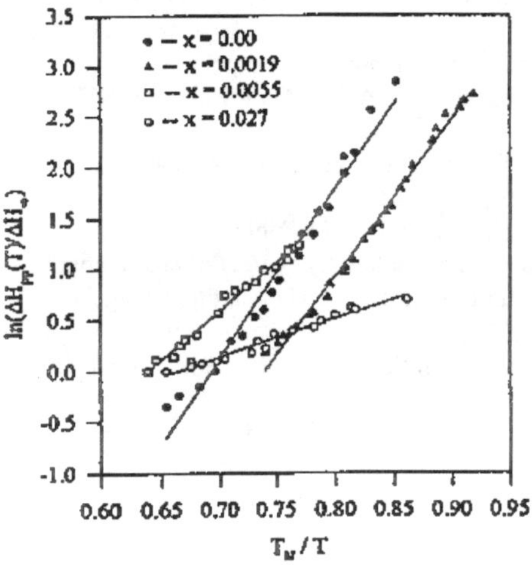

Fig. 9. Straight line fits to the logarithm of normalized linewidth vs. T_N/T yielding soliton energies as slopes.

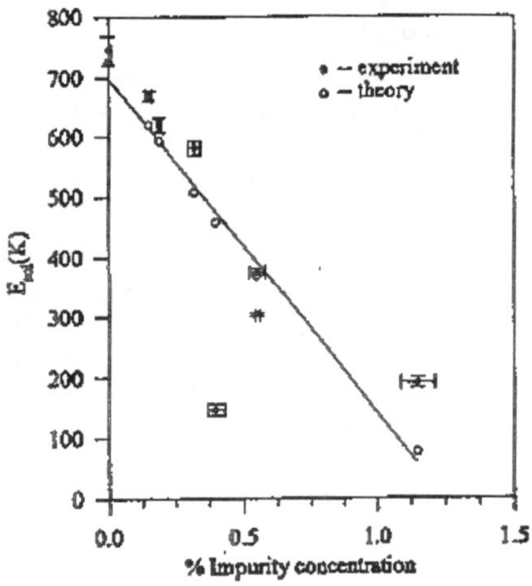

Fig. 10. Cadmiun doped data compared to theory

The dramatic decrease in excitation energy with very small concentration changes strongly suggests a non-linear mechanism. The spinwave spectrum will not noticeably change for impurity concentrations as small as a few per cent.[20] The good qualitative agreement between theory and experiment provides further evidence that solitons play a dominating role in the spin dynamics of the 2D Heisenberg antiferromagnet.

References

1. D. N. Haines and J. E. Drumheller, Bound magnon dominance of the magnetic susceptibility of the 1D Heisenberg ferromagnet cyclohexylammonium trichlorocuprate (II), *Phys. Rev. Letters* **58,** 2702 (1987); K. Ravindran, D. N. Haines, AC magnetic susceptibility of tetramethylammonium copper chloride (TMCuC): S=1/2, Heisenberg one-dimensional ferromagnet, *J. Appl. Phys.* **67**, 5481 (1990); Y. Liu and J. E. Drumheller, AC susceptibility of the 1D spin ½ Heisenberg ferromagnet $(C_6H_{11}NH_3)CuBr_3$, *Phys. Rev.* (submitted) and references therein.

2. R. E. Greeney, C. P. Landee, J. H. Zhang, and W. M. Reiff, One-dimensional Ising ferromagnet [$(CH_3NH]FeCl_3.2H_20$: magnetic properties, crystal stucture and Mössbauer spectroscopy, *Phys. Rev. B***39**, 12 200 (1989).
3. C. P. Landee, R. Keuntzler, and J. J. M. Williams, Specific heat measurements of a one-dimensional Ising ferromagnet FeTAC, *J. Appl. Phys.* **67**, 5604 (1990).
4. M. Date and M. Motokawa, Spin-cluster resonance in $CoCl_2.2H_2O$, *Phys. Rev. Letters* **16**, 1111 (1996).
5. J. B. Torrance, Jr. and M. Tinkham, Magnon bound states in anisotropic linear chains, *Phys. Rev.* **187**, 587 (1969).
6. K. Ravindran, S. Waplak, T. E. Grigereit, J. E. Drumheller, and R. S. Rubins, Spin cluster resonance at 9 GHz in the 1D Ising ferromagnet FeTAC, *Phys. Rev. B***47**, 14568 (1993).
7. R. S. Rubins, T. D. Black, K. Ravindran, T. E. Grigereit, and J. E. Drumheller, Microwave magnetic resonace in the 1D ferromagnet FeTAC, *J. Appl. Phys.* **73**, 6977 (1993).
8. R. S. Rubins, T. D. Black, A. Sohn, and J. E. Drumheller, Spin cluster resonance in an S=2 1D Ising ferromagnet, *Phys. Rev.* **49**, 15 366 (1994).
9. A. Sohn, *M. S. Thesis*, University of Texas, Arlington, pp 13-21 (1994).
10. R. S. Rubins, D K. Smith, T. D. Black, and J. E. Drumheller, The effect of chain length on the temperature dependence of a spin-cluster resonance in a 1D Ising ferromagnet, *IEEE Transactions on Magnetics* **31,** 4148 (1995).
11. T. H. R. Skyrme, Particle states of a quantized meson field, *Proc. Roy. Soc. London* **262**, 237 (1961).
12. A. A. Belavin and A. M. Polyakov, Mteastable states of two-dimensional isotropic ferromagnets, *JETP Letters* **22**, 245 (1975).
13. F. Waldner, Two-dimensional soliton contribution to the ESR linewidth in layered antiferromagnets? *J. Magn. Magn. Mat.* **31-34**, 1203 (1983).
14. F. Waldner, Are Skyrmions (2D solitons) observable in 2D antiferromagnets? *J. Magn. Magn. Mat* **104-107**, 793 (1992)
15. A. M. Kosevich, B. A. Ivanov, A. S. Kovalev, Magnetic solitons, *Phys Reports* **194**, 118 (1990).
16. C. E . Zaspel and J. E. Drumheller, Electron paramagnetic resonance detection of solitons in two-dimensional magnetics, *Int. Jour. Mod. Phys. B***10**, 3649 (1996).
17. C. E. Zaspel, T. E. Grigereit, and J. E. Drumheller, Soliton contribution to the EPR linewidth in the two-dimensional antiferromagnet, *Phys. Rev. Letters* **74**, 4539 (1995).
18. S. Chakravarty and R. Orbach, Electron and nuclear magnetic relaxation in La_2CuO_4 and related cuprates, *Phys. Rev. Letters* **64**, 224 (1990).
19. P. M. Richards and M. B. Salamon, Exchange narrowing of electron spin resonance in a two-dimensional system, *Phys. Rev.* **9**, 32 (1974).
20. R Alben, S. Kirkpatrick, and D. Beeman, Spin waves in random ferromagnets, *Phys. Rev. B***15,** 346 (1977).

Determination of Fractional Dimension by EPR and Spectroscopic Ellipsometry

D. Mo, Z. J. Jiang

Department of Physics, Zhongshan University, Guangzhou 510275, China

Abstract. The line shapes of EPR for fractional-dimensional solids were calculated. We propose a combined, comparative research on the determination of the fractional dimension by EPR and the spectroscopic ellipsometry. The difference between these two methods and the advantage of the comparative measurement are analyzed and discussed.

1 Introduction

Recent progress in low-dimensional condensed matter physics reveals that the dimensionality plays an important role in the properties of condensed matter. The properties of solids are strongly related to their dimension. Some low dimensional systems demonstrate a series of peculiar phenomena. The dimension crossover could happen in some one-dimensional solids under the condition of interchain coupling. The fractional dimension exists in some kinds of solids. In these cases, the dimension may be fractional.

Various experimental techniques have been applied in the past to measure the dimension of solids. Generally speaking, the value of the dimension of a solid is dependent on what kind of the dimension is defined.

We have proposed a definition of fractional dimension to present the anisotropy of the electronic energy band of low-dimensional solids and used it in the investigation of the dimension crossover. On the basis of this definition, a new method using the optical spectroscopy to determine the fractional dimension is suggested. On the other hand, electron paramagnetic resonance(EPR) is a powerful probe of spin localization, so it can also be used for the study of dimensionality.

In this paper, we develop line shape analysis of EPR spectra for the case of fractional dimension. The line shapes for solids with different fractional dimension were calculated. We propose a combined, comparative study of the determination of the fractional dimension of same polymer samples by EPR and spectroscopic ellipsometry(SE). The difference between these two methods and the advantage of the comparative measurements are analyzed and discussed.

2 Line Shape of EPR in Fractional-dimensional Solids

The line shape, as well as the line width on EPR spectrum, is determined by spin-spin relaxation. This relaxation should be different for different dimensional solids. Therefore, the line shape analysis of EPR spectra provides the possibility to determine the values of the dimension of various solids.

In comparison with the line width, we prefer to choose line shape analysis, because the line shape is more sensitive to the dimension of solids and easier for determining the dimension. Therefore, we calculate the line shape of EPR spectra for various dimensional solids.

The basic formulas used for the calculation of the EPR line shape can be found in text-books, such as that by Abragam (1961). These formulas are

$$\phi(t) = \exp[-\int_0^t (t-\tau)\psi(\tau)d\tau]\,, \tag{1}$$

$$I = A\int_0^\infty dt\phi(t)\cos\omega t\,, \tag{2}$$

where $\psi(\tau)$ is a correlation function of the magnetization of the system, as the function of time τ. $\phi(t)$ is a decay function of the transverse magnetization.

In the literature, some authors have used Equation 1 and Equation 2 to calculate the line shape for some specific EPR systems. Most authors calculated the EPR Hamiltonians of some specific solids and obtained the line shape and the corresponding parameters for various structures of the EPR systems. For example, Wang et al. (1992) have analyzed the EPR line shape of polyaniline, and obtained the values of J'/J for various quasi-one-dimensional samples, where J' is the spin interchain coupling and J is the spin in-chain coupling. From another point of view, a different J'/J corresponds to a different dimension. In this work, we perform the calculations of the phenomenological theory, so we do not calculate the specific spin interactions and EPR Hamiltonians, and use the phenomenological expressions of correlation functions.

A full calculation of the correlation function $\psi(\tau)$ is extremely difficult, but in the high-temperature approximation and in a range of enough large τ, $\psi(\tau)$ becomes a simple function. For one-dimensional systems, Hennessy et al. (1973) showed

$$\psi(\tau) = B\tau^{-1/2}\,. \tag{3}$$

And for three-dimensional systems, Gulley et al. (1970) gave

$$\psi(\tau) = B'\tau^{-3/2}\,. \tag{4}$$

We extend above equations to the fractional-dimensional case, and propose the following equation:

$$\psi(\tau) = C_n \tau^{-n/2} , \tag{5}$$

where n is the dimension that can be fractional, and C_n is the constant for the general case. Equation 5 is also a definition of the fractional dimension that is related to the an isotropic property of spin diffusion.

We put Equation 5 into Equation 1 and perform the integral calculation, then obtain the decay of the transverse magnetization:

$$\phi(t) = \exp(-D_n t^{2-n/2}) . \tag{6}$$

From Equation 2 and Equation 6, the line shapes of EPR for various dimensional cases were calculated. The line shapes of different dimensional solids are nearly the same in the linewidth region, but show appreciable differences in the region far away from the resonance line center . Therefore, we wish to study EPR absorption χ'' far from the line center H_0, and prefer to use the quantity $\chi''(H_0)/\chi''(H)$ as the function of $[(H-H_0)/\Delta H_{1/2}]^2$ to express the line shape, where $\Delta H_{1/2}$ is the half line width of a resonance line. Figure 1 shows the line shapes of EPR for various fractional dimension cases.

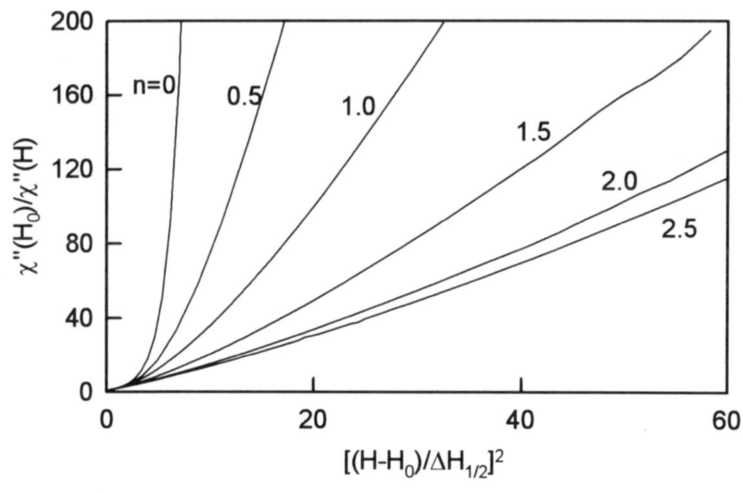

Fig. 1. Theoretical EPR line shapes for various dimensional solids.

3 Shape of Optical Spectra in Fractional-dimensional Solids

He and Mo (1986, 1990) have pointed out that the fractional calculus is useful for spectral analysis and deduced a formula to express the complex dielectric function in optical frequency range with the (n/2)th integral:

$$\varepsilon(E) = i^{r-n} \frac{M_n}{E^2} \frac{d^{-n/2}}{dE_g^{-n/2}} \frac{1}{E_g - E - i\Gamma}, \tag{7}$$

where M_n is a constant, r is the value of the type of critical point, E is the photon energy, E_g is the threshold energy at the critical point, Γ is the broadening parameter, and n is the dimension of solid. The value of dimension n may be a fraction The calculated shapes of imaginary parts of the complex dielectric function spectra are shown in Fig. 2.

From the data measured by the spectroscopic ellipsometry, one will obtain the spectra of the complex dielectric function, and then can determine the value of dimension by data-fitting with Equation 7.

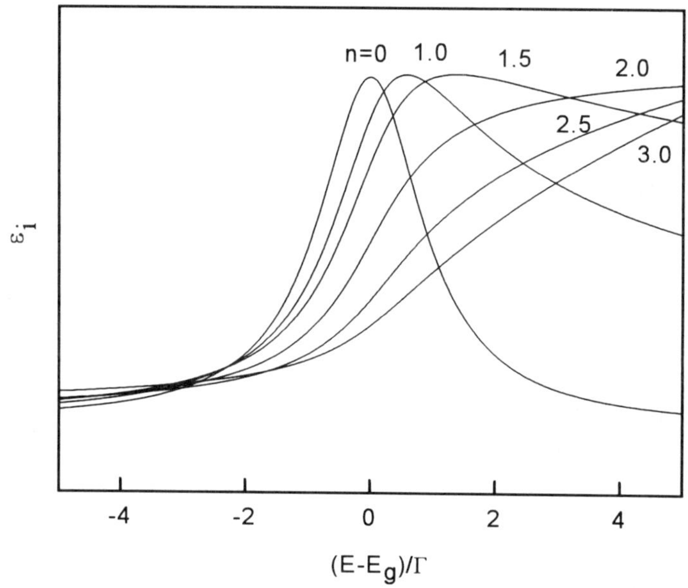

Fig. 2. Theoretical shapes of imaginary parts of the complex dielectric function spectra for various dimensional solids.

4 Combined Study by EPR and Spectroscopic Ellipsometry

We suggest a research project on a combined measurement by EPR and SE for the determination of the fractional dimension. The conducting polymer--polyaniline is one of the available objects for this research. After protonation treatment, the conductivity of polyaniline will greatly increase and the dimension should increase from ~1 to a higher value between 2 and 3. Analyzing the measured EPR and SE data of a series of polyaniline samples with different conductivities, one will obtain two series of fractional dimension values.

Being different from some other dimension-determination methods, which are based on geometric, atomic, lattice-vibrational properties, etc., both EPR and SE are based on the electronic properties. In the EPR case, the line shape analysis is mainly dependent on the exchange interactions of the magnetic moments. In the SE case, the shape of the spectra is mainly dependent on the transitions between the electronic energy levels. Therefore, the general trend of the change of fractional dimension in the samples should be the same, but the exact values determined from two methods may be somewhat different. Because the determination of the fractional dimension is not so direct and the meaning of fractional dimension is dependent on the method used, the comparison of both measured values should be helpful.

References

1. A. Abragam, *Principles of Nuclear Magnetism*, Oxford (1961) pp. 97-103 and 427-441.
2. Z. H. Wang, E. M. Scherr, A. G. MacDiarmid, and A. J. Epstein, *Phys. Rev.* **B45** (1992) 4190-4202.
3. M. J. Hennessy, C. D. McElwee, and P. M. Richards, *Phys. Rev.* **B7** (1973) 930-947.
4. J. E. Gulley, D. Hone, D. J. Scalapino, and B. G. Silbernagel, *Phys. Rev.* **B1** (1970) 1020-1030.
5. X.-F. He, and D. Mo, *Chinese Phys. Lett.* **5** (1986) 565-568.
6. X.-F. He, R.-R. Jiang, and D. Mo, *Phys. Rev.* **B41** (1990) 5799-5805.

An ESR Study of Ce^{3+} in $LiCaAlF_6$

M. Yamaga[1], *T. Yosida*[2], *D. Lee*[3], *T. P. J. Han*[3], *H. G. Gallagher*[3], *B. Henderson*[3], *and K. Kindo*[4]

[1]Department of Electronics, Faculty of Engineering, Gifu University, Gifu, 501-11, Japan
[2]Nakanihon Automotive College, Kamo, 505, Japan
[3]Department of Physics and Applied Physics, University of Strathclyde, Glasgow, G1 1XN, United Kingdom
[4]Research Center for Materials Science at Extreme Conditions, Osaka University, Toyonaka, 560, Japan

Abstract. Electron spin resonance (ESR) spectra of Ce^{3+}-doped $LiCaAlF_6$ (LiCAF) crystals measured at low temperature are reported. The ESR spectra reveal three distinct Ce^{3+} centers in the crystal. The most intense line may be fitted to a spin Hamiltonian with trigonal symmetry and is associated with Ce^{3+} ions that substitute for Ca^{2+} ions in the LiCAF structure. Two other lines have orthorhombic symmetry: they are assigned to Ce^{3+} ions with charge compensating Li^+ ion vacancies in the close vicinity of the Ce^{3+} impurity ion.

1 Introduction

There is much current interest in Ce^{3+}-doped ionic crystals for potential application in UV scintillators and lasers [1,2]. The Ce^{3+} ion in its lowest energy state has a single $4f$ electron outside the closed shells, and the ${}^2F_{5/2}$ ground state is separated from the lowest lying excited state ${}^2F_{7/2}$ by the spin-orbit energy of approximately 2000 cm^{-1}. Higher configurations correspond to the promotion of the $4f$ electron into the $5d$ orbital, some 30000-40000 cm^{-1} higher in energy. The $4f - 5d$ transition is used as a laser transition in the UV region [2].

This paper characterizes the Ce^{3+} dopant in $LiCaAlF_6$ (LiCAF) crystals by electron spin resonance (ESR), an experimental technique which determines the local structure of the rare-earth ion site in such crystals. Such sub-atomic microscopy reveals crystal field distortions intrinsic and extrinsic to the impurity ion. Intrinsic lattice defects are an important source of distortion of the Ce^{3+} environment in fluoride crystals especially when the Ce^{3+} differs in charge from that at the substitutional site: charge compensation then becomes necessary.

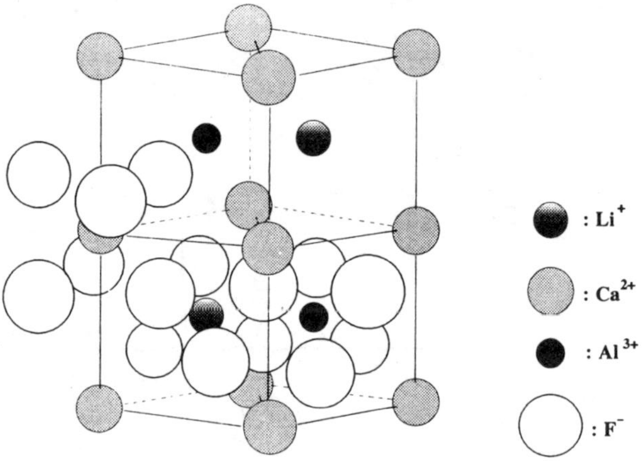

Fig. 1. The crystal structure of $LiCaAlF_6$.

2 Crystal Growth and Experimental Procedure

LiCAF crystallizes trigonal in a space group $P\bar{3}1c$ as shown in figure 1 [3]. The lattice constants are $a = 4.996$ Å and $c = 9.636$ Å. Ca^{2+} ions occupy the center of an octahedron of six fluorine ions. Taking into account the ionic radii of Ca^{2+} (1.06Å), Li^+ (0.78 Å), Al^{3+} (0.57 Å) and Ce^{3+} (1.18 Å) ions, it is expected that Ce^{3+} ions substitute for Ca^{2+} ions. The replacement of Ce^{3+} requires charge compensation, for example, by cation vacancies.

Single crystals of LiCAF doped nominally with 0.5 at.% Ce were grown by the vertical Stockbarger-Bridgman technique using high purity (4N) fluoride components and graphite crucibles. The ESR measurements were made with a Varian X-band spectrometer with 270 Hz field modulation in the temperature range 4.2-70 K.

3 Experimental Results

Figure 2 shows the ESR spectra of Ce^{3+} in LiCAF measured at 4.2 K, a microwave power level of 0.01 mW and a microwave frequency of 9.304 GHz with magnetic field B parallel to the c- and a-axes of the crystal. The ESR signal is saturated even at such low microwave power levels, such that positive signal of the first derivative appears at low magnetic field side when the field is scanned from low to high. There are three ESR lines, denoted by A, B and C in figure 2, when magnetic field is parallel to the c-axis. This result shows there to be

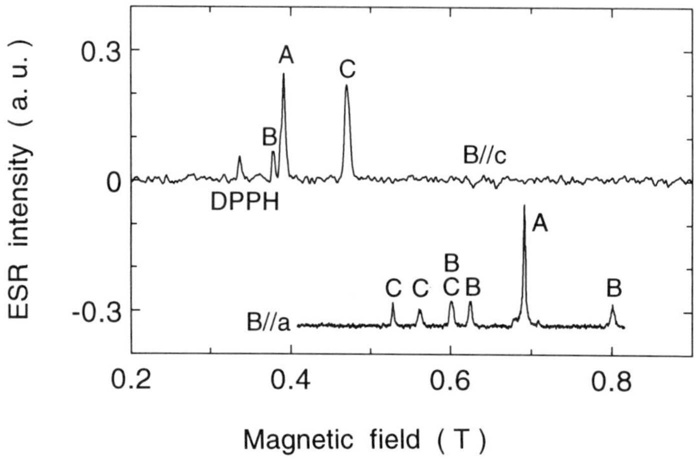

Fig. 2. The ESR spectra of Ce^{3+} in $LiCaAlF_6$ measured at 4.2 K and a microwave frequency of 9.304 GHz and with $B \parallel c,\ a$.

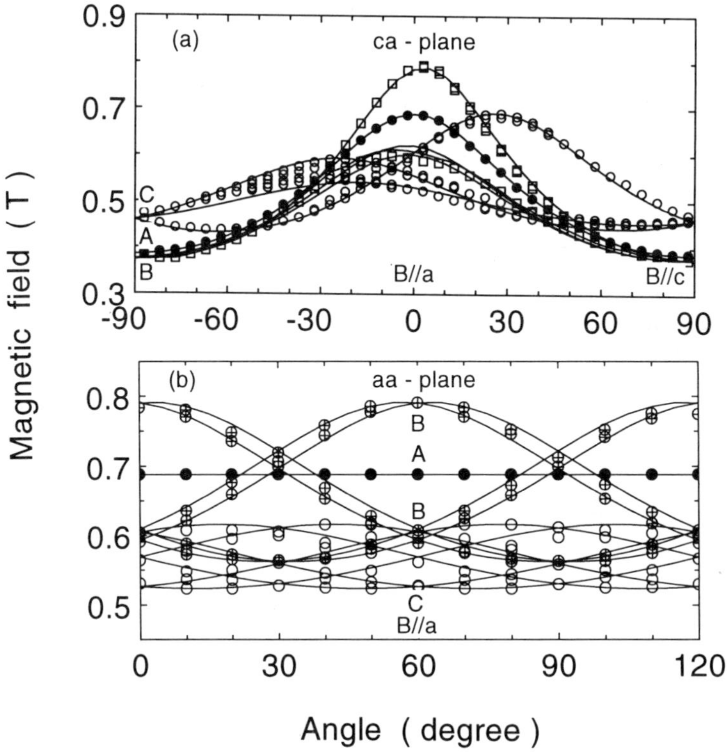

Fig. 3. The orientation dependence of the ESR lines in (a) the *ca*- and (b) *aa*-planes.

Table 1. Spin Hamiltonian parameters and angles of the principal z-axis of Ce^{3+} in $LiCaAlF_6$.

line	A	B	C
g-value	$g_{\parallel}$=1.725(5)	g_z=1.77(1)	g_z=1.54(1)
	$g_{\perp}$=0.965(5)	g_x=0.84(1)	g_x=0.95(1)
		g_y=1.18(1)	g_y=1.27(1)
angle		θ=3(1)	θ=26(2)
		$\phi = \pm$ 4(1)	$\phi = \pm$ 12(2)

three Ce^{3+} centers. On the other hand, the ESR spectrum with $B \parallel a$ consists of more than three lines. Figure 3 shows the angular dependence of the ESR spectra when magnetic field is applied in the *ca*- and *aa*-planes. The intense A line denoted by solid circles does not split at any direction of the magnetic field in these planes. The resonance field is maintained at a constant value in the *aa*-plane. This line is due to a center with trigonal symmetry.

The B and C lines split into four or six lines at any direction in the *ca*- and *aa*-planes: they are due to two different orthorhombic centers. The angular dependence of these ESR spectra may be fitted to a spin Hamiltonian with effective spin $S = 1/2$,

$$\mathcal{H} = \mu_B g_x B_x S_x + \mu_B g_y B_y S_y + \mu_B g_z B_z S_z, \tag{1}$$

where μ_B is the Bohr magneton, the principal z-axis being parallel to the magnetic field direction at which the g value is the largest, taking into account the g values of the A line. The direction of the principal z-axis is defined as angles, θ and ϕ, of polar coordinates, which are rotated from the c- and a-axes, respectively. Solid curves in figure 3(a) and (b) calculated using equation (1), spin Hamiltonian parameters and angles in Table 1 fit the measured data points quite well. The small differences are due to a misalignment of the crystal axes with respect to the magnetic field direction.

4 Discussion

The ${}^2F_{5/2}$ ground state is split into three Kramers doublets by the crystal field. The crystal field Hamiltonian for the case of trigonal symmetry (C_{3v}) [4,5] is given as

$$\mathcal{H}_{cry} = B_2^0 O_2^0 + B_4^0 O_4^0 + B_4^3 O_4^3 + B_6^0 O_6^0 + B_6^3 O_6^3 + B_6^6 O_6^6, \tag{2}$$

where the O_n^m are spin operators and the B_n^m are determined experimentally. The terms O_n^0 $(n = 2, 4, 6)$ and O_6^6 do not mix three wavefunctions $|J, J_z >= |5/2, \pm 1/2 >$, $|5/2, \pm 3/2 >$, and $|5/2, \pm 5/2 >$ in ${}^2F_{5/2}$. The g-values calculated using these pure wavefunctions are not in agreement with the observed

g-values. The terms O_4^3 and O_6^3 mix $|5/2, \pm 1/2>$ and $|5/2, \mp 5/2>$. The mixed wavefunctions are written in the form of

$$|\pm\frac{1}{2}^{'}>= cos\theta|\frac{5}{2}, \pm\frac{1}{2}> \pm sin\theta|\frac{5}{2}, \mp\frac{5}{2}> \tag{3}$$

and the g-values are

$$g_{\|} = \frac{6}{7}|cos^2\theta - 5sin^2\theta|, \tag{4}$$

$$g_{\perp} = \frac{6}{7}(3cos^2\theta). \tag{5}$$

The measured g values of the A line are $g_{\|} = 1.725, g_{\perp} = 0.965$. The $g_{\|}(= g_z)$ and $g_{\perp}(= 1/2(g_x + g_y))$ values of the B and C lines are put approximately to be (1.77, 1.01) and (1.54, 1.11), respectively, which are close to those of the A line. The values ($g_{\|} = 1.99, g_{\perp} = 1.15$) calculated using first-order-theory equations (4) and (5) with θ=48° are slightly different from the experimental data for the A, B and C lines. The differences may be removed by taking into account further admixtures of $|7/2, \pm 7/2>$, $|7/2, \pm 1/2>$, $|7/2, \mp 5/2>$ of $^2F_{7/2}$ by perturbation theory in second-order.

Next consider the environment of Ce^{3+} ions in LiCAF. The A line with trigonal symmetry is associated with Ce^{3+} ions that substitute for Ca^{2+} ions in the LiCAF structure. The B and C lines have orthorhombic symmetry. They are assigned to Ce^{3+} accompanied by Li^+ vacancies. Assuming that there is a single Li^+ ion vacancy in a unit cell in figure 1 and that an Ce^{3+} ion substitutes for one of the twelve Ca^{2+} ions, the Ce^{3+} ion is classified with four groups with different distances between Ca^{2+} and Li^+ ions. The numbers of equivalent structural configurations with the distances of 3.76 Å, 6.25 Å, 7.78 Å, and 9.25 Å are 6, 2, 3, and 1, respectively. The principal z-axis of the C line is rotated through angles of 26° and ±12° from the c- and a-axes, respectively. This line may be assigned to Ce^{3+} ions with the smallest distance (3.76 Å) and highest probability. The configuration with the second smallest distance (6.25 Å) may be associated with the B line because this Ce^{3+} with large difference in g_x and g_y is perturbed strongly in the aa-plane. The resonances from Ce^{3+} with two other configurations may be expected to appear close to the resonance field of the intense A line and produce an inhomogeneous width of the A line.

Acknowledgments

In Japan, the work has been supported by a Grant-in-Aid for Scientific Research on Priority Area "New Development of Rare Earth Complexes" (No 07230233, 08220228) and on Fundamental Research (No 07650049) from The Ministry of Education, Science and Culture, and by joint research (1995-1996) between Japan Society of the Promotion of Science and Royal Society/British Council. At Strathclyde, the research programme has been funded by joint EPSRC/MOD research grants (GR/F/54105 and GR/H/66143).

References

1. G. Blasse and B.C. Grabmaier, *Luminescent Materials*, Springer-Verlag (1994).
2. P. Moulton, *Laser Handbook vol. 5* edited by M. Bass and M.H. Stitch, North-Holland (1985) pp.282-285.
3. W. Viebahan, Untersuchungen an quaternären Fluoriden $LiMe''Me'''F_6$ Die Struktur von $LiCaAlF_6$, *Z. ang. allg. Chem.* **386** (1971) 335-339.
4. A. Abragam and B. Bleaney, *Electron Paramagnetic Resonance of Transition Ions*, Oxford University Press (1970) ch.5.
5. B. R. Judd, The magnetic and spectroscopic properties of certain rare-earth double nitrates, *Proc. Roy. Soc.* **A232** (1955) 458-474.

Spin-lattice Relaxation Times in Amorphous Materials as Effected by Exchange Interactions and TLS Centers

Sushil K. Misra

Physics Department, Concordia University, 1455 de Maisonneuve Boulevard West, Montreal, Quebec H3G 1M8, Canada

Abstract. The mechanisms involving the exchange interaction and TLS centers predominant in effecting electron spin-lattice relaxation at low temperatures in amorphous materials are reviewed. It has been deduced that both of these lead to quadratic temperature dependences at very low temperatures and linear temperature dependences at intermediate temperatures of the relaxation rate in amorphous materials.

1 Introduction

In the last two decades rather unique properties have been reported on amorphous systems in contrast to ionic crystals; the strength and temperature dependence of several parameters have been found to be very unusual. In particular, the homogeneous optical linewidth was observed to be anomalously large, exhibiting a T^2 temperature dependence at low temperatures, which changed into a linear dependence on T at higher temperatures.[1] These dependences were well explained by Lyo and Orbach[1] with the help of interaction with TLS (tunneling level states) centers.

While a large volume of experimental data have been reported on electron spin-lattice relaxation (SLR) in ionic crystals, the reported data are rather scarce on amorphous materials. Some of the interesting studies reported on relaxation-time studies in amorphous materials are as follows: measurements in irradiated ethanol glasses by Bowman and Kevan,[2] in irradiated polymers by Velter-Stefanescu et al.,[3] in amorphous silicon by Gourdon et al.,[4] and by Askew et al.,[5] observation of a linear dependence of T_1^{-1} in DPPH between 1 and 300 K by Swanenburg et al.[6] and by Turkevitch et al.,[7] and the observation of a T^2 followed by T dependence of T_1^{-1} by Kurtz and Stapleton[8] in irradiated ß-alumina at very low temperatures. In addition, rather strong dependence of T_1^{-1} on temperature in glasses doped by Yb^{3+} as reported by Antipin et al.,[9] and by Stevens and Stapleton[10] and by Vergnoux et al.[11] seems to be very unusual in such systems.

For relaxation in amorphous systems, Bowman and Kevan[2] have employed a mechanism involving phonon modulation of electron-nuclear dipolar interaction, the nuclei interacting with the electronic spin constitute TLS (tunneling level states) centers. Kurtz and Stapleton,[8] on the other hand, proposed that the Fermi-contact hyperfine interaction of the electron spin with the cation nuclei in the vicinity modulated by phonons was instrumental in effecting electron relaxation.

Velter-Stefanescu et al.[3] considered that cross relaxation via exchange pairs were responsible for electron spin-lattice relaxation in irradiated polymers.

In the case of amorphous systems doped with paramagnetic centers characterized by partially, or totally, quenched orbital angular momentum (L~ 0), relaxation due to phonon modulation of the crystal field is insignificant, since, in this case, there exists no direct interaction between the paramagnetic center and the crystal field. Thus, for amorphous materials one should consider three principal mechanisms that have been proposed for spin-lattice relaxation in amorphous systems which invoke: (i) Interaction between exchange-coupled pairs, directly modulated by lattice vibrations. (ii) Cross relaxations between an isolated ion and an exchange-coupled electron pair, the latter being perturbed by lattice vibrations. (iii) Electron-nuclear dipolar (END) [2] or Fermi-contact (FC) hyperfine (HF) interaction[8] between electron magnetic moment and LTS centers, the latter being modulated by lattice vibration or phonons.

Mechanisms (i) and (iii) lead to a temperature variation of T_1^{-1} as the temperature is increased. First, T_1^{-1} varies as T^2 at very low temperatures, and then as T at intermediate temperatures. The latter is in contrast to the direct process which only takes place at very low temperatures in ionic crystals, at temperatures less than 4 K. On the other hand, mechanism (ii) in which cross relaxations play an important role, leads to a plateau in the relaxation rate at higher temperatures. The distribution over a wide range of energy (E) of TLS centers and that of strength (J) of exchange interaction responsible for the formation of exchange pairs are characteristics unique to amorphous materials as compared to ionic crystals; in the latter case E and J have rather well-defined sharp values. In amorphous materials, T_1 has been observed to be several orders of magnitude shorter than those reported for single crystals at low temperatures, e.g. in molecular glasses,[12-14] and irradiated ß-alumina.[8]

2. Relaxation via TLS Centers

In the localized-tunneling-level states (TLS) model, applicable specifically to amorphous materials, a group of atoms in amorphous materials resides in asymmetric double-well potentials, characterized by two closely-spaced energy levels. Tunnelling occurs through the barrier separating the two walls. There are two mechanisms of SLR involving TLS centers: electron-nuclear dipolar (END)[2] and Fermi-contact (FC) hyperfine interaction[8] of the relaxing electron with the neighboring nuclei which serve as LTS centers.

2.1 Electron-Spin-Lattice Relaxation: Anderson, Halperin, and Varma and Phillips (AHVP) TLS model

A property unique to amorphous systems is the existence of two level tunneling states (TLS) discussed in detail by Phillips,[15] and Anderson, Halperin and Varma.[16] It follows that all three interactions - FC, END and exhange lead to T^2 followed by a T dependence of the spin-lattice relaxation rate on temperature at low temperatures. From Anderson, Halperin, Varma,[16] and Phillips [15] (AHVP) TLS models, the total average electron- and nuclear-spin-lattice relaxation rates can be expressed by extending the derivation of Estalji, et al [17] as

$$\left(\frac{1}{T_1}\right)_{Tol} = \frac{2}{\hbar^2}\left|\left(\sigma_1\left|H_{T-S}\right|\sigma_2\right)\right|^2 \int_0^{\Delta_m}\int_0^{V_m} \frac{f(\Delta, V)}{\cosh^2\left(\frac{\Delta}{2KT}\right)} \frac{\tau}{1+\omega^2\tau^2} d\Delta\, dV, \qquad (1)$$

where $f(\Delta, V)$ is the mode distribution of the barrier height V between the two wells and of their difference Δ, Δ_M and V_M are the maximum energy difference and maximum barrier height, respectively, of the ADWP model. The correlation time τ_0 is related to the relaxation time of the TLS states. If the TLS states are assumed to be in thermal equilibrium at temperature T, then [15, 18]

$$\tau^{-1} = A\Delta_0^2 E \coth[E/2kT] + \tau_0^{-1}\cosh[\Delta/2kT]\exp[-V/kT]. \qquad (2)$$

The first term describes the one-phonon-assisted tunneling with Δ_0 the ground-state tunnel splitting, $E = \left(\Delta_0^2 + \Delta^2\right)^{1/2}$ and A the strength of the phonon-tunneling coupling, while the second term represents the classical rate process in an asymmetric double-well potential. Different suggestions exist in the literature concerning the distribution function $f(\Delta,V)$.[18-20] Mostly a separation of the form $f(\Delta,V) = \rho(\Delta)g(V)$ has been used, where $\rho(\Delta)$ is assumed to be uniform or weakly dependent on Δ,[20] $\rho(\Delta) \propto \Delta^\gamma$, $0 \le \gamma < 1$, while $g(V)$ is found to decrease rapidly with rising barrier height. Kohler and Friedrich[21] obtained a relation $g(V) \propto 1/\sqrt{V}$ for a number of polymers, while Phillips[22] proposed a distribution $g(V) = \mathrm{sech}(V/V_0)$.

As expected, however, from the integral form of Eq. (1) the results are not very sensitive to the exact form of $f(\Delta,V)$, and the widths of the distribution functions and the cut-off energies Δ_m and V_m are the leading parameters in the equation. For temperature T below V_m/k and Δ_m/k, Eq. (1) can be expressed approximately as a power law,

$$T_1^{-1} \propto T^{1+\alpha}\omega^{-\beta} \qquad (3)$$

with $0<\alpha<1$, and $0.6<\beta<1$, in agreement with the experiments.[25] The spin-lattice relaxation mechanisms via TLS centers explain satisfactorily the magnitude and temperature dependence observed in a number of experiments. There are many factors which determine the relaxation behavior: (i) the distribution function of the disorder modes; (ii) tunneling and activated-disorder-mode relaxation; (iii) coupling between phonons and disorder modes. The nature of the TLS-spin interaction is still an open problem and, admittedly, no information on the detailed mechanism of this interaction can be deduced by most of the available experiments data.

2.2 SLR due to Electron-nuclear Dipolar Coupling (END) to a TLS Center

Bowman and Kevan[2] measured the spin-lattice relaxation rates of trapped electrons in three frozen ethanol glass samples C_2H_5OH, C_2D_5OH, and C_2H_5OD between 5 and 150K. They found that (a) $T_1^{-1} \propto T$ for trapped electrons in ethanol glasses between 7 and 100K, (b) T_1 is sensitive to deuteration of glasses; and (c) the relaxation rate of trapped electrons is about 10^5 times larger for the $T_1^{-1} \propto T$ process in ethanol glasses than in KCl crystals. They were the first to invoke the TLS model to explain the relaxation mechanism. Basically, in this process, an electron spin interacts by END interaction with a neighboring nucleus characterized by TLS. The coupling of this nucleus to lattice vibrations (phonons) effects SLR. As for the resulting temperature dependence in amorphous materials, it is described by an integral of the following form

$$I_{END} = \int_0^\infty \varepsilon \operatorname{cosech}(2\varepsilon / kT)\; n(\varepsilon)\, d\varepsilon . \tag{4}$$

In Eq. (4), 2ε is the energy difference between the two tunnelling states in the double-potential well of the tunnelling particle, and $n(\varepsilon)$ is the density of tunneling modes with an energy splitting between tunneling levels of 2ε. The integral I_{END} is a consequence of the amorphous nature of the sample, i.e. there exists a distribution of the values of ε. I_{END}, as given by Eq. (4), can be shown to exhibit a T^2 dependence at very low temperatures, which changes to a linear (T) dependence as the temperature is increased.

2.3 Electron-spin-lattice Relaxation of Several Simple TLS State Distribution Functions as Discussed by Bowman and Kevan[2]

The electron spin-lattice relaxation is now considered. Using $n(E)dE$ to replace $f(\Delta, V)\, d\Delta\, dV / \cosh^2\left(\frac{\Delta}{2KT}\right)$, and B to replace the constant $\frac{2}{\hbar^2}\left|\left(\sigma_1\middle|H_{T-S}\middle|\sigma_2\right)\right|^2$ in Eq. (1), one obtains

$$\frac{1}{T_1} = B\int_0^{E_m} \frac{\tau_0}{1+\omega^2\tau_0^2}\, n(E)\, dE . \tag{5}$$

Measurements[24] on TLS states indicate that $\omega\tau >> 1$, and at low temperatures $V/kT >> 1$. Thus, considering only the first term of Eq. (2):

$$T_1^{-1} = \langle CD \rangle \int_0^{E_m} E \coth^{-1}(E/2kT)\, n(E)\, dE , \tag{6}$$

where $\langle CD\rangle = AB(\Delta_0^2/\omega^2)$. The actual distribution of TLS-states splitting is not known in glasses. However, from thermal properties of glasses, we know that n(E) must be small for large E and must effectively vanish for E/k to be of the order of 50 K. It is reasonable to replace E_m by ∞. Four different distributions of tunneling-states splitting are considered to examine the effect of the form of n(E) on the temperature dependence of the spin-lattice relaxation rate.

2.3.1 Delta Function Distribution :

$$n(E) = \delta(E-E'),$$

which corresponds to identical tunneling modes. Eq. (2) gives[24, 25]

$$T_1^{-1} = \langle CD\rangle E'\operatorname{csch}(E'/2kT) \approx \langle CD\rangle(2kT/E'),\ \text{for } 2kT >> E', \tag{7}$$

which yields a different high temperature limit from the usual $T_1^{-1} \propto T^2$ limit for most relaxation mechanisms.

2.3.2 Inverse-E Distribution:

$$n(E)=0 \text{ for } E < \Delta_0,\ n(E)=E^{-1} \text{ for } E \geq \Delta_0, \tag{8}$$

which yields

$$T_1^{-1} \propto \langle CD\rangle\int_0^{E_m} E\coth^{-1}(E/2kT)\,n(E)\,dE = \langle CD\rangle 2kT\ln[\coth(\Delta_0/2kT)].$$

For $kT >> \Delta_0$, it becomes

$$T_1^{-1} \propto \langle CD\rangle 2kT\ln(2kT/\Delta_0). \tag{9}$$

In any reasonable temperature range above 4 K, Eq. (9) is very close to being linearly proportional to temperature T.

2.3.3 Step-function Distribution

$$n(E)=0 \text{ for } E<\Delta_0; \quad n(E)=\text{constant for } \Delta_0\leq E\leq E_m; \quad n(E)=0 \text{ for } E > E_m.$$

The average spin-lattice relaxation rate is then given as

$$T_1^{-1} \propto \langle CD \rangle 4kT \sum_{n=0}^{\infty} \frac{\exp\left(-(2n+1)\Delta_0/2kT\right)}{(2n+1)^2} \times$$
$$\left\{1 - \exp\left(-(2n+1)p/2kT\right)\left(kT + \Delta_0(2n+1)\right) - (2n+1)p\exp\left(-(2n+1)^2 p/2kT\right)\right\}, \tag{10}$$

where the csch function has been expanded as an infinite series; $p = E_m - \Delta_0$. When $p << kT$, $T_1^{-1} \propto \mathrm{csch}(\Delta_0/2kT)$. When p is no longer much smaller than kT, which occurs as the temperature is lowered and the range of splitting of tunneling states becomes comparable to the thermal energy, the spin-lattice relaxation rate becomes a much stronger function of temperature.

2.3.4 Low-energy Distribution:

$$n(E) \neq 0, \text{ for } E << kT \quad ; = 0, \text{ otherwise}.$$

Csch(ε/2kT) can be expanded keeping only the first term, 2kT/ε:

$$T_1^{-1} \propto \langle CD \rangle \int_0^{\infty} \mathrm{Exp}(2kT/E)\, n(E)\, dE = \langle CD \rangle 2kT \int_0^{\infty} n(E)\, dE. \tag{11}$$

The integral is just the number of tunneling states, independent of temperature, so that $T_1^{-1} \propto T$.

All of these functions have one feature in common that, above a certain temperature, the spin-lattice relaxation rate is linearly proportional to the temperature of the sample. The first three distribution yield much stronger temperature dependences at very low temperatures. This is in strong contrast to the relaxation mechanisms due to lattice phonons which have been well studied in single crystals doped with paramagnetic transition metal ions.

2.4 SLR due to Fermi-contact (FC) Hyperfine Interaction with a TLS Center

Spin-lattice relaxation measurements on color centers in K, Li, and Na β-alumina[8] revealed expectionally fast relaxation with anomalous temperature and microwave frequency dependences. This behavior was quantitatively described by a mechanism involving the coupling via fermi-contact interaction of a colour center to the phonon-induced relaxation of a nearby localized TLS. This process is discussed well by Kurtz and Stapleton,[8] and is basically the same as that described in Sec. 2.2 above, except that here the interaction is FC instead of END. The temperature dependence for this case in amorphous systems can be shown to be described by the following integral in the notation of Ref. [8]:

$$I_{FC} = \int_0^{\infty} E \operatorname{cosech}(E/kT) n(E) dE \qquad (12)$$

In Eq. (12), E is the energy splitting between the two levels of LTS and n(E) is the density of tunnelling modes. The integral in Eq. (12) arises due to distribution of E values in amorphous materials. Similar to Eq. (4), this leads to a quadratic dependence on temperature at very low temperatures and a linear dependence on temperature of spin-lattice relaxation rate at intermediate temperatures.

3. Influence of Exchange Interaction on Relaxation Rate (T_1^{-1})

3.1 Single crystal

The details of how spin-lattice relaxation of members of exchange-coupled pairs is effected by lattice vibrations, as discussed by Gill,[26] adapted from the theories developed, among others, by Waller[27] and Al'tshuler,[28] are briefly reviewed. The spin Hamiltonian for a pair of spins $\mathbf{S_1}$, $\mathbf{S_2}$ may be written:

$$H = g\mu_B(\mathbf{S_1} + \mathbf{S_2})\cdot\mathbf{H} + D[S_{1z}^2 + S_{2z}^2 - (2/3)S(S+1)] + J\,\mathbf{S_1}\cdot\mathbf{S_2}$$
$$+ (g^2\mu_B^2/r_{12}^3)\,[\mathbf{S_1}\cdot\mathbf{S_2} - (3/r_{12}^2)\,(\mathbf{S_1}\cdot\mathbf{r})(\mathbf{S_2}\cdot\mathbf{r})] + \text{anisotropic exchange terms} \qquad (13)$$

In Eq. (13), S is the magnitude of spins S_1 and S_2, g is the Landé factor, μ_B is the Bohr magneton, $\mathbf{r_{12}}$ is the vectorial distance from spin 1 to spin 2, D is the zero-field splitting parameter and J is the exchange-interaction constant. The last terms are due to exchange splitting in excited orbital levels arising via the spin-orbit coupling; these terms may be ignored since the greatest one of them is of the order $J\lambda/\Delta$, where Δ is the energy of these levels. If J>>D, the energy levels form multiples characterized by effective spins S′. Microwave transitions may be observed between levels belonging to the same multiples S′.

3.1.1 Relaxation of Members of Exchange-coupled Pairs at Low Temperatures as Effected by Modulation of the Parameter J by Lattice Vibrations (phonons) in Single Crystals

This process is similar to that proposed by Van Vleck[29] to account for relaxation of single ions. It is expected to be faster than that of the single ion by a factor ~ $(E_p/E_s)^2$ where E_p, E_s are relevant phonon energies. In this process, lattice vibrations are instrumental in inducing relaxation of an ion coupled to another by exchange interaction at lower temperatures. The probability of relaxation from level i to j is:

$$P_{ij}^{D} = \frac{8\pi^3}{3h^4}\frac{r^2}{\rho v^5}\frac{E_{ij}^{\;3}}{[1-\exp\left(-E_{ij}/kT\right)]}\sum_{q=x,y,z}\left|\left\langle i\left|\frac{dU}{dq}\right|j\right\rangle\right|^2, \qquad (14)$$

where E_{ij} is the energy difference (E_i-E_j), ρ is the density of crystal, v is the average velocity of sound, r is the distance between the two ions of the pair, and U is the interaction whose modulation by displacements in direction q due to lattice vibrations brings about relaxation. If J is sufficiently small for anisotropic exchange to be negligible the interaction dU/dq may be identified with (dJ/dr)($S_1 \cdot S_2$). When J>>D, the largest off-diagonal elements of $S_1 \cdot S_2$ are of the order D/J and couple certain states whose effective spins S′ differ by 2.

It is also noted that at those temperatures for which $E_{ij} >> kT$, Eq. (14) yields a linear dependence of the relaxation rate upon temperature.

3.2 Temperature Dependence of Relaxation Rate in Amorphous Materials

The temperature dependence of the relaxation rate is different in amorphous materials from that in single crystals. Amorphous materials differ from single crystals in that in these materials there exists a distribution of the various parameters characterizing interactions amongst magnetic ions, in particular, the exchange interaction, whose strength depends on the separation between the ions constituting exchange-coupled pairs. A quadratic temperature dependence can be obtained when a three-dimensional uniform distribution of exchange-coupled pair separation, r_{12}, is taken into account. Expression (14) can then be generalized to amorphous samples by appropriately integrating the factor $E_{ij}^3/\left(1-e^{-E_{ij}/kT}\right)$, considering $E_{ij} \sim J$, assuming J to depend on r_{12} as $J = J_0 e^{-\xi r_{12}}$:

$$I_{EX} = \frac{4\pi}{\xi^3}\int_0^{J_{max}} \frac{\left[J\ln\left(J/J_0\right)\right]^2 dJ}{\{\exp(J/kT)-1\}}. \tag{15}$$

The above integral can be evaluated by numerical methods, and for exchange energies less than thermal energies, it leads to a quadratic dependence on temperature at very low temperatures.[30] On the other hand, at relatively higher temperatures for which J >> kT, it is easily seen to yield a linear dependence on temperature. Experimental data on relaxation rates of localized electrons in amorphous silicon[4,5] and those of Cr^{3+} ions in $Cu_{2x}Cr_{2x}Sn_{2-2x}S_4$ [31] are in accordance with those predicted quadratic and linear temperature dependences as analyzed by Misra and Pescia[32] and by Misra,[33] respectively. On the other hand, a linear dependence of relaxation rate upon temperature is found to be true for single crystals, where there does not exist a distribution of J_{12} values, or equivalently that of r_{12}, at both very low and intermediate temperatures.

3.3 Case of Strong Cross Relaxation and Weak Spin-lattice Relaxation of Single ions Situation at Low Temperatures[28]

In this case, a single ion is coupled to an exchange-coupled pair. When a cross relaxation occurs at low temperatures, in which a single ion flips from one energy level to the other accompanied by a simultaneous spin flip of an exchange-coupled pair at an energy Δ' above that of a single ion, a suitably averaged value of temperature-dependent part of Eq. (2) yields an approximate temperature dependence as T^2 for a distribution of values of Δ' as shown by Gill[26] when there are present a variety of clusters. This situation is especially applicable to amorphous samples.

3.3.1 Situation at Higher Temperatures[36]

For the case of strong relaxation of the excited pair, at an energy Δ' above that of a single ion, which becomes more and more populated as the temperature is increased, the temperature dependence turns out to be $T_1^{-1} \propto c_1^2/[1+\exp(\Delta'/kT)]$, where c_1 is the concentration of paramagnetic ions. This leads to a plateau in the relaxation rate at higher temperatures, e.g. that observed in a borate glass sample doped with Fe_2O_3, characterized by a high concentration of paramagnetic ions and a larger antiferromagnetic exchange interaction (J=-0.98 K).[35]

The relaxation rates are much faster when the exchange interaction is the dominant mechanism for SLR as compared to the case when TLS centers, involving electron-nuclear dipolar interaction and Fermi-contact hyperfine interaction, effect spin-lattice relaxation process. In fact, the relaxation rates due to the exchange interaction are of the order of two to three magnitudes higher than those expected on the basis of TLS centres, as evidenced by the measured values of spin-lattice relaxation rates listed in the following table (in units of s^{-1} = Hz) for various samples:

Temperature	**10K**	**20K**	**50K**	**100K**	**200K**	**Effected via (ion)**
Irradiated ethanol glasses[2]	25	60	500	3,000		TLS (END) (trapped electron)
β-alumina irradiated[8]	7	40				TLS (FC) (Color center)
Amorphous Silicon[4]	10^5	5×10^5	2×10^6	10^7	10^8	Exchange (electron)
Amorphous silicon[5]	5×10^4					Exchange (electron)
Borate glass ($0.1\%Fe_2O_3$)[35]		$\sim10^6$	7×10^6	2×10^7	4×10^7	Exchange (Fe^{3+})
Borate glass ($0.5\%Fe_2O_3$)[35]		$\sim10^7$	3×10^7	4.5×10^7	5×10^7	Exchange (Fe^{3+})
$MgO.P_2O_5$ (0.2%Mn)[36]					2.5×10^6	Exchange (Mn^{2+})
Irradiated molecular crystal[30]	10	50	10^3	5×10^3		TLS (trapped electron)
$Cu_{2x}Cr_{2x}Sn_{2-2x}S_4$ (x = 0.8)[33]	5×10^8	8×10^8	10^9	5×10^9	4×10^9 plateau	Exchange (Cr^{3+})

4. Concluding Remarks

The mains features of this review of spin-lattice relaxation in amorphous materials are as follow :

(i) The most effective mechanism of spin-lattice relaxation in amorphous materials is the exchange coupling between two electron spins, being directly perturbed by lattice vibrations (phonons), as judged by the magnitude and temperature dependence of the spin-lattice relaxation rates.

(ii) The relaxation rates are much faster when the exchange interaction is the dominant mechanism for spin-lattice relaxation as compared to the case when TLS centers, involving electron-nuclear dipolar interaction and Fermi-contact hyperfine interaction, effect spin-lattice relaxation process. The relaxation rates due to the exchange interaction are of the order of two to three magnitudes higher than those expected on the basis of TLS centers.

(iii) It has been deduced here that the spin-lattice relaxation rates due to exchange interaction and those effected by TLS centers, exhibit quadratic temperature dependence at very low temperatures and linear temperature dependence at intermediate temperatures.

References

1. S. K. Lyo and R. Orbach, Phys. Rev. B**22**, (1980) 4223.
2. M. K. Bowman and L. Kevan, J. Phys. Chem. **81**, (1977) 456.
3. M. Velter-Stefanescu and R. Grosescu, Rev. Rouman. Phys. **23**, (1978) 369. M. V. Stefanescu, R. Grosecu, I. Ursu and S. Nistor, Proc. 23rd Congress Ampere, Rome, (1986) 234.
4. J. C. Gourdon, P. Fretier and J. Pescia, J. Phys. Lett. **42**, (1981) 21, based on J. C. Gourdon, Ph. D. thesis (Thèse d'Etat, no. 69), Université Paul Sabatier, Toulouse, France (1975).
5. T. R. Askew, P.J. Muench, H.J. Stapleton and K.L. Brower, Solid St. Comm. **49**, (1984) 667.
6. T. J. B. Swanenburg, N. J. Poulis, and W. J. Drewes, Physic **24**, (1963) 713.
7. J. Turkevich, J. Soria, and M. Che, J. Chem. Phys. **56**, (1972) 1463.
8. S. R. Kurtz and H. J. Stapleton, Phys. Rev. B **22**, (1980) 4223.
9. A. A. Antipin, B. I. Kochalev, B. I. Shleklin, J. E. T. P. Lett. **39**, (1984) 182.
10. S. B. Stevens and H.J. Stapleton, Phys. Rev. B**42**, (1990) 9794.
11. D. Vergnoux , P. K. Zinsou, M. Zaripov, G. Ablart, J. Pescia, S. K. Misra and S. Orlinskii, App. Mag. Reson. **11**, (1996) 493.
12. R. Berman, Phys. Rev., **76**, (1949) 315.
13. C. L. Choy, G. L. Salinger, and Y. C. Chiang, J. Appl. Phys., **41**, (1970) 597.
14. J. H. McTaggart and G. A. Slack, Cryogenics, **9**, (1969) 384.
15. W. A. Phillips, J. Low Temp. Phys., **7**, (1972) 351.
16. P. W. Anderson, B. I. Halperin and C. M. Varma, Philosophical Mag., **25**, (1972) 1.
17. S. Estalji, O. Kanert, J. Steiner, H. Jian, and K. L. Ngai, Phys. Rev. B, **43**, (1991) 7481.
18. W. A. Phillips, in *Phonons 89, Proceedings of the Third International Conference on Phonon Physics*, edited by S. Hunklinger, W. Ludwig, and G. Weiss (World Scientific, Singapore, 1990), Vol. **1**, p. 367.

19. W. A. Phillips, *Amorphous Solids, Low-Temperature Properties, Vol. 24* of Topics in *Current Physics* (Springer, Berlin, 1981).
20. S. Hunklinger, Philosophical Mag. B, **56**, (1987) 199.
21. W. Kohler and J. Friedrich, Phys. Rev. Lett., **59**, (1987) 2199.
22. W. A. Phillips, Proc. Roy. Soc. (London), A**319**, (1970) 565.
23. G. B. Jollenbeck, O. Kanert, J. Strinert, and H. Jain, Solid. State Commun. **65**, (1988) 303.
24. W. I. Gamble, I. Miyagawa, and R. L. Hartman, Phys. Rev. Lett., **20**, (1968) 415.
25. J. Murphy, Phys. Rev. B**145**, (1966) 241.
26. J. C. Gill, Proc. Phys. Soc. (London) **79**, (1962) 58.
27. I. Waller, Z. Phys. **79**, (1932) 370. Although this treatment takes into account dipolar interaction only, the calculation can be easily adapted to take into account exchange interaction.
28. S. A. Al'tshuler, Bull. Acad. Sc. URSS **20**, (1956) 1207.
29. J. H. Van Vleck, Phys. Rev. **57**, (1940) 426.
30. L. R. Dalton, A. L. Kwiram and J. A. Owen, Chem. Phys. Letters **17**, (1972) 495.
31. I. Sarda, P. Colombet, G. Ablart, and J. Pescia, C. R. Acad. Sci. Paris, **308**, Series II, (1989) 159.
32. S. K. Misra and J. Pescia, (to be published).
33. S. K. Misra, (to be published).
34. M. B. Schulz and C. D. Jeffries, Phys. Rev. **149**, (1966) 270.
35. P. Zinsou, D. Vergnoux, G. Ablart, J. Pescia, S. K. Misra and R. Berger, App. Mag. Reson. **11**, (1996) 487.
36. D. Vergnoux, J. Pescia and S. K. Misra, Private Communication (1996).

Calculation Method of Spin-lattice Coupling Coefficients and Its Applications

Zheng Wen-Chen

Department of Material Science, Sichuan University, Chengdu 610064, China, and International Center for Material Physics, Academia Sinica, Shenyang 110015, China

Abstract. A uniform and simple method developed by us for calculating the spin-lattice coupling coefficients G_{11} and G_{44} for paramagnetic ions in cubic symmetry is introduced in this paper. From the method, the coefficients G_{11} and G_{44} for d^n ions in 6-fold, 4-fold(or 8-fold) and 12-fold cubic coordinations can be obtained from the distinctive high-order perturbation formulas of zero-field splittings. The uses of this method in solid state physics and material science are reviewed.

1 Introduction

The studies of spin-lattice coupling coefficients G_{ij} of $3d^n$ ions in crystals are of importance and significance from the experimental and theoretical points of view because valuable information on the spin-lattice relaxation time of magnetic centers, the angular variation of EPR linewidths, and the state of internal stress of a crystal, as well as on the microscopic mechanisms of zero-field splitting can be obtained from these studies [1-3]. As is known, experimentally, these coefficients are often measured by the uniaxial stress on the EPR spectra and the ultrasonically modulated EPR methods [1,2], and theoretically, they are usually calculated from the derivatives of zero-field splitting with respect to the strain tensor elements [4]. However, this method is complex and it is sometimes easy to make mistakes when using the high-order perturbation formulas for zero-field splitting. So, in these cases, the analytic expressions of G_{ij} have not been given, and many experimental findings of G_{ij} are also not explained satisfactorily [3,5]. In addition, this method leaves the physical meaning of these coefficients obscure. To improve these weakness, in the previous papers [6-8], we put forward a uniform and simple method of calculating the spin-lattice coupling coefficients for all d^n ions in cubic crystals. From the method, the coefficients G_{11} and G_{44} for d^n ions in 6-fold, 4-fold(or 8-fold) and 12-fold cubic coordinations were obtained from the distinctive high-order perturbation formulas of zero-field splittings and many experimental data were calculated. In the present paper, we will introduce and review this calculation method and its applications to solid state physics and material science.

2 Calculation Method

For d^n ions in 6-fold cubic coordinations, when the stress P is along the [001] axis of crystal, we have [1, 6]

$$dD_{tetra}/dP= (-3/2)G_{11}(S_{11}-S_{12}) \tag{1}$$

where S_{ij} are the elastic constants. The tetragonal distortion due to this stress can be represented by the angle α (note: α is defined by $\tan\alpha=R_{\perp}/R_{/\!/}$, where $R_{\perp}$ and $R_{/\!/}$ stand for the metal-ligand distances perpendicular to and parallel with the tetragonal axis, respectively), and

$$d\alpha/dP=(1/2)(S_{11}-S_{12}) \tag{2}$$

So, we have

$$dD/dP=(\partial D/\partial\alpha)_0(\partial\alpha/\partial P)= (1/2)(\partial D/\partial\alpha)_0(S_{11}-S_{12}) \tag{3}$$

where the zero subscript denotes that the calculation is done for the case of cubic symmetry. Thus

$$G_{11}=(-1/3)(\partial D_{tetra}/\partial\alpha)_0 \tag{4}$$

Similarly, when the stress is along the [111] axis, we can obtain [6]

$$G_{44}= (-\sqrt{2}/6)(\partial D_{tri}/\partial\beta)_0 \tag{5}$$

where β is the angle between the direction of metal-ligand pair and the C_3 axis.

When the stress P is along the [110] axis, it gets [6]

$$G_{11}=(-1/3)(\partial D_{rho}/\partial\alpha)_0, \quad G_{44}= (-1/2)(\partial E_{rho}/\partial\theta)_0 \tag{6}$$

where θ is the rhombic distortion angle (See Fig. 1 of Ref. [6]).

For d^n ions in 4-fold (or 8-fold) coordinations, we have [7]

$$G_{11}= (-2\sqrt{2}/9)(\partial D_{tetra}/\partial\alpha)_0, \quad G_{44}=(4/9)(\partial D_{tri}/\partial \ln R_1)_0-(\sqrt{2}/9)(\partial D_{tri}/\partial\beta)_0 \tag{7}$$

or

$$G_{44}=(1/2)(\partial E_{rho}/\partial\theta)_0 \tag{8}$$

where the bond angles α, β, θ and the bond length R_1 are defined in Fig.1 of Ref. [7].

For d^n ions in 12-fold coordinations, we have [8]

$$G_{11}= (-1/3)(\partial D_{tetra}/\partial\alpha)_0, \quad G_{44}=(1/3)(\partial D_{tri}/\partial \ln R_1)_0+ (\sqrt{2}/6)(\partial D_{tri}/\partial\beta)_0 \tag{9}$$

or

$$G_{44}=(-1/2)(\partial E_{rho}/\partial\theta)_0 \tag{10}$$

where α and β are the angles between the direction of metal-ligand pair and the C_4 and C_3 axes, respectively. R_1 is the bond length defined in Fig. 1 of Ref. [8].

It can be seen that the spin-lattice coupling coefficients G_{11} and G_{44} of d^n ions in cubic crystals denote the derivatives of zero-field splittings with respect to distinctive bond angles and bond lengths. The simple and clear physical meaning of these coefficients enables us to calculate them from the formulae of zero-field splittings D and E in low symmetries very easily. Clearly, the uniform method is simpler than the direct calculation from the strain tensor elements and can be useful to all d^n ions. Considering that the expressions of zero-field splitting are often complex and tedious, the simple method is meaningful.

3 Applications

The uses of the method are as follows:

(1) Checking whether the expressions of G_{11} and G_{44} obtained by other complex methods are correct. For example, we [9] checked the G_{11} and G_{44} for d^5 ions given in Ref.[4] based on various mechanisms and models and found that all their expressions for G_{44} are too large by a factor 2, and hence their ratios G_{11}/G_{44} are half of our values. The reason is perhaps that they confused the actual shear strains e_{ij} ($i \neq j$) with the tensor ones in the calculation. In fact, the former is twice the latter.

(2) By comparing the coefficient G_{44} obtained from two cases, i.e., from the formulas related to D_{tri} and E_{rho}, one can decide whether the formulas of zero-field splitting D_{tri} and E_{rho} are self-consistent. For d^5 ions, we [6,7] found that the formulas D_{tri} and E_{rho} from the spin-orbit coupling mechanism based on two perturbation schemes [4,10] are self-consistent, but those from the covalency and overlap mechanism given in Ref. [11] are not self-consistent.

(3) Calculating the spin-lattice coupling coefficients G_{11} and G_{44} for $3d^n$ ions in some cubic crystals with 6-fold, 4-fold and 8-fold coordinations from the distinctive high-order perturbation formulas of zero-field splitting in low symmetries by using the parameters obtained from optical spectra. The results show reasonable agreement with the observed values (see Table 1).

(4) Studying the local lattice distortions and the local compressibilities (or elastic constants) in the vicinity of $3d^n$ impurity ions in crystals. For $3d^n$ ions in tetragonal crystals with 6-fold coordination, the tetragonal distortion can be characterized by the value of $\alpha-\alpha_0$ (where angle α is defined by $\tan\alpha \approx R_\perp / R_{/\!/}$ and α_0 is the same angle in cubic symmetry). When the tetragonal distortion is small, we have [16]

$$D\approx(\alpha-\alpha_0)\,(\partial D/\partial\alpha)_0 \approx -3G_{11}(\alpha-\alpha_0) \tag{11}$$

So, the local tetragonal distortion can be obtained from the values of G_{11} and D. By using the above formulas, the local tetragonal distortions $(\alpha-\alpha_0)$ for Ni^{2+} and Mn^{2+} in K_2MgF_4 and K_2ZnF_4 crystals were studied [16]. Similarly, for $3d^n$ ions in trigonal crystals, the trigonal distortion $\beta-\beta_0$ (where β is the angle between the direction of metal-ligand pair and the C_3 axis and β_0 is the same angle in cubic symmetry) can be obtained by the formula [16]

$$D\approx(\beta-\beta_0)\,(\partial D/\partial\beta)_0 \approx -3\sqrt{2}G_{44}(\beta-\beta_0) \tag{12}$$

Because [1]

$$dD_{tetra}/dP=(-3/2)G_{11}(S_{11}-S_{12}), \quad dD_{tri}/dP=-G_{44}S_{44} \tag{13}$$

we can study the local elastic constants S_{ij} (and hence the local compressibility) in the vicinity of $3d^n$ ions in crystals from the observed values of dD_{tetra}/dP and dD_{tri}/dP and the calculated values of G_{11} and G_{44}. For example, the local elastic constants S_{ij} for

Cr^{3+} in MgO [12] and $SrTiO_3$ [13] crystals were studied from the above equations.

(5) Determining the substitution position of $3d^5$ ions in crystals. For $3d^5$ ions, the ratio G_{11}/G_{44} in four-coordinated site is obviously different from that in six-coordinated site [7]. So, by comparing the calculated values of G_{11}/G_{44} for various coordination members with the observed value, the substitution site of $3d^5$ ion can be found. For example, for Fe^{3+} in andalusite (Al_2SiO_5) crystal [17], the Fe^{3+} ion replacing the four-fold coordinated site of Al^{3+} was determined from this method.

(6) Determining the internal stress distribution parameters α_1 and β_1 in the vicinity of paramagnetic ions in crystals from EPR linewidth and spin-lattice coupling coefficients. When a crystal contains a high concentration of defects or is poorly annealed, a random distribution of the internal stresses and strains is caused in the crystal. In general, these distributions are Gaussian types and for cubic crystal, they can be written as [1]

$$P(X_{ii})= (2/\pi)^{1/2}\alpha_1^{-1}\exp(-2X_{ii}^2/\alpha_1^2)$$

$$P(X_{ij})= (2/\pi)^{1/2}\beta_1^{-1}\exp(-2X_{ij}^2/\beta_1^2) \qquad (i \neq j) \qquad (14)$$

where X_{ij} are the stress components along the cubic axis. α_1 and β_1 are the internal stress distribution parameters. From the linewidth data of EPR spectra, one can obtain only the parameters $C_{11}\alpha_1$ and $C_{44}\beta_1$ for $3d^n$ ions in a cubic crystal [1], where

$$C_{11}=G_{11}(S_{11}-S_{12}), \qquad C_{44}= G_{44}S_{44} \qquad (15)$$

So, from the spin-lattice coupling coefficients G_{11}, G_{44} and the elastic constants S_{ij}, the parameters α_1 and β_1 can be obtained. We studied the parameters α_1 and β_1 for Fe^{3+} in $Y_3Al_5O_{12}$ (YAG) crystal from this method [18].

(7) Studying the structural phase transitions for the crystals of perovskite structure. Some ABX_3 perovskite compounds show a structural phase transition from cubic to tetragonal (or trigonal) symmetry at a transition temperature Tc. The tetragonal (or trigonal) phase may be viewed as a rotation of BX_6 octahedra around one of their C_4 (or C_3) axis by an angle ϕ(or α_2) which is the order parameter of the phase transition [19]. When a $3d^n$ ion replaces the B ion in ABX_3 crystal, we obtained the zero-field splittings

$$D\approx(3/2)(k-1/2)G_{11}\phi^2 \qquad (16)$$

for the cubic-tetragonal phase transition[20], and

$$D\approx-2(f+1/2)G_{44}\alpha_2^2 \qquad (17)$$

for the cubic-trigonal phase transition[21], where k and f are the release and contraction factors. Thus, by using the paramagnetic ion as the probe, we can study the structural phase transitions for the perovskite compounds from their EPR data and spin-lattice coupling coefficients. For example, based on this method, the phase transitions for $3d^n$ ions in $SrTiO_3$ [20] and $LaAlO_3$ [21] crystals have been studied.

In addition, the above method is also effective in the studies of the spin-lattice coupling coefficients F_{ij} (which are related to the stress dependence of g factor) in cubic crystals [22].

Table 1. The spin-lattice coupling coefficients G_{11} and G_{44} for $3d^n$ ions in some cubic crystals (in units of cm^{-1})

	G_{11}		G_{44}		
	Cal.	Obs.	Cal.	Obs.	Refs.
$MgO:V^{2+}$	0.42	0.42	3.04	3.0	[12]
$MgO:Cr^{3+}$	1.08	0.6	4.86	4.2	[12]
$MgO:Ni^{2+}$	53	57(8)	4.1	36(5)	[5]
$SrTiO_3:Cr^{3+}$	0.68	1.13	4.28	4.73	[13]
$GaP:Fe^{3+}$	–6.0	–6.9	5.2	5.1	[14]
$CaF_2:V^{3+}$	–0.92	–1.2	8.16	7.4	[15]

References

1. E. R. Feher, Effect of uniaxial stresses on the paramagnetic spectra of Mn^{2+} and Fe^{3+} in MgO, *Phys. Rev.* 136 (1964) A145-157.
2. S. D. Devine and W. H. Robinson, Ultrasonically modulated paramagnetic resonance, *Adv. in Magnetic Resonance*, vol. 10 (1982) 53-117.
3. E. B. Tucker, Interaction of phonons with iron-group ions, *Proc. IEEE*, 53 (1965) 1547-1573.
4. W. L. Yu and M. G. Zhao, Spin-Hamiltonian parameters of ^{6}S-state ions, *Phys. Rev.* B37 (1988) 9254-9267.
5. W. C. Zheng, Theoretical calculations of the spin-lattice coupling coefficients G_{11} and G_{44} for $MgO:Ni^{2+}$ crystals, *Phys. Rev.* B40 (1989) 7292-7294.
6. W. C. Zheng, Calculation of the spin-lattice coupling coefficients in cubic symmetry, *Phys. Lett.* A137 (1989) 309-312.
7. W. C. Zheng, Spin-lattice coupling coefficients G_{11} and G_{44} for ^{6}S-state ions in Td symmetry, *Phys. Rev.* B42 (1990) 826-832.
8. W. C. Zheng, Spin-lattice coupling coefficients G_{11} and G_{44} for $3d^5$ ions in 12-fold cubic coordination, *Physics* B 179(1992) 35-42.
9. W. C. Zheng, Spin-lattice coupling coefficients and the method of identifying the dominant mechanism for ^{6}S-state splitting, *Phys. Rev.* B41(1990) 793-794.
10. R. R. Sharma, T. P. Das and R. Orbach, Zero-field splitting of S-state ions, I. Point-Multiple model, *Phys. Rev.* 149 (1966) 257-269.
11. P. Novak and I. Veltrasky, Overlap and covalency contribution to zero-field splitting of S-state ions, *Phys. Status Solidi*, B73 (1976) 575-586.
12. W. C. Zheng, Theoretical explanation of the spin-lattice coupling coefficients G_{11} and G_{44} for d^3 (V^{2+}, Cr^{3+}) ions in MgO crystals, *J. Phys.:Condens. Matter* 1(1989) 8093-8098.
13. W. C. Zheng, Theoretical investigation of the spin-lattice coupling coefficients G_{11} and G_{44} for Cr^{3+} in cubic $SrTiO_3$ crystals, *Phys. Status Solidi*, B154 (1989) K167-170.

14. W. C. Zheng, Calculation of the spin-lattice coupling coefficients G_{11} and G_{44} for GaP:Fe^{3+} crystal, *Nuovo Cimento* D 12 (1990) 1667-1671
15. W. C. Zheng, Investigation of the spin-lattice coupling coefficients G_{11} and G_{44} for CaF_2:V^{3+} crystal, *Phys. Lett.* A 146 (1990) 71-74.
16. W. C. Zheng, Local axial distortion of paramagnetic impurity center in some axial-symmetry crystals, *Solid State Commun.* 94 (1995) 317-319.
17. W. C. Zheng, Determination of the substitution position of the Fe^{3+} ion in an andalusite crystal from the spin-lattice coupling coefficients, *Phys. Lett.* A 166 (1992) 145-147.
18. W. C. Zheng, Determination of the internal stress distribution parameters α and β for Fe^{3+}-doped YAG crystals from the EPR linewidth and spin-lattice coupling coefficients , *J. Phys. : Condens. Matter* 3 (1991) 177-182.
19. K. A. Muller, W. Berlinger and F. Waldner, Characteristic structural phase transition in perovskite-type compounds, *Phys. Rev. Lett.* 21 (1968) 814-817.
20. W. C. Zheng, Investigation of the splitting of $^4A_2$ and ^2E states for Cr^{3+} in a tetragonal $SrTiO_3$ crystal, *J. Phys. :Condens. Matter.* 6 (1994) 2241-2244.
21. W. C. Zheng, Local behavior in the cubic-rhombohedral phase transition for Fe^{3+} ions in $LaAlO_3$ crystal, *Phys. Lett.* A 198 (1995) 248-250.
22. W. C. Zheng, The method of calculating the spin-lattice coupling coefficients F_{ij} in cubic symmetry from the formulas for the g-factor, *J. Phys.:Condens. Matter* 2(1990) 2179-2186.

Lattice Mismatchings in SrF_2-BaF_2 Mixed Crystal Systems Monitored by Eu^{2+} Dopants

K.Kawano and *R.Nakata*

Department of Electronics Engineering,
The University of Electro-Communications, Chofu, Tokyo, Japan 182

Abstracts. Local distortions in mixed crystals or solid solutions $Sr_{1-x}Ba_xF_2$ were investigated with monitoring a doping Eu^{2+} ion by means of ESR measurement. With increasing mixture ratios near x=0 and 1, the resolved hyperfine linewidths broaden in sequence from larger $|M|$ spin magnetic quantum numbers to smaller ones for the $M \rightarrow M-1$ transition. As a source of these characteristic broadening effects, distortion compensation for lattice misfit was considered to be made at local sites in the mixed crystal by adding the 2-nd order zero-field splittings of B_2^0 follwed by B_2^2.

1 Introduction

Mixed fabricated crystals between pairs of isomorphic compounds have attracted many researchers in attempts to find new physical properties and to apply to new devices. Optical spectroscopies take advantage of describing elementary excitations and localized states throughout the overall crystals, but for investigating locally occurred distortions as well as changes in crystal fields, ESR technique becomes a most powerful tool in obtaining the detailed information. Recently, remarkable differences between luminescence properties and local crystal fields (zero-field splittings; ZFS) were found for Eu^{2+} ($4f^7$) ions doped into the mixed crystal systems between SrF_2 and BaF_2 by authors (1993). So far few investigations using ESR have been made on local distortions caused from the lattice mismatching effect in mixed crystals. With increasing mixture ratios near x=0 and 1 in $Sr_{1-x}Ba_xF_2$:Eu systems, the resolved hyperfine linewidths of the ESR spectrum broaden in sequence from larger $|M|$ spin magnetic quantum numbers to smaller ones for the $M \rightarrow M - 1$ transition. It should be emphasized that the line broadenings are neither due to clustering of the doped Eu^{2+} ions nor to inhomogeneous broadening, but originates from lattice distortions occurring in the mixed crystal (1995).

2 Experimental Details

The materials used for this investigation were prepared with a Bridgman crystal growing apparatus in the authors' laboratory. The doping content of the EuF_3 impurities was established as 0.1 mol% to obtain a sufficiently intense ESR signal throughout all the mixed crystals as well as to prevent from broadening of the spectral linewidth due to clustering among the paramagnetic ions. The ESR experiments were carried out with a Varian X-band ESR spectrometer (E3000, USA), and the spectal signals were AD converted (Interface CO. AZI-220, USA) to be sent to a personal computer (NEC PC9801VM, Japan) for the spectral analyses. The magnetic field was measured by a high precision NMR probe (Echo Elec. EFM2000, Tokyo) within $\pm 10^{-6}$ T. The observations were made at room temperature.

3 Experimental Results

The SrF_2 and BaF_2 crystals doped with small amounts of Eu^{2+} ion show well known ESR signals [1969] that have been ascribed to the cation-substitutional Eu^{2+} complexes associated with the surrounding eight F^- ions showing a cubic crystal field symmetry.

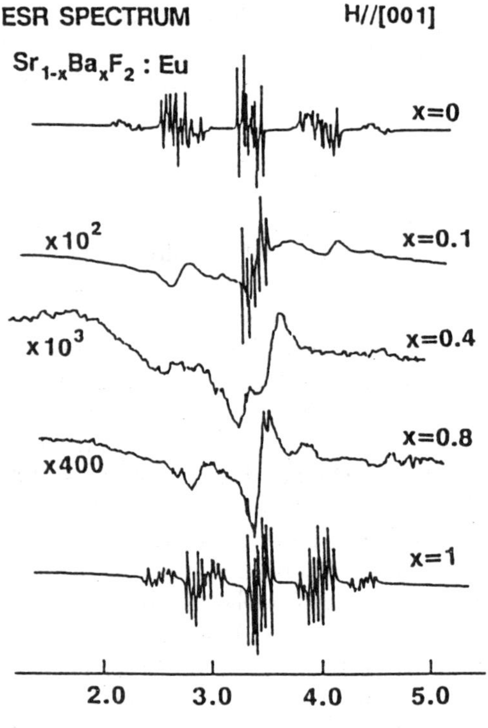

Fig.1. The X-band ESR spectrum for the mixed crystals $Sr_{1-x}Ba_xF_2$:Eu^{2+} with the typical mixture ratio. The magnetic field is applied along the [001] direction.

In Fig.1, typical ESR spectra of the Eu^{2+} monitor (0.1 mol%) are shown when the ratio of mixture was varied from x=0 to 1 in the $Sr_{1-x}Ba_xF_2$ system. As the composition changes from x=0 to 1, the end-to-end field spread of the spectrum is gradually decreased which suggests a continuous weakening of the dominant cubic field in the mixed crystals. Further on increasing the ratio x, the resolved hyperfine linewidths broaden in sequence from the higher $|M|$ value to the lower one in the seven sets of fine structures indicated in Fig.2. Thus the broadening effect contributes more to the transition $M \leftrightarrow M-1$ with larger $|M|$: the line broadening of the transition $M = -\frac{1}{2} \leftrightarrow \frac{1}{2}$ is much smaller as compared with that of the outer field transitions $\pm\frac{7}{2} \leftrightarrow \pm\frac{5}{2}$. Hence even for the larger ratios when x is greater than 0.1, the central $M = -\frac{1}{2} \leftrightarrow \frac{1}{2}$ signals remain with pure resolved hfs lines. Moreover, the broadening rate of each hfs linewidth becomes notable as the ratio of the mixture increases.

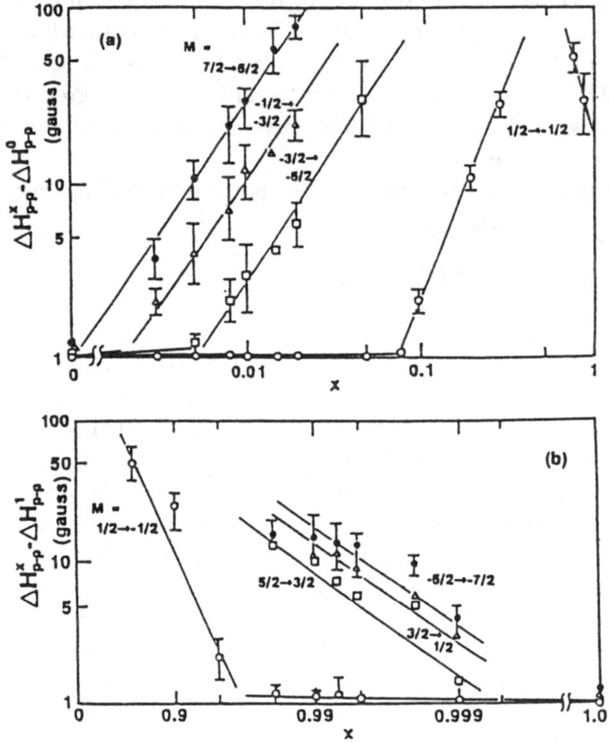

Fig.2. Graphs of the rate of increase of the peak-to-peak hyperfine linewidth for the mixture ratio x; (a) ΔH^x_{p-p}-ΔH^0_{p-p} in the vicinity of x=0 and (b) ΔH^x_{p-p}-ΔH^1_{p-p} in the vicinity of x=1.0 . ΔH^0_{p-p} is the linewidth at x=0 and ΔH^1_{p-p} is that of x=1 .

The rates of increase of peak-to-peak linewidth are graphed in Fig.2 for ΔH^x_{p-p}-$\Delta H^{0/x}_{p-p}$ in the vicinities of x=0 and 1.0. Such plottings of the differeces in these linewidths allow us to avoid the effects of line broadening resulting from properly inhomogeneous origins as well as the clustering of magnetic ions. The values of ΔH^0_{p-p} and ΔH^1_{p-p} are given in Table 1.

Table 1. The peak-peak hfs linewidths ΔH^0_{p-p} and ΔH^1_{p-p} in each fs transition ($M \rightarrow M-1$) for $SrF_2:Eu^{2+}$ (x=0) and $BaF_2:Eu^{2+}$ (x=1), respectively.

M	7/2	-3/2	-1/2	1/2	3/2	5/2	-5/2
ΔH^0_{p-p}(gauss)	9.4	7.5	5.8	5.0	5.1	7.4	9.4
ΔH^1_{p-p}(gauss)	9.5	8.0	5.8	4.5	5.8	7.7	9.2

These values are given by the averaged values of linewidths for m=7/2 and m=-7/2 hfs lines which are at both edges in six hfs lines on each fs transition because these two lines are highly resolved and isolated from other lines. Although the magnitudes of the linewidths are not equal for all seven sets of fs transitions of $M \rightarrow M-1$, the ratio of their integral intensities agree well with the theoretical ratio; 7:12:15:16:15:12:7 [1970]. Every transition line becomes broad on increasing the mixture ratio x, but the broadenings are in sequence starting with from larger $|M|$ values to smaller ones. For a universal interpretation, the following relationship is assumed;

$$\Delta H^x_{p-p} - \Delta H^{0(1)}_{p-p} = \alpha x^n, \tag{1}$$

and the best-fit straight lines are drawn for each transition in Fig.2. The increasing rate for M=±7/2,±5/2 and ±3/2 shows a linear functional change with the powers of n=0.9-1.3 . Otherwise, only the M= 1/2 transition gives more abrupt changes with the powers of n=2.0-2.4 against x. The gradients α of Eq.(1) are derived from the intersections with the x-coordinate which mean the starting mixture ratios of the broadenings because of the log-log plots of the fitted lines. It is noted that the reduced ratio of $|\alpha|$ for each fs transition is very close to 18:12:6:0 when the value of M=1/2 transition is regarded as nil compared with the other transitions. This tendency also applies to Ba-rich region over x=0.8 with minus gradients, in which the spectral changes are related as mirror images for the x and *1-x* crystals. This point is further clarified in contrast to the results of x-dependences of the dominant cubic parameter B_4 of ZFS as well as the hfs constant A-value. The cubic field parameter B_4 can be determined from the separations of the envelope of the fs lines even if the superposed hf lines were unobviously broadened with increasing the mixture ratio x. The variation was investigated by making small changes in the value of x in the vicinity of x=0 and 1, respectively. The experimental values fitted well to a calculated curve in which B_4 was assumed to be a linear function of x. In a similar way, the g-values and A-values showed their linear variations .

4 Discussion

Since the doped Eu^{2+} ion occupies substitutionally the position of either Sr or Ba cations, the spectrum must exhibit cubic field symmetry. Actually, our experimental results show that the crystal structures of the mixed crystal $Sr_{1-x}Ba_xF_2$

system have generally kept the cubic field structure from the viewpoint of Eu^{2+} monitor throughout all the ratios of mixtures because the determined cubic B_4 ZFS parameter varied linearly from CaF_2 to SrF_2. At the same time, conservation of linearity in the g-value supports the continuity of cubic field structures in the overall mixed crystals. Also the linearity in A-value ensures that the influence of change in Sr^{2+} to Ba^{2+} cations does not have an effect on the nuclear position of the Eu monitor through electron-nuclear interaction, but rather that the dominant cubic structure is retained owing to electron clouds bound by constitutional ion cores.

When the Eu^{2+} ion is embedded in the crystal field which is formed by the eight-coordinated F^- ions at the nearest neighbor positions, the energies for spin resonance are determined by solving the energy matrix using the spin Hamiltonian;

$$\mathcal{H} = g\beta\mathbf{HS} + B_4O_4 + B_6O_6 + A\mathbf{IS}. \tag{2}$$

The first term is the Zeeman interaction with isotropic g and Bohr's magneton β; both the second and the third term with the spin operators O_4, O_6 express the 4-th and the 6-th degree ZFS crystal fields. The fourth term in Eq.(2) is caused by the hyperfine (hf) interaction with the hf constant A. The fine structure energies are obtained by solving the 8×8 secular matrix spanned by the electron spin S=7/2 space of Eq.(2) except for the fourth term (1970). If a 2-nd order zero-field splitting term is added to the cubic field of Eq.(2), it is expressed by

$$\mathcal{H}_c = B_2^0O_2^0 + B_2^2O_2^2, \tag{3}$$

and its magnetic field shift $\delta H_{M\leftrightarrow M-1}$ including the 2-nd order perturbation is given by:

$$\delta H_{\pm\frac{7}{2}\leftrightarrow\pm\frac{5}{2}} = 18B_2^0 + \frac{12(B_2^2)^2}{H_0} \quad (4a) \qquad \delta H_{\pm\frac{5}{2}\leftrightarrow\pm\frac{3}{2}} = 12B_2^0 - \frac{3(B_2^2)^2}{H_0} \quad (4b)$$

$$\delta H_{\pm\frac{3}{2}\leftrightarrow\pm\frac{1}{2}} = 6B_2^0 - \frac{12(B_2^2)^2}{H_0} \quad (4c) \qquad \delta H_{\frac{1}{2}\leftrightarrow-\frac{1}{2}} = -\frac{15(B_2^2)^2}{H_0} \quad (4d),$$

where H_0 is the central magnetic field of g=2. From Eqs.(4a)-(4d), it is seen that the broadenings of the linewidth are effected in the ratio 18:12:6:0 times of the linewidth of an elemental B_2^0 for the fs transitions from $|M|$=7/2 to 1/2. As a matter of fact, the observed spectrum results in larger broadenings for the higher $|M|$ transitions, and it is significant that the experimental ratio estimated from the gradient α in Eq.(1) is very close to this ratio 18:12:6:0. This agreement supports a proposition that the observed line broadenings are created by statistical distriburtion of locally additive 2-nd order crystal fields.

As a source of these characteristic broadening effects, distortion compensations for lattice mismatchings were considered to be made at local sites in the mixed crystals by adding the 2nd-order ZFS of B_2^0 follwed by B_2^2. The distortion effects are originated from additions of low-symmetry local crystal fields which are composed of not only the axial but also the orthorhombic fields. Accordingly, the doped Eu^{2+} ion can detect ionic displacements with high sensitivities when the mixed crystals were formed. For the mixed crystal system with insertion of Ba^{2+} ions in place of Sr^{2+} ions, the nearest neighbor eight-coordinated fluorine ions are displaced diagonally by the differences of ionic radii between Sr^{2+} and Ba^{2+} ions. The resultant low symmetry crystal fields produced by the displacement polarization add to the cubic ZFS field that the Eu^{2+} monitor ion originally received from the nearest neighbors. As a result, broadening effects on linewidths are to be observed as the ratio x of the mixture increases.

If we supposed a virtual sphere with a diameter in the fluorite lattice, in which a Eu^{2+} ion is located at the center of the sphere and is involved in N site positions of cations, then we could calculate the contribution of all the displacements of eight fluorines for each replaceable cation over the N cations. For the calculated B_2^n according to the total displacements, the histogram for occupation numbers of the cation positions which gives the same magnitude of total contributions by B_2^0 and B_2^2 does signify the broadening of linewidth caused by locally additive 2-nd order distortions at the mixture ratio of $x = 1/N$. The fundamental idea is on the basis of the polarizable-point-ion (PPI) model that den Hartog [1983] had developed for the analyses of the Gd^{3+} ion. The observed linewidths were compared with the halfwidths of the calculated histograms of locally distributed distortions of B_2^0 and B_2^2. As the results, it was found that the observed line broadening effects would be reasonably explained if the sum of the B_2^0 term and the 2nd-order perturbation term of B_2^2 were distributed over the region from -2.75 to 1.75 gauss. These obtained values correspond to the displacements of ligand fluorine ions to be detectable with the high sensitivity by 0.4 % of the lattice parameter of SrF_2. Considering the difference of ionic radii between Sr^{2+} ion (1.10 Å) and Ba^{2+} ion (1.29 Å), the absolute displacement rate must be estimated to extend to a F-F body diagonal direction by a maximum 3.3 % of the lattice parameter of SrF_2. The obtained small displacement rate below 0.4 %, however, is much smaller, and it suggests that the constituent lattice may play the role of a buffer compensating for the difference of ionic radii instead of reflecting rigid displacement between substitutional ions.

References

1. K.Kawano, R.Nakata and M.Sumita, Proc.9th Int.Conf.Ternary and Multinary Compounds, Yokohama, *Jpn.J.Appl.Phys.*, **32** Suppl.32-3 (1993) 471.
2. Kawano K, Akahane H, Nakata R and Sumita M, *J.Alloys & Compounds* **221** (1995) 218-226.
3. W.R.Hurren, H.M.Nelson, E.G.Larson and J.H.Gardner, *Phys.Rev.*, **185** (1969) 624-629.
4. H.W.den Hartog, *Phys.Rev. B*, **27** (1983) 20-26.
5. P.Dorenbos and H.W.den Hartog, *Phys.Rev. B*, **31** (1985) 3932-3943.
6. A.Abragam and B.Bleaney, *Electron Paramagnetic Resonance of Transition Ions*, Clarendon, Oxford, 1970, p.148.

EPR Investigations of Mn(II) Ions in Cobalt Maleate Tetrahydrate Single Crystals

K.V.Narasimhulu and *J.Lakshmana Rao.*

Department of Physics, S.V.University, Tirupati - 517 502, INDIA.

Abstract. EPR investigations have been carried out on Mn(II) doped cobalt maleate tetrahydrate single cyrstals in the temperature range 123 - 393 K at X-band frequency. The EPR spectrum at room temperature exhibits a group of five fine structure transitions each split into six hyperfine components. The spin-Hamiltonian parameters were evaluated and discussed. It is found that Mn(II) enters the lattice substitutionally and only one type of Mn(II) site is identified. The variation of zero field splitting parameter (D) as a function of temperature is measured. The zero field splitting parameter (D) is found to increase with decrease of temperature. The linewidth is found to increase at low temperatures and the spectrum is broadened out at 123 K. The observation of resolved Mn(II) spectra at room temperature is interpreted in terms of a narrowing mechanism called `host spin-lattice relaxation narrowing'.

1 Introduction

EPR studies are usually done on paramagnetic impurities doped in diamagnetic hosts. The use of paramagnetic hosts for such studies is discouraged because of the expected broadening of impurity resonance lines due to the dipolar interaction between host and impurity ions. However, inspite of this expectation reasonably sharp EPR spectra have been observed in many paramagnetic hosts [1,2]. In these hosts, the effect of dipolar interaction is reduced either by `spin-quenching' or by `motional effects'. In paramagnetic cobalt salts doped with Mn(II) the effect of dipolar interaction between Co(II) and Mn(II) ions are thought to be reduced by host spin-lattice relaxation narrowing and fairly resolved EPR spectra of Mn(II) ions have been observed in some cobalt salts [3-11]. This paper reports the EPR investigations of Mn(II) ions doped in single crystals of cobalt maleate tetrahydrate (hereafter to be referred to as CoMTH).

1.1 Crystal Structure

Cobalt maleate tetrahydrate (CoMTH) belongs to a group of isomorphous series of compounds with general formula $H_2M(C_4H_2O_4)_2.4H_2O$, where M is a divalent cation such as Zn, Co, Mn and Ni [12]. A detailed structure of CoMTH is not available. The unit cell dimensions reported for isomorphous zinc maleate tetrahydrate (ZnMTH) by Saroja and Raman [12] were a = 5.36 Å, b = 7.29 Å, c = 9.73 Å, α =

$63°18'$, $\beta = 110°48'$ $\gamma = 103°36'$ and $z = 1$. The metal ions are octahedrally coordinated by six oxygen atoms, four from water molecules and two oxygen atoms from maleate ligands.

2 Experimental

Single crystals of cobalt maleate tetrahydrate were grown by slow evaporation method. To this compound about 1 mole % of Mn(II) were added as dopant. Transparent crystals, reddish in colour with well developed faces were separated out on concentrating the solution.

The EPR experiments were performed on a JEOL-FE-1X ESR spectrometer operating at X-band microwave frequency equipped with TE_{011} cylindrical cavity with 100 kHz field modulation. The resonance line of DPPH was used as a standard field marker. The angular variation spectra were recorded for every 10° interval. The temperature variation studies were performed using a JES UCI 2AX variable temperature controller.

3 Results and Discussion

The ground state of Mn(II), with a d^5 configuration is unique among d^n configurations in that there is only one state with maximum spin multiplicity ($^6S_{5/2}$). This splits into three Kramer's doublets (± 5/2, ± 3/2 and ±1/2) in an orthorhombic crystalline field, which are further split in the presence of the applied magnetic field. These six levels give rise to five fine structure transitions. Each fine structure transition will be split into six hyperfine components due to ^{55}Mn hyperfine coupling giving in all 30 allowed transitions. The principal axes were found by searching the direction corresponding to the extrema, in the spread of the spectrum. The z axis is defined as the direction of maximal splitting of the EPR spectrum; the y axis is the direction of minimal splitting in the plane perpendicular to the z axis and the x axis is orthogonal to y and z axes. The EPR spectrum recorded when the magnetic field along the direction parallel to z axis in yz plane is shown in Fig. 1. The EPR spectrum consisting of five fine structure components are clearly seen. The angular variation studies of Mn(II) spectra reveal that the total linewidth decreases when the magnetic field is away from the z axis. The occurrence of not more than thirty lines indicate that Mn(II) occupies only one type of site. Fig. 2 shows the angular variation of Mn(II) doped CoMTH single crystals for B in yz plane of the allowed transitions at room temperature.

The observed EPR spectrum can be described by the spin-Hamiltonian [13].

$$\mathcal{H} = \beta\, B.g.S + D\,[S_z^2 - 1/3\, S\,(S+1)] + E\,(S_x^2 - S_y^2) + A.S.I + a/6\,[S_x^4 + S_y^4 + S_z^4 - 1/5\, S(S+1)\,(3S^2 + 3S - 1)]$$

where the symbols have their usual significance and S = I = 5/2 for ^{55}Mn.

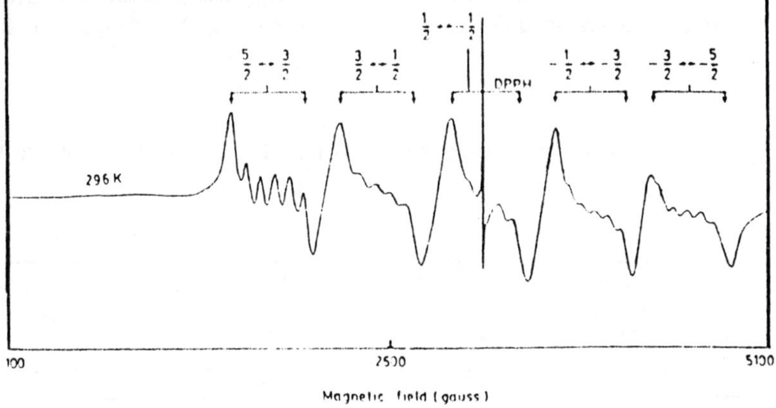

Fig. 1. EPR spectrum of Mn(II):CoMTH single crystals at room temperature for B||z axis.

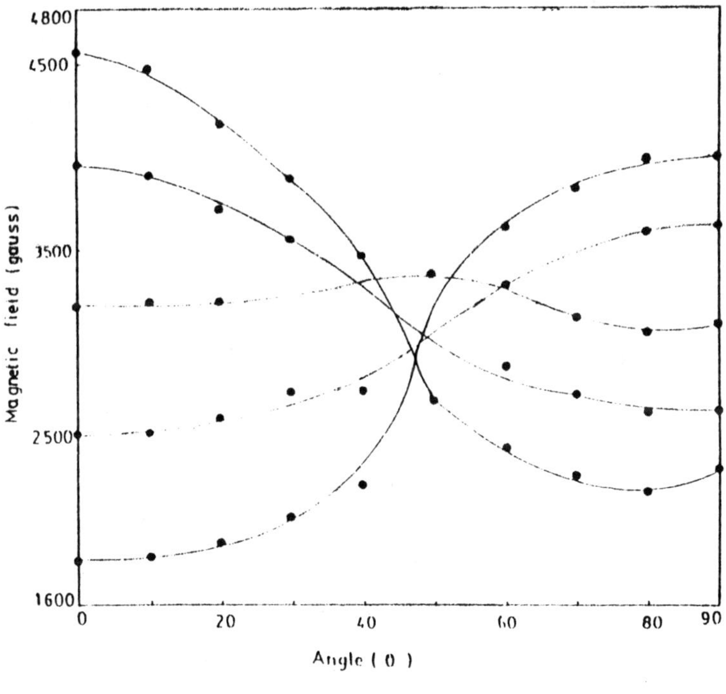

Fig. 2. Angular variation plot of Mn(II):CoMTH single crystals at room temperature for B in yz plane of the allowed transitions.

The spin-Hamiltonian parameters (SHP) obtained for single crystal EPR spectra at 296 K and 393 K are given in Table 1. It was not possible to calculate the spin-Hamiltonian parameters at 123 K, as the spectrum was broadened out at low temperatures.

Table 1. The spin-Hamiltonian parameters of Mn(II):CoMTH single crystals at 296 K and 393 K.

Temperature	*g* ± 0.002	*D (gauss)* ± 3	*A (gauss)* ± 3	*E (gauss)* ±3	*a (gauss)* ± 0.5
296 K	2.029	-350	-90	65	-14.1
393 K	2.030	-316	-90	42	-13.3

From the table, it is clear that the values of zero field splitting parameters are sensitive to the temperature. The value of E/D is consistent with that reported by Rudowicz et al. [14]. The signs of D and E are relative and the value of `A' is always taken to be negative [15].

The zero field splitting parameter (D) is found to increase on lowering the temperature. A best fit of the variation of zero field splitting parameter (D) as a function of temperature is shown in Fig. 3. At low temperatures, the D value increases rapidly. This behaviour of D can be described due to thermal contraction and vibrational mechanism in the lattice [16,17].

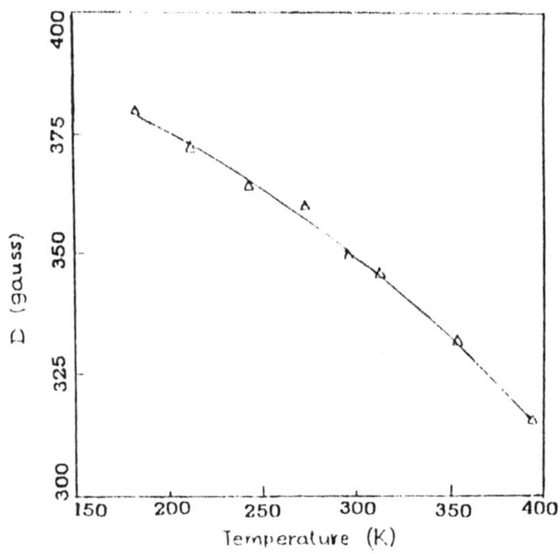

Fig. 3. Variation of zero field splitting parameter (D) as a function of temperature.

The observed g value deviates from free spin value and is slightly anisotropic within the experimental error and the A values remains almost isotropic. A positive shift in g value observed in the present work arises due to the electron-transfer process through spin-orbit interaction from the ligands to the S-state ion [18].

The observed EPR spectral lines of central hyperfine sextet M = 1/2 $\longleftrightarrow$ -1/2 have larger linewidths than those of the outer sextets involving larger quantum numbers. This is true in case of Mn(II) doped paramagnetic hosts [3-5]. Indeed, the same is observed in the present work; at room temperature the authors observed a linewidth of $\approx$ 56 gauss for 1/2 $\longleftrightarrow$ -1/2 transition, while for 5/2 $\longleftrightarrow$ 3/2 transition, it is $\approx$ 49 gauss. This unusual variation of linewidths in the fine structure groups may be due to an unequal influence of exchange interaction between Co(II) and Mn(II) upon the various $\Delta M = \pm 1$ transitions of Mn(II) ions [5].

The EPR spectra were also recorded at different temperatures from 123 K to 393 K for the B || z axis of the crystal. As the temperature of the crystal is lowered from room temperature, the widths of the resonance lines increase and a completely broadened out spectrum is observed at 123 K. On raising the temperature again, the resolved spectrum reappears indicating that the smearing out of the spectrum is due to the increased linewidths. The larger linewidths observed at room and low temperatures are due to the fluctuating magnetic fields created by the cobalt magnetic moments. The spectrum observed at 123 K and 393 K are shown in Fig. 4.

A survey of literature shows that Mn(II) invariably enters the Co(II) host lattices substitutionally [3,5]. This may be attributed to the comparable ionic radii of Mn(II) (0.80 Å) and Co(II) (0.72 Å). So it is reasonable to assume that Mn(II) ions enter substitutionally at Co(II) ion sites in CoMTH crystal.

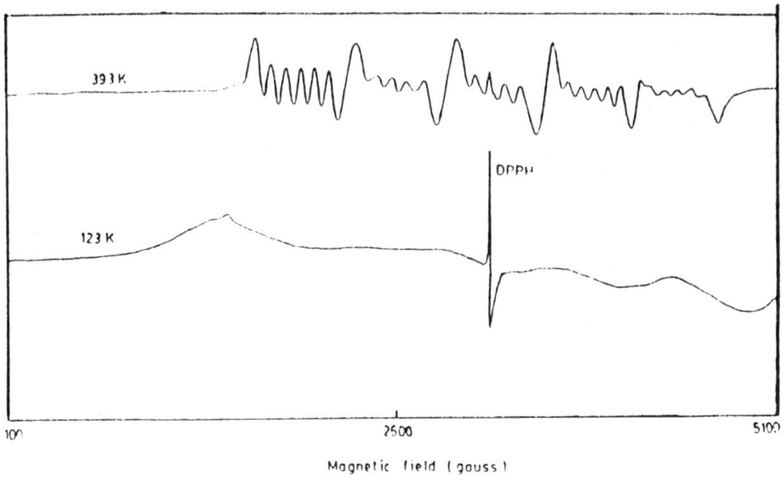

Fig. 4. EPR spectrum of Mn(II):CoMTH single crystals at 393 K and 123 K for B||z axis.

3.1 Estimation of Spin-lattice Relaxation Time (T_1)

Mitsuma [19] proposed that fast relaxation of host can randomly modulate the dipolar and exchange interaction between paramagnetic host and the impurity ion, resulting in what is called "host spin lattice relaxation narrowing".

The impurity linewidth provides information about the host spin-lattice relaxation time (T_1), when the host spin-lattice relaxation mechanism is effective and the formula for the host T_1 is given by [1]:

$$T_1 = (3/20)\,(h/g_h\beta)\,(\Delta B_{imp} / B_d^2)$$

$$\text{with } B_d^2 = 5.1\,(g_h\,\beta n)^2\,S_h\,(S_h + 1)$$

In the above formulae, h is Planck's constant, β is the Bohr magneton, g_h is the host g value, S_h is the host effective spin and is taken to be 1/2. n is the number of host spins per cc. The value of n was calculated from the crystallographic data of isomorphous zinc maleate tetrahydrate (ZnMTH) [12].

The detailed linewidth studies were carried out for M = 5/2 ⟷ 3/2 transition for B || z axis. As the linewidth data for the isomorphous salt Mn(II) doped ZnMTH is not available, the peak to peak linewidth of M = 5/2 ⟷ 3/2 transition in the present work is taken as ΔB_{imp} in the calculation of T_1. The value of g_z = 6.0 for Co(II) ions in most of the hydrated salts [10] is taken to be g_h in the present work.

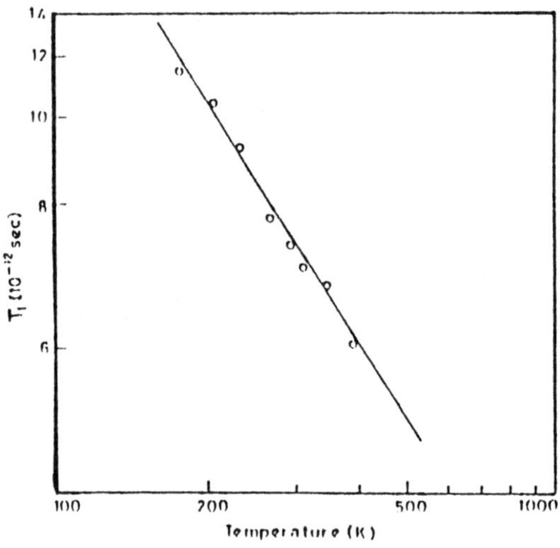

Fig. 5. A logarithmic plot of spin-lattice relaxation time (T_1) as a function of temperature.

The order of spin-lattice relaxation time (T_1) obtained for Co(II) ions at room temperature (7.25 X 10^{-12} sec) in the present work is of the same order reported for other cobalt salts [4, 8-11]. The results for the relaxation time of Co(II) ions obtained in the temperature range 183 - 393 K are shown in Fig. 5, where a logarithmic graph of T_1 as a function of temperature is presented. A slope of which gives information about the spin-lattice relaxation due to two phonon Raman scattering processes [8,11].

4 Conclusions

1. The EPR spectra of Mn(II) in cobalt maleate tetrahydrate single crystals have been studied in the temperature range 123 - 393 K. The detailed studies reveal that Mn(II) ions enter substitutionally at Co(II) sites in CoMTH single crystals.
2. The unusual variation of observed linewidths in Mn(II):CoMTH along the Zeeman field is due to unequal influence of exchange interaction between Co(II) and Mn(II) upon various $\Delta M = \pm 1$ transitions of Mn(II)
3. At low temperatures, the linewidth increases and the EPR spectrum completely broadens out at 123 K. This has been attributed to the fast relaxing of Co(II) ions compared to Mn(II) ions.
4. The spin-Hamiltonian parameters were evaluated. The observed positive shift in g value arises due to the electron transfer process from the ligands to the Mn(II) ion.
5. It is found that the zero field splitting parameter (D) increases on lowering the temperature. This has been attributed to thermal contraction and vibrational mechanism in the lattice.
6. From the observed linewidths, the Co(II) spin-lattice relaxation time has been estimated at different temperatures.

References

[1] G.C.Upreti and R.S.Saraswat, Electron paramagnetic resonance studies of para-magnetic ions in paramagnetic hosts, *Magn. Reson. Rev.*, **7** (1982) 215-237.

[2]V.K.Jain and G.Lehmann, Electron paramagnetic resonance of Mn^{2+} in orthorhombic and higher symmetry crystals, *Phys. Stat. Sol. (b)*, **159** (1990) 495-544.

[3] R.S.Saraswat and G.C.Upreti, A comparative EPR study of Mn^{2+} doped in Magnesium, Cobalt and Nickel acetate tetrahydrate single crystals, *J.Magn. Reson.*, **20** (1975) 39-46.

[4] R.S.Saraswat and G.C.Upreti, Estimation of T_1 of fast relaxing Mn^{2+} and VO^{2+} impurities in some cobalt salt single crystals, *Phys. Stat. Sol. (b)*, **87** (1978) k39-42.

[5] R.S.Saraswat and G.C.Upreti, EPR investigations of Mn^{2+} in $MK_2(SO_4)_2.6H_2O$ Tutton salt single crystals, *Chem. Phys.*, **23** (1977) 97-103.

[6] M.Lynch, G.Kokoszka and P.Szydrik, EPR of a mixed metal two dimensional system: $Mn_{1-f}Co_f(HCOO)_2.2H_2O$, *J.Phys. Chem. Solids*, **45** (1979) 79-83.

[7] V.K.Jain, V.P.Seth, R.S.Bansal and A.Yadav, Electron spin resonance of Mn^{2+} in Tl_2 $Co(SO_4)_2$ $6H_2O$ Single Crystals. *Phys.Stat.Sol.(b)*, **138** (1986) k55-58.

[8] V.K.Jain, V.S.Yadav and J.Singh, Host spin-lattice relaxation narrowing in Rb_2 $Co(SO_4)_2.6H_2O$:Mn^{2+} single crystals, *Solid State Commun.*, **64** (1987) 929-931.

[9] R.Hrabanski, EPR of Mn^{2+} in $[Co(H_2O)_6]PtCl_6$ single crystals: Estimation of spin-lattice relaxation time, *Spectrochim. Acta*, **44A** (1988) 647-650.

[10] V.K.Jain and V.S.Yadav, Estimation of T_1 for Co(II) from the EPR linewidths of Mn(II) in Dibarium Cobalt formate Tetrahydrate single crystals, *Phys. Stat. Sol . (b)*, **148** (1988) k157-160.

[11] J.Leech, A.Slezack and R.Hrabanski, EPR of Mn^{2+} in $[Co(H_2O)](ClO_4)_2$ single crystals: Estimation of spin-lattice relaxation time, *J.Phys.Chem.Solids*, **52** (1991) 685-689.

[12] K.Saroja and S.V.Raman, Space group and unit cell dimensions of Maleate complexes of Zinc and Nickel, *Curr.Sci.*, **41** (1972) 599.

[13] W.Low, *Paramagnetic Resonance in Solids, Solid State Physics*, Academic Prcss, New York, Vol.2 (1960).

[14] C.Rudowicz and R.Bramley, On standardization of the spin Hamiltonian and the Ligand field Hamiltonian for orthorhombic symmetry, *J.Chem. Phys.*,**83** (1985) 5192-5197.

[15] M.Vithal and R.Jagannathan, Electron Spin Resonance Study of Manganese(II) -doped zinc maleate tetrahydrate. *J.Chem.Soc. Dalt. Trans.*, (1988) 983-986.

[16] W.M.Walsh Jr. , Stress and temperature dependence of the Paramagnetic Resonance spectra of Nickel Fluorosilicate, *Phys.Rev.(B)*, **114** (1959) 1473-1485.

[17] Y.Nagaraja Naidu, J.Lakshmana Rao and S.V.J.Lakshman, Electron Paramagnetic Resonance and optical absorption spectra of Mn^{2+} in Magnesium Malate Pentahydrate (MMPH) single crystals, *Solid State Commun.*, **81** (1992) 437-442.

[18] H.Watanabe, g value of Fe^{3+} in II-VI cubic crystals, *J.Phys.Chem.Solids*, **25** (1964) 1471-1475.

[19] T.Mitsuma, Narrowing of resonance lines by spin-lattice relaxation in the system $K_3Co(CN)_6$, *J.Phys.Soc., Japan*, **17** (1962) 128.

Conversions, Standardization and Transformations of the Spin Hamiltonian and the Crystal Field Hamiltonian - Computer Package "*CST*"

C. Rudowicz, I. Akhmadulline and S. B. Madhu*

Department of Physics and Materials Science, City University, Hong Kong
*Present address: E.K. Zavoisky Physical-Technical Institute, Kazan, Russia

Abstract. The abundance of the notations for the operators and parameters hinders efficient comparison of the EPR and optical spectroscopy data, and may lead to misinterpretations. To alleviate this problem a computer package ***CST*** has been developed for various general manipulations of the format of the zero-field splitting parameters [**ZFSP**] (and the crystal-field ones [**CFP**]). The following options are included: (i) **unit conversions** - between several units for ZFSP (CFP), (ii) **notation conversions** - between several major notations for ZFSP (CFP), (iii) **standardization** - confining the ratio b_2^2/b_2^0 [*E/D*] to the standard range (0, 1) [0, 1/3] for the orthorhombic, monoclinic and triclinic ZFSP (CFP), (iv) **transformations** from one axis system into another of ZFSP (CFP) and (module ***GTRANS***) of the Zeeman **g**-matrix. The various options and capabilities of the package are illustrated by applications to facilitate the analysis of the EPR and optical spectroscopy data derived from various sources.

1 Introduction

There is a definite need for a user-friendly computer package for various general manipulations of the format of the experimental zero-field splitting parameters **(ZFSP)** as well as the crystal-field ones **(CFP)**. The package has been developed from the earlier programs: STAND [1], STANDCF and STANDER [2, 3]. The present (FORTRAN 77) package ***'CST'*** can handle ZFSP, CFP, and **g**-matrices of any symmetry. Its capabilities include: **CONVERSIONS: unit conversions** - between several units most often used for ZFSP (CFP) and **notation conversions** - between several major notations for ZFSP (CFP); **STANDARDIZATION** of orthorhombic, monoclinic and triclinic ZFSP (CFP); including the errors of the standardized parameters, and **TRANSFORMATIONS** of ZFSP (CFP) into an arbitrary axis system; including the rotation invariants and the errors of the transformed parameters. A separate ***GTRANS*** module for transformations of the Zeeman **g**-matrix is included. Detailed description of the ***CST*** modules and several examples of applications are provided in the Manual [4]. Here the underlying theory as well as the structure and capabilities of the ***CST*** package are briefly presented.

2 Basic Definitions

2.1 Notations Used for the ZFS Hamiltonian

A number of different operator notations have been employed in the EPR studies [5]. Here we adopt as the *reference* notation the *extended Stevens (ES)* operators O_k^q [6]. Explicit listing of the ES operators and references are provided in [4]. The general ZFS Hamiltonian for a spin $\tilde{S}$ system is given by (see, e.g. [5, 7, 8]):

$$\tilde{H}_{ZFS} = \sum_{kq} B_k^q O_k^q (\tilde{S}_x, \tilde{S}_y, \tilde{S}_z) = \sum_{kq} f_k b_k^q O_k^q \tag{1}$$

The second form with the 'scaling' factors : f_k = 1/3, 1/60 and 1/1260 for k = 2, 4 and 6, respectively, is more often used. For triclinic symmetry all the (2k+1) values of q for each k are allowed. Hence the maximum number of the ZFSP (B_k^q, b_k^q or equivalent) handled by the program is 27 (5 + 9 + 13, for k = 2, 4, 6 respectively), as required for S = 7/2 system. Thus a unification of notations is most important for the high spin and low symmetry cases. In Sec.3.2 the classification of notations and conversions for ZFSP (CFP) are discussed - for details see [4, 5].

2.2 Standardization

2.2.1 Orthorhombic Symmetry

The standardization idea [1] limits, in the ES operator notation, the ratio:

$$\lambda' = B_2^2 / B_2^0 = b_2^2 / b_2^0 \tag{2}$$

to the 'standard' range (0, 1) by a proper choice of the axis system for orthorhombic or lower symmetry. This idea has been extended to the fourth- and sixth-order ZFS terms for orthorhombic [1] and monoclinic symmetry [2]. The standardization transformations S_i (i = 1 to 6) have been defined as follows [1, 9]: S_1 [original] (x,y,z), S_2 (x,z,-y), S_3 (y,x,-z), S_4 (z,x,y), S_5 (y,z,x), S_6(z,y,-x). If the original {λ'} is in the range: (-∞, -3) S_6, (-3, -1) S_4, (-1, 0) S_3, (1, 3) S_2, (3, +∞) S_5, the transformed [λ'] is limitted to (0, 1) by using the transformation S_i as indicated.

2.2.2 Monoclinic Symmetry

For monoclinic symmetry the axis of quantization may be chosen with respect to the monoclinic direction (the C_2 axis for the C_2 and C_{2h}, whereas the direction ⊥ to the reflection plane σ for the C_{1h} group) as: $C_2 \parallel z$, $C_2 \parallel x$, or $C_2 \parallel y$. Each of the three cases corresponds to a different form of the ZFS Hamiltonian as follows [2, 10] :

$$\tilde{H}_{ZFS}(C_2 \parallel z) = \sum_{kq} B_k^q O_k^q \tag{3}$$

with k = 2, q = 0, ±2; k = 4, q = 0, ±2, ±4; k = 6, q = 0, ±2, ±4, ±6,

$$\widetilde{H}_{ZFS}(C_2 \parallel x) = \widetilde{B}_2^0\,\widetilde{O}_2^0 + \widetilde{B}_2^2\,\widetilde{O}_2^2 + \widetilde{B}_2^1\,\widetilde{O}_2^1 + \ldots = \sum_{kq} \widetilde{B}_k^q\,\widetilde{O}_k^q \quad (4)$$

with q = -2, -4, and -6 in (3) replaced by q = -1, -3, and -5, respectively, and

$$\widetilde{H}_{ZFS}(C_2 \parallel y) = \overline{B}_2^0\,\overline{O}_2^0 + \overline{B}_2^2\,\overline{O}_2^2 + \overline{B}_2^1\,\overline{O}_2^1 + \ldots = \sum_{kq} \overline{B}_k^q\,\overline{O}_k^q \quad (5)$$

with q = -2, -4, -6 in (3) replaced by q = 1, 3, 5, respectively. The orthorhombic and monoclinic ZFSP transform separately only into themselves under S_i.

Two suboptions are provided for the monoclinic ZFSP. (1) 'ORTHORHOMBIC' type standardization, which is based only on the ratio λ' of the original ZFSP. The monoclinic ZFSP transform into their counterparts under S_i. This option is suitable for small values of the monoclinic ZFSP as compared with the orthorhombic ones. (2) 'MONOCLINIC' type standardization, where the second-rank monoclinic parameter B_2^{-2}, B_2^{-1} or B_2^1 is first reduced to zero by a suitable rotation α around the z-, x-, and y-axis, respectively. This is equivalent to diagonalization of the second-rank ZFS (CF) terms. The angle α is calculated automatically using the algebraic expressions [2] and then used internally to transform ZFSP (CFP). The value of α is output together with the intermediate and transformed ZFSP (CFP). The option (2) is is suitable for any values of the monoclinic ZFSP.

For triclinic symmetry, both options are suitable only for small values of the triclinic ZFSP as compared with the orthorhombic and monoclinic ones [4].

2.3. Transformations of the ZFS (CF) terms

The transformation relations for the ES operators under a rotation (ϕ/Oz, θ/Oy') [6, 9] have been incorporated into the FORTRAN program. Sequence of two rotations: (ϕ_1, θ_1) followed by (ϕ_2, θ_2) and hence the transformation matrix $R_k = S_k(\phi_1, \theta_1)$. $S_k(\phi_2, \theta_2)$ is used to calculate the transformed ZFSP (CFP) $[B_k^q]$:

$$[B_k^q]^T = \{B_k^q\}^T . R_k \quad (6)$$

in terms of the original ones $\{B_k^q\}$. The norm N_k and the rotation invariants S_k, invariant under arbitrary rotations of the axis system [4], are also calculated.

3. General Structure of the *CST* Package

The package can run either on a mainframe or PC computer. The input form: KEYBOARD or FILE, and the output form: SCREEN or FILE, are provided. The user selects options and/or suboptions from a MENU and can control the flow of the calculations according to the prompts. The general structure of all modules and the list of subroutine files is provided in Appendix 2 of [4]. The opening screen MAIN MENU (Fig. 1) presents the major options. Two lower level menus: the SYMMETRY MENU (Fig. 2) followed by RANK MENU (Fig. 3) will be activated if one of the options 2 - 5 in Fig. 1 is selected. The package can handle ZFSP (CFP) of arbitrary symmetry since it includes all possible point symmetry groups (Fig. 2).

```
                      MAIN    MENU
---------------------------------------------------------------
I      1.      FREE UNIT CONVERSION          I
I      2.      UNIT CONVERSION               I
I      3.      NOTATION CONVERSION           I
I      4.      STANDARDIZATION               I
I      5.      TRANSFORMATION                I
I      Q.      EXIT                          I
---------------------------------------------------------------
              Please choose one option :
```

Fig. 1. Main menu options.

```
        SYMMETRY MENU
--------------------------------------------------------
I  1.  CUBIC I          (O,Td,Oh)                I
I  2.  CUBIC II         (T,Th)                   I
I  3.  HEXAGONAL I      (D6,C6v,D3h,D6h)         I
I  4.  HEXAGONAL II     (C6,C3h,C6h)             I
I  5.  TRIGONAL I       (D3,C3v,D3d)             I
I  6.  TRIGONAL II      (C3,C3i(S6) )            I
I  7.  TETRAGONAL I     (D4,C4v,D2d,D4h)         I
I  8.  TETRAGONAL II    (C4,S4,C4h)              I
I  9.  ORTHORHOMBIC(D2,C2v,D2h)                  I
I 10. MONOCLINIC        (C2,Cs(C1h),C2h)         I
I 11. TRICLINIC         (C1,Ci(S2) )             I
I  M. EXIT TO MAIN MENU                          I
--------------------------------------------------------
        Please choose one option :
```

Fig. 2. Symmetry menu options.

```
                RANK MENU
--------------------------------------------------
I                                                I
I 1. ONLY 2ND ORDER TERMS                        I
I                                                I
I 2. 2ND AND 4TH ORDER TERMS                     I
I                                                I
I 3. 2ND, 4TH AND 6TH ORDER ERMS                 I
I                                                I
I M. EXIT TO MAIN MENU                           I
I                                                I
I                                                I
--------------------------------------------------
        Please choose one option :
```

Fig. 3. Rank menu options.

In order to avoid input of zero values for nonexistent ZFSP (CFP) of an irrelevant higher rank k = 4 or 6, as the case may be, three options are provided (Fig. 3). The package can process several consecutive data sets.

3.1 Unit Conversion

Two options are provided in the MAIN MENU (Fig. 1) for conversion between various units: (1) FREE UNIT CONVERSION, which is not related to any specific symmetry and (2) UNIT CONVERSION, which uses a specified symmetry (Fig. 2) and notation (Sec.3.2) for ZFSP (CFP). The following conversion factors are used to convert to the *standard* unit [10^{-4} cm^{-1}] a value U in a given unit: 8065.54077 * 10^4 * U [eV] , 0.5035 * 10^{20} * U [erg], g * 0.466856 * U [Gauss]; the g-factor must be provided, 6950.38605 * U [K], 0.33356 * U [MHz]. Option (1) and (2) converts an input original unit **A,** selected from the above six options, to two output formats: (i) **"A to B"** - conversion from **A** to any other unit **B** and (ii) **"A to all"** - automatic conversion to all other five unit systems. The number of parameters is limited to 27.

3.2 Notation Conversion

The NOTATION MENU (Fig. 4) enables conversions between major notations used for ZFSP (CFP) as listed in Appendix 1. Selection of an option activates the prompts regarding the symmetry (Fig. 2) and rank (Fig. 3).

```
                    NOTATION   MENU
---------------------------------------------------------------
I          1.        ES    ( bkq )                       I
I          2.        ES    ( Bkq )                       I
I          3.        NS                                  I
I          4.        NCST                                I
I          5.        BST                                 I
I          6.        Ph.M.BST                            I
I          7.        KS/BCS                              I
I          8.        CONVENTIONAL                        I
I          M.        EXIT TO MAIN MENU                   I
---------------------------------------------------------------
                 Please choose one option:
```

Fig. 4. Notation menu options.

The input of the original ZFSP (CFP) must then be done according to the symmetry case chosen. Two output formats are provided: (i) "**A to B**" - conversion from **A** to any other notation **B** selected from the above eight options, and (ii) "**A to all**" - automatic conversion to all other **tensorial** notations (no. 1 to 7). The option (8) is applicable only for some symmetry cases and requires special input format [4]. An example of application for Gd^{3+} in NdP_5O_{14} [11] is provided in Table 1 and 2.

Table 1. Sample input file for the NOTATION CONVERSION option.

```
Ref.  Parrot  and  Canny Phys.Stat.Sol.,(b)110,123 (1982)
393.0,-13.8,13.1,324.0,37.8
0.838,-0.060,0.469,0.621,0.543,-0.097,0.299,0.388,0.890
/
```

Table 2. Sample output file for input in Table 1.

```
      Parrot and Canny. Phys.Stat.Sol.(b),110,123,1982.
       NOTATION CONVERSION
           [ORIGINAL]  (BST)
B20             .3930D+03
B21  Re:       -.1380D+02      Im:      .1310D+02
B22  Re:        .3240D+03      Im:      .3780D+02
B40             .8380D+00
B41  Re:       -.6000D-01      Im:      .4690D+00
B42  Re:        .6210D+00      Im:      .5430D+00
B43  Re:       -.9700D-01      Im:      .2990D+00
B44  Re:        .3880D+00      Im:      .8900D+00
```

```
          [FINAL]          ( ES(bkq) )
B20 = .58950D+03   B21 = .10141D+03   B21M= .96265D+02
B22 = .11905D+04   B22M=-.13889D+03
B40 = .62850D+01   B41 = .40249D+01   B41M= .31461D+02
B42 = .29457D+02   B42M=-.25757D+02  B43 = .17216D+02
B43M= .53067D+02  B44 = .24347D+02    B44M=-.55847D+02
```

3.3 Standardization

Selection of the main option (4), Fig. 1, activates the STANDARDIZATION MENU, which provides two choices: ZFS STANDARDIZATION and CF STANDARDIZATION. For each choice the SYMMETRY and RANK MENU options must be then specified. Note that the standardization procedure deals only with the orthorhombic, monoclinic and triclinic symmetry. An input notation (Fig.4) must be specified since the standardization expressions [1] apply directly only for ZFSP (CFP) in the ES (B_k^q and b_k^q) notation. For notations other than the ES one the conversion to the ES notation and reconversion is carried out automatically. The ZFS (CF) STANDARDIZATION **[ST]** MENU provides three options: (i) AUTOMATIC **ST** - depending on the value of {λ'}, (ii) **ST** TRANSFORMATION - using a specified transformation S_i and (iii) **ST** ERRORS; for details, see [4].

The monoclinic ZFS standardization is applied for Gd^{3+} in $Sm(CH_3COO)_3 .4H_2O$ [12] (Table 3; output file only). The original ES (b_k^q) ZFSP (Z || C_2): {b_2^0, b_2^2, b_2^{-2}..} are non-standard. Since b_2^{-2} is not negligible, the MONOCLINIC type STANDARDIZATION as well as the AUTOMATIC STANDARDIZATION option are selected. The transformation S_4 is automatically chosen to standardize ZFSP. Note that the input values 0.00000001 (.1D-07) are put in place of the unavailable original ZFSP. Consequently the values of the order 'D-07' or smaller in Table 3 mean effectively zero and should be omitted. The angle ALPHA is given in degrees.

Table 3. Sample output file for input in Table 3.

```
      Misra,Bartkowski and Mikolajczak  J.Chem.Phys.,78,5369,1983
MONOCLINIC TYPE STANDARDIZATION (AUTOMATIC:S4 )(MONOCLINIC
SYMMETRY)
                        -------(C2//Z)-------
        ORIGINAL PARAMETERS IN ES(bkq)  NOTATION ARE :
 B20 =-.89700D+00  B22 = .10000D+01    B22M= .94000D-01
 B40 =-.10000D-01  B42 = .10000D-07    B42M= .10000D-07
 B44 = .10000D-07  B44M= .10000D-07

     INTERMEDIATE PARAMETERS IN ES(bkq) NOTATION ARE :
                    ALPHA=    2.685
 B20 =-.89700D+00  B22 = .10044D+01   B22M=-.49670D-17
 B40 =-.10000D-01  B42 = .10892D-07   B42M= .90202D-08
 B44 = .11688D-07  B44M= .79613D-08
```

FINAL PARAMETERS IN ES(bkq) NOTATION ARE :
B20 = .95070D+00 B21 =-.99340D-17 B22 = .84330D+00
B40 =-.37500D-02 B41 = .34512D-08 B42 =-.25000D-01
B43 =-.39532D-07 B44 =-.43750D-01

3.4 Transformations of the ZFS and CF Terms

The transformation relations [8] apply directly to ZFSP (CFP) in the ES notation (B_k^q and b_k^q). To facilitate applications for CFP, the TRANSFORMATION module provides directly two options: the ES and the Wybourne notation [2, 4] for the input parameter notation. The SYMMETRY and RANK MENU options must be specified as well as "PHI" (ϕ - rotation about the original z axis) and "THETA" (θ - about the new y axis) must provided. Input of ϕ and θ may be in the degrees or radians. The norm N_k and rotation invariants S_k are calculated automatically.

4 Summary and Conclusions

The underlying spin Hamiltonian theory as well as the structure and capabilities of the ***CST*** computer package have been presented. The package enables various general manipulations of the zero-field splitting parameters **[ZFSP]** and the crystal-field ones **[CFP]** used in the EPR and optical spectroscopy. In this paper applications to ZFS parameters are only considered. A separate **GTRANS** module for transformations of the Zeeman **g**-matrix is also included. The Manual [4] provides several other examples of applications of the ***CST*** package for the ZFS and CF parameters for various spin systems as well as of the ***GTRANS*** module.

The menu driven structure and the operating instructions displayed on screen during execution of each subprogram make the usage of the package easy and convenient. The ***CST*** package facilitates interpretation and comparison of experimental data derived from various sources. The effort to convert the ZFS and CF parameters expressed in different notations and formats can be substantially reduced by efficient usage of the package. Application of the package may also help clarifying confusion arising from the inadvertent usage by some authors of identical symbols having sometimes different meanings (see Manual [4]). The present package provides an efficient way for bringing various data available in the literature for a given paramagnetic ion and crystal system into a unified form.

Acknowledgements
Technical help with the early versions of the program by Ms Jian Qin is gratefully acknowledged. This work was partly supported by the UGC and the City University of Hong Kong Strategic Research Grant.

Appendix
Apart from the **conventional** notation for ZFSP, which uses various symbols as reviewed in [5], e.g. D and E, the tensor-operator notation can be subdivided into:
(1) the tesseral harmonics operator equivalents:

a) the 'usual' Stevens operators limited to the positive q components [13, 7, 8],
a') the extended Stevens **(ES)** operators including also the negative q components [6],
b) the normalized Stevens **(NS)** operators [14, 15],
c) the normalized combinations of spherical tensor operator **(NCST)** operators [16, 4],
d) five unnamed notations as reviewed in [5].

(2) the spherical harmonics operator equivalents:
a) Buckmaster [17], and Smith & Thornley [18] operators **(BST)**; Wybourne operators [2] are mathematically equivalent to the BST operators,
b) Phase-modified BST operators **(Ph.M.BST)** [5],
c) Koster & Statz [19], and Buckmaster, Chatterjee & Shing [20] **(KS/BCS)**; *cf* also [21],
d) ten unnamed notations as reviewed in [5].

References

1. C.Rudowicz and R. Bramley, *J. Chem. Phys.* **83** (1985) 5192-5202.
2. C. Rudowicz, *J. Chem. Phys.* **84** (1986) 5045-5058.
3. C. Rudowicz, *Mol. Phys.* **74** (1991) 1159-1170.
4. C. Rudowicz, I. Akhmadoulline and S. B. Madhu, *Manual for the computer package "CST" for Conversions, Standardization and Transformations of the spin Hamiltonian and the Crystal field Hamiltonian,* Research Report AP-97-12, Department of Physics and Materials Science, City University of Hong Kong (1997).
5. C. Rudowicz, *Magn. Reson. Rev.* **13** (1987) 1-89.
6. C. Rudowicz, *J. Phys.* **C18** (1985) 1415-1430; ibidem, **C18**, 3837.
7. A. Abragam and B. Bleaney, *Electron paramagnetic resonance of transition ions*, Dover Publications, New York (1970).
8. J. R. Pilbrow, *Transition ion electron paramagnetic resonance*, Clarendon Press, Oxford (1990).
9. C. Rudowicz, *J. Magn. Reson.,* **63** (1985) 95-106.
10. S. K. Misra and C. Rudowicz, *Phys. Stat. Sol. (b)* **147** (1988) 677-683.
11. R. Parrot and B. Canny, *Phys. Stat. Sol. (b)* **110** (1982) 123-131.
12. S. K. Misra, M. Bartkowski and P. Mikolajczak, *J.Chem. Phys.* **78** (1983) 5369-5372.
13. K.W.H. Stevens, *Proc. Phys. Soc.* **65** (1952) 209-215.
14. J.M. Gaite, J. Michoulier, *J. Chem. Phys.* **59** (1973) 488.
15. G. Bacquet, J. Dugas, C. Escribe, J.M. Gaite, J. Michoulier, *J. Phys.* **C7** (1974) 1551.
16. D.G. McGavin and W.C. Tennant, *Mol. Phys.* **55** (1985) 953-964.
17. H.A. Buckmaster, *Can. J. Phys.* **40** (1962) 1670-1677.
18. D. Smith, J.H.M. Thornley, *Proc. Phys. Soc.* **89** (1966) 779-781.
19. G.F. Koster, H. Statz, *Phys. Rev.* **113** (1959) 445-455.
20. H.A. Buckmaster, R. Chatterjee, Y.H. Shing, *Phys. Stat. Sol. (a)* **13** (1972) 9-50.
21. S.A. Marshall, T. Marshall, H.A. Buckmaster, *Magn. Reson. Rev.* **14** (1988) 81-98.

EPR Study of Trivalent Magnetic Ions in Hexagonal $RbZnF_3$ Crystals

Masanori ARAKAWA, Hiroshi EBISU, Hideo TAKEUCHI† *and Masahiro MORI*††

Department of Physics, Nagoya Institute of Technology, Showa-ku, Nagoya, 466, Japan
† School of Engineering, Nagoya University, Chikusa-ku, Nagoya, 464-01, Japan
†† School of Informatics and Science, Nagoya University, Chikusa-ku, Nagoya 464-01, Japan

Abstract. EPR measurements at room temperature have been made on $RbZnF_3$ crystals co-doped with M^{3+} (M = Cr, Gd) and Li^+ using an X-band spectrometer. In some as-grown crystal a new spectrum with trigonal symmetry is observed. The trigonal centre is identified as M^{3+} - Li^+ pair substituted for the Zn^{2+} - Zn^{2+} bond in a Zn_2F_9 unit in hexagonal $RbZnF_3$ (h-$RbZnF_3$), although the hexagonal structure is normally stabilized above about 640 °C. The formation of the hexagonal structure at room temperature is confirmed by X-ray diffraction analysis. This indicates that the M^{3+} - Li^+ pairs in Zn_2F_9 units may be related to the crystal growth of the hexagonal structure in the matrix.

1 Introduction

The crystal structure of $RbZnF_3$ is in a boundary between the cubic perovskite structure and the h-$BaTiO_3$ type hexagonal structure [1]. $RbZnF_3$ crystallizes in the hexagonal phase (h-$RbZnF_3$) from a stoichiometric melt at 730° C. Upon cooling, it normally undergoes a destructive phase transition at about 640° C and becomes cubic (c-$RbZnF_3$) [2]. The hexagonal structure may be regarded as an intermediate one between the $CsNiCl_3$ structure which consists of infinite linear arrays of octahedra sharing faces and the cubic perovskite structure which consists of a three-dimensional network of octahedra sharing corners. That is, two octahedra sharing a face form a $[Zn_2F_9]^{5-}$ unit, which shares corners with six different ZnF_6 octaherda as seen in figure 1.

We have carried out EPR measurements on as-grown single crystals of $RbZnF_3$ co-doped with M^{3+} (M = Cr, Gd) and Li^+ using an X-band spectrometer. Single crystals of $RbZnF_3$ were grown in graphite crucibles by the Bridgman technique. We observed new signals from trigonal M^{3+} centres at room temperature unlike those in cubic perovskite fluorides. This suggests that the hexagonal structure is obtained by co-doping with M^{3+} and Li^+ with a normal cooling procedure.

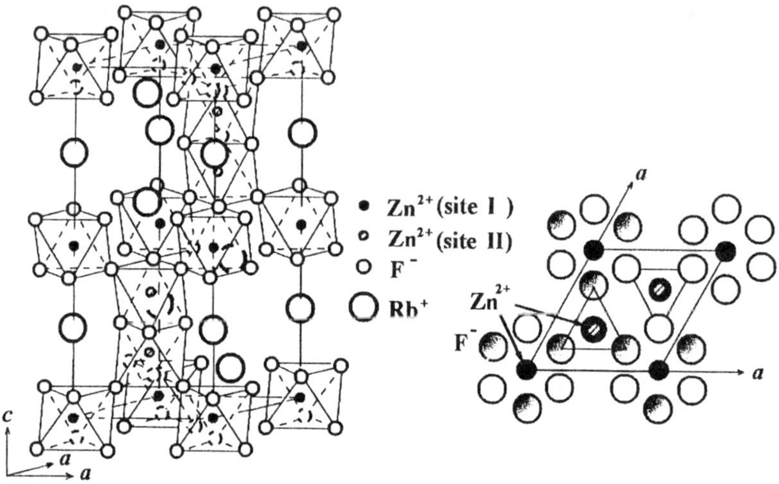

Fig. 1. Unit cell of h-$RbZnF_3$ and projection of a h-$RbZnF_3$ crystal onto the *c*-plane. Two Zn_2F_9 units in unit cell are represented by triangles.

2 Results

2.1 Trigonal Cr^{3+}- Li^+ Centre in H-$RbZnF_3$

In some crystals co-doped with Cr^{3+} and Li^+, we observed signals from a new Cr^{3+} centre at room temperature. The resonance fields of the three fine-structure signals are constant in the *c*-plane. This makes clear that the new centre has an uniaxial symmetry about the *c*-direction. The angular dependences of the signals when H is rotated from the *c*-axis to a direction in the *c*-plane are represented in figure 2. Each signal for the new centre shows no splitting in contrast with the case for the Cr^{3+} - Li^+ centres observed in cubic perovskite fluorides, where Cr^{3+} - Li^+ pairs are formed along the crystalline [100], [010] and [001] axes [3]. This suggests that hexagonal structure may be formed by co-doping with Cr^{3+} and Li^+, although the cubic perovskite structure is normally stabilized at room temperature. The hexagonal structure of the matrix is confirmed by X-ray diffraction analysis. The Zn^{2+} ions in this lattice occupy two crystallographically inequivalent sites (I and II in figure 1), where sites I are in ZnF_6 octaherda and sites II are in Zn_2F_9 units.

In $CsMgCl_3$ crystals with $CsNiCl_3$-like hexagonal structure, the formation of the Cr^{3+} - Li^+ pair substituted for a Mg^{2+}- Mg^{2+} bond in the linear arrays of the octahedra sharing faces is superior to those of other centres [4]. The spectrum observed in h-$RbZnF_3$ strongly suggests that the centre is ascribed to the Cr^{3+} - Li^+ pair in a Zn_2F_9 unit with trigonal symmetry because the excess positive charge of Cr^{3+} is locally just compensated by Li^+ at the other Zn^{2+} site in the Zn_2F_9 unit.

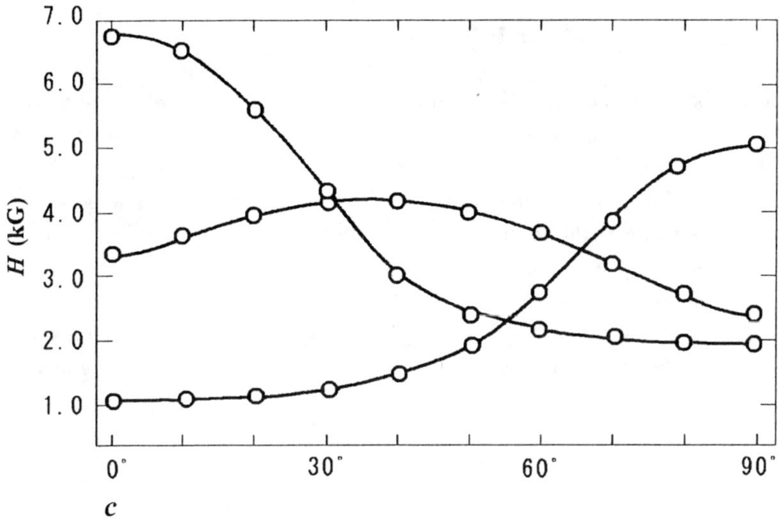

Fig. 2. Angular variation of the EPR spectrum of the trigonal Cr^{3+} centre in h-$RbZnF_3$ when H is rotated from c-axis to the direction in the c-plane. The smooth curves indicate the angular variation computed using the parameters for best fit.

The trigonal spectra for Cr^{3+} - Li^{+} centre can be described by the following spin Hamiltonian:

$$H = g_{\parallel}\,\beta S_z H_z + g_{\perp}\beta (S_x H_x + S_y H_y) + \frac{1}{3}\, b_2^0\, O_2^0 \quad , \tag{1}$$

where the z-axis denotes the trigonal axis parallel to the c-axis. The spin-Hamiltonian parameters obtained by the matrix-diagonalization method are listed in Table 1. The parameters for the trigonal Cr^{3+} - V_K centres in cubic perovskites $KMgF_3$ [5] and $KZnF_3$ [5] are also listed for comparison. The positive sign of b_2^0 is expected from the positive value of $g_{\parallel}$ - $g_{\perp}$ as shown in Table 1. This positive sign was confirmed by the depopulation effect at 4. 2K.

Table 1. Values of the spin Hamiltonian parameters for the trigonal Cr^{3+} centre in h-$RbZnF_3$ at room temperature together with the trigonal Cr^{3+} - V_K centres in $KZnF_3$ and $KMgF_3$ with cubic perovskite structure. The units in 10^{-4} cm^{-1} for b_2^0.

Crystal	T(K)	$g_{\parallel}$	$g_{\perp}$	b_2^0	$g_{\parallel}$ - $g_{\perp}$	Reference
h-$RbZnF_3$	300	1.9762(8)	1.9728(6)	1597.3(5)	0.0034	This work
$KZnF_3$	77	1.9716(5)	1.9725(4)	-1613(5)	-0.0009	[5]
$KMgF_3$	4.2	1.9719(7)	1.9725(4)	-1528(4)	-0.0006	[5]

2.2 Trigonal Gd^{3+} -Li^{+} Centre in H-$RbZnF_3$

In some crystals co-doped with Gd^{3+} and Li^{+}, we observed a Gd^{3+} centre [6] with trigonal symmetry at room temperature, similarly to the case for the crystals co-doped with Cr^{3+} and Li^{+}. The total spread of the seven fine-structure signals shows maximum when H is parallel to the crystalline c-axis. The angular dependence of the signals in the c-plane is represented in figure 3. The seven fine-structure signals observed are almost constant but some ones show slightly dependences of 60° periodicity in this plane. The field directions of the minimum and maximum splittings in figure 3 are called the directions a and b hereafter. The angular dependence of the seven fine-structure signals when

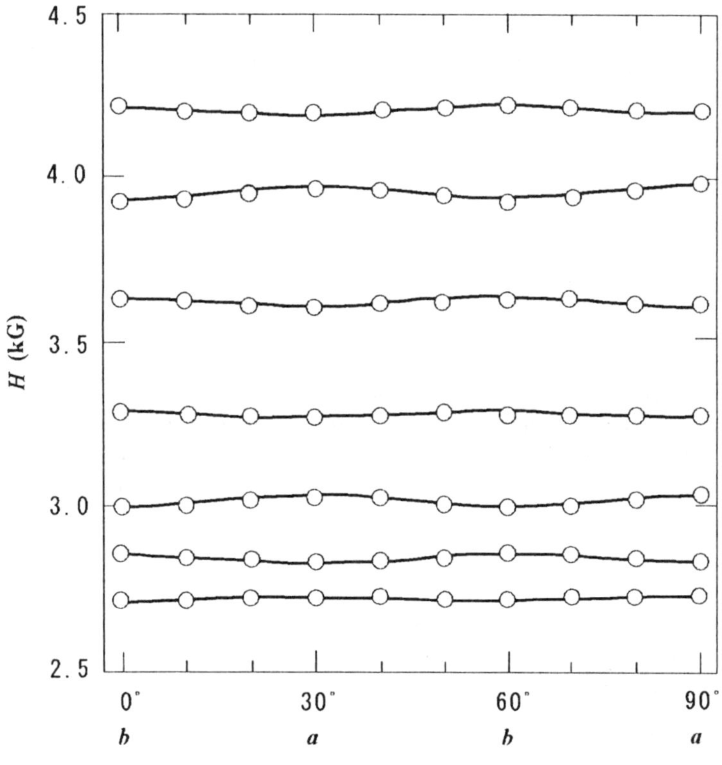

Fig. 3. Angular variation of the EPR spectrum of the trigonal Gd^{3+} centre in h- $RbZnF_3$ with H in the c-plane. The directions a and b are respectively corresponding to those for minimum and maximum splittings in this figure. The smooth curves indicate the angular variation computed using the parameters for best fit.

H is rotated in the plane perpendicular to the direction a is represented in figure 4. Each signal observed in the c-direction further splits into two branches as seen in this figure. The two branches coincide with each other when H is rotated in the plane perpendicular to the direction b. This suggests the presence of only two equivalent trigonal centres with the trigonal axes along the c-direction, which rotate by 60 ° each other about c-axis.

The trigonal spectra for Gd^{3+} - Li^{+} centre can be described by the following spin Hamiltonian:

$$H = g\beta \mathbf{S}\cdot\mathbf{H} + \frac{1}{3}b_2^0 O_2^0 + \frac{1}{60}(b_4^0 O_4^0 + b_4^3 O_4^3) + \frac{1}{1260}(b_6^0 O_6^0 + b_6^3 O_6^3 + b_6^6 O_6^6), \quad (2)$$

where the z-axis denotes the trigonal axis parallel to the c-axis. The x- and y-axes are chosen to be parallel to the directions b and a. The spectra observed were fitted to the spin Hamiltonian by the matrix diagonalization method using a computer. Relative signs among the b_n^m parameters can be uniquely determined. The parameter b_4^3 is related to the cubic component b_{4c} of the fine-structure term as $b_{4c} = -(3/40\sqrt{2})\, b_4^3$, which can be compared with the fine-structure parameters b_4 for cubic Gd^{3+} centres. We select the positive sign of b_4^3 because negative signs of b_4 are obtained for the cubic Gd^{3+} centres in fluorides [7]. The spin Hamiltonian parameters obtained are listed in Table 2.

Table 2. Values of the spin Hamiltonian parameters for trigonal Gd^{3+} centre in h-$RbZnF_3$ at room temperature together with the trigonal Gd^{3+} - V_{Rb} centre in $RbCaF_3$ with cubic perovskite structure.
The Units in $10^{-4}cm^{-1}$ for b_n^m.

Crystal	g	b_2^0	b_4^0	b_4^3	b_6^0	b_6^3	b_6^6
h-$RbZnF_3$ [6]	1.992(1)	-256.3(3)	-9.1(1)	85(6)	-1.7(2)	-26(6)	-17(2)
$RbCaF_3$[7]	1.9918(5)	80.1(1)	8.36(5)	14(2)	1.26(5)	-20(1)	13.0(5)

The parameters for the trigonal Gd^{3+} centre with a vacancy at the nearest Rb^{+} site (the Gd^{3+}- V_{Rb} centre) in cubic perovskite $RbCaF_3$ [8] are listed for comparison.

3 Discussions

In cubic perovskites $KMgF_3$ and $KZnF_3$, Gd^{3+} ions substitute at K^{+} sites in twelve-fold coordinations. This indicates that Mg^{2+} and Zn^{2+} ions are not replaced by Gd^{3+} due to their small ionic radii. The surroundings of a site I in h-$RbZnF_3$ can be considered locally to be those for cubic perovskite structure. The Gd^{3+} in h-$RbZnF_3$ therefore does not substitute for a site I, like that in $KZnF_3$. The

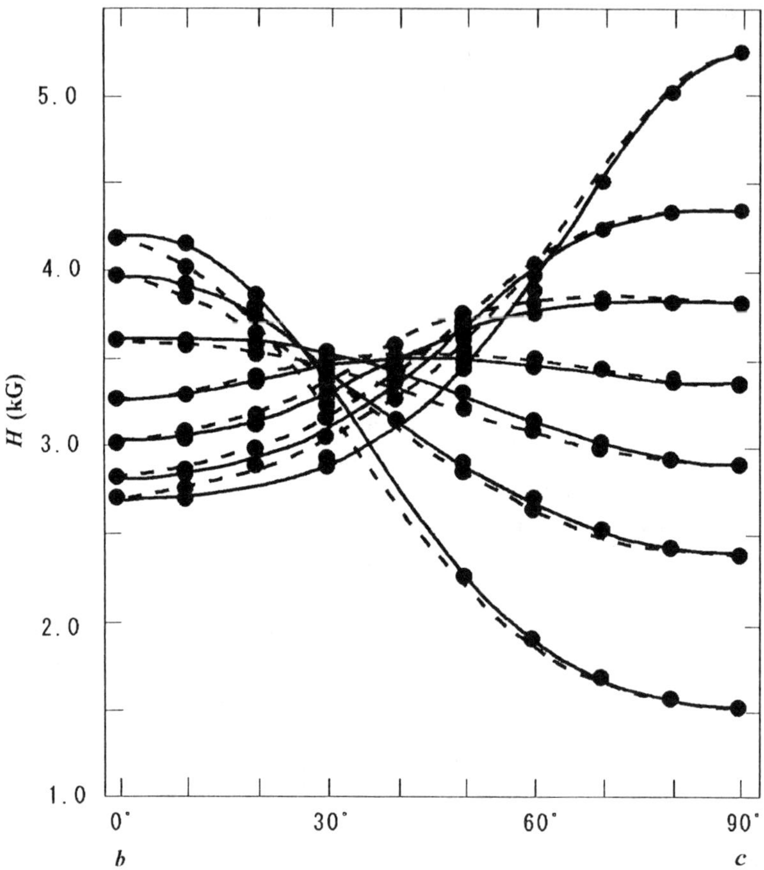

Fig. 4. Angular variation of the EPR spectrum of the trigonal Gd^{3+} centre in h-$RbZnF_3$ with H in the plane perpendicular to the direction a defined in figure 3. The smooth curves indicate the angular variation computed using the parameters for best fit. The full and dotted lines show the calculated branches from the centres, where the principal x-axes differ from each other by 180°.

surroundings of a site II can be considered as a part of $CsNiCl_3$-like hexagonal structure with infinite linear arrays of octahedra sharing faces. In $CsNiCl_3$-like hexagonal crystals such as $RbMgBr_3$ [8], Gd^{3+} ions substitute for Mg^{2+} sites in spite of the small ionic radii of Mg^{2+} ions on substitution of Gd^{3+} ions. In this crystal, excess positive charge of Gd^{3+} ions are compensated locally by forming Gd^{3+} - Li^{+} pairs replaced for Mg^{2+} - Mg^{2+} bonds in the linear arrays of the octahedra [9]. The similar formation of the Gd^{3+} - Li^{+} pair might be possible in h-$RbZnF_3$ to substitute for the Zn^{2+} - Zn^{2+} bond in a Zn_2F_9 unit. In unit cell, there are two Zn_2F_9 units which rotate by 60° each other about the c-axis as seen in figure 1. The two branches in figure 4 are consistent to the Gd^{3+}- Li^{+}

pairs in this two types of Zn_2F_9 units . The direction a in figure 3 is identified to be parallel to the crystalline a-axis. We conclude that the trigonal centres in h-$RbZnF_3$ are ascribed to the M^{3+} - Li^{+} pairs substituted for the Zn^{2+} - Zn^{2+} bonds in Zn_2F_9 units.

In $CsNiCl_3$ - type crystals, octahedra sharing faces make the neighbouring Ni^{2+}-Ni^{2+} distances shorter than those in cubic perovskite crystals. The Coulomb repulsion between the neighbouring divalent cations elongates octahedra along the c-axis. In h-$RbZnF_3$ the octahedra of the Zn_2F_9 unit can be considered to be strongly elongated by the presence of M^{3+} ion for the Zn^{2+} site because of the excess positive charge on the M^{3+} ion. The axial parameters b_2^0 for the trigonal M^{3+} - Li^{+} centres in h-$RbZnF_3$ have the opposite sign with those for the trigonal M^{3+} - V_A centres in ABF_3-type cubic perovskite fluorides as seen in Tables 1 and 2. The signs of b_2^0 for the M^{3+} - V_A centre arise from the deviations of the ligand fluorines away from the trigonal axis by the A^{+} vacancy - that is, ligand octahedron may be considered to be compressed along the trigonal axis. The opposite sign of b_2^0 for the M^{3+} - Li^{+} pair in a Zn_2F_9 unit shows that the elongation of the ligand octahedron is still remained by the presence of a Li^{+} at the other Zn^{2+} site. The result suggests that the M^{3+} ion with a Zn^{2+} vacancy in a Zn_2F_9 unit may be considered to an isolated M^{3+} ion in an almost undistorted octahedron for lack of the repulsive force from the nearest Zn^{2+} ion. The M^{3+} - Li^{+} pairs in Zn_2F_9 units might be expected to be stable such as uncompensated centres in octahedra because of the excess charge on M^{3+} ion being just compensated by Li^{+}. It must be emphasized that crystals of h-$RbZnF_3$ are obtained from one batch co-doped with M^{3+} and Li^{+}. This indicates that the M^{3+} - Li^{+} pairs in Zn_2F_9 units may be related to the crystal growth of the hexagonal structure of the matrix.

References

1. D. Babel, Structural Chemistry of Octahedral Fluorocomplexes of the Transition Elements, *Structure and Bonding*, **3** (1967) 1-87.
2. Ph. Daniel, J.Toulouse , J. Y. Gesland and M. Rousseau, Raman-scattering investigation of the hexagonal perovskite $RbZnF_3$, *Phys. Rev. B*, **52** (1995) 9129-9132.
3. H. Takeuchi and M. Arakawa, EPR Study of Cr^{3+} - Li^{+} centres in Several Perovskite Fluorides, *J. Phys. Soc. Jpn*, **53** (1984) 376-380.
4. G. L. McPherson and K. O. Devaney, EPR spectra of Cr^{3+} and Cr^{+} centres in the linear-chain lattices, $CsMgCl_3$, $CsMgBr_3$ and $CsCdBr_3$, *J. Phys. C : Solid State Phys.*, **13** (1980) 1735-1743.
5. J. L. Patel, J. J. Davies, B. C. Cavenett, H. Takeuchi and K. Horai, Electron spin resonance of axially symmetric Cr^{3+} centres in $KMgF_3$ and $KZnF_3$. *J. Phys. C: Solid State Phys.*, **9** (1976)129-138.
6. M. Arakawa, H. Ebisu and H. Takeuchi, EPR study of Gd^{3+} centres in cubic and hexag-

onal perovskite $RbZnF_3$ crystals, to be published. *J. Phys. : Condens. Matter*, (1997)

7. M. Arakawa, H. Aoki, H. Takeuchi, T. Yosida and K. Horai, Electron Paramagnetic Resonance of Gd^{3+} Centres in Several Perovskite Flourides, *J. Phys. Soc. Jpn*, **51** (1982) 2459-2463.
8. H. Takeuchi, M. Arakawa and H. Ebisu, EPR study of a trigonally symmetric Gd^{3+} centre in $RbCaF_3$, *J. Phys.: Condens. Matter*, 7 (1995)1417-1426.
9. K. H. kirklin and G. L. McPherson, EPR spectra of $RbMgBr_3$ crystals doped with Mn^{2+}, Cr^{3+} and Gd^{3+} : characterisation of two structural phase, *J. Phys. C: Solid State Phys.*, **16** (1983)6539-6550.

ESR Evidence for Radiation Immunity of Mirror symmetry Related Bonds

Usha Tumkurkar, A.Anand, H.L. Bhat and S.V. Bhat

Department of Physics, Indian Institute of Science, Bangalore 560 012, India

Abstract. X-irradiation of Phase III of NH_4NO_3 single crystals leads to NH_3^+ radicals as studied by ESR while in Phase IV no NH_3^+ radicals are produced. The difference in the two structures is that in Phase IV N-H bonds are mirror related while in Phase III they are not. We conclude that mirror related bonds are immune against radiation induced rupture.

1. Introduction

Radiation induced decomposition of molecular ions in the solid state is not yet a fully understood phenomenon. For example, even though each quantum of radiation is more energetic than the bond strengths of the molecular ions, often, selective homolysis is the major outcome, leading to the formation and trapping of only one or two species [1]. Further, the fragments detected are often not the ones expected if the weakest bond had undergone rupture. After a large number of studies, it is now fairly well established that there is no correlation between decomposition yields and the so called 'free space' in the solid, the lattice energy, density, melting point or the free energy of formation of the defects [2]. However, there is some evidence that the crystalline environment and the orientation of the molecular ions in the lattice play important roles in determining the decomposition pathways. Recently, from a detailed examination of bonds ruptured by high-energy radiations in certain ammonium salts, we have found [3,4] evidence for relevance of one important aspect of the crystal structure, namely the symmetry, to the radiation induced decomposition. From ESR studies of NH_3^+ and ClO_3 radicals generated in single crystals of NH_4ClO_4 we have provided evidence that mirror-symmetry related bonds do not break on irradiation even though they are weaker than or as strong as other bonds that get broken. However, these studies could not rule out the possibility of the rupture of weaker bonds and subsequent reorientation of the radical to conform to the lattice symmetry. In the present work we prove conclusively that irradiation is unable to break the mirror symmetry related bonds.

We study by ESR the NH_3^+ free radical generation in two phases of crystalline NH_4NO_3: Phase IV, stable between -18°C to 32.3°C, belongs to the space group P_{mmn} and has the N-H bonds mirror related to each other in pairs [5]. X-irradiation of this phase does *not* produce the NH_3^+ radicals. However, when the same material in phase III (space group P_{nma}, all the N-H bonds are inequivalent [6]) is irradiated with X-rays, there is a production of NH_3^+ radicals.

2 Experimental

Single crystals of NH_4NO_3 were grown by slow evaporation of saturated solution of analar grade NH_4NO_3 in distilled water. Phase III is normally stable in the temperature range 32°C - 84°C. However, it is known [6] that the addition of about three atomic percent KNO_3 to NH_4NO_3 stabilises the phase III at room temperature itself. Therefore we have grown NH_4NO_3 in phase III from the aqueous solution of NH_4NO_3 containing about 5% KNO_3. Powder X-ray diffraction patterns were recorded to confirm that the crystals indeed have Phase III structure.

X-irradiation (40kV 20mA) was carried out using an Enraf Nonius sealed tube X-ray generator.The X-radiation was unpolarised in the plane perpendicular to the beam direction.To eliminate the possibility of relative orientations of the bond directions and the beam direction affecting the bond dissociation, crystals were mounted with their different crystallographic planes perpendicular to the beam direction. Essentially similar results were obtained for these different irradiation geometries. The crystalline axes were determined by measuring the known lattice constants. ESR spectra were recorded at room temperature using a Bruker ER 200D specrometer at X-band, in three mutually orthogonal planes, using a precision goniometer at 10^0 intervals.

3 Results and Analysis

The ESR spectrum of X-irradiated NH_4NO_3 in phase IV is shown in Fig.1. NH_3^+ radical, if present, should give rise to a characteristic 12 line spectrum, made up of a triplet of quartets: the triplet due to the hyperfine interaction of the unpaired electron with the nitrogen nucleus (I=1) and the quartet resulting from the interaction with the three equivalent protons. Such a spectrum is conspicuous by its absence in Fig.1, clearly showing that X-irradiation of NH_4NO_3 in phase IV does not generate NH_3^+ radicals. Irradiation durations were varied between ½ hr to 3 hrs with the same result.

The actual spectrum observed can be analysed in terms of a doublet at g=2.0176 and a single line at g=2.0048. The doublet and the single line show different saturation behavior on increasing the microwave power. The intensity of the doublet goes on increasing while that of the single line first increases and begins to come down around 5 dB. We have not yet identified the paramagnetic center giving rise to the doublet. The g=2.0048 single line is most likely due to the O_2^- center as identified by Jarke and Ashford [7] in γ- irradiated NH_4NO_3 at 77 K.

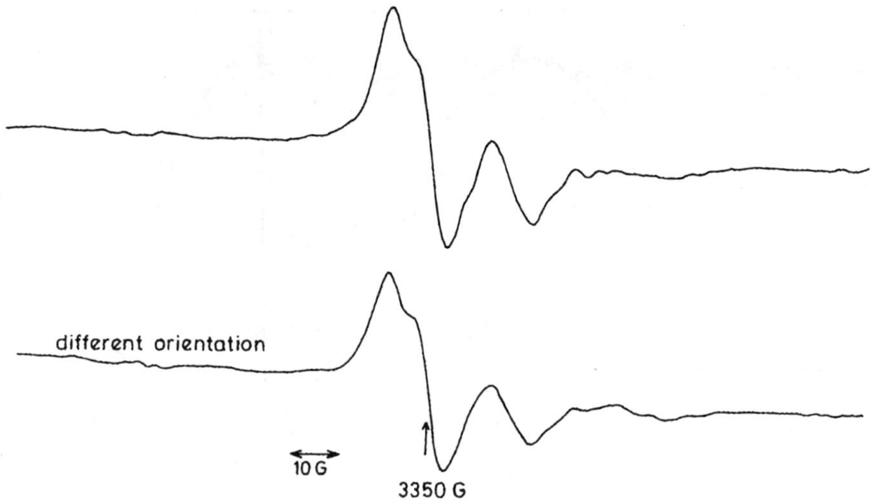

Fig. 1. Room temperature ESR spectrum of X-irradiated NH_4NO_3 single crystals in phase IV. No NH_3^+ signals are observed.

Figure 2 shows ESR spectrum of the X-irradiated single crystal of NH_4NO_3 Phase III; unmistakable presence of NH_3^+ radicals is seen. As indicated in the figure, the spectrum is a result of two magnetically inequivalent NH_3^+ centres. In general, we believe, there are, infact four magnetically inequivalent NH_3^+ radicals, though in each of the three crystallographic planes, we see the evidence of only two sites.

The spectra were fitted to the appropriate spin Hamiltonian

$$H = g\beta \boldsymbol{S}\cdot\boldsymbol{H} + A_H \boldsymbol{S}\cdot\boldsymbol{I}_H + \boldsymbol{S}\cdot\boldsymbol{A}_N\cdot\boldsymbol{I}_N \tag{1}$$

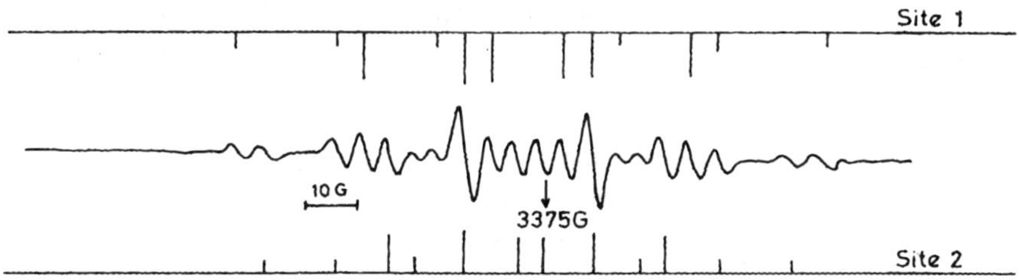

Fig. 2 . Room temperature ESR spectrum of X-irradiated NH_4NO_3 single crystals in phase III for an orientation with H making an angle of 30° with b axis in the bc plane. Characteristic spectra due to two magnetically inequivalent sites are seen.

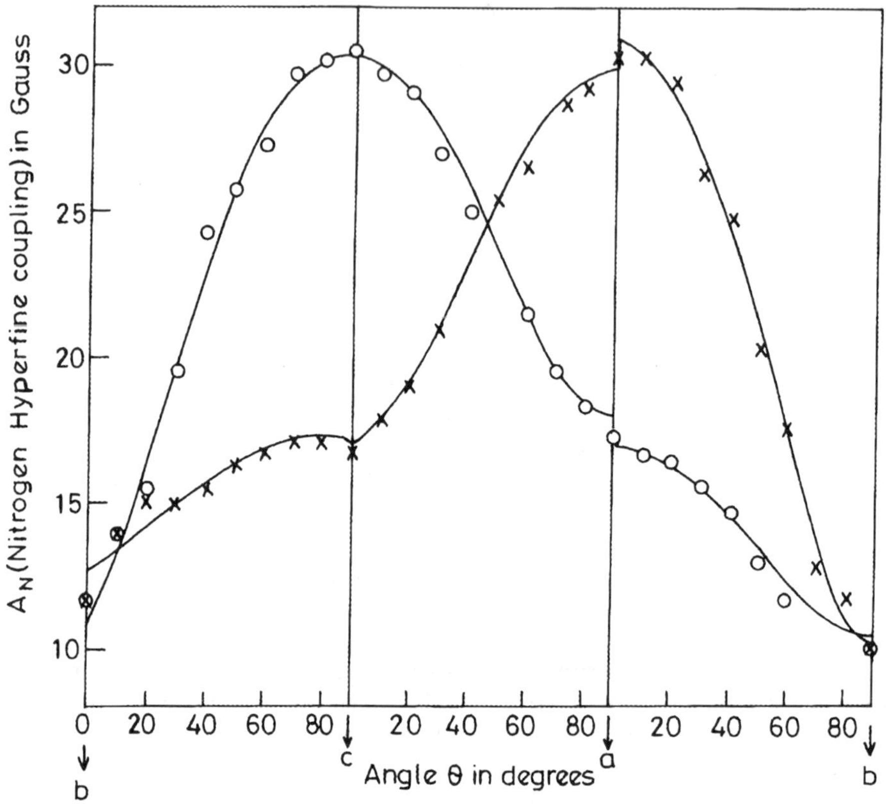

Fig. 3. Orientation dependence of the nitrogen hyperfine splitting in the three crystallographic planes.

with isotropic g and proton hyperfine interaction, but highly anisotropic nitrogen hyperfine interaction. Figure 3 depicts the angular dependence of the nitrogen hyperfine splitting in the three crystallographic planes. We have used the standard method of analysis to obtain the spin-Hamiltonian parameters summarised in Table1. It is seen that the three principal components of the A_N tensor are along the three crystallographic axes. Normally one expects either the largest or the smallest principal components along the line joining the nitrogen to the hydrogen atom which is removed [3]. In the structure of NH_4NO_3 (III) as determined by Choi and Prask [6], none of the N-H bonds lies along any crystallographic axis. Yet we find all the three principal components aligned with the crystallographic axes due to the thermally activated vibrational motion of the ions around the pseudo-four fold axis along the c direction of the crystal. Similar situation was obtained in NH_4ClO_4: NH_3^+ where actual orientation of the principal components of A_N could be obtained only by cooling

Table 1 . Spin Hamiltonian parameters for the NH_3^+ radicals in phase III of NH_4NO_3 single crystals

g = 2.0037	Isotropic
A_H = 25.1 G (proton hyperfine)	Isotropic
A_N (nitrogen hyperfine tensor)	A_{ZZ} = 30.0 G, A_{YY} = 10.7 G and A_{XX} = 17.2 G

the crystals to 153°K when the vibrational time scale becomes slower than the inverse of the hyperfine splitting in frequency units. Keeping this in view it would be useful to record the ESR spectra at low temperatures. Such experiments are in progress at present. In Table 2 we summarise some earlier results available in literature which

Table 2. Anticorrelation of mirror-symmetry and radiolytic decomposition

Compound	Radical	Parent ion	Bond lengths of the parent ion (Å)	Bond broken	Reference Structure	Reference ESR
NH_4ClO_4	NH_3^+	NH_4^+	N-H(1):1.028 N-H(2):1.058 N-H(3):1.031* N-H(4):1.031*	N-H(1)	13	3
$(NH_4)_2SO_4$	NH_3^+	NH_4^+	N(1)-H(1):1.08 N(1)-H(2):1.04 N(1)-H(3):1.09* N(1)-H(4):1.09*	Either N(1)-H(1) or N(1)-H(2)	12	8, 9
			N(2)-H(5):1.05 N(2)-H(6):1.06 N(2)-H(7):1.07* N(2)-H(8):1.07*	Either N(2)-H(5) or N(2)-H(6)		
$(NH_4)_2SO_4$	SeO_3^-	SeO_4^{2-}	S-O(1):1.49 S-O(2);1.49 S-O(3):1.49* S-O(4):1.49*	S-O(1) for $(SeO_3^-)_I$ and S-O(2) for $(SeO_3^-)_{II}$	12	11
$(NH_4)_2SO_4$	CrO_3^-	CrO_4^{2-}	Same as above	S-O(1)	12	10
NH_4ClO_4	ClO_3	ClO_4^-	Cl-O(1):1.449 Cl-O(2):1.440 Cl-O(3):1.446* Cl-O(4):1.446*	Cl-O(2)	14	4

* Mirror-related bonds

provide further confirmation of the stability of the mirror related bonds,although the significance of the mirror plane does not appear to have been recognised except by us. ESR studies of NH_3^+, CrO_3^- and SeO_3^- (the latter two generated respectively by irradiating samples containing CrO_4^{2-} and SeO_4^{2-} doped for SO_4^{2-}) centers in single crystals of $(NH_4)_2SO_4$ have been reported sometime ago [8,9,10,11]. A comparison of the result with the structure determined by neutron diffraction [12] shows (Table 2) that the mirror symmetry related hydrogens of NH_4^+ and oxygens of SO_4^{2-} survive even though they are bound more weakly than or at best as strongly as the atoms that are lost on irradiation. Our recent work [4] on the ESR of ClO_3 centers in single crystals of NH_4ClO_4 included in the same table provides additional confirmation of this conclusion . Considering all these results we come to the inescapable conclusion that mirror symmetry related bonds survive radiation damage.

4 Conclusion

ESR studies of single crystals of NH_4NO_3 show that in phase IV X-irradiation is not able to rupture the N-H bonds since they are related by the mirror symmetry. At the same time, N-H bonds unrelated by any symmetry in phase III, break to form NH_3^+ radicals. We conclude that mirror-symmetry provides radiation immunity to chemical bonds of molecular ions in the solid state.

Acknowledgements

This work was partially supported by the ISRO-IISc Space Technology cell. A. Anand thanks CSIR, India for a fellowship. Professors M.Vijayan and T.N. Guru Row are thanked for the help with irradiation and identification of the crystal axes respectively.

References

1. P.W. Atkins and M.C.R Symons, *The structure of Inorganic Radicals*, Elsevier, Amsterdam, 1967. (Chapter 3)
2. E.R. Johnson, *The radiation-induced decomposition of Inorganic Molecular ions*, Gordon and Breach , New York, 1970. (Chapter 2).
3. S.V. Bhat and M.M.H. Abdel-Gawad, *Curr. Sci*, **61** 397 (1991).
4. B.G. Hegde , A.Rastogi, R.Damle, R.Chandramani and S.V.Bhat, *J.Phys.C.*, **9**, 3214 (1997).
5. C.S. Choi , J.E. Mapes and E. Prince, *Acta Cryst, B* **28** 1357 (1972).
6. C.S. Choi and H.J. Prask, *Acta Cryst, B* **38** 2234 (1982).
7. F.H. Jarke and N. A. Ashford, *J. Chem. Phys.*, **62**, 2923 (1975).
8. C.E.Bailey and R.C.Cuesta-Barro, *J.Chem.Phys.*,**63**, 4120 (1975).
9. M.Fujimoto, L.A. Dressel and T.J.Yu, *J.Phys. Chem. Solids*, **38**,97 (1977).
10. N.C.Mishra and R.Srinivasan, *Chem. Phys. Lett.*, **108**, 288 (1984).
11. N.Shibata, R.Abe and I.Suzuki, *J.Phys.Soc.Jpn.*, **41**, 2011 (1976).
12. E.O.Schlemper and W.C.Hamilton, *J.Chem. Phys.*,**44**, 4498 (1966).
13. C.S.Choi, H.J.Prask and E.J.Prince, *J.Chem.Phys.*,**61**,3523 (1974).
14. H.J.Prask, C.S.Choi, N.J.Chesser, C.J.Rosasco, *J.Chem.Phys.*, **88**, 5106 (1988).

EPR Study of Transition Ions in the $Pb_3(PO_4)_2$ Single Crystals

Y. N. Choi, S. H. Choh and M. S. Jang†

Department of Physics, Korea University, Seoul 136-701, Korea
† Department of Physics, Pusan Nat'l University, Pusan 609-735, Korea

Abstract. Lead phosphate is a ferroelastic crystal, its prototypic phase is rhombohedral, D_{3d}^{5}, and the low temperature phase is monoclinic, C_{2h}^{6}. The first order ferroelastic phase transition was observed at 455.5 K. Below T_c ferroelastic domains are formed, making EPR spectra of impurity ions complicate. EPR spin Hamiltonian parameters of Fe^{3+} are determined and turn out that the principal axes of **D** tensor of $Mn^{2+}(^6S_{5/2})$ ion are different from those of $Cr^{3+}(^4F_{3/2})$ and $Fe^{3+}(^6S_{5/2})$ ions. From the temperature dependence of D value of Cr^{3+}, the local change around the Cr^{3+} ion with temperature is discussed. Although the crystal structure is monoclinic in the ferroelastic phase, evidences of low symmetry effects on the EPR spectra are hardly found.

1 Introduction

Lead phosphate, $Pb_3(PO_4)_2$, is considered as one of the best examples of a pure ferroelastic compound.[1] Since the first observation of ferroelasticity of lead phosphate by Brixner *et al*,[2] numerous studies on the ferroelasticity,[3~4] domain structure[5~8] and phase transition[9~16] have been reported. $Pb_3(PO_4)_2$ is known to undergo a first order phase transition near 453 K (experimental values depend on samples).[9~16] Its low temperature phase has monoclinic structure with space group C_{2h}^{6} and the high temperature phase is rhombohedral with space group D_{3d}^{5}.[17] Our previous study[16] indicates the coexistence of ferro- and paraelastic phase just above T_c.

Electron paramagnetic resonance (EPR) studies of transition metal (TM) ions, Mn^{2+} [9,12] and Cr^{3+} [15~16], and rare earth ions, Gd^{3+} and Eu^{2+} [18], were reported. In this paper we present the EPR parameters of Fe^{3+} and discuss the principal axes (PAs) of the second-order zero-field splitting (ZFS) (represented as **D** tensor) for the Cr^{3+}, Mn^{2+} and Fe^{3+}. From the temperature dependence of second-order ZFS parameter of Cr^{3+} ion, the change of the local environment around the Cr^{3+} ion is discussed. Though there may appear the low symmetry effects (LSEs) on the EPR spectra when the local symmetry around the spin is lower than orthorhombic, but we could not observe the LSEs in $Pb_3(PO_4)_2$.

2 Crystal Structure and Domains

There are two kinds of Pb ions, Pb_I and Pb_{II} in the lead phosphate.[17] In the paraelastic phase, Pb ions and PO_4 groups lie along the ternary axis with three-fold symmetry as shown in Fig. 1(a). In the ferroelastic phase, on the other hand, Pb_I ions keep C_2 symmetry with six-fold coordination of PO_4 groups (Fig. 1(b)) but Pb_{II} ions lost symmetry element and have irregular ten-fold coordination.[9] The nearest Pb-O distances are given in Table 1. There exist three orientation states (or domains) and two kinds of domain walls, W_m (twinned by the pseudo-mirror plane) and W_b (twinned by the binary axis) in the ferroelastic phase.[8] The (100) planes of the three domains are nearly parallel but the binary axes of them make an angle of about 60 degrees.[7]

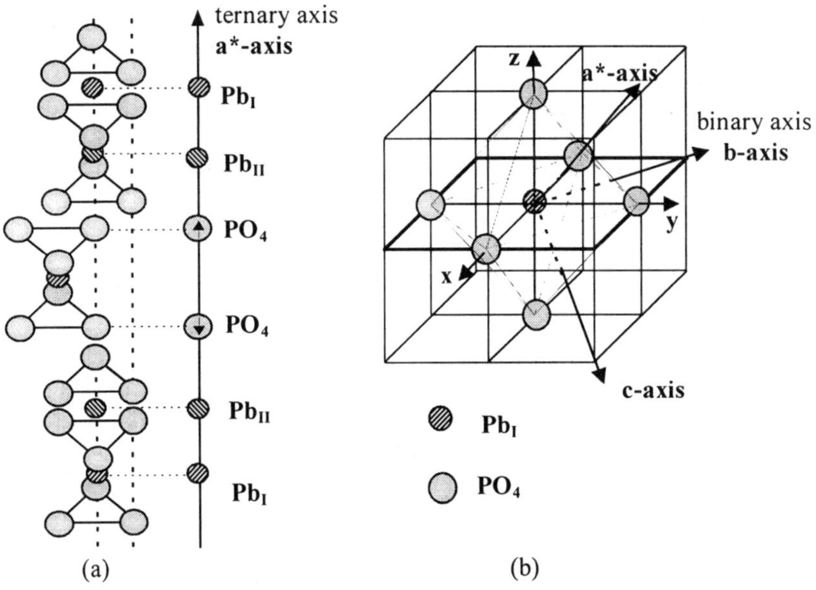

Fig. 1. (a) The stacking order of $Pb_3(PO_4)_2$ in the paraelastic phase. The arrow in the PO_4 indicates one edge of the PO_4 tetrahedron is directed along it. (b) The local structure around the Pb_I ion in the ferroelastic phase. The octahedron of PO_4 is slightly distorted. Principal directions of the octahedron are designated as x, y and z. The crystallographic axes, a*(normal to the bc-plane), b and c, are normal to the octahedral faces.

Table 1. The nearest Pb-O distances of ferro- and paraelastic $Pb_3(PO_4)_2$ crystal at 20℃ and 200℃, respectively.[9]

ferroelastic phase (20℃)	*paraelastic phase* (200℃)
Pb-O : 2.54 Å 2.59 Å 2.73 Å	Pb-O : 2.62 Å

3 Experimental Results and Discussion

3.1 Experiments

Lead phosphate single crystals, doped with 0.03mol% of Cr_2O_3, Mn_2O_3 and Fe_2O_3 respectively, were grown by the Chochralski method.[19] These crystals have many small domains and well developed cleavage planes. EPR measurements were carried out using Bruker ESP300 spectrometers at KBSI Seoul Branch. The rotation patterns of Cr^{3+}, Mn^{2+} and Fe^{3+} EPR spectra and the temperature dependence of Cr^{3+} signals were obtained.

3.2 Principal Axes of D for Mn^{2+}, Cr^{3+} and Fe^{3+}

The spin Hamiltonian parameters (SHPs) and principal axes (PAs) of **D** tensor at room temperature are summarized in Table 2. The possible substitutional cationic site for the impurity ion in $Pb_3(PO_4)_2$ is Pb_I site which lies in the center of distorted octahedron with six PO_4 ligands (Fig. 1(b)). Cr^{3+} and Fe^{3+} ions turn out to have the same PAs of ZFS, but Mn^{2+} has different ones though Mn^{2+} and Fe^{3+} ions have isoelectronic configuration ($[Ar]3d^5$).

In the $^6S_{5/2}$-state ions, five 3d electrons are in two e_g-orbitals, which are oriented toward the ligands, and three t_{2g}-orbitals, oriented to avoid overlap with ligands (Fig. 2). From the arguments of molecular orbital model or angular overlap model,[20] the overlap of e_g-orbital electrons of Mn^{2+} and p-orbital electrons of ligands could make σ-bonds, their overlap is greater than that for the π-bonds. The orientations of σ-bonds are along the x, y and z axes of octahedron and thus the principal directions of spin-orbit coupling are along those directions. Neglecting the contribution of spin-spin coupling to the ZFS, we obtain the PAs of **D** tensor would following those of spin-orbit interaction. Experimentally we observed that the Z-axis of Mn^{2+} rotated by 36° and the direction a*+36° nearly corresponded to the z-axis of octahedron. However, though the same PAs as Mn^{2+} are expected for the Fe^{3+} ion because of the

Table 2. The spin Hamiltonian parameters of Fe^{3+} at room temperature and the principal axes of **D** tensor of TM ions (Mn^{2+}, Cr^{3+} and Fe^{3+})

SHPs of Fe^{3+}	PAs of **D** tensor
$g_X = 1.996(1)$	Mn^{2+} : $Z = a^* + 36^\circ$
$g_Y = 2.001(1)$	$X = a^* - 54^\circ$
$g_Z = 2.006(1)$	$Y = b$
$D = 3B_2^0 = -3.373(7)$ GHz	
$E = B_2^2 = 0.148(7)$ GHz	Cr^{3+} and Fe^{3+} :
$B_4^0 = -0.8(2)$ MHz	$Z = a^*$
$B_4^2 = -0.2(4)$ MHz	$X = c$
$B_4^4 = 1.0(5)$ MHz	$Y = b$
$B_4^{-1} = 2.7(8)$ MHz	
$B_4^{-3} = 12(3)$ MHz	

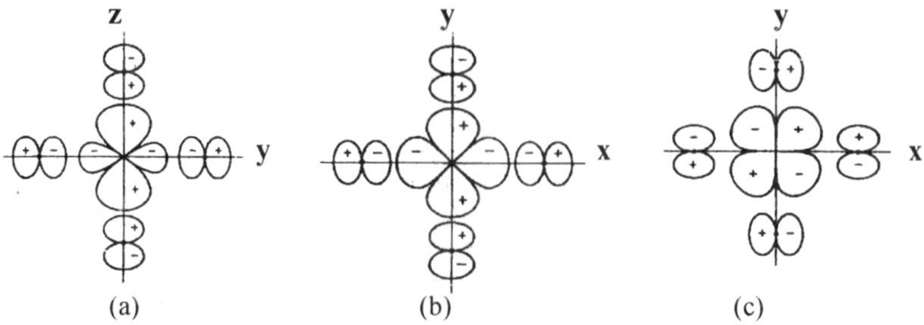

Fig. 2. Representation of the combination of central-ion d orbitals with ligand p orbitals on a plane. (a) $d(z^2)$ and (b) $d(x^2-y^2)$ are e_g orbitals and (c) d(xy) is one of the three t_{2g} orbitals.

same electronic configuration, it has different PAs except the b-axis (Y-axis of **D** tensor). The Z-axis coincides with the a*-axis, normal to the (100) plane, and X-axis with the c-axis. This can be explained by a charge compensator on the pseudo-ternary axis. Due to the new overlap of $d(z^2)$-orbital electron of Fe^{3+} with the charge compensator, possibly it is an electron, the electron clouds may be transferred toward the charge compensator so that the PAs become a*(Z), c(X) and b(Y)-axis.

On the other hand, Cr^{3+} is an F state with $3d^3$ configuration. Three orbitals in the d-shell are in one a_{2g} and two t_{2g} orbitals, of which angular orientations are similar to the t_{2g} orbitals in $3d^5$ configuration. Therefore, all of the 3d orbitals in Cr^{3+} ion are facing the octahedral planes and thus the PAs for the second order ZFS, become a*(Z), b(Y) and c(X)-axis. Similar argument of the orbital overlap of the metal ion with its charge compensator, as in the case of Fe^{3+} ion, could be applied to explain why the PAs of Cr^{3+} are the same as Fe^{3+}. In fact both of the above two mechanisms could elucidate the PAs of Cr^{3+} and may well be in effect simultaneously.

3.3 Temperature Dependence of D for Cr^{3+} Ion

We employed the Cr^{3+} ion as a probe to monitor the phase transition because it is sensitive to the distortions in their local environment.[21] As the temperature of crystal increases, the absolute value of D increases continuously until T reaches T_c and jumps up at T_c as shown in Fig. 3. In the paraelastic phase, there is perfect stacking of Pb and PO_4 along the C_3 axis. As the temperature is lowered, deviations of Pb_{II} and PO_4 groups from the axis occur. Therefore we can explain the temperature dependence of D by considering the displacement of PO_4 ligand. As the temperature increases, PO_4 goes toward the C_3 axis so that it pushes the charge compensator between the Cr^{3+} and PO_4 tetrahedron to the Cr^{3+} ion and consequently the spin-orbit coupling may be increased. Passing through T_c, a sudden increase of D is observed and this can be understood as a sudden lattice displacement.[14]

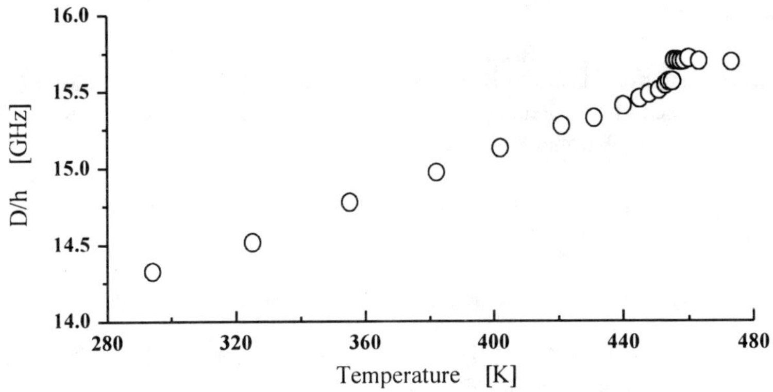

Fig. 3. Temperature dependence of the second order ZFS parameter D of Cr^{3+}.

3.4 Low Symmetry Effects

The low symmetry means monoclinic and triclinic point symmetries. Evidences of the low symmetry effects (LSEs) are known as follows[20]: 1) the non-coincidence of extrema between the fine structures in a magnetic field, 2) the non-coincidence between the tensor **D** and interaction matrix **g** (electron Zeeman splitting) and 3) the appearance of 'forbidden' hyperfine transitions along PAs.

From the EPR spectra of TM ions in the lead phosphate we seldom found the above evidences. The local environment around Pb_I ion is pseudo-trigonal symmetry and thus the monoclinic aspect is not so conspicuous. The determination of accurate PAs of **g** was not available so we took those same as D-tensor, since the angular variation of g-values are very small (see Table 2) with respect to the experimental error. In addition, the forbidden hyperfine transitions of Mn^{2+} are hardly distinguishable from the allowed ones because of the superposition of hyperfine signals from other domains.

4 Summary

We obtained the EPR spin Hamiltonian parameters for the Fe^{3+} and explained why the PAs of Mn^{2+} is different from Cr^{3+} and Fe^{3+}. The first order ferroelastic phase transition was observed at 455.5 ± 0.5 K by Cr^{3+} EPR measurement. The increasing D with temperature can be explained in terms of the displacement of PO_4 ligand toward the C_3 axis. This displacement enhances the repulsion between the PO_4 and charge compensator of Cr^{3+} which strengthen the spin-orbit interaction and hence ZFS. Due to the local pseudo-trigonal symmetry and the domain structure we could hardly find LSEs on the EPR spectra of Cr^{3+}, Mn^{2+} and Fe^{3+} in $Pb_3(PO_4)_2$.

Acknowledgments

This work was supported by the Korea Science and Engineering Foundation (KOSEF) through the Research Center for the Dielectrics and Advanced Matter Physics (RCDAMP) at Pusan National University (1994-97).

References

1. E. Salje, *Phase Transitions in Ferroelastic and Co-elastic Crystals* (Cambridge University Press, Cambridge, 1990).
2. L. H. Brixner, P. E. Bierstedt, W. F. Jeap and J. R. Barkley, Mater. Res. Bull. **8**, 497 (1973).
3. E. Salje and G. Hoppman, Mater. Res. Bull. **11**, 1545 (1976).
4. E. Salje, A. Graeme-Barber and M. A. Carpenter, Acta Crystallogr. **B 49**, 387 (1993).
5. J. Tores, C. Roucau and R. Ayroles, phys. stat. sol. **(a) 70**, 193 (1982).
6. S. Y. Jeong, M. S. Jang, H. J. Kim, C. R. Cho and Y. S. Yu, Ferroelectrics **142**, 121 (1993).
7. A. R. Lim, K. H. Lee, S. H. Choh, S. Y. Jeong and M. S. Jang, J. Korean Phys. Soc. **26**, 384 (1993).
8. S. Y. Jeong, M. S. Jang and A. R. Lim, Phys. Rev. **B 48**, 13340 (1993).
9. H. N. Ng and C. Calvo, Can. J. Phys. **53**, 42 (1975).
10. E. Salje and B. Wruck, Phys. Rev. **B 28**, 6510 (1983).
11. J. P. Chapelle, Cao-Xuan-An and J. P. Benoit, Solid State Commun. **19**, 573 (1976).
12. M. Razeghi, B. Houlier and M. Yuste, Solid State Commun. **26**, 665 (1978).
13. J. M. Kiat, Y. Yamada, G. Chevrier, Y. Uesu, P. Boutrouille and G. Calvarin, J. Phys.: Condens. Matter **4**, 4915 (1992).
14. J. M. Kiat, G. Calvarin and Y. Yamada, Phys. Rev. **B 48**, 34 (1993).
15. M. B. Zapart, W. Zapart and S. Y. Jeong, Ferroelectrics **159**, 233 (1994).
16. Y. N. Choi, S. H. Choh and M. S. Jang, J. Korean Phys. Soc. **29**, S486 (1996).
17. U. Keppler, Z. Kristallogr. **132**, 228 (1970).
18. M. Rageghi, J. P. Busson and B. Houlier, phys. stat. sol. **(b) 95**, 283 (1979).
19. H. J. Kim, M. S. Jang, S. Y. Jeong, C. R. Cho, S. J. Lee and J. H. Ro, Sae mulli (New Phys.) **32**, 862 (1992).
20. J. R. Pilbrow, *Transition Ion Electron Paramagnetic Resonance* (Clarendon Press, Oxford 1992) Chap. 2.
21. F. J. Owens (ed), *Magnetic Resonance of Phase Transitions* (Academic Press, New York 1979) Chap. 2.

Recent 10K EPR Studies of Radiation Induced Defects in Zircon

Rodney F. C. CLARIDGE and W. Craighead TENNANT

Department of Chemistry, University of Canterbury, Private Bag 4800, Christchurch, New Zealand

Abstract X-irradiation of synthetic zirconium silicate at low temperature leads to the formation of a variety of different centres which can be observed by EPR. This paper presents new precise spin Hamiltonian analysis of data on a Ti^{3+} centre. A feature of the analysis is the unequivocal determination of high spin nuclear parameters resulting from inclusion of terms of dimension I^4, BI^3, BI^5, SI^3 and SI^5 .

Introduction

This paper illustrates the power of the EPR technique in elucidating the *microstructure* in the near vicinity of the paramagnetic centre in a single crystal. Modern instrumentation allows such measurements to be carried out by decoupling certain interactions using pulsed or double resonance methods. In this work we carry out an essentially conventional EPR experiment. The measurement is based on many very precisely determined data points (resonant fields) so that interaction matrices (tensors) are considerably over-determined. Data analysis uses "exact" (matrix diagonalisation) numerical methods and a theoretically correct spin-Hamiltonian (SH). Crystal alignments are estimated typically to ±0.01° for the plane of measurement and to the same precision within this plane using the vernier of the cavity-goniometer system. At *each* line position (resonant field) the microwave frequency is measured to ±1kHz and the magnetic field with proton gaussmeter to ±0.002mT. With this precision of measurement the average error of fitting components of the g tensor is ±0.000002, with similar precisions for other parameter matrix components. The science can fairly be described as "EPR crystallography".

There is a large body of data in the literature of rare earth and transition metal ions in natural zircons [1] but the data generally do not have the precision for the type of analysis we wished to undertake. We have undertaken a series of new studies using x-irradiated synthetic zircon crystals. Earlier papers in this work have described in considerable detail X-band 10K EPR studies of a Zr^{3+} centre designated $Zr^{3+}(\alpha)$ [2], two oxygen hole centres [2] and an aluminium compensated oxygen hole centre, designated $[AlO_4]^0$ [3]. These centres are uncontroversial and are not discussed further in this paper.

In this paper we discuss one of the major centres formed by cold x-irradiation and subsequent annealing. This is a d^1 Ti^{3+} centre, labelled herein *B(Ti)*, where the novelty lies in the necessity of having to include high spin nuclear terms (HSN) in the SH in order to describe adequately the spectra.

Experimental

Zircon (zirconium silicate, $ZrSiO_4$) crystallises in the tetragonal space group $I4_1/amd$. The crystal used here was in all cases a high purity single crystal of zircon cut to a parallelepiped of dimensions (3.5x5.5x2.0 mm) with faces parallel to the *a, b,* and *c* axes of the zircon unit cell. The details of the crystal growth and of the instrumentation have been described previously [2]. The single crystal was irradiated at 77K with x-rays from a W tube and transferred cold to the EPR cavity-goniometer system. For the centres here described the crystal was warmed to room temperature (RT) and then cooled to 10K for the EPR measurements.

Results and Discussion

Irradiation of zircon with x-rays at 77K leads to a plethora of paramagnetic defect centres as shown in Fig. 1 of Ref. [2]. Annealing of the crystal to RT releases the electrons resulting in the complete annihilation of the $Zr(\alpha)$ centre and most of the $[AlO_4]^0$ hole centre; other, more thermally stable centres are formed. Fig 1 shows the ***B***||***c*** EPR spectrum at 10K after annealing for 1 hour at RT. The major centres observed are *B(Ti)* [4], a Ti^{3+} electron centre where the Ti^{3+} is in a Zr^{4+} lattice site, and *C* [5], ostensibly part of a triradical centre. We discuss the Ti^{3+} centre in detail below. Three further minor centres are also observed: *Y* [6], a Ti^{3+} electron centre with a Y^{3+} in a nearest neighbour Zr^{4+} site; *P*, possibly a Ti^{3+} centre with a Y^{3+} in a next nearest neighbour Zr^{4+} site; *Q* a further, as yet uncharacterised Ti^{3+} centre.

With ***B***||***c*** the spectrum of *B(Ti)* is simple (see Fig. 1): a strong central line arising from spinless *Ti* nuclei flanked by an octet of lines from $^{47,49}Ti$ nuclei with nuclear spins and abundances 5/2, 7.4% and 7/2, 5.4% respectively. Lines for the two

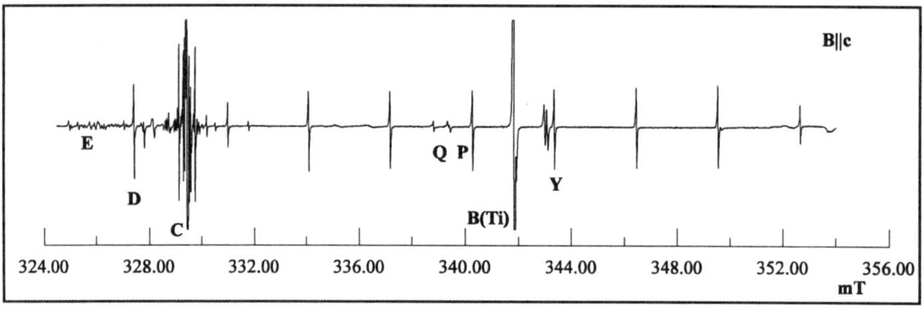

Fig. 1. EPR spectrum of Zircon (***B***||***c***) at 10K after annealing to room temperature. (see text)

isotopes coincide because $A_{||}$ (^{47}Ti) = $A_{||}$ (^{49}Ti). Only allowed transitions are observed. With $\boldsymbol{B}\perp\boldsymbol{c}$ (*ab*-plane) (see Fig. 2) the spectrum is two orders of magnitude weaker and there are observed many “forbidden” lines. In this plane g is strictly isotropic as evidenced by the non-angular dependence of the spinless isotope lines (line 4 of Fig. 3) to within the inherent error of their measurement, ±0.02 mT. However the hyperfine components show complex angular dependence. As an illustration, the angular dependences of five ^{47}Ti lines (marked in Fig. 2) together with the even isotope line (line 4) are plotted in Fig. 3. Plotting as *g*-factors is necessary to obviate errors due to different microwave frequencies at the resonant field. The angular dependence is small but clearly exceeds the error bars for three of the five lines; a feature is the $cos4\varphi$ dependence of each line. Subsequent work has indicated that the angular variation of ^{49}Ti lines in the *ab*-plane is at least an order of magnitude greater.

To explain the angular dependence in the *ab*-plane and to obtain a good fit of the many forbidden lines observed it was necessary to include terms of dimension I^4, BI^3, BI^5, SI^3, and SI^5 . The appropriate SH is then

$$\aleph_S = \beta_e \boldsymbol{S.g.B} + \boldsymbol{S.A.I} + \boldsymbol{I.P.I} - \beta_N \boldsymbol{I.g_N.B} + \text{ terms in } I^4, BI^3, BI^5, SI^3, SI^5.$$

These latter terms are included as tesseral combinations of spherical tensor operators. The terms which are to be considered are restricted by symmetry to those which have the following coefficients. For I^4: $B_{4,m}^{0,0,4}$; for BI^3: $B_{l,m}^{1,0,3}$; for BI^5 : $B_{l,m}^{1,0,5}$; for SI^3: $B_{l,m}^{0,1,3}$ and for SI^5: $B_{l,m}^{0,1,5}$; where the notation $B_{l,m}^{l_B,l_S,l_I}$ is used. For sites of Laue class *4/mmm* only terms with l = *2, 4; m = 0, 4* are allowed. Terms in BSI^2 and BSI^4, which are also symmetry allowed, have not been yet been elucidated in appropriate form for the refinement programme and are not included. With these symmetry restrictions there are 22 ^{47}Ti and 25 ^{49}Ti parameters to be

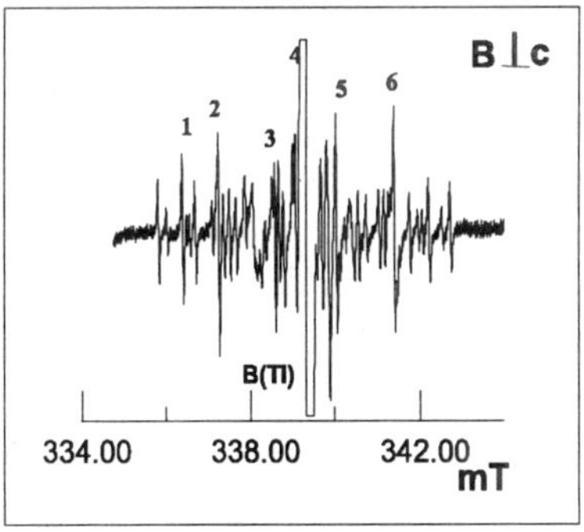

Fig. 2. Epr spectrum in the *ab* plane, along [110], of the B(Ti) centre at 10 K. The numbers label some prominent ^{47}Ti lines which are plotted in Fig. 3. The intense line, labelled 4, is from the even isotopes of titanium.

determined. We ensured that the parameters were well over-determined by setting criteria: Number of data points ≥ 10 times the number of parameters to be iterated upon.

The consequences of including HSN terms are rather subtle but certainly not trivial. We found: (a) a 50% reduction in the root-mean-squared deviation (RMSD) between observed and calculated resonant fields, (b) much improved fit of the "forbidden" hyperfine lines, (c) there is a predicted and observed cos4φ angular dependence (see Fig. 3) in the crystal *ab*-plane. Note that *no* angular dependence is possible for the conventional, uniaxial, SH, (d) The determined ratio $^{47}P_{\parallel}/^{49}P_{\parallel}$ = 2.65±0.05 (*cf.*, 2.54±0.18 as determined from atomic beam magnetic resonance measurements of quadrupole moments), (e) The nuclear Zeeman interaction seems best described by the combination of terms of dimension BI, BI^3, BI^5, all linear in magnetic field, and (f) there is an apparent anisotropy in the nuclear g tensor which is probably to be associated with anisotropy in the chemical shielding tensor.

The SH parameters obtained by best fitting spinless isotope (46 data points), ^{47}Ti (201 data points) and ^{49}Ti (202 data points) data from two crystal planes, (1 0 0) and (1 1 0), are listed in Tables 1, 2.

Table 1. SH parameters for the B(Ti) centre at 10K

	matrix *Y*		Principal values $Y_{\parallel}$	$Y_{\perp}$
even isotopes, I = 0	g		1.92683(1)	1.94067(1)
46 unit weighted data points.			RMSD = 0.0087 mT	
^{47}Ti, *I* = 5/2	g		1.92686(1)	1.94066(2)
	g_N		-0.417(118)	-0.267(25)
	$A/g_e\beta_e$	mT	2.9768(8)	0.8453(30)
	$P/g_e\beta_e$	mT	-0.1698(5)	0.0849(3)
201 unit weighted data points			RMSD = 0.0146 mT	
^{49}Ti, *I* = 7/2	g		1.92683(1)	1.94066(2)
	g_N		FV[b]	FV[b]
	$A/g_e\beta_e$	mT	2.9700(9)	0.8459(5)
	$P/g_e\beta_e$	mT	-0.0667(10)	0.0334(5)
202 unit weighted data points			RMSD = 0.0174 mT	

[a] The parallel values, $Y_{\parallel}$, lie along the c-axis
[b] FV fixed at 1.00028 x ^{47}Ti values

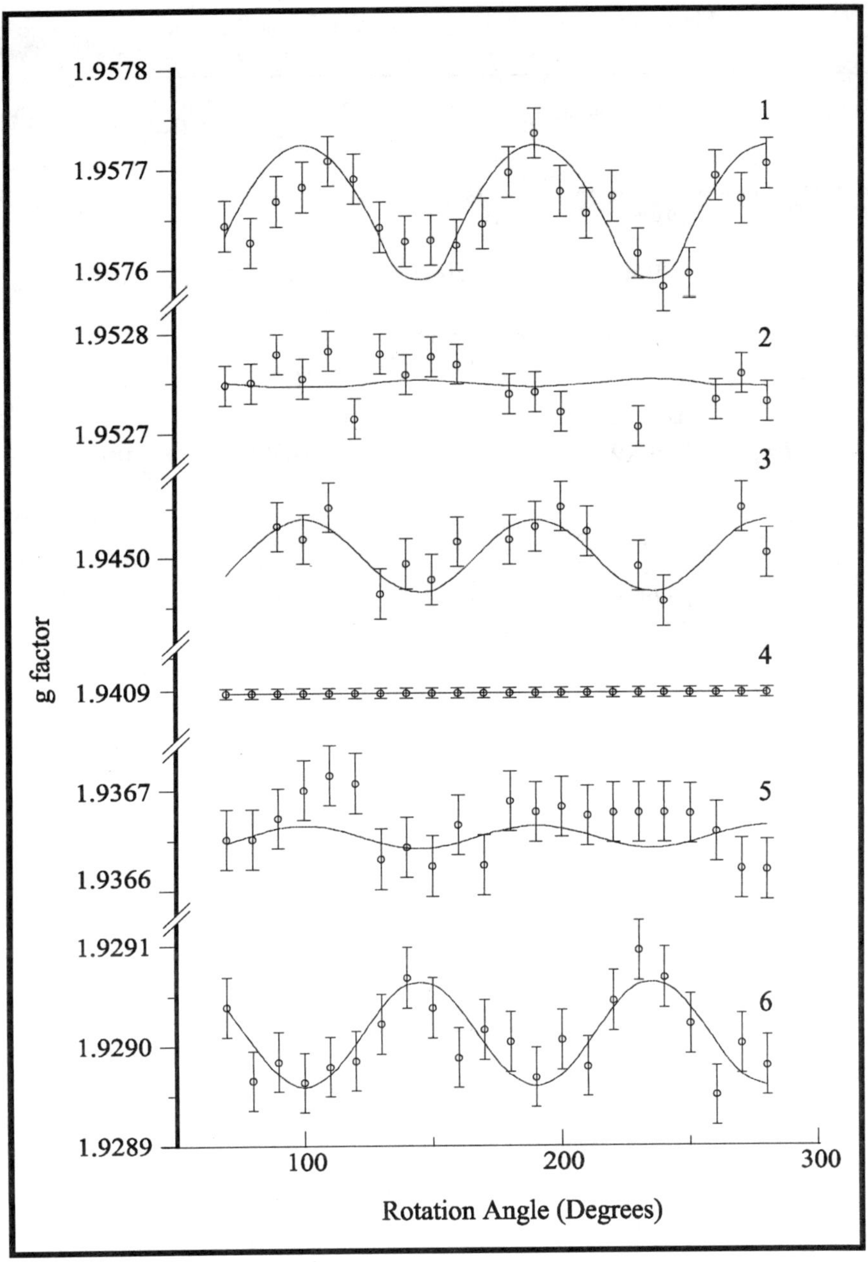

Fig. 3. Angular dependence of selected ^{47}Ti lines in the ab plane. The numbers refer to lines labelled in Fig. 2.

Table 2. Values of high spin parameters $B_{l,m}^{a,b,c}$ for ^{47,49}Ti at 10K

	^{47}**Ti(I=5/2)**		^{49}**Ti(I=7/2)**	
(a) Terms of dimension I^4 (x $1/g_e\beta_e$ (mT))				
	$B_{4,0}^{0,0,4}$	$B_{4,4}^{0,0,4}$	$B_{4,0}^{0,0,4}$	$B_{4,4}^{0,0,4}$
	-0.015(5)	0.001(2)	-0.057(3)	-0.003(1)
(b) Terms of dimension SI^3 (x $1/g_e\beta_e$ (mT))				
	$B_{l,0}^{0,1,3}$	$B_{l,4}^{0,1,3}$	$B_{l,0}^{0,1,3}$	$B_{l,4}^{0,1,3}$
$l = 2$	0.011(9)	--	0.065(8)	--
$l = 4$	0.012(9)	0.19(5)	0.058(7)	0.19(1)
(c) Terms of dimension SI^5 (x $1/g_e\beta_e$ (mT))				
	$B_{l,0}^{0,1,5}$	$B_{l,4}^{0,1,5}$	$B_{l,0}^{0,1,5}$	$B_{l,4}^{0,1,5}$
$l = 4$	0.004(5)	0.09(6)	0.015(1)	0.06(1)
$l = 6$	0.005(6)	0.02(2)	0.011(1)	0.055(6)
(d) Terms of dimension BI^3 (unitless)				
	$B_{l,0}^{1,0,3}$	$B_{l,4}^{1,0,3}$	$B_{l,0}^{1,0,3}$	$B_{l,4}^{1,0,3}$
$l = 2$	-0.055(7)	--	-0.063(3)	--
$l = 4$	-0.016(5)	-0.018(13)	-0.006(3)	-0.042(8)
(e) Terms of dimension BI^5 (unitless)				
	$B_{l,0}^{1,0,5}$	$B_{l,4}^{1,0,5}$	$B_{l,0}^{1,0,5}$	$B_{l,4}^{1,0,5}$
$l = 4$	0.025(6)	-0.023(11)	0.021(1)	-0.012(1)
$l = 6$	0.020(6)	0.052(7)	0.009(1)	0.028(1)

For a d^1 ion in a dodecahedral crystalline electric field, the ground state may be either d_{xy} or $d_{x^2-y^2}$ depending on the choice of x and y axes. Choosing the latter ($^2B_1$ representation) the g and A values, to first order in perturbation are:

$$g_{\parallel} = g_e - 8\zeta/\Delta E_1$$

$$g_{\perp} = g_e - 2\zeta/\Delta_2$$

$$A_{\parallel} = -\wp[4 + 7(g_e - g_{\parallel}) + 3(g_e - g_{\perp}) + 7\kappa]/7$$

$$A_{\perp} = \wp[4 - 11(g_e - g_{\perp}) - 14\kappa]/14$$

where κ is an adjustable parameter and $\wp\kappa$ is the isotropic contact term.

Using the experimental g values and $\zeta = 154\ cm^{-1}$, the free ion spin-orbit coupling constant, we calculate: $\Delta E_1 = 16{,}325\ cm^{-1}$ and $\Delta E_2 = 4{,}997\ cm^{-1}$. Using experimental A values we calculate: $\wp/hc = -24.84\text{x}10^{-4}\ cm^{-1}$ and $\wp\kappa = -25.93\text{x}10^{-4}\ cm^{-1}$. $\wp$ is in good agreement with the theoretical value, $P_{3d} = -23.61\text{x}10^{-4}\ cm^{-1}$ obtained from $P_{3d} = \mu_0 g_e g_N \beta_e \beta_N <r^{-3}>/4\pi$.

Failure to observe site splitting in any crystal orientation indicates that the Ti^{3+} of the $B(Ti)$ centre resides in a site of tetragonal symmetry. This could be either of the Zr^{4+} or Si^{4+} sites of the zircon unit cell each of which possess $\bar{4}2m$ (D_{2d}) point group symmetry. Point charge calculations were made for both the ZrO_8 and SiO_4 dodecahedra using the crystallographic data for the undistorted zircon lattice positions. The calculated energy levels for the former agree rather well with the levels obtained from the experimental g values but the latter do not (see Fig. 4). The designation $[TiO_8]^-$ is therefore suggested [4].

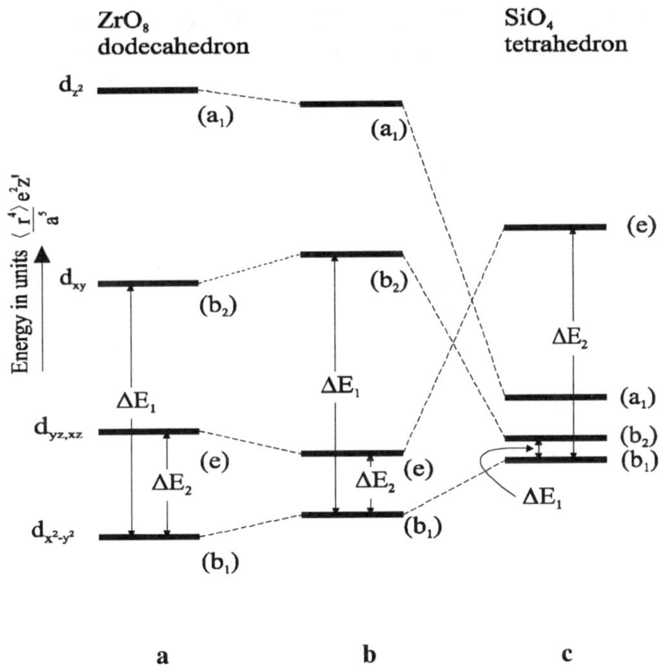

Fig 4. Calculated crystal-field energies in Zircon: (a) ZrO_8 dodecahedron, and (c) SiO_2 tetrahedron calculated using lattice dimensions; (b) ZrO_8 dodecahedron obtained from observed g-values.

Conclusion

By careful measurement and numerical analysis of all "allowed" and "forbidden" lines to a theoretically correct SH, we have obtained parameter matrices and interaction tensors for $^{47,49}Ti$ in zircon with very high precision. We have found, to the best of our knowledge, for the first time, large magnitude HSN terms of dimension BI^3, BI^5, SI^3 and SI^5 and, as a consequence determined the nuclear quadrupole and nuclear Zeeman interaction matrices with precisions normally only attainable by direct nuclear methods. We expect that such measurements will provide very valuable information on chemical shieldings of the *Ti* nucleus and an alternative route to determining nuclear quadrupole moments.

Acknowledgements

We acknowledge the contributions made by Katrina M Mackle, Gordon L A Sutton, Kirsten C Taylor and Charles J Walsby, Dennis G McGavin and Bryce E Williamson to the understanding of these systems. The technical support of the staff of the Mechanical and Electronic Workshops, University of Canterbury is acknowledged. Some equipment was purchased with funding from the New Zealand Lotteries Board.

References

1 V.P. Solntsev and M. Ya Shcherbakova, ESR study of the structural defects in irradiated zircons, *Akad. Nauk. SSSR* **1** (1973) 156-8.

2 R. F. C. Claridge, K. M. Mackle, G. L. A. Sutton and W. C. Tennant,10K EPR of three low symmetry centres in x-irradiated zircon (zirconium silicate), *J. Phys.: Condens. Matter*, **6** (1994) 3429-36.

3 R. F. C. Claridge, K. M. Mackle, G. L. A. Sutton and W. C. Tennant,10K EPR of an oxygen-hole aluminium centre, $[AlO_4]^0$ in x-irradiated zircon (zirconium silicate), *J. Phys.: Condens. Matter*, **6** (1994) 10415-22.

4 R. F. C. Claridge and W. C. Tennant, 10K EPR of a d^1 Ti^{3+} centre in x-irradiated zircon (zirconium silicate), *J. Phys.: Condens. Matter*, **7** (1995) 9049-60.

5 R. F. C. Claridge, G. L. A. Sutton and W. C. Tennant, 10K X-band EPR study of an unusual centre in x-irradiated zircon, *J. Magn. Reson,.* **125** (1997) 107-11.

6 R. F. C. Claridge, K. C. Taylor, W. C. Tennant and C. J Walsby, X-band EPR of a new Ti^{3+} centre in x-irradiated zircon, *J. Phys.: Condens. Matter* **9** (1997) 1-9.

Paramagnetic Defects and Optical Properties of $KTiOPO_4$ Crystals after Various Treatments

V.N.Efimov, M.A.Laruhin, V.A.Nazarova

MRS Laboratory, Physics Department, Kazan State University, Russia

Abstract. The paramagnetic defects were created in KTP crystals with the X-irradiation, vacuum- or H_2 - annealings and also with Nd:YAG laser. Some new hole- and electron-paramagnetic defects have been observed. It has been shown the colour of crystals does not directly correspond to the formation of paramagnetic Ti^{3+} and O^- centers.

1 Introduction

The KTP crystals are used in electrooptics and in a nonlinear optics as converters of laser frequencies near IR range and in high frequency modulators of light. However, damages arising under the laser radiation limit the use of these crystals in laser systems of large and average power. Under the high-power radiation the so-called "gray tracks" or narrow areas of darkening are formed in the crystal. In spite of a great number of papers devoted to a "gray tracks" phenomena this problem is far from the decision yet.

Electron paramagnetic resonance (EPR) is an excellent method for an identification of point defects and their local environment. In this work we describe EPR and optical measurements of hole and Ti^{3+} electron centers in KTP crystals after different treatments.

2 Equipment

The EPR spectra were recorded by X-band THN-251 spectrometer ("Thomson") with a modulation at the frequency of 4.2 kHz. A liquid-helium ESR-9 "Oxford instruments" - cryostat (the temperature range 4.2 - 300 K) with the temperature stabilization accuracy of 0.3 K was used. A BSV-2 (Ag - anticathode) X-ray tube operating at 50 kV and 15 mA during 2 hours at 77 K was used to produce paramagnetic centers. The vacuum annealings were made in pressure of 10^{-3} mbar. Optical spectra were detected with use of Hitachi P7 spectrophotometer.

3 Experiment

3.1 X-irradiation at 77K

After the X-irradiation at 77K the KTP crystal had a light-pink colour and we detected two groups of EPR signals. One of these groups corresponds to the Ti^{3+} electron centers with the typical g-values [1]. Another type of centers has a hole nature because the g-values at any orientations of the external magnetic field are greater than the free-electron g-value. The EPR spectrum of the hole center arising after the X-irradiation (fig.1) is similar with EPR spectrum which had been described by Edwards *et al.* [2].

For an identification of the center we studied the angular dependence of EPR lines in different crystallographic planes. Angular dependence is shown in fig.2 (black triangles), where superhyperfine (shf) splittings are ignored and data points represent the centers of each splitted lines. The least-squares-fitting procedure was used to obtain g-tensor elements. Thus table 1 contains g-matrices, principal values and axes of center 1. As one can see, the principal values of the g-tensor for the hole center 1 and the principal values reported by Edwards *et al.* [2] are not the same. On the one hand the effective g-factor (g_{eff}) and shf splittings along all crystal axes coincide with results of [2]. On other hand the principal values of both g-tensors (i.e. g_a, g_b, g_c) are not the same.

The additional experiments have been carried out to verify our calculation. *First*, minimum and maximum values of g_{eff} were experimentally measured: 2.0034(4) and 2.0442(4), those are comparable with 2.0040(3) and 2.0437(3), obtained from the calculation. *Second*, the angle α from the **a** axis to the maximum g-value principal axis was measured as $65\pm2°$, whereas the calculated angle α=arccos(0.4468)= =63.5° (see table 1).

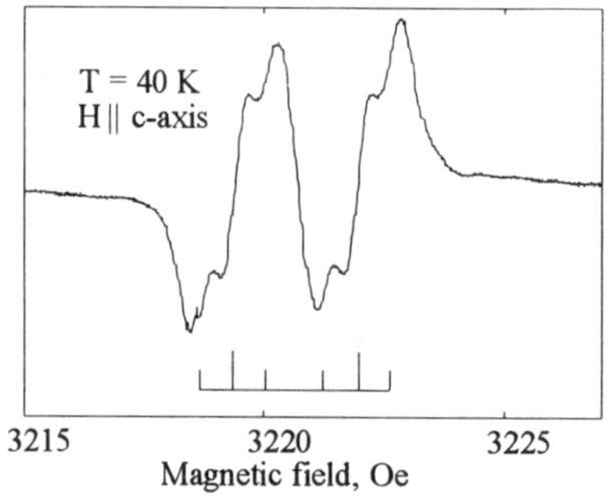

Fig. 1. EPR spectrum of the hole center 1. The data were taken at 40K and the magnetic field aligned along the c-axis of KTP crystal.

The study of the hole center kinetics at different temperatures was the next stage of our researches. Fig.3 indicates such transformation at T=160K. For a few minutes the intensity of the hole center spectrum (labeled as center 1) decreased, and two new lines appeared (center 2 and center 3). The angle dependencies of these hole centers are shown in fig.2 too.

The Ti^{3+} EPR spectra are presented in fig.4. The measurements were made at 40K because the properties of titanium centers allow to observe a strong signal at temperatures below 77K only, whereas the annealing of the KTP crystals was carried out at T=160K. Such transformation of the Ti^{3+} EPR spectrum in fig.4 displays a redistribution of charges in the crystal, and that the Ti^{3+} ions are involved into this process.

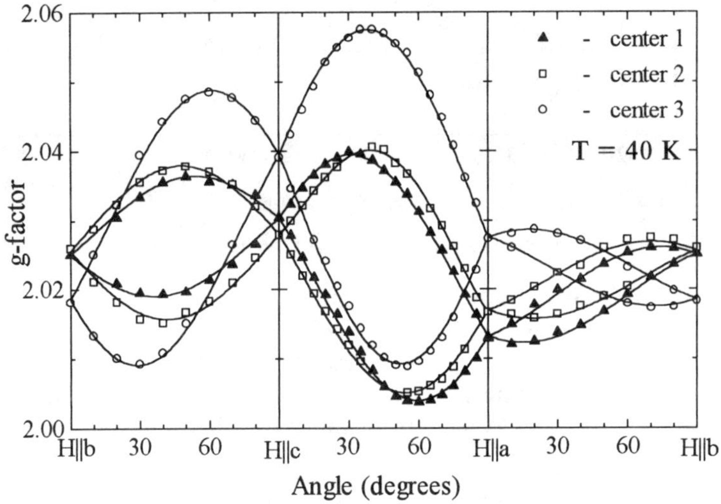

Fig.2. Angular variations of the hole centers 1, 2 and 3. The solid lines were calculated using **g** matrices.

Table 1. Spin-Hamiltonian parameters for the hole-centers produced by X-irradiation at 77K.

	$(\mathbf{g})^2$-matrix			Principal values and axes	θ,°	φ,°
Center ref.[2]	4.0526	.	.	2.0250	79.9	60.4
	.	4.1010	.	2.0008	57.9	156.8
	.	.	4.1209	2.0422	34.1	315.1
Center 1	4.0531	0.0143	0.0626	2.0209:[0.2739,-0.8994,0.3407]	109.9	106.9
	0.0143	4.1011	0.0336	2.0040:[-0.8517,-0.0622,0.5203]	121.4	4.2
	0.0626	0.0336	4.1224	2.0437:[0.4468,0.4327,0.7831]	38.5	44.1

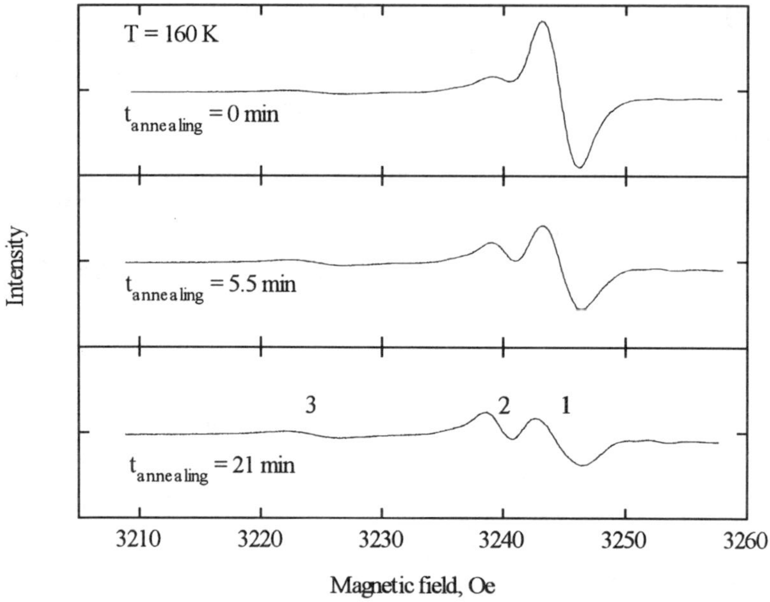

Fig.3. The transformation of the EPR signal of the hole centers in KTP at 160K.

There are four structurally inequivalent Ti^{3+} paramagnetic centers after the annealing of KTP crystals at T=160K. Each signal is tripled by the interaction between Ti^{3+} ions and a couple of approximately equivalent P nuclei, and each line of triplets is splitted by the third P nucleus. The ratio of the centers line intensities, the temperature of their observation, the g-values are almost identical with results of [1].

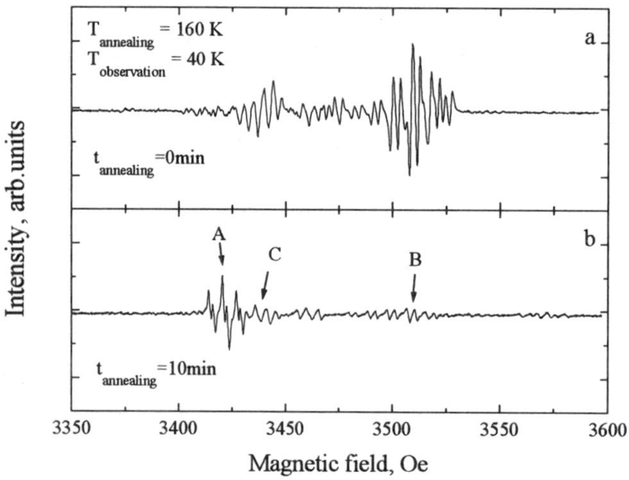

Fig. 4. The EPR of the Ti^{3+} centers in the X-irradiated KTP crystals: a) immediately after the X-irradiation; b) after the heating at 160K. Centers A,B and C are indentical with H2 reduction centers described by Mark G. Roelofs [1].

3.2 X-irradiation at 300K

After X-ray irradiation of crystals at 300K the hole- and electron- centers appeared as well. The titanium spectra are known as identical to spectra of crystals irradiated at 77K (fig.4,b), while the hole center is distinguished. According to their life times, the centers produced by irradiation at 300K are more stable. The colour and EPR spectrum of the crystal after the irradiation at 77K disappears at room temperature, while the crystal, irradiated at 300K, saves the colour and spectrum for several days.

An optical spectrum of samples after irradiation at 300K was taken at room temperature. Although the sample had pink nuance, an absorption spectrum in the visible region could not be detected.

3.3 Vacuum Annealing

The study of KTP crystals after the vacuum annealing was initiated by Martin *et al.* [3]. Samples heated in vacuum at temperatures in the 400-650°C range had a broad optical absorption with maximum at 710nm similar with that observed after H_2 reduction [1]. The Ti^{3+} center had similar features to D center reported by Roelofs [1]. However, such center had slightly different g-values and shf structure.

Vacuum annealing of our crystals was made at 450°C and at 650°C for 1.5-2 hours. After the treatment the samples got a slightly gray colour. An optical absorption in the visible region was not detected. EPR spectrum corresponded to Ti^{3+} centers. We detected the EPR of all four centers and the maximum intensity lines correspond to A (not D) centers [1]. This fact is not agreed with results of Martin *et al.* [3].

3.4 H_2-annealing

The next type of treatments was the reduction annealing in hydrogen atmosphere. During the first stage of procedure the annealing was carried out at 450°C. After the annealing the crystal remained colourless. The EPR spectrum had only Ti^{3+} centers similar with [1]. An optical spectrum of absorption in visible region was not detected. An increase of annealing temperature up to 650°C caused the appearance of gray colour of the crystal. The set of extra Ti^{3+} signals appeared in the EPR spectrum. Optical measurements indicate the presence of weak absorption in the 450-750nm region.

After H_2-annealing at ~800°C the crystal became dark-violet. The broad intensive peak of absorption with 590nm maximum (fig.5,a) appeared in the optical spectrum. In the EPR spectrum the broad lines with strong angular dependence g=0.8-6.5 appeared. The intensity of such lines strongly depends on the magnetic field orientation. Such behavior indicates high-spin state of a paramagnetic center, which could be either the Ti^{2+} or coupled Ti^{3+}-Ti^{3+} centers with electronic spin S=1. After the two years preservation at room temperature the EPR spectrum and the colour of the crystals remained unchanged.

3.5 Air Annealing

The dark colour induced by H_2-treatment could be removed by annealing in air. The ~800°C H_2-reduced sample was annealed isochronally in air at 400-450°. After first 30 min crystal became light-violet. In the EPR spectrum the part of extra lines disappeared. In the optical spectrum the intensity of absorption peak decreased (fig.5,b). During next 30 min the violet nuance disappeared. In the EPR the extra signal and the broad lines with strong angular dependence disappeared. In the optical spectrum the absorption peak became less intensive (fig.5,c). The following annealing resulted in the further enlightenment of the crystal.

3.6 Laser

Attempts to create paramagnetic defects in KTP crystals by YAG:Nd laser radiation at room temperature were made. Crystals, placed into the focus of laser beam (1060nm, 0.5 J/cm^2), have been mechanicaly destroyed without any changes of colour, and any measurements were impossible. One of the samples after the 10 hour irradiation by means of disfocused rays was not destroyed and did not change its colour. But in the EPR spectrum the weak signal from electronic Ti^{3+} center (fig.6) was observed. In a day EPR measurements did not give results. The EPR signal vanished.

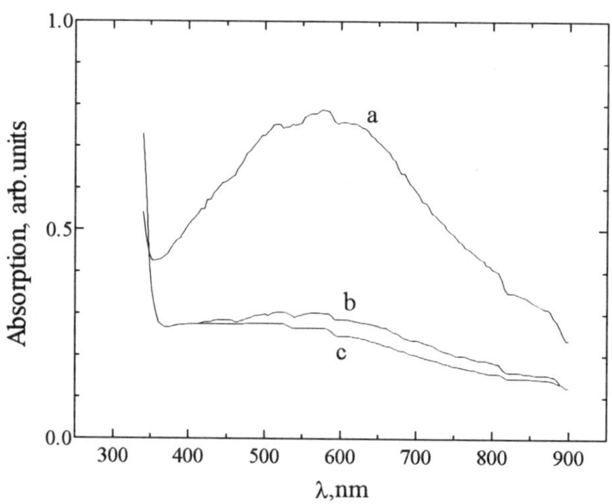

Fig. 5. Optical absorption spectra of KTP crystals (previously annealed at 800°C in H_2 atmosphere) after the oxidizing in air at 400 °C: a) 0 min.; b) 30 min.; c) 60 min.

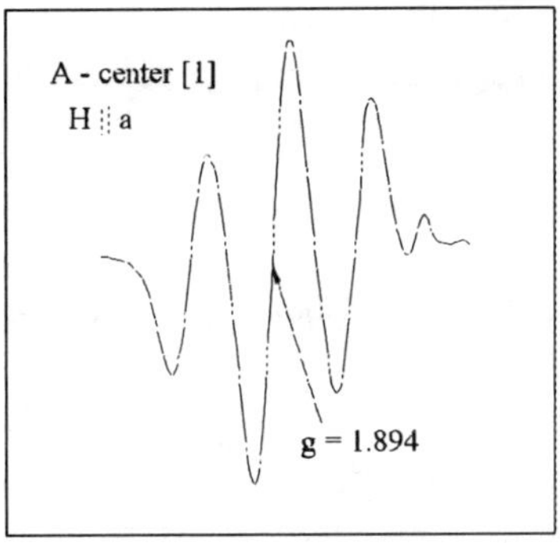

Fig.6. The EPR signal of Ti^{3+} centers in KTP crystal after 10 hour irradiation by the disfocused 1060nm laser beams.

4 Main results

All the results are presented in table 2 for the convenience.

Table 2. The main results of investigations on KTP crystals after various treatments.

Treatment	Time	Temperature	Observed centers	Coloring	Optical absorption	Stability
X-ray	1 hour	77 K	O^-, Ti^{3+}	pink	not detected	below 160 K
X-ray	1 hour	300 K	Ti^{3+}	pink	not detected	at 300 K several hours
X-ray	4 hours	300 K	O^-, Ti^{3+}	pink	not detected	at 300 K several days
laser Nd: YAG	10 hours	300 K	Ti^{3+}	no color	not detected	at 300 K several hours
H_2 annealing	1 hour	400 °C	Ti^{3+}	no color	not detected	a year
H_2 annealing	1 hour	600 °C	Ti^{3+}	gray	590nm	a year
H_2 annealing	1 hour	800 °C	Ti^{3+}, Ti^{2+}?	dark-violet	590nm	a year
Electric field [1]	0.3 hour	200 °C	Ti^{3+}	gray	550nm	
H_2 annealing [1]	10 min.	800-850 °C	Ti^{3+}	blue-gray	700nm	
Vacuum annealing [3]	0.5 hour	400-800 °C	Ti^{3+}	?	710nm	
Vacuum annealing	2.5 hours	650 °C	Ti^{3+}	slightly gray	not detected	

As one can see, the colour of crystals does not directly correspond to the formation of various paramagnetic Ti^{3+} and O^- centers.

References

1. Mark G. Roelofs, Identification of Ti^{3+} in Potassium Titanyl Phosphate and its Possible Role in Laser Damage, *J.Appl.Phys.*, 1989, 65(12), pp.4976-4982.
2. Edwards G.J., Scripsick M.P., Halliburton L.E., Belt R.F., Identification of a Radiation-induced Hole Center in $KTiOPO_4$, *Phys.Rev.B*, 1993, 48(10), pp.6884-6891.
3. M.J.Martin, D.Bravo, R.Sole, F.Diaz, F.J.Lopez, C.Zaldo, Thermal Reduction of $KTiOPO_4$ Single Crystals, *J.Appl.Phys.*, 1994, 76(11), pp.7510-7517.

Electron Spin Relaxation of Cu(II) Ions in Triglycine Selenate $(NH_2CH_2COOH)_3H_2SeO_4$ Single Crystal

Janina Goslar, Stanisław K. Hoffmann, and Wojciech Hilczer

Institute of Molecular Physics, Polish Academy of Sciences, Smoluchowskiego 17, PL-60179 Poznań, Poland

Abstract. Pulsed EPR studies of the electron spin-lattice and phase relaxation of Cu(II) ions were performed in the low temperature range 4.2–100 K. Spin-lattice relaxation is due to direct and Raman type coupling to the acoustic phonons. The relaxation data allowed to determine the Debye temperature $\Theta_D = 168$ K of the crystal. Phase relaxation is governed by molecular librations at low temperatures, and by the T_1-processes above 80 K, with a resonance type minimum in phase-memory time, T_M, around 50 K due to flipping of the glycine GI = $(NH_3^+CH_2COO^-)$ molecules that determine spontaneous polarization of the crystal.

1 Introduction

Triglycine selenate (TGSe) is the order-disorder type ferroelectric crystal with ordering temperature $T_c = 295$ K and belongs to the triglycine sulphate $(NH_2CH_2COOH)H_2SO_4$ = TGS family. Copper(II) ions introduced into the TGSe crystal occupy interstitial positions and coordinate octahedrally with two glycine molecules in the *trans* configuration in the equatorial plane and with SeO_4-groups in the apical positions. Continuous wave EPR spectra and parameters of Cu(II) ions were determined for TGSe by Hoffmann et al. (1996)[1] and reviewed for TGS by Stankowski (1981).[2] These results do not contribute significantly to understanding of molecular mechanism of the ferroelectricity in the crystals since Cu(II) ion is located relatively far from the glycine GI = $(NH_3^+CH_2COO^-)$ responsible for the polar state formation and, moreover, Cu(II) with $S = 1/2$ is rather weakly coupled to the long-distance crystal field components. More pronounced effect can be expected in pulsed EPR studies which allow to explore the spin packets that form inhomogeneously broadened EPR lines. The packets are relatively narrow (order of 10 μT), therefore sensitive to even small changes in crystal lattice dynamics and the second coordination sphere of the Cu(II) ion. In fact, Völkel et al. (1979)[3] observed anomalies of the spin relaxation times around the critical temperature $T_c = 322$ K in TGS due to critical fluctuations. Lattice dynamics of the crystals in ferroelectric state has been not studied by pulsed EPR down to low temperatures where ferroelectric parameterts are still temperature dependent. Anomalous behaviour of TGS crystal was reported by Hadni and Gerbaux (1985).[4] In this paper we present low temperature measurements of the spin-lattice and phase-memory time for low concentration (8×10^{17} spin/gram) Cu(II) ions in TGSe = $(NH_2CH_2COOH)H_2SeO_4$ crystal when spin and spectral diffusion effects are negligible and relaxation is governed by undisturbed relaxation processes and interactions with matrix magnetic nuclei.

2 Spin-lattice Relaxation

The temperature dependence of the Cu(II) spin-lattice relaxation time T_1 is shown in Fig. 1. T_1 was determined by both the saturation recovery and inversion recovery methods. Both the methods gave the same T_1-values in the whole temperature range indicating that Cu(II) concentration is sufficiently low to avoid spectral-diffusion effects. Temperature variations of the relaxation rate $1/T_1$ are well described by the equation:

$$\frac{1}{T_1} = a + bT + cT^9 \int_0^{\Theta_D/T} \frac{x^8 \exp(x)}{[\exp(x) - 1]^2} dx \tag{1}$$

where a results from the zero setting of the detector, bT is the direct one-phonon process, and the third term describes two-phonon Raman process when Debye temperature of a crystal is Θ_D. Using the analytical form of the transport integral of equation (1) published by Hoffmann et al. (1996)[1] the best fit was obtained with parameters: $a = 200\ \mathrm{s^{-1}}$, $b = 204\ \mathrm{s^{-1}K^{-1}}$, $c = 7.5\times10^{-9}\ \mathrm{s^{-1}K^{-9}}$, and $\Theta_D = 168\pm3$ K. The spin-lattice relaxation is entirely determined by coupling of Cu(II) ions to the acoustic lattice phonons via spin-orbit coupling. The local effects as NH_3-group reorientations, molecular librations, and electric polarization fluctuations at low temperatures

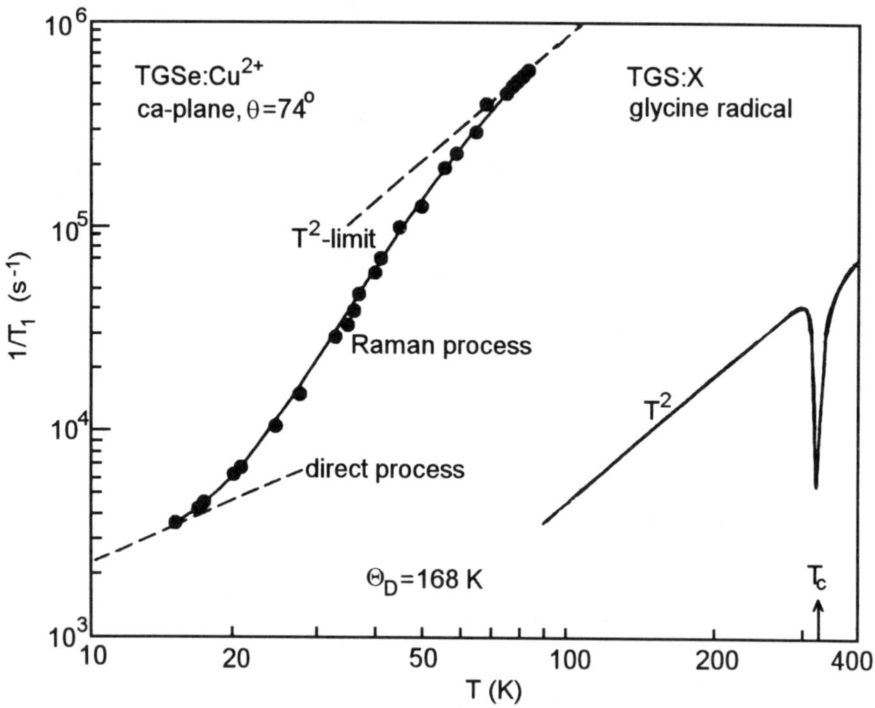

Fig. 1. Temperature variations of the spin-lattice relaxation rate for Cu(II) ion. Solid line is the best fit to the Eq. (1) with $\Theta_D = 168$ K. The Völkel et al. data[3] for glycine radical in TGS are added.

seem to do not contribute to the observed temperature dependence of spin-lattice relaxation time although the linear $T_1^{-1}(T)$ dependence was predicted when phonon-induced relaxation appears by a coupling to nearby localized two-level tunneling state (Murphy (1966)[5], Bownam and Kevan (1977)[6], Kurtz and Stapleton (1980)[7]). This mechanism was found to operate for radicals especially in glassy state, and seems to be of less importance for paramagnetic metal ions in crystals where the direct phonon process dominates at low temperatures.

To the relaxation data of Cu(II) presented in Fig. 1, the data obtained by Völkel et al. (1979)[3] for NH_3^+CHCOO radical in TGS crystal are added. The T_1 times for the radical are much longer that those for Cu(II) ion. It seems to be a general rule that spin-lattice relaxation is much faster for paramagnetic ions than for free radicals because of the most effective relaxation mechanism involve g-factor and hyperfine splitting anisotropy modulation by lattice vibrations (Bowman and Kevan (1979)[8]). A common feature of Cu(II) and glycine radical relaxation in our crystals is quadratic dependence of the relaxation rate $1/T_1$ on temperature in the medium temperature range around 100 K. This is a characteristic dependence for high-temperature limit ($T > \Theta_D/2$) of the Raman process. Such behaviour indicates that relaxation process for Cu(II) and the radical are the same despite the fact that these paramagnetic centres are located at different sites of the crystal unit cell. At high temperature, however, where electron spin echo from Cu(II) ions is nondetectable, an additional mechanisms give measurable contribution to the spin-lattice relaxation rate. They are relaxation via Raman process involving optical phonons ($1/T_1 \propto \mathrm{cosech}^2(\Delta/2kT)$) of energy $\Delta = 49$ cm^{-1} and critical slowing-down of the relaxation rate as observed by Völkel et al. (1982).[9]

The spin-lattice relaxation data for the glycine radical and Cu(II) ion are complementary and allow to determine the relaxation processes and mechanisms in the broad temperature range. Measurements of T_1 for Cu(II) in TGSe by electron spin echo technique is impossible above about 100 K where the echo amplitude becomes too weak as a result of spin-spin relaxation shortening. For free radicals, as a rule, the electron spin echo can be observed at room temperature allowing T_1 measurements at higher temperatures.

3 Phase Relaxation

Phase relaxation of inhomogeneously broadened EPR line is a dephasing process of the spin packets excited by microwave pulse producing simultaneously a coherence in the spin packet precessional motion. This dephasing allows to produce an electron spin echo (ESE) signal after second microwave pulse. When at time τ after the exciting pulse the refocusing pulse is applied, the ESE is formed after the same time τ. The ESE amplitude decreases exponentially with increasing τ with a characteristic time called the phase memory time T_M. This decrease can be due to several mechanisms producing random changes in the phase of the precessional motion. For low concentration of paramagnetic centres, when interactions between unpaired electrons are negligible, the phase relaxation can be due to nuclear spectral diffusion proc-

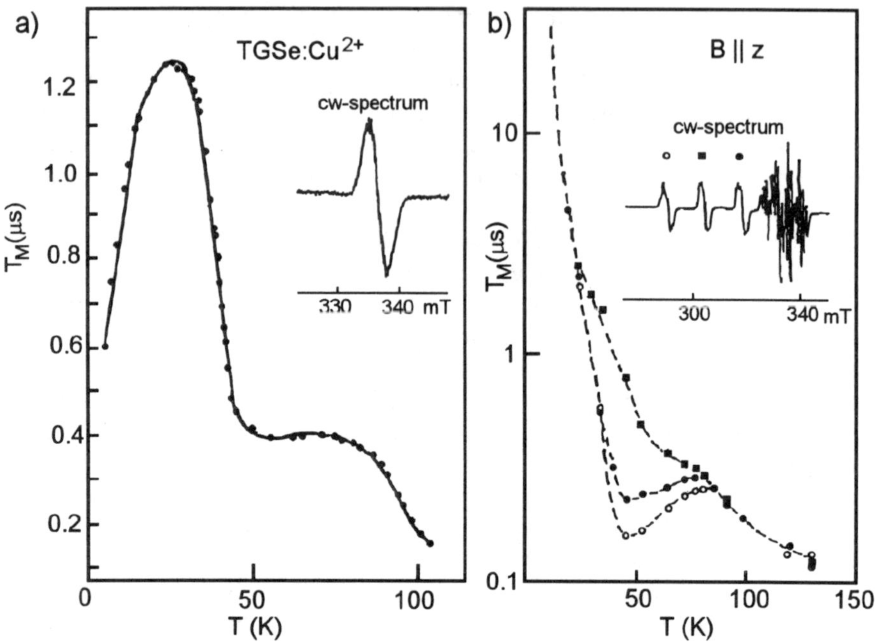

Fig. 2. Temperature variation of the phase memory time, T_M, for two single crystal orientations of TGSe:Cu^{2+}: a) in the *ca*-plane at $\theta = 74°$, b) along the *z*-axis of the $Cu(gly)_2(SeO_4)_2$ complex.

esses, molecular motions and spin-lattice relaxation processes at higher temperature. These processes can determine the ESE decay function $V(2\tau)$ and influence the temperature behaviour of the phase memory time T_M.

We have generated the Hahn type spin echo by two 48 ns pulses with initial interpulse time interval equal to 200 ns. The echo amplitude was modulated by a weak dipolar coupling with the surrounding ^{1}H and ^{14}N nuclei and the decay function was single exponential $V(2\tau) = V_0 \exp(-2\tau/T_M)$ with a good approximation in the whole temperature range, and for all crystal orientations. The temperature variations of T_M for two crystal orientations are shown in Fig. 2. When the external magnetic field was applied in the *ca*-plane at $\theta = 74°$, the microwave pulses excited nearly the whole EPR spectrum consisting of a single resonance line at this crystal orientation (Fig. 2a), whereas for the external field directed close to the local *z*-axis (*bc*-plane, $\theta = 60°$) of the Cu(II) complex the individual hyperfine components were excited separately and different T_M-values were found for the each component. The differences in T_M values and temperature behaviour for different hfs lines will be not discussed in this paper.

A general feature of the T_M above 25 K is a strong decrease on heating with a local minimum in T_M at about $T = 50$ K. This can be more clearly seen in $\log(T_M)$ *vs.* $1/T$ plot presented in Fig. 3 for the data of Fig. 2a. We explain this behaviour as a result of three relaxation processes. Let us consider the spin packets with shape determined by Fourier transform of the decay function and packet width $\Delta B_{packet} = 1/T_M$. Below 25 K the spin packets continuously broaden on heating as a

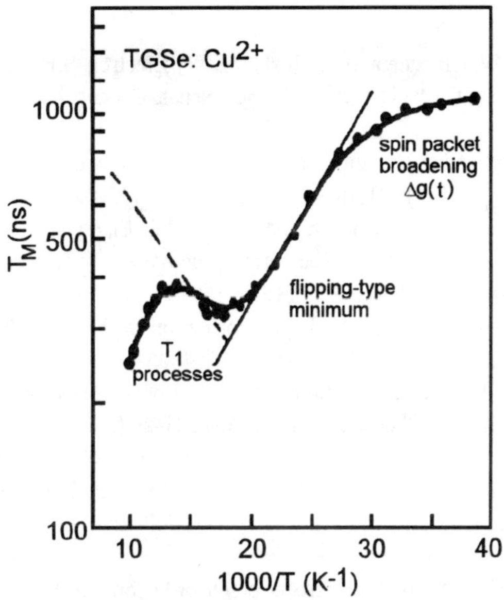

Fig. 3. Temperature variations of the phase memory time T_M of Cu^{2+} in TGSe single crystal (*ca*-plane, $\theta = 74°$) with a flipping-type minimum at 50 K due to glycine molecules motion ($E_a = 1.0$ kJ/mol, $\tau_c = 3.8$ μs). The solid lines is the guide for eyes only.

result of an increase in thermal motions of the coordinated molecules producing *g*-tensor modulations. The resonance type minimum at 50 K is due to an exchange type effect producing merging effect between different spin packets being unresolved hyperfine lines as discussed by Tsvetkov and Dzuba (1990).[10] For slow exchange as compared to the splitting $\Delta\omega$ between spin packets $\Delta\omega\tau_c > 1$, where τ_c is exchange correlation time, the decay function is described as $V(\tau) = V_0 \exp(-2\tau/T_M)$ with $T_M = \tau_c$, whereas in fast exchange limit $\Delta\omega\tau_c < 1$ the decay function has the same form, but $T_M = 2/(\Delta\omega^2\tau_c)$ and in between a minimum of T_M appears. We suppose that the exchange process is a flipping motion of the noncoordinated glycine GI between two positions in TGSe crystal unit cell which is not frozen at low temperatures as indicated by spontaneous polarisation behaviour. From T_M behaviour around $T = 50$ K we found $\tau_c = 3.5\times10^{-8} \exp(E_a/RT)$ with $E_a = 1.0$ kJ/mol and $\tau_c = 38$ μs at T_M minimum. This thermally activated very slow motion with rather small activation energy E_a can not be hindered rotations of NH_3-group which are practically frozen at this temperature range.

At higher temperatures, T_M is influenced by spin-lattice relaxation processes since T_1 becomes one order of magnitude longer than T_M around 100 K. The T_M behaviour below 25 K, visible in Fig. 2a, indicating continuous spin packet broadening on cooling, can be due to the same process that produces optical anomalies as those reported by Hadni and Gerbaux (1985).[4]

This work was supported by Polish Scientific Research Committee under Project KBN-2-P03B-038-08.

References

1. S. K. Hoffmann, W. Hilczer and J. Goslar, EPR, Electron Spin-Lattice Relaxation, and Debye Temperature of Cu(II)-Doped Triglycine Selenate Crystal, *J. Magn. Reson. A* **122** (1996) 37-41.
2. J. Stankowski, Molecular Dynamics of Ferroelectric Crystals of the Triglycine Sulphate Family, *Phys. Reports* **77** (1981) 1-46.
3. G. Völkel, W. Brunner, W. Windsch and D. Steudel, Electron Spin-Echo (ESE) Investigations of Critical Paramagnetic Relaxation Anomalies in X-irradiated Triglycine Sulphate (TGS:X), *Phys. Stat. Solidi* **B95** (1979) 99-105.
4. A. Hadni and X. Gerbaux, Surface Layer Formation at 44 K in Ferroelectric Triglycine Sulphate, *J. Phys C: Solid State Phys.* **18** (1985) 4805-4817.
5. J. Murphy, Spin-Lattice Relaxation Due to Local Vibrations with Temperature-Independent Amplitudes, *Phys. Rev.* **145** (1966) 241-247.
6. M. K. Bowman and L. Kevan, An Electron Spin-Lattice Relaxation Mechanism Involving Tunneling Modes for Trapped Radicals in Glassy Matrices. Theoretical Development and Application to Trapped Electrons in γ-Irradiated Ethanol Glasses, *J. Phys. Chem.* **81** (1977) 456-461.
7. S. R. Kurtz and H. J. Stapleton, Effects of Disorder on Electron-Spin Relaxation in β-Alumina: A Prototype Glass, *Phys. Rev.B* **22** (1980) 2195-2205.
8. M. K. Bowman and L. Kevan, Electron Spin-Lattice Relaxation in Nonionic Solids, in *Time Domain Electron Spin Resonance*, L. Kevan and R. N. Schwartz (eds), Wiley, New York (1979) pp. 67-105.
9. G. Völkel, The Importance of Anharmonic Effects in the Electron Spin-Lattice Relaxation of Ferroelectric Phase Transitions, *Phys. Stat. Solidi* **B114** (1982) K-67-K-69.
10. Yu. D. Tsvetkov and S. A. Dzuba, Pulsed ESR and Molecular Motions, *Appl. Magn. Reson.* **1** (1990) 179-194.

Investigation of the Ground and Metastable State of Nitroprusside Crystals by NMR and EPR Spectroscopy

Jadwiga Tritt-Goc, Narcyz Piślewski, and Stanisław K. Hoffmann

Institute of Molecular Physics, Polish Academy of Sciences, Smoluchowskiego 17, PL-60179 Poznań, Poland

Abstract. The paper presents a brief description of the NMR and EPR results of barium, strontium and sodium nitroprusside crystals. These compounds are very interesting due to a new metastable state generated with laser light. The influence of the phase transitions detected by proton NMR on the temperature below which the new state can be generated is discussed. An evidence of the paramagnetism of the new molecular state in sodium nitroprusside crystal is presented.

1. Introduction

The miniaturization of the classic electronic chips cannot go any further and the next step requires radical new ideas based on molecular electronics. One of the possible solutions is the use of the nitroprusside crystals with the chemical formula $X_n[Fe(CN)_5NO] \times yH_2O$ (X=Na, Sr, K, Ba). This is due to the new metastable electronic states generated with laser light as discovered by Hauser et al. (1977)[1]. The population of the new state depends on the wavelength, intensity, duration of irradiation, polarization of light, with respect to the symmetry axis of nitroprusside ion, and the temperature. A maximum population of 50 % of the new state in sodium nitroprusside or 36 % in barium nitroprusside is obtained with the electric field vector perpendicular to the quasi-fourfold axis (NC-Fe-NO) of the molecule and the temperature below 200 K or 230 K, respectively. The new state is a nearly infinitely long living metastable state (about 10^{10}s) at temperature below 200K. There are two different ways in which the metastable state can be depopulated: via a thermal decay channel by heating the crystal above 190 K or by irradiation of the sample with light in the wavelength range 620-850 nm. The population and depopulation cycle are completely reversible and no changes in the structural or electronic properties of the crystal occur as shown by Rüdlinger et al. (1991)[2]. Therefore, the crystals in the new excited states can be used as optical information storage elements or as fast optical switches. The storage capacity of three-dimensional crystals is extremely high, because the information is localized in the dimension of molecules. Due to this extremely high storage capacity and their technological impact, the nitroprusside crystals have recently attained significant importance.

In our laboratory Tritt-Goc et al. (1986, 1993, 1994, 1995, 1995,1996)[3-8] studied, by ^1H NMR, the structure and dynamics of the water molecules in ground state of the nitroprusside crystal of sodium (SNP), barium (BaNP), strontium (SrNP) and potassium (KNP). The water molecules in the unit cell of these crystals are unusually free, and therefore they undergo different types of motions. We were able to establish the dynamics parameters of the 180^0 flips and oscillations of the water molecules and demonstrate how these librations average the dipolar coupling tensors of the water molecules.

In this paper we will discuss the connection between the phase transition temperatures studied by proton NMR in strontium and barium nitroprusside crystals and the temperatures below which the new state in these crystals is observed. The new molecular excited state was studied in sodium nitroprusside crystals by EPR method and the results will be discussed.

2. Phase Transitions

The phase transitions in strontium $Sr[Fe(CN)_5NO]\times 4H_2O$ (SrNP) and barium $Ba[Fe(CN)_5NO]\times 3H_2O$ nitroprusside crystals were studied by proton spin-lattice relaxation time and proton NMR spectra in a wide range of the temperature. Four phase transitions at T_{c1}=197, T_{c2}=192, T_{c3}=182, and T_{c4}=174 K have been discovered for SrNP and one at 233 K for BaNP.The results of the measurement of the temperature dependence of the proton spin-lattice relaxation time in the strontium nitroprusside compound are presented in Fig. 1. In the temperature range between 340 and 190 K the T_1 dependence on 1/T shows a symmetric V-shape minimum resulting from the 180^0 flips of the water molecules around their quasi-twofold axes. On cooling SrNP undergoes phase transitions. They affect the slope of the T_1 *vs* 1/T curve. The most pronounced change of the slope occurs at T_{c1} and T_{c2} whereas the one at T_{c4} is only slightly visible. The activation energies calculated from the slope of T_1 *vs* 1/T in the ranges I, II, and III are equal to 9.5, 5.1, and 5.3 kJ/mol, respectively. The nature of the phase transitions in SrNP is still unknown. However, the single crystal measurements reported by Tritt-Goc (1995)[7] give evidence that most probably a rearrangement of the water molecules is responsible for the observed transitions because the proton-proton vectors of the water molecules change their orientations in the unit cell of strontium nitroprusside crystal. This conclusion was based on the dipolar coupling tensors measured below and above phase transitions. Similar measurements in barium nitroprusside crystals also reveal the change of the direction of water molecules at phase transition which took place at 233 K. Moreover, Navaza et al. (1992)[9] confirmed the rearrangement of water molecules in BaNP at phase transition by neutron measurements.

We took notice that there is a close connection between the phase transition temperatures and the temperatures below which the new state, after irradiation with argon laser, is observed. The first phase transition in SrNP is observed at 197 K. In BaNP the phase transition occurs at 233 K. The new state is only populated below 190 K and 220 K, in strontium and barium nitroprusside crystals, respectively. Therefore, we suggest that the rearrangement of the water molecules, which takes place at the phase transitions, causes the deformation of the nitroprusside anion

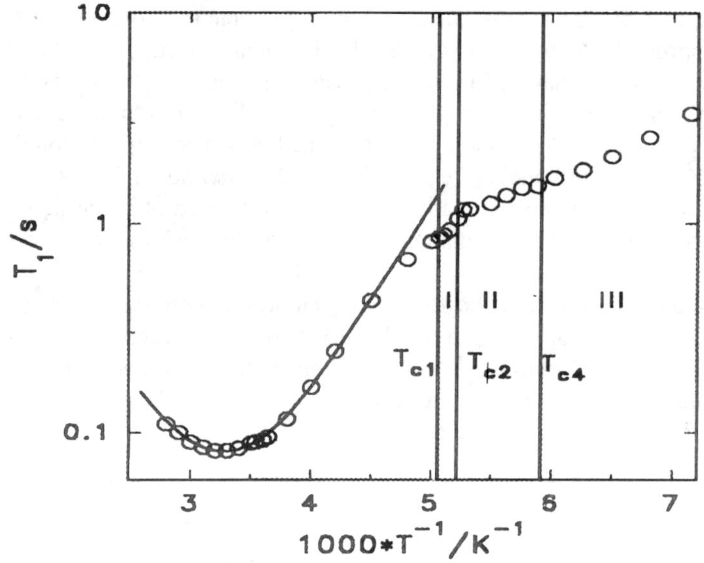

Fig. 1. Temperature dependence of proton spin-lattice relaxation time in polycrystalline sample of SrNP. T_{c1} = 197 K, T_{c2} = 192 K, and T_{c4} = 174 K.

octahedral. Due to this deformation, as it will be discussed further, the degeneracy of the π^*(NO) orbital can be lifted. This is a necessary step in order to transfer the crystal to the new molecular state after irradiation with laser light.

3. Paramagnetism of the Metastable Molecular State

The long lived metastable state was studied by EPR methods in our laboratory in $Na_2[Fe(CN)_5NO]\times 2H_2O$ (SNP) at 77 K. Sodium nitroprusside salt is the first crystal where the metastable state was discovered. The molecular structure of the crystals has been studied at room temperature and in metastable state at 80 K by Rüdlinger et al. (1991)[2]. Physical properties of sodium nitroprusside salt both at the ground and excited states were studied by Mössbauer effect by Hauser et al. (1977)[1] and by Woike et al. (1993)[10], by Raman effect by Woike et al. (1983, 1983, 1984, 1990)[11-14], by

NMR by Gross et al. (1983)[15], by EPR by Kruschel (1984)[16], and DSC spectroscopy by Zöllner et al.(1989)[17] and by Woike et al. (1993)[18]. The molecular-orbital theory and optical-absorption spectroscopy data allowed to calculate a complete scheme of the molecular orbital levels, first done by Manoharan et al. (1965)[19] and propose a model of the new excited molecular state formation by Woike et al. (1983,1984)[12,13]. The energy level diagram of the $[Fe(CN)_5NO]^{2-}$ ion is presented in Fig. 2. The Fe 3d orbital in the ground state of iron in sodium nitroprusside crystal is split into two orbitals: d_{xz}, d_{yz}, and d_{xy} as a consequence of the replacement of one of the CN groups in the nitroprusside ion by an NO group. The degenerate 6e (d_{xz}, d_{yz}) orbital of the central metal atom Fe has mainly metal d character and is occupicd by four electrons. The $2b_2$ (d_{xy}) level is the highest occupied molecular orbital with two electrons and has predominantely metal d character. The lowest unoccupeid orbital in the $[Fe(CN)_5NO]^{2-}$ is the antibonding 7e (π^*NO) ligand orbital about 20 400 cm^{-1} above the ground state. The higher $3b_1$ and $5a_1$ levels are formed of metal $d_{x^2-y^2}$, and d_{z^2} orbitals, respectively. The population of the new state is a two-step process. First a charge transfer transition from 3d orbital of the central Fe atom from $[Fe(CN)_5NO]^{2-}$ molecular ion into the low laying empty antibonding 7e (π^*NO) orbital of the NO ligand occurs, and from this π^*(NO) orbital relaxation into a metastable potential minima (orbital $\pi_2{}^*$(NO)) takes place. In the excited state of the crystal the 3d orbital of iron is also split into two orbitals as indicated in Fig.2 but the $2b_2$ (d_{xy}) is occupied only with one electron. In order to produce a metastable state

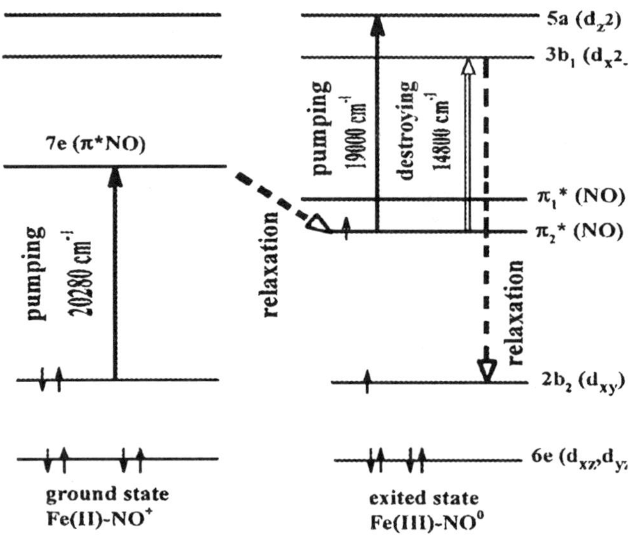

Fig. 2. Energy level diagram on the ground and long-living excited state of $[Fe(CN)_5NO]^{2-}$ complex. The small arrows indicates the spin population in the ground state and after charge transfer by pumping with the gree-blue laser light.

the degeneracy of the 7e (π^*NO) orbital has to be lifted. This can be produced by spin-orbit coupling or by deformation of the nitroprusside anion octahedral. In the light of our NMR results we believe that the degeneracy is lifted due to the deformation.

One can expect that sodium nitroprusside crystals will be paramagnetic in the metastable state, since the charge transfer excitation produces two unpaired electrons as showed by arrows in Fig.2. As a result the iron ion is transformed into a paramagnetic form Fe(III) and paramagnetic nitric NO group appears. However Kruschel's (1984)[16] measurements show that the crystal is EPR silent. On the basis of this paper Zöllner (1989)[17], Rüdlinger (1991)[2], and Woike (1993)[18] claim that the sodium nitroprusside crystal remains diamagnetic in the excited metastable state.

Our EPR results on sodium nitroprusside monocrystal demonstrate unequivocally that the excited state is paramagnetic and the EPR signal of Fe(III) ion can be observed. EPR experiments were performed on a RADIOPAN SE/X-2547 spectrometer operating at about 9.0 GHz with 100 kHz magnetic modulation and TE_{102} microwave cavity. The $Na_2[Fe(CN)_5NO]\times 2H_2O$ single crystal was continuously irradiated at 77 K over 18 hours with polarized light of an air cooled Argon Laser Model 543 (λ = 488 nm) with intensity 130mW/cm^2. The crystal c-axis was parallel to the rotation axis, being perpendicular to the external magnetic field direction and parallel to the magnetic component B_1 of the microwave field. The external magnetic field direction in the crystal ab-plane was chosen to be parallel to the Fe-N-O direction that is expected to be along the principal g - factor axis of $[Fe(CN)_5NO]^{2-}$ complex after excitation with a maximal g-factor value (g_Z - axis). During crystal irradiation the magnetic field was repeatedly swept from 0 to 0.7 T with a scanning time of 5 minutes. EPR spectrum was not observed at room temperature with and without laser irradiation. Also during the first 12 hours of irradiation at 77K no EPR spectrum was observed. Then the dielectric losses in microwave field increased and some spin excitations appeared and disappeared in the form of narrow EPR-type lines during the field sweep. After that a very broad resonance line grew in intensity. This process was completed after about 15 minutes with the stable EPR line of peak-to-peak linewidth ΔB_{pp} = 180 mT located at g = 3.88. During the next 6 hours a slow evolution of EPR spectrum took place. This time evolution of the spectrum is shown in Fig. 3. The line positions described by g - factor as well as the linewidth ΔB_{pp} changed gradually with time, and after about 6 hours this process was completed with the line at g = 2.31, with ΔB_{pp} = 350 mT, although a residual of the starting line was still observed.

From our EPR results some conclusions can be drawn. The measured g - factor has a maximum in the crystal ab - plane and can be assumed as the g_Z - value of the low-spin Fe(III) ion. Our statement is based on the fact that the observed EPR signal is characteristic for S = 1/2 Fe (III) ion in a strong crystal field as shown by Mabbs et al. (1992)[20]. Moreover, a similar signal of $Fe(CN)_5NH_3^{2-}$ ions was observed in single crystal of $Na_2[Fe(CN)_5NO]\times 2H_2O$ by Ezzeh et al. (1974)[21]. There is no chance to mistake the observed signal with that from NO^- radical because of a great difference in the g_z value. It seems that the time beahaviour of the g_Z - factor that we have observed in the excited state of $Na_2[Fe(CN)_5NO]\times 2H_2O$ crystal is due

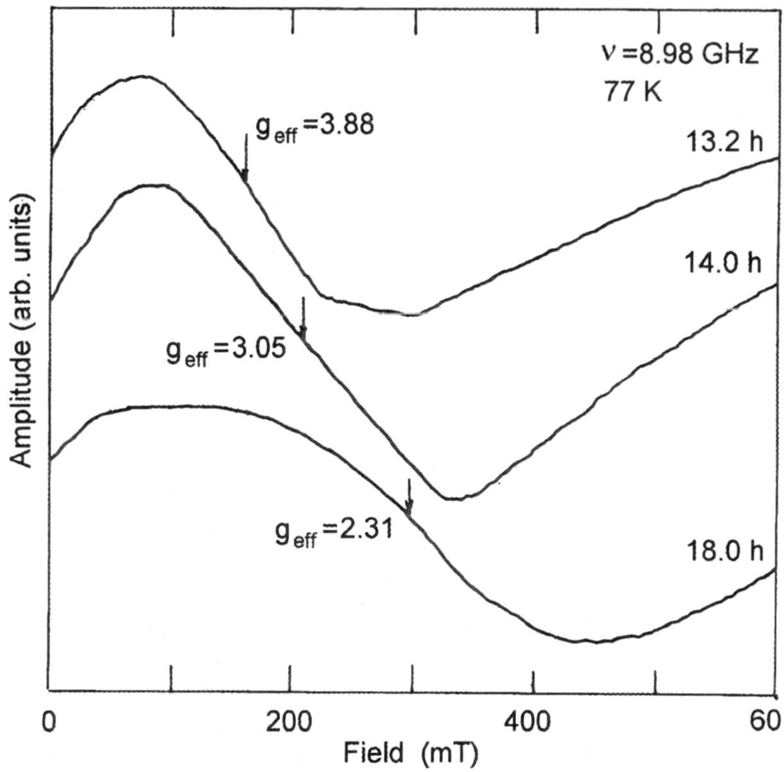

Fig. 3. Spectra of the $Na_2[Fe(CN)_5NO]\times H_2O$ single crystal for various times of the laser light excitation.

to a deformation of the $[Fe(CN)_5NO]$-complex after light excitation. The formation of long-living metastable state is a two-step process. Spin excitations observed as Fe(III) EPR signals appear after about 12 hours of the continuous laser pumping. At this stage the nitroprusside complex still maintains the original geometry, then within the next six hours the crystal slowly tends to an equilibrium state with continuous deformation of the complex geometry. A lowering of the $[Fe(CN)_5NO]$ ion symmetry in the metastable state is confirmed by X - ray diffraction study by Rüdlinger et al. (1991)[2]. When the crystal reaches the equilibrium state, the metastable excited states can exist without light irradiation.

In addition to the EPR signal from unpaired electrons localized on Fe(III) one can expect the signal from NO^- molecule. However, an EPR signal from NO was not observed. One of the possible explanations is that the excited states exist in dynamical equilibrium and an unpaired electron, which excites the (π^*NO) state, can be dynamically nonlocalized. The mobility of this electron can result in a smearing out of the NO^- group EPR signal. Moreover, the situation can be similar to that known for the charge transfer TCNQ and TTF salts where the donor - acceptor electron transfer produces an EPR signal from unpaired electrons localized on donors or acceptors, depending on details of their electronic structure, but hardly ever from both unpaired electrons as discussed by Mabbs et al. (1992)[20].

EPR measurements performed on the sodium nitroprusside single crystal irradiated by laser light at 77K show clearly that the crystal is paramagnetic in the metastable excited state. In the light of thise results it will be possible to explain the disappearing of the ^{14}N NMR lines after irradiation of the sodium nitroprusside crystals at 77K.

This work was supported by National Scientific Reseaech Committee under Project KBN -2-P03B-002-08.

References

1. U. Hauser, V. Oestreich and H. D. Rohrweck, On Optical Dispersion in Transparent Molecular Systems, *Z. Physik A* **280** (1977) 17-25.
2. M. Rüdlinger, J. Schefer, G. Chevrier, N. Furer, H. U. Güdel, S. Haussühl, G. Heger, P. Schweiss, T. Vogt, Th. Woike, and H. Zöllner, Light-induced structural changes in sodiumnitroprusside ($Na_2(Fe(CN)_5NO)\times D_2O$) at 80K, *Z. Phys. B.-Condensed Matter* **83** (1991) 125-130.
3. J. Tritt-Goc, N. Piślewski, NMR Chemical Shift and Asymmetric Dipolar Tensors of Water Protons in Sodium Nitroprusside (SNP), *Chem. Phys.* **102** (1986) 133-140.
4. J. Tritt-Goc, N. Piślewski, Proton NMR Relaxation Study of the Motion of Water Molecules in Hydrated Nitroprussides, *J. Phys. Chem. Solids* **54** (1993) 123-126.
5. J. Tritt-Goc, N. Piślewski, Dynamics of Water Molecules in Barium Nitroprusside Trihydrate Studied at Low Temperature by Proton NMR, *Mol. Phys.* **83** (1994) 949-960.
6. J. Tritt-Goc, N. Piślewski, Proton Dipolar Coupling Tensors in Barium Nitroprusside Trihydrate, *J. Phys. Chem. Solids* **56** (1995) 935-942.
7. J. Tritt-Goc, Motion of the Water Molecules and Phase Transition in $Sr[Fe(CN)_5NO]\times 4H_2O$ Studied by Proton NMR, **86** (1995) 193-200.
8. J. Tritt-Goc, N. Piślewski, The Influence of the Motion of Water Molecules on Proton Dipolar Coupling Tensors in $Sr[Fe(CN)_5NO]\times 4H_2O$, *Mol. Phys.* **87** (1996) 139-149.
9. A. Navaza, G. Chevrier, P. Schweiss, G. Rigotti, Structure of Barium Nitroprusside Trihydrate below the 233-K Phase Transition, Using Neutron Diffraction Data, *J. Solid State Chem.* 99 (1990) 158-165.
10. Th. Woike, W. Kirchner, Hyung-sang Kim, S. Haussühl, V. Rusanov, V. Angelov, Ts., Bonchev, and A. N. F. Schroeder, Mössbauer parameters of two long-lived metastable states in single crystal $Na_2[Fe(CN)_5NO]\times 2H_2O$, *Hyperfine Inter.* **77** (1993) 265-275.
11. Th. Woike, W. Krasser, P. S. Bechthold, and S. Haussühl, Investigation of the Metastable State in $Na_2[Fe(CN)_5NO]\times 2H_2O$ by Optical Spectroscopy. I. Comparison of the Raman Spectra of the Ground and Metastable State, *Solid State Commun.* **45** (1983) 499-502.
12. Th. Woike, W. Krasser, P. S. Bechthold, and S. Haussühl, Investigation of the Metastable State in $Na_2[Fe(CN)_5NO]\times 2H_2O$ by Optical Spectroscopy. II The Electronic Spectra of the Ground and Metastable States, *Solid State Commun.* **45** (1983) 503-506.
13 Th. Woike, W. Krasser, P. S. Bechthold and S. Haussühl, Extremly Long-Living Metastable State of $Na_2[Fe(CN)_5NO]\times 2H_2O$ Single Crystals: Optical Properties, *Phys. Rev. Lett.* **53** (1984) 1767-1770.

14. Th. Woike, H. Zöllner, W. Krasser, and S. Haussühl, Raman-Spesctroscopic and Differential Scanning Calorimetric Studies of the Light Induced Metastable States in $K_2[RuCl_5NO]$, *Solid State Commun.* **73** (1990) 149-152.
15. D. Gross, N. Piślewski, U. Haeberlen, and K. Hausser, The ^{14}N and ^{2}H Quadrupole Coupling Tensors in Sodium Nitroprusside, *J. Magn. Reson.* **54** (1983) 236-253.
16. G. Kruschel, Dissertation, Darmstadt (1984).
17. H. Zöllner , Th. Woike, W. Krasser, and S. Haussühl, Thermal Decay of Laser-Induced Long Living Metastable Electronioc States in $Na_2[Fe(CN)_5NO]\times 2H_2O$ Single Crystal, *Z. Kristallographie* **188** (1989) 139-153.
18. Th. Woike, W. Krasser, H. Zöllner, W. Kirchner and S. Haussühl, Population dynamics of two light induced states in $Na_2[Fe(CN)_5NO]\times 2H_2O$ single crystal *Z. Phys. D* **25** (1993) 351-356.
19. P. T. Manoharan, and H. G. Gray, Electronic Structure of Nitroprusside Ion, *J. Am. Chem. Soc.* **87** (1965) 3340-3348.
20. F. E. Mabbs, and D. Collison, *Electron Paramagnetic Resonance of d Transition Metal Compounds*, Elsevier, Amsterdam, (1992).
21. C. Ezzeh, and B. R. McGarvey, ESR and NMR of $Fe(CN)_5NH_3^{2-}$, *J. Magn. Reson.* **15** (1974) 183-190.

EPR AND LOW FIELD MICROWAVE ABSORPTION IN THE SUPERCONDUCTOR $Hg_{0.9}Tl_{0.2}Ba_2Ca_2Cu_3O_{8+\delta}$

B H HOU[1,2], *Q L XU*[1], *R Y LIANG*[3], *T P ZHANG*[1], *Y X XU*[2], *S X HAN*[2]
S H LIOU[4],

1 Department of Physics, University of Science and Technology of China, Hefei 230026, P.R..China
2 Laboratory of Solid State Microstructure, Nanjing University, Nanjing 210093, P.R..China
3 Laboratory of Structure and Analysis,, University of Science and Technology of China, Hefei 230026, P.R..China
4 Behlen Laboratories of Physics and Center for Material Research and Analysis, University of Nebraska, Lincoln, Nebraska, 68588-0111, USA

Abstract EPR signals in three superconductors $Hg_{0.9}Tl_{0.2}Ba_2Ca_2Cu_3O_{8+\delta}$ represent essentially Cu^{2+} ions from the nonsuperconducting phase. The sample annealed at 300°C for 6h has effective Landé g-factors of ($g_{//}$) 2.217 and ($g_{\perp}$) 2.100; that annealed at 300°C for 12h has that of ($g_{//}$) 2.248 and ($g_{\perp}$)2.048; and an as-synthesized sample has an effective g of about 2.113. Non -resonant low-field microwave absorption (LFA) signals in three samples showed different changes with decreasing temperature, Three *LFA~T* curves indicate that an excess of oxygen "δ" influences the critical temperature and superconducting volume fraction in a very crucial manner.

1 Introduction

An electron paramagnetic resonance (EPR) spectrometer is used to study superconductors in two major respects: electronic configuration and non-resonant low field microwave absorption reduction (LFA). The EPR signal of Cu^{2+} in majority high T_c superconductors (YBaCuO, BiSrCaCuO etc.) is very weak or can not be measured. The weak EPR signal observed is due to some paramagnetic impurity phase in the host, such as *Morimoto et al.*[1] reported, that the EPR spin density is due to a green Y_2BaCuO_5 impurity phase in the host $YBa_2Cu_3O_x$

A non-resonant LFA microwave signal can be detected below the superconducting phase transition temperature Tc with the EPR spectrometer, This is one of the criteria of the superconducting state of a sample[2]. The essence of the LFA signal of a superconducting sample is that, when the sample is in normal state(T>Tc), it absorbs some of the energy from the electromagnetic field due to the skin effect; this loss absorption, which is indicated by the imaginary component of the susceptibility χ'', is independent of the applied magnetic field. Therefore, no signal can be detected with the EPR spectrometer when the sample is in the normal state, because the signal recorded by the EPR spectrometer is a derivative, $d\chi''/dH$. However, in a super-conducting state, due to the Meissner effect, the loss absorption decreases obviously when the applied field H approaches zero, and the absorption increases with increasing H, so a LFA signal is detected near H= 0. It is important to have the idea that the characteristic microwave signal can not be observed when the loss absorption is unchanged with the magnetic field in the normal state of the material, and that signal emerges due to a reduction of the loss absorption in the superconducting state. For many ceramic samples of high Tc superconductors, the LFA signal increases monotonously with temperature. *Portis et al*[3] found experimentally that there was convincing evidence of a linear relationship between the signal intensity and the sample surface area, and the sample was cooled down to liquid helium temperature in zero magnetic field, the signal is maximum, i.e., the absorption is minimum.

In the $HgBa_2Ca_{n-1}Cu_nO_{2n+2}$ series, the $HgBa_2Ca_2Cu_3O_{8+\delta}$ (Hg-1223) oxide superconductor with three CuO_2 layers has become a hot topic, not only because of their high critical temperature T_c that can reach 133K under normal pressure and 157K and 164K under high pressures of about 23.5 and 30GPa respectively, but also due to the fact that the excess oxygen "δ" influences the critical temperature and the superconducting volume fraction in a very crucial manner[4-6]. The highest T_c of Hg-1223 phase caused by the stronger coupling between CuO_2 blocks, and its high critical current density comparable with that of Y-123 and Tl-1223 phase, make it a promising candidate for many important technological applications of microwave and microelectronic devices if the stability of the compound is realized.

It is known that Hg and Tl have a very close electronegtivity and both of Hg^{2+} and Tl^{3+} ions have identical electronic configurations, thus have similar chemical properties. The Hg-1223 compounds are isomorphous with the Tl-1223 phase structure. Therefore Tl is a favorable candidate for stabilizing the Hg-1223 phase structure. Several authors [7-10] have shown that the partial substitution of mercury by thallium introduced extra oxygen atoms into HgO planes and increased the superconducting volume fraction, due to the difference in valence of Hg^{2+} and Tl^{3+} cations (Tl^{3+} has a higher oxygen coordination number). We found that the onset temperature of Hg(Tl)-1223 phase decomposition was higher than that of undoped Hg-system superconductors and the vibration of the apical oxygen shifted to higher frequency which merged with the vibration frequency of interstitial oxygen coordinated with Tl^{3+}. The addition of thallium enhanced the bonding between atoms and increased the structure stability of Hg-1223 phase. All this and the synthesis of a nearly single Hg (Tl)-1223 phase and the study of the thermal stability of the Hg (Tl)-1223 phase are reported in another paper. In this paper, we report on the results of studies of EPR and LFA.

2 Experiments

Superconductor samples were synthesized by a solid state reaction method with a nominal composition $Hg_{0.9}Tl_{0.2}Ba_2Ca_2Cu_3O_{8+\delta}$. After sintering, sample *A* was further annealed at 300°C for 6h in 1atm O_2; sample *B* was further annealed at 300°C for 12h in 1atm O_2; sample *C* was not annealed any more, was as-synthesized.

X-ray powder diffraction (XRD) patterns were measured by using a Rigaku X-ray diffractometer to evaluate the fraction of the Hg(Tl)-1223 and impurity phases. Resistance was measured with a standard four-probe technique and DC magnetization measurements were carried out by using a SQUID magnetometer.

The EPR spectrometer with a temperature controller was used to measure the EPR and LFA signals of the three samples at a frequency of 9.4942GHz, and a method of magnetic field scanning through zero was adopted, within the magnetic field range from -2.2 to 50 millitesla (mT) for the measurement of the LFA signal[11]. The mass of each samples is 5mg.

3 Results and Discussion

3.1 Characterization of Samples

The superconductor samples have the same nominal composition $Hg_{0.9}Tl_{0.2}Ba_2Ca_2$ $Cu_3O_{8+\delta}$. After sintering, there are two superconductive phases, 1223 and 1212, in sample *A*; sample *B* has the pure 1223 phase; oxygen vacancies in sample *C* are more aboundant than in the two other samples. The difference in the preparation of the three samples is only the time of annealing in O_2 atmosphere. It results in a difference of oxygen "δ" in a very crucial manner, and influences significantly the critical temperature, superconducting volume fraction and the other characterizations, such as EPR and LFA.

The XRD peaks in the three samples could be indexed to the Hg-1223 structure which has a tetragonal unit cell (P4/mmm). Only a trace (~5%) of the $CaHgO_2$ impurity phase can be detected from the X-ray diffraction patterns in samples *A* and *B*; and there is some Hg-1212 (~10%) structure[12] in sample *A*. Like other undoped Hg superconductors, transmission electron micrographs of the Hg(Tl)-1223 phase sample display a thick platelet-like crystal. From the very round diffraction pattern dots, we can confirm that there is a relatively complete crystallization and less imperfection or intergrowth existence of other Hg superconducting phases such as Hg-1212 or Hg-1234 in the Hg(Tl)-1223 phase crystal.

The resistance of sample *A* begins to drop at temperature T_c(onset)=140K and goes to zero at T_c(zero)=134K. That of sample *B* is T_c(onset)=132K and T_c(zero)= 130K. That of sample *C* is T_c(onset)=125K and T_c(zero)=122K.

The superconducting properties of the samples are also characterized by DC magnetization measurements using a SQUID magnetometer. The superconducting transition temperatures of samples *A*, *B* and *C* are 133K, 132K and 122K, respectively.

Comparing the characterization of the three samples, the superconducting transition temperature T_c of sample *A* is highest, that of sample *C* is lowest. After sintering, a further annealing of sample *A* results in an increase of impurity phases $CaHgO_2$ and Hg-1212, and an increase of the T_c . The longer annealing time of sample *B* can decrease Hg-1212, but T_c drops slightly.

3.2 EPR Signal of the Cu^{2+}

Figure 1 shows the EPR spectra of the samples at room temperature. The EPR signals represent essentially Cu^{2+} ions from nonsuperconducting phases which are too small to be detected by XRD. EPR has been proved to be a very sensitive, reliable technique for the study of trace paramagnetic phase of less than 10^{-4}. For comparison of actual EPR signal intensity, the signal should be divided by the gain factor indicated in each spectrum in Fig. 1; the stronger the EPR signal is, the smaller the gain is adopted, and the EPR signal intensity indicates the relative amount of impurity in samples with the same mass, so the non-superconducting phase is larger in sample *B*, and smaller in sample C.

It is well known that there is no EPR signal in the superconduc-tive phase and in diamagnetic CuO. The EPR spectra of the three samples have differences not only in intensity but also in peak shape, which arise from different sym-metries of impurity Ba_2CuO_3. The EPR of sample *C* represents a character of cubic symmetry, it arises from cubic phase $BaCuO_2$[12].

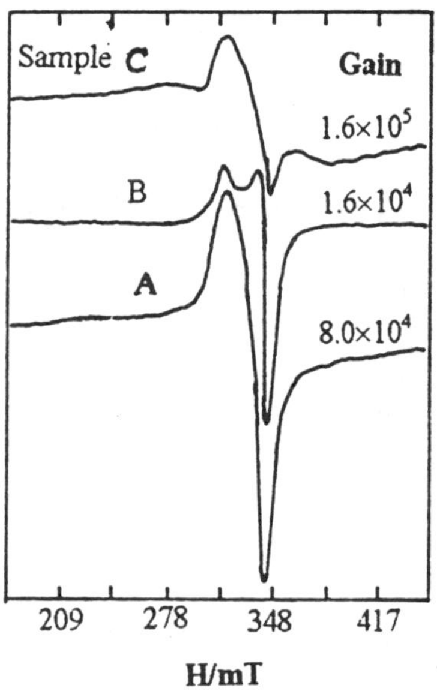

Fig. 1 Cu^{2+} EPR spectra of samples at room temperature

The EPR spectrum of sample *A* is similar to that of a Y_2BaCuO_5 green phase in superconductive YBa_2Cu_3Oy[1], and here it arises probably from a tetragonal distorted $BaCuO_2$ which reflects a distortion of host Hg(Tl)-1223 phase. The EPR of sample *B* arises from distorted $BaCuO_2$ or/and Ca_2CuO_3 whose lattice distortion is more obvious than that of sample *A*. In addition, the EPR of sample *A* has effective Lande g-values of ($g_{//}$) 2.217 and ($g_{\perp}$) 2.100; that of sample *B* has that of ($g_{//}$) 2.248 and ($g_{\perp}$)2.048; that of sample *C* has a *g* of 2.113.

3.3 LFA Signal of the Samples

Figure 2 shows the temperature (*T*) dependence of the non-resonant LFA intensity of the three samples. Generally, the temperature dependence of the LFA signal is similar to that of the diamagnetic susceptibility of a superconducting sample. The T_c got from Fig.2 are 134K (for sample *A*), 132K (for *B*) and 123K (for *C*) respectively. Those are almost identical with the measurements of the R~T curves. The *LFA~T* curve can represent a sensitive feature of a sample below T_c, especially the superconducting volume fraction[13]. The LFA of samples *B* and *C* increases monotonously with decreasing temperature, like that in major bulk samples of high-Tc superconductors; however, the signal of sample *A* shows two peaks at 130K and 125K, and tends to stabilize when the temperature is further reduced; it indicates that there are two superconductive phases with different *Tc* values in sample *A*; in addition, another proof is that an overlap LFA signal has been observed in sample *A* at 127K as shown in Figure 3; a similar overlap LFA signal was observed also in BiPbSrCaCuO:Cr [14].

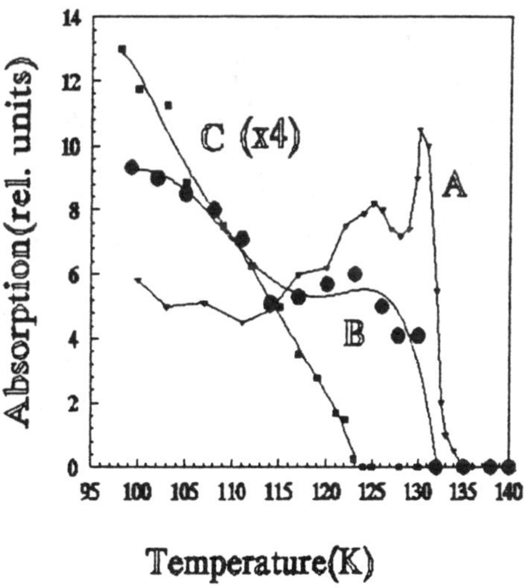

Fig. 2 Temperature dependence of LFA signal intensity of samples.

The phenomenon that the LFA signal shows a peak close to T_c is also observed in BiPbSrCaCuO[15]. It reflects

Maybe it indicates that some of the Hg(Tl)-1223(134K) phase grains are unstable and lose their superconductivity in a very weak magnetic field of about 1mT. Comparing the LFA intensities of the three samples, sample *C* has the largest superconducting volume fraction below 121K.

4 Conclusion

The results of EPR and LFA together with those of XRD and R~T experiments, indicate that various properties in the three superconductors $Hg_{0.9}Tl_{0.2}Ba_2Ca_2Cu_3O_y$ as a function of different annealing times:

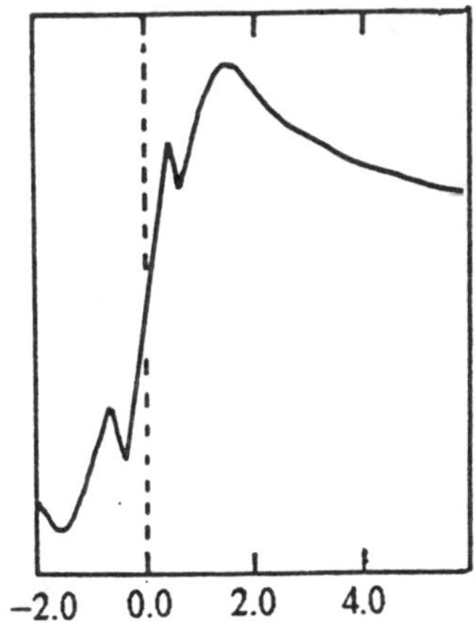

Fig. 3 Overlap LFA signal in sample A at 127K

Sample *A* annealed at 300°C for 6h has an impurity phase $BaCuO_2$ with a little distortion, its g-factors are ($g_{//}$) 2.217 and ($g_{\perp}$) 2.100; and it has two superconductive phases, Hg(Tl)-1223 with T_c of 134K and Hg(Tl)-1212 with T_c 127K. Its LFA signal shows two peaks at 130K and 125K, and an overlap LFA signal appears at 127K.

Sample *B* annealed 300°C for 12h has an impurity phase $BaCuO_2$ with obvious tetragonal distortion, its g-factors are ($g_{//}$) 2.248 and ($g_{\perp}$)2.048. It is almost a single phase Hg(Tl)-1223 with T_c of 132K.

Sample *C* is in an as-synthesized state, there is only little cubic impurity $BaCuO_2$ in it, its effective g is 2.113. It is almost a single phase and a stable Hg(Tl)-1223 phase with T_c of 122K. It has the largest superconducting volume fraction below 121K.

References

1. A. Morimoto, T. Maeda, A. Morto, M. Kumeda and T. Shimizu, EPR and X-ray diffraction Studies on Ba-Y-Cu-O superconductors, *Jpn. J. Appl.Phys.*, **27**, 3 (1988) 56-59.

2. K. W. Blazey, K. A. Mueller, J. G. Bednorz, W. Berliger, G. Amoretti, E. Buluggiu, A.vera and F.C.Matacotta, Low-field microwave absorption in the superconducting copper oxides, *Phys.Rev.B,* **36**, 13 (1987) 7241-7243.
3. A. M. Protis, K. W. Blazey, K. A. Mueller, J. G. Bednorz, Microwave magneto-surface impedance of high-T_c, *Europhysics Letters,***5**,5 (1988) 467-472.
4. S. N. Putilin, E. V. Antipov, O. Chmaissem and M. Marezio, Superconductivity at 94K in $HgBa_2Ca_2Cu_3O_{8+\delta}$, *Nature,* **362** (1993) 226-228.
5. A. Schilling, M. Cantoni, J. D. Guo and H. R. Ott, Superconductivity above 130K in the Hg-Ba-Ca-Cu-O system, *Nature* **363** (1993) 56-58.
6. C. W. Chu, L. Gao, F. Chen, Z. J. Huang, R. L. Meng and Y. Y. Xue, Superconductivity above 150K in $HgBa_2Ca_2Cu_3O_{8+\delta}$ at high pressure, *Nature,* **365** (1993) 323-325.
7. F. Goutenoire, A. Maignam, G. V. Tendeloo, C.M artin, C. Michel, M. Hervieu and B. Raveau, Substitution of Mersury for Thallium in the 2223 cuprate:the 130K superconductor $Tl_{1.6}Hg_{0.4}Ba_2Ca_2Cu_3O_{10-\delta}$, *Solid State Commun.* **90** (1994) 47-50.
8. G. F. Sun, K. W. Wong, B. R. Xu, Y. Xin and D. F. Lu, Tc enhancement of $HgBa_2Ca_2Cu_3O_{8+\delta}$ by Tl substitution, *Phys. Lett. A* **192** (1994) 122-125.
9. P. Dai, B. C. Chakoumakos, G. F. Sun, K. W. Wong, X. Xin and D. F. Lu, Synthesis and nutron powder diffraction study of the superconductor $HgBa_2Ca_2Cu_3O_{8+\delta}$ by Tl substitution, *Physica C,* **246** (1995)201-205.
10. J. Z. Liu, I. C. Chang, M. D. Lan, D. Klavins and R. V. Sheltom, Superconductivity above 130K in $Tl_{1-x}Hg_xBa_2Ca_2Cu_3O_{8+\delta}$, *Physica C,* **246** (1995) 203-206.
11. B. H.Hou, Y. X. Xu, S. Y. Han, Z. M. He, and Z. Z. Wang, Magnetic field scanning through zero of EPR spectrometer, *Chinese Journal of Magnetic Resonance,* **9**, 2 (1992) 201-204.
12. N. H. Hur, N. H. Kim, K. W. Lee, Y. K. Park and J. C. park, Synthesis and characterization of a new mercury-based superconductor $Hg_{0.5}Ba_{0.5}Ba_2Ca_{1-x}Sr_xCu_2O_{6+\delta}$, *Physica C,* **231** (1994) 4-8.
13. B. H.Hou and Z. Z. Wang, Superconducting state volume in high T_c superconductor and non-resonant microwave absorption, *Chinese phys.lett.,* **9**, 3 (1992) 155-158.
14. B. H.Hou, W. L. Cai, S. Z.Jin and Y. H. Zhang, Observation of zero field microwave absorption in superconductor BiPbSrCaCuO with trace Ti. V, Cr,Mn or Mo, *superconductor science and technology,* **3**, (1990) 207-211.
15. B. H.Hou, M. S. Wang, Z.J. Chen, Q. L. Xu, and T. Z. Jin, Low field microwave signal in undoped and Sb_doped Bi-Pb-Sr-Ca-Cu-O, *Acta physica sinica,* **3**, 9 (1994) 708-711.

A CHANGE OF RESONANT MAGNETIC FIELD OF V_k CENTER IN BaF_2

Bihui HOU

1. Department of Physics, University of Science and Technology of China, 230026 Hefei, P.R.China
2. Laboratory of Solid State Microstructure, Nanjing University, 10093 Nanjing, P.R.China

Abstract. The paramagnetic point defects V_k center and nonmagnetic Frenkel exciton were induced by the γ-ray irradiation in BaF_2 at room temperature. It is notable that annihilation of the Frenkel exciton emerges into V_k center when the temperature goes up, and the g-factor of the V_k center changes with temperature. This change relates to a process of transient state in which the orbital angular momentum of an electron in F^0 atom is variable. The g-factor of the V_k center of a steady state is 2.205 ± 0.003.

1 Introduction

Electron spin resonance (ESR) has been proved to be a very sensitive, reliable and important technique for the studies of point defects in solid materials. Barium halide possesses some excellent luminescence properties on which much attention has been paid. Doped with rare earth, they are being studied widely as the γ-ray detection and X-ray image storage materials. The BaF_2 is a useful detection material for high-energy particles, its irradiation effect is of considerable interest both from the research point of view and that of applications. In many years W. Hayes and P. J. Call et al. have devoted themselves to the studies of point defects in alkaline earth fluoride crystals and of defects produced by irradiation[1,2]. We have studied the point defects of color centers in BaF_2 crystallite powder produced by the γ-ray irradiation[3,4]. The point defects in BaF_2 measured with ESR are mainly: F center (an electron trapped by F^- vacancy), V_k center (a F^0 atom at lattice site), H center (a F^0 atom at interstitial site) and M center (adjacency of two F centers). The V_k and H centers are of hole-type defects, whereas the F and M centers are of electron-type defects. The g value of the hole-type defect is larger than 2.0023 that is one of a free electron, that of electron-type defect is smaller than 2.0023. In addition, nonmagnetic point defect, the Frenkel exciton, can also be induced by the γ-ray irradiation at room temperature, that is discovered from the ESR signal of the V_k center enhanced with increasing temperature, because a Frenkel exciton annihilates and changes into a V_k center. The ESR signal of the V_k center enhances when the temperature goes up to 127°C, then weakens and vanishes at about 350°C, and the g-value of the ESR signal of the V_k center also changes with the temperature. Comparing ESR spectra in $BaCl_2$[5] and BaF_2, we found existence of the Frenkel exciton[3,4]. The change of the g-factor of the V_k center is expounded in this paper.

2 Experiments

Three samples are studied. The sample #1 is BaF_2 crystallite powder, the sample #2 is abrasive powder of the BaF_2 single crystal, and the sample #3 is BaF_2 single crystal. The Co^{60} radiation source was used for γ-ray radiation at room temperature, the radiation intensity was 103.3 Gy/min, the irradiated times were 30 min. for sample #1, 5 min. for sample #2, and 30 min. for sample #3, respectively. The ESR experiments were carried out at room temperature. The microwave frequency was 9.4588 GHz. The ESR spectrum of each sample is measured before and after γ-ray irradiation, and is measured after heat treatment, the heated temperatures are from 18°C to 420°C. The values of the g-factor of the ESR signals are determined with a standard sample DPPH (g=2.0037± 0.0003).

The intensity of the ESR signals is found to decay as a function of time (natural recovery) and temperature (thermal recovery). The natural recovery time of the V_k center in sample #1 is shorter than 24 hours.

3 Results and Analysis

Before γ-ray irradiation, no ESR signal can be observed in three samples; so the ESR signals observed after γ-ray irradiation in the sample #1 and #2 stem from irradiation-induced paramagnetic point defects without doubt. The ESR spectra of the sample #1 at 18°C and heated at 97°C are shown in Fig. 1. There are four peaks in the sample #1 with ΔH_{pp} of 2-4mT; there is a wide peak in the sample #2 with ΔH_{pp} of 43.5 mT for γ-ray irradiation. There is still no ESR signal in the sample #3 for 30 min. γ-ray irradiation. The reason might be that the point defects locate often at the surface of grains, and the sample #3 is a perfect crystal, consequently, the surface area per unit mass is much less than that of the sample #1 and #2, and the irradiated defects annihilate below room temperature[1,2]. That is also to say that there are original nonmagnetic point defects in the samples of the BaF_2 powder

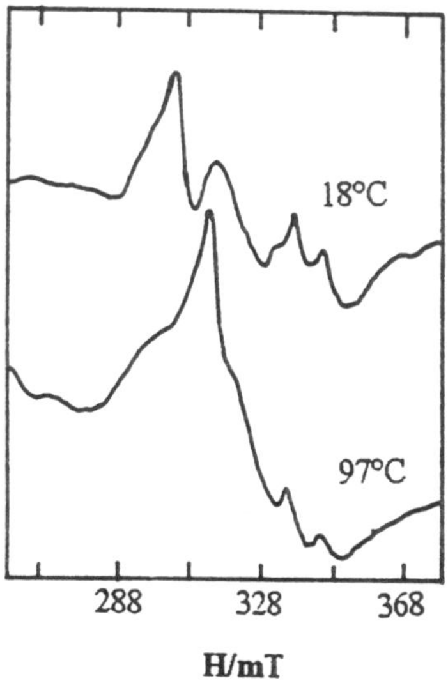

Fig. 1. The ESR spectra of BaF_2 crystallite powder (sample #1) after γ-ray irradiation at 18°C and heated to 97°C.

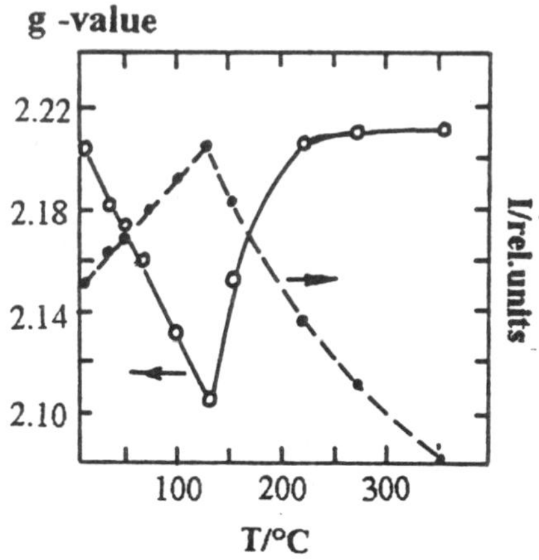

Fig. 2. The dependence of g-factor and ESR intensities of the V_k center in BaF_2 (sample #1) on the heated temperature.

and abrasive powder of the BaF_2 single crystal, such as interstitial and vacant F^- which are located mainly at the surface and boundary plane of crystallite grains. These nonmagnetic point defects can become paramagnetic centers by γ-irradiation at room temperature. There are non-magnetic point defects in the sample #1 and #2, this is a basic condition for appearance of paramagnetic point defects excited by the γ-ray irradiation.

There is not any fine and hyperfine peak in the ESR spectrum, so the g-factor is calculated as

$$g = \frac{h\nu}{\beta H_r},$$

where h is the Planck constant, ν is the microwave frequency , β is the Bohr magneton , and H_r is the resonant magnetic field. And these g values of ESR spectra are compared with g_e (2.0023) of free electron, and with ESR spectra in irradiated $BaCl_2$ crystallite[5], then the four ESR peaks in the sample #1 are affirmed[3], they are V_k (g=2.2022), H (2.0834), F (1.9884) and M [g = (1.9884+1.9426)/2 = 1.9655). The intensities of ESR signals of the H, K and M are monotonously weakened with increasing temperature, and that of H vanishes at 67°C, that of the K and M at 150°C. Because the crystalline grain tends to be perfect, and the defects annihilate with temperature and time[3]. It is found that the ESR signal of V_k center does not decline monotonously with the increasing temperature as of the H, F and M centers do, its intensity I enhances when the temperature goes up to 127°C (it shows the number of the V_k center increases), then weakens, and the V_k center vanishes not higher than 350°C as shown in the Fig.2. It is notable in Fig.1 that the resonant magnetic fields of the V_k center changes when the temperature goes up, so the g-factor of the V_k center varies as shown in Fig.2.

The ESR signal is a wide peak, with ΔH_{pp} about of 43.5mT and g of 2.2212 in the sample #2. The change of intensity of ESR signal with temperature is also not monotonous, the intensity of the signal enhances when the temperature goes up to 100°C, it almost does not change at a range of 100-200°C, then weakens, and vanishes at 420°C. The signal is also of V_k center, the sample #2 is abrasive powder of the BaF_2 single crystal, the anion vacancy is at the surface of the grains, and there is no interstitial anion, so no H, F and M centers. Detail of the experiment and analysis are described in Ref.[3]. The ESR signal of V_k center in $BaCl_2$ is monotonously weakened with increasing temperature, and vanishes at 71°C[5]. The resonant magnetic field of the ESR signal of the sample #2 does not change when the heated temperature goes up, possibly due to the fact that the resonant peak is too wide to distinguish the change of resonant field.

The phenomenon that the ESR signal of the V_k center enhances with increasing temperature in the sample #1 and #2, is related to the Frenkel exciton excited by the γ-ray radiation in BaF_2. We had also found this exciton in BaF_2 with experimentation of absorption spectrum[6], the gap energy is 10.9eV, and the exciton binding energy is 1.1eV. A Frenkel exciton is a tightly bound electron-hole pair, which is called an elementary excitation or an excited state of a single atom. It is induced when an electron of 2p of F^- ion is excited to a higher level in the outer shell orbit, and the electron is coupled with the hole of the 2p orbit. When the electron in the Frenkel exciton gets energy from heating, and leaves the F^- ion, the F^- becomes F^0 (a V_k center), that is to say, the ESR signal of the V_k centers enhances as the temperature increasing, and it shows annihilation of the Frenkel exciton and creation of the V_k center. Owing to the existence of the Frenkel excitons excited by the γ-ray-irradiation, the exciton is nonmagnetic, and it can not be detected with ESR, but can become a V_k center after heated. It is the reason why the ESR signal of the V_k center does not decline monotonously with the increasing temperature as those of the H, F and M centers do. The ESR signal of the V_k center enhances as the temperature increasing, it is a witness to annihilation of the Frenkel exciton. The differentiation between $BaCl_2$ and BaF_2 is due to a disparity in electron affinity, the electron affinity of F atom is 3.9, that of Cl atom is 3.1 and that of Ba atom is 0.9. The electron affinity of Cl atom is smaller than that of F atom, so Frenkel exciton, tightly bound electron-hole pair, can not form in $BaCl_2$. It is found that a temperature range of linear change of the g- factor of the V_k center is from 18°C to 127°C in Fig. 2, and in this range the change of the g- factor of the V_k center relates to a process of transient state of annihilation of the Frenkel exciton. The g-factor of the V_k center is 2.2022 before heated (at 18°C), since then g-factor decreases linear to 2.1058 (heated to 127°C), but is still larger than g_e(2.0023), a thermal recovery of the V_k center is in a temperature range from 127°C to 350°C, and the ESR signal of the V_k center vanishes at 350°C. In this range the g-factor increases, it is 2.2058 heated to 220°C, and is 2.2087 heated to 270°C. As compared with that at 18°C, and in consideration of an experimental error, the g-factor of the V_k center at steady state is about 2.205±0.003.

The change of the g-factor of the V_k center with the temperature is a process of transient state, and this change is related to annihilation of the Frenkel exciton. An electron excited to an outer shell orbit by the γ-ray irradiation is located in a higher level than that of 2p of F^- ion , and the electron is coupled with the hole of the 2p orbit, which is an unsteady state. When the electron in the Frenkel exciton gets energy from heating, it leaves the F^- ion, so the F^- becomes F^0 (a V_k center). In the change the effective electron magnetic moment of the V_k center is unsteady; however when all Frenkel excitons are annihilated, the effective electron magnetic moment of the V_k center becomes steady.

4 Conclusion

Nonmagnetic Frenkel exciton can be induced by the γ-ray irradiation at room temperature. It is noticeable that annihilation of the Frenkel exciton becomes the V_k center, so one should pay much attention to the irradiation character of BaF_2 crystallite powder in its application.

Owing to the existence of the Frenkel excitons excited by γ-ray irradiation, the ESR signal intensity of the V_k center enhances as the temperature increasing, and the g-factor of the V_k center also changes with temperature, both anomalous phenomena are related to annihilation of the Frenkel excitons.

References

1. W. Hayes and R.F. Lambourn, Radiolysis of strontium fluoride, *J.Phys. C:Solid state Phys.*, **6** (1973) 11-26.
2. P.J. Call, W. Hayes and M.N. Kabler, Optical Detection of EPR in Fluorite Crystals, *J.Phys. C:Solid state Phys.*, **8** (1975) L60-62.
3. B.H. Hou, B.X.Yang, X.P.Lu and C.S.Shi, The Υ-ray Radiation Damage in BaF_2, *Acta Physica Sinica*, **44**, 8 (1995) 1296-1301.
4. B.H. Hou and H.B.Yang, Gamma ray irradiation induced point defects in BaF_2, *Nuclear Science and Techniques*, **7**, 2(1996) 105-109.
5. B. H. Hou, Z. Z. Wang, X. P. Lu, B. X. Yang, Y. H .Xu and C. S. Shi, ESR Spectum Analysis of Color Centers in $BaCl_2$ Crystallite, *Bulletin of Analysis & Testing*, **11**, 6 (1992) 14-16.
6. C.S.Shi, T.Kloiber and G.Zimmerer, Special Luminescence Properties of BaF_2 Crystals, *Journal of Luminescence*, **40 & 41** (1988) 189-190.

EPR Study on $BiSrCaCu_2O_y$ High *Tc* Superconductor

Sizhao JIN, Songliu YUAN and Weiyan GUAN

Structure Research Laboratory, University of Science and Technology of China,
Hefei, Anhui 230026, China

Abstract The EPR of the nominal $BiSrCaCu_2O_y$ high *Tc* superconductor at different temperatures is investigated. It is found that there are two EPR signals with different behaviors in this sample. One of them shifts gradually toward higher magnetic field and its Lande factor *g*-value is decreased from 2.78 to 2.24 when the temperature is raised from 115K to 310K. The resonance magnetic field of another signal is almost independent of temperature and its *g*-value is close to one of free electron. There is only zero field signal below *Tc* for this sample and its intensity is increased with decreasing temperature. We propose a possible interpretation for these phenomena based on electron transfer between copper ion and oxygen ion in Cu-O plane.

1 Introduction

The discovery of high *Tc* superconductivity in Bi-Sr-Cu-O system was the first step towards the high *Tc* materials without the need of rare earth elements and superconducting transition temperatures were raised from ~20K to over 100K by the addition of Ca in this system. It is now of general agreement that one important feature of the layered structure is the low-dimensional arrangement of Cu-O bonds. It is therefore very interesting to investigate the properties of copper ions or oxygen ions by various experimental methods. In this paper, we report the results of EPR of the nominal $BiSrCaCu_2O_y$ high *Tc* superconductor at different temperatures.

2 Experiments and Results

The nominal $BiSrCaCu_2O_y$ sample was prepared by solid state reaction. The starting materials Bi_2O_3(>99%), $Sr(NO_3)_2$(>99%), $CaCO_3$(>99%) and CuO(>99%) were completely mixed and ground, then prefired at 860^{0}C for 24h in air. After this, the mixture was reground and pressed into a pellet, 5mm thick and 15mm in diameter. The pellet was sintered in air at 850^{0}C for more than 18h and then brought by furnace cooling to room temperature.

The measurements of resistivity as a function of temperature were performed by the standard four-point method. Also the a.c. susceptibility of the sample was obtained by

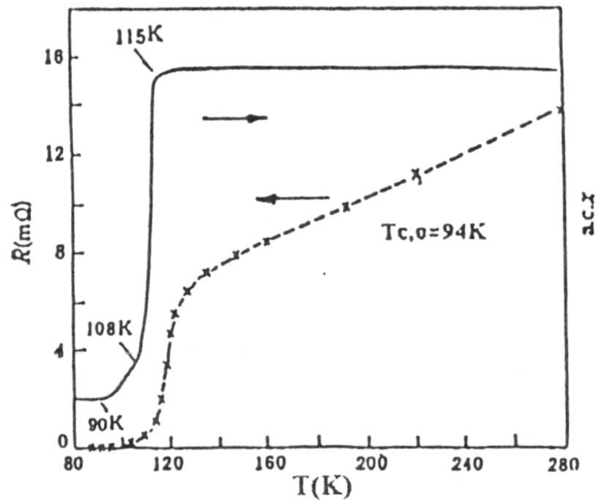

Fig. 1. The R-T and χ-T curves of the nominal $BiSrCaCu_2O_y$ sample.

the mutual inductance measurement. Figure 1 shows the *R-T* and *χ-T* curves of this sample. From *R-T* and *χ-T* curves in Fig.1, it is obvious that the main superconducting phase is the 110K one. Its zero resistance transition temperature is 94K.

X-ray diffraction analysis indicates that there are two phases in this sample. One of them can be identified to an orthorhombic cell of 0.539 × 0.536 × 3.07nm and another one identified to an orthorhombic cell of 0.540 × 0.536 × 3.70nm, which correspond

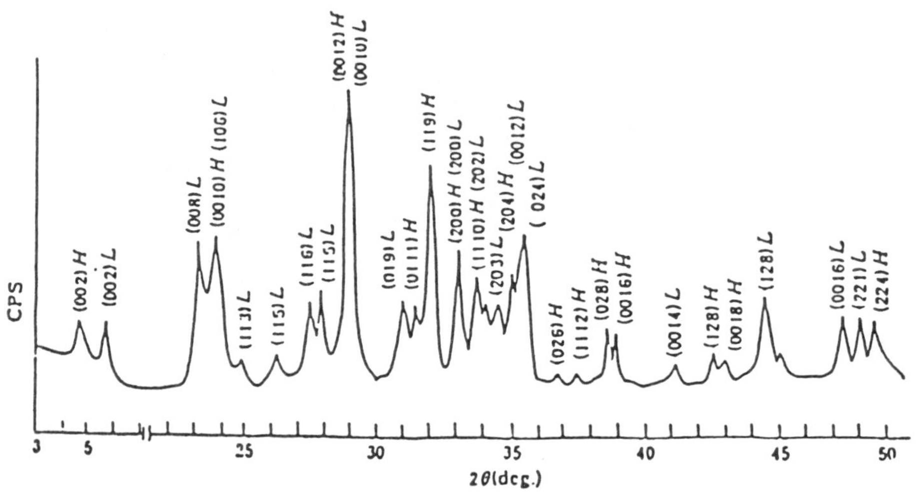

Fig. 2. The X-ray diffraction pattern of the $BiSrCaCu_2O_y$ sample. V ~ 40KV, I ~ 20mA. H and L indicate the peaks of 110K and 85K, respectively.

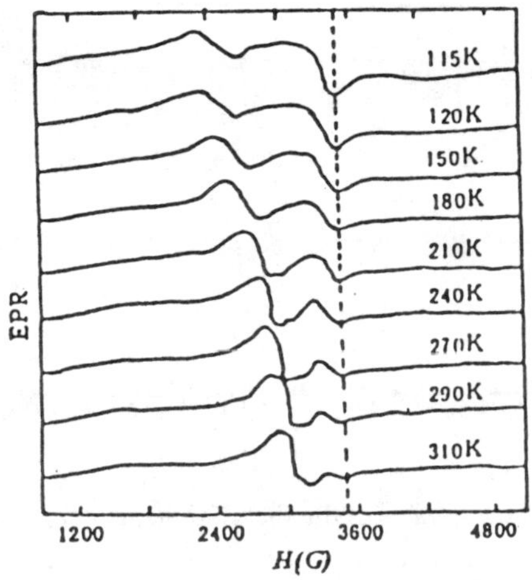

Fig. 3. The characteristic EPR spectra of the $BiSrCaCu_2O_y$ sample in the range of temperatures from 115K to 310K. Microwave frequency and power are 9.50GHz and 20mW, respectively. Field modulation intensity and gain are 3.2Gpp and 5×10^5, respectively. The weight of the sample is 9mg.

to 85K and 110K superconducting phases, respectively. There are no other phases besides the above phases in this sample.(Fig.2)

The EPR experiments were carried out using a Bruker(ER-200D-SRC) reflection X-band type spectrometer equipped with low temperature accessory down to 96K. The frequency and magnetic field were measured by a frequency counter and a proton NMR gauss-meter, respectively. Fig.3 shows the characteristic spectra of the sample at different temperatures before the superconducting transition. It is found that there are two EPR signals with different behaviors in this sample above *Tc*. One of them shifts gradually towards higher magnetic field with raising temperature. Its *g*-value decreases from 2.78 to 2.24 when the temperature increases from 115K to 310K(See Fig.4). The resonance field of another signal is almost independent of temperature and its *g*-value is very close to one of free electron(g_e~2.003). The EPR spectra of the sample below *Tc* at different temperatures are shown in Fig.5. It could be seen that a significant zero field signal appears at temperature lower than superconducting transition temperature and its signal intensity is strengthened when the temperature is decreased below *Tc*.

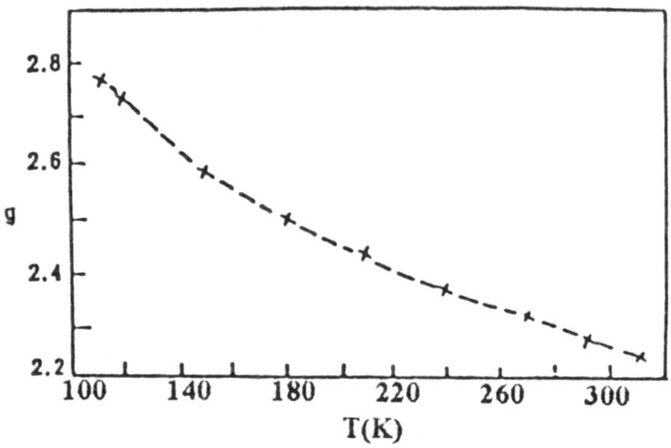

Fig. 4. The dependence of the *g*-value of the first EPR signal in Fig.3 on the temperatures.

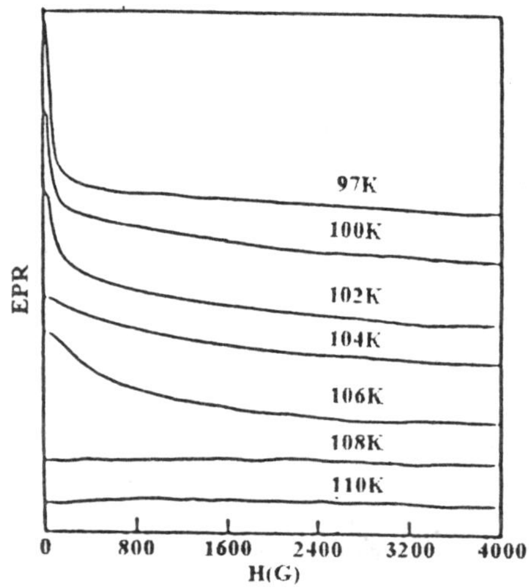

Fig. 5. The EPR spectra of the sample below *Tc*. Microwave frequency and power are 9.50GHz and 8mW, respectively. Field modulation intensity and gain are 0.8Gpp and 3.2×10^3, respectively.

3 Discussion

It is well known that only copper ions in Bi-Sr-Ca-Cu-O system might show paramagnetic behavior. Under the influence of the strong crystal field of oxygen ions around the copper ion, the $d_{x^2-y^2}$ orbital state is lifted above the remaining Cu *d* levels, so that the Cu^{2+} ion ($3d^9$ configuration) has a single electron occupying the $d_{x^2-y^2}$ orbital and a net spin of 1/2, while the Cu^{3+} ion ($3d^8$ configuration) has zero spin which wouldn't show paramagnetic behavior. Therefore the EPR signals observed in the $BiSrCaCu_2O_y$ sample above *Tc* may be related to the Cu^{2+} ions. However, from the observation of EPR in CuO samples at different temperatures, it is found that the EPR signal of CuO is almost independent of temperature except for its signal intensity. It means that the EPR signal observed in the $BiSrCaCu_2O_y$ sample could not only be attributed to the Cu^{2+} ions.

The zero field signal observed in the $YBa_2Cu_3O_{69}$ system was reported by many authors. The first interpretation of the ZFS was in terms of flux-trapping effects due to the interplay between cooling in a magnetic field and the remnant magnetization in a type II superconductor. Different ways of performing the experiment showed that this was not the case.Another possible source for the ZFS could be microwave absorption from Josephson junctions within the sample. However, it is very difficult to detect directly the Josephson radiation, because it is very crucial to the junction's structures in which the Josephson radiation must be well coupled to the resonant cavity formed by the junction. In fact, it is very easy to observe the ZFS below *Tc* for high *Tc* oxide superconductors. Therefore this interpretation may be inappropriate for this case also. In this paper, we propose a possible interpretation for these phenomena based on the electron transfer model.

It is proposed that electrons are transferred from O^{2-} to Cu^{2+} ions due to the shorter Cu-O bonds in Cu-O planes as a result of producing an unpaired electron in the *2p* orbital of the oxygen ion. All of experimental phenomena observed in high *Tc* superconductors could be related to the electron transfer between copper and oxygen ions in Cu-O plane. It is shown in Fig.6. Referring to Fig.6, the electron e_1 is transferred from oxygen ion to copper ion result in an unpaired electron e_2 in the *2p* orbital of oxygen ion. The e_1 electron will not only be subject to the usual Hund rules governing as a result of taking a parallel alignment with respect to the unpaired electron already on the Cu^{2+} ion, but also interacts with the unpaired electron e_2 in the *2p* orbital of oxygen ion. In this case, the e_1 electron doesn't completely transfer into the orbital

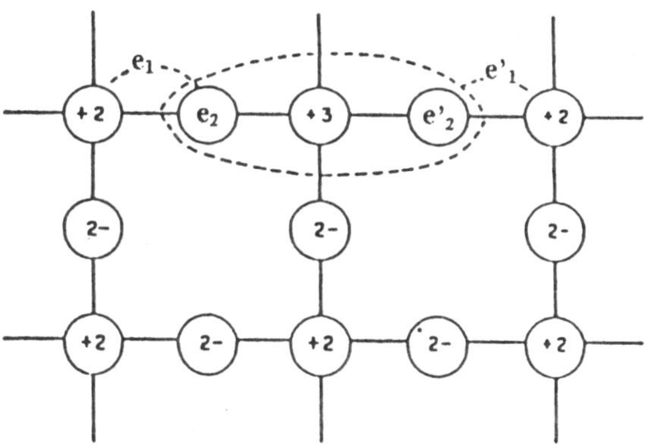

Fig. 6. A possible electron transfer process where the e_1 electron transfer from O^{2-} ion to Cu^{2+} ion and an unpaired electron e_2 is formed in the orbit of the oxygen ion.

of Cu^{2+} ion. It is therefore reasonable to suppose the first EPR signal in Fig.3 could be thought as the contribution of the Cu^{2+} ions and the electrons transferred from oxygen ions. The first EPR signal shifts towards lower magnetic field and its g-value increases from 2.24 to 2.78 when the temperature decreases from 310K to 115K, which could be due to increasing the probability of the electrons from oxygen ions occupying the orbital of Cu^{2+} ions on cooling. The second EPR signal may be related to the unpaired electrons from oxygen ions. Referring Fig.6, it could be seen that the main interaction on the unpaired electrons from oxygen ions comes from the Cu^{3+} ions. In this case, the behavior of the unpaired electrons is similar to one of free electron due to the zero spin of Cu^{3+} ions.

As mentioned above, the probability of the e_1 electron occupying the orbital of Cu^{2+} increases with decreasing temperature. On the other hand, the e_2 electron is paired with the unpaired electron from another oxygen ion by the Coulomb attractive of Cu^{3+} ion below Tc. In this case, the e_1 electron will not interact with the e_2 electron as a result of the e_1 electron indeed occupying the orbital of Cu^{2+} ion. The zero field signal observed below Tc in $BiSrCaCu_2O_y$ high Tc superconductor may be related to this case.

In summary, we investigated the EPR spectra of $BiSrCaCu_2O_y$ high Tc superconductor at different temperatures. The experimental phenomena observed are interpreted based on the electron transfer between copper and oxygen ions in Cu-O planes.

Acknowledgments We thank Prof.C.D.Gong for his useful discussion. We also gratefully acknowledge the support by the Science Foundation of Structure Research Laboratory, University of Science and Technology of China and by the National Natural Science Foundation of China.

References

1 C.W.Chu, J.Bechtold, et al., Phys. Rev. Lett., 60, 941(1988).
2. H.Ihare, et al., Jpn J.Appl.Phys., 26, L460(1987).
3. S.V.Bhat, et al., J.Phys. C: Solid State Phys., 20, L559-L563(1987).

EPR Research of the Nature of Dispersed Systems - Mechanical Activation

M.Kakazey, M.Vlasova, M.Ristich, T.Srechkovich**

Institute for Problems of Material Sciences of NASU,252180 Kyiv, Ukraine
*Joint Laboratory for Advanced Materials of SASA, 11000 Beograd, Yugoslavia

Abstract. Zinc-oxide powder was tribophysically activated by grinding in a vibro-mill in a continual regime in air. Using the method of electron paramagnetic resonance laws were established for defect formation from the viewpoint of the evolution of the fine defect structure of disperse systems. The main focus was on the influence of the length of the tribophysical activation on the process of defect formation. Six signals can be seen on the obtained EPR spectra.

1 Introduction

Tribophysical activation (TA) is used a lot as a method for the modification of physico-chemical properties of disperse systems in technologies for obtaining powders and ceramics [1,2] . However, the nature of this modification, due to the specific multiparameter object of investigation - the disperse system, is still not explained in many cases. In [3] it was suggested that this problem could be dealt with from the viewpoint of the evolution of the fine defect structure (FDS) of the disperse structure that is investigated. The term FDS includes a complete information on the defect type (both the crystallographic lattice and individual particles and their agglomerates) their concentration and space localization. To be able to obtain such a universal information it is necessary to apply a great number of both classical and advanced methods of investigation adapted for polycrystal samples.

In [4-12] EPR signals were investigated on zinc-oxide powders activated in different ways - mechanical, thermal etc. In this research special attention was paid to the influence of the length of the tribophysical activation by grinding on the process of defect formation.

2 Experimental Methods

A zinc-oxide powder (99.99% ZnO) whose specific surface was 3.6 m^2/g was used. Tribophysical activation was performed by grinding in a continual regime in a vibro-mill in air, and the grinding time t_{ml} was from 3 to 300 min. Recording of EPR spectra was performed at room temperature on a radio spectrometer SE/x 2547 with a EVM spectra analyzer.

3 Experimental Results and Discussion

The starting zinc-oxide powder has no EPR signals. EPR signals appear (Fig. 1) with the increase of the grinding time t_{ml} whose intensity does not uniquely depend on the t_{ml} value. Signal appear after grinding starts, while the fields are lower (g > 2.0023), which gradually intensify, and then disappear, to be exchanged for signals with g < 2.0023.

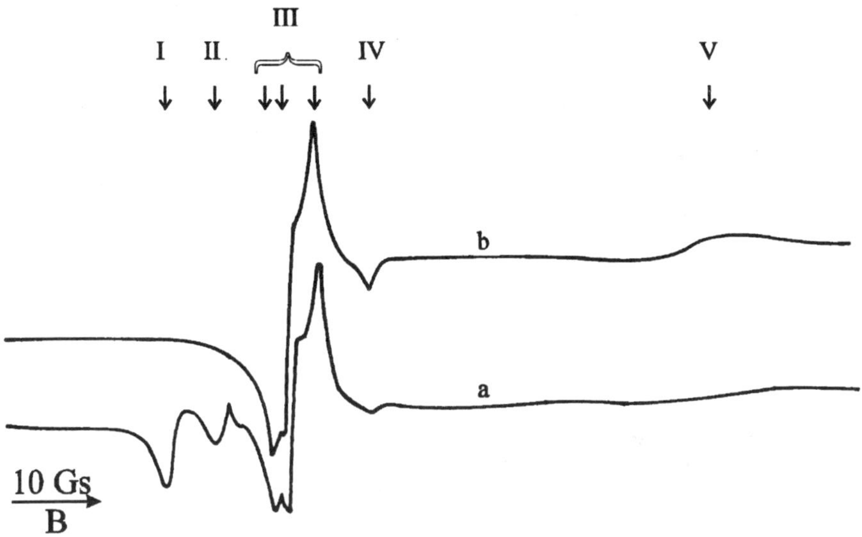

Fig. 1. Obtained EPR signals during TA of zinc-oxide powder for a) t_{ml} = 30 min and b) t_{ml} = 300 min.

We used literature data from EPR investigations of zinc-oxide defects [13-16], which are given in Table I for identifying paramagnetic centers that form during TA. We can see, that the nature of paramagnetic centers in these papers was determined from an analysis of the angular dependence of the position of EPR lines in the magnetic field, the shape and intensity of the lines of the superfine structure, and in some cases (S > 1/2) lines of the fine structure. In EPR spectra of polycrystal samples the shape and size of singularities (sharp peaks) are the result of the calculation of mean values of spectra in single crystal samples [17]:

$$J_{polycrstal}(B) = N / 8\pi^2 \int_0^{\pi} \sin\theta d\theta \int_0^{2\pi} d\varphi \int_0^{2\pi} d\psi \; \Phi\left[B - B_o(\theta,\varphi,\psi)\right] \cdot V(\theta,\varphi,\psi) \qquad (1)$$

where Φ and V are the lines form and intensity of the absorption of the paramagnetic center, respectively; θ, φ and ψ are Euler angles. For centers with S = 1/2, which appear in our case, the positions of singularities in spectra of polycrystal samples correspond to the values of the g-factor of resonant lines in critical points of single crystal samples ($g_{\perp}$ and $g_{\parallel}$ or g_1, g_2 and g_3).

Table 1. Values for obtained parameters for EPR signals in TA zinc-oxide powder and parameters of signals of defects obtained from investigations of single crystals.

Signal	Values for g-factors obtained in this work	Center	Values for g-factors obtained on single crystals	References
I	$g_{\perp}=2.0190$, $g_{\parallel}<g_{\perp}$	$V_{Zn}^{-} : Zn_{i}^{o}$.	$g_{xx}=2.0185$, $g_{yy}=2.0188$, $g_{zz}=2.0040$	13
II	$g_{\perp}=2.0130$, $g_{\parallel}=2.0140$	V_{Zn}^{-}	$g_{\perp}=2.0128$, $g_{\parallel}=2.0142$	14
III	$g_1=2.0075$, $g_2=2.0060$, $g_3=2.0015$	$(V_{Zn}^{-})_2^{-}$	$g_{xx}=2.0077$, $g_{yy}=2.0010$, $g_{zz}=2.0059$	13
IV	$g_{\perp}=1.9965$, $g_{\parallel}=1.9950$	F^{+}	$g_{\perp}=1.9963$, $g_{\parallel}=1.9948$	15
V	$g=1.9640$	SDC	$g\approx1.96$	16

In our case the following signals were obtained and identified:

Signal I: $g_{\perp} = 2.0190\pm0.0005$, $g_{\parallel} < g_{\perp}$. As it can be seen in Table I, such a signal to the greatest degree corresponds to the complex: zinc vacancy with a hole which is localized on one of the four oxygen ions V_{Zn}^{-} and interstitial neutral zinc atom Zn_{i}^{o} - $V_{Zn}^{-} : Zn_{i}^{o}$. The singularity in high fields of such a center (g = 2.0040) was not noted due to its low intensity and also superposition with more intensive lines of other centers.

Signal II: $g_{\perp} = 2.0130\pm0.0005$, $g_{\parallel} = 2.0140\pm0.0005$. This signal completely originates from the V_{Zn}^{-} center and it can form from center I if the interstitial zinc atom moves away from the vacancy at a distance larger than the third coordination sphere. Let us note that when T < 170 K there are two spectra characteristic for the V_{Zn}^{-} centers, which is connected with the inequivalent to oxygen ions that surround the zinc vacancy: a) axial, with a low intensity, caused by the hole, localized on the axial oxygen ion; b) rhombic, which is caused by the localization of the hole on one of the three basic oxygen ions 0^{2-} from the oxygen tetrahedron which surrounds the vacancy. When T >170K, due to the thermally activated hole mobility around the vacancy, a spectrum with parameters of the spin hamiltonian given in Table I is formed.

Signal III: $g_1 = 2.0075\pm0.0005$, $g_2 = 2.0060\pm.0005$ and $g_3 = 2.0015\pm0.0005$. This signal is in accordance with data from Table I caused by the complex which consists of two mutually close zinc vacancies on one plane (001) and which encompass one hole - $(V_{Zn}^{-})_2^{-}$. In accordance with [13] the axis x of the g-tensor complex lies in the plane (1210) and forms an angle of 70° with axis c, while the angle between axes z and c is 54.6°.

Signal IV: $g_{\perp}$ = 1.9965±0.0005 and $g_{||}$ = 1.9950±0.0005. This signal is connected with the oxygen vacancy V_0^{2+} which includes one electron (V_0^{+}) i.e. the F^+ center [15].

Signal V: g = 1.9640±0.0005. This signal is related to small donor centers (SDC) (Table I) and the correlation of its intensity with sample electroconductivity is known [16]. When T = 77 K the intensity of this signal increases more than 50 times.

Signal VI: a wide anisotropy line. This lines is related to the products of the oxidation of the ground material from which the mill is made and its interaction with zinc-oxide. Note that a similar signal appears also after intensive TA, grinding of Al_2O_3 [18], only in this case its intensity is a lot lower, that is understandable having in mind the low abrasiveness of zinc-oxide compared to Al_2O_3.

The quantitative change of the obtained signal with the grinding time requires special attention. Thus, signals I and II are characterized with a linear growth of their amplitude until t_{ml} = 30 min., then it decreases in such a way that in samples with t_{ml} = 300 min., only traces of signal I remain (when this sample is X-rayed this signal increases insignificantly). An increase of the amplitude of signal III is noticed until t_{ml} = 120 min., and then it gradually decreases. Signals IV and V appear only in samples with t_{ml} = 300 min. As the grinding time in our experiments was maximally 300 min., we used data from [6] for estimations with higher grinding times. In [6] EPR data is given for zinc-oxide ground for 10-300 hours. The absence of signal IV in these samples indicates the fact that it appears, grows and disappears in the grinding time interval from 5 to 10 hours. According to [6] the intensity of signal V increases to t_{ml} = 100 h and then decreases, which is accompanied with the growth of the signal with g = 2.003, which authors of [6] attribute to chemo-absorbed molecular O_2^- ions. Signal VI connected with the products of the ground mill material, its oxidation and interaction with zinc-oxide becomes significant in samples with t_{ml} higher than 30 min., when its intensity becomes proportional to the grinding time.

These investigations enabled us to suggest the following scheme of the evolution of defect states in zinc-oxide during tribophysical activation, given in Fig. 2. In the starting period of grinding zinc-oxide, monovacancies in the zinc sublattice are formed $V_{Zn}^-: Zn_i^{\circ}$ and V_{Zn}^- centers. With the increase of the grinding time the defect structure becomes more complex in the metal sublattice -- formation of the di-vacancy complex $(V_{Zn}^-)_2^-$. Further increase of the grinding time leads to the formation of a monovacancy defect structure and F^+ centers in the oxygen sublattice. Further grinding causes the uniting of the cation and anion defect structure, and thus the formation of traps for electrons and contributes to: a) the increase of the samples electroconductivity; b) chemo-absorption of gases (note that at certain grinding stages this process can lead to the decrease of the samples electroconductivity); c) the interaction of the ground material of the mill with zinc-oxide.

This way, it is shown that the evolution of the defect structure in polycrystal zinc-oxide can be followed as part of EPR investigations, during TA, which consists of a gradual transformation of a conditionally ideal lattice in starting samples to the formation and concentration of elementary defects, the formation of various

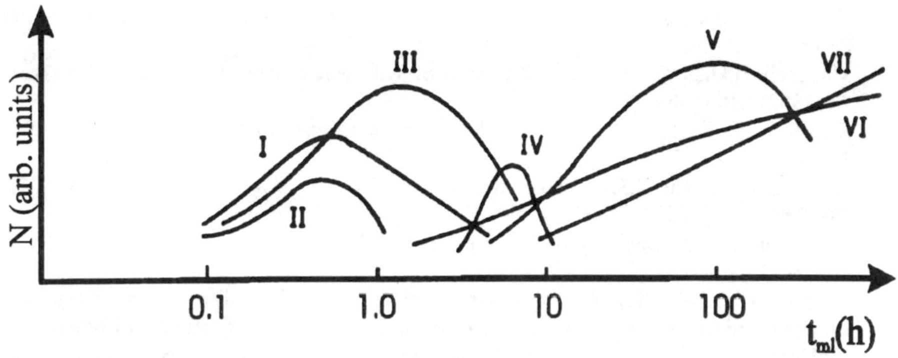

Fig. 2. A descriptive qualitative scheme of the change of the defect structure of zinc-oxide with the length of TA, curve VII is connected with the defect states which form as the result of the interaction of zinc-oxide with the gas environment molecules [12]. Data for t_{ml} >5 h are taken from [6].

interaction of formed defects with the ground mill material and the gas environment, to the disperse system with a modified chemical activity.

4 Conclusion

Six EPR signals were recorded and identified in zinc-oxide samples which were tribophysically activated by grinding an a vibro-mill. It has been shown that they are caused by the formation of defects in the zinc-oxide lattice: $V^{-}_{Zn}:Zn^{\circ}_{i}$ (I), V^{-}_{Zn} (II), $(V^{-}_{Zn})^{-}_{2}$ (III), F^{+} (IV), SDC (V) and the product of the interaction of the ground mill material and zinc-oxide (VI). The noticed dependency of the formation of different centers from the grinding time enabled the suggestion of a scheme of the evolution of the defect structure in polycrystal zinc-oxide samples during tribophysical activation by grinding:

$$\left.\begin{array}{l} Zn^{2+}_{lattice} \to V^{-}_{Zn}:Zn^{+}_{i} \to V^{-}_{Zn}:Zn^{\circ}_{i} \to V^{-}_{Zn}+V^{-}_{Zn} \to (V^{-}_{Zn})^{-}_{2} \\ \qquad\qquad\qquad\qquad O^{2+}_{lattice} \to V^{+}_{0} + O^{-} \end{array}\right\rangle \text{SDC}$$

References

1. G.Heinike, *Tribochemistry*, Akademie-Verlag, Berlin (1984).
2. Ye.G.Avvakumov, *Mechanical Methods of the Activation of Chemical Processes*, Nauka, Novosibirsk (1986).
3. N.G.Kakazey, Evolution of the Defect Structure in Microcrystalline System subjected to Mechanical and Thermal Actions (by EPR Studies), D.Sc. Dissertation Abstract, Riga (1991).
4. G.K.Walters, T.L.Estle, Paramagnetic resonance of defect introduced near the surface of solids by mechanical damage,*J.Appl.Phys*. **32** (1961) 1854 -1859.
5. V.V.Golubev, V.B.Evdokimov, EPR in mechanical activated ZnO, *J.Phys.Chem.* **38** (1964) 477-478 (In Russian).

6. U.Voland, R.Schrader, H.Schneider, Elektron spin resonanz in mechanisch aktivierten Festkörpern, *Z.anorg.und allg.Chem.* **368** (1969) 317-326.
7. M.Codell, H.Gisser, J.Weisberg, ESR study of hydroperoxide on zinc oxide, *J.Phys.Chem.* **72** (1968) 2460-2464.
8. K.M.Sancier, ESR investigation of photodamage to zinc oxide powders, *Surface Sci.* **21**(1970) 1-11.
9. K.Hoffman, D.Hahn, ESR of lattice defects in zinc oxide, *Phys.status solidi* (a) **24** (1974) 637-648.
10. E.B.Baranov, V.E.Holmogorov, Photoinduced EPR signals in ZnO, *Dokl. AS USSR* **146** (1962) 125-128.
11. P.L.Nelson, D.B.Hartley, A comparison of defect centres formed in oxides by explosive shocking and high-energy radiation, *J.Chem.Soc.Dalton Trans.*, N 15 (1972) 1596-1603.
12. I.P.Kuzmina, V.A.Nikitenko, *Zinc Oxide, Production and Optical Properties*, Nauka, Moscow (1984).
13. B.Schallenberger, A.Hausmann, Eigenstörstellen in elektronen bestrahlten zinkoxid, *Z.Physik* **B 23** (1976) 177-181.
14. D.Galland, A.Herve, Temperature dependence of the ESR spectrum of the Zn vacancy in ZnO, *Solid State Commun.* **14** (1974) 953-956.
15. J.M.Smith, W.E.Vehse, ESR of electron irratiated ZnO conformation of the F^- center, *Phys.Lett.* **31A** (1970) 147-148.
16. V.A.Nikitenko, Electronic paramagnetic resonance in ZnO, *J. Appl. Spectroscopy* **57** (1992) 367-385 (In Russian).
17. Ya.G.Klyava, *EPR Spectroscopy Of Disordered Solids*, Zinatne, Riga (1988).
18. M.V.Vlasova, S.I.Gorbachuk, N.G.Kakazey, V.I.Lapshin, V.K.May, L.L.Sukhin, V.I.Chistyakov, EPR study of tribochemical phenomena in active mechanical treatment of powder α-alumina, *Sci.Sintering* **25** (1993) 117-134.

The EPR Investigation of the Distortion of Local Environment of Gd^{3+} at the Structure Transitions in $CsSrCl_3$

V. Vazhenin, M. Artyomov, A. Nikiforov

Institute of Physics and Applied Mathematics, Ural State University, Ekaterinburg, 620083 Russia

Abstract. EPR spectra of the four triclinic centres of Gd^{3+} in the second low-temperature phase of $CsSrCl_3$ have been investigated. The analysis of spin Hamiltonians parameters enabled us to determine the value of the rotation angles of the principal axes fine-structure tensors at phase transitions. In some cases we correlate them with rotations of the nearest chlorine octahedron and evaluate the octahedron distortion.

The EPR investigations of the types of the paramagnetic centres in the single crystal with perovskite structure $CsSrCl_3$: Gd^{3+} have been described in the Refs.1-3. The sequence of crystal phases

$$O_h^1 \xrightarrow{[00\psi]} D_{4h}^5 \xrightarrow{[0\varphi\psi]} D_{2h}^{17} \xrightarrow{[\varphi_2\varphi_1\psi]} C_{2h}^2$$

(ψ and φ-are components of modes M_3 and R_{25}) has been determined on the base of the *qualitative* study of the temperature behaviour of the spectra of cubic centres. Quantitative evaluations of the nearest environment distortion of the paramagnetic ion have been performed in the first low-symmetry phase.

This paper is dedicated to the quantitative investigation of the EPR spectra of the four triclinic centres of Gd^{3+} in the second low-symmetry phase which originate from cubic (type 1) and tetragonal (types $2\|$, $2\perp1$, $2\perp2$ with charge compensation by vacancy of Sr^{2+}) centres of the praphase. Figures 1 and 2 show typical temperature dependences of the resonance fields of the EPR transitions. The measurements of the angular dependences of the EPR spectrum in essentially monodomain (in the first low-symmetry phase) samples have enabled to identify the transitions and to determine spin Hamiltonian parameters (see Table 1). The rotation angles of the principal axes of the fine structure tensors have been determined at the second phase transition (Table 2). To this end, we employed the expressions:

$$\Delta = 1/4 \text{ arctg } (c_{r4}/b_{r4}), \qquad r=4,6 \tag{1}$$

$$\Delta = 1/2 \text{ arctg } (c_{22}/b_{22}), \tag{2}$$

$$\Delta = 1/2 \text{ arctg } \{ b_{21}/(b_{22} - 3\, b_{20}) \}, \tag{3}$$

obtained from the condition that the parameters c_{44}, c_{64}, c_{22} and b_{21} vanish in the coordinate system turned by Δ around axes z and y, respectively.
The average values of the rotation angles of the fourth rank tensor axes are:

$$\psi=\psi_{\|}=7.2(5)^\circ,\quad \varphi=\varphi_{\|}=3.9(5)^\circ,\quad \psi_{\perp}=5.4(5)^\circ,\quad \varphi_{\perp}=1.5(4)^\circ. \tag{4}$$

The symbols ‖ and ⊥ indicate relative orientation of the rotation axis and direction to charge compensator (DCC). We correlate them with the values of the local rotations of the chlorine octahedra. The second rank tensor for determination of rotation angles of the ligand octahedron is unsuitable (see Table 2). The analysis of the second rank parameters of the spin Hamiltonian allowed to determine character and to evaluate local distortion of the octahedron (~10^{-5}nm). The rotation of type ψ in any centres leads to contraction of the ligand octahedron along the rotation axis. The appearance of φ on centres 1 and 2⊥2 leads to contraction of the chlorine octahedron along rotations axis but on centres 2⊥1 and 2‖ leads to extension.

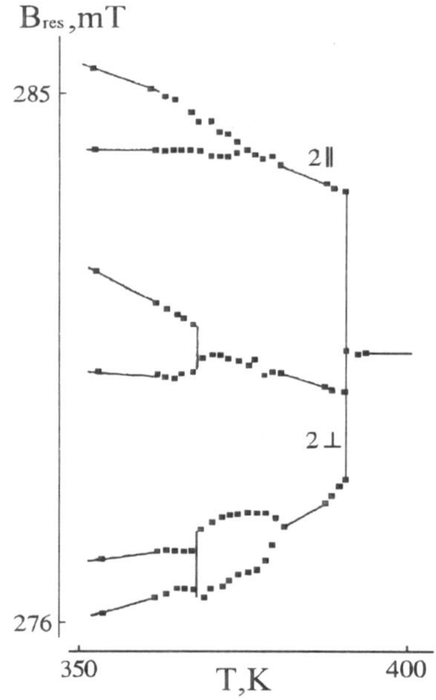

Fig. 1. Temperature dependences of B_{res} in a centre type 2 when C_4‖B⊥DCC, the transition 6↔7.

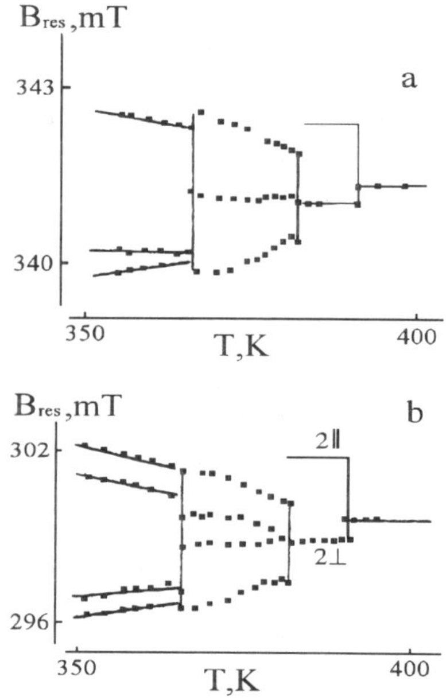

Fig. 2. Temperature dependences of B_{res} for B‖C_2 (a-type 1, the transition 5↔6, b-type 2, the transition 6↔7). Some components have not observed in this sample.

Table 1. Spin Hamiltonian parameters in the second low symmetry phase at 372K.

The values of b_{nm} and c_{nm} are given in MHz.

	1 z‖ψ, y‖φ	2‖ z‖ψ‖DCC, y‖φ	2⊥1 z‖DCC,x‖φ,y‖ψ	2⊥2 z‖DCC‖φ, y‖ψ
g_x	1.9930(15)	1.9920(10)	1.9930(15)	1.9930(15)
g_y	1.9930(15)	1.9930(15)	1.9930(15)	1.9950(10)
g_z	1.9950(20)	1.9950(15)	1.9930(10)	1.9920(15)
b_{20}	29(1)	-573(2)	-638(2)	-631(2)
b_{21}	-7(12)	-56(6)	-274(8)	-276(8)
b_{22}	-4(3)	4(2)	-27(3)	-22(3)
b_{40}	-33.7(4)	-33.0(4)	-32.0(5)	-32.0(5)
b_{41}	-45(6)	18(3)	-61(4)	-63(4)
b_{42}	-1(3)	-1(4)	-5(4)	-5(4)
b_{43}	38(20)	-24(30)	54(18)	55(15)
b_{44}	-153(4)	-171(4)	-188(4)	-185(4)
b_{60}	1.0(6)	0.4(6)	1.3(6)	1.1(6)
b_{61}	-1(8)	-5(6)	7(6)	7(6)
b_{62}	0(7)	-1(6)	-1(5)	-3(6)
b_{63}	5(10)	9(18)	20(18)	26(18)
b_{64}	-24(7)	-20(9)	-26(8)	-22(10)
c_{21}	70(13)	-115(9)	-96(6)	-66(8)
c_{22}	0(6)	-1(6)	33(4)	4(3)
c_{41}	3(6)	-8(4)	-19(4)	2(4)
c_{42}	-2(8)	-6(6)	3(5)	2(4)
c_{43}	0(15)	-37(15)	-22(30)	7(26)
c_{44}	-80(5)	98(3)	4(6)	52(6)
c_{61}	4(8)	-11(9)	-3(6)	1(9)
c_{62}	-3(12)	-17(15)	1(10)	0(10)
c_{63}	-11(10)	-34(21)	17(15)	0(36)
c_{64}	-13(7)	15(9)	-1(10)	8(10)

Table 2. The rotation angles of the principal axes of the fine structure tensors

The symbols ∥ and ⊥ indicate relative orientation of the rotation axis and direction to charge compensator. The angle δ is the result of the two rotations ψ and φ ($\varphi \perp \delta \perp \psi$). All values of rotation angles are given in degree without signs.

vanishing parameter	1	2∥	2⊥1	2⊥2
c_{22}	ψ=0(30)	$\psi_{\parallel}$=7(30)	δ=25(4)	$\varphi_{\parallel}$=5(4)
c_{44}	ψ=6.9(5)	$\psi_{\parallel}$=7.5(4)	δ=0.3(5)	$\varphi_{\parallel}$=3.9(5)
c_{64}	ψ=7(4)	$\psi_{\parallel}$=9(5)	δ=0(5)	$\varphi_{\parallel}$=5(5)
b_{21}	φ=2(5)	$\varphi_{\perp}$=0.8(2)	$\psi_{\perp}$=4.1(2)	$\psi_{\perp}$=4.2(2)
b_{41}	φ=3.8(5)	$\varphi_{\perp}$=1.5(4)	$\psi_{\perp}$=5.3(5)	$\psi_{\perp}$=5.5(5)
c_{21}	δ=20(3)	δ=1.9(2)	$\varphi_{\perp}$=1.6(4)	δ=1.0(1)
c_{41}	δ=0.2(6)	δ=0.7(3)	$\varphi_{\perp}$=1.6(4)	δ=0.2(3)
X-ray	ψ=6.8			

References

1. M.V. Chernitskii, V.A. Vazhenin, A.E. Nikiforov, A.E. Usachev, A.I. Krotkii, M.Yu. Artyomov, Electron spin resonance and structural transformations in $CsSrCl_3$: Gd^{3+}, *Sov. Phys. Solid State.* **33**, (1991) 2011-2015.
2. V.A. Vazhenin, K.M. Starichenko, M.Yu. Artyomov, M.V. Chernitskii, Tetragonal Gd^{3+} centres in ferroelastic $CsSrCl_3$, *Sov. Phys. Solid State.* **34**, (1992) 866-867.
3. V.A. Vazhenin, K.M. Starichenko, M.Yu. Artyomov, A.E. Nikiforov, EPR and the local environment of Gd^{3+} in $CsSrCl_3$, *Sov. Phys. Solid State.* **36**, (1994) 1470-1472.

Electronic Spectra of Cobaltite

*R.V.S.S.N.Ravikumar, A.V.Chandrasekhar, B.J.Reddy, Y.P.Reddy and P.S.Rao**

Department of Physics, S.V.University, Tirupati - 517 502, India
*Dept. of Chemistry, University of Pondicherry, Pondicherry, India

Abstract. Results of the investigations on the Electron Paramagnetic Resonance and Optical absorption spectra of cobaltite mineral are presented in this paper. The EPR spectrum is characteristic of Co^{2+} in cobaltite in distorted octahedra and Spin-Hamiltonian parameters are evaluated. Optical absorption bands are also in accordance with the EPR results. The distortion is trigonal and different crystal field parameters are evaluated.

1 Introduction

Cobaltite is a sulphide of arsenic and cobalt. Its general formula is CoAsS. It is the orthorhombic and cobalt atom is at the centre of an octahedron whose corners are arsenic and sulphur atoms with space group $pca2_1$. Chemical composition and crystal structure of cobaltite have been reported earlier by Bayliss (1982), Mechiling (1921), and Florke (1926). Cobaltite mineral originated from Selwyn, Cloncurry Dist., Queensland, Australia with specimen number BM 1921, 251 procured from The Natural History Museum, London is used for present EPR and optical absorption studies. The present investigation is taken up with a view to ascertain the site symmetry and the probable distortion.

2 Experimental

A polycrystalline sample of cobaltite was used for both EPR and optical absorption spectral recordings. EPR spectrum of the sample taken in a quartz tube was recorded on Varian E-112 EPR Spectrometer operating at X-band frequency (9.44 GHz) and having a 100 kHz field modulation and phase sensitive detection to obtain a first derivative spectrum. The mineral sample was mixed with nujol and the optical absorption spectrum a recorded on Hitachi U 3400 spectrophoto-meter in UV-VIS-NIR regions.

3 Theory

For Co^{2+} ion, both g and A are isotropic in octahedral symmetry. However for Co^{2+} associated with small tetragonal or trigonal distortion g and A values become very anisotropic. In most of the cases $\langle g \rangle = 4.3$, though the range of g-values vary from 1.5 to 7 in distorted octahedral site symmetry as given by Pilbrow (1990). The ^4F term splits into two triplets and a singlet, an orbital triplet being the lowest in energy. In octahedral or near octahedral symmetry the spin-orbit coupling splits the twelve low lying spin states, leaving a doublet state lowest in energy. This Kramer's doublet is split by a magnetic field and the only transitions observed in EPR are between these two states by Abragam and Pryce (1951). Since I = 7/2, the EPR lines exhibit the eight line hyperfine structure. The Kramer's doublet is the ground state.

$Co^{2+}(d^7)$ gives rise to the free ion terms ^4F, ^4P, ^2G and several doublet states. In strong octahedral crystal field (Dq > 1500 cm^{-1}) $^2E_g(G)$ is the ground state, whereas in the intermediate crystal field (Dq = 990 cm^{-1}) $^4T_{1g}(F)$ continues to be the ground state. Accordingly three spin allowed transitions from $^4T_{1g}(F)$ to $^4T_{2g}(F)$, $^4A_{2g}(F)$ and $^4T_{2g}(F)$ are expected. Out of these $^4T_{1g}(F) \rightarrow {}^4A_{2g}(F)$ involves the promotion of two electrons and is expected to be weak according to Koide (1959).

4 Results and Discussion

EPR spectrum of cobaltite is shown in Fig.1. The spectrum clearly indicates an eight line pattern at parallel position and broad line at perpendicular position. Since the nuclear spin of Co is 7/2, the calculated values of parallel and perpendicular hyperfine and g values are obtained as follows:

$$g_{\parallel} = 3.80, \quad A_{\parallel} = 157 \pm 2 \text{ G},$$

$$g_{\perp} = 2.04, \quad A_{\perp} = 75 \pm 2 \text{ G}.$$

It is generally known that most of cobaltous salts show very high anisotropy, if the octahedron has a small tetragonal or trigonal distortion. The structure of cobaltite indicates that the cobalt atom is at the centre of an octahedron whose corners are occupied by arsenic and sulphur atoms. Each (arsenic + sulphur) atom is coordinated by three cobalt atoms and one (arsenic + sulphur) atom in a distorted tetrahedral arrangement. The strong broad line at g = 2.04 also contains a small amount of Fe^{3+} impurity in cobaltite. Cooling the sample to 77 K did not alter the nature of the spectra.

The optical absorption spectrum of cobaltite recorded at room temperature is shown in Fig.2. Two broad bands observed at 1195 and 450 nm and the weak bands at 1100, 575 and 540 nm are characteristic of Co^{2+}. Two intense broad bands at 1195 and 450 nm are attributed to spin allowed transitions $^4T_{1g}(F) \rightarrow {}^4T_{2g}(F)$ and $^4T_{1g}(F) \rightarrow {}^4T_{1g}(P)$ respectively. The weak band at 540 nm is attributed to $^4T_{1g}(F) \rightarrow {}^4A_{2g}(F)$ which involves promotion of two electrons. In octahedral symmetry, the ratio of the energies of the transitions $^4T_{1g}(F) \rightarrow {}^4A_{2g}(F)$: ν_2 and $^4T_{1g}(F) \rightarrow {}^4T_{2g}(F)$: ν_1 is almost invariable at 1.9 to 2.2 as noted by Lever (1984). In the present investigation, the ratio (ν_2 / ν_1) is 18513/8618 = 2.14. This value agrees very well with results reported earlier by Rao et al. (1991), Jorgensen (1963), Narayana et al. (1968), Lakshman Rao and Purandar (1982), Hari et al, (1983), Ramesh et al. (1990), Sivaprasad et al. (1990), Rao et al. (1994), Ravikumar et al. (1995) and Ravikumar et al (1996) and supports present assignment. The remaining bands observed at 1100 and 575 nm are attributed to the transitions $^4T_{1g}(F) \rightarrow {}^2E_g(G)$ and $^4T_{1g}(F) \rightarrow {}^2T_{1g}(H)$ respectively using Tanabe-Sugano (1954) diagrams of d^7 configuration.

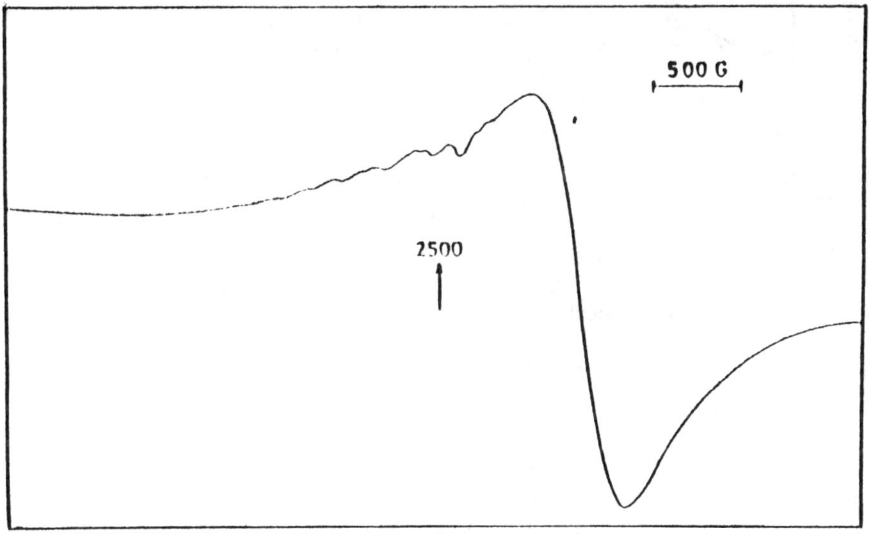

Fig. 1. EPR spectrum of cobaltite mineral (ν = 9.44 GHz , room temperature) .

The observed band positions along with the assignments are presented in Table 1. By solving the cubic field matrices for different sets of interelectronic repulsion parameters (B,C) and crystal field parameters (Dq), the following values are found to give a good fit with the observed data B = 990, C = 3950 and Dq = 990 cm^{-1}. The band attributed to the transition $^4T_{1g}(F) \rightarrow {}^4T_{2g}(F)$ is very broad and its observed value 1195 nm shows considerable deviation from the calculated value. These features might be due to distortion from octahedral symmetry of Co^{2+} site in the mineral.

This distortion might be trigonal or tetragonal as envisaged in EPR spectrum. However in the optical absorption spectrum the band associated with $^2E_g(G)$ is neither split nor broad, but is sharp. Since the E_g term does not split in trigonal symmetry, it is presumed that the deviation from octahedral symmetry in all probability might be trigonal. No bands attributable to Fe^{3+} are observed. Therefore if it is present, it can only be in trace amounts.

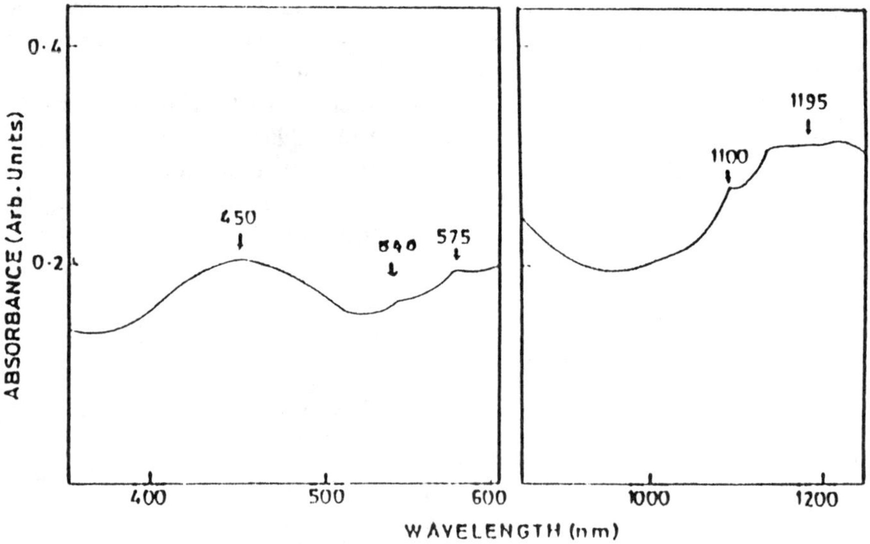

Fig. 2. Optical absorption spectrum of cobaltite mineral.

Table I. Band head data of cobaltite.

Transition from $^4T_{1g}(F)$	Observed band position		Calculated
	Wavelength (nm)	Wavenumber (cm^{-1})	Wavenumber (cm^{-1})
$^4T_{2g}(F)$	1195	8366	8649
$^2E_g(G)$	1100	9088	9082
$^2T_{2g}(H)$	575	17386	17443
$^4A_{2g}(F)$	540	18513	18549
$^4T_{1g}(P)$	450	22216	22247

5 Conclusions

From the EPR results, the site symmetry of Co^{2+} is predicted to be octahedral with either tetragonal or trigonal distortion. In optical absorption spectrum, the bands do not exhibit any splitting indicating the distortion to be too small to cause the splitting. However the band due to ${}^4T_{1g}(F) \rightarrow {}^4T_{2g}(F)$ at 1195 nm is very broad and its observed value shows considerable deviation from the calculated value, whereas the band due to ${}^4T_{1g}(F) \rightarrow {}^2E_g(G)$ is rather sharp and its observed value agrees well with calculated value. This indicates that the distortion has no effect on the E_g state whereas it has a considerable effect on ${}^4T_{2g}$ state. This is possible only in trigonal but not in tetragonal distortion, since of these two distortions trigonal distortion does not effect E_g state. Therefore from both the results of EPR and optical studies it can reasonably be concluded that the minor distortion which causes anisotropy in EPR spectrum is in all probability trigonal.

Acknowledgements

One of the authors (Ravikumar) is thankful to Council of Scientific & Industrial Research, New Delhi for the award of Senior Research Fellowship and foriegn travel grant. The authors gratefully acknowledge Dr.Peter Tandy, Dept. of Mineralogy, The Natural History Museum, London for supplying the desired sample.

References

1. Peter Bayliss, Am. Min., **67** (1982) 1048-1057.
2. M. Mechiling, Zeit. Krist., **1931** (1921) 283-284.
3. W. Florke, Centralb. Mineral. Abt. A., (1926) 337-338.
4. J.R. Pilbrow, Transition ion Electron Paramagnetic Resonance, Clarendon Press, Oxford p 148 (1990).
5. A. Abragam and M.H.I. Pryce, Proc. Roy .Soc., (London) **A206** (1951)173-178.
6. S. Koide, Phil. Mag., **4** (1959) 243.
7. A.B.P. Lever, "Inorganic Electronic Spectroscopy", Elsevier Publishing Co., Amsterdam, 2nd Ed., (1984).
8. S.N. Rao, P. Sivaprasad and Y.P. Reddy, Radiation Effects and Defects in Solids **116** (1991) 369-373.
9. C.K. Jorgensen, Advan. Chem. Phys., **5** (1963) 33.
10. P.A. Narayana, A. Mehra and P. Venkateswarulu, Can. J. Phys., **46** (1968) 325.
11. J. Lakshman Rao and K. Purandar, Acta. Phys. Polonica, **A62** (1982) 337- 342.
12. S. Hari, K. Purandar, J. Lakshman Rao and S.V.J. Lakshman, Phys. Stat. Solids (b) **115** (1983) K45-K50.
13. K. Ramesh, P. Sivaprasad and Y.P. Reddy, Czech. J. Phys., **B40** (1990) 466-471.
14. P. Sivaprasad, K. Ramesh and Y.P. Reddy, Physica. Scripta, **42** (1990) 100-102.
15. S.N. Rao, S. Vedanand, R. Ravikumar, R.V.S.S.N. Ravikumar and Y.P. Reddy, Solid State Comm., **92** (1994) 815-819.
16. R.V.S.S.N. Ravikumar, S.N. Reddy, S.N. Rao, B.J. Reddy and Y.P. Reddy, Cryst. Res. Technol., **30** (1995) 1121-1126.
17. R.V.S.S.N. Ravikumar, S.N. Rao, B.J. Reddy and Y.P. Reddy, Ferroelectrics, in Press. (1996).
18. Y. Tanabe and S. Sugano, J. Phys. Soc., Japan, **9**, (1954) 753-766.

EPR Study of the Thermally-Induced Spin-Crossover in Copper(II)-Doped $[Fe(trz)(Htrz)_2](BF_4)$

C. Cantin, J. Kliava, Y. Servant[1], *L. Sommier and O. Kahn*[2]

[1]Centre de Physique Moléculaire Optique et Hertzienne, Université Bordeaux-I, F-33405 Talence Cedex, France
[2]Laboratoire des Sciences Moléculaires, Institut de Chimie de la Matière Condensée de Bordeaux, F-33608 Pessac Cedex, France

Abstract. Cu^{2+} ions are used as dopant in $[Fe(trz)(Htrz)_2](BF_4)$. The Cu^{2+} EPR spectra in the range 293–403 K indicate a low-spin ($S = 0$) to high-spin ($S = 2$) transition of Fe^{2+} ions with hysteresis depending on Cu^{2+} concentration. A structural phase transition accompanying the spin transition is inferred from the spectroscopic data.

1 Introduction

The non-doped $[Fe(trz)(Htrz)_2](BF_4)$ exhibits a thermally-induced Fe^{2+} spin transition with hysteresis, evidenced by magnetic susceptibility and optical response [1, 2], and by a recent electron paramagnetic resonance (EPR) study [3].

In order to study Fe^{2+} spin transitions by EPR, doping with paramagnetic ions is necessary. A *foreign* EPR probe is able not only to detect a spin transition, but also to yield valuable information on concomitant structural modifications in the material. Moreover, by doping spin transition compounds with a considerable amount of impurity ions, their structure is modified to a certain extent. Thus, one can modify their physical properties, in what concerns the existence, position and broadness, at temperature scale, of the spin transition hysteresis loop. These modifications may be promising from the viewpoint of practical utilisation of these materials.

In this paper, we present a variable temperature EPR study of the compound $[Fe(trz)(Htrz)_2](BF_4)$ containing different concentrations of divalent copper, Cu^{2+}.

2 Samples and Optical Response Results

Three copper-doped $[Fe(trz)(Htrz)_2](BF_4)$ samples have been prepared, containing $x = 1$, 10 and 50 % Cu/Fe. In the 1 and 10 % Cu samples one observes, as in the non-doped compound, a total change of colour from pink in the low-spin (LS) state of Fe^{2+} to white in the high-spin (HS) state of Fe^{2+}, which can be followed with a simple optical device described in [2]. In the 50 % Cu sample the colour contrast between the LS and HS states is weak. The results for the 1 % Cu and 10 % Cu are shown in Figures 1, 2.

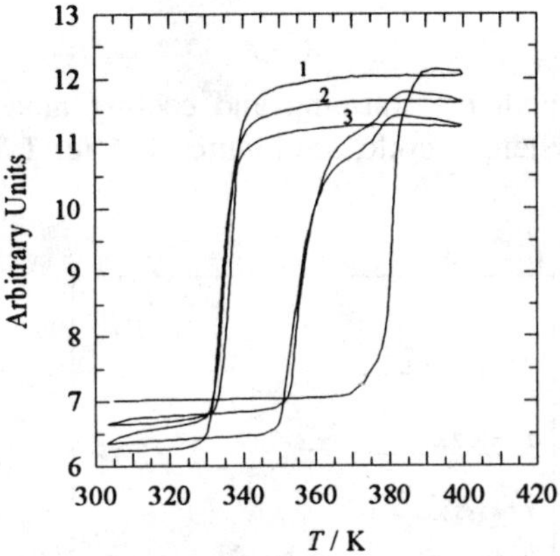

Fig. 1. Optical response of 1 % Cu^{2+}-doped sample. Successive thermal cycles are numbered 1, 2 and 3.

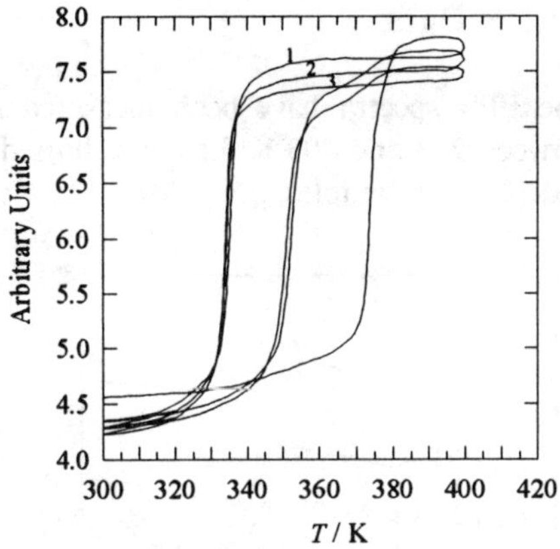

Fig. 2. Optical response of 10 % Cu^{2+}-doped sample. Successive thermal cycles are numbered 1, 2 and 3.

For successive thermal cycles in the 1 % Cu sample, in the cooling mode the transition temperature is constant ($T_c\downarrow$=336 K), while in the warming mode the transition temperature is $T_c\uparrow$=381 K for the first cycle and ~357 K for the following cycles. In the 10 % Cu sample, in the cooling mode the transition temperature is constant ($T_c\downarrow$=334 K), whereas in the warming mode the transition temperature is $T_c\uparrow$=374 K for the first cycle and ~352 K for the following ones.The total width of the stabilised loop, 21 K for 1 % Cu and 18 K for 10 % Cu, is smaller than that obtained for the non-doped compound (29 K) [3] and decreases with doping level. For the 50 % Cu sample, no hysteresis is observed at all; a very gradual LS to HS transition occurs between 200 and 300 K. The critical temperature is $T_c \approx 287$ K, see Figure 3.

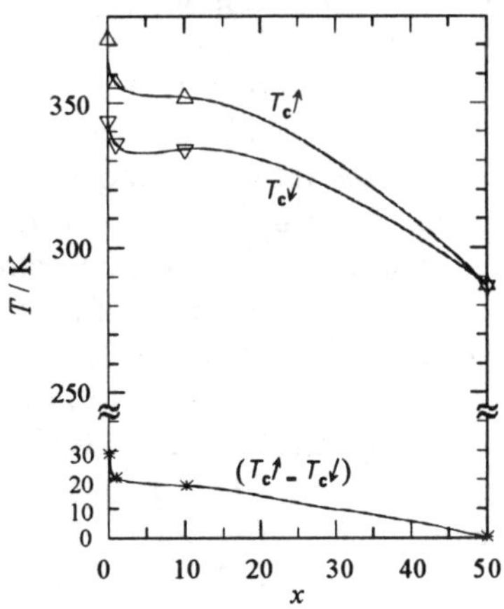

Fig. 3. Critical temperatures in the warming ($T_c\uparrow$) and cooling ($T_c\downarrow$) modes and width of the hysteresis loop ($T_c\uparrow - T_c\downarrow$), versus the concentration of Cu (x).

3 EPR Data

The EPR spectra have been measured in both the warming and cooling modes between 293 and 403 K for the stabilised hysteresis cycle, see Figures 4, 5 for 1 % and 10 % Cu samples.

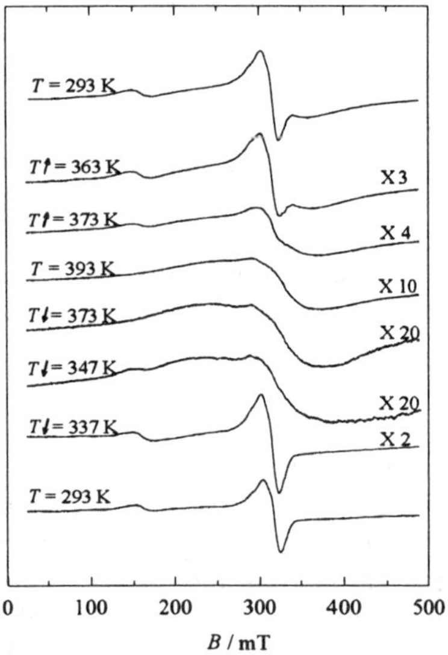

Fig. 4. EPR spectra of the 1 % Cu sample in the warming ($T\uparrow$) and cooling ($T\downarrow$) modes.

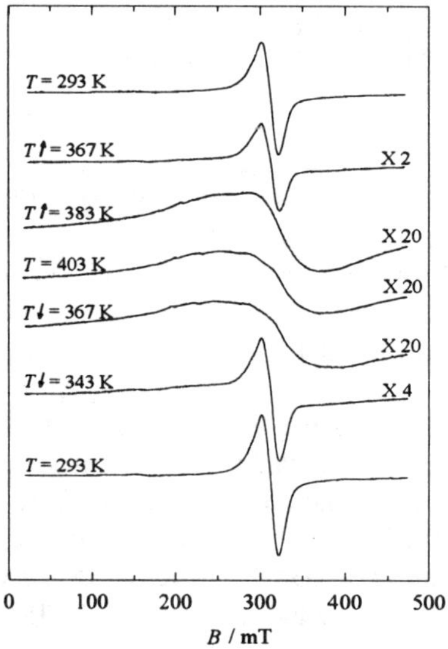

Fig. 5. EPR spectra of the 10 % Cu sample in the warming ($T\uparrow$) and cooling ($T\downarrow$) modes.

At 293 K (LS-state of Fe^{2+} ions), the EPR spectra exhibit a signal due to Fe^{3+} ions: a $g_{ef} = 4.3$ line accompanied by a plateau with a shoulder at $g_{ef} = 9.7$ and a broad line at $g_{ef} = 2.0$. In addition, an EPR signal with poorly resolved hyperfine splitting characteristic of Cu^{2+} ions is observed at $g_{ef} \approx 2$, superimposed with the $g_{ef} = 2.0$ line of Fe^{3+}. The intensity of the Cu^{2+} resonance is roughly proportional to the doping level: in the 1 % Cu sample the contributions of Fe^{3+} and Cu^{2+} ions are comparable, whereas in the 10 and 50 % Cu samples the former one is negligible. As the Cu contents grows, the Cu^{2+} resonance becomes more isotropic.

Lowering the temperature improves the resolution of the Cu^{2+} EPR signal; at 103 K a hyperfine pattern is resolved for both the 1 and 10 % Cu samples. See Figure 6. As the temperature increases above $T_c\uparrow$, the LS to HS transition takes place for the Fe^{2+} ions. The EPR spectrum of Fe^{3+} is transformed to a very broad line at $g_{ef} = 2.0$; the $g_{ef} = 4.3$ line and the plateau at $g_{ef} = 9.7$ disappear. The EPR spectrum of Cu^{2+} broadens, resulting in a structureless broad line with $g_{ef} \approx 2.05$.

In the cooling mode, as the temperature decreases below $T_c\downarrow$, the initial EPR spectrum shape is approximately restored. However, some discrepancies persist for several hours at room temperature. Namely, for the 1 % Cu sample the $g_{ef} = 2.0$ line arising from Fe^{3+} ions is very weak.

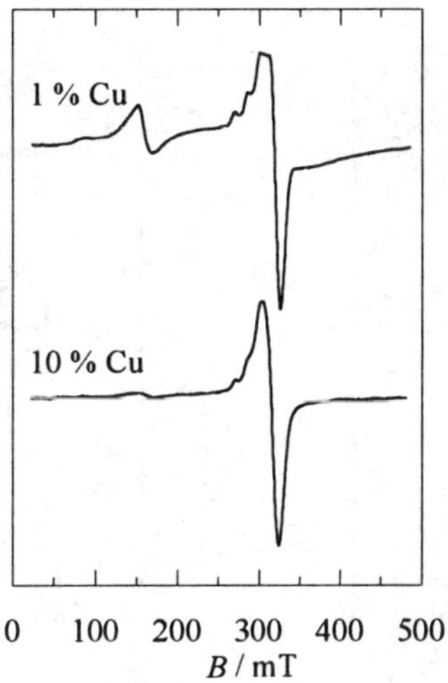

Fig. 6. EPR spectra at 103 K of the 1 and 10 % Cu samples.

4 Computer Simulations and Discussion

At room temperature (LS state of Fe^{2+} ions) the EPR spectra of the Cu^{2+} ions are temperature-broadened. One must go down to liquid nitrogen temperatures in order to get rid of this broadening. At 103 K the Cu^{2+} EPR spectrum (Figure 6) can be described by the spin Hamiltonian:

$$\mathcal{H} = \beta \boldsymbol{S} \cdot \boldsymbol{g} \cdot \boldsymbol{B} + \boldsymbol{S} \cdot \boldsymbol{A} \cdot \boldsymbol{I}, \tag{1}$$

where $\boldsymbol{g}$ and $\boldsymbol{A}$ are the g-tensor and the hyperfine coupling tensor, respectively.

In contrast to some other spin transition molecular complexes (*e.g.*, see [4]), in $[Fe(trz)(Htrz)_2](BF_4)$ no superhyperfine splitting can be resolved.

Computer simulations of this spectrum have been carried out for the 1 % Cu, after subtracting the underlying Fe^{3+} signal (Figure 7). The following Cu^{2+} spin Hamiltonian parameters in the LS state of Fe^{2+} ions are obtained:

$$g_x = 2.045,\ g_y = 2.125,\ g_z = 2.28,\ A_x = -2.5,\ A_y = -0.5,\ A_z = -15.0 \text{ mT}.$$

Theses values characterise a planar square configuration, typical of the Jahn-Teller d^9 electronic configuration of Cu^{2+} ions. (The A_x and A_y values are not very accurate, since the corresponding splittings are not resolved in the EPR spectra.) Note, however, that a considerable rhombic distortion is present.

At room temperature (Figure 8) the computer simulations result in the following Cu^{2+} spin Hamiltonian parameters:

$$g_x = 2.055,\ g_y = 2.135,\ g_z = 2.265,\ A_x = -4.5,\ A_y = -1.0,\ A_z = -12.5 \text{ mT}$$

(with somewhat greater linewidth parameters than at 103 K).

One can see that both the $\boldsymbol{g}$ and $\boldsymbol{A}$ tensors are more isotropic in this case, which can be explained by the dynamic Jahn-Teller effect.

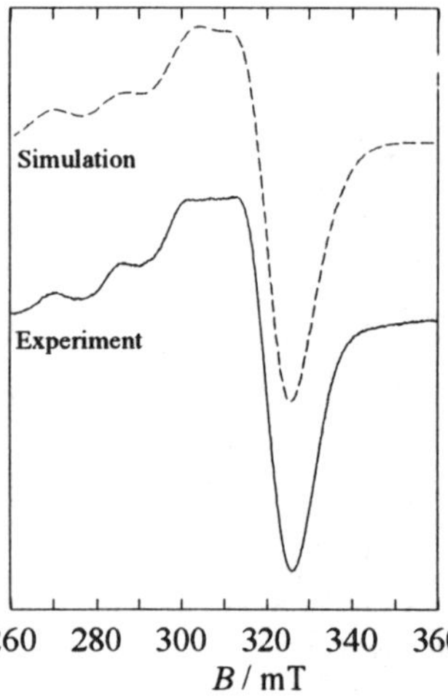

Fig. 7. Computer simulation of the EPR spectrum of the 1 % Cu sample at 103 K.

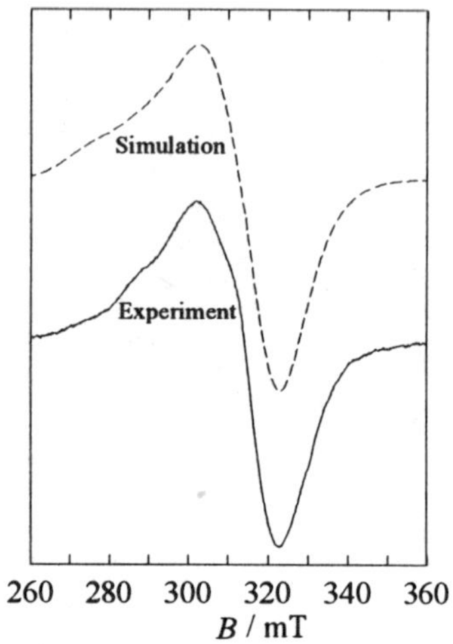

Fig. 8. Computer simulation of the EPR spectrum of the 1 % Cu sample at 293 K.

The Cu^{2+} EPR spectrum in the LS to HS transition of Fe^{2+} is severely broadened. Moreover, it is interesting to note that the *g*-value of the Cu^{2+} EPR line in the HS state of Fe^{2+}, $g_{ef} = 2.05$, is quite different from the *mean* *g*-value of the Cu^{2+} EPR spectrum in the LS state of Fe^{2+}, $\langle g \rangle = (g_x + g_y + g_z)/3 \approx 2.15$. This suggests a structural transformation related to the LS to HS transition, which is also confirmed by the modifications of the EPR spectra of Fe^{3+} and Cu^{2+} ions corresponding to the LS state of Fe^{2+} ions immediately after the HS to LS back transition.

Two mechanisms have been invoked in order to explain why the EPR signal of a spin probe broadens in the course of the LS to HS transition: local magnetic fields generated by the paramagnetic HS Fe^{2+} ions [5] and exchange broadening [4].

If one assumes the broadening of the Cu^{2+} EPR line is due to local magnetic fields, its linewidth can be calculated on the basis of the following considerations. In similar structures copper is supposed to substitute iron [6]. We assume for simplicity that the Cu^{2+} ions occupy at random a number of iron sites, the latter forming an infinite linear equidistant chain of constant *d*. In a *powder* EPR spectrum all orientations of the static magnetic field with respect to the local axes are equally probable. So, the peak-to-peak linewidth ΔB_{pp} due to a local magnetic field distribution generated by the HS Fe^{2+} ions can be calculated by suitably adapting the expression given (for a three-dimensional cubic lattice) in [7], which results in

$$\Delta B_{pp} = \frac{\mu_0}{2\pi} \sqrt{\tfrac{3}{5} S(S+1)}\, \frac{\beta g}{d^3} \left(2 \sum_{n=1}^{\infty} \frac{1}{n^6} \right)^{1/2} . \tag{2}$$

For the HS Fe^{2+} ions $S = 2$ and one gets

$$\Delta B_{pp} / \mathrm{T} = \frac{1.0 \times 10^{-29}}{(d / \mathrm{m})^3} . \tag{3}$$

Using a recent EXAFS evaluation of Fe–Fe distance, $d = 0.363$ nm, results in a peak-to-peak linewidth about 0.21 T, which exceeds the experimentally observed value $\Delta B_{pp} = 0.13\,T$. One can conclude that this mechanism certainly cannot be neglected. On the other hand, at such a distance between the paramagnetic ions the exchange interaction between Cu^{2+} and HS Fe^{2+} ions must also contribute in the EPR linewidth, resulting in a considerable exchange narrowing. Note also that the Cu^{2+} EPR line shape in the HS state of Fe^{2+} ions is rather Lorentzian.

Another mechanism of the Cu^{2+} EPR line narrowing is a fast spin-lattice relaxation of the HS Fe^{2+} ions, which may result in averaging of the local magnetic fields [5].

5 Conclusion

EPR and optical response studies of Cu-doped $[Fe(trz)(Htrz)_2](BF_4)$ show a spin transition of Fe^{2+} ions with hysteresis, the critical temperatures depending on the doping level. Using Cu^{2+} as paramagnetic probe, the spin transition is indicated by a dramatic broadening of its EPR spectrum. The significant change of the mean Cu^{2+} g-value in the course of the spin transition suggests that this transition is accompanied by a structural transformation. Computer simulations of the Cu^{2+} EPR spectra at 103 and 293 K reveal both a static and dynamic Jahn-Teller effect.

References

1. O. Kahn, J. Kröber and C. Jay, Spin transition molecular materials for displays and data recording, *Adv. Mater.* **4** (1992) 718-728.
2. O. Kahn and E. Codjovi, Iron(II)-1,2,4-triazole spin transition molecular materials, *Phil. Trans. R. Soc. Lond.* **A 354** (1996) 359-379.
3. C. Cantin, J. Kliava, Y. Servant, L. Sommier and O. Kahn, Fe^{2+} spin transition in $[Fe(trz)(Htrz)_2](BF_4)$ as evidenced by a variable temperature EPR study, *Applied Magn. Reson.* **12** (1997) 81-86; Variable temperature EPR study of Fe^{2+} spin transition in Cu^{2+}-doped $[Fe(trz)(Htrz)_2](BF_4)$, *ibid.* **12** (1997) 87-93.
4. W. Vreugdenhil, J.G. Haasnoot, O. Kahn, P. Thuéry and J. Reedijk, A copper(II) dope as a detector for the high-spin↔low-spin transition in the two-dimensional compound [*trans*-Bis(thiocyanato)bis(4,4'-bi-1,2,4-triazole)iron]hydrate, *J. Am. Chem. Soc* **109** (1987) 5272-5273.
5. P.S. Rao, A. Reuveni, B.R. McGarvey, P. Ganguli and P. Gütlich, Electron spin resonance and Mössbauer study of the high-spin–low-spin transition in $[Fe(phen)_2(NCS)_2]$ and $[Fe(pic)_3]Cl_2$.EtOH, *Inorg. Chem.* **20** (1981) 204-207.
6. W. Vreugdenhil, J.H. Van Diemen, R.A.G. De Graaff, J.G. Haasnoot, J. Reedijk, A.M. Van Der Kraan, O. Kahn and J. Zarembowitch, High-spin↔Low-spin transition in $[Fe(NCS)_2(4,4'\text{-bis-}1,2,4\text{-triazole})_2](H_2O)$, X-ray crystal structure and magnetic, Mossbauer and EPR properties, *Polyhedron* **9** (1990) 2971-2979.
7. S.A. Al'tshuler and B.M. Kozyrev, *Electron paramagnetic resonance in compounds of transition elements*, John Wiley & Sons, 1974, p.127.

EPR Study of Gd^{3+} Complexes Formed in Layered and Cubic Perovskite Crystals

Hideo Takeuchi, Masanori Arakawa, Hiroshi Ebisu*, Hidekazu Tanaka** and Hisashi Abe*

School of Engineering, Nagoya University, Chikusa-ku, Nagoya, 464-01, Japan
*Department of Physics, Nagoya Institute of Technology, Showa-ku, Nagoya, 466, Japan
**Department of Physics, Tokyo Institute of Technology, Meguro-ku, Tokyo, 152, Japan

Abstract. The basis of spin-Hamiltonian separation, which has been proposed by our group for the analyses of magnetic impurity centres in ionic crystals, are given for the magnetic impurity centre with orthorhombic symmetry. An application of the method to the Cr^{3+} and Gd^{3+} impurity centres in layered perovskite crystals Cs_2CdF_4, Rb_2CdF_4 and Rb_2ZnF_4 are presented. Some EPR results for Gd^{3+} centres in perovskite chlorides $CsCaCl_3$ and $RbCaCl_3$ are also reported.

1 Introduction

$KNiF_3$-type cubic perovskite crystals have advantages in their simple structure to other types of compounds for the study of formation mechanism of impurity centres with six-fold coordination in crystals. The simplicity comes from lack of nearby <110> cation, that may exist in the rock-salt type crystals, and high symmetry of configuration of surrounding farther ions. The structures of various magnetic impurity centres and metal-ligand interaction have been studied in our group for the magnetic ions in cubic perovskite fluorides and related materials by EPR and ENDOR techniques[1-26].

In the crystal growth of fluorides and chlorides doped with trivalent transition-metal or rare-earth ions by the Bridgman technique, usually several kinds of magnetic impurity centres are formed in the as-grown single crystal. This is due to charge misfit of the magnetic ions replacing for the host cations in matrix crystals. As for the cubic perovskite fluorides, transition-metal ions with $3d^n$ electronic configuration have a tendency to substitute for the host divalent cation. Trivalent transition-metal ions such as a Cr^{3+} ion existing in cubic perovskite fluorides are associated with a cation vacancy or a substituting monovalent cation at nearby divalent cation sites. In contrast to the transition-metal ions, rare-earth ions doped in the cubic perovskite fluorides show a variety of magnetic impurity centres. They substitute for both divalent and monovalent host cations in accordance with degree of ionic size fitness. In some cases, therefore, overcharged substitution may be induced by size misfit of magnetic ions for host cations.

K_2NiF_4-type layered perovskite crystals have close structure to the cubic perovskite crystals. They have the perovskite units compressed along the crystalline *c* axis in fluorides and elongated in chlorides and oxides, as shown in figure 1. Perovskite-type chlorides are also interesting compounds by their larger lattice constant and stronger covalency as compared with fluorides. These crystals are therefore expected to show differences in the formation of magnetic impurity centres from fluorides.

Until now, the authors have proposed the method of *spin-Hamiltonian separation* to analyse the experimental data obtained by EPR measurements for the magnetic impurity centres with low symmetry [5, 6, 9, 11-14, 16]. This method is useful to identify the magnetic centres formed in crystals. The low-symmetric fine structure terms in the spin Hamiltonian determined by fitting of EPR spectrum are separated into the terms of 2nd-, 4th- and 6th-rank irreducible tensor operators. It becomes possible by this method to compare the magnetic centres systematically among the same and different types of complexes formed in the same and different matrix crystals. Several other aspects of this method are mentioned elsewhere [16]. In this work we will present the basis of the spin-Hamiltonian separation for the magnetic centre with orthorhombic symmetry. Later, we will show some examples where the spin-Hamiltonian separation will be applied to identify the orthorhombic Cr^{3+} and Gd^{3+} centres observed in the layered perovskite fluorides Cs_2CdF_4, Rb_2CdF_4 and Rb_2ZnF_4 [9, 11, 26]. Some EPR results on the cubic perovskite chlorides $CsCaCl_3$ and $RbCaCl_3$ will be also presented. EPR spectra for chlorides were observed in their cubic phase using X-band spectrometer at the Centre for Instrumental Analysis at Nagoya Institute of Technology.

2 Basis of Spin-Hamiltonian Separation

In cubic perovskite crystals, trivalent transition-metal ions substitute for the host divalent cations. Generally, metal-ligand distance changes from that in the host crystal due to the difference of ionic sizes between the magnetic ion and the host cation.

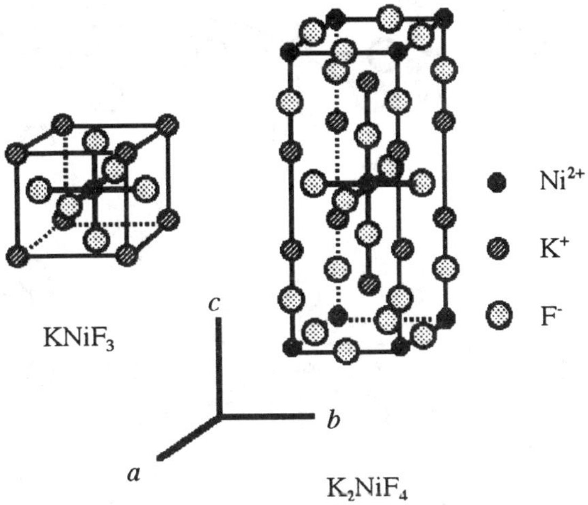

Fig. 1. Unit cells of the $KNiF_3$- and K_2NiF_4-type structure.

Surrounding farther ions except for six ligands make also cubic electric field at the central magnetic ion. Figure 2(a) shows schematically the ligand configuration in a cubic centre. Six metal-ligand distances along the crystalline *a, b* and *c* axes are the same.

When one or two of the ligands on the *c* axis deviate along the metal-ligand bond axis, the complex becomes to have tetragonal symmetry about the *c* axis. Figure 2(b) shows the case where two ligands on the *c* axis deviate toward the central magnetic ion by the same amount Δ_c. Such a tetragonal centre is formed in the K_2NiF_4-type crystals by the substitution for a divalent host cation. Figure 2(c) shows the tetragonal centre about the *b* axis, where two ligands on the *b* axis deviate toward the central magnetic ion by the same amount Δ_b. Another types of tetragonal centres are formed in cubic perovskite crystals when a trivalent magnetic ion substitutes for a host divalent cation with a charge compensator along the cubic axis. It has been found empirically by EPR investigations in our group for the tetragonal centres of Cr^{3+}, Fe^{3+} and Gd^{3+} ions that the 2nd-rank fine structure parameters b_2^0 are negative for tetragonal compression and positive for tetragonal elongation of ligand octahedra [5-14, 16]. Thus, for these ions the tetragonal centres of fig.2(b) and 2(c) may give negative b_2^0 about the *c* and *b* axes, respectively.

When ligand deviation of fig.2(b) and 2(c) happens simultaneously, the resultant ligand configuration becomes that in fig.2(d), where four ligands in the plane perpendicular to the *a* axis deviate toward the central ion with the same amounts $\Delta_c = \Delta_b$. The configuration is that for the tetragonal centre about the *a* axis, where the ligand octahedron distorts to an elongated configuration along the *a* axis in consequence. This centre gives positive b_2^0 corresponding to the elongated octahedron. The third tetragonal centre shown in fig.2(d) can be regarded as a result of two independent tetragonal

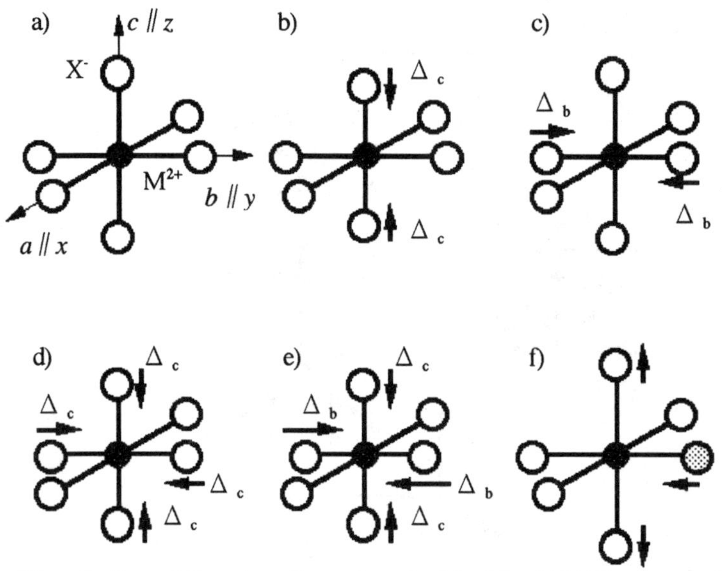

Fig. 2. Several ligand configurations for (a)cubic, (b)(c)(d) tetragonal, and (e)(f)orthorhombic centres.

distortions of the octahedron. The situation is represented mathematically by the spin Hamiltonian as follows:

$$b_2^0\,[3S_x^2-S(S+1)] = -\,b_2^0\,[3S_z^2-S(S+1)] - b_2^0\,[3S_y^2-S(S+1)], \qquad (1)$$

where the x, y and z axes are chosen parallel to the a, b and c crystalline axes, respectively. It should be emphasized that the respective 1st and 2nd terms in the right-hand side of eq.(1) have physical meanings as the uniaxial terms about the z and y axes.

Now, we consider what may happen if the amounts of two deviations are different, that is $\Delta_c \neq \Delta_b$. The resultant ligand configuration makes the complex an orthorhombic centre, where EPR spectra for trivalent magnetic ions are described by the spin Hamiltonian with orthorhombic symmetry. Here, we consider only the positive deviations Δ_c and Δ_b. As for the main quantum axis (ζ axis), three cases are expected. Case [I]; if the deviation along the c axis is comparable to that along the b axis ($\Delta_c \fallingdotseq \Delta_b$), the ζ axis is parallel to the a axis with positive b_2^0. Case [II]; if the deviation along the c axis dominates that along the b axis ($\Delta_c \gg \Delta_b$), the ζ axis is parallel to the c axis with negative b_2^0. Case [III]; if the deviation along the b axis dominates that along the c axis ($\Delta_c \ll \Delta_b$), the ζ axis is parallel to the b axis with negative b_2^0. The three cases are represented mathematically by the spin Hamiltonian terms as follows. For the case [I],

$$b_2^0\,[3S_x^2-S(S+1)] + b_2^2\,[S_y^2-S_z^2] = b_{2a(1)}\,[3S_z^2-S(S+1)] + b_{2a(2)}\,[3S_y^2-S(S+1)], \qquad (2)$$

with

$$b_{2a(1)} = -\,b_2^0 - (1/3)b_2^2 \qquad (3)$$
$$b_{2a(2)} = -\,b_2^0 + (1/3)b_2^2. \qquad (4)$$

For the case [II],

$$b_2^0\,[3S_z^2-S(S+1)] + b_2^2\,[S_x^2-S_y^2] = b_{2a(1)}\,[3S_z^2-S(S+1)] + b_{2a(2)}\,[3S_y^2-S(S+1)], \qquad (5)$$

with

$$b_{2a(1)} = b_2^0 - (1/3)b_2^2 \qquad (6)$$
$$b_{2a(2)} = -\,(2/3)b_2^2. \qquad (7)$$

For the case [III],

$$b_2^0\,[3S_y^2-S(S+1)] + b_2^2\,[S_z^2-S_x^2] = b_{2a(1)}\,[3S_z^2-S(S+1)] + b_{2a(2)}\,[3S_y^2-S(S+1)], \qquad (8)$$

with

$$b_{2a(1)} = (2/3)b_2^2 \qquad (9)$$
$$b_{2a(2)} = b_2^0 + (1/3)b_2^2. \qquad (10)$$

In each case the fine structure terms with orthorhombic symmetry can be separated into two uniaxial terms about the c and b axes, which have physical meanings as axiality of the complex along the respective two axes. In other words, two origins of orthorhombic symmetry can be separated out. Each parameter $b_{2a(1)}$ or $b_{2a(2)}$ of the uniaxial terms is to be compared with those in the other centres in the same or different matrix crystals to clarify the causes of low symmetry or to evaluate the degree of each ligand deviation.

3 Application of Spin-Hamiltonian Separation Analysis

The crystals of Rb_2ZnF_4, Rb_2CdF_4 and Cs_2CdF_4 have the K_2NiF_4-type crystal structure. Here, we analyse the orthorhombic Cr^{3+} and Gd^{3+} centres formed in these crystals.

Centres with orthorhombic symmetry (centre III) were formed in these crystals except for the tetragonal centres (centre I) where Cr^{3+} simply substitute for divalent cations with no charge compensators in their immediate neighbourhood [9]. EPR spectra of the orthorhombic centres III can be described by the spin Hamiltonian in the (*bca*)-coordinate system, where x, y, z axes are parallel to the b, c, a cryatalline axes, respectively. The 2nd-rank fine structure parameters b_2^0 and b_2^2 obtained for Rb_2CdF_4 are listed in table 1. This centre corresponds to the Case I in the preceeding section. Two axial parameters $b_{2a(1)}$ and $b_{2a(2)}$ separated out from b_2^0 and b_2^2 are also listed in table 1. It can be seen that the parameters $b_{2a(1)}$ and $b_{2a(2)}$ are respectively comparable with b_2^0 for the centre I and b_2^0 for the Cr^{3+}-V_{Cd} centre in the cubic perovskite crystal $RbCdF_3$. The centre III may be therefore identified to be the Cr^{3+}- V_{Cd} centre in the layered perovskite crystal. It should be emphasized that the centre III has positive b_2^0 with $|b_2^2| < |b_2^0|$ contrary to the negative b_2^0 for the Cr^{3+}- V_{Cd} centre in $RbCdF_3$. The result shows that the Cr^{3+} ion in the centre III is associated with a nearest Cd^{2+} vacancy on the b axis, but not on the a axis.

The 2nd example of spin-Hamiltonian separation may be presented by the

Table 1. The 2nd-rank fine structure parameters observed and axial parameters calculated for the orthorhombic Cr^{3+} and Gd^{3+} centres. Units in $10^{-4}cm^{-1}$

Matrix	Centre	Coordinate System	b_2^0	b_2^2	$b_{2a(1)}$	$b_{2a(2)}$
Rb_2CdF_4	Cr^{3+} Centre I [9]	*abc*	-666.4	-	-666.4	-
	Cr^{3+} Centre III [9]	*bca*	+554.0	+334.2	-665.4	-442.6
	Gd^{3+} Centre I [11]	*abc*	-549.9	-	-549.9	-
	Gd^{3+} Centre III [11]	*abc*	-400.3	+311.8	-504.2	-207.9
	Gd^{3+} Centre V [26]	*cab*	-2373.1	+788.9	+525.9	-2110.1
Cs_2CdF_4	Gd^{3+} Centre V [26]	*cab*	-2330.6	+808.3	+538.9	-2061.2
Rb_2ZnF_4	Gd^{3+} Centre V [26]	*cab*	-2520.5	+903.2	+602.1	-2219.4
$RbCdF_3$	Cr^{3+}-V_{Cd} [7]	*cab*	-569.3	-	-	-569.3
	Gd^{3+}-V_{Cd} [8]	*cab*	-291.3	-	-	-291.3
$RbCaF_3$	Gd^{3+}-O^{2-} [27]	*cab*	-2614	-	-	-2614

orthorhombic Gd^{3+} centre (Centre III) in Rb_2CdF_4. In this case, the EPR spectrum can be described by the spin Hamiltonian with $|b_2^2|<|b_2^0|$ in the (*abc*)-coordinate system, corresponding to the Case II. The 2nd-rank fine structure parameters b_2^0, b_2^2 and axial parameters $b_{2a(1)}$, $b_{2a(2)}$ are listed in table 1. By comparing the axial parameters with b_2^0 for the centre I in Rb_2CdF_4 and b_2^0 for the Gd^{3+}- V_{Cd} centre in $RbCdF_3$, this orthorhombic centre is identified to be the Gd^{3+}-V_{Cd} centre in the layered perovskite crystal. Although EPR spectra of the centres III for Cr^{3+} and Gd^{3+} have rather different features to each other, these centres are identified to be the same kind of magnetic centres.

The 3rd example is the orthorhombic Gd^{3+} centres V in the layered perovskite fluorides. Some as-grown crystals of Rb_2ZnF_4, Rb_2CdF_4 and Cs_2CdF_4 doped with Gd^{3+} show EPR spectra having a large $|b_2^0|$ with ζ axis along the *a* or *b* axes [26]. It should be pointed out that only the centre V was observed for Rb_2ZnF_4 in contrast to Rb_2CdF_4 and Cs_2CdF_4, where the centres I and III were observed together with the centre V. It has been found in cubic perovskite fluorides that the large Gd^{3+} ion can substitute for large Ca^{2+} and Cd^{2+} at the body centre site of perovskite unit, but do not for small Mg^{2+} and Zn^{2+} ions [4, 25, 28]. When Gd^{3+} ions are doped in $KMgF_3$, $KZnF_3$ and $RbZnF_3$, Gd^{3+} substitute only for K^+ or Rb^+ ions at the 12-fold sites. A Gd^{3+} ion may be too large to substitute for Zn^{2+} ion also in Rb_2ZnF_4. Using the eqs.(8)-(10), axial parameters for the centres V are calculated as shown in table 1. By comparing $b_{2a(2)}$ with b_2^0 for the Gd^{3+}-O^{2-} centre in $RbCaF_3$ [27], these centres V are identified to be the Gd^{3+}-O^{2-} centre, where the Gd^{3+} ion is associated with a O^{2-} ion replacing a ligand fluorine on the *b* axis. It may be interesting that the substitution of a large Gd^{3+} ion for a small Zn^{2+} becomes possible by the simultaneous substitution of Gd^{3+} and O^{2-} ions. These centres have the positive axial parameters $b_{2a(1)}$ along the *c* axis in contrast to the negative values for the Cr^{3+} and Gd^{3+} centres III and I. As seen by comparing $b_{2a(1)}$ for the centres I and V in Rb_2CdF_4, the O^{2-} ion acts strongly on the ligand configuration. If one of the ligand fluorines is replaced by a O^{2-} ion, two ligand fluorines on the *c* axis may be strongly repulsed by the O^{2-} ion apart from the sites in the centre I. As suggested by the positive axial parameter $b_{2a(1)}$, resultant configuration of ligand octahedron is considered to become elongated along the *c* axis, as shown in fig.2(f). This distortion may make the space of the central ion site bigger enough for a Gd^{3+} ion to substitute for the central Zn^{2+} ion even in the matrix of Rb_2ZnF_4.

4 Gd^{3+} Centres in Cubic Perovskite Chlorides

From EPR studies of Gd^{3+} in the perovskite fluorides $CsCaF_3$, $CsCdF_3$, $RbCaF_3$ and $RbCdF_3$, magnitude of the parameter b_4 *increases* in the order $RbCdF_3 \rightarrow RbCaF_3$ and $CsCdF_3 \rightarrow CsCaF_3$ with increasing lattice parameter of the matrix crystal [29]. As found from EPR and ENDOR investigations, magnitude of the parameter b_4 is considered to decrease with increasing Gd^{3+}-ligand distance [23, 30].

For the cubic perovskite crystals $CsCaCl_3$ and $RbCaCl_3$ doped with Gd^{3+}, EPR measurements were made at 340℃ in their cubic phase. The angular variations of the EPR signals showed cubic symmetry with effective spin 7/2. The spectra can be ascribed to Gd^{3+} ions substituted simply for Ca^{2+} ions (hereafter called cubic centre). Determined fine structure parameter b_4 are -12 for $CsCaCl_3$ and -10 for $RbCaCl_3$ in units of $10^{-4}cm^{-1}$. Thus, the magnitude of the b_4 parameter *increases* in the order $RbCaCl_3 \rightarrow CsCaCl_3$, and then decreases in the order $CsCaCl_3 \rightarrow CsSrCl_3 \rightarrow CsPbCl_3$ [28] with increasing lattice parameter of the matrix crystal. This variation of $|b_4|$ in the region of small lattice parameter as chloride crystals is consistent with those found in the series of the perovskite fluorides. It may be conjectured for each kind of ligand that the magnitude of the b_4 parameter for Gd^{3+} in the series of perovskite crystal increases and shows maximum and then decreases with increasing the lattice parameter of matrix crystal.

Acknowledgments

The authors are grateful to Dr Masahiro Mori for useful measurements of x-ray diffraction and x-ray irradiation.

References

1. J. L. Patel, J. J. Davies, B. C. Cavenett, H. Takeuchi, and K. Horai, Electron spin resonance of axially symmetric Cr^{3+} centres in $KMgF_3$ and $KZnF_3$, *J. Phys. C: Solid State Phys.* **9** (1976) 129-138.
2. K. Horai, M. Arakawa, and H. Takeuchi, Evidence for Fluorine Distortion in $KCdF_3$ Doped with Mn^{2+} Derived from Ligand Fluorine ENDOR Measurement, *J. Phys. Soc. Jpn.* **43** (1977) 2095-2096.
3. H. Takeuchi, K. Horai, and M. Arakawa, An Investigation of V^{2+} Impurity Center in $KMgF_3$ and $KZnF_3$ by ^{51}V-ENDOR Measurements, *J. Phys. Soc. Jpn.* **46** (1979) 18-25.
4. M. Arakawa, H. Aoki, H. Takeuchi, T. Yosida, and K. Horai, Electron Paramagnetic Resonance of Gd^{3+} Centres in Several Perovskite Fluorides, *J. Phys. Soc. Jpn.* **51** (1982) 2459-2463.
5. H. Takeuchi, M. Arakawa, H. Aoki, T. Yosida, and K. Horai, EPR and ^{19}F-ENDOR of Cr^{3+} Impurity Centres in K_2ZnF_4 and K_2MgF_4, *J. Phys. Soc. Jpn.* **51** (1982) 3166-3172.
6. H. Takeuchi, and M. Arakawa, EPR of Cr^{3+} Centres in K_2ZnF_4 Crystals Doped with Na^+ and Li^+ Ions, *J. Phys. Soc. Jpn.* **52** (1983) 279-283.
7. H. Takeuchi, and M. Arakawa, EPR Study of Cr^{3+}-Li^+ Centres in Several Perovskite Fluorides, *J. Phys. Soc. Jpn.* **53** (1984) 376-380.
8. M. Arakawa, H. Ebisu, and H. Takeuchi, EPR Study of Gd^{3+}-V_M and Gd^{3+}-Li^+ Centres in Several Perovskite Fluorides, *J. Phys. Soc. Jpn.* **54** (1985) 3577-3583.
9. M. Arakawa, H. Ebisu, and H. Takeuchi, EPR Study of Cr^{3+} Centres in Rb_2ZnF_4, Rb_2CdF_4 and Cs_2CdF_4 Crystals, *J. Phys. Soc. Jpn.* **55** (1986) 2853-2858.
10. H. Takeuchi, M. Arakawa, and H. Ebisu, EPR Study of Fe^{3+}-V_{Cd} and Fe^{3+}-Li^+ Centres in $RbCdF_3$ and $CsCdF_3$ Crystals, *J. Phys. Soc. Jpn.* **56** (1987) 3677-3682.
11. H. Takeuchi, M. Arakawa, and H. Ebisu, Electron Paramagnetic Resonance of Gd^{3+} Centres in Rb_2CdF_4 and Cs_2CdF_4 Crystals, *J. Phys. Soc. Jpn.* **56** (1987) 4571-4580 [Errata, **59** (1990) 2297].
12. M. Arakawa, H. Ebisu, and H. Takeuchi, Host Lattice Effects in the EPR Spectra of Cr^{3+}-Li^+ and Cr^{3+}-V_{Mg} Centres in A_2MgF_4 (A=K, Rb) Crystals, *J. Phys. Soc. Jpn.* **57** (1988) 2801-2804.

13. M. Arakawa, H. Ebisu, and H. Takeuchi, Evidence of Structural Phase Transition in K_2CdF_4 Derived from EPR Measurements for Cr^{3+} Centres, *J. Phys. Soc. Jpn.* **57** (1988) 3573-3579.
14. H. Takeuchi, H. Ebisu, and M. Arakawa, EPR Study of Tetragonal and Orthorhombic Fe^{3+} Centres in Rb_2CdF_4 Crystals, *J. Phys. Soc. Jpn.* **60** (1991) 304-312.
15. T. Yosida, H. Aoki, H. Takeuchi, M. Arakawa, and K. Horai, EPR, ^{19}F-ENDOR and ^{55}Mn-ENDOR of Mn^{2+} Impurity Center in MgF_2 Single Crystal, *J. Phys. Soc. Jpn.* **60** (1991) 625-635.
16. H. Takeuchi, M. Arakawa, and H. Ebisu, EPR study of tetragonal and monoclinic Fe^{3+} centres in Rb_2ZnF_4 crystals, *J. Phys. Condens. Mat.* **3** (1991) 4405-4420.
17. M. Arakawa, T. Hirose, and H. Takeuchi, EPR Study of Local Position for Fe^{3+} in Layer Oxide Bi_2WO_6, *J. Phys. Soc. Jpn.* **60** (1991) 4319-4325.
18. H. Takeuchi, and H. Tanaka, Angular variation of normal-type EPR intensities of V^{2+} in $CsMgCl_3$, *J. Phys. Condens. Mat.* **4** (1992) 10071-10078.
19. T. Hirose, M. Kawaminami, K. Obara, M. Arakawa, and H. Takeuchi, Thermal Anomalies in Bi_2WO_6 and EPR of Doped Gd^{3+} Ions, *J. Phys. Soc. Jpn.* **62** (1993) 1758-1766.
20. H. Takeuchi, H. Tanaka, and M. Arakawa, Direct Matrix Diagonalization of Spin Hamiltonian for Cr^{3+} Pair Centre in $CsMgCl_3$, *J. Phys. Soc. Jpn.* **62** (1993) 2518-2519.
21. H. Takeuchi, H. Tanaka, and M. Arakawa, An EPR study of anisotropies of several Cr^{3+} centres in $CsMgCl_3$, *J. Phys. Condens. Mat.* **5** (1993) 9205-9214.
22. H. Takeuchi, and M. Arakawa, Electrostatic theory and spin-Hamiltonian separation for $3d^3$ ions in orthorhombic and tetragonal centres, *J. Phys. Condens. Mat.* **6** (1994) 3253-3260.
23. H. Takeuchi, H. Ebisu, and M. Arakawa, EPR study of a trigonally symmetric Gd^{3+} centre in $RbCaF_3$, *J. Phys. Condens. Mat.* **7** (1995) 1417-1426.
24. M. Arakawa, H. Ebisu, and H. Takeuchi, EPR Study of Monoclinic Centres in $CsCdF_3$ and $CsCaF_3$ Co-Doped with Gd^{3+} and Na^+ Ions, *J. Phys. Soc. Jpn.* **64** (1995) 1356-1362.
25. M. Arakawa, H. Ebisu, T. Yosida, and K. Horai, Electron Paramagnetic Resonance Studies of Gd^{3+} in $KZnF_3$ and $KCdF_3$, *J. Phys. Soc. Jpn.* **46** (1979) 1483-1487.
26. M. Arakawa, H. Ebisu, and H. Takeuchi, EPR study of Gd^{3+}-O^{2-} centres in Cs_2CdF_4, Rb_2CdF_4 and Rb_2ZnF_4 crystals, *J. Phys. Condens. Mat.* **8** (1996) 11299-11308.
27. J. Y. Buzaré, M. Fayet-Bonnel, and J. C. Fayet, EPR investigations of the superhyperfine interaction and of the zero-field splittings for cubic and tetragonal Gd^{3+} centres at sixfold coordinated sites in AMF_3 crystals, *J. Phys. C : Solid State Phys.* **14** (1981) 67-81.
28. M. M. Abraham, C. B. Finch, J. L. Kolopus, and J. T. Lewis, Electron Paramagnetic Resonance of Several Rare-Earth Impurities in the Cubic Perovskite $KMgF_3$, *Phys. Rev.* **B 3** (1971) 2855-2864.
29. T. Rewaj, M. Krupski, J. Kuriata, and J. Y. Buzaré, Temperature and hydrostatic pressure dependences of the b_4^0 spin-Hamiltonian parameter for Gd^{3+} in fluoroperovskite single crystals, *J. Phys. Condens. Mat.* **4** (1992) 9909-9918.
30. R. A. Allsopp, J. M. Baker, L. J. C. Bluck, and J. Y. Buzaré, ^{19}F ENDOR measurements of transferred hyperfine interactions in $CsCdF_3$: Gd^{3+} and $CsCaF_3$: Gd^{3+}, *J. Phys. C : Solid State Phys.* **20** (1987) 451-462.

An Alternative Parametrization of S-state Ion EPR Spectra

Y.Y. Yeung

Department of Science and Mathematics, Hong Kong Institute of Education
Northcote Campus, 21 Sassoon Road, Hong Kong

Abstract. The EPR spectra of S-state ions doped at some low-symmetry sites are parametrized in an alternative form using the concepts of splitting moments and rotational invariants. In particular, the quadratic moment with magnetic field H is obtained as

$$\sigma_2^2(H) = s_1^2 S(S+1) + \sigma_2^2(O)$$

where $s_1^2 = (1/3)\beta^2 g^2 H^2$ is the rank one rotational for Zeeman interaction, taking account of the fact that g is very isotropic in the ground state of an S-state ion. The second term of the above equation represents the quadratic moment for fine structure splittings. This quadratic moment σ_2^2(H) can be used to determine the g-value directly without the involvement of other spin parameters.

1 Introduction

It is well-known that many parameters are required to describe the zero-field splitting components of the spin Hamiltonian for the EPR spectra of the S-state ion doped at some low-symmetry sites. Nonlinear least squares fits to over ten parameters are notoriously unreliable unless approximate values are known first in order to initiate the fitting procedure. Furthermore, there may exist a continuous minimum in the fitting space because of the indeterminacy of the coordinate axes for low-symmetry systems.

For crystal field in lanthanide systems, Yeung and Newman (1986) have successfully extended and generalized some high order crystal field invariants and worked out their relationships with high order J-multiplet splitting moments. This leads to an alternative scheme of parametrization of the lanthanide crystal field in which the new parameters involved are invariant under any rotation of the coordinates.

For the ground state splitting of the S-state ions, they are traditionally parametrised by a spin Hamiltonian within a model space spanned by the effective spin ***S*** as given below

$$\mathcal{H}_S = \sum_{kq} B_{kq} \mathbf{O}_q^{(k)}(\mathbf{S}) + \mathbf{S} \bullet \mathbf{g} \bullet \mathbf{H} \tag{1}$$

where the first term is known as the zero-field splitting (arising from the crystal field interaction) and the second one as the Zeeman term. All symbols have their usual meanings (see e.g. Newman and Urban 1975). Conventionally, both the g-tensor and the zero-field splitting parameters B_{kq} for low-symmetry sites are coordinate dependent and have to be fitted simultaneously to the observed energy levels, causing much complication and correlation. Hence, it is useful to decouple the two terms by some means.

2 Theory

Firstly, we adopt the Racah's normalization for the spin operator equivalent $\mathbf{O}_q^{(k)}(\mathbf{S})$ (Smith and Thornley 1966) so that

$$\left\langle S \middle\| \mathbf{O}^{(k)} \middle\| S \right\rangle = \frac{1}{2^k}\left(\frac{(2S+k+1)!}{(2S-k)!}\right)^{\frac{1}{2}}$$

and the associated parameters B_{kq} are directly related to the usual B_k^q normalised in Stevens' convention by some constant coefficients given by Kassman (1970). Then, the Zeeman term (second term of (1)) can be rewritten in terms of a rank one spin operator equivalent as

$$H_z = \sum_q B_{lq} \mathbf{O}_q^{(1)}(\mathbf{S}) \tag{2}$$

where

$$\mathbf{O}_0^{(1)} = \mathbf{S}_z \quad , \quad \mathbf{O}_{\mp 1}^{(1)} = \mp \frac{1}{\sqrt{2}} \mathbf{S}_{\pm},$$

$$B_{10} = \beta \sum_\alpha g_{z\alpha} H_\alpha$$

and

$$B_{1,\pm 1} = \mp \frac{1}{\sqrt{2}} \beta \sum_\alpha (g_{x\alpha} \mp i g_{y\alpha}) H_\alpha \cdot$$

On the other hand, the first order perturbation theory gives the energy of the zero-field splitting level Γ_i as

$$E_i = E^0 + \langle \Gamma_i | \mathcal{H}_S | \Gamma_i \rangle \tag{3}$$

where E^0 is the mean energy of the ground state and is usually taken as zero. The nth order moment of all ground state splitting levels is defined by

$$\sigma_n^n \equiv \frac{1}{2S+1}\sum_i (E_i - E^0)^n \tag{4}$$

In the approximation of (3), (4) becomes

$$\sigma_n^n = \frac{1}{2S+1}\sum_i < \Gamma_i|\mathcal{H}_s|\Gamma_i >^n = \frac{1}{2S+1} Tr(\mathcal{H}_s^n) \tag{5}$$

where Tr is the trace operation. The right-hand side of (5) can be evaluated explicitly using diagrammatic angular momentum theory (e.g., see Lindgren and Morrison 1982). Substituting (1) into (5), we obtain the expression for the quadratic moment with magnetic field H as

$$\sigma_2^2(\mathrm{H}) = \frac{1}{2\mathrm{S}+1}\left[\mathrm{s}_1^2 \left\langle \mathrm{S} \| \mathbf{O}^{(1)} \| \mathrm{S} \right\rangle^2 + \sum_{\mathrm{k\ even}} \mathrm{s}_\mathrm{k}^2 \left\langle \mathrm{S} \| \mathbf{O}^{(\mathrm{k})} \| \mathrm{S} \right\rangle^2 \right]$$

$$= \frac{1}{3}\beta^2 g^2 H^2 S(S+1) + \sigma_2^2(O) \tag{6}$$

where we have written $s_1^2 = \frac{1}{3}\beta^2 g^2 H^2$ as the rank one rotational invariant (which is simply the average component) for Zeeman interaction, taking account of the fact that g is very isotropic in the ground state of an S-state ion.

3 Discussion

Immediately, we see from (6) that $\sigma_2^2(H)$ can be used to determine the g-value directly without the involvement of other spin parameters e.g. $\sigma_2^2(2H) - \sigma_2^2(H) = \beta^2 g^2 H^2 S(S+1)$ or the slope of the graph of $\sigma_2^2(H)$ vs H^2 is $\beta^2 g^2 S(S+1)$ and this method is independent of coordinate frame. Moreover, the quadratic invariant s_k for zero-field splitting (k even and < 2S) may be very useful in representing the overall contributions from different rank k whereas the B_{kq} parameters differ by orders of magnitudes. If the ligand positions relative to the S-state ion are known, we may also apply the crystal field invariant method, which was originally proposed by Yeung and Newman (1985) to analyze lanthanide crystal field, for the S-state ion situated in low-symmetry systems because the superposition model is found to be fairly reliable (Newman and Urban 1975).

In conventional EPR experiments, constant microwave frequencies are used with a variable external magnetic field. This means that is $\sigma_2^2(H)$ unobtainable directly from observed EPR transitions. Therefore, the moment approach is applicable only in the EPR spectra measured by the variable frequency method with constant magnetic field (see e.g. Newman and Urban 1975).

4 Concluding Remarks

This paper deals with zero-field splitting and Zeeman splitting together which are physically different from the lanthanide crystal field splitting considered in all the previous papers. The present work applied the rotational invariant and splitting moment concept and the corresponding formulation for the lanthanide crystal field Hamiltonian to the zero-field Hamiltonian with extension to include the magnetic field dependent term - Zeeman interaction. The application was not a straightforward process which actually involved many sophisticated manipulation of some angular momentum tensor operators. Hence, the equation (6) is a piece of new and significant result which gives rise to an alternative method of parametrizing the spin-Hamiltonian parameters of S-state ions doped in some low-symmetry sites. The main advantage of this method is that it is independent of the coordinate frame (i.e. rotational invariant).

References

Kassman, A. J. 1970, J. Chem. Phys. **53**, 4118.

Lindgren, I. and J. Morrison 1982, "Atomic Many-Body Theory" (Springer, Berlin).

Newman, D. J. and W. Urban 1975, *Adv. Phys.* **24**, 793

Smith, D. and J. H. M. Thornley 1966, Proc. Phys. Soc. **89**, 779.

Yeung, Y. Y. and D. J. Newman 1985, *J Chem Phys.* **82**, 3747

Yeung, Y. Y. and D. J. Newman 1986, *J. Chem. Phys.* **84**, 4470.

The g and D tensor of Gd^{3+} in $BiVO_4$ Single Crystal

Tae Ho Yeom,[†] *Sung Ho Choh,*[‡] *and Czeslaw Rudowicz*[¶]

[†] Department of Physics, Chongju University, Chongju 360-764, Republic of Korea
[‡] Department of Physics, Korea University, Seoul 136-701, Republic of Korea
[¶] Department of Physics and Materials Science, City University of Hong Kong, Hong Kong

Abstract. Electron paramagnetic resonance of the Gd^{3+} ion in a ferroelastic $BiVO_4$ single domain crystal has been investigated at room temperature using a Q-band spectrometer. The rotation patterns of the resonance fields measured on the crystallographic planes are analyzed and the monoclinic spin Hamiltonian parameters of Gd^{3+} ion are determined. The principal Z-axis of the second-order zero-field splitting tensor **D** is found to coincide with the crystallographic b-axis. The principal values g_{ii} and D_{ii} as well as the principal axes indicate that the local site symmetry of Gd^{3+} ion in $BiVO_4$ crystal is monoclinic.

1 Introduction

Electron paramagnetic resonance (EPR) as well as NMR studies have been performed to investigate various properties of $BiVO_4$ [1-7]. In this preliminary report, we study the rotation patterns of the resonance fields of Gd^{3+} impurity on the crystallographic ab, bc^*, and c^*a planes at Q-band and room temperature. EPR spectra of Gd^{3+} (S = 7/2) ion are analyzed using the monoclinic spin Hamiltonian. The principal axes X, Y, and Z of the second-order zero-field splitting (ZFS) tensor **D**, and **g**-tensor are determined.

2 Experimental Aspects

Single domain $BiVO_4$ crystals were grown by the Czochralski method [8]. The crystallographic principal axes were determined by the x-ray Laue method. The crystallographic data [9] show that the monoclinic C_2 axis is parallel to the crystallographic b axis with the choice a = 51.966 nm, b = 117.044 nm, c = 50.935 nm. Q-band EPR spectra at room temperature were recorded using a 100 kHz field modulation. The rotation patterns of the resonance fields on the crystallographic ab-, bc^*-, and c^*a-plane were obtained at 1° or 4° intervals at room temperature. The c^*-axis was chosen to be perpendicular to the ab-plane, but it is nearly parallel to the c-axis, since β is slightly different from 90.0°. The accuracy of the sample alignment is estimated to be within ± 0.5°.

3 Spin Hamiltonian Analysis

The experimental resonance fields for Gd^{3+} (S = 7/2) ion can be analyzed using the monoclinic spin Hamiltonian with the axis y//C_2 [10]:

$$H = \mu_B \mathbf{B}\cdot\mathbf{g}\cdot\mathbf{S} + B_2^0 O_2^0 + B_2^1 O_2^1 + B_2^2 O_2^2 + B_4^0 O_4^0 + B_4^1 O_4^1 + B_4^2 O_4^2 + B_4^3 O_4^3 + B_4^4 O_4^4 . \quad (1)$$

The "laboratory" axes are labeled (x, y, z) and are chosen to coincide with (a, b, c^*). Magnetic axes are labeled (X, Y, Z). The direction of the magnetic field was measured with a precision better than 0.1°. The computer program (EPR.FOR version 5.13) yields the final best fit of the **g** and **D** components as listed in Table 1. The corresponding Stevens parameters : B_2^0 = 177.4, B_2^1 = -91.4, B_2^2 = 385.9 (in 10^{-4} cm^{-1}) agree with the earlier ones [1]. The sum of the final root mean squares of weighted differences between the observed and calculated values is found to be 34 MHz for the transition frequencies and 1.2 mT for the resonance fields.
The estimated statistical uncertainties in the last significant digit are given in parentheses. The principal Z-axis of the **D**-tensor is found to lie along the crystallographic b-axis. The other principal directions are in the plane normal to the b-axis. One principal axis of the **g**- tensor and that of the **D**-tensor coincides with the monoclinic direction, whereas the non-coincidence of the two remaining principal axes of **g** and **D** reveals the monoclinic symmetry effects for Gd^{3+} in $BiVO_4$ crystal. This EPR study confirms that the Gd^{3+} site symmetry in the ferroelastic $BiVO_4$ is monoclinic. Full analysis, including the fourth- and the sixth-order ZFS parameters, will be published in a forthcoming paper.

Table 1. The principal values and axes of matrices Y = **g**, **D** for Gd^{3+} in $BiVO_4$ single crystal at room temperature. The y-axis is taken as parallel to the C_2 axis.

Y			k	Y_k	θ_k (deg)	φ_k (deg)
			g			
1.9757(2)	0	0.0005(2)	1	1.9812(4)	90	90
	1.9812(4)	0	2	1.9764(2)	143(9)	180
		1.9760(2)	3	1.9753(2)	53(9)	180
			D (10^{-4} cm^{-1})			
208.4(2)	0	-45.7(2)	1	368.0(2)	164.0(1)	0
	-563.3(2)	0	2	195.3(2)	106.0(1)	180
		354.8(2)	3	-563.3(2)	90	90

Acknowledgements: This work was supported by the KOSEF through the RCDAMP at Pusan Nat. Univ. (1994-97) and UGC /CityU grant. The program EPR.FOR ver. 5.13 (authors: D. G. McGavin, M. Y. Mombourquette and J. A. Weil; Dept. of Chemistry, Univ. of Saskatchewan, Canada) has been employed.

References

1. N. P. Baran, V. I. Barchuk, V. G. Grachev and B. K. Krulikovskii, Spectra of paramagnetic probes of Gd, Er, and Mn in the paraelastic and ferroelastic phases of $BiVO_4$, *Sov. Phys. Crystallogr.* **30** (1985) 410-414.
2. N. P. Baran, V. I. Barchuk, V. G. Grachev and B. K. Krulikovskii, Critical temperature dependences of the characteristic ESR spectra of Gd^{3+} and Mn^{2+} in $BiVO_4$ paraelastic and ferroelastic phases, *Sov. Phys. Solid State*, **28** (1986) 485-488.
3. E. Y. Moon, S. H. Choh, Y. H. Park, H. Y. Yeom and M. S. Jang, A new twin structure of ferroelastic $BiVO_4$, *J. Phys. C: Solid State Phys.* **20** (1987) 1867-1873.
4. T. H. Yeom, S. H. Choh, and M. S. Jang, Further study of the Mn^{2+} EPR center in ferro- and para-elastic $BiVO_4$: spin Hamiltonian parameters and temperature dependence, *J. Phys. Condens. Matter* **4** (1992) 587-598.
5. T. H. Yeom, A. R. Lim, S. H. Choh and K. S. Hong, Temperature dependence of the ^{51}V NMR spectrum in a ferroelastic $BiVO_4$ single crystal, *J. Korean Phys. Soc.* **26** (1993) 670-674.
6. T. H. Yeom, S. H. Choh, M. L. Du, and M. S. Jang, EPR study of Fe^{3+} center in $BiVO_4$ single crystal: experimental and theoretical investigation of spin Hamiltonian parameters, *Phys. Rev. B* **53** (1996) 3415-3421.
7. T. H. Yeom, C. Rudowicz, S. H. Choh, and D. G. McGavin, Monoclinic spin Hamiltonian analysis of EPR spectra of Mn^{2+} in $BiVO_4$ single crystals, *phys. stat. sol. (b)* **198** (1996) 839-851.
8. M. S. Jang, H. L. Park, J. N. Kim, J. H. Ro and Y.H. Park, Raman spectrum in monoclinic $BiVO_4$, *Jpn. J. Appl. Phys.* **24-2** (1985) 506-507.
9. L. P. Avakyants, A. M. Antonenko, E. F. Dudnik, D. F. Kiselev, I. E. Mnushkina, and M. M. Firsova, Elastic constants of ferroelastic $BiVO_4$, *Sov. Phys. Solid State* **24** (1982) 1411-1412.
10. S. K. Misra and C. Rudowicz, Effect of monoclinic symmetry on the EPR spectra of Gd^{3+}-doped hydrated single crystals of rare-earth trichlorides, *phys. stat. sol. (b)* **147** (1988) 677-684.

FMR of Co/Cu Multilayers

Zhong J. Wang, S. Mitsudo, K. Watanabe, K. Saito, H. Fujimori and M. Motokawa

Institute for Materials Research, Tohoku University, Katahira, Sendai 980-77, Japan

Abstract. Angular dependent ferromagnetic resonance (FMR) experiments have been performed on $[Co(15Å)/Cu(d_{Cu})]_{20}$ multilayer systems. In the strong FM-IEC region (d_{Cu}<7Å), bulk spin wave modes were found. While in the weak IEC region (d_{Cu}>9Å), besides the main modes the surface modes were also observed. With increasing the Cu layer's thickness the oscillation effect of the FMR spectra was found. The oscillation period was the same as that of the saturation field and the giant magnetoresistance.

1 Introduction

Magnetic multilayers have attracted significant attention since 80's because of their various fascinating properties[1]. So far unique properties have been found in magnetic multilayer systems, for example, the interlayer exchange coupling(IEC) and giant magnetoresistance effect (GMR) which are the two of the most important properties found recently and are being intensively studied now. Because of the IEC bridge the multilayer system is connected together, and the excitation of the crossing-multilayer collective spin waves can be expected. Investigation of spin wave excitation in multilayer systems will give valuable information on IEC. Now FMR has been successfully used in the investigation of the interlayer coupled sandwich systems[2], but the FMR study on the real multilayer structures, especially on magnetic/non-magnetic multilayer systems are not conclusive yet. In the previous work we observed a series of spin wave modes[3] in the strong FM-IEC region and found that the experimental results could be well explained by van Stapele's theory[4]. Here we have performed an FMR experiment on $[Co(15Å)/Cu(d_{Cu})]_{20}$ multilayer systems (5.2Å≤d_{Cu}≤80Å), and an oscillation effect of FMR spectra was observed.

2 Experimental Method

All of the samples were prepared on the single crystal Si substrates using ion beam sputtering technique under the same deposition conditions. The first Co layers were deposited onto Si substrates directly and the last Co layers were capped with Cu layers. A total of 20 Co/Cu bilayers were prepared. In order to analyze the FMR results, the structures and the magnetic properties of the samples were also examined by X-ray diffraction, SQUID and GMR measurements.

Usually FMR experiment of the multilayers is more difficult than that of the sandwiches. In the case of the sandwich structures, the optical mode, which directly reflects the interlayer coupling, can be easily observed by parallel-field FMR. In multilayers, usually only the uniform mode can be observed in the parallel-field FMR spectra[5]. Wigen et al. have found that the perpendicular-field FMR can provide more information than the parallel-field resonance[5]. But because of the strong demagnetizing field, a higher magnetic field, sometimes even higher than 2 T, is required to perform the perpendicular field X-band FMR, which is difficult to be achieved by a traditional electromagnet. To overcome this difficulty, new X-band and K-band FMR spectrometers were developed, and a Helium-free superconducting magnet (upper field is 6 T) was used. In addition the angular dependence and temperature dependence of FMR spectra can be easily investigated.

3 Results and Discussions

The large and small angle X-ray diffraction showed that both of Co layers and Cu layers grow along the cubic (111) direction and the samples have as good multilayer structures as the designed ones. In the SQUID and GMR measurements the oscillation effects of IEC and GMR were observed. For example, Fig. 1 (a) shows the Cu thickness dependence of the saturation field Hs, the higher Hs means the AF-IEC, while the lower Hs means the FM-IEC. The results of structure and magnetic measurements showed that we obtained the high quality of coupled multilayer systems.

All of the samples were investigated using the X-band FMR experiments at room temperature. The FMR results for the perpendicular and parallel configurations were summarized in Fig.1 (b) and (c). Two resonance modes, larger one and smaller one, were observed in the FMR spectra. The larger mode (main mode) was always observed and showed little changes with varying Cu thickness, so it can be easily determined as the uniform modes. The smaller mode showed a very complex but regular variations, implying that the resonance field of the smaller mode was dependent with the interlayer coupling. Thus the smaller mode must be some kind of spin wave modes.

In order to determine these smaller modes the angular dependent FMR spectra were measured for all of the samples. It was found that with varying thickness of the Cu layers the field angular dependence of FMR spectra was also changed. Generally speaking, the field angular dependence of FMR spectra had two kinds of patterns. The two typical examples was shown in Fig. 2 and Fig. 3.

As shown in Fig. 2 for the samples in the strong FM-IEC region (d_{Cu}<7Å), when the external field was rotated away from the normal direction and towards the parallel direction, the small peak became smaller and eventually disappeared. The larger modes are always the main mode. When $0° \leq \theta \leq 86°$ only the main mode could be observed. These results were very similar to that for strongly ferromagnetic

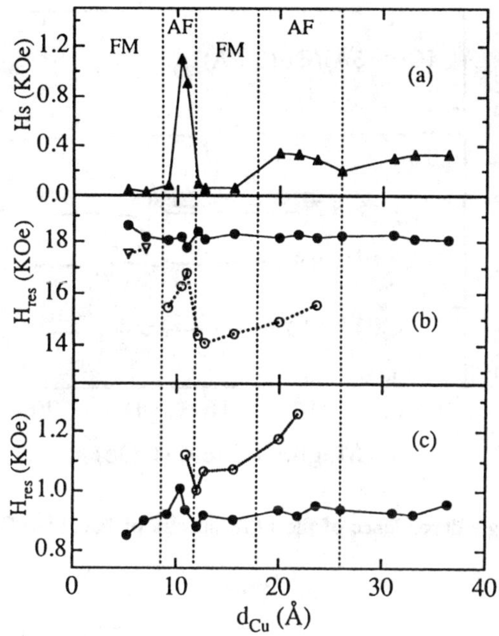

Fig. 1. The Cu thickness dependence of the parallel saturation fields (a), the perpendicular and parallel resonance fields (b) (c). ● o ∇ indicate the main modes, the surface modes and the spin wave modes, respectively. All of the experiments were made at room temperature.

coupled $[Co(50Å)/Cu(4Å)]_{20}$ multilayers [3], which implied that the smaller mode was the bulk spin wave mode. In addition, when the field was rotated away the normal direction the main mode became more larger, that means the main mode was not a complete uniform mode but a surface-like mode[5].

When $9Å<d_{Cu}<25Å$, for ether anti-ferromagnetic or ferromagnetic IEC multilayer systems the field angle dependence of the FMR spectra showed the similar variation patterns, as shown in Fig. 3. When the field got away the normal direction the larger mode (at the higher field side) became smaller while the smaller one (at the lower field side) became larger and turn to the main mode at last. For the strong AF-IEC sample $[Co(15Å)/Cu(10.4Å)]_{20}$, when $\theta^2 85°$ the higher field mode disappeared absolutely. For the other samples the higher field modes were observable until that the parallel direction was reached.

Wigen et al. supposed that in FMR the single magnetic layer behaves like a big moment. Using this model Wigen et al. derived the perpendicular resonant dispersion relation and predicted that the first bulk mode and the surface mode should be easily observed in the perpendicular resonance spectra[6]. With this theory Wigen et al. successfully explained the FMR spectra of FM-IEC Co/Pt multilayers[5]. According to Wigen's theory we assumed the smaller modes would be the surface spin wave modes. In order to confirm this conclusion we calculated

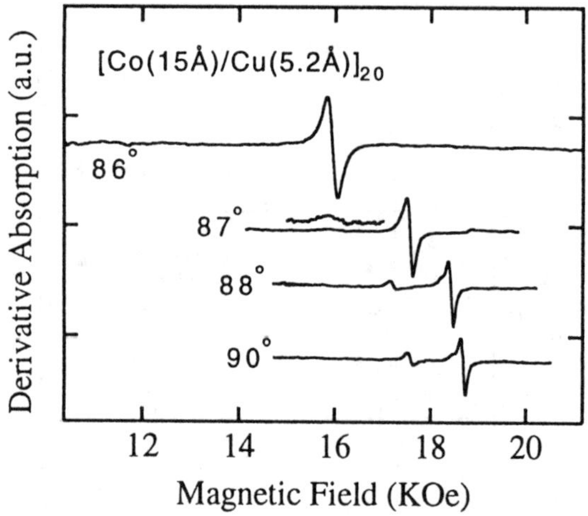

Fig. 2. The field angle dependence of the FMR spectra of [Co(15Å)/Cu(5.2Å)]$_{20}$. (at room temperature)

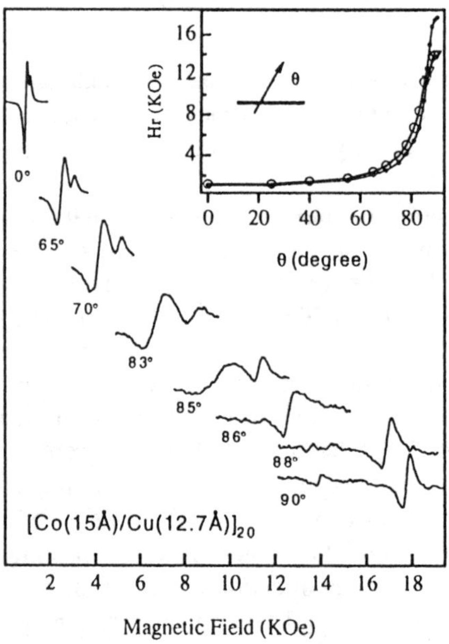

Fig. 3. The field angle dependence of the FMR spectra and the resonant fields of [Co(15Å)/Cu(12.7Å)]$_{20}$. (at room temperature)

the field angular dependence of the dispersion relations using Wigen's method. The calculated results were consistent with the experimental results. We also found that in the weak IEC region the surface mode can be observed even in the parallel resonance, while if IEC is not very weak the surface mode can not be observed in the parallel resonance, just as in the sample [Co(15Å)/Cu(10.4Å)]$_{20}$. The calculated results confirmed our analysis on the smaller modes.

In Fig. 1 we also found that the surface mode oscillated with the thickness of Cu layers. The oscillation period was the same as that of Hs and it indicated that the resonance field of the surface mode was dependent on the interlayer coupling. The oscillation effect of the surface mode can be easily explained by the calculated dispersion relation.

When 25Å<d_{Cu} , the main mode became weak and the surface mode disappeared. The weakness may arise from the interface roughness effect. With increasing the thickness of Cu layers the interface roughness would also increase, and the inter layer roughness would result in a wide distribution of anisotropy. As a result the resonance peak would become broader. This effect has been proved by the measurement of the peak-width. In addition, in this region the IEC was very weak and the surface mode in fact only included the surface layer, so it will be very weak and difficult to be observed.

We know that different from other Co layers which were sandwiched by the Cu layers, the first Co layer was deposited onto the Si substrate directly. According to the other experimental results, the anisotropy of the first Co layer may have a smaller easy-plane anisotropy than other Co layers. The surface spin wave modes just resulted from this anisotropy differences and located at the area near to the bottom of multilayer systems.

References

1. M.N. Baibich, J.M. Broto, A. Fert. F.Nguyen Van Dau, F. Petroff, P. Eittenne, G. Creuzet, A. Friederich and J. Chazelas, *Phys. Rev. Lett.*, 61 (1988) 2472.
2. B. Heinrich, S. T. Purcell, J. R. Dutcher, K. B. Urquhart, J. F. Cocvhran and A. S. Arrott, *Phys. Rev. Lett.*, 38 (1988) 12879.
3. Z. J. Wang, S. Mitsudo, K. Watanabe, S. Awaji, K. Saito, H. Fujimori and M. Motokawa, Spin Wave Resonance in Ferromagnetic Coupled Co/Cu Multilayers (submitted to *J. Magn. Magn. Mat.*).
4. R. P. van Stapele, F. J. A. M. Greidanus and J. M. Smits, J. Appl. Phys. 57 (1985) 1282
5. Z. Zhang, P. E. Wigne and T. Suzuki, *J. Magn. Soc. Jpn.* Vol. 17, Supplement No. S1 (1993) pp119.
6. P. E. Wigen, Resonance in Coupled Ferromagnetic Layer Structures, *Magnetic Multilayers*, edited by L. H. Bennett and R. E. Watson, (World Scientific publishing Co. Ptc. Ltd. 1994).

EPR Studies of Phase Transitions and Structural Solitons in Improper Ferroelastic $MgTiF_6\,6H_2O$ Doped with 3d - ions.

A.M. Ziatdinov, V.G. Kuryavyi and T.F. Antohina

Institute of Chemistry, Far Eastern Division of the RAS,Vladivostok, Russia.

Abstract. Electron paramagnetic resonance (EPR) of Mn^{2+} and Ni^{2+} in magnesium fluotitanate hexahydrate ($MgTiF_6\,6H_2O$) crystals has been studied at X - and Q - band frequency in the temperature interval 100 ÷ 400 K. At cooling of crystals with natural abundance of 3d - ions, at $T_c \approx 300$ K they undergo an improper ferroelastic first-order phase transition with the unit cell doubling. Above T_c a succession of step-wise changes in the slope of Mn^{2+} line shape splitting curve is observed. These step-wise discontinuities were interpreted as a result of structural phase transitions of the incommensurate phase - incommensurate phase type. In incommensurate phases the Mn^{2+} line shape was analyzed using the structural soliton model. In mixed ferroelastics $Mg_{1-x}Zn_xTiF_6\,6H_2O$ at $x < 0.75$ (>0.75) the Mn^{2+} EPR spectrum parameters and the phase transition temperatures are close to those in $MgTiF_6\,6H_2O$ ($ZnTiF_6\,6H_2O$) crystals.

1 Introduction

Magnesium fluotitanate hexahydrate (MFTH) belongs to isomorphous compounds of the type $ABF_6\,6H_2O$ (where A and B are di- and fourvalent metals, respectively) in which two complex ions $A(H_2O)_6$ and BF_6 can be distributed between two orientations around the 3 - fold axis [1,2] . The presence of superstructure reflections first observed by X-ray diffraction of Mg, Mn and Fe compounds is inconsistent with the structure models proposed before (space group: P3m) [1-5]. Therefore, Chevrier [6-8] had suggested new structure models: the crystals consist of domains (space group: P3) with two different octahedra orientations related by pseudomirrors (110), and with equal volume parts. In the case of Mg and Mn fluosilicates (T > 300 K [6]; T > 240 K [7]), these domains are of great extension, whereas in Fe - fluosilicates (T > 240 K [8]), their size is a function of temperature. A number of compounds are characterized by an improper ferroelastic phase transition from a rhombohedral modification to a low-temperature monoclinic phase [3, 5-11]. The latter is stable at low temperatures. According to the literature data at $T \approx 300$ K MFTH crystals undergo the 1-st order improper ferroelastic phase transition, which was attributed to a direct transition of the type R3m → P2 $_1$ /c [12] . In this paper we present experimental EPR results of MFTH crystals containing i) natural abundance of 3d - ions (<10 at.%), ii) ≈0.1 at. % Mn^{2+} and iii) ≈3 at.% Ni^{2+}. Analisys of EPR data suggests that the present point of view of high temperature crystal structure organization is incomplete and indicates the presense an unknown succession of phase transitions and regions with the lattice displacement modulation (structural solitons) in these crystals at $T>T_c$.

2 Experimental Procedure

The EPR measurements were carried out using X - and Q - band spectrometers in three mutually perpendicular crystal planes.The studies were performed in the temperature range 100 ÷ 400 K. Cooper-constantan thermocouples were used to monitor temperatures with one junction attached to the outside of the wave -guide close to the sample. The accuracy in the measuring temperature was ≈ 0.3 K. Single crystals of MFTH were prepared from aqueous solutions at 300 K.

3 Result

The EPR investigations of MFTH crystals containing natural abundance of Mn^{2+} and other ions show (Figure 1) that at cooling at T_c = 300 ± 1 K they undergo the 1st order improper ferroelastic phase transition. (The T_c is defined as the temperature at which intensities of Mn^{2+} EPR low field hyper-fine structure (HFS) lines from different crystal phases are equal.) The value of the temperature hysteresis of transition is ~ 10 K . Below T_c , the EPR spectra of ~ 0.1 at. % Mn^{2+} and ~ 3 at. % Ni^{2+} correspond to six spatially inequivalent rhombohedral centres, two of which came from orientetional domain. The principial axis of a single centre for both kinds of ions and the crystal trigonal axis are rotated by $(8 \pm 2)^{o}$. The angular dependence of a single magnetic centre spectrum for Mn^{2+} (Ni^{2+}) is in good agreement with the spin Hamiltonian of a rhombohedral symmetry with the following parameters at 300 (250) K: g=2.0009 ± 0.0003, D = - 203 ± 1, |E| = 28 ± 0.5, a = 9 ± 1, $A_\perp$ = $A_\parallel$ = - 89 ± 1 (g= 2.25 ± 0.01, D = 2.20 ± 0.02, |E| = 0.22 ±0.03), where D, E, A and a are presented in units of 10^{-4} cm^{-1}. The above type of $3d^n$ ions EPR spectra is characteristic for the monoclinic phase $P2_1$ /c in studied crystals. At $T > T_c$ and at a constant external magnetic field (**H**) parallel to **c** - axis the Mn^{2+} EPR spectrum consists of 6x5 HFS lines. Outlines of Mn^{2+} HFS lines depend on electron

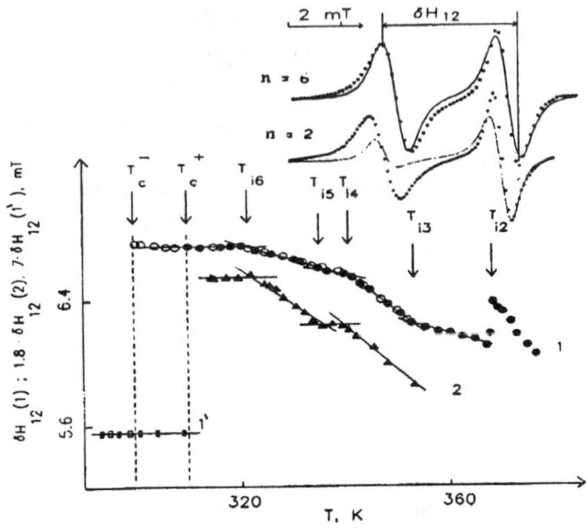

Fig 1. Temperature dependence of the value of ~0.1 at.% Mn^{2+} EPR low field HPS line splitting δH_{12} in MFTH crystals at angles θ=0° (1,1') and 55° (2). The black and white points correspond to parameters at heating and cooling of the crystals, respectively. Comparison between the experimental (pointed line) and simulated (solid line) (~0.1 at %) Mn^{2+} EPR low field HFS line shapes were obtained using the following values of modulation parameters: h_1=3.0 (3.7) mT, h_2=-3.25 (-3.0) mT, δW=1.3 (1.3) mT, Δ = 0.36 (0.6) mT, b= 0.2 (0.1) mT, Φ_o=0° (1°), n = 6(2) . The Q-band **H**||c.

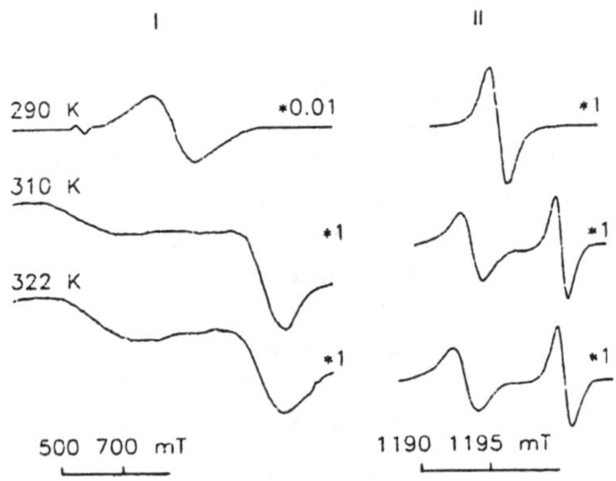

Fig 2. Temperature evolution of the (~5 at%) Ni^{2+} EPR line shape in MFTH crystals (I). At 290 K (in ferroelastic phase), the low field line with small intensity corresponds to the transitions between the M_s=±1 doublet. For comparison, the evolution of the Mn^{2+} (natural abudence) EPR low field HFS line shape of the same crystal is shown (II). The Q-band **H**|| c.

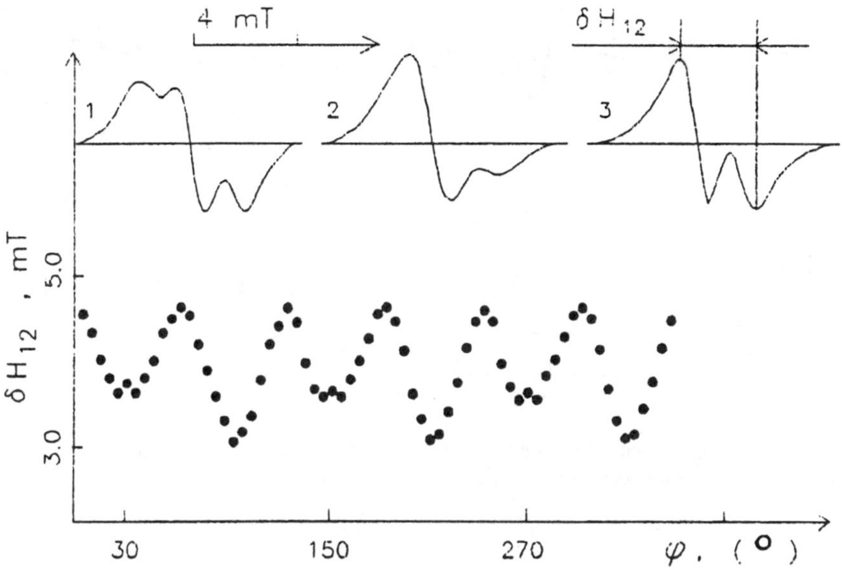

Fig 3. Angular dependence of the value of(~0.1 at%) Mn^{2+} EPR low field HFS line splitting δH_{12} in MFTH crystals at rotating of the sample around the **c**- axis (along the φ angle), which makes the angle θ=55° with H_o.1-ϕ_n = n×60° ; 2- φ_n=30° +n×120°, φ_n=90° +n×120° (n=0,1,2.,...). The X-band **H**||**c**. T=320 K.

quantum number M_s , however all except for the slightly asymmetric HFS lines of the central group have a two peak form (see Figure 1 and 2).
The Ni^{2+} spectrum at Q-band and **H**||**c** consist of a broad two peak line of fine structure (FS) (see Figure 2). Up to ≈ 400 K, the Mn^{2+} and Ni^{2+} spectrum axial axes are preserved along **c**-axis. At $T_{i2} > T_c$, all spectrum and line shape parameters of Mn^{2+} undergo small and close to step-wise changes with the temperature hysteresis of ~ 1 K . Between T_{i2} and T_c , a succession of step-wise changes in the slope of Mn^{2+} line shape parameter curves which are less than those at T_{i2} is observed (see Figure 1). The temperatures of these step-wise discontinuities T vary from sample to sample within the limit of ~ 4 K , but whatever and independently on temperature variation direction they occurs at practically the same amounts of line shape parameters. Having larger FS line width and, as a consequence, a larger value of experimental error,Ni^{2+} parameters do not fix these step-wise changes. At rotating the crystals around **c**-axis which makes a certain angle θ = 0 , 90° with **H** , Mn^{2+} HFS lines do undergo complicated changes which are repeated every 120° (see Figure 3). The amplitude of the HFS line two-peak splitting value variation vs. φ angle increases at T → T_c . In mixed crystals $Mg_{1-x}Zn_xTiF_6$ $6H_2O$ containing natural abundance of Mn^{2+} ions, up to x < 0.75, the Mn^{2+} EPR spectrum is qualitatively similar to its spectrum

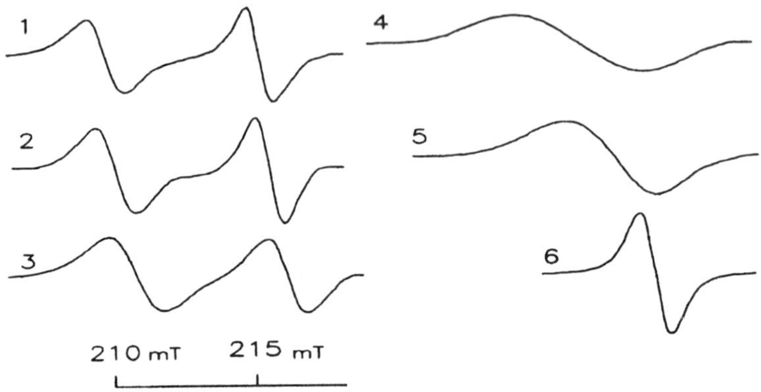

Fig 4. Shape and position of the (~0.1 at%) Mn^{2+} EPR low field HFS line in $Mg_{1-x}Zn_xTiF_6\ 6H_2O$ crystals vs Zinc concentration (x). Spectra 1,2,3,4,5 and 6 correspond to x=0.0,0.33,0.75,0.8,0.9 and 1.0, respectively. The X-band **H**||**c**.

in MFTH crystals (see Figure 4). In crystals with $x > 0.75$ Mn^{2+} EPR spectrum is qualitatively analogous to its spectrum in crystals $ZnTiF_6\ 6H_2O$ (see Figure 4). It is interesting to note that in all crystals investigated only one of the above-mentioned type of Mn^{2+} EPR spectra was observed. In crystals with $x < 0.75$ and $x > 0.75$ the critical temperature regions, in which the improper ferroelastic phase transition is observed, were close to those in MFTH and in $ZnTiF_6\ 6H_2O$ crystals, respectively (see Table 1).

Table 1. The ferroelastic phase transition parameters in crystals $Mg_{1-x}Zn_xTiF_6\ 6H_2O$

x	T_c^+	T_c^-	$\delta T(°)$
0.00	310 ± 1	300 ± 1	~3
0.13	312 ± 1	299 ± 1	~3
0.33	315 ± 1	299 ± 1	~6
0.45	310 ± 2	296 ± 2	~10
0.64	304± 2	292 ± 2	~16
0.75	300 ± 2	290 ± 2	~20
0.81	199 ± 2	194± 2	~24
0.92	190 ± 2	187 ± 2	~10
1.00	183 ± 2	180 ± 1	~ 5

δT is the temperature interval near T_c in which the Mn^{2+} EPR low field HFS line intensity decreases by 90 %.

4 Discussion

The two-peak form of Mn^{2+} HFS lines and Ni^{2+} FS lines indicates on magnetically inequivalent centres of these ions in high temperature phase of MFTH crystals. This fact is incompatible with attributing MFTH crystals to the space group R3m with Z=1. On the other hand, at $T > T_c$ the Mn^{2+} and Ni^{2+} EPR line shapes, their temperature, φ and θ angular dependences are qualitatively identical to the characteristics of this ions in incommensurate (IC) phases of dielectric crystals, in particular, in IC phases of $MgSiF_6\,6H_2O$ [13, 15] and $MgGeF_6\,6H_2O$ [14, 15] crystals. The aforesaid allows to suggest that in high temperature phase the MFTH crystals have also an IC structure. As for the IC phase-paraelastic (P) phase transition which is observed in the latter two crystals at $T_{i1} > T_{i2}$, it may by suggusted that temperature of such transition in MFTH crystals is higher than temperature of its dehydration. Therefore, Mn^{2+} EPR line shape in the high temperature phase of MFTH crystals have been interpreted in terms of the model analogous to that of Blinc [16] for the interpretation of magnetic resonance spectra in crystal IC phases. The scheme of calculations is as follows. The calculations of the Mn^{2+} HFS line shape in high temperature phase are carried out in the assumption of the one-dimensional lattice modulation which is given by

$$U(x) = A(x)\cos[\Phi(x) + \Phi_o] = A(x)\cos\alpha(x),$$

where $\alpha(x)$ is the phase of the local modulation in the direction of the one-dimensional modulation x, Φ_o is the phase shift, $A(x) = A_o + \delta A(x)$ is the amplitude of the lattice displacement, assumed at calculations as $\delta A(x) << A_o$. The resonance field of a given paramagnetic centre can be expanded in powers of $\cos\alpha(x)$

$$H = H_o + h_1 \cos\alpha(x) + h_2 \cos^2 \alpha(x) + ...$$

where H_o indicates the resonance field corresponding to the structure without modulation, parameters h_1 and h_2 are certain functions of x.

The overall line shape F(H) can then be given by analogous to Blinc[16] by the distribution density of the lines

$$f(H) = \mathrm{const}/(|-\sin\alpha[\Delta^2 + \cos(n\alpha)]^{1/2} + b\cos\alpha\cos(n\alpha)\sin(n\alpha)|\, dH/dU) \quad (1)$$

and the Lorenzian line shape of the individual lines L(H) with full width at half maximum height δW, as follows,

$$F(H) = \int L(H-H_R)\, f(H_R)\, dH_R .$$

In equation (1) b is the numerical parameter and Δ connected with the soliton density n_s as $n_s = (1+\Delta^2)^{1/2}/2K$, where K is the complete elliptic integral of the 1-st kind. The line shape computer simulation results are given in Figure 1. As shown, the experimental and calculation results are in good agreement. Thus, it is reasonable to admit that the trigonal distorsion of an aquacomplex which determines the value of FS line parameter of Mn^{2+} and Ni^{2+} EPR spectra are modulated at $T > T_c$.

The fact that in IC phase of MFTH crystals, $n_s < 1$ implies that the crystal consist of two types of areas: one with a lattice displacement modulation (solitons), the other without it (domains).

The existance of angular dependence on φ for Mn^{2+} line shape in high temperature phase allows to drow the conclusion that in the IC phase the cubic crystalline field axes for different Mn^{2+} ions do not coincide due to the rotation of complex ions around the crystal c-axis. We assumed [13, 15] that the similar disorientation of the cubic crystalline field axes in IC phase of $MgSiF_6\,6H_2O$ and $MgGeF_6\,6H_2O$ crystals must be connected with the modulation of the rotation angle of the complex around the c-axis relative to the position in the P-phase and that this angle is an order parameter of P $\rightarrow$IC phase transition and it, therefore, determines the value of the trigonal distorsion modulation. Similar suggestions may be made according to an origin of the disorientation of the cubic crystal line field axes in high-temperature phase of MFTH crystals. The calorimetric investigations of MFTH crystals in temperature interval 120 $\div$ 400 K recorded two new phase transition[17]. Temperatures of these transitions are near T_{i4} and T_{i6}. So, one can conclude that other T_{in} are also the phase transition temperatures. In this case, these transitions formed previously unknown in MFTH crystals succession of IC phase $\leftrightarrow$ IC phase type phase transitions.

References

1. W. Hamilton, Bond distances and thermal motion in ferrous fluosilicate hexahydrate: neutron diffraction study, *Acta Cryst.* 15,(1962) 353-360.
2. D.C. Price, Static and dynamic crystal-field effects in ferrous fluosilicate, *Can.J.Phys.*. 65, (1987) 1280-1293.
3. G.Jehanno and F.Varret, Desorde structural et transformation de phase dans les fluosilicates de Fe, Mg, Mn, *Acta Cryst.* A31,(1975) 857.
4. S.Syoyama and K.Osaki, An X-ray study of the low temperature form of $NiSiF_6\,6H_2O$ and the relation between the crystal lattices of low- and high-temperature forms, *Acta Cryst.* B28, (1972) 2626-2628.
5. S. Ray, A. Zalkin and D. H. Templton, Crystal structures of the fluosilicate hexahydrates of cobalt, nickel and zinc, *Acta Cryst.* **B29**, (1973) 2741-2747.
6. G. Chevrier and G. Jehhano, Antiphase peridique orientationnelle dans le fluosilicate de magnesium, *Acta Cryst.* **A35**, (1979) 912-916.

7. G. Chevrier, A new example of an Arrangement of antiphase Domains: Investigation by Neutron Diffraction of Deuterated Manganese Fluosilicate, *Acta Cryst.* **B47**, (1991) 224-228.
8. G. Chevrier, A. Hardy and G. Jehhano, Antiphase periodique orientannelle et transformation de phase dans le fluosilicate de fer, *Acta Cryst.* **A37**, (1981) 578-583.
9. R. S. Rubins, Bobby C. Griffin and Burris R., Electron paramagnetic resonan- ce in nickel fluotitanate, *J. Chem. Phys..* **64**, (1976) 3349-3352.
10. Yu. V. Yablokov, A. M. Zaripov, A. M. Ziatdinov and R. L. Davidovich, The phase transition in $ZnGeF_6$ $6H_2O$, *Chem. Phys. Lett.* **48**, (1977) 443-446.
11. A. M. Ziatdinov, V. G. Kuryavyi and R.L. Davidovich, Angular dependence EPR line shape Mn^{2+} in incommensurate phase of $MgTiF_6$ $6H_2O$ crystals, *Sov. Phys.: Solid State* **29**, (1987) 120-123.
12. R. S. Rubins and Kian-Kwet Kwee, Electron paramagnetic resonance of divalent nickel in the low temperature phase of magnesium fluosilicate and magnesium fluotitanate, *J. Chem. Phys.* **66**, (1977) 3948-3951.
13. A. M. Ziatdinov and V. G. Kuryavyi, EPR studies of phase transitions and incommensurate states in $3d^n$ - ions doped $MgSiF_6$ $6H_2O$ crystals, *Ferroelectrics* **143**, (1993) 99-107.
14. A. M. Ziatdinov and V. G. Kuryavyi, EPR studies of phase transitions and incommensurate states in ferroelastic $MgGeF_6$ $6H_2O$ doped with 3d - ions, *Ferroelectrics*, **156**, (1994) 395-400.
15. A. M. Ziatdinov, Materials with electronic and structural incommensurate states, including intercalation compounds, *Ferroelectrics* **155,** (1994) 383-390.
16. R. Blinc, P. Prelovsek and R. Kind, Amplitude-fluctuation effects on magnetic-resonance line shape and soliton density in incommensurate systems, *Phys. Rev.* **B27**, (1983) 5401-5408.
17. I. N. Flerov, M. V. Gorev, S. V. Melnikova, M. L. Afanas'ev and K..S.Alexandrov, Investigation of phase transitions in ABF_6 $6H_2O$ crystals, *Sov. Phys.. : Solid State* **33**, (1991) 1921.

Conduction ESR and Current Carriers Injection Phenomenon at Incommensurate Crystallization of "Guest" Molecules in Acceptor Graphite Intercalation Compounds

A.M. Ziatdinov and N.M. Mishchenko

Institute of Chemistry, Far-Eastern Division of the RAS, Vladivostok, Russia

Abstract. The investigations of the effect of temperature on the conduction electron spin resonance (CESR) spectra of graphite intercalation compounds (GICs) $C_{10}HNO_3$ were realized with the 2D-CESR line shape analysis procedure developed by the authors. Conductivities along σ_a and perpendicular σ_c to GICs basal planes have also been measured. As a result, we established an earlier unknown phenomenon: when layers of "guest" molecules undergo an aggregate phase transition (at T_c ~ 250 K), σ_a and σ_c increase in spite of the decrease of the charge carrier mobility. CESR measurements of charge carrier concentration N vs. temperature show that this unusual "nonmetallic" behavior of GICs results from the increase of the hole concentration in π-bands. The increase of N is directly connected with the partial localization of the conduction π-electrons (in terms of GICs tight binding model). Because the localized electrons are scattering centers for the conduction electrons, the increase in N must obviously be accompanied by a decrease in conduction electron mobility and an increase in CESR line width.

1 Introduction

The temperature dependence of the conduction electron spin resonance (CESR) line width (ΔB) in metals has been successfully explained [1] in terms of the Elliot's theory [2] according to which $\Delta B = \alpha(\Delta g)^2/\tau$, where α is a constant, Δg is the g-shift from the free electron value, and τ is the conductivity relaxation time. The frequency of the electron-electron and electron-phonon collisions decreases with temperature and, as a result, τ increases, while the concentration (N) of the current carriers remains practically constant because of large values of Fermi temperature [3]. As a result, when the temperature lowers, the conductivity of pure metals $\sigma = Ne^2\tau/m^* = Ne\mu$ (m^*, e and μ are the effective mass, charge and mobility of charge carriers, respectively) increases, and simultaneously, if Δg is temperature independent, the CESR line width decreases.

In graphite intercalation compounds (GICs) of the acceptor type $C_{10}HNO_3$ we found a relation between σ and ΔB other than in pure metals. In this synthetic metal during and after intercalate crystallization Δg is unchanged, while both the electroconductivity and the CESR line width increase simultaneously. In this paper we report the results of an investigation of the CESR line width "nonmetallic" temperature dependence origin in GIC $C_{10}HNO_3$

2 Experimental

GICs comprise a wide class of synthetic metals and consist of an alternating sequence of n hexagonal graphite monolayers (n is the stage index) and a monolayer of "guest" atoms or molecules (intercalate) [4]. The compounds $C_{10}HNO_3$ investigated belong to the second stage of α - modification of GICs with nitric acid of the general formula $C_{5n}HNO_3$ (n = 1, 2, 3,...) [4,5]. According to the data of various physical methods [4], in these GICs the two-dimensional liquid-like layers of HNO_3 are ordered and form a two-dimensional crystal at $T \leq 250$ K. In $C_{10}HNO_3$ layers of HNO_3 are incommensurable with a carbon net and they undergo a structural phase transition of the incommensurate phase-commensurate phase type at T~210 K [6].

CESR spectra of GIC $C_{10}HNO_3$ plates were registered at X-band (ν=9.5 GHz) in the temperature range 100 ÷ 300 K using the rectangular resonator with the TE_{102} mode and 2.5 kHz modulation of an external constant magnetic field $\mathbf{B}_0$. At the conventional setting of the resonator, the electrical component of the microwave field is parallel to $\mathbf{B}_0$; in the geometric center of the resonator the magnetic component of this field is parallel to the vertical axis of the resonator.

Highly oriented pyrolitic graphite plates required for GIC synthesis were cut out of a single bar with the basal plane conductivity equal to $(1.2 \pm 0.2)\times 10^4$ $Ohm^{-1}\cdot cm^{-1}$. These plates had a shape of a rectangular parallelepiped with dimensions: width (l) × height (h)× thickness (d), where $l\times h$ is the square of a basal plane. Accuracy in determining the size of plates was $\sim 5\times 10^{-4}$ cm. Synthesis of GIC $C_{10}HNO_3$ was carried out in nitric acid with the density ρ=1.562 $g\cdot cm^3$. The GIC stage was controlled by the diffraction method. The effect of temperature on CESR line shape parameters was investigated using the plate of size: $0.40\times 0.40\times 0.02$ cm^3.

3 Results and Discussion

In the entire temperature range of investigations a single asymmetric CESR signal of $C_{10}HNO_3$ is observed. Down to ~ 250 K the asymmetry parameter of the first derivative of the absorption line A/B (which is defined as the ratio of the maximum peak height (A) to the minimum peak height (B), both measured with respect to the zero line of the resonance derivative) and the width of the CESR line at half height of the peak A are independent of temperature and are equal to 3.0±0.1 and 2.2×10^{-5} T, respectively. At all temperatures, the spectrum is axial with respect to the **c**- axis and is characterized by $g_{\parallel} = 2.0023 \pm 0.0001$ and $g_{\perp} = 2.0028 \pm 0.0001$. The line width increases as temperature decreases below 250 K (Figs 1 and 2). Simultaneously, the A/B value decreases, and it is equal to 2.6 ± 0.1 at 96 K (Fig. 2). The derivative $d(\Delta B)/dT$ undergoes the step-wise changes at T_{c1^-} =250 K, T_{c2^-} =249.8 K, T_{c3^-} =248 K, T_{c4^-} =247 K, T_{c5^-} =243 K (Fig 1). As the temperature increases ΔB and A/B values change inversely but with a "global" temperature hysteresis (Figs. 1 and 2). Temperatures T_{ci^-} (T_{ci^+}) (i=1÷5) are close to the temperature of the intercalate crystallization (melting) in $C_{10}HNO_3$ (T_c ~ 250 K) known from literature [4]. This allows us to treat the CESR line shape transformations observed in $C_{10}HNO_3$ at T_{ci^-} (T_{ci^+}) (i=1÷5) as a result of the multi-step change in $C_{10}HNO_3$ the aggregate state of the "guest" molecules. The CESR line shape transformation at T_{c3}*, is probably due to

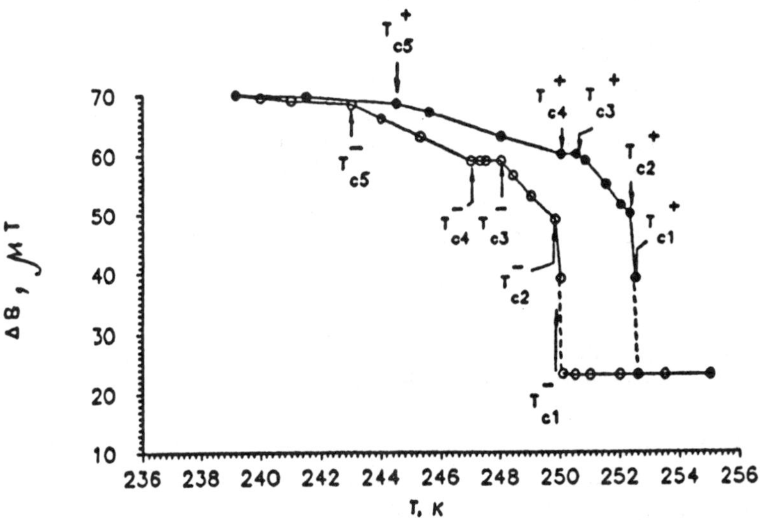

Fig. 1. Temperature dependence of CESR linewidth ΔB of $C_{10}HNO_3$ plate with dimensions: 0.4 × 0.4 × 0.02 cm³ during crystallization (unfilled symbols) and melting (filled symbols) of intercalate. T_{ci^-} (T_{ci^+}) are the temperatures of the step-wise changes of the derivative $d(\Delta B)/dT$ at cooling (heating) of the sample, respectively. $\mathbf{B}_0 \perp \mathbf{c}$, ν = 9.52 GHz.

the incommensurate-commensurate phase transition. At powers of the microwave field far from saturation of the CESR signal and at the same temperature, ΔB values in Q- and X- bands coincide indicating that the CESR line is homogeneously broadened.

The basal plane electroconductivity (σ_a) of $C_{10}HNO_3$ plates was measured by the contactless induction method using a device analogous to that described by Pendrys *et al.* [7]. The results of σ_a measurements are

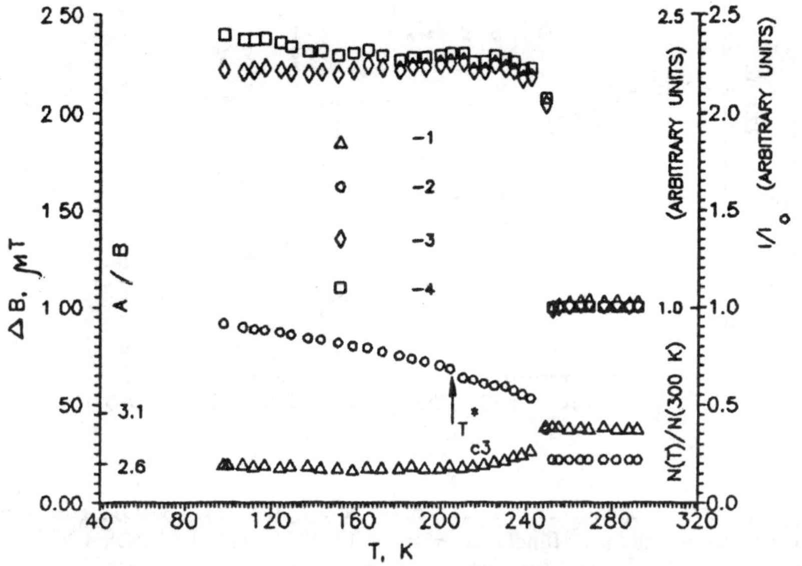

Fig. 2. Temperature dependence of CESR line parameters of $C_{10}HNO_3$ plate with dimensions: 0.4 × 0.4 × 0.02 cm³, i.e., $A/B(T)$ (curve 1), $\Delta B(T)$ (curve 2), $I(T)/I_o$ (curve 3), where $I = (A+B)\,\Delta B^2$ and I_o is the intensity of the Mn^{2+} ESR line in a standard sample $ZnS{:}Mn^{2+}$, and the relative spin concentration $N(T)/N(300\text{ K})$ (curve 4), determined by the expression (1). Unfilled symbols correspond to experimental values of parameters at cooling of the GIC. T_{c3}* is the temperature of the peak of the derivative $d(\Delta B)/dT$. $\mathbf{B_o} \perp \mathbf{c}$, $\nu = 9.52$ GHz.

presented in Fig. 3, and indicate that the σ_a increases as temperature decreases.

In the microwave field of a given configuration only the conduction plate regions adjacent to the vertical faces $(h \times l)$ and $(h \times d)$, situated approximately within the skin-depth, contribute to the CESR [8,9]. In $C_{10}HNO_3$ one may neglect the contribution to CESR from regions adjacent to the basal planes $(h \times l)$ due to the high conduction anisotropy (~ 10^5) [4]. This peculiarity of $C_{10}HNO_3$ plates allows us to analyze their CESR line shape, in particular, at contactless determination of the **c**-axis electroconductivity (σ_c) (see below), using the one-dimensional Dyson expression [10] for CESR line shape in isotropic metals and well known Feher and Kip [11] or Kodera [12] nomograms calculated on the basis of this expression.

The σ_c measurement is not a trivial problem. The contactless method to determine the σ_c of acceptor GICs was suggested by Saint-Jean and McRae [13]. They recommended the use of the well known A/B vs l/δ_c nomograms [8,9,12,13] at interval $l/\delta_c < 2.5$ for this purpose. At this interval the ratio of A/B does not depend on spin carriers mobility [8,9,12,13]. This enables one to determine the value of δ_c and, consequently, the value of σ_c unambiguously, by measuring the values of A/B and l. However, in narrow

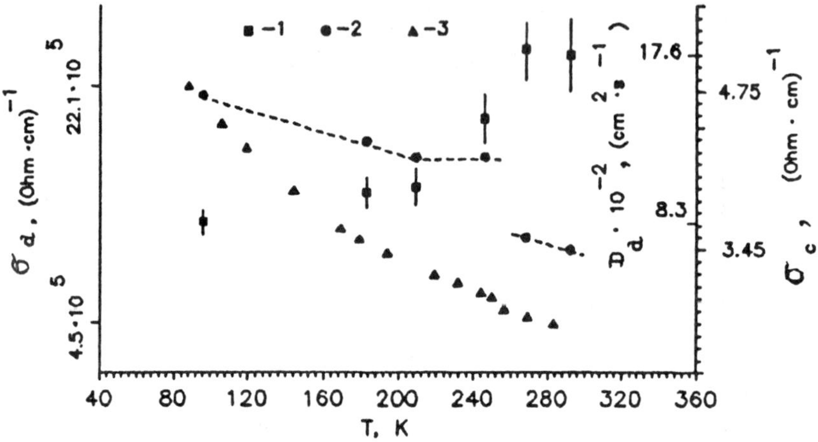

Fig. 3. Temperature dependence of D_a (1), σ_c (2) and σ_a (3) in $C_{10}HNO_3$. The dashed lines correspond to the linear functions $\sigma_c = [(-4.17\times10^{-3}/\ K)T + 4.7\ ]$ (Ohm cm)$^{-1}$, σ_c =4.21 (Ohm cm)$^{-1}$ and $\sigma_c = [(-5.86\ \times10^{-3}/\ K)T+5.28]$ (Ohm cm)$^{-1}$ for $T > T_{cl'}$, 209 K $<T$ <246 K and 96 K $< T <$ 209, respectively.

samples of GICs with HNO_3 [14] as well as in narrow samples of GICs with AsF_5 [13] the values of σ_c and spin carriers diffusion constant differ from values of this parameters in wide samples. This unusual result for narrow GIC samples is likely connected with the fact that the intercalate layers composition and organization depend on a sample size [14]. That is why another contactless method was used to determine the value of σ_c in our study. The principles of this method are as follows.

It follows from the one-dimensional Dyson expression for CESR line shape in metals [10] that the coordinate of the maximum (l_{max}/δ_c) of the $A/B(l/\delta_c)$ dependence and the value of asymmetry parameter on "plateau" at $(l/\delta_c > 10)$ of this dependence $[(A/B)_{plat}]$ are unambiguously related to each other. The corresponding nomogram calculated by the authors is presented in Fig.4. The theoretical value of l_{max}/δ_c can be found from this nomogram, if the experimental value of $(A/B)_{plat}$ is known. Then, at a given frequency of the microwave field, the value of σ_c can be easily calculated, if the experimental value of l_{max} is known as well. It must be emphasized that the given procedure of σ_c -determination is applicable only in the case when (1) one may neglect the contribution to CESR from regions adjacent to the basal planes ($h \times l$) of GIC plates and (2) the CESR line shapes of plates with $l \sim l_{max}$ as well as plates with l from "plateau" of the $A/B(l)$ dependence should be described by the one-dimensional Dyson expression for CESR line shape [10] with the same value of σ_c and the in-plane diffusion time across the skin -depth δ_c (T_{Da}) determined by the σ_c .Because of this, before the procedure under consideration

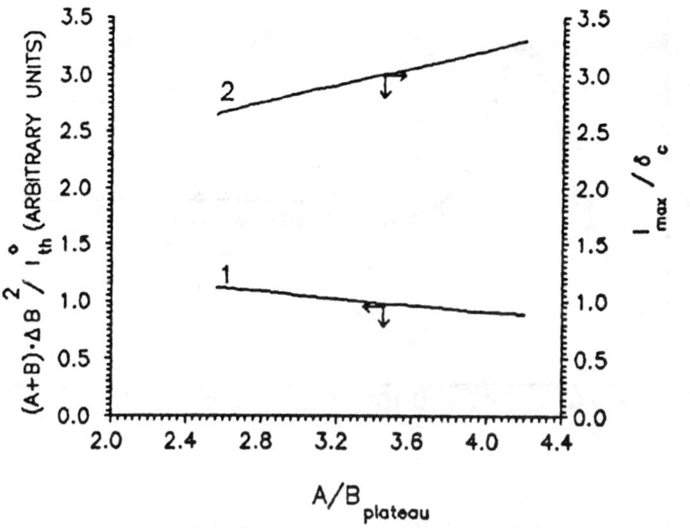

Fig. 4. The relation of theoretical value of intensity $I = (A+B)\cdot\Delta B^2 / I_{th}^o$ (curve 1) and coordinate of the maximum (l_{max} / δ_c) of the theoretical dependence $A/B(l / \delta_c)$ (curve 2) plotted against the value of A/B on the "plateau" of the $A/B(l / \delta_c)$ dependence at $(l / \delta_c > 10)$. I_{th}^o is the value of the I_{th} at A/B = 3.0 (in the quasi-liquid phase of the intercalate).

was applied to determine the σ_c-value in $C_{10}HNO_3$, the experimental $A/B(l)$ dependence was studied in more detail at fixed temperature both before (at 292 K and 268 K) and after (at 246 K, 209K, 183 K and 96 K) the phase transition. Then this dependence was approximated by the corresponding theoretical curve calculated from the one-dimensional Dyson expression for CESR line shape [10] with the value of σ_c determined from the experimental values of l_{max} and $(A/B)_{plat}$ and the value of T_{Da} determined by the standard procedure [11,12] (using the values of ΔB and A/B of GIC plates with l from "plateau" of the $A/B(l)$ dependence). As seen from Fig. 5, the theoretical and experimental dependences $A/B(l)$ are in good agreement at $l > l_{max}$. Taking it into consideration , the values of σ_c at other temperatures were determined using the experimental values of l_{max} and $(A/B)_{plat}$ only. By this procedure of σ_c -determination we established that σ_c, as well as σ_a increases with decreasing the temperature (Fig. 3). Hence in $C_{10}HNO_3$, during and after intercalate crystallization the ΔB increases simultaneously with the conductivity in spite of the fact that the g tensor values of the spin carriers are temperature independent. We have found a possible explanation for this uncommon property of the CESR line width temperature dependence by investigating the temperature effect on N and on the in-plane spin diffusion constant (D_a).

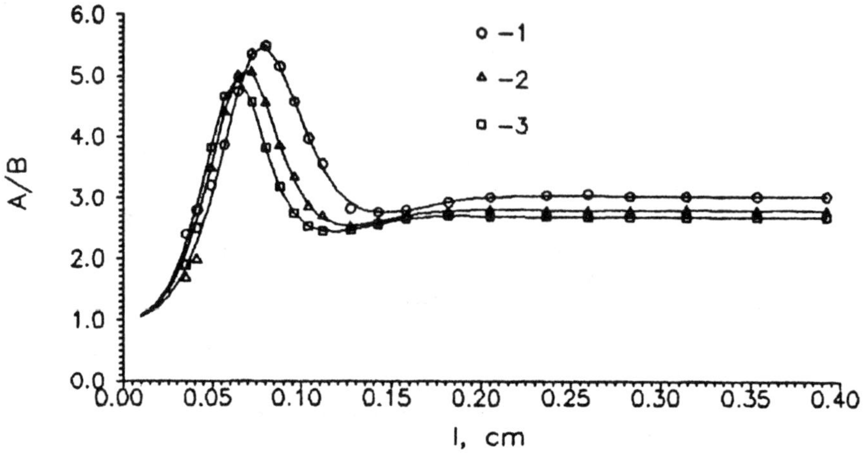

Fig. 5. The experimental (open symbols) and theoretical (solid lines) values of the line asymmetry parameter *A/B* of the $C_{10}HNO_3$ plates with dimensions $l \times 0.40 \times 0.02$ cm³ as a function of sample width (l). The experimental curves 1, 2 and 3 correspond to 296 K , 209 K and 96 K, respectively. $\mathbf{B_0} \perp \mathbf{c}$, $\nu = 9.52$ GHz. The theoretical curves at 296 K, 209 K and 96 K are one-dimensional Dyson's [10] curves with σ_c = 3.45, 4.2 and 4.75 (Ohm cm)$^{-1}$, T_{Da} = 21.7, 30.5, and 33.5 in the units of 10^{-8} s, respectively.

At a given geometry of the GIC sample the CESR signal integral intensity $I=(A+B)\Delta B^2$ is proportional to $N\delta_c$. Because of this, the increase in I simultaneously with the decrease in δ_c , which takes place in $C_{10}HNO_3$ as the temperature decreases from T_{c1}- to T_{c3}* (Fig. 2), suggests that N increases in this temperature range. This conclusion does not change once the 7.5 % increase of I (relative to its value in liquid phase of intercalate) along with the decrease of *A/B* at this temperature range has been taken into consideration with the help of the corresponding nomogram in Fig. 4. Further, using the relation $I \sim N\delta_c$ and approximating the experimental values of σ_c before and after phase transition by the linear functions (see Fig. 3) we have calculated the temperature dependence of the relative spin carriers concentration (Fig. 2):

$$N(\mathrm{T})/N(300\ \mathrm{K}) = [I(\mathrm{T})/I(300\ \mathrm{K})][\sigma_c(\mathrm{T})/\sigma_c(300)]^{1/2}, \qquad (1)$$

where N(300 K), I(300 K) and σ_c(300 K) are the values of N, I and σ_c at 300 K, respectively. As should be expected from the results of the qualitative analysis of the experimental data, N increases as the temperature is lowered in the crystal phase of the intercalate (Fig. 2). Concerning the behaviour of N in the quasi-liquid phase of the intercalate, all its variations occur within the experimental error (Fig. 2). It follows from above that in synthetic metal $C_{10}HNO_3$ in the solid phase of the intercalate the increase of σ_a is at least partially due to the increase in N.

With knowledge of the experimental values of A/B and ΔB of GIC plates with l from " plateau " of the $A/B(l)$ dependence and using the Feher and Kip [11] nomograms one can easily determine T_{Da}. According to Dyson [10], in the approximation of independent electrons, the value T_{Da} is connected with the value of D_a by the relationship $D_a = \delta_c^2/2T_{Da}$. The estimation of D_a according to this expression has shown (Fig. 3) that its value during and after intercalate crystallization decreases.

The diffusion constant is equal to

$$D_a = (1/3)\,\Lambda_a v_a = (1/3)\,\tau_a v_a{}^2 \qquad (2)$$

by definition [3]. In the expression (2) Λ_a, τ_a and v_a are the mean free path, the electroconductivity relaxation time and the spin carriers velocity, all along the basal plane. If in expression (2) v_a is identified with Fermi velocity v_F and it is taken into account that in metals v_F is proportional to the square of current carriers concentration [3] then it may be concluded, from this expression, that τ_a in $C_{10}HNO_3$ decreases during and after intercalate crystallization (since at this process D_a decreases simultaneously with the increase of N). This suggests that "nonmetallic" broadening of the CESR line in $C_{10}HNO_3$ is due to the decrease of the in-plane current carriers mobility.

The broadening of the CESR line is homogeneous. The g-tensor values of spin carriers are independent of the changes in aggregate states of the intercalate and they are close to the g-tensor value of the free electron. These facts testify that the density of charge carriers on intercalate molecules is small and does not change at phase transition. Above, it is shown that it is possible to analyze the transformations of the electronic properties of $C_{10}HNO_3$ induced by phase transition in terms of the rigid electronic band model of GICs [4,15]. In terms of this model, in acceptor GICs related to the conductors with the hole Fermi surface [4] the increase of N as the temperature decreases is possible only at lowering of the Fermi energy, for instance, due to the transfer of the additional electrons from carbon layers to intercalate layers [4,15], or due to the localization of spin carriers in graphite layers, preferentially at the edges of the Dumas-Herold islands [16]. In this case, the increase in N must obviously be accompanied by the decrease in τ_a, and consequently, by the increase in ΔB, because the localized electrons are the perturbation centers for the conduction electrons. That is, in the crystal phase of the "guest" molecules subsystem, the rigid electronic band model predicts qualitatively the correct temperature effect on these experimental values. Note that several recent experimental results confirm the existence in GICs of the localized moments and traps. For instance, Davidov *et al.* [17] observed the presence of localized moments in the graphite layers by spin-echo techniques performed on AsF_5 GICs and suggested the existence of traps. A precise analysis of the spin-relaxation time [13] and the anisotropy of the line width with the orientation of $\mathbf{B}_0$ [18] confirms this assumption. At the present time, the exact nature of these traps remains to be determined.

Acknowledgments

This work was supported by Russian Foundation for Basic Research grant (N 97-03-33346 a). The authors are grateful to L.B. Nepomnyashchii (State Research Institute of Graphite, Moscow) for the HOPG plates used in the synthesis of GIC.

References

1 P. Monod and F. Beuneu, Conduction electron spin flip by phonons in metals: Analysis of experimental data, *Phys.Rev.B* **19**, 12 (1979) 911-916.
2. R.J. Elliot, Theory of the effect of spin-orbit coupling on magnetic resonance in some semiconductors, *Phys.Rev.* **96** (1954) 266.
3. N.W. Ashcroft and D. Mermin, Solid State Physics, Cornel University, Holt, Rinehard and Winston, N.Y. (1976).
4. M.S. Dresselhaus and G. Dresselhaus, Intercalation compounds of graphite, *Adv.Phys.* **30** (1981) 139.
5. M.J. Bottomly, G.S. Parry and A.R. Ubbelohde, Thermal expansion of some salts of graphite, *Proc.R.Soc.London* **A279**, 1378 (1964) 291-301.
6. F. Batallan, I. Rosenman, A. Magerl and H. Fuzellier, Two-dimensional molecular diffusion and phase transition in HNO_3 - intercalated graphite studied by quasielastic neutron scattering, *Phys.Rev.B* **32** (1985) 4810.
7. L.A. Pendrys, C. Zeller and F.L. Vogel, Electrical transport properties of natural and synthetic graphite, *J.Mat.Sci.* **15**, 11 (1980) 2103-2112.
8. A.M. Ziatdinov and N.M. Mishchenko, Electron spin resonance line shape and kinetic parameters of the conduction electrons in highly anisotropic conductors: highly oriented pyrolytic graphite *Phys. Solid State*, **36**, 8 (1994) 1280-1289.
9. J. Blinowski, P. Kacman, C. Rigaux and M. Saint-Jean, Effect of conduction anisotropy of GIC on ESR line shape. *Synth.Met.* **12** (1985) 419-423.
10. F.J. Dyson, Electron spin resonance absorption in metals. II. Theory of electron diffusion and the skin effect. *Phys.Rev.* **98** (1955) 349-359.
11. G. Feher and A.F. Kip, Electron spin resonance absorption in metals I. Experimental *Phys.Rev.* **98**, 2 (1955) 337-348.
12. H. Kodera , Dyson effect in the electron spin resonance of phosphorus doped silicon, *J.Phys.Soc.Jpn.* **28**, 1, (1970) 89-98.
13. M. Saint-Jean and E. McRae, Planar diffusion constant D_a in acceptor graphite intercalation compounds, *Phys.Rev.* **43**, 5 (1991) 3969-3974.
14. A.M. Ziatdinov and N.M. Mishchenko, unpublished data.
15. S.Y. Leung and G. Dresselhaus, Dispersion relations in graphite intercalation compounds: Electronic energy band, *Phys.Rev.B* **24** (1981) 3490-3504.
16. A.M. Ziatdinov, N.M. Mishchenko and Yu.M. Nikolenko, Phase transitions and incommensurate states in GIC C_{5n} HNO_3, *Synth.Met.* **59**, 2 (1993) 253-258.
17. D. Davidov, A. Grupp, H. Kass and P. Hoffer, Pulsed ESR studies of graphite intercalation compounds, *Synth.Met.* **23** (1988) 291-295.
18. S. Shimamura, A model for c-axis electrical conduction in graphite intercalation compounds, *Synth. Met.* **12** (1985) 365-370.

Contributions to the Ground State Splitting of d^5 ^6S-State Ions: Mn^{2+}

Guei-Gu SIU

Department of Physics and Materials Science, City University of Hong Kong, Kowloon, Hong Kong

Abstract. Expressions are obtained for several significant contributions to the spin-Hamiltonian parameters which describe the ground state splitting of Mn^{2+}. Calculations are consistent with the superposition model analysis of experimental data. Numerical estimates provide a qualitative explanation of the experimental results.

1. Introduction

There has been lasting interest in the experimental and theoretical studies of the zero-field splitting (ZFS) of the S-state ions, e.g. Gd^{3+}, Eu^{2+} and Mn^{2+}. Progress of experimental works on ZFS of Mn^{2+} (the bibliography in [1] and [2-6]), e.g. the ligand- and coordination number- dependence of its intrinsic parameter $\bar{b}_2$, entails a more detailed theoretical study for suitable interpretation. Results of various approaches are still far from being satisfactory. The over-simplified electrostatic model could not give reasonable results [7]. Others approaches [8-10] are not able to comply with the newly discovered mechanisms [11-13] and the agreement with experimental data seems fortuitous. Overall *ab initio* calculations relate the ZFS parameters to the underlying quantum mechanical processes but fail to explain the relationship between the spin-Hamiltonian (SH) and the crystal field (CF) parameters.

We have discussed contributions to the ground S-state splitting of Gd^{3+} (paper I [13]) and proposed a theoretical framework for studying the ground S-state splitting of Mn^{2+} (paper II [14]). Both studies aim at explaining how the ground S-state is split by the action of the crystalline environment. This work extends the studies of I and II by considering contributions of CF and relativistic CF (RCF) to ZFS of Mn^{2+}.

Because the conventional CF cannot split the S-ground state of half-filled electronic shell in the first order, the ZFS theory requires the consideration of usually ignored mechanisms, e.g. the high order perturbations, intermediate coupling, spin-spin interaction and so on. Investigating these secondary mechanisms are in the direction of the extension of the usual CF theory. Here contributions to ZFS from of CF and RCF, together with those from the spin-correlated CF (SCCF) and the electrostatically correlated CF [15-17] are studied. New effects such as the convalency effects on RCF are also considered. Numerical estimates for the complex Mn^{2+}/F^{-}-ligand are made.

2. Theoretical framework

The contribution of crystalline anisotropy to ZFS for the d^5 ^6S state is expressed by

$$H_s = \frac{1}{3}\sum_m b_m^2 O_m^2 + \frac{1}{60}\sum_m b_m^4 O_m^4 \qquad (1.1)$$

where operators O_m^k are defined in [7]. Conventionally a cubic parameter $a = (2/5)\, b_4^4$ and non-cubic ones $F = 3b_0^4 - (3/5)b_4^4$, $D = b_0^2$ and $E = (1/3)\, b_2^2$ are used [7].

However, it is essential to separate physical and geometrical aspects of ZFS or CF before a meaningful calculation can be carried out. For example, ZFS values for Mn^{2+} in isomorphous $MgCl_2$ and $CdCl_2$ are -136×10^{-4} cm^{-1} and $+2 \times 10^{-4}$ cm^{-1} respectively [18] and the ZFS parameter D varies from 36×10^{-4} cm^{-1} for Mn^{2+} in K_2ZnF_4 and 107 $\times 10^{-4}$ cm^{-1} for Mn^{2+} in K_2MgF_4 to negative values for Rb_2ZnF_4 and Rb_2MgF_4 [19]. Hence, a common quantity for the same type of ligands is needed. The superposition model (SM) aims at deriving expressions and estimating the magnitude of single-ligand contributions to the *intrinsic parameters*, which depend only on the type and number of the nearest ligands and are applicable for different crystals with similar structures and the same ligands. SM deals with a single ligand in the principal axis of the coordinates and only axially symmetric operators O_0^k are left

$$H_s = \frac{1}{3}\, b_0^2 O_0^2 + \frac{1}{60}\, b_0^4 O_0^4$$

and $\bar{b}_k = b_0^k$ for a single ligand are the intrinsic parameters. This work aims to determine contributions to $\bar{b}_{2,4}$ from CF and RCF in so far as this is possible.

The conventional CF constitutes only one of the perturbing mechanisms accounted for by

$$<^6S|V_{CF}|^6S> = 0$$
$$<^4L_1|V_{CF}|^4L_2><^4L_2|\Lambda|^6S>/(E_{6S}-E_{4L1})(E_{6S}-E_{4L}) << 1$$

where $^4L_1$ and $^4L_2$ are the excited states and Λ is one of the other perturbation mechanisms. Hence, in ZFS, CF has no priority in perturbation calculations and is similar to other perturbations such as the compound mixing mechanism

$$\Lambda^{(0)} = H_{SO} + H_{SS} + H_{SOO}$$

where H_{SO}, H_{SS} and H_{SOO} are respectively the spin-orbit, spin-spin and spin-other-orbit interactions. $\Lambda^{(0)}$ is considered first. Then contributions to $\bar{b}_{2,4}$ from CF and other effects to are calculated. The ground $3d^5$ ^6S state is expressed in terms of LS-coupled states as follows:

$$|^6S> = s\,|^6S_{5/2}> + p\,|^4P_{5/2}> + d\,|^4D_{5/2}> + f\,|^4F_{5/2}> + g\,|^4G_{5/2}>$$
$$+ d_1\,|^2D1_{5/2}> + d_2\,|^2D2_{5/2}> + d_3\,|^2D3_{5/2}> \qquad (2.1)$$

where the conditions $\delta(JJ')$ and $\delta(M_J M_{J'})$ of the $\Lambda^{(0)}$ matrix elements limit J to 5/2. The following values of coefficients in Eq.(2.1) are obtained [II]:

$$s = 0.99975; \quad p = 2.2230 \times 10^{-2}; \quad d = -4.1405 \times 10^{-4};$$
$$f = 7.632 \times 10^{-6}; \quad g = -2.220 \times 10^{-7}; \quad d_1 = 2.5439 \times 10^{-4};$$
$$d_2 = -4.639 \times 10^{-6}; \quad d_3 = -3.9500 \times 10^{-4} \qquad (2.2)$$

according to the expansion

$$|^6\mathcal{S}> = |^6S> + \sum \frac{<{}^4L|\Lambda^{(0)}|^6S>}{E_{6S} - E_{4L}} \,|^4L> + \sum \frac{<^4L|\Lambda^{(0)}|^4L_1><^4L_1|\Lambda^{(0)}|^6S>}{(E_{6S} - E_{4L})(E_{6S} - E_{4L_1})} \,|^4L>$$

where the summations are over L_1, M_{L1}, M_S of the excited states.

This approach is contrary to the Blume-Orbach procedure. It separates CF into cubic and non-cubic parts and then inserts spin-orbit coupling in between. The separation of the crystal field into a large part and a small part is undoubtedly a proper procedure of ordinary perturbation theory, but it is inappropriate that another perturbation (spin-orbit interaction) is considered beforc the CF effect has been fully studied, since it spoils the integrity of the crystal field.

Linear phenomenological parametrization of the spectra of $3d^n$ ions is chosen [20]. There are only two sets of parameters available for the conventional CF [21-22]. For secondary mechanisms, models are used to calculate their parameter values, e.g. the overlap model for RCF and SCCF. The processes involving admixtures of states into excited configurations are separated from those which contribute to matrix elements of the LS-coupling states within the $3d^5$ configuration. The former are treated as providing effective Hamiltonian within $3d^5$ configuration while the latter are used explicitly to determine perturbation contributions to ZFS. Successful applications of SM analysis to SH and CF parameters for the iron group ions show that the interactions between the metal ion and the crystalline environment are predominantly short-range and linear.

3. CF and RCF

The conventional CF potential can be expressed as a sum of single-electron tensor operators with parameters

$$V_{CF} = \sum_i \sum_{k,q} B_q^k c_q^{(k)}(i) = \sum_{k,q} B_q^k C_q^{(k)} \tag{3.1}$$

and the linear CF contributions to ZFS may be schematically represented as

$$<^6S|\Lambda^{(0)}|^4L><^4L|V_{CF}|^4L'><^4L'|\Lambda^{(0)}|^6S>. \tag{3.2}$$

RCF results from the application of relativistic Hartree-Fock wavefunctions [23,24] and generates effective operators w.r.t. non-relativistic LS coupled states

$$V_{RCF} = \sum_{k,q} \sum_{\kappa,\kappa'} b_q^k(\kappa,\kappa') W_q^{(\kappa,\kappa')k} \tag{3.3}$$

where $W_q^{(\kappa,\kappa')k}$ are the double tensor operators of n-equivalent electrons defined by Judd [24]. RCF contributions to ZFS are given by $<^6S|V_{CF}^{(R)}|^4L><^4L|\Lambda^{(0)}|^6S>$. In SM analysis, only axially symmetric operators with q = 0 are left in Eqs.(3.1) and (3.3).

The contributions of CF and RCF to $\bar{b}_{2,4}$ are respectively

$$\bar{b}_2(CF) = \frac{\overline{B}_2}{49\sqrt{105}}\left(\frac{63}{5}\sqrt{21}pd + 3\sqrt{6}df + \frac{135}{2}fg\right) = -1.059\times10^{-6}\,\overline{B}_2 = -99.61\times10^{-4}\,cm^{-1}$$

$$\bar{b}_4(CF) = \frac{10\overline{B}_4}{21\sqrt{210}}\left(\frac{5\sqrt{3}}{7}df - \sqrt{7}pg - \frac{33\sqrt{2}}{14}fg\right) = 3.008\times10^{-10}\,\overline{B}_4 = 1.470\times10^{-6}\,cm^{-1} \tag{3.4}$$

using values of the intrinsic CF parameters: $\bar{B}_2 = 9{,}420$; $\bar{B}_4 = 4{,}888$ (cm^{-1}) [13] and

$$\bar{b}_2(RCF) = -\frac{6\sqrt{5}}{125}\Delta\bar{B}_2\left(sp - \frac{6\sqrt{14}}{49}sf\right) = -2.385\times10^{-3}\Delta\bar{B}_2 = -2{,}201\times10^{-4}\,cm^{-1}$$

$$\bar{b}_4(RCF) = \frac{8\sqrt{5}}{245}\Delta\bar{B}_4\left(\frac{\sqrt{14}}{4}sf - \frac{8}{15}pd\right) = 8.796\times10^{-7}\Delta\bar{B}_4 = 0.2973\times10^{-4}\,cm^{-1} \quad (3.5)$$

with the RCF parameters $\Delta\bar{B}_2$ = 92.3 cm^{-1} and $\Delta\bar{B}_4$ = 33.8 cm^{-1} [15]. These results alone are sufficient to explain the sign of $\bar{b}_2$ of Mn^{2+} with F^- ligand [3] but $\bar{b}_2$ (CF + RCF) = −0.23 cm^{-1} is one order of magnitude larger than the experimental value (~ −0.03 cm^{-1}). This value will be reduced to the similar order of magnitude to the experimental one by cancellation effects due to, e.g. SCCF [16] and electrostatically correlated CF (ECCF) [17]. However, RCF should be first improved. The correlation of the ZFS parameter D with the trigonal deformation of an octahedron surrounding the S-state ion in garnet [24] and the strong ligand-dependence of b_2 in magnitude and sign [1] indicate that overlap and covalency might be important.

4. Covalency effect in RCF

RCF is based on the model of wavefunction expansion, allowing $nd \rightarrow n'd$ inter-configuration interaction due to the spin-orbit coupling and essentially concerns the large component of the Dirac wavefunction [15]. The matrix elements of the spin-orbit operator between *3d* and *nd* are −3/2 and 1 for total angular momen-tum j = 3/2 (−) and j = 5/2 (+) respectively so that j-dependent relativistic wave-functions are constructed and matrix elements of any operator are j-dependent. RCF parameters are determined by calculating the differences between energies E^+ and E^-. Improvement on RCF can be achieved from the complex model [25], where the energy shift of an $n\ell^N$ state due to virtual charge-transfer excitations of a ligand electron into the unfilled $n\ell$ state can be expressed as a negative contribution to the total energy in the form of $-|N|^2/D$, where N is the cross matrix element of the total Hamiltonian for all electrons in the metal ion-ligand complex between the many-electron ground state and excited state and D is the energy difference between these states (not the axial ZFS parameter D in Section 2). The wavefunction expansion model deals with only the j-dependence of N. Now the j-dependence of D should be taken care of. It can be derived that

$$E_{covalent} = -E_0 + \Sigma|N^j{}_m|^2/D_j \qquad (4.1)$$

where the summation is over all occupied orbitals and $E_0 = -\Sigma_m(|N^+{}_m|^2/D_+ + |N^-{}_m|^2/D_-)$ is the m-independent negative contribution, m is the magnetic quantum number and D is effectively independent of all orbital quantum numbers. In (4.1), the j-dependence is emphasised and the energy shift Δ of E^+ from E^- is

$$\Delta = |N^+{}_m|^2/D_+ - |N^-{}_m|^2/D_-. \qquad (4.2)$$

Wave expansion ensures that $|N^-{}_m|^2 < |N^+{}_m|^2$ and $|D_-| < |D_+|$. The competition between these two processes determines the sign of Δ.

The covalency contributions due to the j-dependence of denominator D is calculated in a way similar to the case of Gd^{3+} [26]. The excitation of a ligand electron into the ground ^{6}S state of Mn^{2+} results in intermediate coupling states of $3d^6$, the same electronic configuration of Fe^{2+} with the ground states ^{5}D. These states split up into energy levels under the spin-orbit interaction. The energy levels $|d^5(^6S)d, {}^5D_J\, M>$ are calculated to estimate $D_+ - D_-$ for Mn^{2+}. The states can be transformed to *jj*-coupling by applying the 9-*j* symbols:

$$|d^5(^6S)d;\, {}^5D_J\, M> = \sum_j (-1)^{J+1}[5(2j+1)]^{1/2} \begin{Bmatrix} j & J & 5/2 \\ 2 & 1/2 & 2 \end{Bmatrix} |d^{5\,6}S_{5/2}d; JM>. \quad (4.3)$$

The matrix elements of the spin-orbit interaction are then

$$<\boldsymbol{H}_{SO}> = <d^5(^6S)d;\, {}^5D_J\, M|\boldsymbol{H}_{SO}|\, d^5(^6S)d;\, {}^5D_J\, M> = (-1)^{J+1}\times 30\zeta \begin{Bmatrix} 2 & 2 & J \\ 2 & 2 & 1 \end{Bmatrix} \quad (4.4)$$

based on the LHS of Eq. (4.3) and from the RHS

$$<\boldsymbol{H}_{SO}> = 180\zeta \left[\begin{Bmatrix} 5/2 & J & 5/2 \\ 2 & 1/2 & 2 \end{Bmatrix}^2 - \begin{Bmatrix} 3/2 & J & 5/2 \\ 2 & 1/2 & 2 \end{Bmatrix}^2 \right] \quad (4.5)$$

where ζ is the radial integral of the spin-orbit operator whose value is taken as $\zeta = 437$ cm^{-1} [25]. It is noticed that Eqs. (4.4) and (4.5) are identical and satisfy

$$\Sigma_J(2J+1) <d^5(^6S)d;\, {}^5D_J\, M|\boldsymbol{H}_{SO}|\, d^5(^6S)d;\, {}^5D_J\, M> = 0, \quad (4.6)$$

based the orthonormality of the 6-*j* symbols. The total contribution to the spin-orbit matrix elements owing to transfer of a $j = 3/2$ electron is -90ζ and that owing to a $j = 5/2$ electron is 90ζ. There are totally $\Sigma^4_{J=0}(2J+1) = 25$ states. The mean difference of 180ζ over 25 states is $36/5\zeta$, taken as an estimate of $D_+ - D_-$ for d^5 system, i.e. $D_+ - D_- = 0.0143$ au. This result holds for other parents of d^6 ^{5}D state, which can be readily shown using a recoupling scheme similar to Eq. (4.3). Thus for the whole d^6 ^{5}D state the estimate is applicable.

For calculating the covalency contributions, the SCCF results [16] are used. The charge transfer energy of $F^-(2p$ or $2s) \rightarrow Mn^{2+}(3d)$ is $D_p = 0.4084$ au and $D_s = 1.3034$ au, which may be regarded as the mean of the energy denominators D_+ and D_- for *p* and *s*-electron covalency respectively. The cross matrix elements due to the charge transfer are $N_s = -0.653\times10^{-2}$, $N_{p\sigma} = -1.84\times10^{-2}$ and $N_{p\pi} = 1.18\times10^{-2}$ (all in unit of au). Then the *j*-dependent parameters $\overline{B}_k$ are obtained from their general relation with *E* (Eqs.(3.5-6) of [16]) and RCF parameters are then obtained

$$\Delta\overline{B}_2(\text{cov.}) = -9.1\ \text{cm}^{-1} \qquad \Delta\overline{B}_4(\text{cov.}) = -5.4\ \text{cm}^{-1}$$

which have opposite signs to those from the wavefunction expansion model (c.f. the parameters in Section 3). It follows that the total RCF parameters are reduced to

$$\Delta\overline{B}_2(total) = 83.2\ \text{cm}^{-1}; \quad \Delta\overline{B}_4(total) = 28.4\ \text{cm}^{-1}.$$

The contributions of RCF are reduced to

$$\overline{b}_2(RCF) = -0.1984\ \text{cm}^{-1};\ \overline{b}_4(RCF) = 0.254\times10^{-4}\ \text{cm}^{-1}.$$

Its rank-2 contribution is still much larger than the experimental value. Together with Eq.(3.4) of CF, $\overline{b}_2(SCCF) = 83.7\times10^{-4}$ cm^{-1}, $\overline{b}_4(SCCF) = 9.35\times10^{-6}$ cm^{-1} [16] and $\overline{b}_2(ECCF) = 599.04\times10^{-4}$ cm^{-1}, $\overline{b}_4(ECCF) = -1.92\times10^{-4}$ cm^{-1} [17], we have

$$\overline{b}_2(total) = -0.1401\ \text{cm}^{-1}; \quad \overline{b}_4(total) = -1.56\times10^{-4}\ \text{cm}^{-1}.$$

Comparison of theoretical and experimental values suggests that there are more cancellation effects.

In summary, we show that the contributions from CF, RCF, SCCF, ECCF mechanisms to ZFS can explain the sign of experimental $\bar{b}_2$ of Mn^{2+} with F^- ligand but $\bar{b}_2$ (theo.) ~ 4.7 $\bar{b}_2$ (expt.). Further study is needed to improve the agreement between theory and experiment.

References

1. M. Heming and G. Lehmann, Correlation of zero-field splittings and site distortions. Variation of $\bar{b}_2$ for Mn^{2+} with ligand and coordination number, *Chem. Phys. Lett.* **80** (1981) 235-7
2. M. Heming and G. Lehmann, Correlation of zero-field splittings and site distortions. III. $MnBr_4^{2-}$ and crystal structure refinements for Rb_3ZnBr_5 and Cs_2ZnBr_4, *Z. Narurforsch* **36A** (1981) 286-93
3. M. Heming, G. Lehmann, K. Recker and F. Wallrafens, EPR and luminescence of Mn^{2+} in distorted tetrahedral symmetry in $CaZnF_4$ crystals, *Phys. Stat. Sol. (b)* **117** (1983) 271-5
4. M. Heming, S. Remme and G. Lehmann, Experimental and theoretical investigations of the second-order zero field splitting parameters of Mn(II) with Cl^- as ligand, *Ber. Bunsenges. Phys. Chem.,* **88** (1984) 946-53
5. M. Heming and G. Lehmann, Correlation of zero field splittings and site distortions. Interstitial incorporation of Mn^{2+} in urea-ammoniumchloride $(H_2N)_2CO$-NH_4Cl, *Phys.* Stat. *Sol. (b)* **124** (1984) 315-23
6. M. Heming, S. Remme and B. Lehmann, Correlation of zero-field splittings and site distiortions. X. Complexes of Mn^{2+} with mixed ligands, *J. Magn. Res.* **69** (1986) 134-43
7. D.J. Newman and W. Urban, Interpretation of ***S***-state ion E.P.R. spectra, *Adv. Phys.* **24** (1975) 793-844
8. P.R. Sharma, T.P. Das and R. Orbach, Zero-field splitting of S-state ions. I. Point-multipole model, *Phys. Rev.* **149** (1966) 257-69
9. P.R. Sharma, T.P. Das and R. Orbach, Zero-field splitting of S-state ions. II. Overlap and covalency model, *Phys. Rev.* **155** (1967) 338-52
10. P.R. Sharma, T.P. Das and R. Orbach, Zero-field splitting of S-state ions. III. Corrections to parts I and II and application to distorted cubic crystals, *Phys. Rev.* **171** (1968) 378-88
11. A. van Heuvelen, *J. Chem. Phys.* **46** (1967) 4903-5
12. R. Chatterjee, M.R. Smith and H.A. Buckmaster, *Can. J. Phys.* **54** (1976) 1224-7
13. G.G. Siu, Contributions to the ground-state splitting of S-state ions: II. A theoretical framework for the $3d^5$ ^6S ground-state splitting of Mn^{2+}, *J. Phys. C: Solid State Phys.* **21** (1988) 3927-37
14. G.G. Siu and D.J. Newman, Contributions to the ground state splitting of S-state ions: I. The $^8S_{7/2}$ state of Gd^{3+}, *J. Phys. C: Solid State Phys.* **15** (1982) 6753-64

15. G.G. Siu, The relativistic effects in the S ground-state splitting of Mn^{2+} ions, *J. Phys. C: Solid State Phys.* **21** (1988) 3917-25
16. G.G. Siu, The effect of spin polarisation on the crystal field and the S ground-state splitting of Mn^{2+}, *J. Phys. C: Solid State Phys.* **21** (1988) 5385-92
17. G.G. Siu and Y.M. Poon, Effects of the electrostatically correlated crystal field in S ground-state splitting of Mn^{2+}, *J. Phys.: Condens. Matter* **1** (1989) 1911-7
18. A. Edgar, E. Siegel and W. Urban, *J. Phys. C: Solid State Phys.* **13** (1980) 6649-57
19. P.A. Narayana, Overlap contribution to the axial field of Mn^{2+} in K_2MgF_4, *Phys. Rev.*B **10** (1974) 2676-80
20. D.J. Newman, B. Ng and Y.M. Poon, Parametrisation and interpretation of paramagnetic ion spectra, *J. Phys. C: Solid State Phys.* **17** (1984) 5577-84
21. M.G. Zhao, G.R. Bai and H.C. Jin, d-orbital theory for Mn^{2+}-complex ions in crystal, *J. Phys. C: Solid State Phys.* **15** (1982) 5959-64
22. K. König and R. Schnakig, *Phys. Stat. Sol. B* **78** (1976) 543-51
23. C.A. Hutchinson, B.R. Judd and D.E.F. Pope, *Proc. Phys. Soc. B* **70** (1957) 514-20
24. P. Novak and I. Veltrusky, Overlap and covalency contribution to the zero-field splitting of S-state ions, *Phys. Stat. Sol.b* **73** (1976)575-86
25. D.J. Newman, *Adv. Phys.* **20** (1971) 197-256
26. G.G. Siu, Covalency effects in the relativistic crystal field, *J. Phys. C: Solid State Phys.* **21** (1988) L1001-5
27. J.E. Hansen and A.J.J. Raassen, *Physica* **111** C (1981) 76-101

Comparative Analysis of EPR Data for Low Symmetry Paramagnetic Centers - Application to Cr^{3+}:$KTiOPO_4$

C. Rudowicz, I. Akhmadoulline and S. B. Madhu*

Department of Physics and Materials Science, City University, Hong Kong
* Present address: Zavoisky Physical-Technical Institute, Kazan 420029, Russia

Abstract. Comparison of data sets for low symmetry paramagnetic centers, comprising the principal values as well as the orientations of the principal axes of the second-rank spin Hamiltonian terms, e.g., **D, g, A** and **P,** is often difficult. The data sets ascribed to supposedly distinct centers by some authors may actually belong to the physically same centers. A quantitative method for comparative analysis of EPR data for low symmetry paramagnetic centers is proposed. Parameters S_A and S_D are introduced to describe the closeness of the orientation of any two axis systems and the principal values of any two tensors, respectively. A computer program based on this method has been developed. The data sets for Cr^{3+} centers in $KTiOPO_4$ consisting of the principal values of **D** and **g** as well as the orientations of their principal axes are considered. Various other applications of the program are briefly discussed.

1 Introduction

In the course of our recent work on the low symmetry Cr^{3+} centers in $KTiOPO_4$ (*KTP*) we have encountered problems when comparing and identifying the supposedly distinct centers determined by different authors. Khasanova et al. [1] have identified six Cr^{3+} centers in KTP : five centers located at Ti(1) sites and one at Ti(2) sites. The study [2] have revealed twelve Cr^{3+} centers - the six already identified [1] centers as well as new centers: two located at Ti(1) and four at Ti(2) sites. Recently, two Cr^{3+} centers in KTP [3] and from careful reanalysis of weaker intensity lines two additional ones [4] have been identified. The principal values of **D** and **g** as well as the orientations of their principal axes were determined [3, 4] indicating two Cr^{3+} centers located at Ti(1) and two at Ti(2) sites. The question arises which of the centers identified in [3, 4] correspond to the centers observed in [1, 2] as well as which data sets ascribed to supposedly distinct centers in [1-4] may actually belong to the physically same centers.

In this paper, we propose a quantitative method for comparison of various data sets comprising the principal values and/or the orientations of the principal axes of the second-rank spin Hamiltonian terms, e.g., **D, g, A** and **P**. Numerical parameters S_A and S_D are introduced to describe the closeness of the orientation of any two axis systems as well as of the principal values of any two 'tensors', respectively. A computer program based on this method has been worked out. Usefulness of the program is illustrated by quantitative analysis of the data sets consisting of the principal values of **D** and **g** as well as the orientations of their principal axes for Cr^{3+} centers in KTP. The closeness of several data sets for the supposedly distinct centers identified so far [1-4] indicates that some of them may actually belong to the physically same centers. Various other applications of the program are briefly discussed.

2 Background Theory

EPR spectra of the centers located at triclinic symmetry sites are analysed using a spin Hamiltonian (*SH*):

$$H_{SH} = \mu_B \vec{B} \bullet \mathbf{g} \bullet \vec{S} + \vec{S} \bullet \mathbf{D} \bullet \vec{S} + \{\text{higher order ZFS terms}\} \quad (1)$$

with the zero field splitting (ZFS) expressed in the conventional notation [5, 6] or

$$H_{SH} = \mu_B \vec{B} \bullet \mathbf{g} \bullet \vec{S} + \sum_{q=-2}^{2} B_2^q O_2^q + \sum_{q=-4}^{4} B_4^q O_4^q + \sum_{q=-6}^{6} B_6^q O_6^q \quad (2)$$

in the extended Stevens operator notation [5], expressed in the laboratory axis system (X, Y, Z). In general the principal axes of the second-rank 'tensors' **g** and **D** are not coincident for triclinic symmetry [6]. A full SH data set for the spin S=3/2 consists of the principal values of **g** and **D** in Eq.(1) (or equivalently B_2^0 and B_2^2 in Eq.(2)) as well as the orientation of the principal axis system (*PAS*) for each SH term with respect to (X, Y, Z). Hence, physically different centers can be distinguished by either significantly different principal axis systems and/or principal values of **g** and **D** (as compared with the respective experimental errors). When comparing two data sets of the same type for a given type of substitutional impurity with valence different from that for the host ion two cases may arise. The case (i): the PASs are nearly identical, whereas the principal values are different. This would mean that an additional charge compensator located along one of the principal axes exists, which makes the two "orientationally" equivalent centers physically distinguishable. The case (ii): the PASs are different, whereas the principal values are nearly identical. Then it can be expected that one deals with two (or more; in the case of KTP - four) magnetically equivalent sites connected by proper symmetry operations. In the latter case the experimental data points for such sites should be jointly taken into account in the final fitting procedure.

For triclinic symmetry a direct comparison of the often apparently close numerical values of the respective components in the SH data sets from various authors does not provide any global conclusions. For this purpose, a quantitative method, based on the geometrical properties of scalar product, is proposed as follows. Consider, for a given SH term, two PASs independently determined from EPR spectra as (supposedly) corresponding to two distinct paramagnetic centers. Each PAS is given by nine either angles or direction cosines, every three of which (α, β, γ) or (*l*, *m*, *n*) characterise the orientation of a particular principal axis relative to the reference frame; equivalently six polar angles (θ, ϕ) are also used. Therefore, the angles or the direction cosines for the three principal axes can be considered as a "vector" in a 9-dimensional space. In general, the closeness of the mutual orientation of two vectors can be measured in terms of the normalized scalar product. Extending this idea, the closeness of the relative "orientation" of two 3×3 matrices (a_{ij}, b_{ij}) representing two PASs can be defined as

$$S_A = \frac{\sum_{i,j} a_{ij} \cdot b_{ij}}{\sqrt{\sum_{i,j} a_{ij}^2} \cdot \sqrt{\sum_{i,j} b_{ij}^2}} \cong 1. \tag{3}$$

Note that the matrices dealt with are real and symmetric, hence $b^*_{ji} \equiv b_{ij}$ in Eq.(3). The parameter S_A is an analogue of the cosine of an angle between two vectors and for two identically oriented PASs $S_A \cong 1$. Eq.(3) provides a convenient quantitative measure of the proximity of the orientation of any two PASs expressed in the same reference frame. This algorithm is used in the computer program, which besides S_A calculates also the angles θ_1, θ_2, θ_3, and their average (labelled in the program as T1, T2, T3; AvT), between the pairs of the corresponding axes (X_a, X_b), (Y_a, Y_b), (Z_a, Z_b), respectively, of the two PASs *a* and *b* defined by

$$\cos\theta_1 = \vec{X}_a \bullet \vec{X}_b \ , \quad \cos\theta_2 = \vec{Y}_a \bullet \vec{Y}_b \ , \quad \cos\theta_3 = \vec{Z}_a \bullet \vec{Z}_b \tag{4}$$

where $\vec{X}_{a,b}$, $\vec{Y}_{a,b}$, $\vec{Z}_{a,b}$ are the unit vectors in the direction of the respective principal axes of the PAS *a* and *b*.

Similar approach can be used for comparison of the principal values of any two tensors of the same nature, e.g., $\mathbf{D_a}$ and $\mathbf{D_b}$ ascribed to (supposedly) different centers '*a*' and '*b*'. The three principal values (A_{11}, A_{22}, A_{33}) of any second-rank SH tensor **A** can be considered as a "vector" and the relative "orientation" of two such "vectors" $\vec{A}$ and $\vec{B}$ can also be defined in terms of the normalized scalar product. However, since each PAS is a 'local' system, first the components of each vector in the common reference frame (X, Y, Z), i.e. (A_X, A_Y, A_Z)= $\vec{A}_r$, must be calculated by projecting each principal value A_{ii} onto the axes (X, Y, Z) using the values of (α, β, γ) or (*l*, *m*, *n*). Then the quantity S_D defined as

$$S_D = \frac{\vec{A}_r \bullet \vec{B}_r}{A \cdot B} = \cos\theta_{ab} \tag{5}$$

represents the angle θ_{ab} between the two vectors $\vec{A}$ and $\vec{B}$. The magnitudes *A* and *B* in Eq.(5) are also useful quantities, since they enable direct comparison of the 'lengths' of $\vec{A}$ and $\vec{B}$. For the ZFS parameters, the norm N_k [8, 4], which is independent of the axis system used, can alternatively be used for this purpose. In the PAS of **D** the norm N_2 is given by

$$N_2 = (B_2^0)^2 + \tfrac{1}{3}(B_2^2)^2 = \tfrac{1}{4}(D_{zz})^2 + \tfrac{1}{12}(D_{xx} - D_{yy})^2 \ . \tag{6}$$

It should be noted that when comparing data sets for two centers (especially from different authors) both the parameter S_D, Eq.(5), describing the relative "orientation" of two "principal values vectors", and S_A, Eq.(3), describing the relative "orientation" of two "PAS matrices", should be taken into account. It must be ensured for S_A that both PASs are referred to the same reference frame, whereas for S_D that the convention used for the principal values of a given tensor is the same. The **standard** convention [7, 8] for the principal values of **D**, i.e. $|D_{xx}| \le |D_{yy}| \le |D_{zz}|$, means that the PAS of **D** has been chosen so that the ratio λ defined by

$$\lambda = \frac{E}{D} = \frac{D_{xx} - D_{yy}}{3 \cdot D_{zz}} \tag{7}$$

satisfies the standardization condition [7]: $0 \le \lambda \le 1/3$. If one or both initial data sets contain nonstandard principal values (D_{xx}, D_{yy}, D_{zz}), the standardization [7, 8] should be performed first, since only comparison of standardized data sets, i.e. expressed using the same convention, is meaningful. The present program detects the nonstandard data sets and marks them by an asterisk (*). Then it is necessary to perform an appropriate standardization transformation. This is the case of some data sets in [2]. The recently developed computer package ***CST*** for conversions, standardization and transformations of SH (and crystal field Hamiltonian) [9] facilitates standardization transformations for orthorhombic [7] as well as monoclinic and triclinic symmetry [8]. An example of application of Eq.(5) to nonstandardized data sets, which yields meaningless results, is considered in [10].

The program calculates also the errors of S_A and S_D (denoted as ErrSA and ErrSD) induced by the experimental errors $\delta\phi$ and $\delta\,\mathrm{Vec}$ in determination of the angles describing the orientation of the principal axes and the principal values, respectively, distributed according to the normal law:

$$ErrSA = K_A \cdot \delta\phi \;, \qquad ErrSD = K_D \cdot \delta\,Vec \tag{8}$$

Full description of the program ***COMPARE*** as well as the definition of the coefficient K_A and K_D in Eq.(8), which represent the relative error of S_A and S_D, respectively, will be given elsewhere [10].

3 Results and Discussion

Here two illustrative cases of application of the program for analysis of data for Cr^{3+} ions in KTP [3, 4] are presented. In a similar way, the data sets for the six [1] and twelve [2] Cr^{3+} centers in $KTiOPO_4$ are analysed in [10]. Each of the two crystallographically inequivalent Ti positions in KTP crystal: Ti(1) and Ti(2) gives rise to four crystallographically equivalent but magnetically inequivalent Ti sites. Both Ti sites lie in slightly distorted oxygen octahedra with local site symmetry C_1. Cr^{3+} ions can substitute Ti^{4+} ions at either position [1].

Firstly, we consider the data sets [3, 4] consisting of the orientation of the PAS of **g** and its principal values for the centers A and C identified as $(Cr^{3+})_{Ti(2)}$ and the $(Cr^{3+})_{Ti(1)}$ centers B and D. For simplicity the errors are not considered (dPHI= 0, dVec= 0). Data are taken from Table 2 of [3] for the center A (SA1) and B (SB1), and Table 2 of [4] for the center C (SC1) and D (SD1). The magnitude of |g| for the four centres is: (A) 3.412578, (B) 3.413454, (C) 3.424591 and (D) 3.425771. The computer generated output, with proper labelling of the centers (*a;b*) and the repeating lines (*b;a*) omitted, is given in Table 1 (with angles in degrees and S_A or S_D in the double (D+00) precision).

Table 1. Comparison of the **g**-tensor data sets for the four $(Cr^{3+})_{Ti}$ centers in KTP of [3, 4].

PAIR	T1	T2	T3	AvT	SA	SD
A;A	.0	.0	.0	.0	.1000D+01	.9984D+00
A;B	87.2	133.6	88.2	103.0	-.2033D+00	-.2252D+00
A;C	124.0	151.3	140.1	138.5	-.7349D+00	-.7710D+00
A;D	83.6	122.3	66.8	90.9	-.9631D-02	-.7455D+00
B;B	.0	.0	.0	.0	.1000D+01	.1000D+01
B;C	145.6	54.9	59.0	86.5	.8834D-01	.4335D+00
B;D	11.5	30.2	149.1	63.6	.3285D+00	.7933D+00
C;C	.0	.0	.0	.0	.1000D+01	**.7544D+00**
C;D	152.0	47.3	150.9	116.8	-.3597D+00	.7930D+00
D;D	.0	.0	.0	.0	.1000D+01	.9926D+00

The pairs of the centers in Table 1 appear to be physically distinct. An interesting point is, however, revealed. For any (*a;a*) pair the value of S_D should be close to 1. It is the case for the pairs (A;A), (B;B) and (D;D) in Table 1, whereas not for the pair (C;C) - $S_D \cong 0.7544$ indicates that for the centre C the angles (θ, ϕ) are ill-defined, i.e. the axes do not form an orthogonal set. This has been confirmed by calculating the scalar products $\mathbf{a}_i.\mathbf{a}_j$ defined by $\mathbf{a}_i = (l_i, m_i, n_i)$ for each pair of axes $(i \neq j)$. The non-zero values $\mathbf{a}_1.\mathbf{a}_2 = 0.193$ and $\mathbf{a}_1.\mathbf{a}_3 = 0.671$ indicate that the respective two axes are not perpendicular. Hence for the centre C in [4] either the angles (θ, ϕ) are misprinted or the fitting has been incorrect. This means that the values of S_A and S_D in the lines (A:C) (B:C) and (C:D) in Table 1 are unreliable. Thus the present program serves as a *quality assurance measure* for low symmetry EPR data sets. Calculation of the products $\mathbf{a}_i.\mathbf{a}_j$ and verification of the orthogonality of the axes will be included in the next version of the program ***COMPARE***.

Secondly, similar analysis is carried out for the **D**-tensor data sets taken from Table 3 of [3] for the center A (SA1) and B (SB1), and Table 3 of [4] for the center C (SC1) and D (SD1). The norm N_2 for the four centers is (in $10^{-4}cm^{-1}$): (A) 3097760, (B) 2337789, (C) 2724093, and (D) 2088504. A part of the computer generated output is given in Table 2, where the lines for the (*a;a*) pairs are omitted since the values of S_A and S_D are very close to 1 for all cases.

Table 2. Comparison of the **D**-tensor data sets for the four $(Cr^{3+})_{Ti}$ centers in KTP of [3, 4].

PAIR	T1	T2	T3	AvT	SA	SD
A;B	103.0	38.6	92.5	78.0	.1712D+00	-.2177D+00
A;C	17.5	16.7	172.9	69.0	.3064D+00	.9250D+00
A;D	101.3	41.0	90.4	77.6	.1839D+00	-.2568D+00
B;C	116.4	43.7	80.4	80.2	.1485D+00	-.4089D+00
B;D	**3.5**	**2.6**	**2.4**	**2.8**	**.9987D+00**	**.9979D+00**
C;D	115.4	45.5	82.5	81.1	.1345D+00	-.4271D+00

Apart from the pair (B;D), other pairs of the centers in Table 2 appear to be physically distinct as the values S_A are far from 1 and the angles between their PASs are large (but not comparable with $\pi/2$). For the pair (B;D), S_A=0.9987≅1 and the very small angles T1, T2 and T3 between the respective principal axes of **D** indicate that the PASs of the two centers are nearly coincident. Moreover, their "principal vectors" **D** are nearly "parallel" (S_D = 0.9979). Hence the center B and D appear to be nearly identical. Would the Cr^{3+} be an isovalent impurity in KTP host crystal, this closeness would indicate possibility of some errors in the experimental or fitting procedure. However, since Cr^{3+} ions are not isovalent with Ti^{4+}, it may mean that a kind of charge compensator must exist in the lattice. The most common charge compensator in oxide crystals is the F^+-center (oxygen vacancy with trapped one electron). If a F^+ -center is placed close to the Cr^{3+} ion, its wave function would overlap considerably (the F^+-center wave function usually spreads over the next-nearest neighbours in oxide crystals) with that of Cr^{3+}, resulting in screening of the crystal field and therefore reducing the strength of the ZFS parameters. Since it is not the case, this type of charge compensator is presumably placed beyond the first coordination sphere, so the wave functions of the F^+ center and Cr^{3+} ion do not overlap. In this case, however, due to the ferroelectric nature of $KTiOPO_4$ host, its dielectric constant is large and one can expect significant screening of the charge compensators, so their effect on the paramagnetic impurity would be comparatively weak. This can explain the closeness of the **D**-tensor data sets for the center B and D obtained from recognisably distinct lines in the EPR spectra [3, 4]. This conclusion is supported by the slightly different values of N_2: (B) 23402446 and (D) 2088504. In other respects the two centers are equivalent, i.e. both have the same environment.

Cross-comparison of Table 1 and 2 reveals that the **g**- and **D**-tensor data sets have different characteristics. This might be because of the ill defined values of the angles for the center C in Table 1. If the angles (θ, ϕ) are well defined for the center C then probably the two tables may show similar characteristics for the comparable pairs. This would not effect the pair (B:D) and hence in spite of the closeness of the center B and D indicated by the **D**-tensor data sets, these two centers may be considered as physically different.

We have also considered the interrelationships between the four magnetically equivalent (i.e., symmetry related) sites for each center A and B [3] using the **D** and **g** data sets in Table 2 and 3 of [3]. This analysis reveals [10] an interesting general characteristics of the symmetry related data sets as well as some inconsistencies in the results [3]. More extensive comparative analysis of the **g**- and **D**-tensor data sets for Cr^{3+} in KTP taken from several sources (including [1, 2]) as well as other systems will be carried out in [10]. In general, such analysis do indicate that some supposedly distinct centers in [1, 2] should rather be considered as physically same centers. Moreover, the proposed method helps also to identify the ill-defined data sets.

4 Summary and Conclusions

A quantitative method for comparative analysis of EPR data for low symmetry paramagnetic centers is proposed. A computer program has been worked out and applied to in-depth analysis of data for the Cr^{3+} centers in $KTiOPO_4$ consisting of the principal values of **D** and **g** as well as the orientations of their principal axes.

In general the method and the program can also be applied for the following problems: a) determination of the pairs of data sets which may likely belong to the same physical center, but had been identified as two separate centers, b) assessment of the closeness of the orientation of any two axis systems in terms of well defined numerical parameters, c) estimation of the goodness of the match between the principal axis systems of **D** and any other axis systems, e.g., the axis system related to the bonds between the paramagnetic ion and the ligands, d) estimation of the closeness of any two principal axis systems, e.g. of the **D**, **g**, **P** and/or **A** spin Hamiltonian terms. These applications of the method will be dealt with elsewhere [10] using examples from recent EPR studies of low symmetry paramagnetic centers.

Acknowledgements. We thank University Grants Committee and the City University of Hong Kong Strategic Research Fund for financial support.

References

1. N. M. Khasanova, N. M. Nizamutdinov, A. S. Izugzon, G. R. Bulka, V. M. Vinokurov, N.I. Pavlova, and I. S. Rez, in *Physics of Minerals and their Synthetic Analogs,* Kazan Univ. Press (1988) pp. 87-110 (in Russian).
2. A. A. Galeev, N. M. Nizamutdinov, G. R. Bulka, N. M. Khasanova, N. I. Pavlova, G. A. Ermakov, and V. M. Vinokurov, in *Spectroscopy, Crystal Chemistry and Real Structure of Minerals and their Analogs*, Kazan Univ. Press (1990) pp. 140-148 (in Russian).
3. S. W. Ahn, S. H. Choh, and J. N. Kim, *J. Phys.: Cond. Matter* **7** (1995) 667-674.
4. S. W. Ahn, S. H. Choh, and C. Rudowicz, *Appl. Magn. Reson.* (1997) - in press.
5. C. Rudowicz, *Magn. Reson. Rev.* 13 (1987) 1-89.
6. J. R. Pilbrow, M. R. Lowrey, *Rep. Prog. Phys.* 43 (1980) 433-495.
7. C.Rudowicz and R. Bramley, *J. Chem. Phys.* 83 (1985) 5192-5202.
8. C. Rudowicz, *J. Chem. Phys.* 84 (1986) 5045-5058.
9. C. Rudowicz and I. Akhmadoulline, *CST program manual,* Res. Rep. AP-97-xx, Dep. of Physics and Materials Science, City University of Hong Kong (1997) - to be published.
10. C. Rudowicz, S. B. Madhu, and I. Akhmadoulline, (1997) - in preparation.

Chapter 4

New Developments and Cross Disciplinary Areas

Electrical Detection of EPR and ENDOR: New Possibilities for the Study of Point Defects in Semiconductors

J.-M. Spaeth

Fachbereich Physik, Universität-GH Paderborn, 33095 Paderborn, Germany

Abstract. Electron paramagnetic resonance (EPR) and electron nuclear double resonance (ENDOR) of defects in semiconductors can be detected as a microwave (EPR)- and microwave- and radio-frequency-induced (ENDOR) change of the electrical conductivity caused by excess charge carriers, the latter being either generated by above-band-gap light (photoconductivity) or injection (pn-junctions). Hyperfine and fine structure interactions can be resolved as in conventional EPR and ENDOR, but the sensitivity is several orders of magnitude higher both for EPR and ENDOR. The present understanding of the mechanism and the experimental conditions to observe electrical detection of EPR and ENDOR are reviewed and new possibilities for the study of defects in semiconductors available in this detection scheme are discussed.

1 Introduction

The study of point defects in semiconductors often suffers from the fact that there are only samples available in small volumes such as in semiconductor devices or in thin epitaxial layers, while their concentrations are in the usual range of interest, i.e. between 10^{15}-10^{17} cm^{-3}. In such cases the conventional detection of electron paramagnetic resonance (EPR) or electron nuclear double resonance (ENDOR) is not possible because these methods are not sensitive enough. Although optical detection of EPR and ENDOR is generally more sensitive and can overcome this problem, as was successfully demonstrated in many cases, this method is not generally applicable. The use of the magneto-optical method via the magnetic circular dichroism of the optical absorption (MCDA-detected EPR/ENDOR) usually requires layer thicknesses of about 50-100 μm or more and an optical transition associated with the defect is needed, while the detection via donor-acceptor-recombination luminescence requires the donor-acceptor-recombination to be radiative, which is often not the case, in particular in indirect semiconductors such as silicon. There are only very few ENDOR experiments via luminescence-detected EPR (for an overview see e.g. [1]). Therefore, a more sensitive method is needed to detect EPR and ENDOR, which, in principle, are very powerful tools to study the microscopic and electronic structures of point defects (see e.g. [1]).

Attempts to measure EPR at room temperature as microwave-induced changes of the electrical conductivity of Si-diodes have, in principle, been successful, but gave in most cases not much information, except for an unstructured line at g=2 [2-16]. Only in a few cases a hyperfine (hf) structure was resolved [5,17], allowing the identification of the defect involved. It was also a matter of luck whether or not an ED-EPR signal could be detected. Most authors explain their observations by assuming a spin-dependent recombination (SDR) of excess charge carriers at a paramagnetic recombination defect as the responsible process [2-9] and a number of specific mechanisms has been proposed [4,15,18-20]. The understanding of the experimental observations and of the conditions to observe electrical detection of EPR (EDEPR) remained quite poor such that no 'ED-EPR' spectroscopy has yet been developed. Only recently some progress was made with a systematic study of the conditions to observe EDEPR of shallow and deep-level defects in silicon for low or moderate concentrations of such defects using microwave-induced changes of the photoconductivity. It was shown that a donor-acceptor pair recombination process can explain satisfactorily most experimental observations [21] and that also ENDOR can be detected via the photoconductivity [22]. In this paper these results are reviewed and the present understanding is discussed. Furthermore, some additional features for the study of point defects are pointed out which are not available with conventional detection of EPR/ENDOR. Finally, open questions are mentioned, which require further theoretical and experimental investigations.

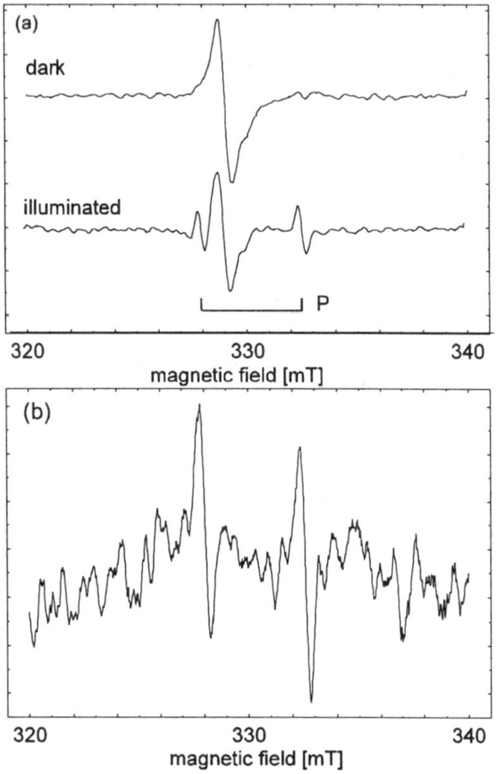

Fig. 1. (a) Conventional EPR spectrum of a compensated Si sample containing shallow P donors and shallow B acceptors, both in a concentration of about $10^{16}cm^{-3}$. ν_{EPR}=9.23GHz, P_{MW}=20μW, T=10K, B_0||[100]. Upper trace: measurement in the dark, lower trace: measurement with above-band-gap light using a halogen lamp. (b) Electrical detection of the EPR of the sample described in fig. 1(a).

2 Results and Discussion on Electrical Detection of EPR

2.1 Experimental Observation of EDEPR

Various defects were investigated in donor and acceptor-doped Czochralski (Cz)-grown silicon. In order to generate deep-level defects, the samples were irradiated with 2MeV electrons at a fluence of $10^{17}cm^{-2}$ at room temperature producing oxygen-vacancy pair defects (so-called A centres [23]). Annealing of Cz-grown Si generated thermal double donors (NL8 [24]). The photoconductivity was generated by above-band-gap light and measured via electrical contacts. It was carefully checked that the contacts were ohmic. A constant current source was used and the conductivity changes were measured as voltage changes across the sample. To induce microwave transitions, a standard homodyne X-band EPR spectrometer operating between 10K and 300K was used.

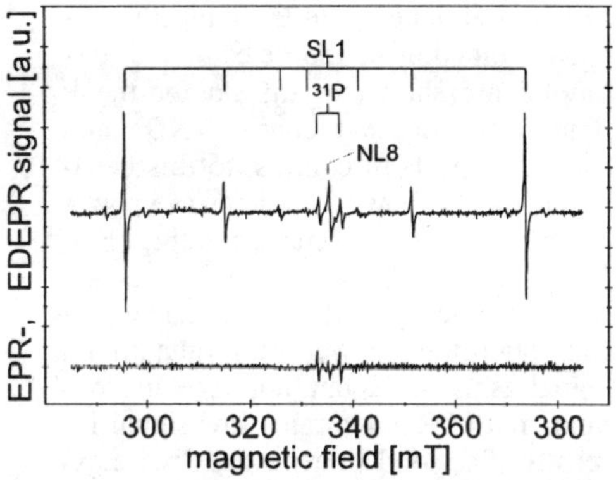

Fig. 2. Conventional EPR (lower trace) and ED-EPR (upper trace) of an Si sample containing shallow P donors, thermal double donors (NL8) and SL1 centres under above-band-gap illumination. T=20K and $B_0||[011]+2°$, ν_{EPR}=9.3GHz

In a silicon sample containing only shallow phosphorus donors (n-type, Fermi-level near the conduction band) no EDEPR effect was detected, whereas, of course, in conventional EPR the P hyperfine (hf) doublet was measured. However, upon compensation of the sample with boron the ED-EPR signal could be seen. This is shown in fig. 1. Fig. 1(a), upper trace, shows the conventional EPR spectrum in the compensated P-doped sample in the dark. The signal of an unknown defect is detected. The phosphorus is not seen, since the Fermi-level is below the 0/+ level of the P donors. Upon illumination with above-band gap light, the paramagnetic state of P^0 is occupied, the P hf doublet is seen. Fig. 1(b) shows the EDEPR spectrum, in which the ^{31}P hf doublet is clearly seen, however, not that other EPR line seen in the dark and under illumination with conventional EPR. Analogous experiments using, for example, thermal double donors (TDDs) as shallow donors generated by the annealing of the oxygen-containing silicon at 450°C gave the same result: if the sample contained B acceptors in such a concentration that in the dark the Fermi-level was below the +/++ level of TDD^+ (i.e. the number of B acceptors exceeded that of both TDDs and TDD^+s), then the EDEPR signal was seen. Many experiments confirmed that the condition for the observation of an EDEPR signal is that a donor-acceptor pair recombination takes place, whereby the electrons and holes must recombine from separate paramagnetic donors and acceptors [21,25]. In a phosphorus and TDDs containing sample, irradiated with 2MeV electrons at room temperature, the spectrum of fig. 2 was measured. The lower trace represents the conventional EPR spectrum measured under above-band-gap illumination, the upper trace the EDEPR spectrum. The signal-to-noise ratio of the EDEPR spectrum is about the same as that for the conventional EPR spectrum for the P donors, but greatly enhanced for the so-called SL1 centre, an excited triplet state of the oxygen-vacancy pair defect (A-centre) [26] for various orientations of the defect. In the EDEPR spectrum the line width is the same as that in the conventional EPR spectrum and the structural information resolved is also identical: hyperfine (hf), fine structure and superhyperfine (shf) structure of ^{29}Si ligands are resolved. A clear difference between conventional EPR and EDEPR is the dependence of the signal intensity on the microwave power measured at the same temperature [21]. The conventional EPR spectrum saturates and has a maximum signal intensity at relatively low microwave power following the law

$$S_{EPR} \propto (1+S)^{-3/2}(P_{MW})^{1/2} \tag{1}$$

where S is the saturation factor

$$S = \tfrac{1}{4}\gamma^2 B_1^2 T_1 T_2 \tag{2}$$

and P_{MW} is the microwave power. At high microwave power an inverse relation $S_{EPR} \propto (P_{MW})^{-1}$ is observed. Eq. 1 describes the saturation of the ground state polarisation and takes into account the response function of the detector [8]. Examples are shown in fig. 3a for the P donors, the thermal donors TDD^+ and the SL1 defect. In contrast to this behaviour, fig. 3b shows the microwave power dependence of the corresponding EDEPR signals. The saturation curves are of a completely different character: at high microwave power a high maximum signal is reached but not like in conventional EPR a saturated low signal intensity. The solid lines in fig. 3b are experimental fits using the relation

$$S_{EDEPR} = C\frac{\beta P_{MW}}{1+\beta P_{MW}}. \qquad (3)$$

β and C are parameters. For low microwave power there is a linear relationship between S_{EDEPR} and P_{MW}, and for high P_{MW} values there is constant signal C.

Interestingly, a very similar power dependence has been measured recently for a luminescence-detected EPR signal of the residual donor in GaN [27,28]. The basis for the optical detection of EPR via luminescence is a radiative donor-acceptor pair recombination (see e.g. [1]). Thus, this observation together with the experimental results described above point to a donor-acceptor pair recombination mechanism to explain the occurrence of EDEPR effects, at least for low and moderate defect concentrations (see below). It should be noted that the B acceptors in silicon are difficult to observe with EPR. Their lines are very broad, and they consequently escape detection [29].

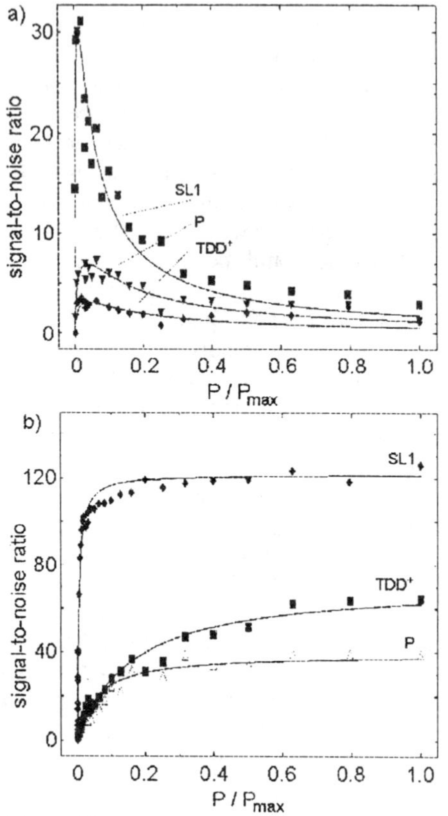

Fig. 3. EPR microwave power dependence of the lines of SL1, P and TDD^+ (NL8) defects shown in fig. 2. The various symbols are the experimental signal-to-noise ratios of the EPR lines. The solid curves are fitted according to theory. P_{max}=200mW (ν_{EPR}=9.3GHz, $B_0 \|$ [011], T=20K). (a) Microwave power dependence of the conventionally detected EPR lines, fits are according to eq. (1) and (2). (b) Microwave power dependence of the EDEPR lines, fits are according to eq. (8).

The recombination of weakly exchange-coupled neutral donors and acceptors, be it radiative or non-radiative, is a spin-dependent process and can thus be influenced by magnetic dipole transitions as is well known from ODEPR [1]. It was proposed to be the mechanism which can quantitatively explain the observation of EDEPR [21]. At low concentration single defects, such as P donors alone, cannot be detected by EDEPR in contrast to earlier models [15,19,20]. The concept that electron-hole recombination is responsible for EDEPR has been suggested previously [18], however, with the difference that the electron-hole recombination was thought to occur at one defect and not between separated donors and acceptors. At least, this mechanism of separated but weakly exchange-coupled pairs of D^0 and A^0 defects was not thought to be the decisive mechanism. If the electron-hole recombination could occur at one paramagnetic defect alone, P donors alone would have been seen or e.g. deep chalcogen paramagnetic centres independent of the compensating B concentration. This was not the case [21]. The situation is different when the concentration of defects increases. Then, for instance, when hopping conductivity between donors can occur, other mechanisms seem to operate (see below).

2.2 The Donor-Acceptor Recombination Model

A general description of the EDEPR signal with the donor-acceptor pair recombination model leads to a complicated set of rate equations by which the temporal change of the excess charge carriers is calculated as a function of their generation (e.g. by above-band-gap light) and recombination via the spin-dependent exchange coupled donor-acceptor pair states. For the concentration n of the conduction band electrons one obtains for example [25,41]

$$\frac{d}{dt}n = g_0 - c_n\ n(N - n_D - n_{DA}) + e_n(n_D + n_{DA}) \qquad (4)$$

where g_0 is the generation rate of the charge carriers and N is the number of possible D^0-A^0 pairs, c_n the capture cross section of the donors, e_n the emission rate, n_D the concentration of D^0-A^0 pairs having captured an electron and n_{DA} the concentration of D^0-A^0 pairs having captured an electron and a hole. An analogous expression holds for the concentration of holes in the valence band. For n_A, n_D, n_{DA} also rate equations hold containing recombination rates, microwave transition probability, spin-lattice relaxation, etc. Such a complicated system of rate equations cannot be solved analytically. Furthermore, non-spin-dependent recombination processes (shunt processes) are not yet taken into account in eq. (4) and many quantities are not well enough known, such as the capture cross sections and emission rates (for a detailed discussion see [25]). For a certain set of parameters a numerical solution of the rate equation systems can be calculated. Some examples are found in ref. [25].

However, many of the essential features of the experimental observations can be explained within a simplified donor-acceptor pair recombination model, in which the capture and emission of electrons and holes by the donors and acceptors (which may have different concentrations) is neglected, i.e. the details of the connection between the charge carrier concentration in the bands and the single and coupled donors and acceptors are not taken into account explicitly as in eq. (4). The temporal development of the charge carrier concentration is assumed to be

$$\frac{dn}{dt} = g_0 - Rn \qquad (5)$$

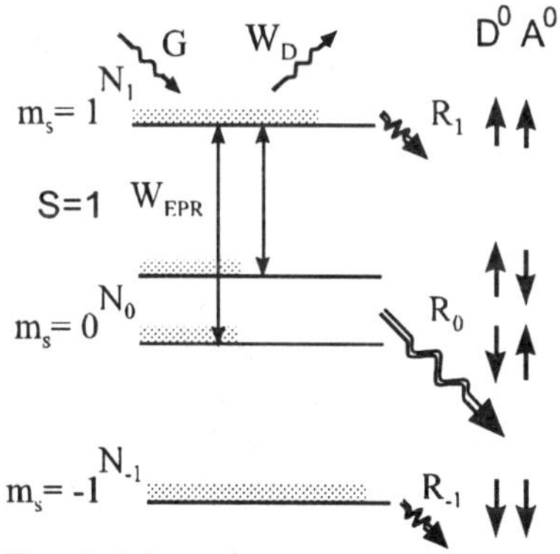

Fig. 4. Schematic representation of donor-acceptor D^0-A^0 pair recombination. Spin-lattice relaxation is neglected. The levels with m_s=1,0,-1 are occupied with a pair generation probability G. $N_{1,0,-1}$ are the occupation numbers of the m_s states. W_{EPR} is the EPR transition probability, W_D denotes a dissociation probability and $R_{1,0,-1}$ are the (non-radiative) recombination probabilities of the m_s states.

where R is the total recombination rate of the charge carriers determined by the D^0-A^0 pair recombination process. Since only the stationary solution is needed (dn/dt=0) and one is only interested in the difference effect with and without microwave transitions, one obtains

$$S_{EDEPR} \propto \Delta n_{EPR} = \frac{\partial n}{\partial R}\Delta R = -\frac{g_0}{R}\frac{\Delta R}{R} = -n\frac{\Delta R}{R} \tag{6}$$

The calculation of ΔR requires the solution of a system of rate equations describing the temporal evolution of the occupancy of the exchange-coupled D^0-A^0 system according to fig. (4), which is a schematic representation of their states and of the recombination to the singlet ground state. Spin-lattice relaxation is neglected. The levels with m_s=1,0,-1 are occupied with a pair generation probability G. N_1,N_0 and N_{-1} are the occupancies of the m_s states. Exchange coupling is assumed to be very weak; since the ground state is a singlet state, the recombination probability R_0 is expected to be much larger than those of the other two states R_1 and R_{-1} so that the occupancy of the $m_s=\pm1$ levels under stationary above-band-gap illumination conditions is much higher than that with m_s=0. W_{EPR} is the transition probability between the $m_s=\pm1$ levels and the m_s=0 levels. W_D denotes the dissociation probability describing a thermally induced dissociation of electron-hole pairs. N is the number of states which can be occupied. For this scheme one obtains the following rate equations for the occupancies N_i(i=1,0,-1) [21,25]

$$\frac{dN_i}{dt} = G(N - \sum_{i=-1}^{+1} N_i) - N_i(W_{EPR} + W_D + R_i) + \sum_{i \neq i} N_i W_{EPR} \tag{7}$$

The simplest case would be to assume $R_0 \neq 0$, $R_1=R_{-1}=0$ and W_D=0. The EDEPR signal is proportional to the difference of the occupancies with and without inducing EPR transitions. In this approximation with a stationary solution of eq. (7) one obtains

$$S_{EDEPR} \propto \sum_i (N_i(EPR) - N_i(0)) = C\frac{W_{EPR}\ T_R}{1 + W_{EPR}\ T_R} \tag{8}$$

with

$$C = \frac{2GN}{4G + R_0} \quad \text{and} \quad T_R = \frac{4G + R_0}{GR_0} \tag{9}$$

The result of eq. (8) qualitatively describes the experimental saturation curves of fig. 3b. The parameter β in eq. (3) is the parameter T_R of eq. (8). It was fitted to the experimental curves in fig. 3b and reflects the different recombination probabilities R_0. G is probably also not exactly the same in all experiments. For low microwave power the signal is proportional to W_{EPR}, which in turn is proportional to the microwave power P_{MW}:

$$W_{EPR} = \frac{1}{4\hbar} g\mu_B B_1^2 \propto P_{MW} \tag{10}$$

where g is the g-factor of either the donor or the acceptor, μ_B the Bohr magneton, and B_1 the amplitude of the microwaves. The maximum effect depends on the pair generation probability G as well as on the number of defects. The former is given by the electron hole capture probability into the D^+-A^- states as well as by the above-band-gap light intensity producing electron-hole pairs. The results are in qualitative agreement with the experimental observation within the approximation of eq. (5), i.e. regarding only the recombination of the coupled pairs. Inclusion of spin-lattice relaxation or W_D does not change the form of eq. (8). Only the expressions for C and T_R become more complicated [25].

An increase of the exchange coupling between donors and acceptors D^0 and A^0 would result in a relative shift of the level position in fig. 4 and probably results in a line broadening compared to the conventional EPR lines because of a statistical distribution of couplings due to the statistical distribution of the separations between donors and acceptors. Such a broadening has not yet been observed. The reason probably is that due to the increase in the recombination rates upon increasing the exchange coupling, the EPR transitions cannot be induced anymore within the life times of the spin states because of the limited microwave power of 100-200 mW available. The observation of the line broadening or line shift probably requires higher microwave powers. In the limiting case of strong coupling one would deal with a triplet system. The SL1 centre provides such a triplet system for which ED-EPR could be observed very well (see fig. 2). Its recombination mechanism is discussed in detail in ref. [25] .

A discussion of limiting cases, such as $W_D \rightarrow \infty$, $N \rightarrow 0$, $G \rightarrow \infty$, 0 showed that eq. (6) and (8) do not provide an adequate description of the EDEPR effects. The reason is not that the D^0-A^0 pair mechanism is not correct, but that the model discussed so far does not adequately consider the connection between the charge carrier concentrations and the valence and conduction bands, respectively, and the generation, dissociation and recombination of the D^0-A^0 pairs. For details, the reader is referred to [25].

Recently, Xiong and Miller [30] presented rate equations for the occupation and recombination of weakly exchange coupled electron-hole pairs, in which they also included spin-lattice relaxation and the electron-hole dissociation probability W_D. Their expression for S_{EDEPR} generally has the same form as eq. (8), however, in their work C is proportional to W_D. For vanishing W_D the whole EDEPR effect vanishes, which does not seem compatible with the D^0-A^0 recombinations observed [21] in EDEPR as well as in ODEPR at very low temperature (6K). At very low temperature W_D is certainly small considering the ionization energies of the defects involved. The temperature dependence of EDEPR shows (see below) that the EDEPR effect increases still below 6K [25]. The explanation for the discrepancy probably is that in the rate equations of ref. [30] the occupation rate is constant for all four levels of fig. 4, which is not true, if there is only a finite number of electron-hole pairs available of the same order of magnitude as levels N_i which can be distributed over the four weakly coupled electron-hole pair states.

2.3 Concentration and Temperature Dependence of the EDEPR Signals

The concentration dependence of EDEPR is not yet well understood and needs further study. High defect concentrations seem unfavourable. When producing $10^{14}cm^{-3}$ SL1 centres in Si by electron irradiation, no EDEPR signals are observed. It was qualitatively observed that signals appear after annealing out the SL1 centres. The signals have a rather broad maximum for concentrations where the conventional EPR signal is hardly detectable any more (see fig. 2 in ref. [31]) until it decreases for very low concentrations again. In a series of experiments the concentration of thermal donors TDD^+ was varied by variation of the annealing time of Cz-Si at 450°C.

Surprisingly, it was observed that EDEPR signals were still measured with a very good signal-to-noise ratio for concentrations well below $10^{12}cm^{-3}$ at which IR control measurements could not detect TDDs any more. It sufficed to quench the Cz samples from high temperature (770°C, 1200°C) in order to observe the EDEPR signals of TDDs, a new experience [31,25]. Experimentally it seems that there is a broad range of concentrations of 'distant' donor and acceptor pairs, in which EDEPR can be observed with very good signal-to-noise values. As mentioned above, if the recombination times become too fast compared to the inverse of the EPR transition probability, as they probably do for closer pairs, then EDEPR is not observable any more. The limit for R_0^{-1} seems to be of the order of 100ns for a microwave power of 100-200mW, usually available in conventional EPR spectrometers. However, a quantitative relation between defect concentration and EDEPR signal intensity must yet be established. It will probably also depend on the energy level of the defects in the gap, i.e. whether the recombination involves only shallow or also deep level defects.

First experiments with high concentration of N donors in 6H-SiC revealed interesting different features to what was discussed so far. For a concentration of about $10^{18}cm^{-3}$ N donors a single line about g=2 was observed as an increase of conductivity and not as a decrease upon enhanced recombinations as in the case of the D^0-A^0 pair recombination mechanism. When investigating the EDEPR of N donors in a B compensated 6H-SiC sample with lower concentration, the N hf triplet ($I(^{14}N)=1$) was observed [32]. At high concentration the mechanism seems to be different from the one discussed so far. The effect is probably connected with hopping conductivity occurring at higher donor concentrations. The loss of hf splitting points to a rapid motion of the donors [1]. From the observed temperature dependence according to our experiments it does not seem to be clear whether the mechanism proposed earlier by Morigaki and Onda [33] on the basis of observations in germanium is correct, according to which electrons are excited to an impurity conduction band by a thermal effect due to the microwaves. Further investigations are under way [32].

It is well possible that many observations made in diodes, in which usually only a structureless line about g=2 is observed without hf splittings, are also the result of donor concentrations which were too high, such that hopping effects dominated. It will be important in the future to measure whether the conductivity increases or decreases upon microwave transitions and which is the size of the effect in order to learn more about the underlying processes (the relative change of the conductivity typically is of the order of 10^{-3}-10^{-4} for D^0-A^0 pair recombinations in silicon).

Little is known about the temperature dependence of the EDEPR effects. It was reported that signals measured in diodes at room temperature disappeared upon cooling [17]. The general observation for the D^0-A^0 pair recombination as studied in Si [21,25] was that upon cooling the signals increased. Most experiments were done between 6 and 20K. For thermal donors in Si an investigation between 6 and 16K revealed a temperature dependence following a T^{-2} law: $S_{EDEPR} \propto T^{-2}$ [25]. This can be explained by assuming that the dominant influence is the temperature dependence of the spin-lattice relaxation time T_1 for which a direct process, including the phonon bottleneck effect, was assumed. For the latter process $T_1 \propto T^{-2}$ is expected [34]. Surprisingly it was found that above 16K the EDEPR effect increased again, possibly due to increasing D^0-A^0 pair dissociation effects restoring the concentration of carriers in the conduction band [25]. However, more systematic studies of the temperature dependence are needed.

3 Electrical Detection of Electron Nuclear Double Resonance (ENDOR)

In a Cz-grown P-doped sample, which contained apart from shallow P donors also the SL1 centre after electron irradiation, electron nuclear double resonance (ENDOR) of the P donors could be detected in the photoconductivity [22]. Fig. 5 shows the ENDOR spectrum for $B_0\|[100]$ measured in the low-field EDEPR line as an rf and microwave-induced change of the photoconductivity. The best signals were obtained with frequency modulation of 1kHz and at 4.5K. The spectrum of fig. 5a contains two ENDOR lines at about 52.60 and 64.93MHz for $m_s=\pm 1/2$ due to the well known ^{31}P hf interaction separated by $2\nu_n(^{31}P)$ = 12.33MHz (see e.g. [1]), ν_n being the Larmor frequency of the free ^{31}P nucleus. At $2\nu_n(^{29}Si)$ = 2.84MHz the ^{29}Si 'distant' ENDOR line is observed. Up to 6MHz well-known ^{29}Si shf lines could be measured [35,36] the frequency positions of which are indicated by a stick diagram in fig. 5c. The P concentration was chosen to be small enough that EDEPR and EDENDOR could be measured, but high enough for conventional ENDOR measurements. Fig. 5b shows the conventional stationary ENDOR spectrum measured with the same sample as in fig. 5a at the same orientation, using the same illumination and frequency modulation. The ^{31}P hf lines are detected and the ^{29}Si distant ENDOR line, but the concentration was not sufficient to detect the ^{29}Si ligands. Their signals seem to be particularly intense in electrical detection.

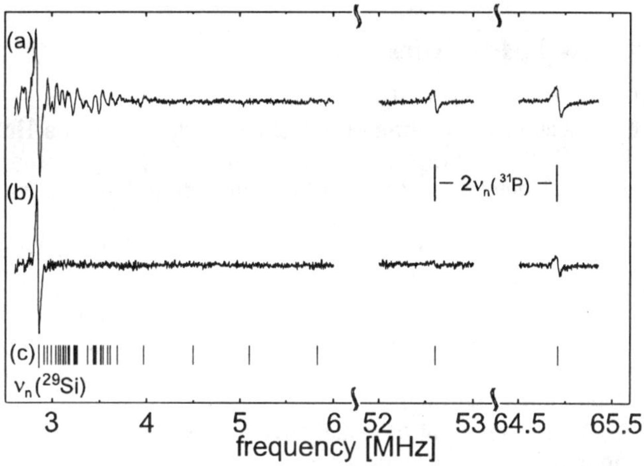

Fig. 5. (a) EDENDOR spectrum of the shallow P donors in Si measured with frequency modulation and with above-band-gap illumination in the low field EDEPR line (see fig. 1(b)) at B=333.3mT and at T=4.5K (P_{MW}=1mW). (b) Conventional stationary ENDOR spectrum of the same sample as in (a) measured with above-band-gap illumination. (c) Stick spectrum of the ^{31}P and ^{29}Si ENDOR lines calculated from the known hf and shf interaction after ref. [36].

A systematic study of the dependence of the EDENDOR signals on the microwave power P_{MW} and on the rf power P_{HF} qualitatively gave the result that the signals increased both with P_{MW} and P_{HF} (see e.g. fig. 3 in [22]). The experimental results could be described with a quantitative relation for the EDENDOR signals, which was also derived from rate equations similar to eq. (7), however, expanded by terms comprising the rf NMR transition probabilities and the nuclear Zeeman levels [25]:

$$S_{EDENDOR} \propto \frac{W_{EPR}\,T_R}{1+W_{EPR}\,T_R}\;\frac{W_{NMR}\,T_{HF}}{1+W_{NMR}\,T_{HF}} \tag{11}$$

W_{EPR} and W_{NMR} are the transition probabilities of the EPR and NMR transitions, respectively, and T_R and T_{HF} are parameters depending on the details of the D^0-A^0 pair recombination model assumed. In the simplest case T_R is given by the expression of eq. (9). T_{HF} also depends on W_{EPR}:

$$T_{HF} \propto \frac{(W_{EPR} + \frac{1}{T_R})T_{R'}}{1 + W_{EPR}\, T_{R'}} \tag{12}$$

which, for large values of W_{EPR} becomes an expression independent of W_{EPR} depending only on the recombination rates W_D, G etc. (for details see [25]).

Thus, the EDENDOR signal depends on the microwave power similarly as the EDEPR signals and on the rf power in an analogous way. This latter result is very different from conventional ENDOR, where upon high rf power the ENDOR signals saturate and decrease (see e.g. [1]).

4 New Possibilities

In order to see an EDEPR effect it is not necessary to illuminate the whole sample. Even when illuminating only a small spot, e.g. with a HeNe laser (632nm, 50mW), the EDEPR spectrum can be detected (see e.g. fig. 2 and 3 of ref. [21]). Thus, with electrical detection via the photoconductivity, it is possible to measure EPR with spatial resolution. In the experiment described in ref. [21] the light spot had a diameter of 0.6mm. The light beam can be scanned throughout the sample surface and a two-dimensional distribution can, in principle, be measured, whereby in this kind of experiment the sample is integrated over the penetration depth of the light, in Si about 5μm. One must be sure though that no other diamagnetic recombination processes take place simultaneously. Further work is needed to study in detail such kind of mapping experiments.

The sensitivity of EPR and EDENDOR can be several orders of magnitude higher compared to conventional detection. In fig. 2, for example, for the SL1 centre the signal-to-noise ratio was higher than in conventional EPR [37], while that for the P donors was about equal. Yet, one has to consider that in conventional EPR [37] the whole sample is measured, while in EDEPR only a layer of the thickness of the above-band-gap light penetration depth is measured. Thus, even for the P donors in Si in fig. 2 the sensitivity gain is about a factor of 600 (sample thickness 3mm, light penetration 5μm). Since it is possible to illuminate only a spot, defects in small volumes can be measured. For P and the SL1 centre containing Si sample as described in ref. [21] (in fig. 2) it was estimated that $6 \cdot 10^7$ P atoms were detectable. Similarly, in the EDENDOR experiment about $6 \cdot 10^9$ P nuclei were detectable, which is a sensitivity gain of 10^4-10^5

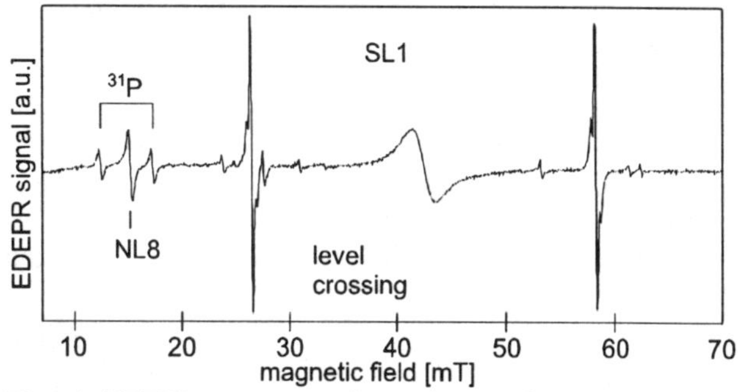

Fig. 6: EDEPR spectrum of an electron irradiated Cz-Si:P crystal containing also thermal double donors NL8 measured with ν_{EPR}=280GHz at 11K (B_0||[110]) using a low-frequency EDEPR spectrometer of special design (see fig. 7) after [39].

compared to conventional ENDOR. As seen from fig. 5, the sensitivity gain is particularly high for the ligand ENDOR lines.

A further interesting feature is the fact that the EDEPR signal intensities do not depend on the size of the magnetic field. This is a direct consequence of the D^0-A^0 pair recombination model and in contrast to earlier models for the effect by Lepine [4], who claimed that the effect of EDEPR is due to spin-dependent carrier capture and should be proportional to thc spin polarization of the conduction electrons and proportional to the polarization of the Zeeman ground state of the defect. According to this model, the decrease of the photoconductivity, i.e. the EDEPR effect, should vary with $(B/T)^2$. However, it was observed that given the B_1 amplitudes at the sample is the same, the signal-to-noise ratio of the EDEPR signal does not depend on the resonance frequency (or field), respectively. Fig. 6 shows the EDEPR spectrum of a sample containing P shallow donors, thermal donors and the SL1 centre measured at about 280MHz. The signal-to-noise ratio is excellent and - only depends on $P_{MW} \propto B_1^2$ and the temperature (see also ref. [21]). This failure of the Lepine model was also observed previously [38]. In fig. 6 there is also a level crossing of the SL1 centre seen at about 40mT, also previously observed by Vlasenko [39]. The angular dependence measured at 280MHz shows the typical features of a low-field spectrum with loops rather than open branches (see e.g. fig. 4 of ref. [21]). The possibility to measure at low field or low frequency opens up the possibility to construct inexpensive and small spectrometers to measure EDEPR. Fig. 7 shows the cross section of such a low frequency spectrometer recently built at the University of Paderborn. The spectrum of fig. 6 was measured with it [40]. At low frequency one is not restricted to a small microwave cavity. For frequencies such as 300MHz a system of solenoids can be used instead of a cavity with which also larger samples can be investigated, e.g. for mapping purposes using the feature of spatial resolution. All this new potential has not yet been exploited except for a few first steps.

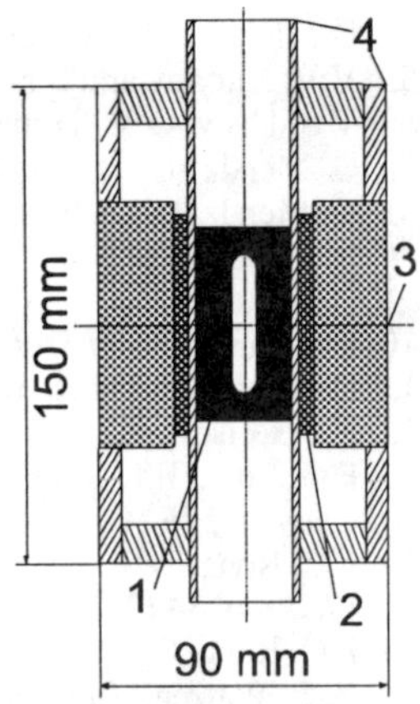

Fig. 7. Cross section of a low-frequency EDEPR spectrometer [40]: 1 rf solenoids 2 modulation coils 3 magnetic field coils 4 holder for cryo-system.

Acknowledgement

The author is indebted to Dr. S. Greulich-Weber and Dr. B. Stich for helpful discussions. The work was supported by the Deutsche Forschungsgemeinschaft.

References

1. J.-M. Spaeth, J.R. Niklas, and R.H. Bartram, Structural Analysis of Point Defects in Solids: An introduction to multiple magnetic resonance spectroscopy, *Springer* (1992) vol. 43
2. J. Schmidt and I. Solomon, *C.R. Acad.Sci.* Paris **263**, (1966) 169
3. D.J. Lepine and J.J. Prejean, *Proceedings of the 10th International Conference on the Physics of Semiconductors* (1970) Cambridge, 805

4. D.J. Lepine, *Phys. Rev. B* **6**, 2 (1972) 436
5. V.V. Kurylev and S.N. Karyagin, *phys. stat. sol.* (a) **21** (1974) K127
6. V.S. L'vov, O.V. Tretyak, and I.A. Kolomiets, *Sov. Phys. Semicond.* **11**, 6 (1977) 661
7. G. Mendz and D. Haneman, *J. Phys. C: Solid State Phys.* **11** (1978) L197
8. G. Mendz, D.J. Miller, and D. Haneman, *Phys. Rev. B* **20** (1979) 5246
9. L.S. Vlasenko, M.P. Vlasenko, V.N. Lomasov, and V.A. Khramtsov, *Sov. Phys. JETP* **64** (1986) 612
10. I. Solomon, *Solid State Commun.* **20** (1976) 215
11. D. Kaplan and M. Pepper, *Solid State Commun.* **34** (1980) 803
12. F.I. Borisov, V.I. Strikha, and O.V. Tretyak, *Sov. Phys. Semicond.* **15** (1981) 1149
13. F.C. Rong, E.H. Poindexter, M. Harmatz, W.R. Buchwald, and G.J. Gerardi, *Solid State Commun.* **76** (1990) 1083
14. F.C. Rong, W.R. Buchwald, E.H. Poindexter, W.L. Warren, and D.J. Keeble, *Solid State Electron.* **34** (1991) 835
15. F.C. Rong, G.J. Gerardi, W.R. Buchwald, E.H. Poindexter, M.T. Umlor, D.J. Keeble, and W.L. Warren, *Appl. Phys. Lett.* **60** (1992) 610
16. P. Christmann, C. Wetzel, B.K. Meyer, A. Asenov, and A. Endrös, *Appl. Phys. Lett.* **60** (1992) 1857
17. B. Stich, S. Greulich-Weber,and J.-M. Spaeth, *Semicond. Sci. Technol.* **8** (1993) 1385
18. D. Kaplan, I. Solomon, and N.F. Mott, *Le Journal de Physique Lettres* **39** (1978) L51
19. A. Honig, *Phys. Rev. Lett.* **17** (1966) 186
20. M. Lannoo, D. Vuillaume, D. Deresmes, and D. Stiévenard, *Microelectron. Eng.* **22** (1993) 143
21. B. Stich, S. Greulich-Weber, and J.-M. Spaeth, *J. Appl. Phys.* **77** (1995) 1546
22. B. Stich, S. Greulich-Weber, and J.-M. Spaeth, *Appl. Phys. Letters* **68** (1996) 1102
23. G.D. Watkins and J.W. Corbett, *Phys. Rev.* **121** (1960) 1001
24. S.H. Muller, M. Sprenger, E.G. Sieverts, and C.A.J. Ammerlaan, *Solid State Commun.* **25** (1978) 987
25. B. Stich, Untersuchungen zu den Mechanismen und den Anwendungsmöglichkeiten der elektrisch detektierten magnetischen Resonanz, Edition Wissenschaft, Reihe Physik, Band 44, Tectum Verlag, Marburg, 1997
26. K.L. Brower, *Phys. Rev. B* **4** (1971) 1968
27. F.K. Koschnick and J.-M. Spaeth, *Phys. Rev. B* **54** (1996) 11042
28. F.K. Koschnick, private communication
29. A. Neubrand, *phys. stat. sol.* (b) **86** (1978) 269
30. Z.Xiong and D.J. Miller, *Appl. Phys. Lett.* **63** (1993) 352
31. S. Greulich-Weber, B. Stich, and J.-M. Spaeth, *Mat. Sci. Forum* **196-201** (1995) 1509
32. I. Grasa-Molina, S. Greulich-Weber, and J.-M. Spaeth -to be published-
33. K. Morigaki and M. Onda, *J. Phys. Sci. Japan* **33** (1972) 1031
34. A. Abragam and B. Bleaney, Electron Paramagnetic Resonance of Transition Ions, *Dover Publications Incorp., N.Y.* (1986)
35. G. Feher, *Phys. Rev.* **114** (1959) 1219
36. E.B. Hale and R.L. Mieher, *Phys. Rev.* **184** (1969) 739
37. L.S. Vlasenko, *Phys. Rev. B* **52** (1995) 1144
38. L.S. Mima, V.I. Strikha, and O.V. Tretyak, *Sov. Phys. Semicond.* **14** (1980) 1328
39. L.S. Vlasenko and V.A. Khramtsov, *Sov. Phys. JETP Lett.* **42** (1985) 38
40. C. Pinkawa, Diplomarbeit 1996, University of Paderborn
41. X. de le Broise, M. Lannoo and M.D. Stiévenard, Rapport de project, Univ. des Sciences et Techniques de Lille (1994), unpublished

Interdisciplinary Applications of ESR and ESE - Dating, Dosimetry and Microscopy -

Motoji Ikeya

Department of Earth and Space Science, Graduate School of Science, Osaka University,
1-1 Machikaneyama, Toyonaka, Osaka 560, Japan

Abstract. Interdisciplinary applications of electron spin resonance (ESR) and electron spin echo (ESE) to radiation dosimetry, environmental assessment and geological and archaeological dating based on natural radiation effects have been reviewed with emphasis on outer planetary science in future as well as A-bomb and accident radiation dosimetry. Computer tomography (CT-) ESR and ESE microscopes were made by inserting field gradient wires into a TE_{011} cylindrical cavity, while the scanning ESR microscope was made using a TE_{111} cavity with an aperture over which a flat surface sample is scanned mechanically using an X-Y stage.

1 Introduction

A new field of "quantum geophysics" has emerged in an inter-disciplinary field between ESR and the geosciences (Ikeya 1993, Marfunin 1979). Some knowledge of quantum physics as well as physics and chemistry of radiation effects in solid materials are needed in ESR applications. Geologists sampling in field have no such training and find it difficult to develop the method. Objectives of applied quantum physics are shifting from physics to chemistry and finally to geophysics. A new field of "Quantum Geophysics" related with EPR will be a fruitful field requiring collaboration in an interdisciplinary field.

The signal intensity of radiation-induced paramagnetic centers can be used in radiation dosimetry and also in dating archaeological and geological materials. Electron spin resonance (ESR) spectroscopy has thus been used as a method of radiation dosimetry and dating (Ikeya, 1993). ESR signal intensity due to radiation-induced centers can be used as indicative of radiation dose and the ages.

Carbonate stalactites and stalagmites from limestone caves at the Akiyoshi cave were dated two decades ago (Ikeya, 1975) and then, fossil bones and teeth in anthropology (Ikeya and Miki, 1980), fossil shells and corals in marine geology (Ikeya and Ohmura, 1981), volcanic ash (Imai *et al.* 1984), and geothermal silica and geological fault gouge (Ikeya *et al.*, 1982) followed. The ESR dating is thus becoming a useful dating technique in geophysics with measurements of natural radiation effects in minerals and fossils and the *First International Symposium on ESR Dating* including ESR radiation dosimetry was held (Ikeya and Miki ed 1985).

ESR dosimetry of teeth was performed for A-bomb radiation at Hiroshima and Nagasaki (Ikeya et al, 1984), for personnels relevant with A-bomb test and with the reactor accident at Chernobyl and other nuclear facility accidents in former Soviet Union. Human dosimetry using teeth enamel and bones has also been performed for several radiation accidents.

In this review, the principles and procedures of ESR dating based on natural radiation effects are described, together with examples of ESR microscopy based on my book "New Applications of ESR".

2 Principle and Procedures of ESR Dating and Dosimetry

2.1 Principle of ESR and ESR Dosimetry

2.1.1 Basic Principle Based on Natural Radiation Effects

ESR dating described utilizes accumulated radiation effects in crystalline and amorphous materials, while that in forensic science and history is based on chemical reaction of radical formation or oxidation of impurities (Ikeya and Miki 1980a, Ikeya 1993). A variety of materials that could not be dated by other methods have been dated with ESR.

An ESR detects unpaired electrons either created by natural radiation as shown by one hundred times used cartoon in Figure 1 or formed by chemical reaction. The direction of the unpaired electron spin (self-rotation), represented by a dancing girl and a boy is changed by microwave absorption under a magnetic field as shown in (c) and (d). The concentration of the unpaired electrons (simply called defects) can be determined directly by microwave absorption as the ESR signal intensity.

a b c d

Fig. 1. One thousand times used cartoon illustrating the principle of ESR dating and dosimetry, (a) Paired electrons are spinning in opposite directions, (b) ionization by α, β, and γ-rays knocks off one of the paired electrons (girl) leaving the other (boy) in a hole, (c) the unpaired electron (girl) is trapped by an impurity (gangster), and (d) microwave absorption occurs for unpaired electrons.

2.1.2 Events Which Set the Clock to Zero: Recent Fault and Lithic Tool Dating

Crystallization has been dated on a slice of a stalactite in the Akiyoshi limestone cave in concordance with the growth in stratigraphic order (Ikeya 1975). Biocarbonates of shells and corals (Ikeya and Ohmura, 1981) and biophosphates of bones and teeth (Ikeya and Miki, 1980) belong to crystallization in living organisms. Partial recrystallization of these materials can, however, give misleadingly young ages.

Temperature rise also set the clock time to zero. Artifacts at the time of firing and geothermal events can thus be determined. The age of a burn lithic tool in the norther part of Japan, Kamitakamori, was determined as old as 650,000 years recently (Tani et al 1996) though there is some controversy over heating by ancient man .

Mechanical annealing was proposed for dating a geological fault. Deformation produces movement of dislocations and introduces new dislocations in solids. It also creates a local temperature rise by frictional heat. Spin relaxation times of E' centers measured with spin echo are different for ordinary and crushed quartz (Ikeya *et al.* 1993). Recent results in this laboratory on the bore core gouge of the Nojima Fault which moved in the Kobe Earthquake on January 17, 1995 clearly indicates that the intensities of both E' centers and Al hole centers decreases at the fault plane suggesting that they are annealed thermally. The basis of dating a fault was confirmed.

Sunlight bleaching effect on the decay of the signal intensity from quartz grains (SiO_2) for $g = 1.997$ (Ge center) can be observed (Sato 1985). An attempt to date sediments by ESR based on this mechanism has been made (Tanaka *et al*, 1985).

2.2. Procedures and Materials for Dating and Dosimetry

2.2.1 Additive Dose Method : A New Extrapolation Method Considering Lifetime

An ESR signal intensity does not indicate the radiation dose unless it is calibrated to the dose. A technique to use artificial irradiation by γ-rays from a ^{60}Co or a ^{137}Cs source is called the "additive dose method". The intensity is enhanced with

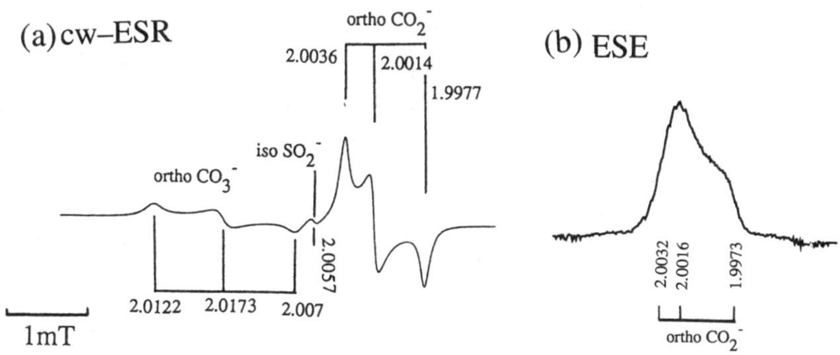

Fig. 2 (a) ESR spectrum of He^+-irradiated coral showing signal of orthorhombic g factor due to CO_2^-. (b) ESR and electron spin echo (ESE) spectra of CO_2^- in the orthorhombic symmetry in a fossil shell (Kohno et al., 1994).

the absorbed dose of artificial irradiation, Q' giving the equivalent dose by the used additive radiation. The γ-ray equivalent dose (E_d) is determined from the enhancement, assuming a linear relationship between the intensity and dose as $I(Q) = I_o(1 + Q / E_d)$, where I_o and $I(Q)$ are the observed signal intensities before and after irradiation, respectively. The growth curve may be fitted to a saturation curve.

New extrapolation method was suggested and described (Ikeya 1993) and developed in a program including the estimation of errors in E_d (Takaki, 1997). The estimation of *ED* by the additive dose method has also been used for ESR dosimetry.

Dose to age conversion is made by assessing the annual dose rate, *D*, in mGy/a by a separate method in nuclear measurements. An age, *T*, in [ka] is obtained as T [ka] $= E_d$ [Gy] / D [mGy/a]. The unit "Gy" (gray) is the SI unit of absorbed radiation dose equivalent to the energy of 1 Joule dissipated in 1 kg of material (1Gy = 1J/kg). The equivalent dose (E_d) and the dose rate (D) estimated for geological and archaeological materials constitutes the main part of ESR dating. However, ESR dating is said to be an exercise of various radiation assessments and measurements.

2.2.2 Materials, Models of Defects and Time Range of ESR Dating

ESR signals are identified with spin Hamiltonian parameters such as the *g* factor hyperfine *A* value etc. ESR parameters for defects in minerals are tabulated in text books and reviews (Poole and Farach 1986, Marfunin 1979, Ikeya 1993). The following are models of defects used for ESR dating and for environmental assessment. Table 1 summarizes defects which are stable for a sufficiently long time and some geological events. ESR signals produced by chemical processes rather than by natural radiation are indicated by * . Models of defects are only and always tentative and might be changed as further studies are carried out. Applications are limited to typical ones given in the textbook (Ikeya 1993) and to some recently developed in this laboratory. Recent works on E' centers indicated that there are several species with different linewidth and thermal stability. Applications to dating geological fault may be made using some siganal intensity of E' centers.

Chemical ESR dating, like dating bloodstain in days in forensic science, is a different method from the ESR dating based on the dosimetry of natural radiation. Preparation in the laboratory, cleaning and magnetic separation, light etching, sample size selection, , X-ray or microscope studies (Ikeya 1993). These conditions are always tentative as such are in research and will never be eternal as such is in life. Improved procedures may be found even after standard procedures are established.

The Arrhenius plot of lifetimes, τ at various annealing temperatures, *T*, gives a straight line from which τ, at ambient temperatures, can be estimated. A geothermal temperature assessment can be made using the presence/absence of ESR signals.

The upper limitation of ESR dating is imposed from the lifetime of defects and from the saturation of the defect formation. The lower limit comes from the signal to noise (*S/N*) ratio and interfering impurity signals , especially for young samples, practically, one thousand years (*ED* = 1 Gy) for common materials, a few hundred years (*ED*=0.1Gy and *D*=0.05 mGy/y)) for corals.

Table 1. Materials and paramagnetic defects used for ESR dating, dosimetry and microscopy [1]

Materials	Chemical Form (Model of Defects)	Applications
ESR DATING		
Terrestrial Materials		
Carbonates	**$CaCO_3$ (CO_2^-, SO_3^-)**	stalactites, stalagmites, travertines, corals, shells, foraminifers, egg
	$NaHCO_3$ (CO_3^{3-}-H\ or CO_2^--H)	nahcholite (dessert deposits)
Sulfates	**$CaSO_4$ (SO_3^-)**	
	$CaSO_4$ $2H_2O$ (CO_3^- or O_2^{3-}, SO_3^-)	gypsum (desert & cave deposits)
Phosphates	**$Ca_{10}(PO_4)_6X_2$** (CO_2^-, O$^-$]	tooth enamel, bone
	X = F$^-$, $CaHPO_4$	fluorapatite, phosphate nodule
Silica SiO_2	(E', peroxy, Al (Al^{3+} h), (Ge^{4+}+e), (Ti^{4+}+e), O-vacancy)	geological fault, volcanic rock altered rock
Silicates	**$ZrSiO_4$**, Plagioclase (E', peroxy, O$^-$)	volcanic ash & rock
Molecular Ices	**(Future dating of outer planets with a low ambient temperature)**	
Ice	**H_2O** (OH, CO_2^-, O_2H)	future dating of comets
Dry ice	**CO_2** (CO_2^-, CO_3^-, O_3^-)	CO_2 in polar area of the Mars
Organics	(peroxy radicals, Fe^{3+}, Cu^{2+})	coal, petroleum, foods, crops leather, paper, blood, mummy
ESR DOSIMETRY		
Organics	Alanine ($CH_3CHCOOH$)	rod with polystyrene or paraphine
	Sugar	Chernobyl accident dosimetry
Carbonate	Shell (CO_2^-)	shell buttun of clothes (A-bomb)
H-apatire	Tooth (CO_2^-)	A-bomb at Hiroshima & Nagasaki Chernobyl residents, Accident
Granite [3]	Quartz (E', (Ge^{4+}+e))	Bridge pillar at Hiroshima

[1] ESR imaging have also been made using these signals in addition to the signal intensity of paramagnetic ions.

[2] Chemical reaction of radical formation or valency change is used in chemical ESR dating of organics.

[3] Natural radiation dose is subtracted from the depth profile using signal intensity geologically not so stable but stable for dosimetric purpose.

3 ESR Dosimetry and Its Applications

3.1 ESR Dosimeters and Apparatus : A Portable ESR

The principle of ESR dosimetry is based on measurements of concentration of paramagnetic centers produced by radiation, which is detected as an ESR signal intensity proportional to the absorbed radiation dose. The sensitivity heavily depends on the production yield of unpaired spins, so called G-values (the number of radicals per 100 eV) and also on the linewidth of the ESR signal since we generally detect a signal as a derivative spectrum using magnetic field modulation.

The number of radicals produced by the dose of 1 Gy is about 6.3×10^{16} / kg for a material with G = 1.0 . Thus, the number of radicals, n, for 1 gram of materials is given as $n = 6.3 \times 10^{13}$ G in $[\mathrm{Gy}^{-1}\mathrm{g}^{-1}]$. The minimum detectable dose is evaluated with the sensitivity of a commercial X-band ESR spectrometer, ~ 10^{10} spins / 0.1 mT. Additional requirement is that the lifetime of centers must be longer than a few years, i.e., the annealing temperature is above 200 °C for inorganic solids. The detection of a few mGy exposure is possible using newly developed sulfates as dosimeter substance.

ESR spectrometers have increased in the sensitivity and reproducibility dramatically. A compact and small portable ESR spectrometer is being developed using the rare earth magnet of Nd-B-Fe weighting less than 2 kg (Ikeya 1993). The principle of CT- and scanning ESR microscopes are demonstrated using the small equipment.

3.2 A-bomb ESR Dosimetry

Accident dosimetry of direct exposure to human beings can be made by measuring human tooth enamel and shell buttons. Tooth enamel is the best biological dosimeter (Ikeya et al. 1984). ESR of tooth enamel has been applied to dosimetry of cancer patients under radiation therapy, to the cumulative dose of dentists and to human dosimetry of radiation accidents. Since signals of CO_2^- in tooth enamel is sufficiently stable, the lifetime cumulative radiation dose of personnel working in nuclear facilities will be determined using teeth.

The ESR signal of bone tissue and rat tail has been used to determine the radiation dose (Swartz 1965). The sensitivity (minimum measurable dose) was about 1 Gy for canine, human and bovine bones, now reduced to 50 mGy though the subtraction of a broad organic signal is needed. Sensitivity enhancement was made by detecting rapid passage out-of-phase (second harmonic) signal at 77K (Galtsev *et al*. 1993).

Shells and corals have been used for ESR dating. Shells, especially shell buttons are ideal dosimeters. A-bomb radiation dosimetry using a piece of shell button was made in the early work on A-bomb dosimetry at Nagasaki (Ikeya et al. 1984). Plastic buttons did not show a stable signal after irradiation.

ESR dosimetry with natural rocks was attempted in A-bomb dosimetry. The total dose of natural radiation (T_d) is higher than artificial dose of an A-bomb or accident radiation (A_d). However, the A_d for natural rocks can be determined considering the difference of defect formation. The absorption effect of A-bomb radiation in the bridge pillar granite rock at Hiroshima is clear.

The intensity of the Ge center at the surface is relatively smaller than that in slightly deeper layers. On the other hand, the intensity of E'_1 centers is large at the surface, suggesting a charge transfer from the Ge to E'_1 center. A similar tendency of the intense E'_1 signal at the surface was observed in the building wall granite of the Fukoku Insurance Company: the granite was exposed to radiation from the A-bomb.

3.3 Accident ESR Dosimetry: Chernobyl Resident and Environmental Studies

Sugar samples collected from a hotel, schools and polyclinic at Pripyat city, 3 km north-west of the Chernobyl nuclear reactor were measured with ESR (Nakajima *et al.* 1991). External radiation to evacuees at the Chernobyl accident was estimated from the sugar samples to be 55 ~ 91 mGy in good agreement with the results of teeth dosimetry (Ikeya 1993) except for a few percent who might have been exposed to much higher radiation from presumably hot spots. *Sugar-coated drug tablets* have been used for dosimetry (Regulla and Deffner 1989, Kai and Ikeya 1988).

Accidental dosimetry with ESR has been successfully performed on workers in Ebgland and the result was used as a legal proof. The absorbed doses to legs of two workers in a radiation accident at a San Salvador 60Co-irradiation facility (Desrosiers 1991) and to a piece of tooth and finger bones in British radiation facilities (Lloyd et al. 1994) were determined.

Radiation environment study has recently been initiated in this laboratory. Pieces of living shells, snail shells and egg shells was suggested to use as indicators for wild-life radiation exposure around A-bomb test site and the contaminated area. Dose limitation of $CaCO_3$ of about tenth of Gy is possible. We have visited Semipalatinsk nuclear bomb test site in Kazafstan Republic to get samples.

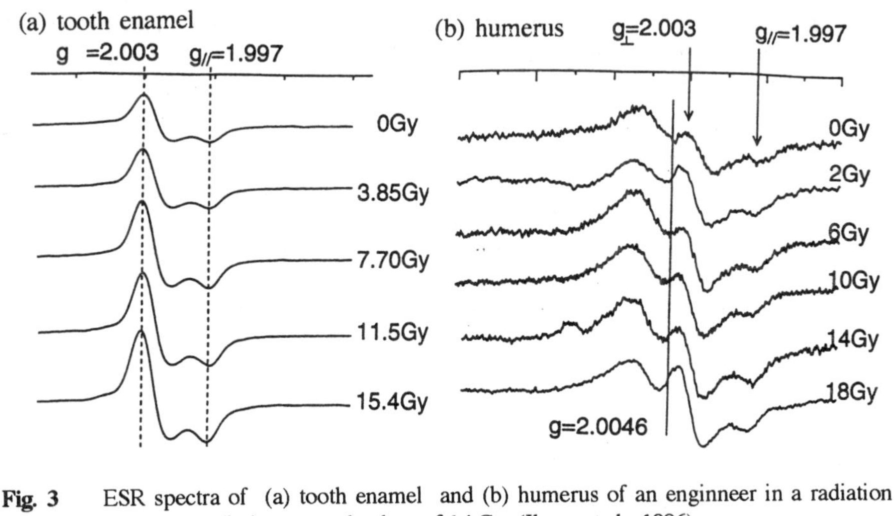

Fig. 3 ESR spectra of (a) tooth enamel and (b) humerus of an enginneer in a radiation accident. Additive irradiation gave the dose of 14 Gy (Ikeya et al., 1996).

4. ESR (EPR) Microscope

4.1 Field Gradient Wires inside a Cavity for CW- and Pulsed ESR

Medical use of magnetic resonance imaging (MRI) or CT-NMR employing the magnetic field gradient for spatial resolution is common in hospitals. The condition of the resonance transition by the microwave absorption is given aa $g\,\beta\,H_o = h\,\nu$. The intensity of the magnetic field, H, and the frequency of the electromagnetic wave, ν, are two physical quantities that can be used to obtain spatial distribution. High spatial distribution of ESR signal intensities may be obtained by spatial variation of H through adding an intense field gradient with a high quality (linearity, magnitude and the available space). Straight wires were inserted into a TE_{011} mode cavity to generate an intense field gradient with no appreciable microwave loss. Two-dimensional cw-ESR and electron spin echo (ESE) imaging were obtained as in Fig. 5 (Yamamoto and Ikeya 1994) though the sensitivity is not so high at the moment.

4.2 Scanning ESR Microscope and Its Applications: Decoding Cyclic Pattern

Spatial localization and scanning of this physical quantity leads to a scanning microscopy. A localized alternating modulation field and a localizedstatic resonance field are scanned to obtain the surface image of the spin distribution. Most ESR imagings are concerned with the distribution of spin labeled chemicals to demonstrate the resolution. Attractive example of applications are unfortunately few.

Catalytic chemistry involves some rare earth and transition metal ions. The mechanism of a catalytic reaction and appropriate design including efficiency, size, proper preparation, poisoning effects and lifetime can be studied from the spatial spin distribution of the valence and states of paramagnetic ions before and after reactions. Catalytic centers in tablets formed by drying a mixture of $H_3PMo_{12}O_{40}$, H_2O and silica gel, were studied by heating in a stream of N_2 and H_2 gases. The intensity of V^{4+} is high at the lower surface and the outer edge of the upper surface: the reduction proceeded from the bottom and from the edge.

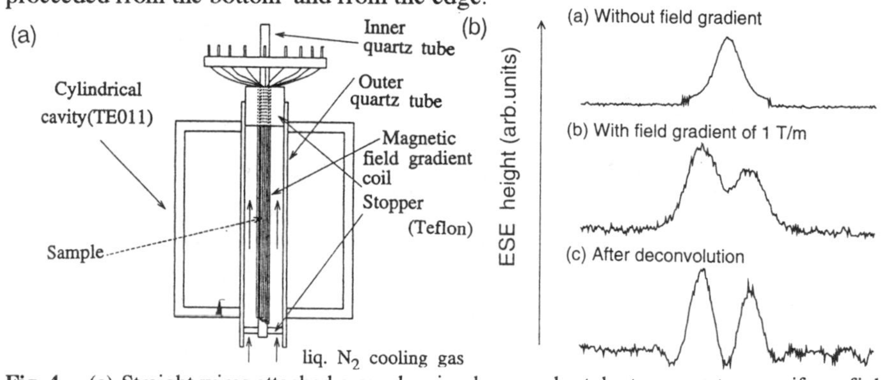

Fig. 4 (a) Straight wires attached around a circular sample tube to generate an uniform field gradient for CT-ESR imaging. (b) Field-sweep electron spin echo (ESE) spectra: a. without field gradient, b. splitting with field gradient and c. spin distribution by deconvolution.

Crystal growth, mineralogy and paleontology are the next field of applications. A synthetic diamond grown in Fe-Co-Ni solvent under high temperature and high pressure using a temperature gradient method gave the ESR signal intensity of nitrogen (N) and nickel (Ni), which were mapped with a scanning ESR microscope.

Distribution of paramagnetic ion impurities and radical species in natural minerals and fossils can be determined using an ESR microscope. Cyclic variation or stripes can tell the changes of the growth environment such as the climate changes. Decoding the stripe pattern may be made with ESR as was made for corals.

Dental radiology and medical application are vast area left to ESR microcopy. The distribution of radicals has been imaged in human teeth exposed to gamma rays and X-rays. Radicals are observed in the enamel for a gamma- rayed tooth, while only the surface of the enamel showed an ESR signal in the case of X-ray exposure.

Semiconductor and ceramic studies are potential field of ESR imaging until further breakthrough is made in medical use of ESR imaging. Dangling bonds on chemical vapor deposition (CVD) films, especially their annealing properties have been studied with scanning ESR. Thermal gradient allows to determine the annealing at each pixel leading to determination of annealing parameters. Dangling bonds and its effect to light exposure were also imaged as given in this volume by Fututa *et al*.

5 Summary

An inter-disciplinary research with geosciences lead to ESR dating with applications in various fields. A vast area for the application of ESR dating and dosimetry is opening and awaits further research. Extension of ESR dating to space science is anticipated in the next century by launching a light weight spectrometer to planets, satellites and comets and measuring radicals in solid H_2O, CO_2 and CH_4.

A-bomb radiation doses at Nagasaki and Hiroshima and accident doses to people close to Chernobyl were evaluated with ESR using tooth enamel extracted by dental treatments. Human radiation measurement by ESR is practical using tooth enamel,

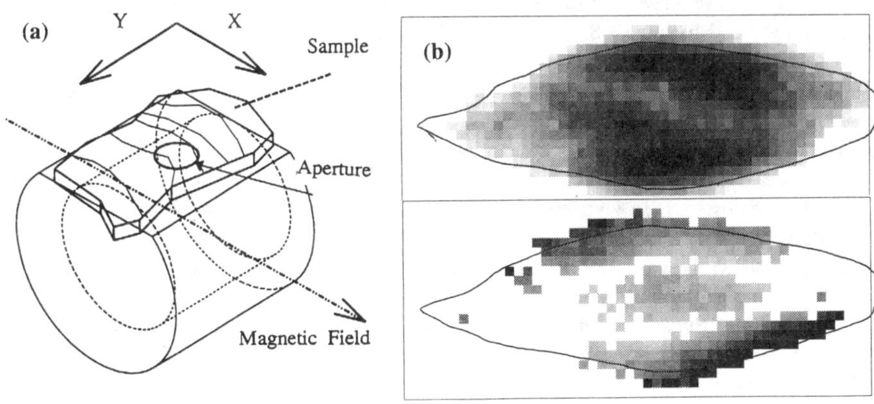

Fig. 5 (a) A TE_{111} cavity for the scanning ESR microscope. (b) An ESR image for an area of 10 mm x 25 mmm of a piece of natural gypsum crystal and (c) the equivalent dose image by additive dose (Omura and Ikeya 1995).

bone and shell buttons. Thus, ESR is useful both for accident dosimetry and for radiation monitoring. Efforts to obtain high sensitive ESR dosimeter have resulted in a minimum detectable dose of 10 mGy at present.
The principles of CT-ESR and of several types of scanning ESR microscopy will be useful in material sciences such as ceramics and semiconductor studies as well as in clinical diagnosis, chemistry, biology and mineralogy. Detailed references in ESR dating and dosimetry should be seen in the book (Ikeya 1993).

References

1. M. Ikeya, *New Applications of Electron Spin Resonance - Dating, Dosimetry and Microscopy,* World Scientific, Singapore (1993).
2. M. Ikeya, Dating a stalactite by electron paramagnetic resonance, *Nature.* **255** (1975) 48-50.
3. M. Ikeya and T. Miki ed, *ESR Dating and Dosimetry,* Ionics, Tokyo (1985).
4. A. S. Marfunin *Spectroscopy, Luminescence and Radiation centers in Minerals*, Springer-Verlag, Berlin (1979).
5. C. P. Poole and H. A. Farach Electron spin resonance of minerals. *Magn. Reson. Rev.* **4** (1986) 225-289.
6. R. Grün Electron spin resonance (ESR) dating. *Quat. Intern.* **1**, (1989) 65-109.
7. L. V. Bershov, V. O. Martirosyan, A. S. Marfuninand A. V. Speranskii, EPR and structure models for radical ions in anhydrite crystals. *Fortschr. Mineral.* **52** (1975) 591-604.
8. R. Debuyst, M. Bidiamambuand F. Dejehet An EPR study of g and -irradiated synthetic powdered calcite labeled with 13C. *Nucl. Tracks.* **18**, (1991) 193-201.
9. H. Kohno, C. Yamanaka and M. Ikeya Pulsed and cw-ESR studies of CO_2^- in $CaCO_3$ irradiated by He+ and g-rays for ESR dating, *Jpn. J. Appl. Phys.* **33** (1994) 5743-5746.
10. M. Ikeya and T. Miki Electron spin resonance dating of animal and human bones, *Science* **207** (1980) 977-979.
11. M. Ikeya, T. Miki and K. Tanaka Dating of a fault by ESR on intrafault materials. *Science.* **215** (1982) 1392-1393.
12. M. Ikeya, J. Miyajima, S. Okajima, ESR dosimetry for atomic bomb survivors using shell buttons and tooth enamel., *Jpn. J. Appl. Phys.* **23** (1984) L699-L701.
13. H. M. Swartz, Long-lived electron spin resonance in rats irradiated at room temperature. *Radia. Res.* **24** (1965) 579-586.
14. M. Ikeya, H. Sumitomo, C. Yamanaka, D.C.Lloyd and A.A. Edwards, ESR dosimetry of a deceased radiation worker, *Appl. Radiat. Iso*. **47** (1996) 1341-1344.
15. M. F. Desrosiers , In vivo assessment of radiation exposure. *Health Phys.* **61** (1991) 859-861.
16. A. Kai and M. Ikeya, ESR accident dosimetry using medicine tablets coated with sugar. *Radiat. Protect. Dos.* **34** (1989) 307-310.
17. T. Nakajima Sugar as an emergency populace dosimeter for radiation accident. *Health Phys.* **55** (1988) 951-955.
18. D. F. Regulla and U. Deffner, Progress in alanine/ESR transfer dosimetry in high dose dosimetry. *Trends in Radiation Dosimetry* (ed. W..L. Mclaughlin, Pergamon Press, Oxford) *Appl. Radiat. Isot.* **33 (1982)** 101-1114.
19. T. Omura and M. Ikeya Evaluation of the ambient temperature of mineral (gypsum ($CaSO_42H_2O$) growth by ESR microscope. *Geochem. J.* **29** (1995) 317-324.

ESR Characterization of Silicon Oxides and Nitrides for VLSI Failure Analysis Applications

K.Kawano, K.Ishida and †*Khine Nyunt*

Department of Electronics Engineering,
The University of Electro-Communications, Chofu, Tokyo, Japan 182
†Institute of Microelectronics, National University of Singapore,
11 Science Park Road, Science Park II, Singapore 117685

Abstract. **In the requirement of higher integrations of VLSI, formations of defects on the processed wafers were investigated by ESR techniques from the concept of built-in-reliability. For some ion-implanted and RIE (Reactive Ion Etching) treated specimens, new ESR centers other than the well-known E′ and P_B centers were observed with narrower linewidths and were assigned to new defects created on fabrication of LSI.**

1 Introduction

To meet the challenge of todays' microelectronics market, new designs must do more than the basic system function. They must also offer novel, sophisticated features to customers as well as the highest quality and reliability. For the requirement of higher integrations of VLSI, however, we have faced a few significant problems to be resolved for lattice defects produced by process treatment; firstly defects introduced into the Si substrate as the result of ion implantation and chemical etching process, secondly defects left over after the annealings, thirdly knocked-on atomic defects from SiO_2 to the substrate. First studies of these defects by ESR were made by Izumi and Matsumori [1]. Characterization using ESR is very useful to detect high-sensitively and non-destructively the lattice defects accompanied with lone-pair electrons as reported by Jousse et al. [2], Murakami et al. [3], Evans et al. [4,5] and Notake et al. [6]. The photo emission microscopy and TEM observation by Shivanandan and Khine Nyunt [7] have clarified them to correspond to the defects created in ion implanted n^+-layers near the gate oxide. In addition, element-analytical studies accompanied with SEM have examined ESR characterization obtained.

2 Experimental Details

All the tested chips were prepared by the Institute of Microelectronics (IME), National University of Singapore, with cooperation of Siemens Co. These involve a wide variety of Si-based VLSI devices under different conditions such as etching rate/time of RIE (Reactive Ion Etching) and various chemical wet methods like BOE (Buffer Oxide Etching). For ion-implanted specimens, a Ga^+ ion beam in FIB (Focused Ion Beam: Model 611, FEI, USA) was used in the area size

150×150 μm^2 under the conditions; 25 kV, 100 pA, 5 min. All the FIB treatments were made under implantation perpendicular to the surface through SiO_2 with the thickness of 0.5 μm on Si layers. In the present study, the measured specimens were microprocessor chips with topside polyimide layer on <100> Si substrate. The VLSI chips were measured by X-band ESR (Bruker, ESP300E) with the very-low-temperature control attachment (Oxford, ER4112HV). ESR signals were accumulated 20-30 times to achieve high sensitivities. For the measurements in ESR cavity, specimens were in early works cut out from the standard size of 12×12 mm^2, and thereafter specimen sizes were arranged in 6×3 mm^2. No differences of the signals between the sample sizes were observed.

3 Experimental Results and Discussion

For a LSI specimen with topside polyimide etched by N_2O gas plasma (200 mTor, 150 W and 117 V) in RIE, two spieces of defects: (1) E′ center (g = 2.0013, ΔH = 5G) and (2) Si-dangling bond (g = 2.0061, ΔH = 8G) were clearly observed. The E′ center is assigned to be a hole-trapped in O^{2-}-vacancy in SiO_2, and were confirmed in terms of γ-irradiation to SiO_2 by Evans et al. [4,5] and Notake et al.[6].

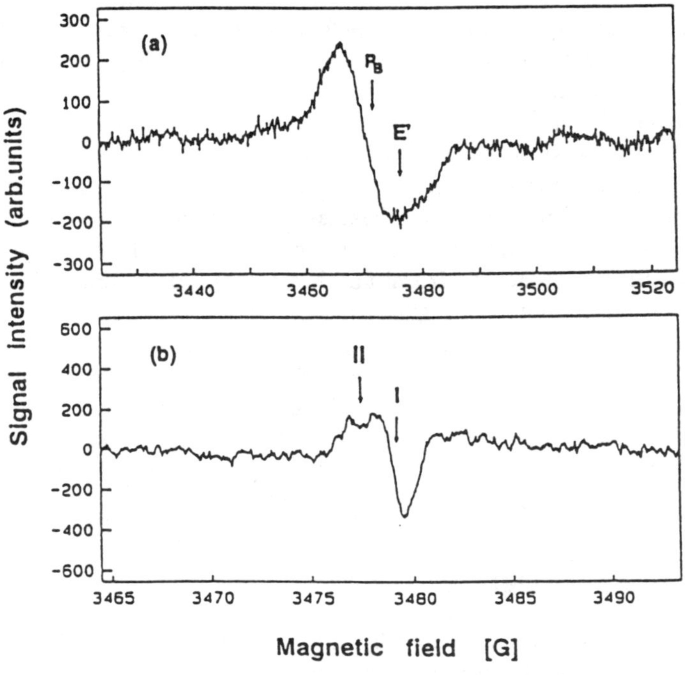

Fig. 1. The spectra of (a) E′ center in Si and P_B center in SiO_2 for VLSI specimen under RIE etched by N_2O plasma and (b) new centers observed for specimen under FIB by Ga^+. Note that scanning magnetic field widths are different between (a) and (b) by 3.5 times.

Table 1. List of ESR centers observed in fabricated VLSI specimen.

Conditions	g-factor	p-p width(G)	ESR Centers
RIE by N_2O plasma	2.0013	5	E' Center in SiO_2
	2.0061	8	P_B Center in Si
FIB by Ga^+	2.0031	1.1	New I
	2.0041	0.8	New II
RIE by CF_4/O_2 plasma	2.0032	1.2	New I
	2.0039	0.9	New II

These centers are well known as defects created by ion implantation and subsequent annealings from the results of modeling SiO_2/Si samples by Izumi and Matsumori [1], Murakami et al. [3] and Evans et al. [4,5]. For FIB-treated specimen by Ga^+, however, two kinds of new signals were observed with much narrower widths. Similar centers, which are designated by New I and New II, appear in also RIE etched specimen by CF_4/O_2 plasma (8 min). All the observed ESR spectra are shown in Fig.1 and the determined parameters are summed up in Table I. As preliminary experiments, Si_3N_4 layered specimen as well as the chemically wet-etched ones had very weak signals of new I and II, too. For the origin of new centers, some ambiguities remained about common fabrication conditions to determine definite structures because the new centers are observed by elaborate measurements even in N_2O plasma-etched samples.

So far many researches of defects for simple materials like SiO_2 on Si have been made from the view of BIR, but few works for defects on real VLSI devices were approached owing to complicated factors involved. A typical structure of NMOS LSI element that appears often in microprocessor chips is shown in Fig.2.

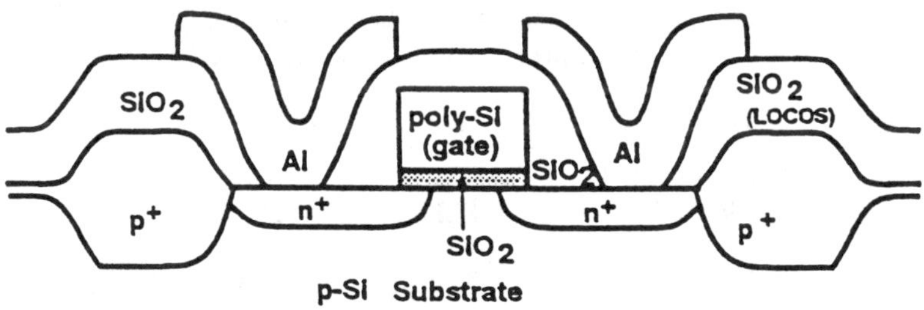

Fig. 2. Typical cross-section structure of NMOS-type LSI element.

The basic parts that produce the defects with lone-pair electrons have been confirmed to be in SiO_2 on Si, but a lot of possibilities of newly created defects exist in boundary layers among metals (Al, Poly-Si), SiO_2 and n^+ source and drain parts including p^+ isolation. New centers do not depend on LSI fabrication processes and they may be electron-trapped centers judging from near-free electron's (g=2.0023) g-factors. There are two origins for proper defects which create some charge-trapped centers in SiO_2: extrinsic impurities and intrinsic O^{2-}-vacancies, proposed by Miki et al. [8] and Kaneta [9]. Taking into account the differences of g-factors and relative signal intensities, we believe that the new center I may be caused by an impurity center like a monovalent OH, H, F or Cl ion replacing for O^{2-} ion, and the new center II may be assigned as O^{2-}-vacancy. Models of the new centers are shown in Fig.3 by the molecular bond diagram in SiO_2.

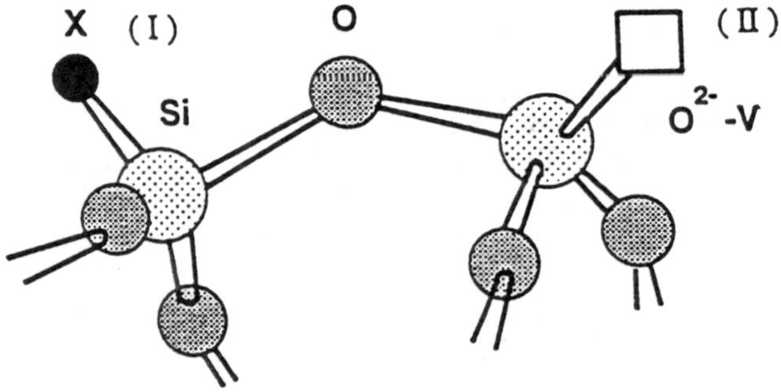

Fig. 3. Models of new centers I and II in SiO_2 molecular bonds.

For Si_3N_4 film, a diffusive O^{2-} ion out of the film of LOCOS (Selective Oxidation) mask may be another candidate for the new II center. Precise experiment at low temperatures, which is now underway, will decide on clear assignments of the new observed defects.

Acknowledgment

This research is the international collaborative research project between UEC in Japan and IME in Singapore. The authors wish to thank Dr.George T.T.Sheng for guidance in TEM specimen preparations and Dr.G.P.Goh for FIB at IME.

References

1. T.Izumi and T.Matsumori, Jpn.J.Appl.Phys, **14** (1975) 1067-1076.
2. D.Jousse, J.Kanicki, D.T.Krick and P.M.Lenahan, Appl.Phys.Lett. **52** (1988) 445-447.
3. K.Murakami, S.Fujita and K.Masuda, Mat.Sci.Forum. **83** (1992) 75-80.
4. H.L.Evans, R.K.Lowry, W.L.Schultz, T.J.Morthorst, P.M.Lenahan and J.F.Conley, IEEE/IRPS (1994) 410-418.
5. H.L.Evans, R.K.Lowry, W.L.Schultz, T.J.Morthorst, P.M.Lenahan and J.F.Conley, Proc.20th Int.Symp.Testing and Failure Analysis (1994) 33-39.
6. Y.Notake, K.Harada, H.Kitahara, K.Tomita and H.Hazama, ISTFA '95, 21st Int.Symp. Testing and Failure Analysis (1995) 15-18.
7. K.Shivanandan and Khine Nyunt, *Ibid* (1995) 69-71.
8. H.Miki, M.Noguchi, K.Yokogawa, B.Kim, K.Asada and T.Sugano, IEEE Trans,Electron Devices **35** (1988) 2245-2251.
9. C.Kaneta, Jpn.J.Appl.Phys. **35** (1996) 1540-1543.

ESR Study of Electrons and Holes Localized in Amorphous Si_3N_4

Vladimir A. Gritsenko

Institute of Semiconductor Physics, 630090 Novosibirsk, Russia
Present address: City University of Hong Kong, Tat Chee Avenue, Kowloon, Hong Kong

Alexandr A. Milov

Institute of Chemical Kinetics and Combustion, 630090, Novosibirsk, Russia

Hei Wong

City University of Hong Kong, Tat Chee Avenue, Kowloon, Hong Kong

Abstract. This work studies the localization effect of electrons and holes in amorphous silicon nitride (Si_3N_4) using ESR measurements. Although the number of localized electrons and holes are two orders larger than that reported in others works, no ESR signal of localized carriers is observed. We explain this phenomenon with antiferromagnetic ordering effect. Because of a high density of localized states, the localized electrons can interact with each other by resonance tunneling through unoccupied quantum wells. Antiferromagnetic ordering will occur if the exchange interaction energy is greater than the Zeeman splitting of the level of a localized electron. Si-Si bonds are proposed to be the responsible trap origins for electrons and holes localization in Si_3N_4.

1 Introduction

Amorphous Si_3N_4 is one of the major insulators for silicon device applications. As it can localize injected charge carriers and the delocalization time is very long (about 100 years at 300K), the nitride film has been widely used in silicon EEPROM (Electrically Erasable Read Only Memory) devices. The nature of localization and delocalization has been studied for more than a quarter of a century. However, many fundamental problems are still unresolved. Recently, more completed studies on the localized states in nitride film have been conducted experimentally by using ESR measurements [1] and a theoretical explanation has been proposed [2]. This work provides a further discussion based on the recent works [1,2].

2 Experiments

To block electrons and holes discharge, Silicon-Oxide-Nitride-Oxide (SONO) structures were prepared (see Fig.1). The nitride layer, with thickness of about 2100 Å, was prepared by pyrolysis of a mixture of SiH_2Cl_2 and NH_3 at 800 °C. The thermal oxide layer between the silicon and nitride is 200 Å thick and the thickness of the top CVD oxide is 500 Å. The electron and holes injection made use of corona discharge which induced a high electric field on the surface and enables the carrier tunneling from silicon substrate into the insulator [3-4]. After polarization, the injected charge density is measured using capacitance-voltage method and the absolute number of localized carriers was about 5×10^{14}. Since the trap density in SiO_2 is low, most of the injected carriers should be trapped in the silicon nitride. The X band ESR spectra were recorded using a Bruker ESP-300 spectrometer at 300 K with a 100 kHz modulation field. A small amount Mg^{++} was used as a reference for calibrating the g-values.

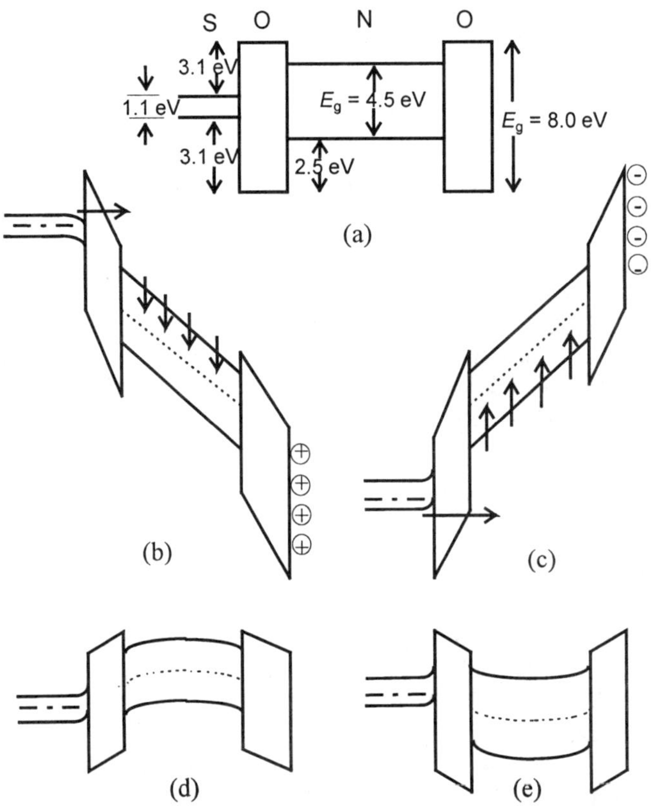

Fig. 1. Energy diagram of silicon-oxide-nitride-oxide structures at different conditions. (a) at flat-band; (b) during polarization for electron injection by accumulating positive ions on the surface using corona discharge; (c) during polarization for hole injection by accumulating negative ions on the surface; (d) with localized electrons; and (e) with localized holes.

3 Theory

3.1 Wigner Crystallization

When an electron is trapped in a trap, the second electron with kinetic energy E_{ph} will keep at a distance, $2R_q$, with this trap.

$$2R_q = \frac{e^2}{4\pi\varepsilon\varepsilon_0 E_{ph}}, \tag{1}$$

where e, ε, and ε_0 are the electronic charge, the relative permittivity of silicon nitride and vacuum permittivity, respectively.

The occupied trap density, n_t, can be estimated to be

$$n_t = (2R_q)^{-3} = \frac{(4\pi\varepsilon\varepsilon_0 E_{ph})^3}{e^6}. \tag{2}$$

The two-dimensional square quasi-periodic lattice with period $2R_q$. The lattice is formed by hard spheres with radius R_q. A gap of magnitude $N^{-1/3}$ exists between the spheres. This quantity gives the characteristic scale of the variance in the lattice period. A phase transition will occur when the Coulomb interaction energy is greater than the binding energy (E) of an electron in a trap, *i.e.* $e^2 n_t^{1/3} / 4\pi\varepsilon\varepsilon_0 >> E$ [5-6]. Another obvious criterion is $E >> kT$. This means that the ionization probability of the traps is low. In addition, R_q should be greater than the distance between the traps: $R_q > N^{-1/3}$. The trapping cross section (σ) is determined not by the individual properties of a trap or the mechanism of energy loss, but rather by the square of R_q:

$$\sigma = \pi R_q^2 = \frac{\pi e^4}{(8\pi\varepsilon\varepsilon_0 E_{ph})^2}. \tag{3}$$

Hence, the relation between density of filled traps and the trapping cross section is:

$$n_t^{-1/3} / 2 = (\sigma / \pi)^{1/2} = R_q. \tag{4}$$

3.2 Antiferromagnetic Ordering of Localized Electrons due to Resonance Exchange Interaction

When all traps in the insulator have the same energy and are located at the same distance, antiferromagnetic pairing of localized spins will occur (see Figs. 2 and 3). Consider the "superlattice" formed by traps distributed periodically (see Fig. 2 (b)), the exchange integral J for electrons 1 and 2 at the sites A and B is given by

$$J = \iint \psi_A(1)\psi_B(2)\frac{e^2}{|\mathbf{r}_1 - \mathbf{r}_2|}\psi_B(1)\psi_A(2)d\mathbf{r}_1 d\mathbf{r}_2\,, \tag{5}$$

where $\mathbf{r}_1$ and $\mathbf{r}_2$ are measured from the sites A and B, respectively. It has been estimated that direct exchange interaction is too low to cause a spin flip [2].

However, exchange interaction between localized electrons through unoccupied nearby well is highly possible [2]. Figure 3 shows the potential energy diagram for a pair of nearest localized electrons. The barrier between the two traps is lowered because of the overlapping of the Coulomb polaron potentials. This potential can be modeled by

$$V(z) = -\frac{e^2}{4\pi\varepsilon_p\varepsilon_0}\left(\frac{1}{z} + \frac{1}{N^{-1/3} - z}\right), \tag{6}$$

where $\varepsilon_p^{-1} = \varepsilon_\infty^{-1} - \varepsilon^{-1}$ is the inverse polaron insulator permittivity; ε_∞ and ε are the high and low-frequency permittivities, respectively.

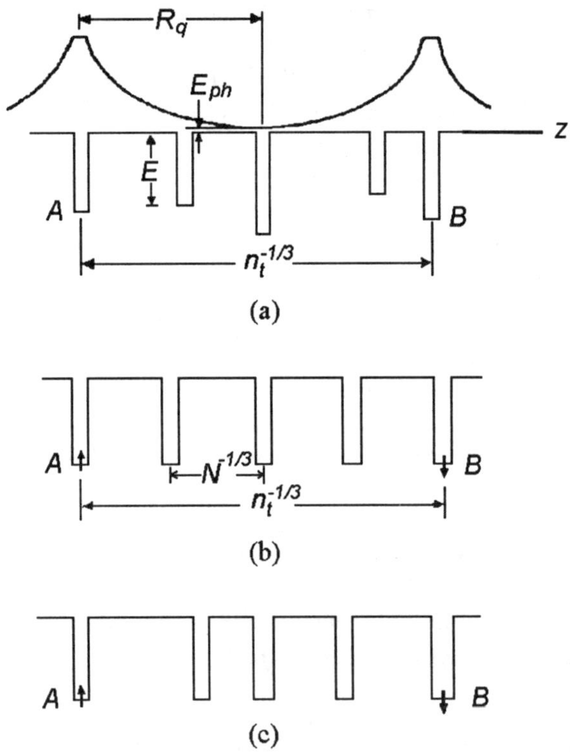

Fig. 2. (a) Potential energy diagram illustrating the forming of the quasi-periodic lattice due to Coulomb repulsion localized electron. Antiferromagnetic ordering in a periodic lattice due to the resonance exchange interaction in shown in (b) and aperiodic lattice is indicated in (c).

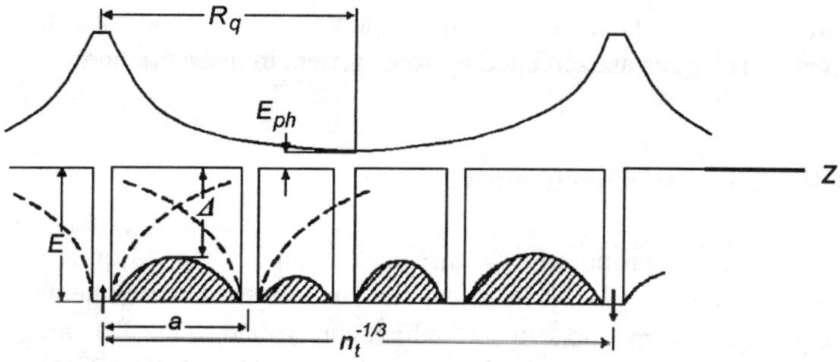

Fig. 3. Potential energy diagram illustrating spin pairing as a result of an exchange interaction with tunneling via unfilled states.

The probability of finding a tunneling electron in a well can be calculated using the WKB approximation. If only the potential of the nearest traps is considered, the tunneling probability between two neighbor traps is

$$P = \exp\left(-\frac{2}{\hbar} \int_{z_1}^{z_2} \sqrt{2m^*\left(E + V(z)\right)} dz \right). \tag{7}$$

Antiferromagnetic ordering will occur if the exchange interaction energy is greater than the Zeeman splitting of the level of a localized electron, i.e.

$$\frac{e^2 n_t^{1/3} P}{4\pi\varepsilon\varepsilon_0} > g\mu H , \tag{8}$$

where μ and H are the Bohr magnetron and the intensity of the magnetic field, respectively.

A few rules for the exchange interaction can be proposed here. First, when the barrier transmittances are substantially different, the total tunneling probability will be determined by tunneling between the most widely separated pair of traps (the widest barrier rule). Second, an electron should choose a chain of traps with the minimum barrier width (minimum barrier width rule). Third, binary pairing of spins should take place in neighbor filled traps which is proposed to resolve the percolation problem. Finally, it is noted that an exchange interaction can also occur when the traps distribute aperiodically in space but have same energy (see Fig. 2(c)). In that case, the estimation of the electron tunneling probability is more complicated.

Although Wigner crystallization in the insulator in a MIS structure could occur in three-dimensional case, antiferromagnetic ordering of the spins can only occur in a plane parallel to the surface of the semiconductor. In the direction normal to the surface, the electric field will result in the displacement of a trap energy which prevents tunneling exchange in a direction normal to the surface. In the direction parallel to the plane of the semiconductor, there is no electric field because the localized charges are screened by free carriers in the semiconductor and the metal.

4 Results and Discussion

It was proposed that an electron can be trapped on a diamagnetic defects, ≡Si-Si≡, in Si_3N_4 [7-10]. In direct Raman scattering experiments, feature lines with frequency 475 $cm^{-1}$and 296 cm^{-1} which correspond to the TO and LO mode vibration of the Si-Si bond, respectively, were recorded [11]. MINDO/3 quantum-mechanical modeling further found that the Si-Si defect in Si_3N_4 can trap both electrons and holes [12]. These trapping effects should lead to the formation of electron spin resonance (ESR).

Figure 4 shows the ESR spectra for a SONO structure. Two ESR signals, *i.e.* g = 2.0055 and g = 2.00055, are observed. The wide signal with g =2.0055 (Δ =0.5 mT) observed in samples without polarization is due to the defects on the silicon polished wafer surface. This signal is also observed in amorphous silicon and is usually attributed to the silicon triply coordinated silicon atoms with an unpaired electron, i.e. Si_3Si•. Whereas the signal with g =2.00055 (Δ= 0.14 mT) is due to the well-known E' center or O_3Si• in SiO_2 [13,14]. An E' center arises when a hole is trapped at a Si-Si bond in SiO_2 near Si-SiO_2 interface. Similar experiments in silicon nitride were conducted, no ESR signal is observed in silicon nitride with localized electrons and holes, however. This phenomenon can be explained with the newly proposed weak electron interaction-resonance exchanging interaction of electrons on account of tunneling via localized states given below.

These experiments reveal that hole trapping on Si-Si defect will be accompanied by the formation of paramagnetic defect (O_3Si•) in SiO_2. However, no ESR signal is found in Si_3N_4. The expected ESR signal in Si_3N_4 is K center with g = 2.003 (N_3Si• defect) [15]. The absence of an ESR signal in Si_3N_4 with localized electrons and holes can be explained with pairing of localized spins so that the sum of the spins equals zero.

Since the trap density in silicon nitride could be very high, the average spacing $N^{-1/3}$ between traps may be less than the effective trapping radius $(\sigma / \pi)^{1/2}$. If only a fraction of traps are filled with injected electrons and the electron potential energy exceeds the kinetic energy, pairing of spins can occur on neighboring filled traps via the exchange interaction. The localized electrons can interact with each other by resonance tunneling through unoccupied traps. In addition, antiferromagnetic ordering will occur if the exchange interaction energy is greater than the Zeeman splitting of the level of a localized electron. It occurs when most of the traps have the same energy and the trap distribution is uniform in the insulator. The Si-Si bond traps in amorphous Si_3N_4 meet these requirements. It was found that the Si-Si bond concentration in silicon nitride is in the order of 10^{21} cm^{-3} and the mean distance between two traps can be estimated to be about 10 Å.

However, due to Coulomb repulsive effect, only a small part of traps can be occupied by the localized electrons (holes) (Figs.2 and 3). The Coulomb repulsive radius in Si_3N_4 can be calculated as $R_q \approx 50$ Å. These localized electrons result in the formation of a two-dimensional Wigner quasi-periodic crystal (Fig.2). Assuming $N = 10^{21}$ cm^{-3}, $m^* \approx 0.4\ m_e$ [16], and $E = 2.0$ eV, the exchange interaction energy is about 5×10^{-5} eV which is greater than the typical Zeeman splitting energy for electrons localized in a solid (about 6×10^{-5} eV). It indicates that the presence of a high density of unfilled traps would result in the exchange interaction between localized electrons and hence the magnetic ordering of the spin subsystem.

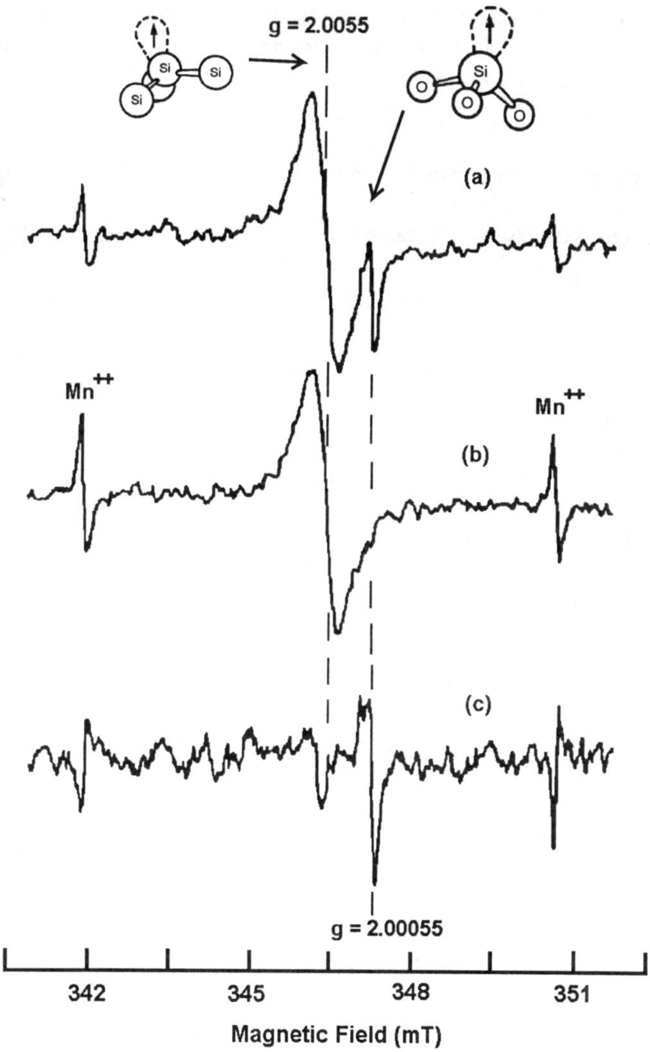

Fig. 4. ESR spectra of a SONO structure: (a) sample with localized holes; (b) nonpolarized sample. Trace (c) shows the difference between the two spectra.

5 Conclusion

In this work, we attempt to study the localized electrons and holes in Si_3N_4 using ESR measurements. However, no ESR of localized electrons and holes was observed. This observation has been interpreted in terms of two new phenomena: (a) Wigner quasi-crystal formed due to filled traps; and (b) antiferromagnetic ordering of localized electrons due to resonance exchange through unoccupied traps. These phenomena occur only in those materials that have a high-enough trap density. We proposed that the Si-Si bond, which can capture both electrons and holes, is the origin for those observations in amorphous silicon nitride. The absence of ESR in a system with localized electrons can serve as a criterion of a resonance exchange interaction.

Acknowledgments. The authors would like to thank Yu. N. Morokov, V. I. Belinicher, A. V. Chaplik, and M. V. Entin for their helpful discussions, T. I. Nedosekin and A. G. Klirnenko for preparing the samples for ESR measurements.

References

1. V. A. Gritsenko and A. D. Milov, JETP Lett. **64** (1996) 531.
2. V. A. Gritsenko, JETP Lett. **64** (1996) 525.
3. P. M. Lenahan, D. J. Krick, and J. Kanicki, Appl. Surf. Sci. **39** (1989) 392.
4. L. Warren, J. Kinicki, J. R. Robertson E. H. Poindexter, and P. J. McWhorter, J. Appl. Phys. **74** (1993) 4034.
5. E. Wigner, Phys. Rev. **46** (1934) 1002.
6. T. Ando, A. Fowler, and F. Stern, Rev. Mod. Phys. **54** (1992) 2.
7. V. A. Gritsenko and P. A. Pundur, Sov. Phys. Solid State **28** (1986) 1829.
8. V. A. Gritsenko, Silicon Nitride in Electronics, Materials science Monograph **34**, Elsevier (1988).
9. V. A. Gritsenko, Structure and Electronic Structure of Amorphous Insulators in Silicon MIS Structures [in Russian], Novosibirsk, Nauka (1993).
10. P. A. Pundur, J. G. Shavalgin, and V. A. Gritsenko, Phys. State Solid, **a94** (1986) K107.
11. Y. G. Chen and F. Zhang, Phys. Status Solidi B **152** (1989) 73.
12. V. A. Gritsenko, A. D. Milov, Y. N. Morokov, Y. N. Novikov, H. Wong, and Y. C. Cheng, in Proceeding of 1996 Materials Research Symposium, Boston, to be published.
13. E. Griscom, J. Non-Cryst. Solids **73** (1985) 51.
14. S. Withinan and P. M. Lenahan, Appl. Phys. Lett. **51** (1987) 1007.
15. W. L. Warren and P.M. Lenahan, Phys. Rev. B **42** (1990) 1773.
16. V. A. Gritsenko and E. E. Meerson, Mikroelektronika (Sov.) **17** (1988) 249.

Fabrication, Spatial and Spin Chain Structure of Langmuir-Blodgett Films of Copper tetra-4-tert-butyl Phthalocyanine

Tomohiro KOMATSU, Takumi TAKAMURA † and Yuhei SHIMOYAMA

Department of Physics, Hokkaido University of Education, Hakodate 040, Japan.
† Hokkaido Industrial Technology Center, Hakodate 041, Japan.

Abstract. Formation of the Langmuir monolayer, and structure and magnetic properties of the LB films of copper tetra-4-tert-butyl phthalocyanine (Cu-Pc) were investigated by surface pressure-area isotherms and electron paramagnetic resonance (EPR) spectroscopy, respectively. Upon lowering the solute concentration of developing solution, we found that the limiting area increases up to 0.63 nm^2 at the 5 μM of the Cu-Pc solution. Dilution of the developing solution enables formation of a well-defined Cu-Pc monolayer on the subphase surface. For the LB films with one to five layers, EPR revealed that molecular axis of Cu-Pc holds the perpendicular position against the substrate plane; the film molecules tilt in the layers beyond six. The gradience of the tilt angles along thickness of layer indicated a clear evidence for surface force of substrate onto the film molecules. We revealed that a spin chain system in the Cu-Pc LB films forms two-dimensional order.

1 Introduction

Langmuir (L) monolayer and Langmuir-Blodgett (LB) films have opened a new horizon for technological breakthrough of molecular electronics [1]. Fabrication techniques for well-oriented ultra thin films are very crucial for LB methodology [1]. Thus far, we have been developing a fabrication methods of a defect-free LB film for various organic molecules [2]. Copper tetra-4-tert-butyl phthalocyanine (Cu-Pc) is a highly colored macrocycle having a copper ion at the center, which has been widely known as a material for solar cell [3]. Although Cu-Pc has been extensively studied in the forms of the evaporated thin film, until recently, few researches have been studied in L or LB films [4]. Electron paramagnetic resonance (EPR) provides an elucidation method of structure of materials containing π-electron and transition metal ions. For the organic thin film, EPR has been used to study molecular ordering in the LB films of fatty acid-transition metal salts [2, 5].

In the present study, we report on the formation process of Langmuir monolayer of Cu-Pc molecule on the subphase by π-A isotherms. We then elucidate the static structure of LB films of Cu-Pc by EPR. Special efforts are paid on the magnetic properties, in particular, spin chain system in the LB films by EPR spectroscopy.

2 Materials and Methods

Chemicals used in the present study were copper tetra 4-tert-butyl phthalocyanine (Tokyo Kasei kogyo Co.) and benzene of spectral grade (Kishida kagaku Co.). Benzene solution of Cu-Pc was used for the formation of monolayer. Concentration was varied from 2.5 μM to 1 mM. For the subphase, we used ultra pure water at ca. 17 ℃. The benzene solution was deposited onto the subphase. After evaporation of benzene, the film was compressed by an automatic barrier to obtain the solid phase. Measurement of the π-A isotherm was done by the Wilhelmy plate method. Langmuir-Blodgett films were fabricated by the Kuhn type (Kyowa Kaimenkagaku Co.). Quartz (SiO_2) slides were used as the substrate and washed several times in a ultrasonic bath containing ethyl alcohol. The surface pressure of compressions was fixed at 20 mN/m. The monolayer was transferred onto the substrate by the dipping/lifting strokes. The stroke speed was 2.5 mm/min. LB films were prepared by the stacking of one to nineteen layers. EPR measurements were carried out by a spectrometer (Jeol FEX-1H) with an automatic phase controller and digitizing systems that allowed the second harmonic detection. All the spectra were measured at room temperature and at the first harmonic mode. Five LB films were prepared with different number of layers. The angular variations were made by a homemade goniometer from 0 to 360 degrees. The sample plate was located in the applied magnetic field of the EPR spectrometer and rotated with respect to the dipping direction.

The experimental g-values were fitted by Gaussian distribution function. Considering the tilt angle θ, and the deviation angle σ, we may define an extended distribution function of ϑ, an angle between the surface of substrate and the applied magnetic field as following:

$$f(\vartheta)=\frac{1}{\sqrt{2\pi}\,\sigma}\exp\left(\frac{-(\vartheta-\theta)^2}{2\sigma^2}\right) \tag{1}$$

Using the least square fitting procedure, we evaluated the tilt and deviation angle from the angular variations of g-value.

The overall angular dependence of the EPR linewidth demonstrates the prominent role of the null frequency as the dimensionality decreases [6, 7]. This is observed through the minimum at the magic angle at low dimensions. Due to the nonstandard time decay of the correlation function at one-dimensional (1D) and two-dimensional (2D), the power law of the secular dipolar contribution becomes different in 1D and 2D. When the spectral density of the spin correlation function exhibits a flat frequency dependence for exchange narrowed EPR line in the three-dimension (3D), the peculiarities are not observed. The angular variation of linewidth, ΔB is given by the equation below. For the 1D system, the half-width at half-height obeys an angular variation of the following expression [6]:

$$\Delta B \propto \left|1-3\cos^2\vartheta\right|^{\frac{4}{3}} \tag{2}$$

The following equation depicts the angular variation in the 2D system:

$$\Delta B \propto \left(1-3\cos^2\vartheta\right)^2 \tag{3}$$

The 3D spin system yields a relationship of linewidth versus angle [7]:

$$\Delta B \propto \left(1+\cos^2\vartheta\right) \tag{4}$$

3 Results

Figure 1A shows π-A isotherms of the L-monolayer formed by developing solutions at various concentrations. By lowering concentration of the developing solution, we found that the limiting area of single Cu-Pc molecule increases. Figure 1B summarizes the relationship between the limiting area versus concentration of the solution. As the concentration lowers from 1 mM to 5 μM, the limiting area varies from 0.40 to 0.63 nm^2, respectively. However, the limiting area remained constant at the lower concentrations below 5 μM. The lowest value for the limiting area, 0.63 nm^2 is almost equal to the cross section of the edge of phthalocyanine ring. Therefore, Cu-Pc molecules orient perpendicular to the substrate surface.

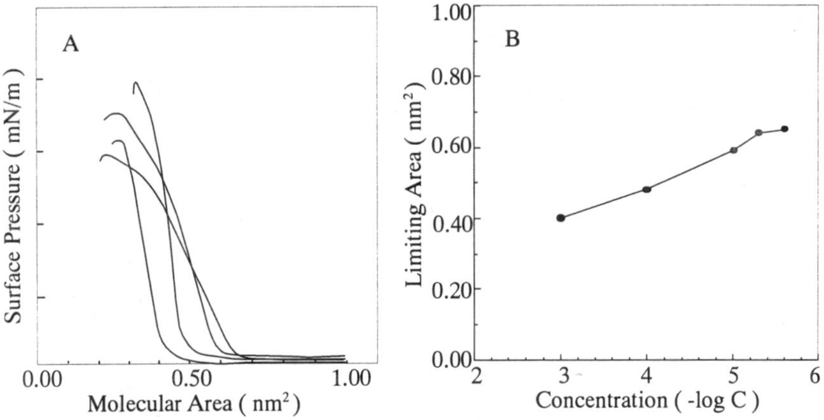

Fig. 1. A) Surface pressure versus molecular area isotherms at various concentrations of developing solution of copper phthalocyanine. B) A correlation between concentration of developing solution of copper phthalocyanine and limiting area.

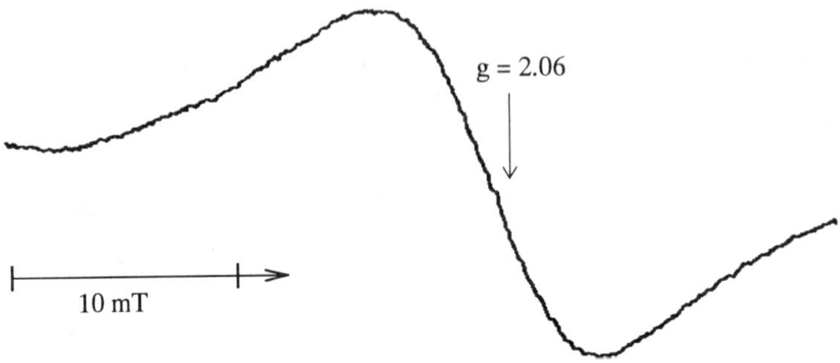

Fig. 2. An EPR spectrum of LB films of copper phthalocyanine.

Figure 2 shows an EPR spectrum of the Cu-Pc LB films, suggesting a broad signal at g = 2.06. The peak around the g-value is originated from the strong spin-spin interaction among copper ions. Figure 3A shows the angular variation of the g-value. The g-value anisotropy reflects the in-plane ordering. Angular variation of the g-value (open circles in Fig. 3A) shows a characteristic shape for molecules in a layer, with a symmetric shape at the substrate normal axis. This implies that the Cu-Pc plane tilts with a certain angle. The angular distribution of the Cu-Pc plane orientation obeys a Gaussian distribution function (Eq. 1) of ϑ, an angle between the magnetic field and the surface direction. We found that for the small number of layers, i.e., one to five layers, the molecular axis holds almost the perpendicular position (85 °) against the substrate plane. On the other hand, the film molecules in the layers beyond six tilt with twelve degrees from the layer normal. From deviation angle, we found a constant width of distribution of film molecules despite the number of layers.

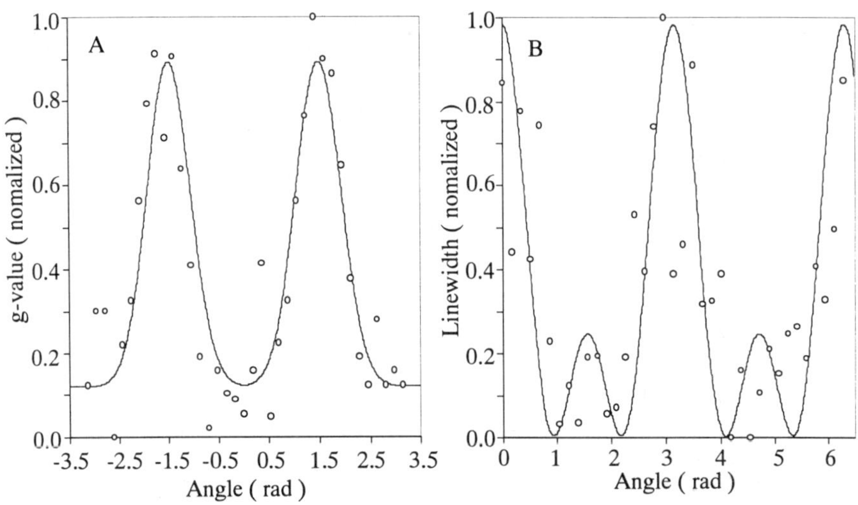

Fig. 3. Angular variation of A) the g-value, and B) the linewidth of EPR spectra from LB films of copper phthalocyanine.

Figure 3B shows the angular variation of the linewidth (open circle). For the fitting to the linewidth equations (2-4), we normalized the linewidth data for the technical convenience. We immediately recognized that the linewidth hits the minimum value at the magic angle (55.4°). This is because that the dipolar interactions vanished at this point. In fact, Eq. (3) gave a best-fit for the angular variation of the linewidth. The EPR spectrum recorded at the magic angle shows Lorentzian lineshape, suggesting that the secular contribution be dominant. Based on these facts, we may attribute the spin chain system possessing the 2D order.

4 Discussion

4.1 Structure of Monolayer and LB Films

We successfully formed the Cu-Pc monolayer on the water surface by the dilution of the developing solution. Upon lowering the solute concentration of the developing solution, we found that the limiting area increases up to 0.63 nm^2 at the 2.5 μM of the Cu-Pc solution. The limiting value corresponds to the area of the edge of a Cu-Pc molecule. Cu-Pc molecule is reported to be difficult to develop on the water surface, because of the aperiodic arrangement due to the steric hindrance of the tert-butyl moieties. Therefore, the dilution procedure enables formation of a well-defined Cu-Pc monolayer on the subphase surface. It is likely that strong cohesion force among Cu-Pc molecules forms multilayers on the water surface.

It has been suggested for a long time that in-plane ordering be a matter of interest and importance, because it controls the physical properties of the LB films [1]. We found that the EPR method detected the same tilting phenomena. Using the EPR spectroscopy, Cook *et al*. reported that the Cu-Pc molecules align with the plane of the phthalocyanine ring at 80 $\pm 10°$ to the subphase surface [8]. They employed the spectral simulation methods to reveal the tilt phenomena. Very recently, Pasimeni *et al.* reported the tilt phenomena of ca. 86° of Cu-Pc molecule on the quartz substrate by means of UV-visible spectroscopies [9]. It should be noted that pinning aggregates were found in the in-plane organization of the LB films by means of EPR method. Crystalline order of Cu-Pc LB films was reported to be ill-defined at the in-plane direction, while periodicity was well defined along the layer normal [4]. Periodicity along the layer normal of the LB films was formed to be better than that of vacuum deposition.

A remarkable correlation between the number of layers and tilting angle was found in the LB film. The tilt angle of the LB film increases with the number of layers up to the maximum angle given by XRD of the powder crystals of Cu-Pc. We interpret this as the tilt gradience along the layer normal axis. It provides clear evidence of the surface force between the film molecules and the substrate at the molecular level. The tilting phenomenon has been first reported in the LB film of

Mn-arachidate [2, 10]. The molecules near the substrate are perpendicular to the substrate. In the layers far from the substrate, the molecules tilt to the in-plane directions. Thus, we may conclude that the interaction force within the layer plane and from the substrate induces the tilt orientation.

4.2 Spin Chain Organization

New information about dimensionality of spin-chain was obtained from the angular variation of the EPR linewidth of the Cu-Pc LB films. Fitting with the Henessy's formula Eq. (3) suggests existence of the spin chain of 2D order (Fig. 3) [7]. Using the Dietz's formula, we have reported that the LB films of diacetylene manganese salt exhibits the spin chain of the 1D order [6]. It is worth mentioning that these relations are not strictly correct at 1D and 2D. This is due to the angular dependence of the secular contribution, which is dominant. However, a difference between low dimensionality and 3D is observed as the non-secular contributions increases at the Larmor frequency of electron (ω_e) and $2\omega_e$. Efficiency of the in-plane or interlayer interactions enhances the difference, as the temperature decreases.

We obtained direct evidence of the low dimensional magnetism using EPR linewidth measurements. It is remarkable to find pure 2D magnetism in the Cu-Pc LB films at the molecular level. This is due to the anisotropic alignment in the in-plane space. Major magnetic interactions among Cu ions of Cu-Pc in the LB films can be classified into the spin-spin dipolar and exchange, and the hyperfine interactions. The spin-spin interactions give a broad band spectrum and the hyperfine interaction a well-resolved quintet hyperfine line. Since these interactions are anisotropic and sensitive to the magnetic field, the angular variation reflects the Cu ions being in-plane order in the LB film. Theoretically, the linewidth dependence upon Eq. (3) is due to the spatial average of secular contribution to the dipolar interactions over the 2D space [5]. Pomerantz *et al.* carried out the elucidation of the 2D magnetism in LB film of Mn-stearate both by EPR spectroscopy and magnetic susceptibility [11]. They found the weak ferromagnetism in the LB film. They further suggested the possibility of 2D magnet in the LB film system. Haseda *et al.* reported the magnetic susceptibility measurements of Mn-stearate powder in a wide temperature range [12]. They explored the ferromagnetism at the temperature of 0.5 K, and further showed a novel magnetic ordering, the pseudo-one dimensional or 2D magnetism. These two groups suggested the possibility of the low dimensional magnetic system in both the powder and the LB films of the same sample, i.e., Mn-stearate.

5 Conclusions

We investigated the formation of the Langmuir monolayer and structures of the LB films of copper tetra-4-tert-butyl phthalocyanine. Dilution of developing solution enables a well-defined Cu-Pc monolayer on the subphase surface. EPR method revealed that molecular axis holds the perpendicular position against the substrate plane for the LB films with one to five layers. The film molecules tilt in the layers beyond six. Angular variation of EPR linewidth of Cu-Pc LB films suggests a spin chain of 2D order. Measurements of magnetic susceptibility of the LB films of Cu-Pc are now undertaken to perform the more detailed quantitative analysis.

References

[1] A. Ulman, *An Introduction to Ultrathin Organic Films from Langmuir-Blodgett to Self-Assembly*, Academic Press Boston (1991).

[2] T. Takamura, K. Matsushita, Y. Shimoyama and K. Mukasa, A Structure Analysis of Langmuir-Blodgett Magnetic Films by X-ray Diffraction, *J. Surf. Sci. Soc. Jpn.* **15** (1994) 615-620.

[3] A. L. Thomas, *Phthalocyanine research and applications*, CRC Press Boca Raton (1990).

[4] S. Baker, M. C. Petty, G. G. Roberts and M. V. Twigg, The Preparation and Properties of Stable Metal-free Phthalocyanine Langmuir-Blodgett Films, *Thin Solid Films* **99** (1983) 53-59.

[5] C. P. Poole, Jr., *Electron Spin Resonance*, Wiley-Interscience (1983)

[6] P. Turek, Spin Correlations in Organic Radical Magnet, *Mol. Cryst. Liq. Cryst.* **233** (1993) 191-207.

[7] M. J. Hennesy, C. D. McElwaa and P. M. Richards, Effect of Interchain Coupling on Electron-Spin Resonance in Nearly One-Dimensional System, *Phys. Rev. B* **7** (1973) 930-946.

[8] M. J. Cook, A. J. Dunn, A. A. Gold and A. J. Thomson, Association and Orientation of Copper (II) Tetra-t-butylphthalocyaninate in Multilayer Langmuir-Blodgett Films as determined by Electron Paramagnetic Resonance Spectroscopy, *J. Chem. Soc. Dalton Trans* **1988** (1988) 1583-1589.

[9] L. Pasimeni, M. Meneghetti, R. Rella, L. Valli, C. Granito and L. Troisi, Structure and spectroscopic characterization of Cu(II) [tetrakis-(3,3-dimethyl-1-butoxycarbon yl)] phthalocyanine thin films deposited by the Langmuir-Blodgett technique, *Thin Solid Films* **265** (1995) 58-65.

[10] T. Takamura, K. Matsushita and Y. Shimoyama, X-ray Diffraction study of Langmuir-Blodgett Films of Mn-Arachidate, *Jpn. J. Appl. Pyhs.* **35** (1996) 5831-5835.

[11] M. Pomerantz, F. H. Dacol and A. Segmüier, Preparation of Literally Two-Dimensional Magnets, *Phys. Rev. Lett.* **40** (1978) 246-249.

[12] T. Haseda, H. Yamada, M. Ishizuka, Y. Okuda, T. Kubita, M. Hara and K. Amaya, Measurements of magnetic susceptibility of Mn stearate powder, *Solid State Commun.* **24** (1977) 599-603.

Simultaneous ESR-CT Imaging of Plural Radicals Using Spectral-Spatial Imaging Techniques

Ken-ichiro Matsumoto and Hideo Utsumi*

Faculty of Pharmaceutical Sciences, Kyushu University, Maidashi, Higashi-ku, Fukuoka 812, JAPAN

*Present address: Showa College of Pharmaceutical Sciences, Higashi-Tamagawagakuen, Machida, Tokyo 194, JAPAN

Abstract. The ESR-CT imaging by the phantom involving two radical species giving different signals is reported. Spectral-spatial images were obtained along four directions on X-Z plane. Spatial information of each signal species was separated from the corresponding spectral-spatial image. The spatial information was obtained by two methods; 1) using peak height, and 2) using peak area. Two sets of CT images of two different spin probes were reconstructed. In the former method, separation was not complete. CT images had good agreement with the arrangement of phantom. This technique was applied to imaging of evaluation of drug-delivery-system. Images were obtained from both spin-labeled compounds encapsulated in and released from liposome. The other application of this technique was simultaneous imaging of spin trapped nitric oxide (•NO) and hydroxyl radical (•OH) as one of reactive oxygen species (ROS). 2D distributions of both radicals were obtained using phantom.

1 Introduction

Since first papers on ESR(EPR) imaging were independently reported by two groups, Karthe and Wehrsdorfer,[1] and Hoch and Day[2] in 1979, several developments have been done for the technique of ESR imaging.[3-6] After the development of L-band ESR spectroscopy, ESR imaging was applied to *in vivo* experiments. Berliner and Fujii[7] reported, for the first time, L-band ESR imaging of nitroxide in a celery stem. Ishida et al.[8] reported the imaging of the distribution of nitroxide radical in the head of a rat. Alecci et al.[9] described the three dimensional imaging in rat tail. We also reported the image of nitroxide radical in mouse lungs,[10] and in abdomen.[11] In the methods used for ESR-CT in the above reports, it is essential to deconvolute the spectrum under field gradient with that under zero field gradient to obtain spatial information. Deconvolution with Fourier transformation is, however, based on the assumption that the only one spectral component is existing in all space observed. When spectrum involves more than two components, ESR-CT with Fourier transformation produces the wrong image.

In the living organism, it is very often that more than two signal components are coexisting. Therefore, it should be necessary to distinguish the spectral components spatially at each space along desired spatial direction. The following three techniques have been reported on that purpose. Katsuda et al.[4] have developed localized ESR measurement, while this technique requires a special detector such as pin hole cavity and requires accurate stepping motors to scan the sample. Modulated Field Gradient technique that has been developed by Herrling et al.[5] also requires a special field gradient coil. Recently, Maltempo et al.[6,12,13] developed the spectral-spatial ESR image reconstruction. Spectral-spatial image reconstruction has large advantage since most devices used for ESR-CT are available and any other special device is not necessary.

In the present paper, spectral-spatial image reconstruction was used. The combination of spectral-spatial image reconstruction with CT image reconstruction may make the image construction from plural signal components possible. The authors developed algorithm to separate ESR spectra spatially and obtain the 2D distributions of following samples. 1) Two different spin probes, PTIO and C-PROXYL. 2) Liposomal high concentration and free low concentration aqueous solution of a spin probe called CAT-1.[14] 3) Nitric oxide (•NO) and hydroxyl radical (•OH) spin trapped with MGD_2-Fe^{2+} and DMPO, respectively.[15]

2 Two Different Spin Probes

2.1 Experimental

Materials. PTIO (2-phenyl-4,4,5,5,-tetramethylimidazoline-3-oxide-1-oxyl) and C-PROXYL (3-carbamoyl-2,2,5,5,-tetramethylpyrrolidine-1-oxyl) were purchased from Tokyo Kasei Kogyo Co., Ltd. (Tokyo, Japan) and Aldrich Chemical Co., Inc. (Milwaukee, WI), respectively. These compounds were dissolved in distilled water at 5 mM (PTIO) and 0.5 mM (C-PROXYL), and then transferred into capillaries (i.d. 1 mm). The phantom was made of four capillaries, two PTIO and two C-PROXYL, the location of which were shown in Fig. 1a, and was set in a TE mode ESR cavity.

ESR Measurement. ESR spectra were obtained with an X-band ESR spectrometer (JES-RE3X, JEOL Ltd., Tokyo, Japan) equipped with a field gradient coil unit which involved three dimensional field gradient coils and rapid scan coils. Control of field gradient coils and data sampling were perfomed with a personal computer. Conditions for ESR measurement were as follows; Modulation frequency was 100 kHz (first derivative), modulation amplitude was 0.1 mT, microwave frequency was 9.42 GHz, microwave power was 10 mW.

Spectral-Spatial Image Reconstruction. Spectral-spatial imaging data were obtained with four directions shown in Fig. 1a. For each spectral-spatial image, 29 projection spectra were collected with angular increments of 5° between -70° and +70° over 5 cm / 5 mT plane. Spectra were manipulated as described by Maltempo et al.[12,13] The spectral-spatial image was reconstructed by FBP (filtered back-projection) at first, followed by ART (Algebraic Reconstruction Technique) algorism with three iterations. Spectral-spatial images were constructed onto a 128 × 128 points image matrix.

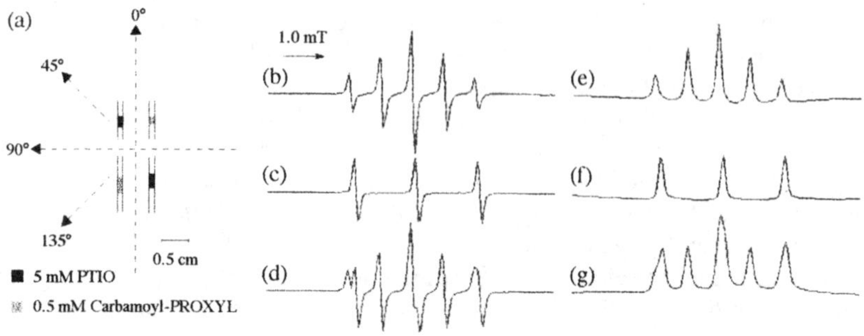

Fig. 1. Phantom and X-band ESR spectra of components. a) Phantom. b) Spectra of 5 mM aqueous solution of PTIO, c) 0.5 mM aqueous solution of C-PROXYL and d) overlappeng of b) and c). e, f) and g) are integrated spectra of b), c) and d), respectively.

CT (2D distribution) Image Reconstruction. Spatial information of PTIO and C-PROXYL were obtained from the spectral-spatial images with two methods described as below. With four projection sets of spatial information of PTIO and C-PROXYL obtained with method 1 and 2, 2D distribution images were reconstructed. Each image was reconstructed on 128 × 128 matrix using also combination of ART with FBP.

2.2 Results and Discussion

The X-band ESR spectra of two sample solutions, PTIO and C-PROXYL, under no field gradient are shown in Fig. 1b and 1c, respectively. Aqueous solution of PTIO gave five lines (hfs = 0.77 mT, peak height ratio = 1:2:3:2:1) and C-PROXYL gave a triplet of lines (hfs = 1.52 mT, peak height ratio = 1:1:1). Fig. 1d shows the spectrum of a phantom containing both radicals. Fig. 1e-g show the integrated spectra of Fig. 1b-d, respectively. The second or fourth lines of PTIO are spectrally separated from that of C-PROXYL, while the other three lines are overlapping with the lines of C-PROXYL.

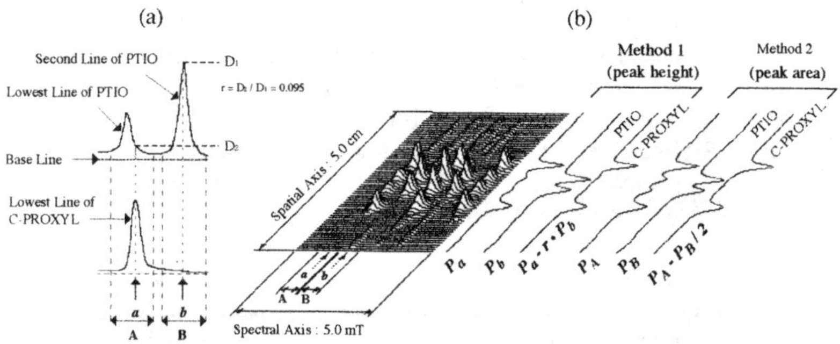

Fig. 2. Detail of signal superimposing and spectral-spatial image. a) Expansion of lowest two lines of PTIO and a lowest line of c-PROXYL. b) Spectral-spatial image on 135° direction of the phantom.

Fig. 3 shows a spectral-spatial image obtained on a 135° direction in the Fig. 1a. Another three spectral-spatial images were also observed with 0°, 45° and 90° directions. Spatial information was obtained by two methods. Method 1; Spatial information were obtained from peak height. Method 2; Spatial information were obtained from peak area.

Method 1. The point **a** in Fig. 2 is corresponding to the peak of the lowest line of C-PROXYL and point **b** is that of the second line of PTIO. The spatial profile obtained from point **b** that is indicated as $\mathbf{P_b}$ must be the spatial information of PTIO. On the other hand, the spatial information from point **a** indicated as $\mathbf{P_a}$ includes amount of not only the lowest line of C-PROXYL but also a small amount of superimposing PTIO. The contribution of PTIO at the point **a** was calculated to be $r = D_2 / D_1 = 0.095$ of the second line of PTIO in this case. The spatial information of the C-PROXYL was obtained by subtraction of 0.095 fold of $\mathbf{P_b}$ from $\mathbf{P_a}$.

Method 2. The range **A** in Fig. 2 divides the whole line shape of both lowest line of C-PROXYL and lowest line of PTIO. The range **B** divides the second line of PTIO. The profile $\mathbf{P_B}$ that was obtained from range **B** was used as spatial information of PTIO. The profile $\mathbf{P_A}$ obtained from range **A** is including an amount of the lowest line of C-PROXYL and also the lowest line of PTIO. The lowest line of PTIO has half the area of the second line of PTIO. Therefore, spatial information of C-PROXYL was obtained by subtraction of half of the area of $\mathbf{P_B}$ from $\mathbf{P_A}$.

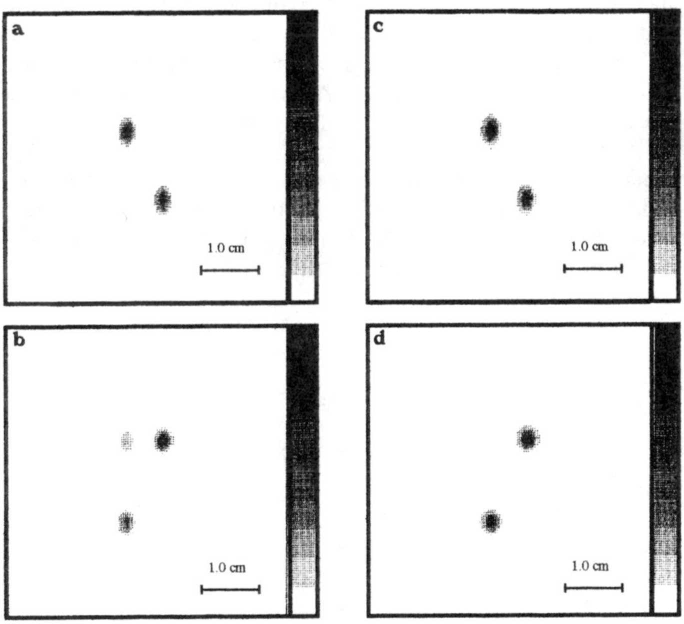

Fig. 3. CT images (2D distributions) of each radical species on 5 cm × 5 cm space on X-Z plane, a) and b) were obtained by the separation method 1. c) and d) were obtained by the method 2. a), c) were of PTIO and b), d) were of C-PROXYL.

The spatial information for the other three directions was obtained by the same procedure, and then two sets of CT images of two components were reconstructed independently. Fig. 3a and 3b are reconstructed from the spatial information obtained by method 1, and Fig. 3c and 3d are those of method 2. These CT images show good agreement with the arrangement of phantom. However, spatial information of PTIO was contaminated in the Fig. 3b. This means that point subtraction is not suitable and to obtain spatial information exactly, use of peak area is much reliable.

3 Liposomal and Free Spin Probe

Liposomes are good candidates as carriers for drug delivery systems. When spin probes are encapsulated into liposomes, pharmacodynamics of liposomes can be estimated in the living mouse using L-band ESR spectroscopy. The free spin probes released from liposomes that contain high concentration of spin probes can also be distinguished by the line shape analysis of ESR spectra, since ESR spectra depend strongly on the concentration of spin probe. The authors applied the above imaging method to the phantom including liposomal high concentration and free low concentration of spin probe solution.

3.1 Experimental

Materials. CAT-1 (4-trimethylammonium-2,2,6,6-tetramethylpiperidine-1-oxyl, iodide) was purchased from Molecular Probes Inc (Junction City, OR). Presome (type C-1; hydrogenated egg PC / cholesterol = 1 / 1) was a gift from Nippon Fine Chemical Co. Ltd. (Osaka, Japan). An aqueous solution of 150 mM CAT-1 was encapsulated into liposomes of Presome C-1. A Solution of 2 mM CAT-1 was used as the free probes. The phantom was made as shown in Fig. 4a.

ESR Measurement. Imaging data were collected with an L-band ESR spectrometer (JEOL Ltd., Tokyo, Japan) equipped with field gradient coils. The conditions of ESR measurements were as follows; modulation frequency was 100 kHz, modulation width was 0.16 mT, microwave frequency was 1223 MHz, microwave power was 2 mW.

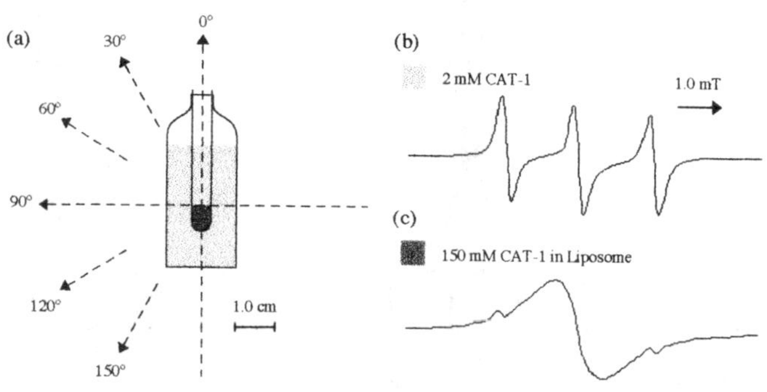

Fig. 4. Phantom and L-band ESR spectra of components. a) Phantom. b) the spectrum of 2 mM CAT-1. c) the spectrum of liposomal 150 mM CAT-1.

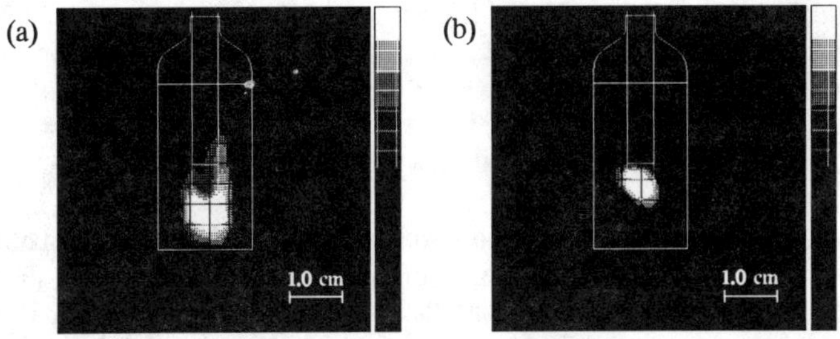

Fig. 5. 2D distribution reconstructed. a) Free CAT-1. b) Liposomal CAT-1.

Spectral-Spatial Image Reconstruction. Six spectral-spatial images were obtained with six directions in Fig. 4a. Each spectral-spatial image was reconstructed with 13 projections between -72° and +72° over 6 cm / 6 mT plane. The spectral-spatial images were reconstructed using the combination of ART with FBP.

CT Image Reconstruction. Spatial information of the liposomal and the free CAT-1 were obtained from the spectral-spatial images. Each CT image was reconstructed from six spatial information by the combination of ART with FBP.

3.2 Results and Discussion

The 150 mM of CAT-1 gave a broad single line (Fig. 4c), but the 2 mM gave sharp triple lines (Fig. 4b). In this case, the single line was overlapping with the all lines of triplet. The peak area of whole magnetic width was corresponding to the amount of whole CAT-1. To obtain the amount of free CAT-1, the area of the lowest line of the triplet was evaluated by subtracting a trapezoid area from the observed peak. The trapezoid area was an approximated back ground and a portion originated from a single broad line. Then, the amount of free CAT-1 was multiplied by 3. The amount of liposomal CAT-1 was obtained by subtraction of that of free CAT-1 from whole CAT-1.

2D distributions shown in the Fig. 5 were roughly coincided with the phantom. This imaging technique is applicable for the liposomal drug delivery system, and be

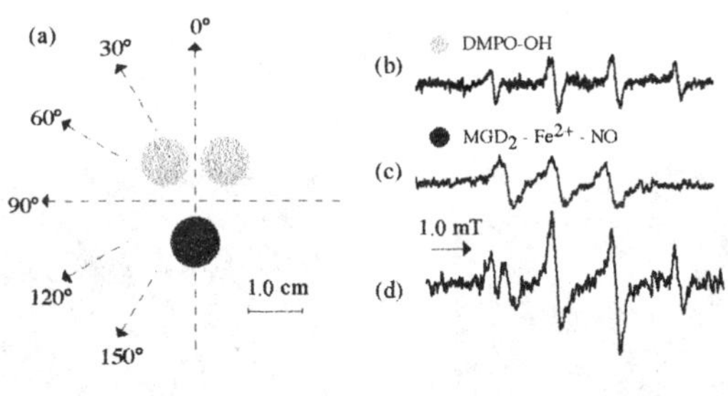

Fig. 6. Phantom and L-band ESR spectra of components. a) Phantom. b) a spectrum of DMPO-OH. c) a spectrum of MGD_2-Fe^{2+}-NO. d) superimposing b) and c).

able to obtain the pharmacokinetics of liposomes and release of encapsulated drugs simultaneously.

4 Nitric oxide (•NO) and hydroxyl radical (•OH)

The •NO and the most of reactive oxygen species (ROS) are free radicals. They appear to transact and have some important roles in living organisms. Thus, the authors tried simultaneous imaging of the •NO and the •HO that was one of major ROS.

4.1 Experimental

Materials and Method. The •NO gas was bubbled into MGD_2-Fe^{2+} (Fe^{2+}-*N*-methyl-D-glucamine dithiocarbamate) solution to prepare MGD_2-Fe^{2+}-NO complex. The •OH were synthesized by Fenton reaction and trapped with DMPO (5,5-dimethyl-1-pyrroline-*N*-oxide, Dojindo Laboratories, Kumamoto, Japan) to obtain DMPO-OH. The phantom was made as Fig. 6a. The ESR and the spectral-spatial imaging conditions were the same as those described in the section 3.1. The spatial information of DMPO-OH and MGD_2-Fe^{2+}-NO were separated from the spectral-spatial images. The CT images were reconstructed by the same way described above.

4.2 Results and Discussion

The L-band ESR spectra of DMPO-OH and MGD_2-Fe^{2+}-NO, under no field gradient are shown in Fig. 6b and 6c, respectively. The aqueous solution of DMPO-OH gave four lines (g = 2.007, peak height ratio = 1:2:2:1) and MGD_2-Fe^{2+}-NO gave a triplet of lines (g = 2.04, peak height ratio = 1:1:1). Fig. 6d shows the spectrum of the phantom containing both radicals. The fourth line of DMPO-OH was spectrally separated from MGD_2-Fe^{2+}-NO. Thus spatial information of DMPO-OH was obtained from the peak area of the fourth line of DMPO-OH. The spatial information of MGD_2-Fe^{2+}-NO should be obtained with a subtraction of superimposing DMPO-OH signal from the lowest lines. 2D distributions of two radicals were independently reconstructed with six projections.

The position and scale of reconstructed images of DMPO-OH and MGD_2-Fe^{2+}-NO were well fitted with those of the phantom (Fig. 7). The result suggests that this technique makes simultaneous assessment of distribution of •NO and ROS possible.

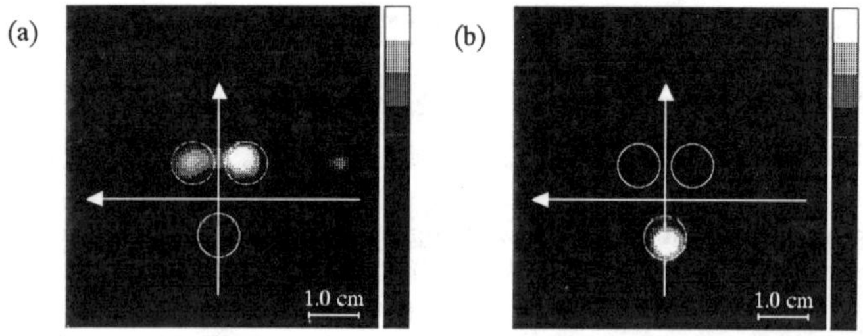

Fig. 7. 2D distribution reconstructed. a) DMPO-OH. b) MGD_2-Fe^{2+}-NO.

5 Conclusion

The present experiment demonstrates the possibility of ESR-CT for plural signal components. The separation of the spatial information used in this paper is the easiest technique. It may be not available for complicated spectra, for example, several broad components which are distributing inhomogeneously and overlapping each other. In that case, other calculation procedures such as deconvolution of overlapping spectra should be introduced for the separation of plural signals. This could be feasible using spectral simulation and line fitting.

References

1. W. Karthe, E. Wehrsdorfer, The Measurenent of Inhomogeneous Distributions of Paramagnetic Centers by Means of EPR, *J. Magn. Reson.* **33** (1979) 107-111.
2. M. J. R. Hoch, A. R. Day, Imaging of Paramagnetic Centres in Diamond, *Solit State Commun.* **30** (1979) 211-213.
3. K. Ohno, A Method of EPR Imaging: Application to Spatial Distributions of Hydrogen Atoms Trapped in Sulfuric Acid Ices, *Jpn. J. Appl. Phys.* **20** (1981) L179-L182.
4. H. Katsuda, T. Kobayashi, H. Saito, T. Matsunaga, M. Ikeya, Electron Spin Resonance Imaging of Mouse B16 Melanoma, *Chem. Pharm. Bull.* **38** (1990) 2838-2840.
5. T. Herrling, N. Klimes, W. Karthe, U. Ewert, B. Ebert, EPR Zeugmatography with Modulated Magnetic Field Gradient, *J. Magn. Reson.* **49** (1982) 203-211.
6. M. M. Maltempo, Differentiation of Spectral and Spatial Components in EPR Imaging Using 2-D Image Reconstruction Algorithms, *J. Magn. Reson.* **69** (1986) 156-161.
7. L.J. Berliner, H. Fujii, Magnetic Resonance Imaging of Biological Specimens by Electron Paramagnetic Resonance of Nitroxide Spin Labels, *Science* **227** (1985) 517-519.
8. S. Ishida, S. Matsumoto, H. Yokoyama, N.Mori, H. Kumashiro, N. Tsuchihashi, T. Ogata, M. Yamada, M. Ono, T. Kitajima, et. al. An ESR-CT imaging of the head of a living rat receiving an administration of a nitroxide radical, *Magn. Reson. Imaging* **10** (1992) 109-14.
9. M. Alecci, S. Colacicchi, P. L. Indovina, F. Momo, P. Pavone, A. Sotgiu, Three-dimensional in vivo ESR imaging in rats, *Magn. Reson. Imaging* **8** (1990) 59-63.
10. K. Takeshita, H. Utsumi, A. Hamada, ESR measurement of radical clearance in lung of whole mouse, *Biochem. Biophys. Res. Commun.* **177** (1991) 874-80.
11. T. Masumizu, T. Tatebe, E. Mutou, H. Utsumi, A. Hamada, Trial manufacture of ESR-CT apparatus, *Magn. Reson. Medicine* **2** (1991) 63-68.
12. M. M. Maltempo, S. S. Eaton, G. R. Eaton, Spectral-Spatial Two-Dimensional EPR Imaging, *J. Magn. Reson.* **72** (1987) 449-455.
13. M. M. Maltempo, S. S. Eaton, G. R. Eaton, Reconstruction of Spectral-Spatial Two-Dimensional EPR Images from Incomplete Sets of Projections without Prior Knowledge of the Component Spectra, *J. Magn. Reson.* **77** (1988) 75-83.

14 K. Matsumoto, A. Hamada, H. Utsumi, Simultaneous ESR Imaging of Spin Probes Encapsulated into and Released from Liposomes, *Magn. Reson. Medicine* **6** (1995) 71-73.

15. K. Matsumoto, H. Utsumi, Simultaneous ESR Imaging of •NO and •OH, *Magn. Reson. Medicine* **8** (1997) 87-90.

Magnetic Resonance in Intermetallides Co_5Zn_{21} and Mn_5Zn_{21}

Yu.A. Ryabikin, O.V. Zashkvara, V.D. Melichov

Physical-Technical Institute, MS-AS RK , 480082 Almaty, Republic of Kazakstan

Abstract. Υ-phases of solid solutions on basis of intermetallides Mn_5Zn_{21} and Co_5Zn_{21} were studied with the EPR-method irradiated at the temperature range 4 - 300K. Irradiated by neutron fluence $3.25 \cdot 10^{18}$ $n \cdot cm^{-2}$ Mn_5Zn_{21} at low temperatures demonstrates spin-glass behaviour. For Co-Zn solutions the EPR-spectra consist of single asymmetric lines for nonstoichiometric compositions, and for stoichiometric composition of the solution the EPR-spectrum is characterized by triaxial anisotropy. It was established that in this case ions Co^{2+} are in the field of slightly oblate tetrahedron.

1 Introduction

At the last time the intermetallides with ordered structure and alloys with their inclusions are considered as perspective construction materials working under extremal conditions [1].

In present report the results are given of investigation with EPR method of the γ-range alloys of Mn-Zn (13-18wt.%Mn) and Co-Zn (13-24wt.%Co) system, forming the solid solutions of manganese and zinc, and also cobalt and zinc on the basis of intermetallides Mn_5Zn_{21} and Co_5Zn_{21}, respectively.

The alloys Mn-Zn and Co-Zn with stoichiometric composition have complex cubic structure with 52 atoms in elementary cells disposed at lattice points in ordered manner. Known literature data [2] about X-rays diffraction investigations of Fe_5Zn_{21}, Ni_5Zn_{21} , possessing the same type crystal lattice ($D8_{1-3}$), permit certainly to suggests that from 10 atoms manganese and cobalt, 7 ones occupy tetrahedral positions, two ones are in octahedral surroundings and one - in cubic surroundings. In such a way for stoichiometric composition the more part of manganese and cobalt ions in the field of tetrahedral symmetry.

Magnetic behaviour of these systems is in general due to ions Co^{2+} and Mn^{2+}. In spite of that though crystall structure of studied intermetallides is practically identical, their EPR behaviour, though having some similar features, significantly differs one from another.

2 Experiment

EPR measurements were carried out by homodyne type spectrometer working in

X-band. Irradiation of monolittic samples Mn_5Zn_{21} by neutrons was performed in WWRK-reactor by fluences: $9.06 \cdot 10^{14}$; $1.37 \cdot 10^{16}$; $8.8 \cdot 10^{17}$; $3.25 \cdot 10^{18}$; $8.8 \cdot 10^{18} n \cdot cm^{-2}$ (E>2.35 Mev) at the temperature not higher than 353K.
To obtain EPR spectra fine-dispersive (~50μm) powder of intermetallide Mn_5Zn_{21}, prepared from irradiated plates and sealed up in quartz capillaries with 1.0mm diameters was put to spectrometer resonator.

3 EPR in Mn_5Zn_{21}.

The EPR method is sufficient widely used for study of intermetallides with various composition and structure [3]. For example, in paper [4] results were given of study of more simple, than we use, compound Mn-Zn with monoclinic lattice and 28 atoms in elementary cell. In this alloy the EPR signal was observed only at the temperatures lower than 120K. For Mn_5Zn_{21} alloy samples sufficiently intensive EPR signal is registered at room temperature too. On fig.1 (curve 1) there is experimental dependence of the EPR signal intensity for the samples of γ-phase of solid solutions on Mn_5Zn_{21} intermetallide basis on manganese concentration in the alloy. Region of γ-phase homogeneity is limited by the interval 14-18 wt.% within which there is the alloy of stoichiometric composition Mn_5Zn_{21} (16,7 wt.%). One can see from fig. 1 (curve 1) that EPR signal intensity firstly increases with manganese concentration growth achieving maximal value in stoichiometric composition region, and then, with further increase of manganese concentration, begins to decrease. Further secondary growth of the signal intensity at manganese concentrations higher 18 wt% is connected with the appearance in the solution the particles of ferromagnetic β-phase, and below it will not be discussed.

At consideration of increasing part of curve 1 attention is attracted to the fact that the signal increase occures significantly faster than it may be expected only due to manganese concentration increase. Calculating straight line at fig.1 (curve 2) demonstrates possible increase of the EPR signal for the alloy due to only rise of manganese concentration. We suppose that complementary increment of the EPR signal intensity with concentration increase may be conditioned by the arising of exchange ferromagnetic interaction between separate pairs of manganese ions. One could expect that at the region «higher» than alloy stoichiometric composition the increase of manganese atom numbers in intermetallide lattice leads that supplemented atom spin forms antiferromagnetic connection with one of spins of pair which, in result, is destroyed. Sequence of this is the decrease of cluster number in which spins were connected by ferromagnetic interaction and, respectively, decrease of the EPR signal intensity. The signal intensity decrease with manganese concentration growth can be also explained by suppose that again formed clusters demonstrate the antiferromagnetic interaction with already available clusters. The presence of ferromagnetic interaction at room temperature between separate manganese atoms in the alloy doesn't mean that the whole sample is in ferromagnetic state. Really, with temperature decrease as for irradiated samples as for non-irradiated ones it is occurred the growth of the EPR signal intensity, the lines

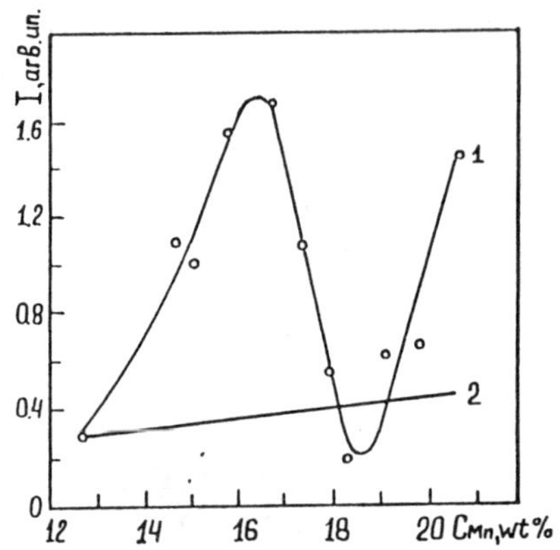

Fig. 1. Dependence of signal intensity in Mn-Zn compound on manganese concentration . 1 - experimental curve; 2 - calculating dependence.

became some times as broad and move to region of weak fields. Such a behaviour of EPR lines with temperature decrease is typical for the system of magnetic moments undergone the ferromagnetic ordering [4]. Let us consider given on fig.2 (curves 2-5) experimental dependences of the EPR line width on irradiation fluence for some temperature values. On fig. 2 (curve1) it is also given lattice parameter change with irradiation fluence. One could see that at room temperature the EPR line width at maximal fluence $8.8\cdot10^{18}n\cdot cm^{-2}$ increases almost twofold. Practically linear dependence of line width on irradiation fluence tells that localized moments of manganese ions interact with point radiation defects. It could be expected that at initial irradiation dozes the experimental EPR line width is conditioned by portional contributions of line widths from alloy's γ- and ε-phases, as from constantly increasing, with doze, component related to radiation defects. Particles of ε-phase arising in γ-phase of Mn-Zn alloy as a result of radiative stimulated transition $\gamma\rightarrow\varepsilon$ at irradiation fluences $9.06\cdot10^{14}$-$8.8\cdot10^{17}n\cdot cm^{-2}$ were discovered with X-ray diffraction method. At more high irradiation fluences it occures dissolution of particles of ε-phase. Since the width of the EPR line of ε-phase is equal 43mT, and of γ-phase - 28mT, that small deviation to the decreasing direction from linear growth of the line width at irradiation fluence $3.25\cdot10^{18}$ $n\cdot cm^{-2}$ may be conditioned only by dissolution of ε-phase. Upon that again appearing particles of γ-phase, as adjoining to them regions, can be considerably disordered due to this transitions. It is confirmed, as one could see on fig.2.(curve 2), by remarkably change of intermetallide lattice parameter at irradiation by fluence $3.25\cdot10^{18}$ $n\cdot cm^{-2}$. At the increase of irradiation fluence the lattice parameter almost completely returns to initial value. At temperature decrease on doze dependences of line width with fluence $3.25\cdot10^{18}$ $n\cdot cm^{-2}$ a minimum appears which value increases with temperature decrease. Such a behaviour of doze dependences of line width permits to suppose that magnetic behaviour of investigated alloy irradiated by fluence

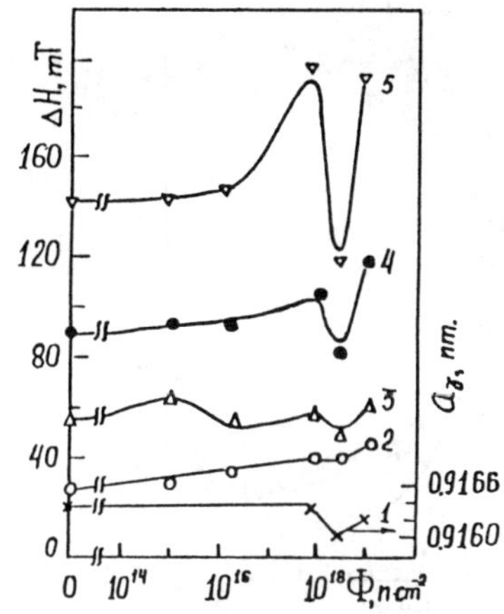

Fig. 2. Dependence of line width (left ordinate) at some temperatures and lattice parameter (right ordinate) on irradiation fluence. 1 - lattice parameter; 2 - 300 K; 3 - 130 K; 4 - 70 K; 5 - 30 K.

$3.25 \cdot 10^{18}$ n·cm^{-2} at low temperatures remarkably differs from its behaviour at another irradiation dozes.

Really, presented on fig. 3 the temperature dependences of the EPR line width for fluence $3.25 \cdot 10^{18}$ n·cm^{-2} and $8.8 \cdot 10^{18}$ n·cm^{-2} have essential differences. If the EPR line width of the sample irradiated by fluence $8.8 \cdot 10^{18}$ n·cm^{-2} as well as non-irradiated, increases monotonically with temperature decrease, the line width of the sample irradiated by fluence $3.25 \cdot 10^{18}$ n·cm^{-2} at the temperature 20K achieves a maximal value and then decreases with temperature decrease. Similar behaviour of the EPR line width was observed for alloys AuFe [5] and FeMn [6], which was interpreted as transition of these alloys to cluster spin glass state at the temperatures corresponding to the EPR line width maximum. One could consider the temperature behaviour of signal intensity, presented on fig.3 (curve 1), as some complementary confirmation of the transition of Mn_5Zn_{21} alloy to cluster spin glass state at the temperature T_g=20K. If maximum at T~70K can be explained by partial transition to antiferromagnetic state, the maximum at 20K shows the transition to cluster spin glass state as a result of competition between the ferromagnetic interaction and antiferromagnetic one. Sharp fall of the EPR signal intensity below 20K indicates the disordering of magnetic moments of clusters and their freezing with random orientation. Thus the irradiation of Mn_5Zn_{21} alloy by neutron fluence $3.25 \cdot 10^{18}$ n·cm^{-2} leads to its strong structural disordering which promotes the transition to spin glass state at the temperature T_g=20K. As far as we known, it is first when the transition to spin state due to radiation effect on metallic alloy is observed.

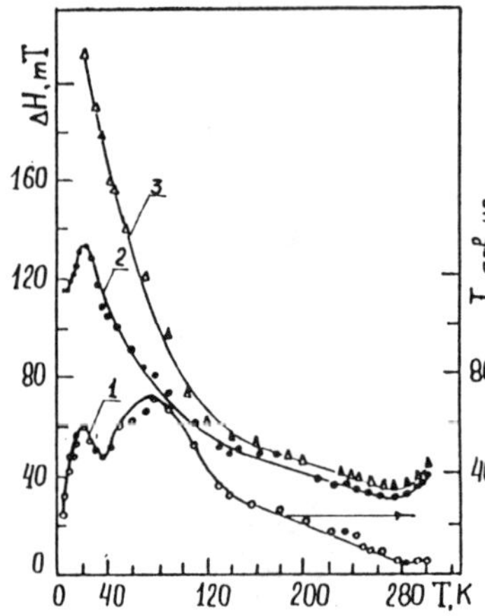

Fig. 3. Dependence of line width on temperature (left ordinate). 2 - sample irradiated by fluence $3.25 \cdot 10^{18}$ n·cm^{-2}; 3 - sample irradiated by fluence $8.8 \cdot 10^{18}$ n·cm^{-2}; and 1 - signal intensity at sample irradiated by fluence $3.25 \cdot 10^{18}$ n·cm^{-2} (right ordinate).

4 EPR in Co_5Zn_{21}.

In investigated system Co-Zn, as in some other cobalt-containing metallic compounds [7-8], the EPR is observed at room temperature.

The EPR spectra of investigated compound Co-Zn at some cobalt concentrations are given on fig.4. To the exclusion of spectrum of the sample with 17.4wt.% cobalt content, all rest spectra in the region of cobalt concentration change 13 - 20wt.% consist of single asymmetry lines. And in the stoichiometry region (17,4wt.% of cobalt) the spectrum has a complex form on which basis it can be conclude that Co^{2+} ions are in the field with triaxial anysotropy. The form of experimental spectrum of system Co-Zn in the region of stoichiometric composition is well described by spin-Hamiltonian

$$H = g_x' \beta H_x S_x' + g_y' \beta H_y S_y' + g_z' \beta H_z S_z' \tag{1}$$

with effective spin $S'=1/2$ and g'-tensor component experimentally defined from the spectrum: $g'_x=5.63$; $g'_y=4.90$; $g'_z=1.95$. Effective spin values indicate that observed spectrum can be conditioned by the transitions within lower Kramers doublet which appeares as a result of influence to level ^4F of tetrahedral field, of spin-orbital interaction and low-symmetry component of crystalline field. True spin of bivalent cobalt ion is $S=3/2$ and total Hamiltonian, describing the lower doublet behaviour, is

$$H = g_x \beta H_x S_x + g_y \beta H_y S_y + g_z \beta H_z S_z + D[S_z^2 - \frac{1}{3} S(S+1)] + E(S_x^2 - S_y^2) \tag{2}$$

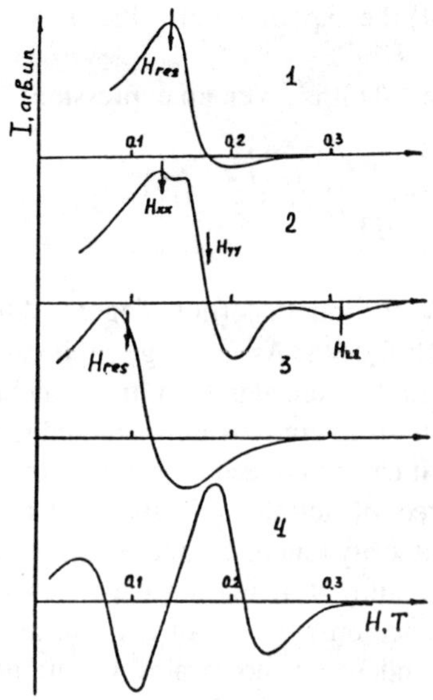

Fig. 4. EPR spectra of Co-Zn compound at some cobalt concentration: 1 - 13 wt.%; 2 - 17.4wt%; 3 - 18 wt%; 4 - 24 wt%.

Theory, describing the behaviour of cobalt ion in the tetrahedral symmetry field, is sufficiently worked out in papers [9-11].

Connection between g-tensor components in Hamiltonians (1) and (2) is described by expressions:

$$\begin{aligned} g_x' &= g_x(C_2^2 + \sqrt{3}C_1C_2) \\ g_y' &= g_y(C_2^2 - \sqrt{3}C_1C_2) \\ g_z' &= g_z(C_2^2 - 3C_1^2) \end{aligned} \tag{3}$$

The anisotropy of experimentally observed g'-factor is determined by intermixing of spin states to most degree than by anisotropy of orbital contribution which can lead to various values of g_x and g_y. This permits, with rather good approximation, to suppose that $g_x = g_y = g_\perp$.

In these conditions for C_2 from (3) it can be obtained the followed expression:

$$C_2^2 = \frac{3(g_x'+g_y')^2}{(g_x'-g_y')^2 + 3(g_x'+g_y')^2} \tag{4}$$

which permits to evaluate degree of mixing between higher and lower Kramers' doublets . By substituting to (4) the experimental values g'_x=5.63; g'_y=4.90 we obtain that C_2^2=0,9984.

From [9-11] for relation E/D it is given an expression:

$$\frac{E}{D} = \pm\frac{1}{\sqrt{3}}\{[1/(2C_2^2-1)^2]-1\}^{1/2} \tag{5}$$

which permits to evaluate tetrahedral distortion degree. After substitution to (5) obtained value C_2^2 we get E/D=0,0448. As far as $g_\perp > g_\parallel$ it can be asserted that there is a small compression of tetrahedron. Thus in the stoichiometric region when intermetallide Co_5Zn_{21} lattice is structurally wholly ordered Co^{2+} ions are in the field of slightly oblate tetrahedron. It can be noticed that X-ray structural analysis does not permit to fix so little degree of tetrahedral distortions. On fig. 5 there are dependences of EPR signals intensity (curve 1) and asymmetry parameter (curve 3) on cobalt concentration in the alloy. Calculating dependence shows the change of EPR line intensity in dependence only on cobalt concentration in the sample. It is obvious that experimental dependence demonstrates deviation from this straight line at 16wt.% Co. At 20wt.% Co the experimental intensity exceeds almost eightfold the calculated one and at concentration 23-24wt.% this exceeding is already about two orders. Such a behaviour of signal intensity for Co^{2+} ions essentially differs from concentration dependence for ions Mn^{2+} on fig.1. The dependence of signal intensity on cobalt concentration can be explained by supposing that at C=16wt.% it begins to form the clusters in which ions Co^{2+} are connected by ferromagnetic interaction. At cobalt concentration C=20wt.% and higher in the alloy it becomes apparent the ferromagnetic interaction already between clusters. The spectrum for cobalt concentration C=24wt.% as it is shown from fig.4 consists of two lines. This permits to conclude that in the region of high cobalt concentration it occures formation of two magnetic sublattices within which ions interact ferromagnetically.

Presented on fig.5 (curve3) dependence of asymmetry parameter on cobalt concentration shows the change of this parameter in wide range: from 8 at 13wt% Co to 2 at 20wt% Co . It is known that for paramagnetic ions in metals this parameter must be close to 2.55. This parameter value at cobalt concentration 13wt%-14wt.% is more characteristic for conduction electrons diffusing through skin depth during ~10^{-10}sec. It is difficult to suppose that the Co^{2+}ions could migrate with such a speed from point to point of lattice but there is information about anomalously high diffusion of cobalt atoms , for example, in its alloy with zirconium [12].

As possible explanation of large value of asymmetry parameter one could suppose that as a result of d-s spin exchange the fixed cobalt ion becomes as if «migrating» through skin depth. It can be also expected that on the signal asymmetry due to mixture of absorption and dispersion signals because of skin depth influence it is superimposed the asymmetry because of axial distortion. However, all these assumptions require additional investigations.

Low temperature investigations show that for the stoichiometric composition the spectrum form of Co-Zn alloy is significantly dependent on temperature change direction at getting of the dependence. At temperature decrease all components of

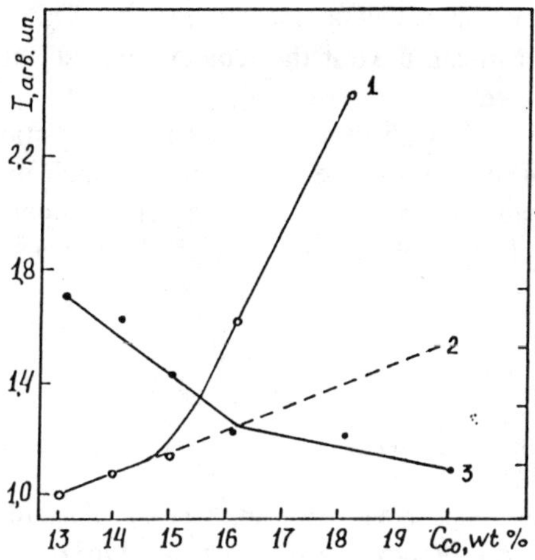

Fig. 5. Dependence on cobalt concentration in Co-Zn compound : 1 - of experimental intensity of EPR lines; 2 - of calculating intensity of EPR lines; 3 - of asymmetry parameter of EPR line.

the spectrum begin to broaden and move to weak field range, peak corresponding to g_x' moving faster than g_y' peak and at temperature lower 150K it becomes non-registered.

The peak, corresponding to g_z', at temperature decrease strongly broadens too and moves in direction to g_y' peak. Below 150K it disappeares. When spectra are got at temperature increase their forms is repeated up to 30-40K. At more high temperatures it is apparently affected some delay in reconstruction of magnetic moment system that shows in the change of spectra forms in comparison with those which were registered at temperature increase. Restoration of spectrum form at the room temperature occures approximately trough round the clock. The cooling seems to give unusual changes in intermetallide structure and the restoration of its structure requires prolonged anneal.

5 Conclusion

Concentration dependences of solid solutions were investigated by the EPR method in γ-phase range on the basis of intermetallides Mn_5Zn_{21} and Co_5Zn_{21}. For compound Mn-Zn in the region of manganese concentration change up to stoichiometric composition it is found the exhibition of the exchange ferromagnetic interaction between separate atoms, and after stoichiometry with manganese concentration increase the antiferromagnetic interaction becomes apparent. As for Co-Zn alloys, in whole region of cobalt concentration increase it occures the raising of ferromagnetic interaction. At maximal cobalt concentration, two sublattices of ferromagnetically interacting cobalt atoms are formed. On irradiated by neutron fluence $3.25\cdot10^{18} n\cdot cm^{-2}$ Mn-Zn alloy at the temperature 20K it was discovered the transition to spin glass state.

The EPR spectrum of cobalt in stoichiometric composition of Co-Zn alloys is characterized by triaxial anisotropy, the exhibition of which in metallic system has been found first. The change of this spectrum form with temperature decreasing tells about low- temperature changes of Co_5Zn_{21} intermetallide structure parameters.

References.

1. J. Tobolski, Intermetallische phasen als basis fur neue structurwerkstoffe, *Metall,* **41** W.-Berlin (1987) 134-138
2. Anders Johansson, Hans Ljang, Sven Westman, X-rays and neutron diffraction studies on Γ-Ni,Zn and Γ-Fe,Zn, *Acta Chemica Scandinavica.* **22** (1968) 2743-2753.
3. R.H. Coles Electron spin-resonance studies of the onset of magnetic order in intermetallic compounds, *J.Phys.F.* **5** (1975) 121-142.
4. B.R. Coles, B.V.B. Sarkissian, R.H. Taylor, The role of finite magnetic cluster in Au-Fe alloys near percolation concentration, *Phil. Mag.**B*** **37** (1978) 489-498.
5. A.D. Caplin, J.B. Dunlop, Dilute intermetallic compounds: property of zinc transition metal phases, *J.Phys.F.* **3** (1973) 1621-1647.
6. I.F. Rosenbaum, L.W. Rupp, G.A. Thomas, W.M. Walh, H,S, Chen, J.R. Banavar, Electron spin resonance indication of the ferromagnet-spin glass transition in amorphous FeMn alloys, *Solid State Comm.* **42** (1982) 725-727.
7. R.C. Barnes, D.A. Cornel, D.R. Torgenson Paramagnetic Resonance of Cobalt in Intermetallic Compounds, *Phys. Rev.Lett.* **16** (1966) 233-234.
8. V.P. Kalantarian, V.I. Chechernikov, Interpretastiya Magnitnykh svoistv splavov i soedinenii perekhodnykh metallov na osnove ikh zonnoi structury, *Electronnay structura perekhodnykh metallov, ikh splavov i intermetalicheskikh soedinenii,*Naukova Dumka, Kiev (1979) 85-89.
9. M.C. Chen, Y.O. Artman, Electron Paramagnetic Resonanse of Co Doped $PbWO_4$, *Phys.Rev.* **187** (1969) 723-732.
10. Yu.V. Yablokov, V.K. Voronkova, V.Ph. Shishkov, A.V. Ablov, J.Yu. Vaisbein, Paramagnitnyi resonans dvukh valentnogo Cobalta v tetraedricheskom okrushchenii, *Fisika Tverdogo Tela* **13** (1971) 997-1002.
11. V.K. Voronkova, M.M. Zaripov, Yu.V. Yablokov, A.V. Ablov, M.A. Ablova, Paramagnitnyi rezonans Co(II) iskashchennoij tetraedricheskoi koordinastii, *Doklady* AN *USSR* **214** (1974) 377-380.
12. W. Petry, G. Voge, A. Keigemann, K.H. Streinmetz, Anomalously fast diffusion of Cobalt in β - zirconium., *Phylosoph. Magaz. A* **55** (1987) 183-201.

Minima in Temperature Variations of the Electron Spin Phase Memory Time T_M Due to Molecular Motions – ESE Studies in Single Crystals

Stanisław K. Hoffmann, Wojciech Hilczer, Janina Goslar, and Maciej Gramza

Institute of Molecular Physics, Polish Academy of Sciences, Smoluchowskiego 17, PL-60179 Poznań, Poland

Abstract. Temperature variations of the phase memory time T_M, describing a dephasing of the spin packets of inhomogeneously broadened EPR lines, are reviewed and discussed for molecular crystals containing a small amount of Cu(II) ions or free radicals. The variations due to the radical motions, librational and rotational motion of a matrix nuclei, cross-relaxation and spin lattice relaxation processes are considered. Activation energy and correlation times of thermally activated motions are calculated. Resonance type minima produced by hindered rotations of CH_3 and NH_3 groups, and flipping motion of the glycine molecules are discussed. It is concluded that T_M temperature behaviour is an analogue of the behaviour of nuclear spin-lattice relaxation time T_1 in studies of local motions in crystals.

1 Phase Relaxation and Phase Memory Time in the Spin Packets Model

EPR lines of magnetically diluted dielectric solids are inhomogeneously broadened and consist of unresolved spin packets of different Larmor precession frequencies (Fig. 1a). Spin relaxation is not described by Bloch equations in such case. The free induction decay (FID) signal after the line excitation by a microwave pulse decreases exponentially with characteristic time T_2^* related to the line width $\Delta B_{line} = 1/T_2^*$ (Fig. 1b).

For inhomogeneous lines an electron spin echo (ESE) signal can be formed by a series of microwave pulses. In a two-pulse experiment the Hahn-type ESE can be

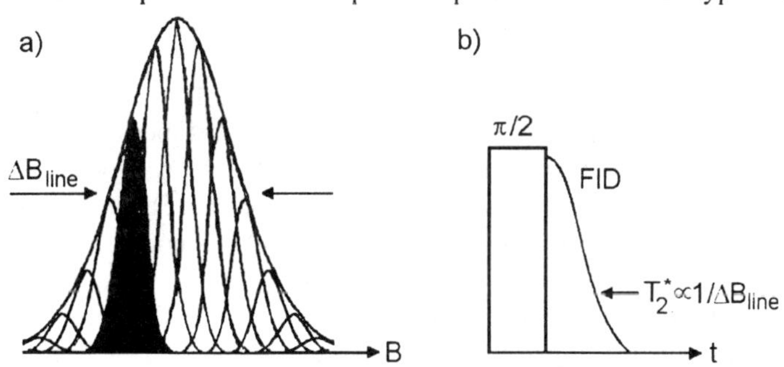

Fig. 1. Inhomogeneously broadened EPR line and its FID signal after $\pi/2$ pulse.

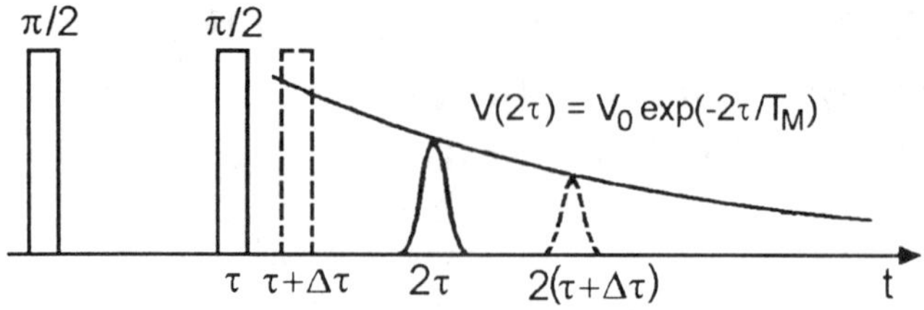

Fig. 2. Electron spin echo of Hahn-type excited in two pulse experiment. The echo amplitude decays exponentially with characteristic time T_M.

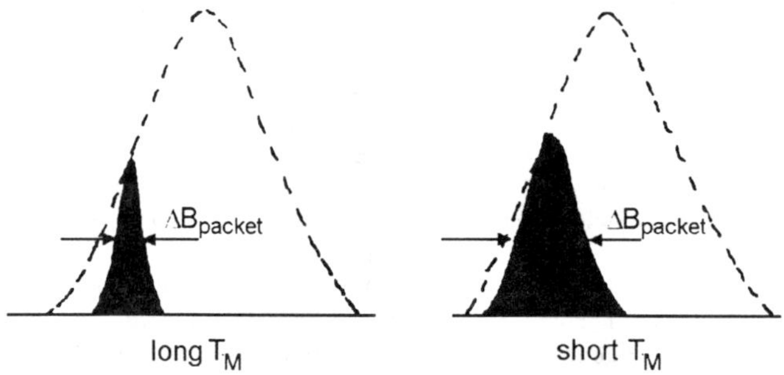

Fig. 3. Spin packet width and the phase memory time.

created by exciting and refocusing π/2 pulses with interpulse time interval τ. In such case at time 2τ ESE is formed (Fig. 2). When τ increases the ESE amplitude $V(2\tau)$ decreases. In most cases the decrease is exponential $V(2\tau) = V_0 \exp(-2\tau/T_M)$ with characteristic time T_M called the phase memory time. The Fourier transform of the decay function $V(2\tau)$ gives the spin packet shape of the width $\Delta B_{packet} = 1/T_M$ (Fig. 3). The spin packets are very narrow ($\Delta B_{packet} = 10$ μT) thus they are very sensitive to the spin and molecular dynamics as is reviewed in this paper.

2 Phase Relaxation Mechanism and Processes Causing ESE Amplitude Decay.

Two-pulse echo amplitude decay is exponential in many cases and can be written in a general form:

$$V(2\tau) = V_0 \exp\left(-a\tau^n\right) \tag{1}$$

where $n = 0.5$-3, and a-coefficient depends on a processes and mechanisms which govern the ESE decay. For $n = 1$ and $a = 2/T_M$ a single exponential decay is obtained as shown in Fig. 2. The processes and mechanisms are described in a book of Salikhov et al. (1976).[1]

In most cases the spectral width of a microwave pulse is not sufficient to excite the whole EPR line. Thus only part of the line is excited (spins A) and non-excited spins (spins B) exist in the sample. Only spins A take part in ESE signal formation because only their precessional motion can be refocussed. The number of spins A which are able to create ESE signal decreases in time because:

– the number of spins A decreases as a result of the spin-lattice relaxation processes or a spin diffusion converting spins A into B;
– a random fluctuations of the phase of the precessional motion through instantaneous and spectral diffusion producing fluctuating local magnetic fields in a crystal.
The phase relaxation processes are mutually independent and the decay function can be written as:

$$V(2\tau) = V_{ID} \cdot V_{SD} \cdot V_0 \tag{2}$$

where V_{ID} describes a contribution from instantaneous diffusion produced by the refocusing pulse (Kurshev, Ichikawa (1992)[2]), V_{SD} is a contribution from spectral diffusion within unpaired electron system (Brown (1979)[3], Salikhov and Tsvetkov (1979)[4]). The V_{ID} and V_{SD} depend on the paramagnetic centres concentration whereas the last term V_0 contains contributions from mechanisms which are concentration independent (Fig. 4) and dominate when interactions between unpaired electrons are negligible.

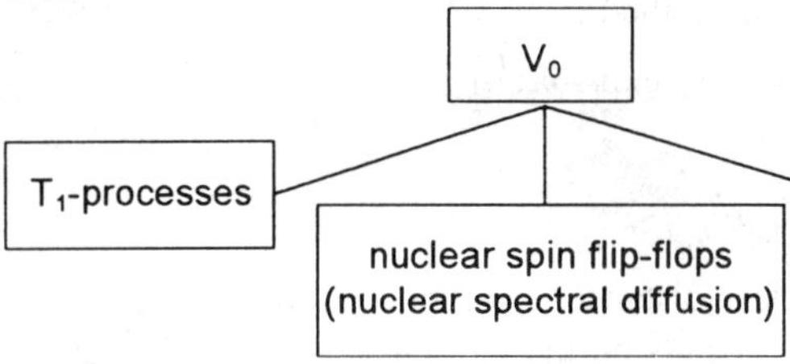

Fig. 4. Concentration independent mechanisms of the phase relaxation dominating in sufficiently diluted magnetic systems

In this paper the examples of crystals where V_{ID} and V_{SD} are overdominated by V_0 will be presented, allowing to observe an influence of the matrix structure and dynamics on T_M.

The decay function $V(2\tau)$, as a rule, is strongly modulated by a weak dipolar coupling of paramagnetic centre to the surrounding magnetic nuclei. The proton ^1H and nitrogen ^{14}N modulations are shown in Fig. 5. The relaxation function has to be subtracted from experimental modulated ESE amplitude decay, and beacause of this T_M can only be determined with a limited accuracy.

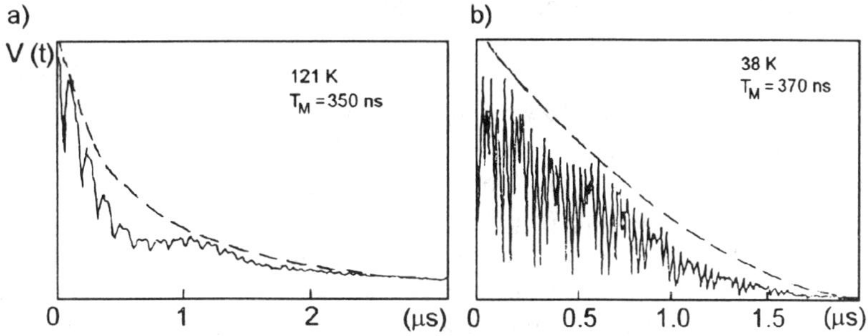

Fig. 5. Modulated experimental ESE amplitude decays for a) SeO_3^- radical in $(NH_4)_3H(SeO_4)_2$ crystal, b) Cu^{2+} ions in triglycine selenate crystal. The "pure" decay functions are plotted by dashed lines.

3 Magnetic Matrix Nuclei Contribution to the ESE Amplitude Decay

For low concentration of paramagnetic centres, lower than about 10^{18} spins/cm^3 or 10^{-4} M/dm^3, diffusion within electron spin system does not contribute to the phase relaxation. Then nuclear spins play the role of spins B and phase relaxation can be governed by nuclear spectral diffusion processes. Only nuclei weakly coupled to the unpaired electron can take part in the spectral diffusion processes. These are the nuclei in shells outside a radius r_s defined as a electron-nuclear distance at which electron-nuclear interactions equal the nucleus-nucleus interactions. This diffusive barrier r_s is of about 1 nm and can be calculated from the equation:

$$\Delta\omega_{nn} r_s^{\,3} = \tfrac{1}{2}\hbar\gamma_n\gamma_e\left|1 - 3\cos^2\theta\right| \tag{3}$$

where $\Delta\omega_{nn}$ is related to the NMR linewidth ($\Delta\omega_{nn} = 1/T_2$ nuclear). Nuclear spin diffusion causes two-pulse ESE decay, described by:

$$V(2\tau) = V_0 \exp\left(-m\tau^k\right) \tag{4}$$

where

$$m = 2\pi W_n c_n \left(3 r_n \gamma_n \gamma_e \hbar\right)^{3/4}. \tag{5}$$

In Eq. (5), r_n – average internuclear distance, c_n – concentration of nuclei, W_n – nuclear flip-flop rate. The index k depends on the interpulse distance τ, equals to $k = 3$ for short, and $k = 7/4$ for long τ (Salikhov et al. (1976)[1]).

4 Temperature Dependence of the Phase Memory Time T_M

The nuclear spectral diffusion is practically temperature independent as can be seen from equations (4) and (5). Pulsed EPR experiments on magnetically diluted crystals show, however, that T_M can vary with temperature, due to dynamical changes of the spin-packet width.

4.1 Motional Narrowing or Broadening of Spin Packets

Thermal *motions of paramagnetic centres* (radical molecule) having anisotropic g-factor and hyperfine splitting A_{hf} will produce time-dependent shifts $\Delta g(t)$ and $A_{hf}(t)$ which determine the spin-packet linewidth (Fig. 6a). An increase of the motion amplitude with temperature results in a continuous spin-packet broadening, and T_M will continuously *shorten* on heating (Fig. 6b).

When the spin packet width is determined by the local magnetic field fluctuations due to nuclear spectral diffusion, the *matrix molecular motions* can average out the local magnetic fields producing spin packet *narrowing*. As a result, a continuous increase in T_M can be observed (Fig. 6b).

Such behaviour was reported by Hilczer et al. (1993)[5] for SeO_3^- radical in $(NH_4)_3H(SeO_4)_2$ single crystal (Fig. 7). A decrease in T_M on heating at low tempera-

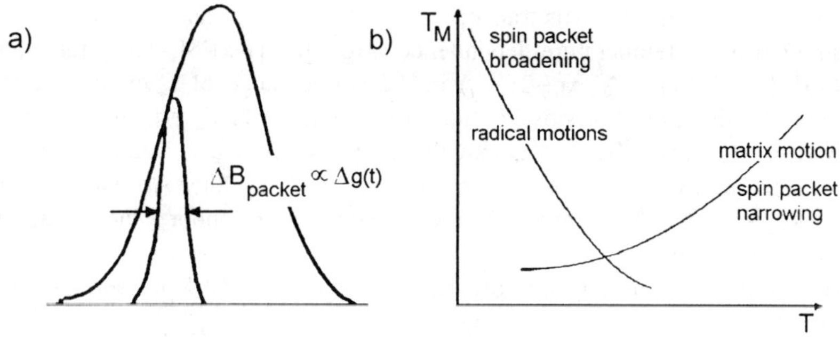

Fig. 6. Spin packet width ΔB_{packet} broadening and narrowing effect.

tures is due to an increase in the amplitude of the radical motion. This process is dominated by thermally activated (with small activation energy $E_a = 1.2$ kcal/mol) matrix motion above about 50 K. Above $T = 170$ K, spin-lattice relaxation time becomes very short and influences phase relaxation.

4.2 Spin Packet Merging Effect

When transitions with average rate $1/\tau_c$ are possible (τ_c – correlation time) between two spin packets of Larmor frequencies ω_1 and ω_2 then the merging effect can be observed, similar to that produced by quantum-mechanical exchange in cw-EPR of paramagnetic systems (Fig. 8a). When the transition rate increases, the packets broaden, then coalesce into a single broad one which continuously narrows with further in-

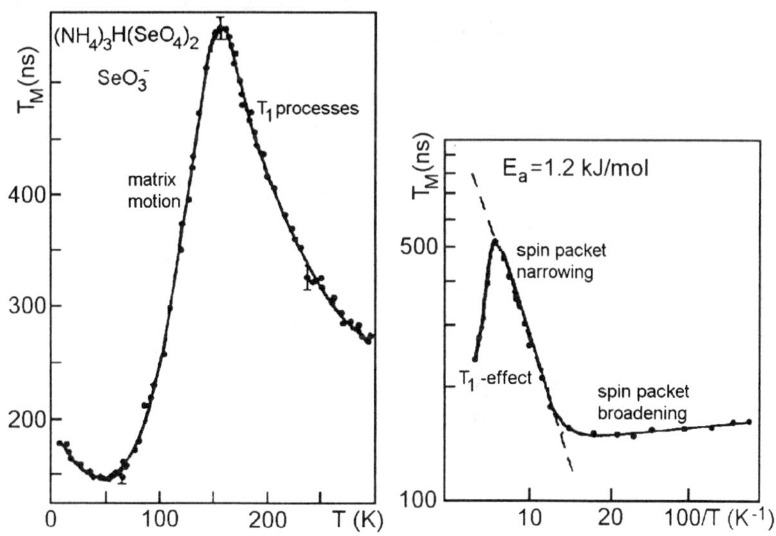

Fig. 7. The phase memory time temperature variations in linear and logarithmic scales. Radical motions determines T_M at low temperatures (below 50 K). For higher temperatures the matrix molecules motions continuously narrow spin packet. Above 170 K the T_1-processes dominate.

crease in the transition rate. This transition from slow to fast motion limit is observed as a minimum in T_M temperature dependence (Fig. 8b). The ESE decay function can be described as $V(2\tau) = V_0 \exp(-2\tau/T_M)$ in the whole range of τ_c except the region around the T_M minimum. For slow motion limit ($\Delta\omega\tau_c >> 1$) $T_M = \tau_c$, whereas for fast limit ($\Delta\omega\tau_c << 1$) $T_M = 2/(\Delta\omega^2\tau_c)$ as calculated by Tsvetkov and Dzuba (1990).[6] If a motion is thermally activated and τ_c is described by the Arrhenius equation $\tau_c = \tau_0 \exp(E_a/kT)$, $T_M(T)$-minimum allows to determine E_a and τ_c since at the minimum $T_M = \tau_c$ approximately.

Various molecular motions can produce minima in $T_M(T)$ dependencies. Minima

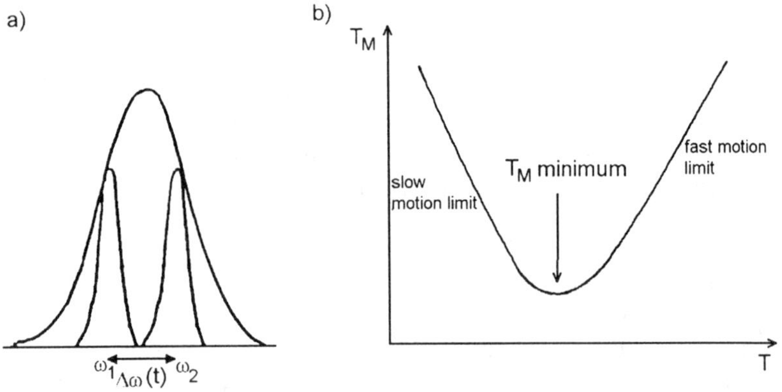

Fig. 8. „Exchange" process between spin packets produces a merging effect and T_M minimum in temperature dependence.

due to *hindered rotation of CH_3 groups* have been identified in many solids. Tsvetkov and Dzuba (1990)[6] observed that for nitroxyl radicals the depth of the T_M minimum is correlated with the number of methyl groups bonded to the radical molecule. The CH_3-group related minima were reported for Cr(V) complexes by Nakagawa et al. (1992)[7] and for Cu(II) complexes by Du et al. (1994).[8] As a rule such minima appear for temperatures higher than about 100 K, and their parameters can be verified by a comparison with proton spin-lattice relaxation data of NMR experiments.

We have found (Hoffmann et al. (1995)[9]) a minimum in $T_M(T)$ for X-ray generated glycine radical in diglycine nitrate (DGN) crystal (Fig. 9a). The minimum at $T = 170$ K is due to a hindered rotation of the NH_3 groups of the glycine molecule adjacent to the glycine radical with activation energy $E_a = 12$ kJ/mol and $\tau_c = 70$ μs at 170 K.

The motional type minimum of T_M was observed by Hoffmann et al. (1996)[10] for Cu(II) ions in triglycine selenate (TGSe) crystal at $T = 50$ K (Fig. 9b). The shape of the minimum is disturbed by spin-lattice relaxation processes (T_1-processes) for $T > 80$ K where T_1 is only one order of magnitude longer than T_M. For $T < 35$ K the effect of spin-packet broadening is visible due to a modulation of the g-tensor anisotropy by librations of the $Cu(gly)_2$ complex. We have assumed that the motion responsible for the T_M minimum is a flipping of the noncoordinated glycine molecule GI in the crystal unit cell. This is a relatively slow motion with $\tau_c = 38$ μs at 50 K and with a low activation energy $E_a = 1.0$ kJ/mol.

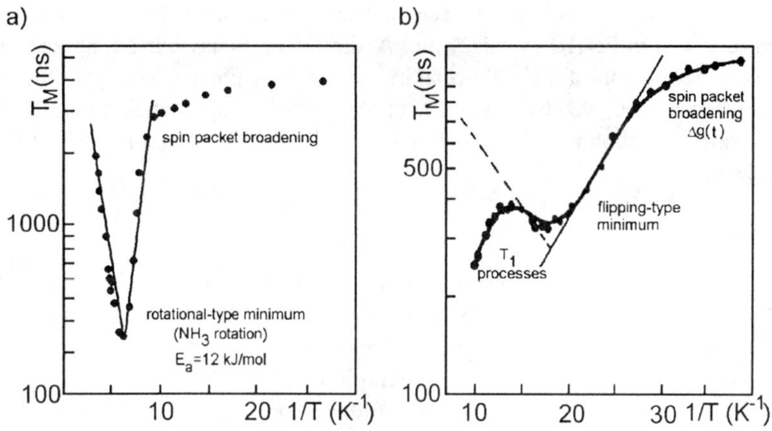

Fig. 9. T_M vs. $1/T$ for a) NH_3CH_2COO-radical in diglycine nitrate, b) Cu(II) ion in triglycine selenate crystal.

In the above described glycine crystals we have discovered that T_M depends on the crystal orientation and is different for different hyperfine lines.

4.3 Cross-relaxation Effects

Cross-relaxation type minimum was identified by Hilczer et al. (1995)[11] for T_M of the spin label covalently bonded to a cytochrome *c* molecule containing heme centre. When heme iron was paramagnetic Fe(III) a very deep minimum at $T = 40$ K was observed. After a chemical reduction of the iron ion to the diamagnetic form Fe(II) the minimum disappears (Fig. 10). It indicates that T_M minimum is due to a cross-

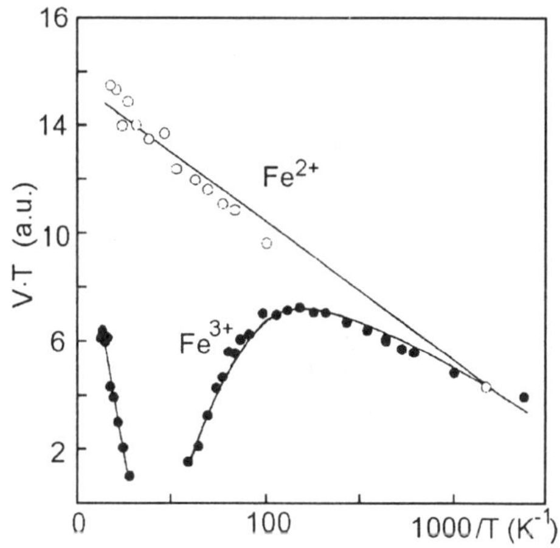

Fig. 10. Temperature variations of the product of ESE amplitude V and temperature T being proportional to T_M (Novoibirsk method) for nitroxide radical in oxidized (Fe^{3+}) and reduced (Fe^{2+}) cytochrome *c*

relaxation by dipolar coupling between unpaired electron of the spin label and an electron of low-spin Fe(III) ($S = 1/2$) ion. A similar T_M behaviour i.e. an acceleration of the phase relaxation of a slowly relaxing centre by dipolar coupling to the fast relaxing centre was observed by Rakowsky et al. (1995)[12] in spin labeled porphyrins.

This paper was supported by Polish Scientific Research Committee, Grant KBN-2-P03B-038-08.

References

1. K. M. Salikhov, A. G. Semenov, and Yu. D. Tsvetkov, *Electron Spin Echoes and Their Applications*, Nauka, Novosibirsk 1976 (in Russian).
2. V. V. Kurshev and T. Ichikawa, Effect of Spin Flip-Flop on Electron Spin Echo Decay due to Instantaneous Diffusion, *J. Magn. Reson.* **96** (1992) 563-573.
3. I. M. Brown, Electron Spin Echo Studies of Relaxation Processes in Molecular Solids, in *Time Domain Electron Spin Resonance*, edited by L. Kevan and R. N. Schwartz, Wiley, New York 1979, pp. 196-229.
4. K. M. Salikhov and Yu. D. Tsvetkov, Electron Spin Echo Studies of Spin-Spin Interaction in Solids, in *Time Domain Electron Spin Resonance*, edited by L. Kevan and R. N. Schwartz, Wiley, New York (1979), pp. 231-277.
5. W. Hilczer, S. K. Hoffmann, J. Goslar, J. Tritt-Goc, and M. A. Augustyniak, Electron Spin Echo Studies of Spin-Lattice and Spin-Spin Relaxation of SeO_3^- Radical in $(NH_4)_3H(SeO_4)_2$ Crystal, *Solid State Commun.* **85** (1993) 585-587.
6. Yu. D. Tsvetkov and S. A. Dzuba, Pulsed ESR and molecular Motions, *Appl. Magn. Reson.* **1** (1990) 179-194.
7. K. Nakagawa, M. B. Candelaria, W. W. Chik, S. S. Eaton, and G. R. Eaton, Electron-Spin Relaxation Times of Chromium (V), *J. Magn. Reson.* **98** (1992) 81-91.
8. J. L. Du, G. R. Eaton, and S. S. Eaton, Effect of Molecular Motion on Electron Spin Phase Memory Time for Copper (II) Complexes in Doped Solids, *Appl. Magn. Reson.* **6** (1994) 373-378.
9. S. K. Hoffmann, M. Gramza, and W. Hilczer, Molecular Dynamics of Diglycine Nitrate Studied by Phase Memory Relaxation Time of Glycine Radical, *Ferroelectrics* **172** (1995) 431-435.
10. S. K. Hoffmann, W. Hilczer, and J. Goslar, Electron Spin Echo Studies of Flipping Type Minimum in Phase Memory Time of Cu(II) Ions in Triglycine Selenate Crystal at Low Temperature, *Solid State Commun.* **100** (1996) 449-452.
11. W. Hilczer, J. Goslar, M. Gramza, S. K. Hoffmann, W. Blicharski, A. Osyczka, B. Turyna, and W. Froncisz, A Resonance Enhancement of the Phase Relaxation in the Electron Spin Echo of Nitroxide Covalently Attached to Cytochrome *c*, *Chem. Phys. Lett.* **147** (1995) 601-606.
12. M. H. Rakowsky, K. M. More, A. Y. Kulikov, G. R. Eaton, and S. S. Eaton, Time-Domain Electron Paramagnetic Resonance as a Probe Electron-Electron Spin-Spin Interaction in Spin-Labeled Low-Spin Iron Porphyrins, *J. Am. Chem. Soc.* **117** (1995) 2049-2057.

EPR and optical absorption spectra of x $R_2B_4O_7$ - (1-x) $VOSO_4$ (R = Li, Na or K) glasses

A.Murali and J.Lakshmana Rao

Department of Physics, S.V.University, Tirupati - 517 502, INDIA.

Abstract Electron paramagnetic resonance (EPR) and optical absorption studies have been made for VO(II) ions in glasses with composition x $R_2B_4O_7$ - (1-x) $VOSO_4$ (R = Li, Na or K). The EPR spectrum shows eight hyperfine lines characteristic of VO(II) ions. The spin-Hamiltonian parameters, Fermi contact interaction parameter K, and dipolar hyperfine coupling parameter P have been evaluated from the spectra. The EPR spectra of 0.9 $Li_2B_4O_7$ - 0.1 $VOSO_4$ and 0.98 $Na_2B_4O_7$ - 0.02 $VOSO_4$ glasses have been studied at various temperatures. The increase in $\Delta g_\perp$ value (g_e - $g_\perp$) with temperature is attributed to an increase in octahedral symmetry around V^{4+} ion with temperature. From the optical absorption spectra, a shift in band positions towards lower energy side was observed with change in alkali content (i.e., from Li to K). The theoretical values of the optical basicity (Λ_{th}) have been calculated. By correlating the EPR and optical data, the molecular orbital coefficients have been evaluated.

1 Introduction

Over the last twenty years, glass materials have found their way into a wide variety of applications. Glasses were found very useful for numerous applications such as soda-lime, silica for displays, ferrites for storage, molybdenum for printers etc. [1]. In addition, today glass is used as a densifying aid to sinter metals. Glass for electronic communication is perhaps the fastest growing area. Glassy materials with a wide variety of electrical, thermal, magnetic and mechanical properties are the challenging needs of the industry.

Oxide glasses containing transition metal (TM) ions in two different valence states are electronic conductors. The current carriers are highly localized in the glass, and the basic conduction mechanism is a transfer of electrons between TM ions in different valence states [2].

Transition metal ions doped sodium and lithium tetraborate glasses attracted the attention of many researchers [3-11] for their electrical, spectroscopic properties and cluster and dimer formations.

The present study describes the site symmetry around VO(II) ions in $xR_2B_4O_7$ - (1-x)$VOSO_4$ glasses as revealed from its electron paramagnetic resonance and optical absorption spectra.

2 Experimental

The starting materials used in the present work are analar grade $Li_2B_4O_7$, $Na_2B_4O_7$, $K_2B_4O_7$ and $VOSO_4$. The glass samples were prepared by following the conventional melt quenching technique. The glass samples thus prepared were transparent and green in colour.

EPR spectra were recorded at room temperature on a JEOL-FE-1X ESR spectrometer operating in the X band frequencies (9.205 GHz) and with a field modulation of 100 kHz. 100 mg of coarsely powdered glass sample was taken in a fused quartz sample tube for EPR measurements.

EPR spectra of 0.90 $Li_2B_4O_7$ - 0.10 $VOSO_4$ and 0.98 $Na_2B_4O_7$ - 0.02 $VOSO_4$ glass samples were recorded at different temperatures (123 K to 393 K) using a variable temperature controller (JES - UCT - 2AX).

Optical absorption spectra for 0.98 $R_2B_4O_7$ - 0.02 $VOSO_4$ (R = Li, Na or K) glass samples were recorded at room temperature on a Perkin Elmer 551 spectrophotometer in the wavelength range of 450 nm to 900 nm.

3 Results and Analysis

No EPR signal was detected in x $R_2B_4O_7$ - (1-x) $VOSO_4$ (R = Li, Na or K) (x = 1) glass samples. The EPR spectra of all other glass samples (x $\neq$ 1) show eight hyperfine lines, which are characteristic of a hyperfine interaction arising from an unpaired electron with the ^{51}V nucleus whose nuclear spin is 7/2 and which is present in 99.76 % abundance. Figure 1 shows the EPR spectrum of 0.98 $Li_2B_4O_7$ - 0.02 $VOSO_4$ glass sample at room temperature. The EPR spectra of x $R_2B_4O_7$ - (1-x) $VOSO_4$ (R = Na or K) glass samples at room temperature are very similar and are not shown.

The observed EPR spectra of all glass samples are analysed by using an axial spin-Hamiltonian [12]:

$$= g_{\parallel}\beta H_z S_z + g_{\perp}\beta (H_x S_x + H_y S_y) + A_{\parallel} S_z I_z + A_{\perp} (S_x I_x + S_y I_y), \qquad (1)$$

where the symbols have their usual meaning; the quadrupole and nuclear Zeeman interaction terms are ignored. The spin-Hamiltonian parameters were evaluated by an iterative method as suggested by Muncaster and Parke [13]. The spin-Hamiltonian parameters thus obtained along with those values reported in literature [14-17] for VO(II) ions in glassy materials are presented in Table 1.

EPR spectra of 0.90 $Li_2B_4O_7$ - 0.10 $VOSO_4$ and 0.98 $Na_2B_4O_7$ - 0.02 $VOSO_4$ glasses were recorded at different temperatures (123 K to 393 K). Figure 2 shows a

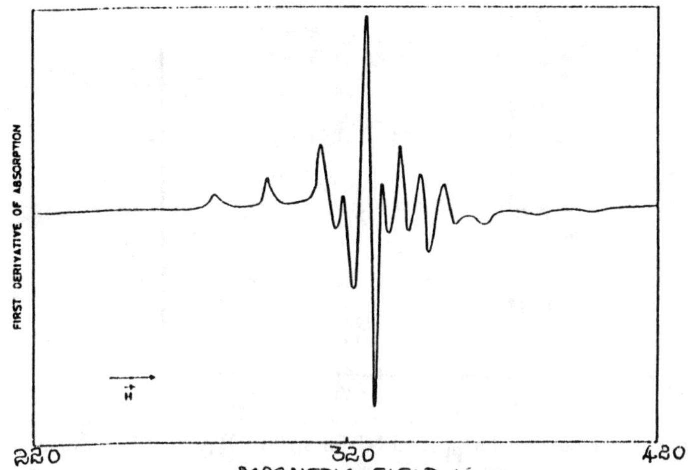

Fig. 1 EPR spectrum of 0.98 $Li_2B_4O_7$ - 0.02 $VOSO_4$ glass sample at room temperature.

Table 1. Spin-Hamiltonian parameters for VO^{2+} ions reported in literature and present results.

VO^{2+} ions in	$g_{\|\|}$	$g_{\perp}$	$\|A_{\|\|}\|$ in $10^{-4}cm^{-1}$	$\|A_{\perp}\|$ in $10^{-4}cm^{-1}$	$\Delta g_{\|\|}/\Delta g_{\perp}$	Ref
$ZnO-B_2O_3$	1.923	1.961	165	59	1.920	14
$Li_2O-Na_2O-B_2O_3$	1.931	1.956	171	50	1.540	15
$Na_2SO_4-ZnSO_4-B_2O_3$	1.936	1.982	181	73	3.266	16
$Cs_2SO_4-ZnSO_4-B_2O_3$	1.934	1.984	180	72	3.732	16
$CoO-BaO-B_2O_3$	1.935	1.958	165	55	1.519	17
$Li_2B_4O_7$	1.948	1.966	158	55	1.496	*
$Na_2B_4O_7$	1.944	1.972	162	52	1.924	*
$K_2B_4O_7$	1.947	1.971	165	50	1.767	*

* Present work

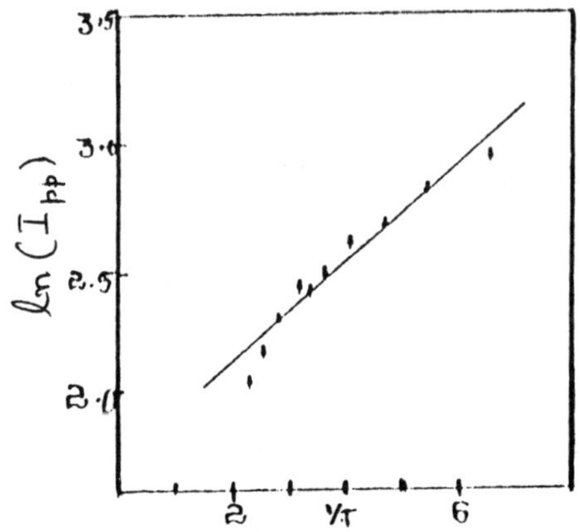

Fig. 2 Graph between $\ln(I_{pp})$ and 1/T for 0.90 $Li_2B_4O_7$ - 0.10 $VOSO_4$ glass sample.

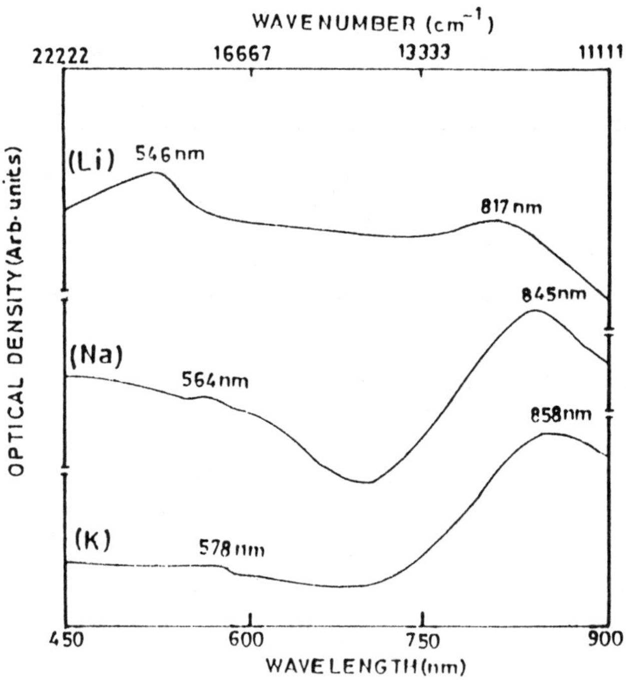

Fig. 3 Optical absorption spectra of 0.98 $R_2B_4O_7$ - 0.02 $VOSO_4$ (R = Li, Na or K) glass sample at room temperature.

graph between ln (I_{pp}) [I_{pp} is the peak-to-peak height of the $m_I = 3/2 \longleftrightarrow m_I = 3/2$ transition of the perpendicular component] and reciprocal of absolute temperature for the 0.90 $Li_2B_4O_7$ - 0.10 $VOSO_4$ glass sample. The variation of Q value of the cavity with the temperature is small and hence not considered.

The optical absorption spectra of 0.98 $R_2B_4O_7$ - 0.02 $VOSO_4$ (R = Li, Na or K) glass samples were recorded at room temperature and are shown in Figure 3. The observed spectra consist of two bands and are assigned to the d-d transition $^2B_{2g} \longrightarrow {}^2B_{1g}$ and $^2B_{2g} \longrightarrow {}^2E_g$ in the decreasing order of energy respectively.

4 Discussion

The EPR spectrum of V^{4+} ions in glasses is rich in hyperfine structure due to the ^{51}V nucleus with a nuclear spin of I = 7/2 and is easily observable in most systems [14-17] at room temperature.

The observed EPR spectra are satisfactorily explained in terms of an axial spin-Hamiltonian with S = 1/2 and I = 7/2. Therefore, the resonances are due to vanadium with a single unpaired electron in an axially distorted site. However, the values of $g_{\|}$ and $g_{\perp}$ depend critically on the type of distortion.

An octahedral site with a tetragonal compression would give $g_{\|} < g_{\perp} < g_e$. The g values obtained in the present work agree with this relationship. Therefore, it is confirmed that the V^{4+} ions in $R_2B_4O_7$ (R = Li, Na or K) glasses exist as VO^{2+} ions in octahedral coordination with tetragonal compression and belong to C_{4v} symmetry.

The deviations of $g_{\|}$ and $g_{\perp}$ values from the free electron g_e value ($g_e = 2.0023$) are generally denoted by $\Delta g_{\|}$ and $\Delta g_{\perp}$ respectively. The value of $\Delta g_{\|}/\Delta g_{\perp}$ is the measure of tetragonality of V^{4+} site. The values of $\Delta g_{\|}/\Delta g_{\perp}$ are calculated for 0.98 $R_2B_4O_7$ - 0.02 $VOSO_4$ (R = Li, Na or K) glasses and are given in Table 1. A decrease in $\Delta g_{\|}/\Delta g_{\perp}$ value shows that the octahedral symmetry around V^{4+} ions is improved[17]. From Table 1, we can observe that $\Delta g_{\|}/\Delta g_{\perp}$ value is less in $Li_2B_4O_7$ glass sample as compared to $Na_2B_4O_7$ and $K_2B_4O_7$ glass samples. Thus we can assert that the vanadyl ions in $Li_2B_4O_7$ glasses are less tetragonally distorted than those in $Na_2B_4O_7$ and $K_2B_4O_7$ glasses.

Also it is observed that both the $\Delta g_{\|}$ and $\Delta g_{\perp}$ values for 0.98 $Na_2B_4O_7$ - 0.02 $VOSO_4$ glasses are increasing with temperature. It is known that a decrease in $\Delta g_{\perp}$ value corresponds to a more tetragonal distortion of the V^{4+} site [18]. In the present study, since $\Delta g_{\perp}$ value increases with temperature, we can assume an improved octahedral symmetry around the V^{4+} ion in $Na_2B_4O_7$ glasses with temperature.

Figure 2 shows the Arrhenius plot of $\log_e$ (peak to peak intensity) [ln I_{pp}] versus 1/T. The anisotropy energy can be calculated from the slope of the graph. The anisotropy energy, thus calculated in the present work is 2.62 X 10^{-14} erg ($\approx$ 16.4 meV). The anisotropy energy calculated in the present work is in the same order expected for paramagnetic centers [19].

The values of dipolar hyperfine coupling parameter P and Fermi contact interaction parameter K were evaluated for the glasses using the expressions given by Muncaster and Parke [13] and are given in Table 2.

The theoretical values of the optical basicity of the glass can be estimated using the formula :

$$\Lambda_{th} = z_i r_i / |z_O| \gamma_i , \quad (2)$$

where the symbols have their usual meaning given by Duffy and Ingram [20]. The theorctical values of optical basicity (Λ_{th}) calculated for glass samples were given in Table 2. From these values, we can observe that the changes in optical basicity are independent of the changes in spin-Hamiltonian parameters. Also, it can be observed that the theoretical values of Λ increases from Li to K as has been reported by Rao et al [21].

For vanadyl ions in octahedral crystal field with tetragonal distortion, we can expect three bands corresponding to the transitions: $^2B_{2g} \longrightarrow {}^2E_g$, $^2B_{2g} \longrightarrow {}^2B_{1g}$ and $^2B_{2g} \longrightarrow {}^2A_{1g}$. In the present case we observed the first two transitions only.

EPR and optical data can be correlated to calculate the molecular orbital coefficients α^2 and γ^2 from the following equations [13]:

$$g_{\parallel} = g_e [1-(4\alpha^2\beta^2\lambda/E_1)]$$

$$(3)$$

$$g_{\perp} = g_e [1- (\gamma^2\beta^2\lambda/E_2)]$$

Table 2. Band positions transitions and the molecular orbital coefficients for VO^{2+} ions in $R_2B_4O_7$ glasses (R = Li, Na or K).

Glass	$^2B_{2g}\rightarrow$ $^2B_{1g}$ cm^{-1}	$^2B_{2g}\rightarrow$ ^2E_g cm^{-1}	*P in* 10^{-4} cm^{-1}	*K*	α^2	γ^2	Λ_{th}
$Li_2B_4O_7$	18310	12240	115	0.737	0.499	0.891	0.512
$Na_2B_4O_7$	17725	11830	125	0.658	0.518	0.719	0.533
$K_2B_4O_7$	17295	11650	128	0.653	0.480	0.732	0.564

where E_1 and E_2 are the crystal field splitting energies of the transitions $^2B_{2g} \longrightarrow {}^2B_{1g}$ and $^2B_{2g} \longrightarrow {}^2E_g$ respectively; λ is the spin orbit coupling constant; its value is 249 cm^{-1} for V^{4+} [13]. The terms α^2 and γ^2 are measures of degree of σ and π bondings with the equatorial ligands and equal to unity if the bonding is purely ionic, while the term β^2 represent the in-plane π bonding with the equatorial ligands.

The values of α^2 and γ^2 are given in Table 2. From the values, it can be concluded that the in plane σ bonding is significantly covalent in nature while the out of plane π bonding is more ionic one.

Table 2 also shows the band positions along with their assignments for the glass samples. From the table we can observe that all the bands are shifted towards longer wavelength (i.e., red side) in 0.98 $K_2B_4O_7$ - 0.02 $VOSO_4$ glasses as compared to those of 0.98 $Li_2B_4O_7$ - 0.02 $VOSO_4$ and 0.98 $Na_2B_4O_7$ - 0.02 $VOSO_4$ glasses. This confirms that the heavier the molecule, the more the bands are shifted towards lower energy side [21].

5 Conclusions

- The EPR and optical absorption spectra of vanadyl ions in $R_2B_4O_7$ (R = Li, Na or K). glasses have been studied.
- The results show that $g_{\parallel} < g_{\perp} < g_e$ (= 2.0023) and $|A_{\parallel}| > |A_{\perp}|$. Therefore, it is confirmed that V^{4+} ions in $R_2B_4O_7$ (R = Li, Na or K) glasses exist as VO^{2+} ions in octahedral coordination with tetragonal compression.
- The octahedral symmetry around V^{4+} ion in $Na_2B_4O_7$ glasses improves with temperature. The anisotropy energy was calculated from the temperature variation of EPR intensity.
- From the optical absorption spectra, it is observed that the bands are shifted towards lower energy side as the alkali content is changed from lithium to potassium. This confirms that the heavier the molecule, the more the bands are shifted towards the lower energy side.
- The theoretical values of optical basicity were calculated for all glasses.
- By correlating EPR and optical data, the molecular orbital coefficients α^2 and γ^2 have been evaluated. From these values, it is concluded that the out of plane π bonding is ionic and in-plane σ bonding is covalent in nature for all the glasses studied.

Acknowledgments

One of the authors (JLR) thanks University Grants Commission (New Delhi) for providing financial assistance.

References

[1] R.Thummala Rao, Glasses in microelectronics, *J.Non-Cryst. Solids*, **73** (1985) 409 -411.

[2] F.P.Koffyberg and F.A.Benko, Conductivity and optical absorption of semiconducting WO_3 glasses, *J.Non-Cryst. Solids*, **40** (1980) 7-17.

[3] I.Ardelean, O.Cozar and G.Ilonca, EPR and magnetic susceptibility studies of xCuO.(1-x) ($2B_2O_3.Li_2O$) glasses, *Solid State Commun.*, **50** (1984) 87-90.

[4] A.A.Kutub, A.E.Mohamed Osman and C.A.Hogarth, A spectroscopic study of some sodium tetraborate glasses containing lead and copper, *J.Mater. Sci.*, **22** (1987) 2199-2202.

[5] A.A.Kutub and M.S.Elman Haraway, Effect of gamma ray irradiation on some optical properties of sodium tetraborate glasses containing iron, *J.Non-Cryst. Solids*, **104** (1988) 31-37.

[6] E.E.Khawaja and F.F.Al Adel, Optical absorption of vanadium in sodium diborate glasses, *J.Mater. Sci.*, **23** (1988) 1391-1394.

[7] M.V.Dirke, S.Muller, K.Barner and H.Rager, Cluster formation of WO_3 in $Li_2B_4O_7$ glasses, *J.Non-Cryst. Solids*, **124** (1990) 265-270.

[8] A.A.Kutub and M.S.Elman Haraway, Gamma ray irradiation induced effects in sodium diborate glasses doped with some transition metals, *J.Mater. Sci.*, **26** (1991) 3964-3968.

[9] A.A.Kutub, Electrical conductivity studies of sodium tetraborate glasses containing copper, *J.Mater. Sci. Lett.*,**13** (1994) 1157-1158.

[10] A.A.Kutub, Optical, infrared and DSC studies of sodium tetraborate glasses containing copper oxide, *J.Mater. Sci.*, **30** (1995) 724-728.

[11] E.Culea and I.Milea, Gd-O-Gd dimer formation in x Gd_2O_3 - (1-x) $Na_2O.2B_2O_3$ glasses, *J.Non-Cryst. Solids*, **189** (1995) 246-250.

[12] A.Abragam and B.Bleaney, *Electron Paramagnetic Resonance of Transition Ions*, Clarendon Press, Oxford (1970) p.175.

[13] R.Muncaster and S.Parke, ESR spectra of glasses in the system V_2O_5-TeO_2 , *J.Non-Cryst. Solids*, **24** (1977) 399-412.

[14] A.Yadav and V.P.Seth and S.K.Gupta, Electron spin resonance and optical spectra of VO^{2+} and Cu^{2+} in ZnO-B_2O_3 and PbO-B_2O_3 glasses, *J.Non-Cryst. Solids*, **101** (1988) 1-7.

[15] D. Suresh Babu, M.V. Ramana, S.G. Sathyanarayan and G.S. Sastry, Electron paramagnetic resonance studies of VO^{2+} in lithium-sodium borate glasses, *Phys. Chem. Glasses*, **31** (1990) 80-82.

[16] B.Sreedhar, J.Lakshmana Rao and S.V.J.Lakshman, Electron spin resonance spectra of VO^{2+} ions in alkali zinc borosulphate glasses, *J.Non-Cryst. Solids*, **116** (1990) 111-114.

[17] V.P.Seth, S.P.Yadav and S.K.Gupta, EPR in CoO.BaO.B_2O_3 glasses containing two transition elements, *Radiat. Effects Def. Solids*, **132** (1994) 187-191.

[18] G.Hochstrasser, Detection of VO^{2+} in glass by electron spin resonance, *Phys. Chem. Glasses*, **7** (1966) 178-182.

[19] E.Guedes de Sousa, S.K.Mendiretta and J.M.Machado da Silva, EPR studies of magnetic interactions in glasses containing iron, *Portugal Phys.*, **17** (1986) 203-214.

[20] J.A.Duffy and M.D.Ingram, Optical basicity-IV. Influence of electronegativity on the Lewis basicity and solvent properties of molten oxyanion salts and glasses, *J.Inorg. Nucl. Chem.*, **37** (1975) 1203-1206.

[21] A.S.Rao, J.L.Rao and S.V.J.Lakshman, Electron paramagnetic resonance and optical absorption spectra of VO^{2+} ions in alkali cadmium borosulphate glasses, *Phys. Chem. Glasses*, **33** (1992) 1-4.

EPR of Fullerene Ions and Superconductivity in Potassium Intercalated Fullerene

*Wojciech Kempiński, Lidia Piekara-Sady, Jan Stankowski and Peter Scharff**

Institute of Molecular Physics, Polish Academy of Sciences, Smoluchowskiego 17, PL–60179 Poznań, Poland
*Institut für Anorganishe Chemie, TU Clausthal, Clausthal-Zellerfeld, Paul Ernst-Strasse 4, D–3392, Germany

Abstract. The process of potassium intercalation into C_{60} fcc lattice has been studied in detail. EPR has been applied to show the appearance and evolution of spins while superconductivity was demonstrated by Magnetically Modulated Microwave Absorption method. The process of phase separation, $K_1C_{60} \rightarrow \alpha\text{-}C_{60} + K_3C_{60}$, has been studied in terms of C_{60}^{-1} further reduction to C_{60}^{-3} as the result of potassium redistribution from octahedral do tetrahedral sites of C_{60} lattice. In the course of doping process a new, metastable superconducting phase has been identified with $T_c = 21$K, whereas K_3C_{60} superconducting transition occurs at $T_c = 18.5$K.

1 Introduction

The application of EPR technique in the studies of alkali metal-intercalated fullerenes provides information about the number and localisation of spins related to the charge carriers involved in the phenomenon of superconductivity. An EPR spectrometer with the magnetic field switched off is a perfect instrument to determine the onset temperature of a transition to the superconducting phase. This comprises Magnetically Modulated Microwave Absorption method (MMMA) [1] in which the absorption of microwave field energy by intergrain Josephson junctions formed upon the transition to the superconducting state is recorded. Both EPR and MMMA techniques allows to follow the evolution of superconductivity at already low doping levels [2-4]. We will review our EPR results on fullerene paramagnetic centres and present the detailed study on potassium intercalation into C_{60} lattice leading to the formation of superconducting phases.

2 Fullerene Ions

2.1 C_{60}^{+} Defect Centre

C_{60} single crystal is diamagnetic. However, polycrystalline C_{60} exhibits weak paramagnetism characterised by EPR as a single narrow line (≈ 2G) at $g \cong 2.0026$. It

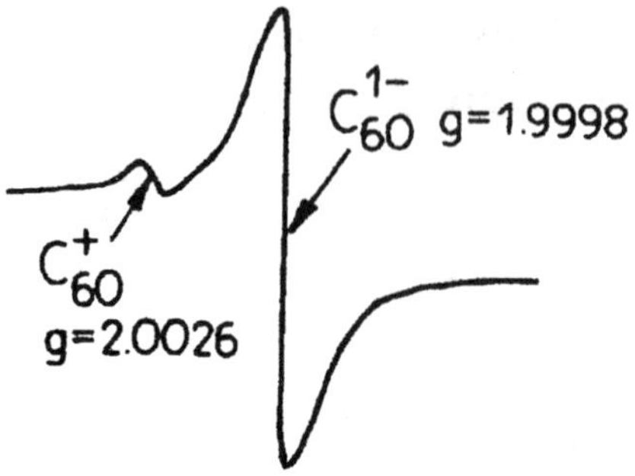

Fig. 1. EPR spectrum of C_{60} and potassium after 5 minutes of heating at 500 K

was expected that large molecular ionisation potential rules out the possibility of C_{60} ionisation. It has been already demonstrated that it is possible after all. C_{60} in a fumic sulphuric acid gives at least one oxidised state with g value greater than free electron g–factor [5]. In a C_{60} solid, the appearance of an electron in the conduction band results in the formation of a hole in the fundamental band, localised on one C_{60} molecule. We believe that the localised hole type excitation which gives EPR at $g \cong 2.0026$ [6]. The hole nature of that centre can be proved by K-intercalation of C_{60}. Intercalation of fullerene with alkali metals is accompanied by an electron transfer from potassium to C_{60}. Immediately after C_{60} powder was added to K, the C_{60}^{+} line decreased its amplitude several times. After the first 10 minutes of heating C_{60} and potassium, this signal has been already vanished. Polycrystalline sample of C_{60} intercalated with K exhibited C_{60}^{+} signal for a prolonged time of the heat treatment. The occurrence of C_{60}^{-1} signal (Fig. 1) while C_{60}^{+} has not been neutralised yet is the result of a larger value of C_{60} granules. It proves that the $g = 2.0026$ resonance is a bulk effect, it does not come from any surface defects.

2.2 C_{60}^{-1} and C_{60}^{-3}

In the fcc C_{60} structure there are octahedral and tetrahedral sites big enough to accommodate alkali metal atoms. Intercalation of fullerene with alkali metals is accompanied by an electron transfer from the metal to C_{60} molecule, thus fullerides are composed of C_{60}^{-n} anions. The methodology of K-intercalation into C_{60} has been already described in detail [2-4]. At the beginning of the doping process, strong C_{60}^{-1} and weak C_{60}^{-3} EPR signals are observed (Fig. 2a). If the sample is left long enough at room temperature, the signal due to C_{60}^{-3} radicals dominates over that from C_{60}^{-1} (Fig. 2a'). Time dependent changes in the integral intensities of both C_{60}^{-1} and C_{60}^{-3} lines at room temperature correspond to the phase separation of K_1C_{60} into α-C_{60} and K_3C_{60} as a redistribution of potassium from octa- to tetrahedral sites. The α-C_{60} phase (i.e. K_xC_{60}, where $x \cong 0.1$ [7,8]) is composed of C_{60}^{-1} also. The integral intensity of the EPR signal can be a measure of the reaction progress α in Avrami model [9]. The reaction progress is given by:

$$\alpha = 1 - \exp\left(-(kt)^n\right) \qquad (1)$$

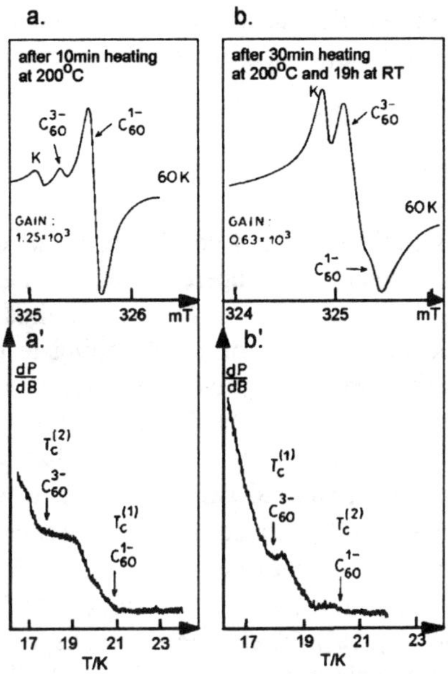

Fig. 2. EPR spectrum of C_{60} and potassium: **a)** and **a')** after annealing the sample at room temperature; MMMA signals **b)**: from the sample characterised by the EPR spectrum a), **b')**: from the sample characterised by the EPR spectrum a'). Signal K comes from potassium impurity.

where k is the rate constant of the reaction. The value of the exponent n depends on the mechanism of the reaction. The solid lines in Fig. 3 are the fit of experimental points to the Avrami equation (Eq. (1)). Both, decreasing C_{60}^{-1} and growing in C_{60}^{-3} signals gave $n \approx 2.3$ which is characteristic for a diffusion controlled reaction. The process of phase separation could easily be reversed by heating the sample again

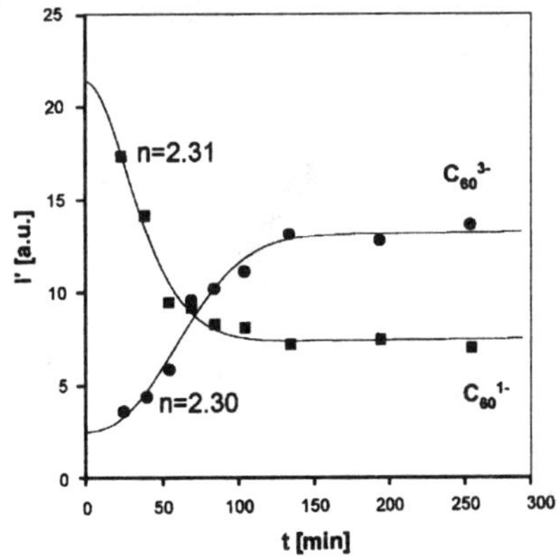

Fig. 3. Integral intensity of C_{60}^{-1} and C_{60}^{-3} EPR signals at the very low doping level versus the annealing time at room temperature. The solid lines represent Avrami equation fitted to the experimental data.

above 323 K [10] to further populate octahedral sites by potassium ions. Then, C_{60}^{-1} signal dominates again over the C_{60}^{-3} one in the EPR spectrum [2, 3].

3 Superconductivity

Much of the interest in fullerides comes from the existence of superconductivity in A_3C_{60} (where A is alkali metal) with superconducting transition temperatures as high as 33 K for Rb_3C_{60}. The application of MMMA method allows us to study the formation of superconducting phases at different doping levels. On examining the microwave absorption at the very low doping level (Fig. 4), after only 5 minutes of heating at 500K, we found a very distinct $T_c = 21$K which has not been reported before for K-intercalated fullerenes. When C_{60}^{-3} signal has been already detectable (Fig. 2a), microwave absorption showed two different and well separated onset T_c temperatures: $T_c^{(1)}$=(20.5 ± 0.5) K related to the presence of C_{60}^{-1} signal and $T_c^{(2)}$ = (18.5 ± 0.5) K related to the presence of C_{60}^{-3} line (Fig.2b). After annealing the sample at room temperature further reduction of C_{60}^{-1} to C_{60}^{-3} gives strong C_{60}^{-3} line relative to C_{60}^{-3} line (Fig.2a'). That process results in a very low microwave absorption at $T_c^{(1)}$ and a strong absorption at $T_c^{(2)}$ (Fig.2b'). The final result of doping is single C_{60}^{-3} line, slightly asymmetric, characteristic for metallic state, and only one T_c, i.e. $T_c^{(2)}$, that of the bulk K_3C_{60} in microwave absorption [3].

To make sure that two-step MMMA signal comes from two superconducting phases, microwave absorption dependence on magnetic field was examined at a doping level characterised by both EPR lines and MMMA signals well separated. As shown in Fig. 5, further separation of the two signals takes place with the increase of the external magnetic field. At 300 mT there is almost no microwave ab-

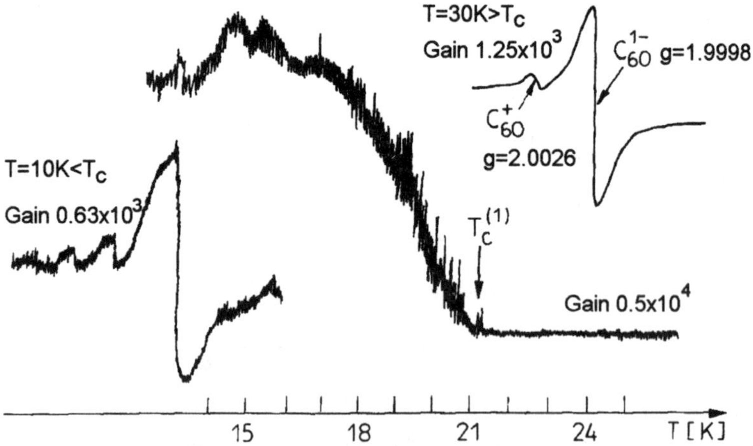

Fig. 4. EPR and MMMA signals at a very low doping level. The EPR spectrum below T_c shows Josephson dissipation

sorption at $T_c^{(2)} = 18$ K (K_3C_{60}) while that at $T_c^{(1)} = 21$ K has still significant amplitude. Since microwave absorption signals respond in a different way to the increase of the external magnetic field, we have indeed two superconducting phases with different magnetic field critical values.

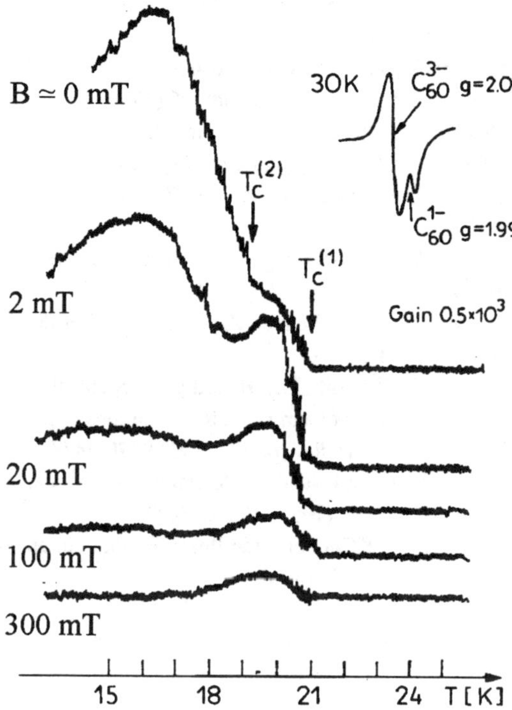

Fig.5 Dependence of microwave absorption intensity on the external magnetic field.

4 Conclusions

The process of K-intercalation in C_{60} powdered samples first leads to neutralise the hole-type paramagnetic centres. The experiments performed for two C_{60} samples of different granularity proved the bulk nature of the defect centres which give EPR at $g \cong 2.0026$. The evolution of EPR lines upon C_{60} doping with potassium showed suppression of C_{60}^{+} line and formation of C_{60}^{-1} and C_{60}^{-3} radical ions. It was shown that the process of potassium redistribution in phase separation: $K_1C_{60} \rightarrow \alpha\text{-}C_{60} + K_3C_{60}$ is controlled by solid state diffusion. C_{60} exhibits superconductivity upon doping at the very low doping levels. Two superconducting phases were found: K_3C_{60} phase with $T_c = 18.5$ K and metastable superconducting phase with $T_c = 21$ K.

This work was supported by Polish Scientific Research Committee under Project KBN-2-P03B–100–11.

References

1. J. Stankowski, P.K. Kahol, N.S. Dalal, and J.S. Moodera, Possible Josephson Oscillations Spectra and Electron Paramagnetic Resonance of Cu^{2+} in Y-Ba-Cu-O, *Phys. Rev. B* **36** (1987) 7126-7128.
2. W. Kempiński, and J. Stankowski, EPR and MMMA Study of C_{60} upon K-doping, *Acta Phys. Polonica* **88** (1995) 549-552.
3. W. Kempiński, and J. Stankowski, Two Superconducting Phases Related to Different Stoichiometry of K-doped C_{60}, *Solid State Commun.* **95** (1996) 1079–1083.
4. W. Kempiński, P. Scharff, J. Stankowski, L. Piekara-Sady, and Z.Trybuła, EPR of Fullerene Ions in K-fullerides at Low Doping Level, *Physica C,* in press.
5. P. Scharff, J. Patyk, and F. Rozpłoch, Electronic Properties of C_{60} Derivatives with Oleum, to be published.
6. J. Stankowski, W. Kempiński, A. Koper, and J. Martinek, Valence State of Paramagnetic Centres in Fullerenes, *Appl. Magn. Reson.* **6** (1994) 145–154.
7. D.M. Poirier, D.W. Owens, and J.H. Weaver, Alkali-metal-fullerides Phase Equilibrium, *Phys.Rev. B,* **51** (1995) 1830–1843.
8. D.M. Poirier, T.R. Ohno, G.H. Kroll, P.J. Benning, F. Stepniak, J.H. Weaver, L.P.F. Chibante, and R.E. Smalley, X-ray Fotoemission Investigations of Binary and Ternary C_{60} Fullerides of Na, K, Rb and Cs, *Phys.Rev. B,* **47** (1993) 9870-9877.
9. J. Stankowski, L. Kevan, B. Czyżak, and B. Andrzejewski, Model of Rb Diffusion in Rb_xC_{60} fullerides, *J. Phys. Chem.* *97* (1993) 10430–10432.
10. D.M. Poirier, and J.H. Weaver, KC_{60} Fulleride Phase Formation: an X-ray Fotoemission Study, *Phys. Rev. B,* **47** (1993) 10959–10962.